3701987340

MODERN FOOD
MICROBIOLOGY

MODERN FOOD MICRO- BIOLOGY

Fourth Edition

JAMES M. JAY
Wayne State University

CHAPMAN & HALL
New York • London

This edition published by
Chapman & Hall
One Penn Plaza
New York, NY 10119

Published in Great Britain by
Chapman & Hall
2-6 Boundary Row
London SE1 8HN

Copyright © 1992 by Van Nostrand Reinhold
Library of Congress Catalog Card Number 91-32527
ISBN 0-442-00733-7
Printed in the United States of America

Library of Congress Cataloging-in-Publication Data
Jay, James M. (James Monroe), 1927–
    Modern food microbiology/James M. Jay.—4th ed.
        p.   cm.
    Includes bibliographical references and index.
    ISBN 0-442-00733-7
    1. Food—Microbiology.   I. Title.
QR115.J3   1991
576'.163—dc20   91-32527

British Library Cataloguing in Publication Data available

Please send your order for this or any **Chapman & Hall book to Chapman & Hall,
29 West 35th Street, New York, NY 10001, Attn: Customer Service Department.**
You may also call our Order Department at 1-212-244-3336 or fax your purchase order
to 1-800-248-4724.

For a complete listing of Chapman & Hall's titles, send your requests to **Chapman &
Hall, Dept. BC, One Penn Plaza, New York, NY 10119.**

CONTENTS

PREFACE

This fourth edition of *Modern Food Microbiology* is written primarily for use as a textbook in a second or subsequent course in microbiology. The previous editions have found usage in courses in food microbiology and applied microbiology in liberal arts, food science, food technology, nutritional science, and nutrition curricula. Although organic chemistry is a desirable prerequisite, those with a good grasp of biology and chemistry should not find this book difficult. In addition to its use as a textbook, this edition, like the previous one, contains material that goes beyond that covered in a typical microbiology course (parts of Chaps. 4, 6, and 7). This material is included for its reference value and for the benefit of professionals in microbiology, food science, nutrition, and related fields.

This edition contains four new chapters, and with the exception of Chapter 15, which received only minor changes, the remaining chapters have undergone extensive revision. The new chapters are 17 (indicator organisms), 18 (quality control), 21 (listeriae and listeriosis), and 24 (animal parasites). Six chapters in the previous edition have been combined; they are represented in this edition by Chapters 12, 13, and 14.

In the broad area of food microbiology, one of the challenges that an author must deal with is that of producing a work that is up to date. While recent research and advances in some of the areas covered have been relatively minor, others are experiencing an explosion of activity and advances. Since much of this has been presented in other books and monographs, I have opted to cite many of these sources in the introductions to six of the seven parts. Readers are referred to these introductory statements for an overview of the respective chapters.

Numerous references have been consulted in the preparation of this edition, and I thank the many investigators whose findings I have drawn upon so heavily. I have been assisted by the following individuals, who critiqued various sections of my drafts: D. A. Corlett, Jr., C. D. Jeffries, M. J. Loessner, M. C. Johnson-Thompson, C. R. Sterling, H. M. Wehr, N. B. Wehr, and D. J. Wilson. Those who assisted me with the previous three editions are acknowledged in the respective editions.

My thanks also go to the diligent editors of Beehive Production Services, especially Beverly Miller, copyeditor, for their professional work on the manuscript.

MODERN FOOD
MICROBIOLOGY

PART I
Historical Background

The material in this chapter provides a glimpse of some of the early events that ultimately led to the recognition of the significance and role of microorganisms in foods. Food microbiology as a defined subdiscipline does not have a precise beginning. Some of the early findings and observations are noted, along with dates. The selective lists of events noted for food preservation, food spoilage, food poisoning, and food legislation are meant to be guideposts in the continuing evolution and development of food microbiology.

1

History of Microorganisms in Food

Although it is extremely difficult to pinpoint the precise beginnings of human awareness of the presence and role of microorganisms in foods, the available evidence indicates that this knowledge preceded the establishment of bacteriology or microbiology as a science. The era prior to the establishment of bacteriology as a science may be designated the prescientific era. This era may be further divided into what has been called the **food-gathering period** and the **food-producing period**. The former covers the time from human origin over 1 million years ago up to 8,000 years ago. During this period, humans were presumably carnivorous, with plant foods coming into their diet later in this period. It is also during this period that foods were first cooked.

The food-producing period dates from about 8,000 to 10,000 years ago and, of course, includes the present time. It is presumed that the problems of spoilage and food poisoning were encountered early in this period. With the advent of prepared foods, the problems of disease transmission by foods and of faster spoilage caused by improper storage made their appearance. Spoilage of prepared foods apparently dates from around 6000 B.C. The practice of making pottery was brought to Western Europe about 5000 B.C. from the Near East. The first boiler pots are thought to have originated in the Near East about 8,000 years ago. The arts of cereal cookery, brewing, and food storage were either started at about this time or stimulated by this new development. The first evidence of beer manufacture has been traced to ancient Babylonia, as far back as 7000 B.C. (7). The Sumerians of about 3000 B.C. are believed to have been the first great livestock breeders and dairymen and were among the first to make butter. Salted meats, fish, fat, dried skins, wheat, and barley are also known to have been associated with this culture. Milk, butter, and cheese were used by the Egyptians as early as 3000 B.C. Between 3000 B.C. and 1200 B.C., the Jews used salt from the Dead Sea in the preservation of various foods. The Chinese and Greeks used salted fish in their diet, and

3

the Greeks are credited with passing this practice on to the Romans, whose diet included pickled meats. Mummification and preservation of foods were related technologies that seem to have influenced each other's development. Wines are known to have been prepared by the Assyrians by 3500 B.C. Fermented sausages were prepared and consumed by the ancient Babylonians and the people of ancient China as far back as 1500 B.C. (7).

Another method of food preservation that apparently arose during this time was the use of oils such as olive and sesame. Jensen (6) has pointed out that the use of oils leads to high incidences of staphylococcal food poisoning. The Romans excelled in the preservation of meats other than beef by around 1000 B.C. and are known to have used snow to pack prawns and other perishables, according to Seneca. The practice of smoking meats as a form of preservation is presumed to have emerged sometime during this period, as did the making of cheese and wines. It is doubtful whether people at this time understood the nature of these newly found preservation techniques. It is also doubtful whether the role of foods in the transmission of disease or the danger of eating meat from infected animals was recognized.

Few advances were apparently made toward understanding the nature of food poisoning and food spoilage between the time of the birth of Christ and A.D. 1100. Ergot poisoning (caused by *Claviceps purpurea*, a fungus that grows on rye and other grains) caused many deaths during the Middle Ages. Over 40,000 deaths due to ergot poisoning were recorded in France alone in A.D. 943, but it was not known that the toxin of this disease was produced by a fungus. Meat butchers are mentioned for the first time in 1156, and by 1248 the Swiss were concerned with marketable and nonmarketable meats. In 1276 a compulsory slaughter and inspection order was issued for public abattoirs in Augsburg. Although people were aware of quality attributes in meats by the thirteenth century, it is doubtful that there was any knowledge of the causal relationship between meat quality and microorganisms.

Perhaps the first person to suggest the role of microorganisms in spoiling foods was A. Kircher, a monk, who as early as 1658 examined decaying bodies, meat, milk, and other substances and saw what he referred to as "worms" invisible to the naked eye. Kircher's descriptions lacked precision, however, and his observations did not receive wide acceptance. In 1765, L. Spallanzani showed that beef broth that had been boiled for an hour and sealed remained sterile and did not spoil. Spallanzani performed this experiment to disprove the doctrine of the spontaneous generation of life. However, he did not convince the proponents of the theory since they believed that his treatment excluded oxygen, which they felt was vital to spontaneous generation. In 1837 Schwann showed that heated infusions remained sterile in the presence of air, which he supplied by passing it through heated coils into the infusion. While both of these men demonstrated the idea of the heat preservation of foods, neither took advantage of his findings with respect to application. The same may be said of D. Papin and G. Leibniz, who hinted at the heat preservation of foods at the turn of the eighteenth century.

The event that led to the discovery of canning had its beginnings in 1795, when the French government offered a prize of 12,000 francs for the discovery of a practical method of food preservation. In 1809, a Parisian confectioner, François (Nicholas) Appert, succeeded in preserving meats in glass bottles that had been kept in boiling water for varying periods of time. This discovery was made public in 1810, when Appert was issued a patent for his process. Not being a scientist, Appert was probably unaware of the long-range significance of his discovery or why it worked. This, of course, was the beginning of canning as it is known and practiced today. This event occurred some 50 years before L. Pasteur demonstrated the role of microorganisms in the spoilage of French wines, a development that gave rise to the rediscovery of bacteria. A. Leeuwenhoek in the Netherlands had examined bacteria through a microscope and described them in 1683, but it is unlikely that Appert was aware of this development since he was not a scientist and Leeuwenhoek's report was not available in French.

The first person to appreciate and understand the presence and role of microorganisms in food was Pasteur. In 1837 he showed that the souring of milk was caused by microorganisms, and in about 1860 he used heat for the first time to destroy undesirable organisms in wine and beer. This process is now known as pasteurization.

HISTORICAL DEVELOPMENTS

Some of the more significant dates and events in the history of food preservation, food spoilage, food poisoning, and food legislation are listed below.

Food Preservation

1782—Canning of vinegar was introduced by a Swedish chemist.
1810—Preservation of food by canning was patented by Appert in France.
    —Peter Durand was issued a British patent to preserve food in "glass, pottery, tin or other metals or fit materials." The patent was later acquired by Hall, Gamble, and Donkin, possibly from Appert.
1813—Donkin, Hall, and Gamble introduced the practice of postprocessing incubation of canned foods.
    —Use of SO_2 as a meat preservative is thought to have originated around this time.
1825—T. Kensett and E. Daggett were granted a U.S. patent for preserving food in tin cans.
1835—A patent was granted to Newton in England for making condensed milk.
1837—Winslow was the first to can corn from the cob.
1839—Tin cans came into wide use in the United States.
    —L. A. Fastier was given a French patent for the use of brine bath to raise the boiling temperature of water.
1840—Fish and fruit were first canned.
1841—S. Goldner and J. Wertheimer were issued British patents for brine baths based on Fastier's method.
1842—A patent was issued to H. Benjamin in England for freezing foods by immersion in an ice and salt brine.
1843—Sterilization by steam was first attempted by I. Winslow in Maine.
1845—S. Elliott introduced canning to Australia.
1853—R. Chevallier-Appert obtained a patent for sterilization of food by autoclaving.

1854—Pasteur began wine investigations. Heating to remove undesirable organisms was introduced commercially in 1867–1868.

1855—Grimwade in England was the first to produce powdered milk.

1856—A patent for the manufacture of unsweetened condensed milk was granted to Gail Borden in the United States.

1861—I. Solomon introduced the use of brine baths to the United States.

1865—The artificial freezing of fish on a commercial scale was begun in the United States. Eggs followed in 1889.

1874—The first extensive use of ice in transporting meat at sea was begun.

—Steam pressure cookers or retorts were introduced.

1878—The first successful cargo of frozen meat went from Australia to England. The first from New Zealand to England was sent in 1882.

1880—The pasteurization of milk was begun in Germany.

1882—Krukowitsch was the first to note the destructive effects of ozone on spoilage bacteria.

1886—A mechanical process of drying fruits and vegetables was carried out by an American, A. F. Spawn.

1890—The commercial pasteurization of milk was begun in the United States.

—Mechanical refrigeration for fruit storage was begun in Chicago.

1893—The Certified Milk movement was begun by H. L. Coit in New Jersey.

1895—The first bacteriological study of canning was made by Russell.

1907—E. Metchnikoff and coworkers isolated and named one of the yogurt bacteria, *Lactobacillus bulgaricus*.

—The role of acetic acid bacteria in cider production was noted by B. T. P. Barker.

1908—Sodium benzoate was given official sanction by the United States as a preservative in certain foods.

1916—The quick freezing of foods was achieved in Germany by R. Plank, E. Ehrenbaum, and K. Reuter.

1917—Clarence Birdseye in the United States began work on the freezing of foods for the retail trade.

—Franks was issued a patent for preserving fruits and vegetables under CO_2.

1920—Bigelow and Esty published the first systematic study of spore heat resistance above 212°F. The "general method" for calculating thermal processes was published by Bigelow, Bohart, Richardson, and Ball; the method was simplified by C. O. Ball in 1923.

1922—Esty and Meyer established $z = 18°F$ for *C. botulinum* spores in phosphate buffer.

1928—The first commercial use of controlled-atmosphere storage of apples was made in Europe (first used in New York in 1940).

1929—A patent issued in France proposed the use of high-energy radiation for the processing of foods.

—Birdseye frozen foods were placed in retail markets.

1943—B. E. Proctor in the United States was the first to employ the use of ionizing radiation to preserve hamburger meat.

1950—The D value concept came into general use.

1954—The antibiotic nisin was patented in England for use in certain processed cheese to control clostridial defects.

1955—Sorbic acid was approved for use as a food preservative.

—The antibiotic chlortetracycline was approved for use in fresh poultry (oxytetracycline followed a year later). Approval was rescinded in 1966.

1967—The first commercial facility designed to irradiate foods was planned and designed in the United States.

Food Spoilage

1659—Kircher demonstrated the occurrence of bacteria in milk; Bondeau did the same in 1847.

1680—Leeuwenhoek was the first to observe yeast cells.

1780—Scheele identified lactic acid as the principal acid in sour milk.

1836—Latour discovered the existence of yeasts.

1839—Kircher examined slimy beet juice and found organisms that formed slime when grown in sucrose solutions.

1857—Pasteur showed that the souring of milk was caused by the growth of organisms in it.

1866—L. Pasteur's *Étude sur le Vin* was published.

1867—Martin advanced the theory that cheese ripening was similar to alcoholic, lactic, and butyric fermentations.

1873—The first reported study on the microbial deterioration of eggs was carried out by Gayon.

—Lister was first to isolate *Lactococcus lactis* in pure culture.

1876—Tyndall observed that bacteria in decomposing substances were always traceable to air, substances, or containers.

1878—Cienkowski reported the first microbiological study of sugar slimes and isolated *Leuconostoc mesenteroides* from them.

1887—Forster was the first to demonstrate the ability of pure cultures of bacteria to grow at 0°C.

1888—Miquel was the first to study thermophilic bacteria.

1895—The first records on the determination of numbers of bacteria in milk were those of Von Geuns in Amsterdam.

—S. C. Prescott and W. Underwood traced the spoilage of canned corn to improper heat processing for the first time.

1902—The term *psychrophile* was first used by Schmidt-Nielsen for microorganisms that grow at 0°C.

1912—The term *osmophilic* was coined by Richter to describe yeasts that grow well in an environment of high osmotic pressure.

1915—*Bacillus coagulans* was first isolated from coagulated milk by B. W. Hammer.

1917—*Bacillus stearothermophilus* was first isolated from cream-style corn by P. J. Donk.

1933—Olliver and Smith in England observed spoilage by *Byssochlamys fulva*; first described in the United States in 1964 by D. Maunder.

Food Poisoning

1820—The German poet Justinus Kerner described "sausage poisoning" (which in all probability was botulism) and its high fatality rate.

1857—Milk was incriminated as a transmitter of typhoid fever by W. Taylor of Penrith, England.

1870—Francesco Selmi advanced his theory of ptomaine poisoning to explain illness contracted by eating certain foods.

1888—Gaertner first isolated *Salmonella enteritidis* from mean that had caused 57 cases of food poisoning.

1894—T. Denys was the first to associate staphylococci with food poisoning.

1896—Van Ermengem first discovered *Clostridium botulinum*.

1904—Type A strain of *C. botulinum* was identified by G. Landman.

1906—*Bacillus cereus* food poisoning was recognized. The first case of diphyllobothriasis was recognized.

1926—The first report of food poisoning by streptococci was made by Linden, Turner, and Thom.

1936–

1937—Type E strain of *C. botulinum* was identified by L. Bier and E. Hazen.

1937—Paralytic shellfish poisoning was recognized.

1938—Outbreaks of *Campylobacter* enteritis were traced to milk in Illinois.

1939—Gastroenteritis caused by *Yersinia enterocolitica* was first recognized by Schleifstein and Coleman.

1945—McClung was the first to prove the etiologic status of *Clostridium perfringens (welchii)* in food poisoning.

1951—*Vibrio parahaemolyticus* was shown to be an agent of food poisoning by T. Fujino of Japan.

1955—Similarities between cholera and *Escherichia coli* gastroenteritis in infants were noted by S. Thompson.

—Scombroid (histamine-associated) poisoning was recognized.

—The first documented case of anisakiasis occurred in the United States.

1960—Type F strain of *C. botulinum* identified by Moller and Scheibel.

—The production of aflatoxins by *Aspergillus flavus* was first reported.

1965—Foodborne giardiasis was recognized.

1969—*C. perfringens* enterotoxin was demonstrated by C. L. Duncan and D. H. Strong.

—*C. botulinum* type G was first isolated in Argentina by Gimenez and Ciccarelli.

1971—First U.S. foodborne outbreak of *Vibrio parahaemolyticus* gastroenteritis occurred in Maryland.

—First documented outbreak of *E. coli* foodborne gastroenteritis occurred in the United States.

1975—*Salmonella* enterotoxin was demonstrated by L. R. Koupal and R. H. Deibel.

1976—First U.S. foodborne outbreak of *Yersinia enterocolitica* gastroenteritis occurred in New York.

—Infant botulism was first recognized in California.

1978—Documented foodborne outbreak of gastroenteritis caused by the Norwalk virus occurred in Australia.

1979—Foodborne gastroenteritis caused by non-01 *Vibrio cholerae* occurred in Florida. Earlier outbreaks occurred in Czechoslovakia (1965) and Australia (1973).

1981—Foodborne listeriosis outbreak was recognized in the United States.

1982—The first outbreaks of foodborne hemorrhagic colitis occurred in the United States.

1983—*Campylobacter jejuni* enterotoxin was described by Ruiz-Palacios et al.

Food Legislation

1890—The first national meat inspection law was enacted. It required the inspection of meats for export only.

1895—The previous meat inspection act was amended to strengthen its provisions.

1906—The U.S. Federal Food and Drug Act was passed by Congress.

1910—The New York City Board of Health issued an order requiring the pasteurization of milk.

1939—The New Food, Drug, and Cosmetic Act became law.

1954—The Miller Pesticide Chemicals Amendment to the Food, Drug, and Cosmetic Act was passed by Congress.

1957—The U.S. Compulsory Poultry and Poultry Products law was enacted.

1958—The Food Additives Amendment to the Food, Drug, and Cosmetics Act was passed.

1962—The Talmadge-Aiken Act (allowing for federal meat inspection by states) was enacted into law.

1963—The U.S. Food and Drug Administration approved the use of irradiation for the preservation of bacon.

1967—The U.S. Wholesome Meat Act was passed by Congress and enacted into law on December 15.

1968—The Food and Drug Administration withdrew its 1963 approval of irradiated bacon.

—The Poultry Inspection Bill was signed into law.

1969—The U.S. Food and Drug Administration established an allowable level of 20 ppb of aflatoxin for edible grains and nuts.

1973—The state of Oregon adopted microbial standards for fresh and processed retail meats. They were repealed in 1977.

1984—A bill was introduced in the U.S. House of Representatives that would promote the irradiation of foods by, among other things, defining irradiation as a process rather than as a food additive.

References

1. Bishop, P. W. 1978. Who introduced the tin can? Nicolas Appert? Peter Durand? Bryan Donkin? *Food Technol.* 32(4):60–67.

2. Brandly, P. J., G. Migaki, and K. E. Taylor. 1966. *Meat Hygiene*, 3d ed., ch. 1. Philadelphia: Lea & Febiger.

3. Farrer, K. T. H. 1979. Who invented the brine bath?—The Isaac Solomon myth. *Food Technol.* 33(2):75–77.

4. Goldblith, S. A. 1971. A condensed history of the science and technology of thermal processing. *Food Technol.* 25(12):44–50.

5. Goldblith, S. A., M. A. Joslyn, and J. T. R. Nickerson. 1961. *Introduction to Thermal Processing of Foods*, vol. 1. Westport, Conn.: AVI.

6. Jensen, L. B. 1953. *Man's Foods*, chs. 1, 4, 12. Champaign, Ill.: Garrard Press.

7. Pederson, C. S. 1971. *Microbiology of Food Fermentations*. Westport, Conn.: AVI.

8. Schormüller, J. 1966. *Die Erhaltung der Lebensmittel.* Stuttgart:Ferdinand Enke Verlag.
9. Stewart, G. F., and M. A. Amerine. 1973. *Introduction to Food Science and Technology*, ch. 1. New York: Academic Press.
10. Tanner, F. W. 1944. *The Microbiology of Foods*, 2d ed. Champaign. Ill.: Garrard Press.
11. Tanner, F. W., and L. P. Tanner. 1953. *Food-borne Infections and Intoxications*, 2d ed. Champaign, Ill.: Garrard Press.

PART II

Taxonomy, Sources, Types, Incidence, and Behavior of Microorganisms in Foods

Many changes in bacterial taxonomy have occurred since the ninth edition of *Bergey's Manual* was published in 1984–1986, and those that were approved through 1990 are noted in Chapter 2 for most groups of interest in foods. Similarly, many recent changes have been made in the taxonomy of yeasts and molds, and some of these are noted in Chapter 2. In addition to the references cited in the chapter, the monograph on yeasts edited by Rose (listed below) may be consulted.

The intrinsic and extrinsic parameters that affect microbial growth are presented in Chapter 3, and additional general information can be found in several of the references noted below.

Chapter 4 lists the numbers and types of microorganisms reported for a variety of foods and some food processing technologies. Except for the latter, this chapter is meant to be a compilation of findings for reference use.

The following sources are recommended for the reader interested in additional details:

G. J. Banwart. 1989. *Basic Food Microbiology*, 2d ed. New York: Van Nostrand Reinhold. This is a general food microbiology text.

W. C. Frazier and D. C. Westhoff. 1988. *Food Microbiology*, 4th ed. New York: McGraw-Hill. This is a general food microbiology text.

International Commission Microbiological Specification Foods (ICMSF). 1980. *Microbial Ecology of Foods*, vol. 1. New York: Academic Press. This volume gives a detailed treatment of the intrinsic and extrinsic parameters that affect microbial growth in foods.

T. J. Montville, ed. 1987. *Food Microbiology*, vol. 1. Boca Raton, Fla.: CRC Press. This book provides in-depth coverage of intrinsic parameters of microbial growth.

A. H. Rose, ed. 1987. *Biology of Yeasts*. New York: Academic Press. A thorough treatment of the general biology and taxonomy of yeasts is given in this volume.

2

Taxonomy, Role, and Significance of Microorganisms in Foods

Since human food sources are of plant and animal origin, it is important to understand the biological principles of the microbial flora associated with plants and animals in their natural habitats and respective roles. While it sometimes appears that microorganisms are trying to ruin our food sources by infecting and destroying plants and animals, including humans, this is by no means their primary role in nature. In our present view of life on this planet, the primary function of microorganisms in nature is self-perpetuation. During this process, the heterotrophs carry out the following general reaction:

All organic matter
(carbohydrates, proteins, lipids, etc.)
↓
Energy + Inorganic compounds
(nitrates, sulfates, etc.)

This, of course, is essentially nothing more than the operation of the nitrogen cycle and the cycle of other elements (Fig. 2-1). The microbial spoilage of foods may be viewed simply as an attempt by the food flora to carry out what appears to be their primary role in nature. This should not be taken in the teleological sense. In spite of their simplicity when compared to higher forms, microorganisms are capable of carrying out many complex chemical reactions essential to their perpetuation. To do this, they must obtain nutrients from organic matter, some of which constitutes our food supply.

If one considers the types of microorganisms associated with plant and animal foods in their natural states, one can then predict the general types of microorganisms to be expected on this particular food product at some later stage in its history. Results from many laboratories show that untreated foods may be expected to contain varying numbers of bacteria, molds, or yeasts, and

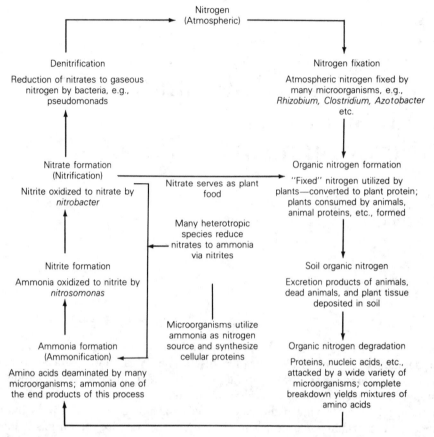

FIGURE 2-1.   Nitrogen cycle in nature is here depicted schematically to show the role of microorganisms.
From Microbiology *by M. J. Pelczar and R. Reid (25); copyright © 1965 by McGraw-Hill Book Company, used with permission of the publisher.*

the question often arises as to the safety of a given food product based upon total microbial numbers. The question should be twofold: What is the total number of microorganisms present per g or ml? and what *types* of organisms are represented in this number? It is necessary to know which organisms are associated with a particular food in its natural state and which of the organisms present are not normal for that particular food. It is therefore of value to know the general distribution of bacteria in nature and the general types of organisms normally present under given conditions where foods are grown and handled.

BACTERIAL TAXONOMY

Many changes have taken place in the classification or taxonomy of bacteria since the ninth edition of *Bergey's Manual* was published in 1984. Many of the new taxa have been created as a result of the employment of molecular genetic methods, alone or in combination with some of the more traditional methods:

- DNA homology and moles% G + C content of DNA
- 23S, 16S, and 5S rRNA sequence similarities
- Oligonucleotide cataloging
- Numerical taxonomic analysis of total soluble proteins or of a battery of morphological and biochemical characteristics
- Cell wall analysis
- Serologic profiles
- Cellular fatty acid profiles

While some of these have been employed for many years (e.g., cell wall analysis and serologic profiles), others (e.g., rRNA sequence similarity) have come into wide use only during the 1980s. The methods that are the most powerful as bacterial taxonomic tools are outlined and briefly discussed below.

rRNA Analyses

Taxonomic information can be obtained from RNA in the production of nucleotide catalogs and the determination of RNA sequence similarities. First, the prokaryotic ribosome is a 70S (Svedberg) unit, which is composed of two separate functional subunits: 50S and 30S. The 50S subunit is composed of 23S and 5S RNA in addition to about 34 proteins, while the 30S subunit is composed of 16S RNA plus about 21 proteins.

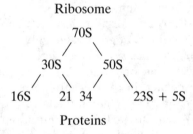

The 16S subunit is highly conserved and is considered to be an excellent chronometer of bacteria over time (39). By use of reverse transcriptase, 16S rRNA can be sequenced to produce long stretches (about 95% of the total sequence) to allow for the determination of precise phylogenetic relationships (23). Because of its smaller size, 5S rRNA has been sequenced totally.

To sequence 16S rRNA, a single-stranded DNA copy is made by use of reverse transcriptase with the RNA as template. When the single-stranded DNA is made in the presence of dideoxynucleotides, DNA fragments of various sizes result that can be sequenced by the Sanger method. From the DNA sequences, the template 16S rRNA sequence can be deduced. It was through studies of 16S rRNA sequences that led Woese and his associates to propose the establishment of three kingdoms of life forms: Eukaryotes, Archaebacteria, and Prokaryotes. The last include the cyanobacteria and the eubacteria, with the bacteria of importance in foods being eubacteria. Sequence similarities of 16S rRNA are widely employed, and some of the new foodborne taxa were created primarily by its use along with other information. Libraries of eubacterial 5S rRNA sequences also exist, but they are fewer than for 16S.

Nucleotide catalogs of 16S rRNA have been prepared for a number of organisms, and extensive libraries exist. By this method, 16S rRNA is subjected to digestion by RNAse T1, which cleaves the molecule at G(uanine) residues. Sequences (−mers) of 6 to 20 bases are produced and separated, and similarities S_{AB} between organisms can be compared. While the relationship between S_{AB} and percentage similarity is not good below SAB values of 0.40, the information derived is useful at the phylum level. The sequencing of 16S rRNA by reverse transcriptase is preferred to oligonucleotide cataloging, since longer stretches of rRNA can be sequenced.

Analysis of DNA

The moles % G + C of bacterial DNA has been employed in bacterial taxonomy for several decades, and its use in combination with 16S and 5S rRNA sequence data makes it even more meaningful. By 16S rRNA analysis, the gram-positive eubacteria fall into two groups at the phylum level: one group with moles % G + C >55, and the other <50 (39). The former includes the genera *Streptomyces*, *Propionibacterium*, *Micrococcus*, *Bifidobacterium*, *Corynebacterium*, *Brevibacterium*, and others. The group with the lower G + C values includes the genera *Clostridium*, *Bacillus*, *Staphylococcus*, *Lactobacillus*, *Pediococcus*, *Leuconostoc*, *Listeria*, *Erysipelothrix*, and others. The latter group is referred to as the Clostridium branch of the eubacterial tree. When two organisms differ in G + C content by more than 10%, they have few base sequences in common.

DNA-DNA or DNA-RNA hybridization has been employed for some time, and this technique continues to be of great value in bacterial systematics. It has been noted that the ideal reference system for bacterial taxonomy would be the complete DNA sequence of an organism (38). It is generally accepted that bacterial species can be defined in phylogenetic terms by use of DNA-DNA hybridization results, where 70% or greater relatedness and 5°C or less Tm defines a species (38). When DNA-DNA hybridization is employed, phenotypic characteristics are not allowed to override except in exceptional cases (38). While a genus is more difficult to define phylogenetically, 20% sequence similarity is considered to be the minimum level of DNA-DNA homology (38).

Even if there is not yet a satisfactory phylogenetic definition of a bacterial genus, the continued application of nucleic acid techniques, along with some of the other methods listed above, should lead ultimately to a phylogenetically based system of bacterial systematics. In the meantime, changes in the extant taxa may be expected to continue to occur.

Some of the important genera known to occur in foods are listed below in alphabetical order. Some are desirable in certain foods; others bring about spoilage or cause gastroenteritis.

Bacteria

Acinetobacter	*Enterobacter*	*Pediococcus*
Aeromonas	*Erwinia*	*Proteus*
Alcaligenes	*Escherichia*	*Pseudomonas*
Alteromonas	*Flavobacterium*	*Psychrobacter*

Bacillus	*Hafnia*	*Salmonella*
Brochothrix	*Lactococcus*	*Serratia*
Campylobacter	*Lactobacillus*	*Shewanella*
Carnobacterium	*Leuconostoc*	*Shigella*
Citrobacter	*Listeria*	*Staphylococcus*
Clostridium	*Micrococcus*	*Vagococcus*
Corynebacterium	*Moraxella*	*Vibrio*
Enterococcus	*Pantoea*	*Yersinia*

Molds

Alternaria	*Cladosporium*	*Mucor*
Aspergillus	*Colletotrichum*	*Penicillium*
Aureobasidium	*Fusarium*	*Rhizopus*
Botrytis	*Geotrichum*	*Trichothecium*
Byssochlamys	*Monilia*	*Wallemia*
		Xeromyces

Yeasts

Brettanomyces	*Issatchenkia*	*Schizosaccharomyces*
Candida	*Kluyveromyces*	*Torulaspora*
Cryptococcus	*Pichia*	*Trichosporon*
Debaryomyces	*Rhodotorula*	*Zygosaccharomyces*
Hanseniaspora	*Saccharomyces*	

Protozoa

Cryptosporidium parvum	*Giardia lamblia*
Entamoeba histolytica	*Toxoplasma gondii*

PRIMARY SOURCES OF MICROORGANISMS FOUND IN FOODS

The genera and species previously listed are among the most important normally found in food products. Each genus has its own particular nutritional requirements, and each is affected in predictable ways by the parameters of its environment. Eight environmental sources of organisms to foods are listed below, and these, along with the 36 genera of bacteria and four protozoa noted, are presented in Table 2-1 to reflect their primary food-source environments.

1. Soil and Water. These two environments are placed together because many of the bacteria and fungi that inhabit both share a lot in common. Soil organisms may enter the atmosphere by the action of wind and later enter water bodies when it rains. They also enter water when rainwater flows over soils into bodies of water. Aquatic organisms can be deposited onto soils through the actions of cloud formation and subsequent rainfall. This common cycling results in soil and aquatic organisms being one in the same to a large degree. Some aquatic organisms, however, are unable to persist in soils, especially those that are indigenous to marine waters. *Alteromonas* spp. are

TABLE 2-1.   Relative Importance of Eight Sources of Bacteria and Protozoa to Foods

Organisms	Soil & Water	Plants/Products	Utensils	Gastrointestinal Tract	Handlers	Animal Feeds	Animal Hides	Air & Dust
Bacteria								
Acinetobacter	XX	X	X				X	X
Aeromonas	XX[a]	X						
Alcaligenes	X	X	X	X			X	
Alteromonas	XX[a]							
Bacillus	XX[b]	X	X		X	X	X	XX
Brochothrix		XX	X					
Campylobacter				XX	X			
Carnobacterium	X	X	X					
Citrobacter	X	XX	X	XX				
Clostridium	XX[b]	X	X	X	X	X	X	XX
Corynebacterium	XX[b]	X	X		X		X	X
Enterobacter	X	XX	X				X	
Enterococcus	X	X	X	XX	X	X	X	X
Erwinia	X	XX	X					
Escherichia	X	X		XX	X			
Flavobacterium	X	XX					X	
Hafnia	X	X		XX				
Lactococcus		XX	X	X			X	
Lactobacillus		XX	X	X			X	
Leuconostoc		XX	X	X			X	
Listeria	X	XX		X	X	X	X	
Micrococcus	X	X	X		X	X	X	XX
Moraxella	X	X					X	
Pantoea	X	X		X				
Pediococcus		XX	X	X			X	
Proteus	X	X	X	X	X		X	
Pseudomonas	XX	X	X			X	X	
Psychrobacter	XX	X	X				X	
Salmonella				XX		XX		
Serratia	X	X	X	X		X	X	
Shewanella	X	X						
Shigella				XX				
Staphylococcus				X	XX		X	
Vagococcus	XX			XX				
Vibrio	XX[a]			X				
Yersinia	X	X		X				
Protozoa								
C. parvum	X			X	X			
E. histolytica	XX[a]			X	X			
G. lamblia	XX[a]			X	X			
T. gondii		X		XX				

Note: XX indicates a very important source.
[a] Primarily water.
[b] Primarily soil.

aquatic forms that require seawater salinity for growth and would not be expected to persist in soils. The bacterial flora of seawater is essentially gram negative, and gram-positive bacteria exist there essentially only as transients.

2. Plants and Plant Products.   It may be assumed that many or most soil and water organisms contaminate plants. However, only a relatively small number find the plant environment suitable to their overall well-being. Those that persist on plant products do so by virtue of a capacity to adhere to plant surfaces so that they are not easily washed away and because they are able to obtain their nutritional requirements. Notable among these are the lactic acid bacteria and some yeasts. Among others that are commonly associated with plants are bacterial plant pathogens in the genera *Corynebacterium, Curtobacterium, Pseudomonas,* and *Xanthomonas* and fungal pathogens among several genera of molds.

3. Food Utensils.   When vegetables are harvested in containers and utensils, one would expect to find some or all of the surface organisms on the products to contaminate contact surfaces. As more and more vegetables are placed in the same containers, a normalization of the flora would be expected to occur. In a similar way, the cutting block in a meat market along with cutting knives and grinders are contaminated from initial samples, and this process leads to a build-up of organisms, thus ensuring a fairly constant level of contamination of meat-borne organisms.

4. Intestinal Tract of Humans and Animals.   This flora becomes a water source when polluted water is used to wash raw food products. The intestinal flora consists of many organisms that do not persist as long in waters as do others, and notable among these are pathogens such as salmonellae. Any or all of the Enterobacteriaceae may be expected in fecal wastes, along with intestinal pathogens, including the four protozoal species already listed.

5. Food Handlers.   The microflora on the hands and outer garments of handlers generally reflects the environment and habits of individuals, and the organisms in question may be those from soils, waters, dust, and other environmental sources. Additional important sources are those that are common in nasal cavities and the mouth and on the skin and those from the gastrointestinal tract that may enter foods through poor personal hygienic practices.

6. Animal Feeds.   This continues to be an important source of salmonellae to poultry and other farm animals. In the case of some silage, it is a known source of *Listeria monocytogenes* to dairy and meat animals. The organisms in dry animal feed are spread throughout the animal environment and may be expected to occur on animal hides.

7. Animal Hides.   In the case of milk cows, the types of organisms found in raw milk can be a reflection of the flora of the udder when proper procedures are not followed in milking and of the general environment of such animals. From both the udder and the hide, organisms can contaminate the general environment, milk containers, and the hands of handlers.

8. Air and Dust.   While most of the organisms listed in Table 2-1 may at times be found in air and dust in a food-processing operation, the ones that can persist include most of the gram-positive organisms listed. Among fungi, a number of molds may be expected to occur in air and dust along with some yeasts. In general, the types of organisms in air and dust would be those that are constantly reseeded to the environment. Air ducts are not unimportant sources.

SYNOPSIS OF COMMON FOODBORNE BACTERIA

These synopses are provided to give readers glimpses of bacterial groups that are discussed throughout the textbook. They are not meant to be used for culture identifications. For the latter, one or more of the cited references should be consulted.

Acinetobacter (A · ci · ne'to · bac · ter; *Gr. akinetos, unable to move*).   These gram-negative rods show some affinity to the family Neisseriaceae, and some that were formerly achromobacters and moraxellae are placed here. Also, some former acinetobacters are now in the genus *Psychrobacter*. They differ from the latter and the moraxellae in being oxidase negative. They are strict aerobes that do not reduce nitrates. Although rod-shaped cells are formed in young cultures, old cultures contain many coccoid-shaped cells. They are widely distributed in soils and waters and may be found on many foods, especially refrigerated fresh products. The moles % G + C content of DNA for the genus is 39–47. (See Chap. 9 for further discussion relative to meats.) It has been proposed, based on DNA-rRNA hybridization data, that the genera *Acinetobacter*, *Moraxella*, and *Psychrobacter* be placed in a new family (Moracellaceae), but this proposal has not been approved.

Aeromonas (ae · ro · mo'nas; *gas producing*).   These are typically aquatic gram-negative rods formerly in the family Vibrionaceae but now in the family Aeromonadaceae (24). As the generic name suggests, they produce copious quantities of gas from those sugars fermented. They are normal inhabitants of the intestines of fish, and some are fish pathogens. The moles % G + C content of DNA is 57–65. (The species that possess pathogenic properties are discussed in Chap. 25.)

Alcaligenes (al · ca · li'ge · nes; *alkali producers*).   Although gram-negative, these organisms sometimes stain gram positive. They are rods that do not, as the generic name suggests, ferment sugars but instead produce alkaline reactions, especially in litmus milk. Nonpigmented, they are widely distributed in nature in decomposing matter of all types. Raw milk, poultry products, and fecal matter are common sources. The moles % G + C content of DNA is 58–70, suggesting that the genus is heterogeneous.

Alteromonas (al · te · ro · mo'nas; *another monad*).   These are marine and coastal water inhabitants that are found in and on seafoods; all species require seawater salinity for growth. They are gram-negative motile rods that are strict aerobes. The species once classified as *A. putrefaciens* and *A. colwelliana* have been transferred to the genus *Shewanella*. The moles % G + C content of DNA is 43.2–48.

Bacillus (ba · cil'lus).   These are gram-positive sporeforming rods that are aerobes in contrast to the clostridia, which are anaerobes. Although most are mesophiles, psychrotrophs and thermophiles exist. The genus contains only two pathogens: *B. anthracis* (cause of anthrax) and *B. cereus*. While most strains of the latter are nonpathogens, some cause foodborne gastroenteritis (further discussed in Chap. 20). The moles % G + C content of DNA of 32–62 suggests heterogeneity, and it may be expected that the genus as now constituted will be altered. With 16S rRNA and DNA-DNA hybridization data overriding features such as morphology and endospores, some of the species will undoubtedly be placed with nonsporeformers.

Brochothrix (bro · cho · thr'ix: *Gr. brochos, loop; thrix, thread*).   These gram-positive nonsporeforming rods are closely related to the genera *Lactobacillus* and *Listeria* (30), and some of the common features are discussed in Chapter 21. Although they are not true coryneforms, they bear resemblance to this group. Typically exponential-phase cells are rods, and older cells are coccoids, a feature typical of coryneforms. Their separate taxonomic status has been reaffirmed by rRNA data, although only two species are recognized: *B. thermosphacta* and *B. campestris*. They share some features with the genus *Microbacterium*. They are common on processed meats and on fresh and processed meats that are stored in gas-impermeable packages at refrigerator temperatures. In contrast to *B. thermosphacta*, *B. campestris* is rhamnose and hippurate positive (32). The moles % G + C content of DNA is 36.

Campylobacter (cam · py' · lo · bac · ter; *Gr. campylo, curved)*.   Although most often pronounced "camp'lo · bac · ter," the technically correct pronunciation should be noted. These gram-negative, spirally curved rods were formerly classified as vibrios. They are microaerophilic to anaerobic. The genus has been restructured since the 1984 publication of *Bergey's Manual* (22). The once *C. nitrofigilis* and *C. cryaerophila* have been transferred to the new genus *Arcobacter*; the once *C. cinnaedi* and *C. fenneliae* are now in the genus *Helicobacter*; and the once *Wolinella carva* and *W. recta* are now *C. curvus* and *C. rectus* (33). The moles % G + C content of DNA is 30–35. (They are further discussed in Chap. 23.)

Carnobacterium (car · no · bac · terium; *L. carnis, of flesh-meat bacteria)*.   This genus of gram-positive, catalase-negative rods was formed to accommodate some organisms previously classified as lactobacilli. They are phylogenically closer to the enterococci and vagococci than to lactobacilli (3, 36). The following four species are now recognized, along with the former taxon designation:

C. divergens (*Lactobacillus divergens*)
C. piscicola (*Lactobacillus piscicola*)
C. gallinarum
C. mobile

They are heterofermentative, and most grow at 0°C and none at 45°C. Gas is produced from glucose by some species, and the moles % G + C for the genus is 33.0–37.2. They differ from the lactobacilli in being unable to grow on acetate medium and in their synthesis of oleic acid. They are found on vacuum-packaged meats and related products, as well as on fish and poultry meats (6, 16, 36).

Citrobacter (cit · ro · bac'ter).   These enteric bacteria are slow lactose-fermenting, gram-negative rods. All members can use citrate as sole carbon source. *C. freundii* is the most prevalent species in foods, and it and the other species are not uncommon on vegetables and fresh meats. The moles % G + C content of DNA is 50–52.

Clostridium (clos · tri'di · um; *Gr. closter, a spindle*).   These anaerobic spore-forming rods are widely distributed in nature, as are their aerobic counterparts, the bacilli. The genus contains many species, some of which cause disease in humans (see Chap. 20 for *C. perfringens* food poisoning and botulism). Mesotrophic, psychrotrophic, and thermophilic species/strains exist; their importance in the thermal canning of foods is discussed in Chapter 14.

Corynebacterium (co · ry · ne · bac · ter' · um; *Gr., coryne, club*).   This is one of the true coryneform genera of gram-positive, rod-shaped bacteria that are sometimes involved in the spoilage of vegetable and meat products. Most are mesotrophs, although psychrotrophs are known, and one, *C. diphtheriae*, causes diphtheria in humans. The genus has been reduced in species with the transfer of some of the plant pathogens to the genus *Clavibacter* and others to the genus *Curtobacterium*. The moles % G + C of DNA content is 51–63.

Enterobacter (en · te · ro · bac'ter).   These enteric gram-negative bacteria are typical of other Enterobacteriaceae relative to growth requirements, although they are not generally adapted to the gastrointestinal tract. They are further characterized and discussed in Chapter 17. *E. agglomerans* has been transferred to the genus *Pantoea*.

Enterococcus (en · te · ro · coc'cus).   This genus was erected to accommodate some of the Lancefield serologic group D cocci. It has since been expanded to 16 species of gram-positive ovoid cells that occur singly, in pairs, or in short chains. They were once in the genus *Streptococcus*. At least three species do not react with group D antisera. The genus is characterized more thoroughly in Chapter 17, and its phylogenetic relationship to other lactic acid bacteria can be seen in Figure 21-1 (Chap. 21).

Erwinia (er · wi'ni · a).   These gram-negative enteric rods are especially associated with plants, where they cause bacterial soft rot (see Chap. 8). The moles % G + C content of DNA is 53.6–54.1.

Escherichia (esch · er · i'chi · a).   This is clearly the most widely studied genus of all bacteria. Those strains that cause foodborne gastroenteritis are discussed in Chapter 22, and *E. coli* as an indicator of food safety is discussed in Chapter 17.

Flavobacterium (fla · vo · bac · te'ri · um).   These gram-negative rods are characterized by their production of yellow to red pigments on agar and by their association with plants. Some are mesotrophs, and others are psychrotrophs, where they participate in the spoilage of refrigerated meats and vegetables.

Hafnia (haf'ni · a).   These gram-negative enteric rods are important in the spoilage of refrigerated meat and vegetable products; *H. alvei* is the only species at this time. It is motile and lysine and ornithine positive, and it has a moles % G + C content of DNA of 48–49.

Lactobacillus (lac · to · ba · cil'lus).   Taxonomic techniques that came into wide use during the 1980s have been applied to this genus, resulting in some of those in the ninth edition of *Bergey's Manual* being transferred to other genera. Based on 16S rRNA sequence data, three phylogenetically distinct clusters are revealed (4), with one cluster encompassing the *Leuconostoc paramesenteroides* group. In all probability, this genus will undergo reclassification. They are gram-positive, catalase-negative rods that often occur in long chains. Although those in foods are typically microaerophilic, many true anaerobic strains exist, especially in human stools and the rumen. They typically occur on most, if not all, vegetables, along with some of the other lactic acid bacteria. Their occurrence in dairy products is common. A recently described species, *L. suebicus*, was recovered from apple and pear mashes; it grows at pH 2.8 in 12–16% ethanol (20). Many fermented products are produced, and these are discussed in Chapter 16. Those that are common on refrigerator-stored, vacuum-packaged meats are discussed in Chapter 9.

Lactococcus (lac · to · coc'cus).   The nonmotile Lancefield serologic group N cocci once classified in the genus *Streptococcus* have been elevated to generic status. They are gram-positive, nonmotile, and catalase-negative spherical or ovoid cells that occur singly, in pairs, or as chains. They grow at 10°C but not at 45°C, and most strains react with group N antisera. L-lactic acid is the predominant end product of fermentation. The following four species and three subspecies are recognized (12, 27):

Present Taxa	Former Taxa
L. lactis subsp. *lactis*	*Strep. lactis* subsp. *lactis* and *Lactobacillus xylosus*
L. lactis subsp. *cremoris*	*Strep. lactis* subsp. *cremoris*
L. lactis subsp. *hordniae*	*Lactobacillus hordniae*
L. garvieae	*Strep. garvieae*
L. plantarum	*Strep. plantarum*
L. raffinolactis	*Strep. raffinolactis*

Leuconostoc (leu · co · nos'toc; *colorless nostoc*).   Along with the lactobacilli, this is another of the genera of lactic acid bacteria. They are gram-positive, catalase-negative cocci that are heterofermentative. By 16S rRNA analysis, some lactobacilli group with leuconostocs to form a "leuconostoc branch" of the lactobacilli (40), with *L. oenos* being more distant than the other leuconostocs studied. *L. oenos* is the only acidophile in this genus, and it is important in wines (10). Leuconostocs may be expected to be found on the same types of foods as the lactobacilli, although their presence in vacuum-packaged foods is not as constant.

Listeria (lis · te'ri · a).   This genus of seven species of gram-positive, non-sporing rods is closely related to *Brochothrix*. The seven species show 80% similarity by numerical taxonomic studies; they have identical cell walls, fatty acid, and cytochrome composition. They are more fully described and discussed in Chapter 21.

Micrococcus (mi · cro · coc'cus).   These cocci are gram positive and catalase positive, and some produce pink to orange-red to red pigments, while others are nonpigmented. Most can grow in the presence of high levels of NaCl, and most are mesotrophs, although psychrotrophic species/strains are known. This is a large genus, and they are widely distributed in nature in food preparation environments, on the person of food handlers, and naturally in many foods. The moles % G + C content of DNA is 66–75.

Moraxella (mo · rax · el'la).   These short gram-negative rods are sometimes classified as *Acinetobacter*. They differ from the latter in being sensitive to penicillin and oxidase positive and having a moles % G + C content of DNA of 40–46. The newly erected genus *Psychrobacter* includes some that were once placed in this genus. Their metabolism is oxidative, and they do not form acid from glucose.

Pantoea (pan · toe'a).   This genus consists of gram-negative, noncapsulated, nonsporing straight rods, most of which are motile by peritrichous flagella. Some are yellow pigmented, and all are oxidase negative. They are found on plants and seeds and in soil, water, and human specimens; their moles % G + C ranges from 55.1 to 60.6. The genus consists of two species, most notable of which is *P. agglomerans* (which encompasses the former *Enterobacter agglomerans*, *Erwinia herbicola*, and *E. milletiae*); and *P. dispersa*. They do not grow at 44°C (13).

Pediococcus (pe · di · o · coc'cus; *coccus growing in one plane*).   These homofermentative cocci are lactic acid bacteria that exist in pairs and tetrads resulting from cell division in two planes. *P. acidilactici*, a common starter species, caused septicemia in a 53-year-old male (15). Their moles % G + C content of DNA is 34–44; they are further discussed in Chapter 16.

Proteus (pro'te · us).   These enteric gram-negative rods are aerobes that often display pleomorphism; hence, the generic name. All are motile and typically produce swarming growth on the surface of moist agar plates. They are typical of enteric bacteria in being present in the intestinal tract of humans and animals. They may be isolated from a variety of vegetable and meat products, especially those that undergo spoilage at temperatures in the mesophilic range.

Pseudomonas (pseu · do'mo · nas; *false monad*).   These gram-negative rods constitute the largest genus of bacteria that exists in fresh foods. The moles % G + C content of their DNA of 58–70 suggests that it is a heterogeneous group, and this has been verified. They are typical of soil and water bacteria and are widely distributed among foods, especially vegetables, meat, poultry, and seafood products. They are by far the most important group of bacteria that bring about the spoilage of refrigerated fresh foods since many species and strains are psychrotrophic. Some are notable by their production of water-soluble, blue-green pigments, while many other food spoilage types do not.

Psychrobacter (psy·chro' bac·ter).   This genus was created primarily to accommodate some of the nonmotile gram-negative rods that were once classified in the genera *Acinetobacter* and *Moraxella*. They are plump coccobacilli that occur often in pairs. Also, they are aerobic, nonmotile, and catalase and oxidase positive, and generally they do not ferment glucose. Growth occurs in 6.5% NaCl and at 1°C, but generally not at 35°C or 37°C. They hydrolyze between 80, and most are egg-yolk positive (lecithinase). They are sensitive to penicillin and utilize γ-aminovalerate, while the acinetobacters do not. They are distinguished from the acinetobacters by being oxidase positive and amino-valerate users and from nonmotile pseudomonads by their inability to utilize glycerol or fructose. Because they closely resemble the moraxellae, they have been placed in the family Neisseriaceae. The genus contains some of the former achromobacters and moraxellae, as noted. They are common on meats, poultry, and fish, and in waters (18, 28). One report has been made of a nosocomial ocular infection in a 12-day-old infant (14).

Salmonella (sal·mon·el'la).   All members of this genus of gram-negative enteric bacteria are considered to be human pathogens, and they are the subject of Chapter 22. The moles % G + C content of DNA is 50–53.

Serratia (ser·ra'ti·a).   These gram-negative rods that belong to the family Enterobacteriaceae are aerobic and proteolytic, and they generally produce red pigments on culture media and in certain foods, although nonpigmented strains are not uncommon. *S. liquefaciens* is the most prevalent of the food-borne species; it causes spoilage of refrigerated vegetables and meat products. The moles % G + C content of DNA is 53–59.

Shewanella (she·wa·nel'la).   The bacteria once classified as *Pseudomonas putrefaciens* and later as *Alteromonas putrefaciens* have been placed in this new genus as *S. putrefaciens*. They are gram-negative, straight or curved rods, nonpigmented, and motile by polar flagella. They are oxidase positive and have a moles % G + C of 44–47. The other three species in this genus are *S. hanedai*, *S. benthica*, and *S. colwelliana*. All are associated with aquatic or marine habitats, and the growth of *S. benthica* is enhanced by hydrostatic pressure (8, 24).

Shigella (shi·gel'la).   All members of this genus are presumed to be human enteropathogens; they are further discussed in Chapter 22.

Staphylococcus (staph·y·lo·coc'cus; *grapelike coccus*).   These gram-positive, catalase-positive cocci include *S. aureus*, which causes several disease syndromes in humans, including foodborne gastroenteritis. It and other members of the genus are discussed further in Chapter 19.

Vagococcus (vago·coccus; *wandering coccus*).   This genus was created to accommodate the group N lactococci based upon 16S sequence data (5). They are motile by peritrichous flagella, gram positive, catalase negative, and grow at 10°C but not at 45°C. They grow in 4% NaCl but not 6.5%, and no growth occurs at pH 9.6. The cell wall peptidoglycan is Lys-D-Asp, and the moles % G + C is 33.6%. At least one species produces H_2S. They are found on fish, in feces, and in water and may be expected to occur on other foods (5, 36). Information on the phylogenetic relationship of the vagococci to other related genera is presented in Chapter 21 (Fig. 21-1).

Vibrio (vib'ri · o).   These gram-negative straight or curved rods are members of the family Vibrionaceae. Several former species have been transferred to the genus *Listonella* (24). Several species cause gastroenteritis and other human illness; they are discussed in Chapter 23. The moles % G + C content of DNA is 38–51. (See ref. 7 for environmental distribution.)

Yersinia (yer · si'ni · a).   This genus includes the agent of human plague, *Y. pestis*, and at least one species that causes foodborne gastroenteritis, *Y. enterocolitica*. All foodborne species are discussed in Chapter 23. The moles % G + C content of DNA is 45.8–46.8. The sorbose positive biogroup 3A strains have been elevated to species status as *Y. mollaretti* and the sorbose negative strains as *Y. bercovieri* (37).

SYNOPSIS OF COMMON GENERA OF FOODBORNE MOLDS

Molds are filamentous fungi that grow in the form of a tangled mass that spreads rapidly and may cover several inches of area in two to three days. The total of the mass or any large portion of it is referred to as **mycelium**. Mycelium is composed of branches or filaments referred to as **hyphae**. Those of greatest importance in foods multiply by ascospores, zygospores, or conidia. The **ascospores** of some genera are notable for their extreme degrees of heat resistance. One group forms pycnidia or acervuli (small, flask-shaped, fruiting bodies lined with conidiophores). **Arthrospores** result from the fragmentation of hyphae in some groups.

There were no radical changes in the systematics of foodborne fungi during the 1980s. The most notable changes involve the discovery of the sexual or perfect states of some well-known genera and species. In this regard, the **ascomycete** state is believed by mycologists to be the more important reproductive state of a fungus, and this state is referred to as the **teleomorph**. The species name given to a teleomorph takes precedence over that for the **anamorph**, the imperfect or conidial state. **Holomorph** indicates that both states are known, but the teleomorph name is used. (More information on foodborne molds can be obtained from refs. 1 and 26.)

The taxonomic positions of the genera described are summarized below. (Consult refs. 1 and 26 for identifications; see ref. 17 for the types that exist in meats.)

Division: Zygomycota

    Class: Zygomycetes (nonseptate mycelium, reproduction by sporangiospores, rapid growth)

Order: Mucorales
   Family: Mucoraceae
      Genus: *Mucor*
         Rhizopus
         Thamnidium

Divison: Ascomycota
   Class: Plectomycetes (septate mycelium, ascospores produced
   in asci usually number 8)
      Order: Eurotiales
         Family: Trichocomaceae
            Genus: *Byssochlamys*
               Eupenicillium
               Emericella
               Eurotium

Division: Deuteromycota (the "imperfects", anamorphs; perfect stages are
unknown)
   Class: Coelomycetes
      Genus: *Colletotrichum*
   Class: Hypomycetes (hyphae give rise to conidia)
      Order: Hyphomycetales
         Family: Moniliaceae
            Genus: *Alternaria*
               Aspergillus
               Aureobasidium (*Pullularia*)
               Botrytis
               Cladosporium
               Fusarium
               Geotrichum
               Helminthosporium
               Monilia
               Penicillium
               Stachybotrys
               Trichothecium

The genera are listed below in alphabetical order.

Alternaria.   Septate mycelia with conidiophores and large brown conidia are
produced. The conidia have both cross and longitudinal septa and are variously
shaped (Fig. 2-2*A*). They cause brown to black rots of stone fruits, apples,
and figs. Stem-end rot and black rot of citrus fruits are also caused by species/
strains of this genus. This is a field fungus, which grows on wheat. Additionally,
it is found on red meats. Some species produce mycotoxins (see Chap. 25).

Aspergillus.   Chains of conidia are produced (Fig. 2-2*B*). Where cleistothecia
with ascospores are developed, the perfect state of those found in foods is
Emericella, *Eurotium*, or *Neosartorya*. *Eurotium* (the former *A. glaucus* group)
produces bright-yellow cleistothecia, and all species are xerophilic. *E. herbariorum* has been found to cause spoilage of grape jams and jellies (31).

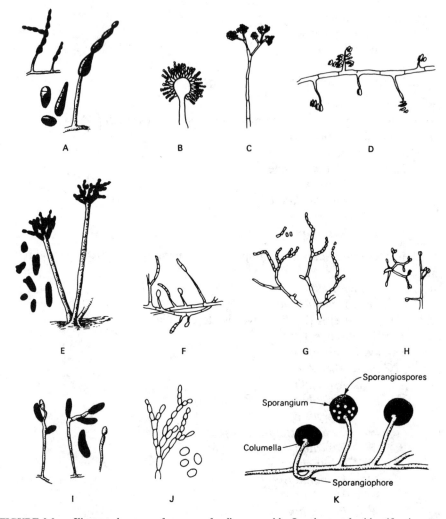

FIGURE 2-2.   Illustrated genera of common foodborne molds. See the text for identification.

Emericella produces white cleistothecia, and *E. nidulans* is the teleomorph of *Aspergillus nidulans*. *Neosartorya* produces white cleistothecia and colorless ascospores. *N. fischeri* is heat resistant, and resistance of its spores is similar to those of *Byssochlamys* (26).

The aspergilli appear yellow to green to black on a large number of foods. Black rot of peaches, citrus fruits, and figs is one of the fruit spoilage conditions produced. They are found on country-cured hams and on bacon. Some species cause spoilage of oils, such as palm, peanut, and corn. *A. oryzae* and *A. soyae* are involved in the shogu fermentation and the former in koji. *A. glaucus* produces katsuobushi, a fermented fish product. The *A. glaucus–A. restrictus* group contains storage fungi that invade seeds, soybeans, and common beans. *A. niger* produces β-galactosidase, glucoamylase, invertase, lipase, and pectinase, and *A. oryzae* produces α-amylase. Two species produce aflatoxins, and others produce ochratoxin A and sterigmatocystin (see Chap. 25).

Aureobasidium (*Pullularia*).   Yeastlike colonies are produced initially. They later spread and produce black patches. *A. pullulans* (*Pullularia pullulans*) is the most prevalent in foods. They are found in shrimp, are involved in the "black spot" condition of long-term-stored beef, and are common in fruits and vegetables.

Botrytis.   Long, slender, and often pigmented conidiophores are produced (Fig. 2-2C). Mycelium is septate; conidia are borne on apical cells and are gray in color, although black, irregular sclerotia are sometimes produced. *B. cinerea* is the most common in foods. They are notable as the cause of gray mold rot of apples, pears, raspberries, strawberries, grapes, blueberries, citrus, and some stone fruits (see Chap. 8).

Byssochlamys.   This genus is the teleomorph of certain species of *Paecilomyces*, but the latter does not occur in foods (26). The ascomycete *Byssochlamys* produces open clusters of asci, each of which contains eight ascospores. The latter are notable in that they are heat resistant, resulting in spoilage of some high-acid canned foods. In their growth, they can tolerate low Eh values. Some are pectinase producers, and *B. fulva* and *B. nivea* spoil canned and bottled fruits. These organisms are almost uniquely associated with food spoilage, and *B. fulva* (Fig. 2-3F) possesses a thermal D value at 90°C between 1 and 12 min with a z value of 6–7°C (26).

Cladosporium.   Septate hyphae with dark, treelike budding conidia variously branched, characterize this genus (Fig. 2-2E). In culture, growth is velvety and olive colored to black. Some conidia are lemon shaped. *C. herbarum* produces "black spot" on beef and frozen mutton. Some spoil butter and margarine, and some cause restricted rot of stone fruits and black rot of grapes. They are field fungi that grow on barley and wheat grains. *C. herbarum* and *C. cladosporiodes* are the two most prevalent on fruits and vegetables.

Colletotrichum.   They belong to the class Coelomycetes and form conidia inside acervuli (Fig. 2-3G). Simple but elongate conidiophores and hyaline conidia that are one celled, ovoid, or oblong are produced. The acervuli are disc or cushion shaped, waxy, and generally dark in color. *C. gloeosporioides* is the species of concern in foods; it produces anthracnose (brown/black spots) on some fruits, especially tropical fruits such as mangos and papayas.

Fusarium.   Extensive mycelium is produced that is cottony with tinges of pink, red, purple, or brown. Septate fusiform to sickle-shaped conidia (macro-conidia) are produced (Fig. 2-2F). They cause brown rot of citrus fruits and pineapples and soft rot of figs. As field fungi, some grow on barley and wheat grains. Some species produce zearalenone and trichothecenes (see Chap. 25).

Geotrichum (once known as *Oidium lactis* and *Oospora lactis*).   These yeast-like fungi are usually white. The hyphae are septate, and reproduction occurs by formation of arthroconidia from vegetative hyphae. The arthroconidia have flattened ends. *G. candidum*, the anamorph of *Dipodascus geotrichum*, is the most important species in foods. It is variously referred to as dairy mold, since it imparts flavor and aroma to many types of cheese, and as machinery mold, since it builds up on food-contact equipment in food-processing plants, especially tomato canning plants. *G. albidum* is shown in Figure 2-2*G*. They cause sour rot of citrus fruits and peaches and the spoilage of dairy cream. They are widespread and have been found on meats and many vegetables. Some participate in the fermentation of gari.

Monilia.   Pink, gray, or tan conidia are produced. *M. sitophila* is the conidial stage of *Neurospora intermedia*. *Monilia* is the conidial state of *Monilinia fructicola*. *M. americana* is depicted in Figure 2-2*J*. They produce brown rot of stone fruits such as peaches. *Monilina* sp. causes mummification of blueberries.

Mucor.   Nonseptate hyphae are produced that give rise to sporangiophores that bear columella with a sporangium at the apex (Fig. 2-2*K*). No rhizoids or stolons are produced by members of this large genus. Cottony colonies are often produced. The conditions described as "whiskers" of beef and "black spot" of frozen mutton are caused by some species. At least one species, *M. miehei*, is a lipase producer. It is found in fermented foods, bacon, and many vegetables. One species ferments soybean whey curd.

Penicillium.   When conidiophores and conidia are the only reproductive structures present, this genus is placed in the Deuteromycota. They are placed with the ascomycetes when cleistochecia with ascospores are formed as either *Talaromyces* or *Eupenicillium*. Of the two teleomorphic genera, *Talaromyces* is

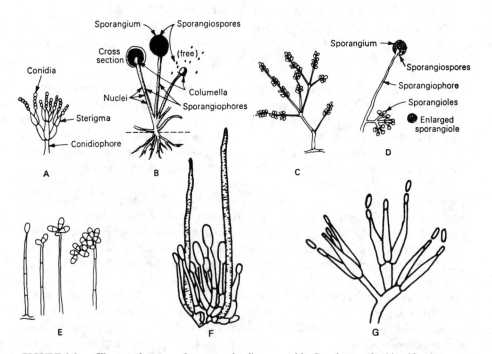

FIGURE 2-3.   Illustrated genera of common foodborne molds. See the text for identification.

the more important in foods (26). *T. flavus* is the teleomorph of *P. dangeardii*, and it has been involved in the spoilage of fruit juice concentrates (19). It produces heat-resistant spores.

When conidia are formed in the penicillus, they pinch off from **phialides** (Fig. 2-3*A*). Typical colors on foods are blue to blue-green. Blue and green mold rots of citrus fruits and blue mold rot of apples, grapes, pears, and stone fruits are caused by some species. One species, *P. roqueforti*, produces blue cheese. Some species produce citrinin, yellow rice toxin, ochratoxin A, rubra-toxin B, and other mycotoxins (see Chap. 25).

Rhizopus.   Nonseptate hyphae are produced that give rise to stolons and rhizoids. Sporangiophores typically develop in clusters from ends of stolons at the point of origin of rhizoids. (Fig. 2-3*B*). *R. stolonifer* is by far the most common species in foods. Sometimes referred to as "bread molds," they produce watery soft rot of apples, pears, stone fruits, grapes, figs, and others. Some cause "black spot" of beef and frozen mutton. They may be found on bacon and other refrigerated meats. Some produce pectinases, and *R. oligosporus* is important in the production of oncom, bonkrek, and tempeh.

Thamnidium.   These molds produce small sporangia borne on highly branched structures (Fig. 2-3*D*). *T. elegans* is the only species, and it is best known for its growth on refrigerated beef hindquarters where its characteristic growth is described as "whiskers." It is less often found in decaying eggs.

Trichothecium.   Septate hyphae that bear long, slender, and simple conidio-phores are produced (Fig. 2-3*E*). *T. roseum* is the only species, and it is pink and causes pink rot of fruits. It also causes soft rot of cucurbits and is common on barley, wheat, corn, and pecans. Some produce mycotoxins (see Chap. 25).

Other Molds.   Two categories of organisms are presented here, with the first being some miscellaneous genera that are found in some foods but are generally not regarded as significant. These are *Cephalosporium*, *Diplodia*, and *Neurospora*. *Cephalosporium* is a deuteromycete often found on frozen foods (Fig. 2-2*D*). The microspores of some *Fusarium* species are similar to those of this genus. *Diplodia* is another deuteromycete that causes stem-end rot of citrus fruits and water tan-rot of peaches. *Neurospora* is an ascomycete, and *N. intermedia* is referred to as the "red bread" mold. *Monilia sitophila* is the anamorph of *N. intermedia*. The latter is important in the oncom fermen-tation and has been found on meats.

The second category consists of xerophilic molds, which are very impor-tant as spoilage organisms. In addition to *Aspergillus* and *Eurotium*, Pitt and Hocking (26) include six other genera among the xerophiles: *Basipetospora*, *Chrysosporium*, *Eremascus*, *Polypaecilum*, *Wallemia*, and *Xeromyces*. These molds are characterized by the ability to grow below a_w 0.85. They are of significance in foods that owe their preservation to low a_w. Only *Wallemia* and *Xeromyces* are discussed further here.

Wallemia produces deep-brown colonies on culture media and on foods. *W. sebi* (formerly *Sporendonema*), the most notable species, can grow at an a_w of 0.69. It produces the "dun" mold condition on dried and salted fish.

Xeromyces has only one species, *X. bisporus*. It produces colorless cleistothecia with evanescent asci that contain two ascospores. This organism has the
lowest a_w growth of any other known organisms (26). Its a_w high is <0.97,
optimum is 0.88, and minimum is 0.61. Its thermal D at 82.2°C is 2.3 min. It
causes problems in licorice, prunes, chocolate, syrup, and other similar types
of products.

SYNOPSIS OF COMMON GENERA OF FOODBORNE YEASTS

Yeasts may be viewed as being unicellular fungi in contrast to the molds, which
are multicellular; however, this is not a precise definition, since many of what
are commonly regarded as yeasts actually produce mycelia to varying degrees.

Yeasts can be differentiated from bacteria by their larger cell size and their
oval, elongate, elliptical, or spherical cell shapes. Typical yeast cells range
from 5 to 8 µm in diameter, with some being even larger. Older yeast cultures
tend to have smaller cells. Most of those of importance in foods divide by
budding or fission.

Yeasts can grow over wide ranges of acid pH and in up to 18% ethanol.
Many grow in the presence of 55–60% sucrose. Many colors are produced by
yeasts, ranging from creamy to pink to red. The **asco-** and **arthrospores** of
some are quite heat resistant. (Arthrospores are produced by some yeastlike
fungi.)

Regarding the taxonomy of yeasts, newer methods have been employed in
the past decade or so consisting of 5S rRNA, DNA base composition, and
coenzyme Q profiles. Because of the larger genome size of yeasts, 5S rRNA
sequence analyses are employed more than for larger RNA fractions. Many
changes have occurred in yeast systematics, due in part to the use of newer
methods but also to what appears to be a philosophy toward grouping rather
than splitting taxa. The most authoritative work on yeast systematics is that
edited by Kreger-van Rij and published in 1984 (21). In this volume, the
former *Torulopsis* genus has been transferred to the genus *Candida*, and
some of the former *Saccharomyces* have been transferred to *Torulaspora* and
Zygosaccharomyces. The teleomorphic or perfect states of more yeasts are
now known, and this makes references to the older literature more difficult.

The taxonomy of 14 foodborne genera is summarized below. For excellent
discussions on foodborne yeasts, the publications by Deak and Beuchat (9),
Beneke and Stevenson (1), and Pitt and Hocking (26) should be consulted. For
identification, Deak and Beuchat (9) have presented an excellent simplified
key to foodborne yeasts. See ref. 11 for those present in dairy products, and
ref. 17 for the species found in meats.

Division: Ascomycotina
   Family: Saccharomycetaceae (ascospores and arthrospores formed;
   vegetative reproduction by fission or budding)
      Subfamily: Nadsonioideae
         Genus: *Hanseniaspora*
      Subfamily: Saccharomycotoideae

            Genus: *Debaryomyces*
                   Issatchenkia
                   Kluyveromyces
                   Pichia
                   Saccharomyces
                   Torulaspora
                   Zygosaccharomyces
        Subfamily: Schizosaccharomycetoideae
            Genus: *Schizosaccharomyces*

Division: Deuteromycotina
    Family: Cryptococcaceae (the "imperfects"; reproduce by budding)
        Genus: *Brettanomyces*
               Candida
               Cryptococcus
               Rhodotorula
               Trichosporon

The above genera are listed below in alphabetical order.

Brettanomyces.   These asporogenous yeasts form ogival cells, terminal budding, and produce acetic acid from glucose only under aerobic conditions. *B. intermedius* is the most prevalent, and it can grow at a pH as low as 1.8. They cause spoilage of beer, wine, soft drinks, and pickles, and some are involved in after-fermentation of some beers and ales.

Candida.   This genus was erected in 1923 by Berkhout and has since undergone many changes in definition and composition (see 34). It is regarded as being a heterogenous taxon that can be divided into 40 segments comprising three main groups, based mainly on fatty acid composition and electrophoretic karotyping (35). The generic name means "shining white," and cells contain no carotenoid pigments.

   The ascomycetous imperfect species are placed here, including the former genus *Torulopsis* as follows:

Candida famata (*Torulopsis candida*; *T. famata*)
Candida kefyr (*Candida pseudotropicalis*, *T. kefyr*; *Torula cremoris*)
Candida stellata (*Torulopsis stellata*)
Candida holmii (*Torulopsis holmii*)

Many of the **anamorphic** forms of *Candida* are now in the genera *Kluyveromyces* and *Pichia* (see 9). *Candida lipolytica* is the anamorph of *Saccharomycopsis lipolytica*.

   Members of this genus are the most common yeasts in fresh ground beef and poultry, and *C. tropicalis* is the most prevalent in foods in general. Some members are involved in the fermentation of cacao beans, as a component of kefir grains, and in many other products, including beers, ales, and fruit juices.

Cryptococcus.   This genus represents the anamorph of *Filobasidiella* and other Basidiomycetes. They are asporogenous, reproduce by multilateral budding, and are nonfermenters of sugars. They are hyaline and red or orange, and they may form arthrospores. They have been found on plants and in soils, strawberries and other fruits, marine fish, shrimp, and fresh ground beef.

Debaryomyces.   These ascosporogenous yeasts sometimes produce a pseudomycelium and reproduce by multilateral budding. They are one of the two most prevalent yeast genera in dairy products. *D. hansenii* represents what was once *D. subglobosus* and *Torulaspora hansenii*, and it is the most prevalent foodborne species. It can grow in 24% NaCl and at an a_w as low as 0.65. It forms slime on wieners, grows in brines and on cheeses, and causes spoilage of orange juice concentrate and yogurt.

Hanseniaspora.   These are apiculate yeasts whose anamorphs are *Kloeckera* spp. They exhibit bipolar budding, and consequently lemon-shaped cells are produced. The asci contain two to four hat-shaped spores. Sugars are fermented, and they can be found on a variety of foods, especially figs, tomatoes, strawberries, and citrus fruits, and the cacao bean fermentation.

Issatchenkia.   Members of this genus produce pseudomycelia and multiply by multilateral budding. Some species once in the genus *Pichia* have been placed here. The teleomorph of *Candida krusei* is *I. orientalis*. They typically form pellicles in liquid media. They contain coenzyme Q-7 and are prevalent on a wide variety of foods.

Kluyveromyces (Fabospora).   These ascosporeforming yeasts reproduce by multilateral budding, and the spores are spherical. *K. marxianus* now includes the former *K. fragilis*, *K. lactis*, *K. bulgaricus*, *Saccharomyces lactis*, and *S. fragilis*. *K. marxianus* is one of the two most prevalent yeasts in dairy products. *Kluyveromyces* spp. produce β-galactosidase and are vigorous fermenters of sugars, including lactose. *K. marxianus* contains coenzyme Q-6 and is involved in the fermentation of kumiss. It is also used for lactase production from whey and as the organism of choice for producing yeast cells from whey. They are found on a wide variety of fruits, and *K. marxianus* causes cheese spoilage.

Pichia.   This is the largest genus of true yeasts. They reproduce by multilateral budding, and the asci usually contain four spheroidal, hat- or saturn-shaped spores. Pseudomycelia and arthrospores may be formed. Some of the hat-shaped sporeformers may be *Williopsis* spp., and some of the former species are now classified in the genus *Debaryomyces*. *P. guilliermondii* is the perfect state of *Candida guilliermondii*. The anamorph of *P. membranaefaciens* is *Candida valida*. *Pichia* spp. typically form films on liquid media and are known to be important in producing indigenous foods in various parts of the world. Some have been found on fresh fish and shrimp, and they are known to grow in olive brines and to cause spoilage of pickles and sauerkraut.

Rhodotorula.   These yeasts are anamorphs of Basidiomycetes. The teliospore producers are in the genus *Rhodosporidium*. They reproduce by multilateral budding and are nonfermenters. *R. glutinis* and *R. mucilaginosa* are the two most prevalent species in foods. They produce pink to red pigments, and most are orange or salmon-pink in color. The genus contains many psychrotrophic species/strains that are found on fresh poultry, shrimp, fish, and beef. Some grow on the surface of butter.

Saccharomyces.   These ascoporogenous yeasts multiply by multilateral budding and produce spherical spores in asci. They are diploid and do not ferment lactose. Those once classified as *S. bisporus* and *S. rouxii* are now in the genus *Zygosaccharomyces*, and the former *S. rosei* is now in the genus *Torulaspora*. All bakers', brewers', wine, and champagne yeasts are *S. cerevisiae*. They are found in kefir grains and can be isolated from a wide range of foods, such as dry-cured salami and numerous fruits, although *S. cerevisiae* rarely causes spoilage. *S. bailii* is a spoilage yeast and is well known for its activities in tomato sauce, mayonnaise, salad dressing, soft drinks, fruit juices, ciders, and wines. It is resistant to benzoate and sorbate preservatives.

Schizosaccharomyces.   These ascosporogenous yeasts divide by lateral fission or cross-wall formation and may produce true hyphae and arthrospores. Asci contain from four to eight bean-shaped spores, and no buds are produced. They are regarded as being only distantly related to the true yeasts. *S. pombe* is the most prevalent species; it is osmophilic and resistant to some chemical preservatives.

Torulaspora.   Multilateral budding is the method of reproduction with spherical spores in asci. Three haploid species formerly in the genus *Saccharomyces* are now in this genus. They are strong fermenters of sugars, and contain coenzyme Q-6. *T. delbrueckii* is the most prevalent species.

Trichosporon.   These nonascosporeforming oxidative yeasts multiply by budding and by arthroconidia formation. They produce a true mycelium, and sugar fermentation is absent or weak. They are involved in cacao bean and idli fermentations and have been recovered from fresh shrimp, ground beef, poultry, frozen lamb, and other foods. *T. pullulans* is the most prevalent species, and it produces lipase.

Zygosaccharomyces.   Multilateral budding is the method of reproduction, and the bean-shaped ascospores formed are generally free in asci. Most are haploid and they are strong fermenters of sugars. *Z. rouxii* is the most prevalent species, and it can grow at a_w of 0.62, second only to *Xeromyces bisporus* in ability to grow at low a_w (26). Some are involved in shoyu and miso fermentations, and some are common spoilers of mayonnaise and salad dressing, especially *Z. bailii*, which can grow at pH 1.8 (26).

References
1. Beneke, E. S., and K. E. Stevenson. 1987. Classification of food and beverage fungi. In *Food and Beverage Mycology*, 2d ed., ed. L. R. Beuchat, 1–50. New York: Van Nostrand Reinhold.
2. Beuchat, L. R., ed. 1987. *Food and Beverage Mycology*, 2d ed. New York: Van Nostrand Reinhold.
3. Champomier, M.-C., M.-C. Montel, and R. Talon. 1989. Nucleic acid relatedness

studies on the genus *Carnobacterium* and related taxa. *J. Gen. Microbiol.* 135: 1391–1394.

4. Collins, M. D., U. Rodrigues, C. Ash, M. Aguirre, J. A. E. Farrow, A. Martinez-Murcia, B. A. Phillips, A. M. Williams, and S. Wallbanks. 1991. Phylogenetic analysis of the genus *Lactobacillus* and related lactic acid bacteria as determined by reverse transcriptase sequencing of 16S rRNA. *FEMS Microbiol. Lett.* 77:5–12.

5. Collins, M. D., C. Ash, J. A. E. Farrow, S. Wallbanks, and A. M. Williams. 1989. 16S ribosomal ribonucleic acid sequence analyses of lactococci and related taxa. Description of *Vagococcus fluvialis* gen. nov., sp. nov. *J. Appl. Bacteriol.* 67:453–460.

6. Collins, M. D., J. A. E. Farrow, B. A. Phillips, S. Ferusu, and D. Jones. 1987. Classification of *Lactobacillus divergens*, *Lactobacillus piscicola*, and some catalase-negative, asporogenous, rod-shaped bacteria from poultry in a new genus, *Carnobacterium*. *Int. J. System. Bacteriol.* 37:310–316.

7. Colwell, R. R., ed. 1984. *Vibrios in the Environment.* New York: Wiley.

8. Coyne, V. E., C. J. Pillidge, D. D. Sledjeski, H. Hori, B. A. Ortiz-Conde, D. G. Muir, R. M. Weiner, and R. R. Colwell. 1989. Reclassification of *Alteromonas colwelliana* to the genus *Shewanella* by DNA-DNA hybridization, serology and 5S ribosomal RNA sequence data. *System. Appl. Microbiol.* 12:275–279.

9. Deak, T., and L. R. Beuchat. 1987. Identification of foodborne yeasts. *J. Food Protect.* 50:243–264.

10. Dicks, L. M. T., H. J. J. van Vuuren, and F. Dellaglio. 1990. Taxonomy of *Leuconostoc* species, particularly *Leuconostoc oenos*, as revealed by numerical analysis of total soluble cell protein patterns, DNA base compositions, and DNA-DNA hybridizations. *Int. J. System. Bacteriol.* 40:83–91.

11. Fleet, G. H. 1990. Yeasts in dairy products. *J. Appl. Bacteriol.* 68:199–211.

12. Garvie, E. I., J. A. E. Farrow, and B. A. Phillips. 1981. A taxonomic study of some strains of streptococci which grow at 10°C but not at 45°C, including *Streptococcus lactis* and *Streptococcus cremoris*. *Zentralbl. Bakteriol. Parasitenkd. Infektionskr. Hyg. Abt. I Orig. Reihe* C2:151–165.

13. Gavini, F., J. Mergaert, A. Beji, C. Mielcarek, D. Izard, K. Kersters, and J. de Ley. 1989. Transfer of *Enterobacter agglomerans* (Beijerinck 1888) Ewing and Fife 1972 to *Pantoea* gen. nov. as *Pantoea agglomerans* comb. nov. and description of *Pantoea dispersa* sp. nov. *Int. J. Syst. Bacteriol.* 39:337–345.

14. Gini, G. A. 1990. Ocular infection caused by *Psychrobacter immobilis* acquired in the hospital. *J. Clin. Microbiol.* 28:400–401.

15. Golledge, C. L., N. Stingemore, M. Aravena, and D. Joske. 1990. Septicemia caused by vancomycin-resistant *Pediococcus acidilactici*. *J. Clin. Microbiol.* 28: 1678–1679.

16. Holzapfel, W. H., and E. S. Gerber. 1983. *Lactobacillus divergens* sp. nov., a new heterofermentative *Lactobacillus* species producing L(+)-lactate. *System. Appl. Microbiol.* 4:522–534.

17. Jay, J. M. 1987. (Fungi in) meats, poultry, and seafoods. In *Food and Beverage Mycology*, 2d ed., ed. L. R. Beuchat, 155–173. New York: Van Nostrand Reinhold.

18. Juni, E., and G. A. Heym. 1986. *Psychrobacter immobilis* gen. nov., sp. nov.: Genospecies composed of gram-negative, aerobic, oxidase-positive coccobacilli. *Int. J. System. Bacteriol.* 36:388–391.

19. King, A. D., Jr., and W. U. Halbrook. 1987. Ascospore heat resistance and control measures for *Talaromyces flavus* isolated from fruit juice concentrate. *J. Food Sci.* 52:1252–1254, 1266.

20. Kleynmans, U., H. Heinzl, and W. P. Hammes. 1989. *Lactobacillus suebicus* sp. nov., an obligately heterofermentative *Lactobacillus* species isolated from fruit mashes. *System. Appl. Bacteriol.* 11:267–271.

21. Kreger-van Rij, N. J. W., ed. 1984. *The Yeasts, A Taxonomic Study*. Amsterdam: Elsevier.

22. Krieg, N. R., and J. G. Holt, eds. 1984. *Bergey's Manul of Systematic Bacteriol*, vol. 1. Baltimore: Williams & Wilkins.

23. Lane, D. J., B. Pace, G. J. Olsen, D. A. Stahl, M. L. Sogin, and N. R. Pace. 1985. Rapid determination of 16S ribosomal RNA sequences for phylogenetic analyses. *Proc. Nat'l. Acad. Sci. USA* 82:6955–6959.

24. MacDonell, M. T., and R. R. Colwell. 1985. Phylogeny of the Vibrionaceae, and recommendation for two new genera, *Listonella* and *Shewanella*. *System. Appl. Microbiol.* 6:171–182.

25. Pelczar, M. J., Jr., and R. D. Reid. 1965. *Microbiology*, 2d ed., ch. 36, app. B. New York: McGraw-Hill.

26. Pitt, J. I., and A. D. Hocking. 1985. *Fungi and Food Spoilage*. New York: Academic Press.

27. Schleifer, K. H., J. Kraus, C. Dvorak, R. Kilpper-Balz, M. D. Collins, and W. Fischer. 1985. Transfer of *Streptococcus lactis* and related streptococci to the genus *Lactococcus* gen. nov. *System. Appl. Microbiol.* 6:183–195.

28. Shaw, B. G., and J. B. Latty. 1988. A numerical taxonomic study of non-motile non-fermentative gram-negative bacteria from foods. *J. Appl. Bacteriol.* 65:7–21.

29. Smibert, R. M. 1978. The genus *Campylobacter*. *Ann. Rev. Microbiol.* 32:673–709.

30. Sneath, P. H. A., and D. Jones. 1976. *Brochothrix*, a new genus tentatively placed in the family Lactobacteriaceae. *Int. J. Syst. Bacteriol.* 26:102–104.

31. Splittstoesser, D. F., J. M. Lammers, D. L. Downing, and J. J. Churey. 1989. Heat resistance of *Eurotium herbariorum*, a xerophilic mold. *J. Food Sci.* 54:683–685.

32. Talon, R., P. A. D. Grimont, F. Grimont, F. Gasser, and J. M. Boeufgras. 1988. *Brochothrix campestris* sp. nov. *Int. J. System. Bacteriol.* 38:99–102.

33. Vandamme, P., E. Falsen, R. Rossau, B. Hoste, P. Segers, R. Tytgat, and J. deLey. 1991. Revision of *Campylobacter*, *Helicobacter*, and *Wolinella* taxonomy emendation of generic descriptions and proposal of *Arcobacter* gen. nov. *Int. J. System. Bacteriol.* 41:88–103.

34. Viljoen, B. C., and J. L. F. Kock. 1989a. The genus *Candida* Berkhout nom. conserv.—A historical account of its delimitation. *System. Appl. Microbiol.* 12:183–190.

35. Viljoen, B. C., and J. L. F. Kock. 1989b. Taxonomic study of the yeast genus *Candida* Berkhout. *System. Appl. Microbiol.* 12:91–102.

36. Wallbanks, S., A. J. Martinez-Murcia, J. L. Fryer, B. A. Phillips, and M. D. Collins. 1990. 16S rRNA sequence determination for members of the genus *Carnobacterium* and related lactic acid bacteria and description of *Vagococcus salmoninarum* sp. nov. *Int. J. Syst. Bacteriol.* 40:224–230.

37. Wauters, G., M. Janssens, A. G. Steigerwalt, and D. J. Brenner. 1988. *Yersinia mollaretti* sp. nov. and *Yersinia bercovieri* sp. nov., formerly called *Yersinia enterocolitica* biogroups 3A and 3B. *Int. J. System. Bacteriol.* 38:424–429.

38. Wayne, L. G., D. J. Brenner, R. R. Colwell, P. A. D. Grimont, O. Kandler, M. I. Krichevsky, L. H. Moore, W. E. C. Moore, R. G. E. Murray, E. Stackebrandt, M. P. Starr, and H. G. Truper. 1987. Report of the ad hoc committee on reconciliation of approaches to bacterial systematics. *Int. J. System. Bacteriol.* 37:463–464.

39. Woese, C. R. 1987. Bacterial evolution. *Microbiol. Rev.* 51:221–271.

40. Yang, D., and C. R. Woese. 1989. Phylogenetic structure of the "Leuconostocs": An interesting case of a rapidly evolving organism. *System. Appl. Microbiol.* 12: 145–149.

3

Intrinsic and Extrinsic Parameters of Foods That Affect Microbial Growth

Since our foods are of plant and animal origin, it is worthwhile to consider those characteristics of plant and animal tissues that affect the growth of microorganisms. The plants and animals that serve as food sources have all evolved mechanisms of defense against the invasion and proliferation of microorganisms, and some of these remain in effect in fresh foods. By taking these natural phenomena into account, one can make effective use of each or all in preventing or retarding the microbial spoilage of the products that are derived from them.

INTRINSIC PARAMETERS

The parameters of plant and animal tissues that are an inherent part of the tissues are referred to as **intrinsic parameters** (56). These parameters are

1. pH
2. moisture content
3. oxidation-reduction potential (Eh)
4. nutrient content
5. antimicrobial constituents
6. biological structures

Each of these is discussed below, with emphasis placed on their effects on microorganisms in foods.

pH

It has been well established that most microorganisms grow best at pH values around 7.0 (6.6–7.5), while few grow below 4.0 (Fig. 3-1). Bacteria tend to

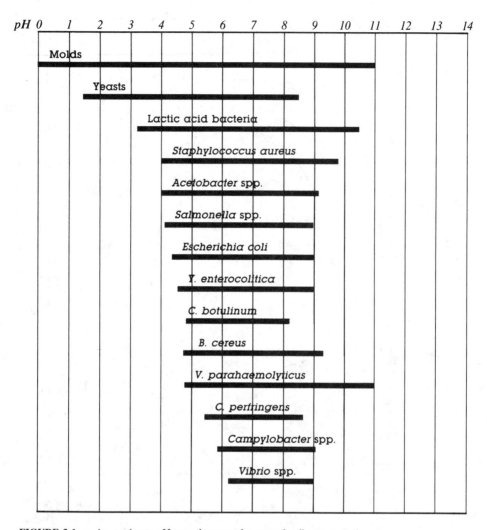

FIGURE 3-1.   Approximate pH growth ranges for some foodborne organisms.

be more fastidious in their relationships to pH than molds and yeasts, with the pathogenic bacteria being the most fastidious. With respect to pH minima and maxima of microorganisms, those represented in Figure 3-1 should not be taken to be precise boundaries, since the actual values are known to be dependent on other growth parameters. For example, the pH minima of certain lactobacilli have been shown to be dependent on the type of acid used, with citric, hydrochloric, phosphoric, and tartaric acids permitting growth at lower pH than acetic or lactic acids (49). In the presence of 0.2 M NaCl, *Alcaligenes faecalis* has been shown to grow over a wider pH range than in the absence of NaCl or in the presence of 0.2 M sodium citrate (Fig. 3-2). Of the foods presented in Table 3-1, it can be seen that fruits, soft drinks, vinegar, and wines all fall below the point at which bacteria normally grow. The excellent keeping quality of these products is due in great part to pH. It is a common observation that fruits generally undergo mold and yeast spoilage, and this is

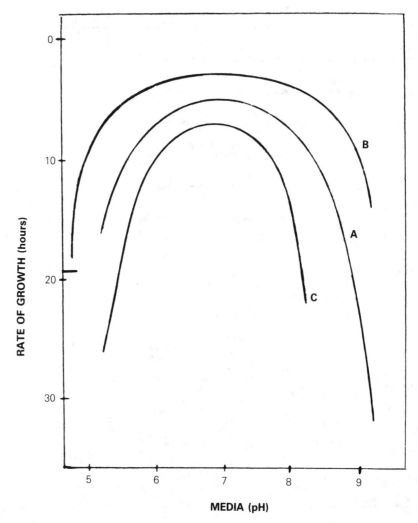

FIGURE 3-2.   Relationship of pH, NaCl, and Na citrate on the rate of growth of *Alcaligenes faecalis* in 1% peptone: A = 1% peptone; B = 1% peptone + 0.2 M NaCl; C = 1% peptone + 0.2 M Na citrate.
Redrawn from Sherman and Holm (75); used with permission of the publisher.

due to the capacity of these organisms to grow at pH values <3.5, which is considerably below the minima for most food spoilage and all food poisoning bacteria. It may be further noted from Table 3-2 that most of the meats and seafoods have a final ultimate pH of about 5.6 and above. This makes these products susceptible to bacterial as well as to mold and yeast spoilage. Most vegetables have higher pH values than fruits, and consequently vegetables should be subject more to bacterial than fungal spoilage.

With respect to the keeping quality of meats, it is well known that meat from fatigued animals spoils faster than that from rested animals, and that this is a direct consequence of final pH attained upon completion of rigor mortis. Upon the death of a well-rested meat animal, the usual 1% glycogen

TABLE 3-1.   Approximate pH Values of Some Fresh Fruits and Vegetables

Product	pH	Product	pH
Vegetables		**Fruits**	
Asparagus (buds and stalks)	5.7–6.1	Apples	2.9–3.3
Beans (string and Lima)	4.6 & 6.5	Bananas	4.5–4.7
Beets (sugar)	4.2–4.4	Figs	4.6
Broccoli	6.5	Grapefruit (juice)	3.0
Brussels sprouts	6.3	Limes	1.8–2.0
Cabbage (green)	5.4–6.0	Melons (honeydew)	6.3–6.7
Carrots	4.9–5.2; 6.0	Oranges (juice)	3.6–4.3
Cauliflower	5.6	Plums	2.8–4.6
Celery	5.7–6.0	Watermelons	5.2–5.6
Corn (sweet)	7.3	Grapes	3.4–4.5
Eggplant	4.5		
Lettuce	6.0		
Olives	3.6–3.8		
Onions (red)	5.3–5.8		
Parsley	5.7–6.0		
Parsnip	5.3		
Potatoes (tubers & sweet)	5.3–5.6		
Pumpkin	4.8–5.2		
Rhubarb	3.1–3.4		
Spinach	5.5–6.0		
Squash	5.0–5.4		
Tomatoes (whole)	4.2–4.3		
Turnips	5.2–5.5		

TABLE 3-2.   Approximate pH Values of Dairy, Meat, Poultry, and Fish Products

Product	pH	Product	pH
Dairy products		**Fish and shellfish**	
Butter	6.1–6.4	Fish (most species)[a]	6.6–6.8
Buttermilk	4.5	Clams	6.5
Milk	6.3–6.5	Crabs	7.0
Cream	6.5	Oysters	4.8–6.3
Cheese (American mild and cheddar)	4.9, 5.9	Tuna fish	5.2–6.1
		Shrimp	6.8–7.0
Meat and poultry		Salmon	6.1–6.3
		White fish	5.5
Beef (ground)	5.1–6.2		
Ham	5.9–6.1		
Veal	6.0		
Chicken	6.2–6.4		

[a] Just after death.

is converted into lactic acid, which directly causes a depression in pH values from about 7.4 to about 5.6, depending on the type of animal. Callow (17) found the lowest pH values for beef to be 5.1 and the highest 6.2 after rigor mortis. The usual pH value attained upon completion of rigor mortis of beef is around 5.6 (6). The lowest and highest values for lamb and pork, respectively, were found by Callow to be 5.4 and 6.7, and 5.3 and 6.9. Briskey (9) reported that the ultimate pH of pork may be as low as approximately 5.0 under certain conditions. The effect of pH of this magnitude upon microorganisms, especially bacteria, is obvious. With respect to fish, it has been known for some time that halibut, which usually attains an ultimate pH of about 5.6, has better keeping qualities than most other fish, whose ultimate pH values range between 6.2 and 6.6 (66).

Some foods are characterized by inherent acidity; others owe their acidity or pH to the actions of certain microorganisms. The latter type is referred to as biological acidity and is displayed by products such as fermented milks, sauerkraut, and pickles. Regardless of the source of acidity, the effect on keeping quality appears to be the same.

Some foods are better able to resist changes in pH than others. Those that tend to resist changes in pH are said to be **buffered**. In general, meats are more highly buffered than vegetables. Contributing to the buffering capacity of meats are their various proteins. Vegetables are generally low in proteins and consequently lack the buffering capacity to resist changes in their pH by the growth of microorganisms (see Table 8-1 for the general chemical composition of vegetables).

The natural or inherent acidity of foods may be thought of as nature's way of protecting the respective plant or animal tissues from destruction by microorganisms. It is of interest that fruits should have pH values below those required by many spoilage organisms. The biological function of the fruit is the protection of the plant's reproductive body, the seed. This one fact alone has no doubt been quite important in the evolution of present-day fruits. While the pH of a living animal favors the growth of most spoilage organisms, other intrinsic parameters come into play to permit the survival and growth of the animal organism.

pH Effects.   Adverse pH affects at least two aspects of a respiring microbial cell: the functioning of its enzymes and the transport of nutrients into the cell. The cytoplasmic membrane of microorganisms is relatively impermeable to H^+ and ^-OH ions. Their concentration in the cytoplasm therefore probably remains reasonably constant despite wide variations that may occur in the pH of the surrounding medium (69). The intracellular pH of resting baker's yeast cells was found by Conway and Downey (25) to be 5.8. While the outer region of the cells during glucose fermentation was found to be more acidic, the inner cell remained more alkaline. On the other hand, Peña et al. (61) did not support the notion that the pH of yeast cells remains constant with variations in pH of medium. It appears that the internal pH of almost all cells is near neutrality. Bacteria such as *Sulfolobus* and *Methanococcus* may be exceptions, however. When microorganisms are placed in environments below or above neutrality, their ability to proliferate depends on their ability to bring the environmental pH to a more optimum value or range. When placed in acid environments, the cells must either keep H^+ from entering or expel H^+ ions as rapidly as they enter. Such key cellular compounds as DNA and ATP require neutrality (10). When most microorganisms grow in acid media, their metabolic activity results in the medium or substrate's becoming less acidic, while those that grow in high pH environments tend to effect a lowering of pH. The amino acid decarboxylases that have optimum activity at around pH 4.0 and almost no activity at pH 5.5 cause a spontaneous adjustment of pH toward neutrality when cells are grown in the acid range. Bacteria such as *Clostridium acetobutylicum* raise substrate pH by reducing butyric acid to butanol, while *Enterobacter aerogenes* produces acetoin from pyruvic acid to raise the pH of its growth environment. When amino acids are decarboxylated, the increase in pH occurs from the resulting amines. When grown in the alkaline range, a group of amino acid deaminases that have optimum activity at about pH 8.0 cause the spontaneous adjustment of pH toward neutrality as a result of the organic acids that accumulate.

With respect to the transport of nutrients, the bacterial cell tends to have a residual negative charge. Nonionized compounds therefore can enter cells, while ionized cannot. At neutral or alkaline pH, organic acids do not enter, while at acid pH values, these compounds are nonionized and can enter the negatively charged cells. Also, the ionic character of side chain ionizable groups is affected on either side of neutrality, resulting in increasing denaturation of membrane and transport enzymes.

Among the other effects that are exerted on microorganisms by adverse pH is that of the interaction between H^+ and the enzymes in the cytoplasmic membrane. The morphology of some microorganisms can be affected by pH. The length of the hyphae of *Penicillium chrysogenum* has been reported to decrease when grown in continuous culture where pH values increased above 6.0. Pellets of mycelium rather than free hyphae were formed at about pH 6.7 (69). Extracellular H^+ and K^+ may be in competition where the latter stimulates fermentation, for example, while the former represses it. The metabolism of glucose by yeast cells in an acid medium was markedly stimulated by K^+ (70). Glucose was consumed 83% more rapidly in the presence of K^+ under anaerobic conditions and 69% more under aerobic conditions.

Other environmental factors interact with pH. With respect to temperature, pH of substrate becomes more acid as temperature increases. Concentration of salt has a definite effect on pH growth rate curves, as illustrated in Figure 3-2, where it can be seen that the addition of 0.2 M NaCl broadened the pH growth range of *Alcaligenes faecalis*. A similar result was noted for *Escherichia coli* by these investigators. When salt content exceeds this optimal level, the pH growth range is narrowed. An adverse pH makes cells much more sensitive to toxic agents of a wide variety, and young cells are more susceptible to pH changes than older or resting cells.

When microorganisms are grown on either side of their optimum pH range, an increased lag phase results. The increased lag would be expected to be of longer duration if the substrate is a highly buffered one in contrast to one that has poor buffering capacity. In other words, the length of the lag phase may be expected to reflect the time necessary for the organisms to bring the external environment within their optimum pH growth range. Analysis of the substances that are responsible for the adverse pH is of value in determining not only speed of subsequent growth but also minimum/maximum pH at which an organism may grow. In a study of the minimum pH at which salmonellae would initiate growth, Chung and Goepfert (23) found the minimum pH to be 4.05 when hydrochloric and citric acids were used but 5.4 and 5.5 when acetic and propionic acids were used, respectively. This is undoubtedly a reflection of the ability of the organisms to alter their external environment to a more favorable range in the case of hydrochloric and citric acids than with the other acids tested. It is also possible that factors other than pH come into play in the varying abilities of organic acids as growth inhibitors. For more information on pH and acidity, see Corlett and Brown (26).

Moisture Content

One of the oldest methods of preserving foods is drying or desiccation, and precisely how this method came to be used is not known. The preservation of foods by drying is a direct consequence of removal or binding of moisture, without which microorganisms do not grow. It is now generally accepted that the water requirements of microorganisms should be defined in terms of the **water activity** (a_w) in the environment. This parameter is defined by the ratio of the water vapor pressure of food substrate to the vapor pressure of pure water at the same temperature—$a_w = p/p_0$, where p = vapor pressure of solution and p_0 = vapor pressure of solvent (usually water). This concept is related to relative humidity (R.H.) in the following way: R.H. = $100 \times a_w$ (20). Pure water has an a_w of 1.00, a 22% NaCl solution (w/v) has a_w of 0.86, and a saturated solution of NaCl is 0.75 (Table 3-3).

The a_w of most fresh foods is above 0.99. The minimum values reported for the growth of some microorganisms in foods are presented in Table 3-4 (see also Chap. 15). In general, bacteria require higher values of a_w for growth than fungi, with gram-negative bacteria having higher requirements than gram positives. Most spoilage bacteria do not grow below a_w 0.91, while spoilage molds can grow as low as 0.80. With respect to food-poisoning bacteria, *Staphylococcus aureus* has been found to grow as low as 0.86, while *Clostridium botulinum* does not grow below 0.94. Just as yeasts and molds grow over a wider pH range than bacteria, the same is true for a_w. The lowest reported

TABLE 3-3.   Relationship Between Water Activity And Concentration of Salt Solutions

Water Activity	Sodium Chloride Concentration	
	Molal	Percent, w/v
0.995	0.15	0.9
0.99	0.30	1.7
0.98	0.61	3.5
0.96	1.20	7
0.94	1.77	10
0.92	2.31	13
0.90	2.83	16
0.88	3.33	19
0.86	3.81	22

Source: From *The Science of Meat and Meat Products*, by the American Meat Institute Foundation (36). W. H. Freeman and Company, San Francisco; copyright © 1960.

TABLE 3-4.   Approximate Minimum a_w Values for Growth of Microorganisms Important in Foods

Organisms	a_w	Organisms	a_w
Groups		**Groups**	
Most spoilage bacteria	0.9	Halophilic bacteria	0.75
Most spoilage yeasts	0.88	Xerophilic molds	0.61
Most spoilage molds	0.80	Osmophilic yeasts	0.61
Specific Organisms		**Specific Organisms**	
Clostridium botulinum, type E	0.97	*Candida scottii*	0.92
Pseudomonas spp.	0.97	*Trichosporon pullulans*	0.91
Acinetobacter spp.	0.96	*Candida zeylanoides*	0.90
Escherichia coli	0.96	*Staphylococcus aureus*	0.86
Enterobacter aerogenes	0.95	*Alternaria citri*	0.84
Bacillus subtilis	0.95	*Penicillium patulum*	0.81
Clostridium botulinum,		*Aspergillus glaucus*[a]	0.70
types A and B	0.94	*Aspergillus conicus*	0.70
Candida utilis	0.94	*Aspergillus echinulatus*	0.64
Vibrio parahaemolyticus	0.94	*Zygosaccharomyces rouxii*	0.62
Botrytis cinerea	0.93	*Xeromyces bisporus*	0.61
Rhizopus stolonifer	0.93		
Mucor spinosus	0.93		

[a]Perfect stages of the *A. glaucus* group are found in the genus *Eurotium*.

value for bacteria of any type is 0.75 for halophilic (literally, "salt-loving") bacteria, while xerophilic ("dry-loving") molds and osmophilic (preferring high osmotic pressures) yeasts have been reported to grow at a_w values of 0.65 and 0.61, respectively. When salt is employed to control a_w, an extremely high level is necessary to achieve a_w values below 0.80 (Table 3-3).

Certain relationships have been shown to exist among a_w, temperature, and nutrition. First, at any temperature, the ability of microorganisms to grow is reduced as the a_w is lowered. Second, the range of a_w over which growth occurs is greatest at the optimum temperature for growth; and third, the presence of nutrients increases the range of a_w over which the organisms can survive (54). The specific values given in Table 3-4, then, should be taken only as reference points, since a change in temperature or nutrient content might permit growth at lower values of a_w.

Effects of Low a_w.   The general effect of lowering a_w below optimum is to increase the length of the lag phase of growth and to decrease the growth rate and size of final population. This effect may be expected to result from adverse influences of lowered water on all metabolic activities since all chemical reactions of cells require an aqueous environment. It must be kept in mind, however, that a_w is influenced by other environmental parameters such as pH, temperature of growth, and Eh. In their study of the effect of a_w on the growth of *Enterobacter aerogenes* in culture media, Wodzinski and Frazier (85) found that the lag phase and generation time were progressively lengthened until no growth occurred with a lowering of a_w. The minimum a_w was raised, however, when incubation temperature was decreased. When both the pH and temperature of incubation were made unfavorable, the minimum a_w for growth was higher. The interaction of a_w, pH, and temperature on the growth of molds on jam was shown by Horner and Anagnostopoulos (45). The interaction between a_w and temperature was the most significant.

In general, the strategy employed by microorganisms as protection against osmotic stress is the intracellular accumulation of compatible solutes. Those employed by bacteria include K^+ ions, glutamate, glutamine, proline, γ-aminobutyrate, alanine, glycinebetaine, sucrose, trehalose, and glucosylglycerol. Gram negatives tend to accumulate proline by the mechanism of enhanced transport. Halotolerant and xerotolerant fungi tend to produce polyhydric alcohols, such as glycerol, erythritol, and arabitol. More specific information on this phenomenon is presented below for individual organisms and varying solutes. For more information, the thorough review by Csonka (27) should be consulted, along with reviews by Sperber (77) and Troller (80).

With regard to specific compounds used to lower water activity, results akin to those seen with adsorption and desorption systems (see Chap. 15) have been reported. In a study on the minimum a_w for the growth and germination of *Clostridium perfringens*, Kang et al. (50) found the value to be between 0.97 and 0.95 in complex media when sucrose or NaCl was used to adjust a_w but 0.93 or below when glycerol was used. In another study, glycerol was found to be more inhibitory than NaCl to relatively salt-tolerant bacteria but less inhibitory than NaCl to salt-sensitive species when compared at similar levels of a_w in complex media (52). In their studies on the germination of *Bacillus* and *Clostridium* spores, Jakobsen and Murrell (47) observed strong inhibition of spore germination when a_w was controlled by NaCl and $CaCl_2$ but less inhibition when glucose and sorbitol were used, and very little inhibition when glycerol, ethylene glycol, acetamide, or urea were used. The germination of clostridial spores was completely inhibited at a_w 0.95 with NaCl, but no inhibition occurred at the same a_w when urea, glycerol, or glucose was employed. In another study, the limiting a_w for the formation of mature spores by *B. cereus* strain T was shown to be about 0.95 for glucose, sorbitol, and NaCl but about 0.91 for glycerol (48). Both yeasts and molds have been found to be more tolerant to glycerol than to sucrose (45). Using a glucose minimal medium and *Pseudomonas fluorescens*, Prior (63) found that glycerol permitted growth at lower a_w values than either sucrose or NaCl. It was further shown by this worker that the catabolism of glucose, sodium lactate, and DL-arginine was completely inhibited by a_w values greater than the minimum for growth when a_w was controlled with NaCl. The control of a_w with glycerol allowed catabolism to continue at a_w values below that for growth on glucose. In all cases where NaCl was used by this investigator to adjust a_w, substrate catabolism ceased at an a_w greater than the minimum for growth, while glycerol permitted catabolism at lower a_w values than the minimum for growth. In spite of some reports to the contrary, it appears that glycerol is clearly less inhibitory to respiring organisms than agents such as sucrose and NaCl.

There are some definite effects of lowered a_w on microorganisms. Some bacteria accumulate proline as a response to low a_w, and increases in some "pool" amino acids have been reported to occur in some salt-tolerant *Staphylococcus aureus* strains (13). Using a defined medium, Christian (19) found that *Salmonella oranienburg* at a_w of <0.97 had the added requirement for the amino acid proline, and Christian and Waltho (22) showed that proline stimulated respiration at reduced values of a_w. With *S. aureus* MF-31 in 10% NaCl, proline was shown to accumulate by transport while glutamine accumulated via synthesis (2). As to the trigger for proline synthesis, K^+ accumulates inside cells as a_w is lowered and catalyzes the formation of proline precursors (81). However, proline synthesis is not stimulated by osmotic stress among enteric bacteria but instead by transport from the culture medium. The accumulation of proline in *E. coli* and *S. typhimurium* has been shown to be mediated by two transport systems, PPI and PPII, with the activity of the latter being elevated under osmotic stress (41).

Osmophilic yeasts accumulate polyhydric alcohols to a concentration commensurate with their extracellular a_w. According to Pitt (62), the xerophilic fungi accumulate compatible solutes or osmoregulators as a consequence of the need for high internal solutes if growth at low a_w is to be possible. In a comparative study of xero-tolerant and nonxero-tolerant yeasts to water stress, Edgley and Brown (31) found that *Zygosaccharomyces rouxii* responded to low a_w controlled by polyethylene glycol by retaining within the cells increasing proportions of glycerol. However, the amount did not change greatly, nor did the level of arabitol change appreciably by a_w. On the other hand, a nontolerant *S. cerevisiae* responded to a lowering of a_w by synthesizing more glycerol but retaining less. The *Z. rouxii* response to low a_w was at the level of glycerol permeation/transport, while that for *S. cerevisiae* was metabolic. It appears from this study that a low a_w forces *S. cerevisiae* to divert a greater proportion of its metabolic activity to glycerol production accompanied by an increase in the amount of glucose consumed during growth. In a more recent study, it was noted that up to 95% of the external osmotic pressure exerted on *S. cerevisiae*, *Z. rouxii*, and *Debaryomyces hansenii* may be counterbalanced by an increase in glycerol (67). *Z. rouxii* accumulates more glycerol under stress while ribitol remains constant.

While yeasts concentrate polyols as "osmoregulators" and enzyme protectors (11), halophilic bacteria operate under low a_w conditions by virtue of their ability to accumulate KCl in the same general manner. In the case of halophilic bacteria, KCl is a requirement, while osmophilic yeasts have a high tolerance for high solute concentrations (12).

It is known that the growth of at least some cells may occur in high numbers at reduced a_w values, while at the same time certain extracellular products are not produced. For example, reduced a_w results in the cessation of enterotoxin B production by *S. aureus* even though high numbers of cells are produced at the same time (79). In the case of *Neurospora crassa*, low a_w resulted in nonlethal alterations of permeability of the cell membrane, leading to loss of several essential molecules (18). Similar results were observed with electrolytes or nonelectrolytes.

Overall, the effect of lowered a_w on the nutrition of microorganisms appears to be of a general nature where cell requirements that must be mediated through an aqueous milieu are progressively shut off. In addition to the effect on nutrients, lowered a_w undoubtedly has adverse effects on the functioning of the cell membrane, which must be kept in a fluid state. The drying of internal parts of cells would be expected to occur upon placing cells in a medium of lowered a_w to a point where equilibrium of water between cells and substrate occurs. Although the mechanisms are not entirely clear, all microbial cells may require the same effective internal a_w. Those that can grow under extreme conditions of low a_w apparently do so by virtue of their ability to concentrate salts, polyols, and amino acids (and possibly other types of compounds) to internal levels not only sufficient to prevent the cells from losing water but that may allow the cell to extract water from the water-depressed external environment. For more information, see Christian (21).

Oxidation-Reduction Potential

It has been known for many years that microorganisms display varying degrees of sensitivity to the oxidation-reduction potential (O/R, Eh) of their growth medium (43). The O/R potential of a substrate may be defined generally as the ease with which the substrate loses or gains electrons. When an element or compound loses electrons, the substrate is said to be oxidized, while a substrate that gains electrons becomes reduced:

$$Cu \xrightleftharpoons[\text{reduction}]{\text{oxidation}} Cu + e.$$

Oxidation may also be achieved by the addition of oxygen, as illustrated in the following reaction:

$$2\,Cu + O_2 \rightarrow 2\,CuO$$

Therefore, a substance that readily gives up electrons is a good reducing agent, and one that readily takes up electrons is a good oxidizing agent. When electrons are transferred from one compound to another, a potential difference is created between the two compounds. This difference may be measured by use of an appropriate instrument, and expressed as millivolts (mv). The more highly oxidized a substance is, the more positive will be its electrical potential, and the more highly reduced a substance is, the more negative will be its electrical potential. When the concentration of oxidant and reductant is equal, a zero electrical potential exists. The O/R potential of a system is expressed by the symbol Eh. Aerobic microorganisms require positive Eh values (oxidized) for growth, while anaerobes require negative Eh values (reduced) (Fig. 3-3). Among the substances in foods that help to maintain reducing conditions are—SH groups in meats and ascorbic acid and reducing sugars in fruits and vegetables.

According to Frazier (38), the O/R potential of a food is determined by

1. The characteristic O/R potential of the original food
2. The **poising capacity**, that is, the resistance to change in potential of the food
3. The oxygen tension of the atmosphere about the food
4. The access that the atmosphere has to the food

With respect to Eh requirements of microorganisms, some bacteria require reduced conditions for growth initiation (Eh of about $-200\,mv$), while others require a positive Eh for growth. In the former category are the anaerobic

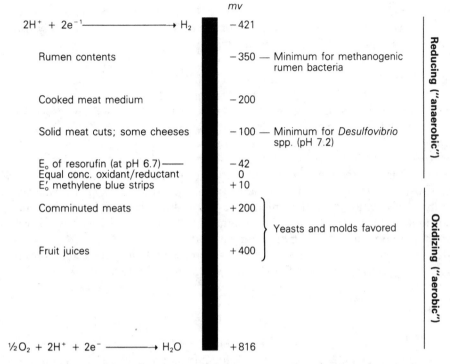

FIGURE 3-3.   Schematic representation of oxidation-reduction potentials relative to the growth of certain microorganisms.

bacteria such as the genus *Clostridium*; in the latter belong aerobic bacteria such as the genus *Bacillus*. Some aerobic bacteria actually grow better under slightly reduced conditions, and these organisms are often referred to as **microaerophiles**. Examples of microaerophilic bacteria are lactobacilli and streptococci. Some bacteria have the capacity to grow under either aerobic or anaerobic conditions. Such types are referred to as **facultative anaerobes**. Most molds and yeasts encountered in and on foods are aerobic, although a few tend to be facultative anaerobes.

In regard to the Eh of foods, plant foods, especially plant juices, tend to have Eh values of from +300 to +400. It is not surprising to find that aerobic bacteria and molds are the common cause of spoilage of products of this type. Solid meats have Eh values of around −200 mv; in minced meats, the Eh is generally around +200 mv. Cheeses of various types have been reported to have Eh values on the negative side, from −20 to around −200 mv.

With respect to the Eh of prerigor as opposed to postrigor muscles, Barnes and Ingram (4, 5) undertook a study of the measurement of Eh in muscle over periods of up to 30 h postmortem and its effect on the growth of anaerobic bacteria. These authors found that the Eh of the sternocephalicus muscle of the horse immediately after death was +250 mv, at which time clostridia failed to multiply. At 30 h postmortem, the Eh had fallen to about −130 mv in the absence of bacterial growth. When bacterial growth was allowed to occur, the Eh fell to about −250 mv. Growth of clostridia was observed at Eh values of −36 mv and below. These authors confirmed for horse meat the finding for whale meat: that anaerobic bacteria do not multiply until the onset of rigor mortis because of the high Eh in prerigor meat. The same is undoubtedly true for beef, pork, and other meats of this type.

Eh Effects.   Microorganisms affect the Eh of their environments during growth just as they do pH. This is true especially of aerobes, which can lower the Eh of their environment while anaerobes cannot. As aerobes grow, O_2 in the medium is depleted, resulting in the lowering of Eh. Growth is not slowed, however, as much as might be expected due to the ability of cells to make use of O_2-donating or hydrogen-accepting substances in the medium. The result is that the medium becomes poorer in oxidizing and richer in reducing substances (54). The Eh of a medium can be reduced by microorganisms by their production of certain metabolic by-products such as H_2S, which has the capacity to lower Eh to -300 mv. Since H_2S reacts readily with O_2, it will accumulate only in anaerobic environments (10).

Eh is dependent on the pH of the substrate and the direct relationship between these two factors is the rH value defined in the following way:

$$Eh = 2.303 \frac{RT}{F} (rH - 2pH),$$

where $R = 8.315$ joules, $F = 96,500$ coulombs, and $T =$ absolute temperature (58). Therefore, the pH of a substrate should be stated when Eh is given. Normally Eh is taken at pH 7.0 (expressed Eh'). When taken at pH 7.0, 25°C, and with all concentrations at 1.0 M, Eh = Eh_0' (simplified Nernst equation). In nature, Eh tends to be more negative under progressively alkaline conditions.

Among naturally occurring nutrients, ascorbic acid and reducing sugars in plants and fruits and—SH groups in meats are of primary importance. The presence or absence of appropriate quantities of oxidizing/reducing agents in a medium is of obvious value to the growth and activity of all microorganisms.

While the growth of anaerobes is normally believed to occur at reduced values of Eh, the exclusion of O_2 may be necessary for some anaerobes. When *Clostridium perfringens, Bacteroides fragilis,* and *Peptococcus magnus* were cultured in the presence of O_2, inhibition of growth occurred even when the medium was at a negative Eh of -50 mv (83). These investigators found that growth occurred in media with an Eh as high as $+325$ mv when no O_2 was present.

With regard to the effect of Eh on lipid production by *Saccharomyces cerevisiae,* it has been shown that anaerobically grown cells produce a lower total level, a highly variable glyceride fraction, and decreased phospholipid and sterol components than aerobically grown cells (65). The lipid produced by anaerobically grown cells was characterized by a high content (up to 50% of total acid) of 8:0 to 14:0 acids and a low level of unsaturated fatty acid in the phospholipid fraction. In aerobically grown cells, 80–90% of the fatty acid component was associated with glyceride, and the phospholipid was found to be 16:1 and 18:1 acids. Unlike aerobically grown cells, anaerobically grown *S. cerevisiae* cells were found to have a lipid and sterol requirement. For more on Eh, see Brown and Emberger (14).

Nutrient Content

In order to grow and function normally, the microorganisms of importance in foods require the following:

1. water
2. source of energy
3. source of nitrogen
4. vitamins and related growth factors
5. minerals

The importance of water to the growth and welfare of microorganisms was presented earlier in this chapter. With respect to the other four groups of substances, molds have the lowest requirement, followed by yeasts, gram-negative bacteria, and gram-positive bacteria.

As sources of energy, foodborne microorganisms may utilize sugars, alcohols, and amino acids. Some few microorganisms are able to utilize complex carbohydrates such as starches and cellulose as sources of energy by first degrading these compounds to simple sugars. Fats are used also by microorganisms as sources of energy, but these compounds are attacked by a relatively small number of microbes in foods.

The primary nitrogen sources utilized by heterotrophic microorganisms are amino acids. A large number of other nitrogenous compounds may serve this function for various types of organisms. Some microbes, for example, are able to utilize nucleotides and free amino acids, while others are able to utilize peptides and proteins. In general, simple compounds such as amino acids will be utilized by almost all organisms before any attack is made on the more complex compounds such as high-molecular-weight proteins. The same is true of polysaccharides and fats.

Microorganisms may require B vitamins in low quantities, and almost all natural foods tend to have an abundant quantity for those organisms that are unable to synthesize their essential requirements. In general, gram-positive bacteria are the least synthetic and must therefore be supplied with one or more of these compounds before they will grow. The gram-negative bacteria and molds are able to synthesize most or all of their requirements. Consequently, these two groups of organisms may be found growing on foods low in B vitamins. Fruits tend to be lower in B vitamins than meats, and this fact, along with the usual low pH and positive Eh of fruits, help to explain the usual spoilage of these products by molds rather than bacteria.

Antimicrobial Constituents

The stability of some foods against attack by microorganisms is due to the presence of certain naturally occurring substances that have been shown to have antimicrobial activity. Some spices are known to contain essential oils that possess antimicrobial activity. Among these are eugenol in cloves, allicin in garlic, cinnamic aldehyde and eugenol in cinnamon, allyl isothiocyanate in mustard, eugenol and thymol in sage, and carvacrol (isothymol) and thymol in oregano (74). Cows' milk contains several antimicrobial substances, including lactoferrin, conglutinin, and the lactoperoxidase system (see Chap. 11). Casein, as well as some free fatty acids that occur in milk, have been shown to be antimicrobial. Eggs contain lysozyme, as does milk, and this enzyme, along with conalbumin, provides fresh eggs with a fairly efficient antimicrobial system. The hydroxycinnamic acid derivatives (p-coumaric, ferulic, caffeic, and chlorogenic acids) found in fruits, vegetables, tea, molasses, and other plant sources all show antibacterial and some antifungal activity. (See Chap. 11 for more specific information on the above and other related antimicrobials.)

Biological Structures

The natural covering of some foods provides excellent protection against the entry and subsequent damage by spoilage organisms. In this category are such structures as the testa of seeds, the outer covering of fruits, the shell of nuts, the hide of animals, and the shells of eggs. In the case of nuts such as pecans and walnuts, the shell or covering is sufficient to prevent the entry of all organisms. Once cracked, of course, nutmeats are subject to spoilage by molds. The outer shell and membranes of eggs, if intact, prevent the entry of nearly all microorganisms when stored under the proper conditions of humidity and temperature. Fruits and vegetables with damaged covering undergo spoilage much faster than those not damaged. The skin covering of fish and meats such as beef and pork prevents the contamination and spoilage of these foods, partly because it tends to dry out faster than freshly cut surfaces.

Taken together, these six intrinsic parameters represent nature's way of preserving plant and animal tissues from microorganisms. By determining the extent to which each exists in a given food, one can predict the general types of microorganisms that are likely to grow and, consequently, the overall stability of this particular food. Their determination may also aid one in determining age and possibly the handling history of a given food.

EXTRINSIC PARAMETERS

The extrinsic parameters of foods are those properties of the storage environment that affect both the foods and their microorganisms. Those of greatest importance to the welfare of foodborne organisms are (1) temperature of storage, (2) relative humidity of environment, and (3) presence and concentration of gases in the environment.

Temperature of Storage

Microorganisms grow over a very wide range of temperatures. Therefore, it would be well to consider at this point the temperature growth ranges for organisms of importance in foods as an aid in selecting the proper temperature for the storage of different types of foods (see Fig. 3-4).

The lowest temperature at which a microorganism has been reported to grow is $-34°C$; the highest is somewhere in excess of $90°C$. It is customary to place microorganisms into three groups based on their temperature requirements for growth. Those organisms that grow well at or below $7°C$ and have

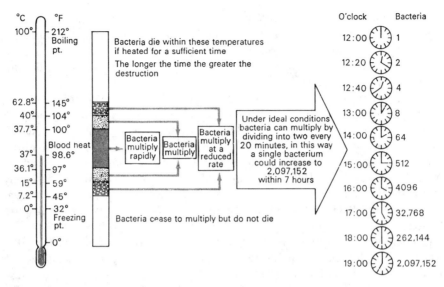

FIGURE 3-4.   Effect of temperature and time on the growth of bacteria. Safe and dangerous temperatures for foodstuffs.
From Hobbs (44), reproduced with permission of the publisher.

their optimum between 20° and 30°C are referred to as **psychrotrophs** (see Chap. 13). Those that grow well between 20°C and 45°C with optima between 30° and 40°C are referred to as **mesophiles,** while those that grow well at and above 45°C with optima between 55° and 65°C are referred to as **thermophiles**. (Physiological properties of these groups are treated in Chaps. 13 and 14.)

With regard to bacteria, psychrotrophic species and strains are found among the following genera of those presented in Chapter 2: *Alcaligenes, Shewanella, Brochothrix, Corynebacterium, Flavobacterium, Lactobacillus, Micrococcus, Pseudomonas, Psychrobacter, Enterococcus,* and others. The psychrotrophs found most commonly on foods are those that belong to the genera *Pseudomonas* and *Enterococcus* (see Chap. 13). These organisms grow well at refrigerator temperatures and cause spoilage of meats, fish, poultry, eggs, and other foods normally held at this temperature. Standard plate counts of viable organisms on such foods are generally higher when the plates are incubated at about 7°C for at least seven days than when incubated at 30°C and above. Mesophilic species and strains are known among all genera presented in Chapter 2 and may be found on foods held at refrigerator temperature. They apparently do not grow at this temperature but do grow at temperatures within the mesophilic range if other conditions are suitable. It should be pointed out that some organisms can grow over a range from 0° and 30°C or above. One such organism is *Enterococcus faecalis*.

Most thermophilic bacteria of importance in foods belong to the genera *Bacillus* and *Clostridium*. Although only a few species of these genera are thermophilic, they are of great interest to the food microbiologist and food technologist in the canning industry.

Just as molds are able to grow over wider ranges of pH, osmotic pressure, and nutrient content, they are also able to grow over wider ranges of temperature than bacteria. Many molds are able to grow at refrigerator temperatures, notably some strains of *Aspergillus, Cladosporium,* and *Thamnidium,* which may be found growing on eggs, sides of beef, and fruits. Yeasts grow over the psychrotrophic and mesophilic temperature ranges but generally not within the thermophilic range.

The quality of the food product must also be taken into account in selecting a storage temperature. While it would seem desirable to store all foods at refrigerator temperatures or below, this is not always best for the maintenance of desirable quality in some foods. For example, bananas keep better if stored at 13°–17°C than at 5°–7°C. A large number of vegetables are favored by temperatures of about 10°C, including potatoes, celery, cabbage, and many others. In every case, the success of storage temperature depends to a great extent upon the R.H. of the storage environment and the presence or absence of gases such as CO_2 and O_3.

Temperature of storage is the most important parameter that affects the spoilage of highly perishable foods, and this fact has been emphasized by the work of Olley and Ratkowsky and their co-workers. According to these investigators, spoilage can be predicted by a spoilage rate curve (58). The general spoilage curve has been incorporated into the circuitry of a temperature function integrator which reads out the equivalent days of storage at 0°C and thus makes it possible to predict the remaining shelf life at 0°C. It has been shown that the rate of spoilage of fresh poultry at 10°C is about twice that at 5°C and that at 15°C is about three times that at 5°C (29). Instead of using the Arrhenius law equation, the following was developed to describe the relationship between temperature and growth rate of microorganisms between the minimum and optimum temperatures (64):

$$\sqrt{r} = b(T - T_0)$$

where r = the growth rate, b = the slope of the regression line, and T_0 = a conceptual temperature of no metabolic significance. The linear relationship has been shown to apply to spoilage bacteria and fungi when growing in foods or when utilizing amino acids (64).

Relative Humidity of Environment

The R.H. of the storage environment is important both from the standpoint of a_w within foods and the growth of microorganisms at the surfaces. When the a_w of a food is set at 0.60, it is important that this food be stored under conditions of R.H. that do not allow the food to pick up moisture from the air and thereby increase its own surface and subsurface a_w to a point where microbial growth can occur. When foods with low a_w values are placed in environments of high R.H., the foods pick up moisture until equilibrium has been established. Likewise, foods with a high a_w lose moisture when placed in an environment of low R.H. There is a relationship between R.H. and temperature that should be borne in mind in selecting proper storage environments for the storage of foods. In general, the higher the temperature is, the lower is the R.H., and vice versa.

Foods that undergo surface spoilage from molds, yeasts, and certain bacteria should be stored under conditions of low R.H. Improperly wrapped meats such as whole chickens and beef cuts tend to suffer surface spoilage in the refrigerator much before deep spoilage occurs, due to the generally high R.H. of the refrigerator and the fact that the meat spoilage flora is essentially aerobic in nature. While it is possible to lessen the chances of surface spoilage in certain foods by storing under low conditions of R.H., it should be remembered that the food itself will lose moisture to the atmosphere under such conditions and thereby become undesirable. In selecting the proper environmental conditions of R.H., consideration must be given to both the possibility of surface growth and the desirable quality to be maintained in the foods in question. By altering the gaseous atmosphere, it is possible to retard surface spoilage without lowering R.H.

Presence and Concentration of Gases in the Environment

The storage of food in atmospheres containing increased amounts of CO_2 up to about 10% is referred to as "controlled atmosphere" or modified atmosphere (MA) storage. The effect of MA storage on plant organs has been known since 1917 and was first put into commercial use in 1928. The use of MA storage for fruits is employed in a number of countries, with apples and pears being the fruits most commonly treated. The concentration of CO_2 generally does not exceed 10% and is applied either from mechanical sources or by use of dry ice (solid CO_2). Carbon dioxide has been shown to retard fungal rotting of fruits caused by a large variety of fungi. While the precise mechanism of action of CO_2 in retarding fruit spoilage is not known, it is probable that it acts as a competitive inhibitor of ethylene action. Ethylene seems to act as a senescence factor in fruits (71), and its inhibition would have the effect of maintaining a fruit in a better state of natural resistance to fungal invasion.

It has also been known for many years that ozone added to food storage environments has a preservative effect on certain foods. At levels of several parts per million (ppm), this gas has been tried with several foods and found to be effective against spoilage microorganisms. It is effective against a variety of microorganisms (16). Since it is a strong oxidizing agent, it should not be used on high-lipid-content foods, since it would cause an increase in rancidity. Both CO_2 and O_3 are effective in retarding the surface spoilage of beef quarters under long-term storage.

Effect of CO_2 and O_2.   The use of CO_2 atmospheres to extend the storage life of meats received a lot of attention during the 1980s. During the 1930s, the meat storage lockers of ships transporting meat carcasses were enriched with CO_2 as a means of increasing storage life (39, 73). New interest in this practice seems to have developed along with the increasing use of vacuum packaging of meats. A large number of workers have found that hyperbaric CO_2 increases the shelf life of a variety of meats, and some of these findings are summarized below. For reviews, see Daniels et al. (28), Genigeorgis (39), and Dixon and Kell (30). More information on vacuum packaging is presented in Chapters 4 and 9.

In general, the inhibitory effects of CO_2 increase with decreasing temperature, due primarily to the increased solubility of CO_2 at lower temperatures, and the pH of meats stored in high CO_2 environments tends to be slightly lower than that of air-stored controls due to carbonic acid formation. Gram-negative bacteria are more sensitive to CO_2 than gram positives (76, 79), with pseudomonads being among the most sensitive and the lactic acid bacteria and anaerobes being among the most resistant.

With steaks stored in 100% CO_2, significantly lower counts were seen 16–27 days after slaughter, compared to steaks stored in 100% N_2 or 100% O_2 or in air (46). Steaks stored at 1°C in 15% CO_2 + 85% air, or in 15% CO_2 + 85% O_2 for up to 20 days had consistently lower counts (about one-third) than those stored in air (3). The CO_2 + O_2 mixture was slightly more effective than the CO_2 + air mixture. When lamb chops were stored at −1°C in 80% O_2 + 20% N_2, or 80% air or O_2 + 20% CO_2, or 80% N_2 or H_2 + 20% CO_2, psychrotrophic organisms decreased successively when compared to those stored in air (57). Using slices of lean beef inoculated with pseudomonads and *Moraxella-Acinetobacter* spp. and stored at 5°C in 70–85% O_2 + 15% CO_2, shelf life was increased by 9 days compared to storage in air (24). In a study of pork stored at 4° or 14°C, the time needed for the APC at 4°C to reach 5×10^6 organisms/cm^2 was about three times longer when stored with 5 atm CO_2, and about 15 times longer than in air (7). With poultry stored in air, APC of drip after 16 days at 10°C was log 9.40, while in 20% CO_2 log APC was 6.14 (82). Carbon dioxide was less effective against coliforms in the latter study with log numbers of 3.28 and 3.21 for air and 20% CO_2 storage respectively.

A gas mixture of 10% CO_2, 5% O_2, and 85% N_2 was found to be more effective than two other mixtures for preserving steaks in zero barrier film (1). The other mixtures used were 15% CO_2, 40% O_2, and 45% N_2; and 60% CO_2 and 40% O_2. No lactics were identified, and *B. thermosphacta* was found, but they did not dominate. *Serratia liquefaciens* increased under all treatments. Meat sandwiches stored in 50% CO_2 and 50% air at 4°C remained acceptable up to 35 days, with roast beef being acceptable longer (28 to 35 days) than hamburger (14 days) (55). The film had the following transmission properties: 12.1 cc O_2/m^2/day/23°C. Treatment of raw whole milk with CO_2 was shown to extend the generation time of pseudomonads when stored at 7°C (68). Time for APC to increase tenfold was extended by at least 24 h by the CO_2 treatment.

Carbon dioxide atmosphere storage is effective with fish. Using 80% CO_2 + air, log numbers after 14 days at 35°C were approximately 6.00/cm^2 compared to air controls with numbers >10.5 cm^2. The pH of CO_2-stored products after 14 days decreased from around 6.75 to around 6.30, while controls increased to around 7.45 (60). The shelf life of rockfish and salmon at 4.5°C was extended by 20–80% CO_2 (15). At least 1 log difference in bacterial counts over controls was obtained when trout and croaker were stored in CO_2 environments at 4°C (42). When fresh shrimp or prawns were packed in ice in an atmosphere of 100% CO_2, they were edible for up to 2 weeks, and bacterial counts after 14 days were lower than air-packed controls after 7 days (53). When cod fillets were stored at 2°C, air-stored samples spoiled in 6 days, with aerobic plate counts (APC) of $\log_{10}$ 7.7, while samples stored in 50% CO_2 + 50% O_2 or 50% CO_2 + 50% N_2 or 100% CO_2 did not show bacterial spoilage

TABLE 3-5.   Effect of Storage on the Microflora of Two Meats Held from 48 to 140 Days at 4°C

		Smoked Pork Loins		
	0 Day	Vacuum 48 d	CO_2 48 d	N_2 48 d
Log APC/g	2.5	7.6	6.9	7.2
pH	5.8	5.8	5.9	5.9
Dominant flora (%)	Flavo (20)	Lactos (52)[a]		
	Arthro (20)		Lactos (74)[b]	
	Yeasts (20)			Lactos (67)[c]
	Pseudo (11)			
	Coryne (10)			

		Frankfurter Sausage		
	0 Day	Vacuum 98 d	CO_2 140 d	N_2 140 d
Log APC/g	1.7	9.0	2.4	4.8
pH	5.9	5.4	5.6	5.9
Dominant flora (%)	Bac (34)	Lactos (38)	Lactos (88)[d]	Lactos (88)[e]
	Coryne (34)			
	Flavo (8)			
	Broch (8)			

Note: Percent flora represented by *Lactobacillus viridescens:* [a] 40; [b] 72; [c] 50; [d] 22; [e] 35.
Flavo = *Flavobacterium*; Arthro = *Arthrobacter*; Pseudo = *Pseudomonas*; Coryne = *Corynebacterium*; Bac = *Bacillus*; Broch = *Brochothrix*.
Source: Adapted from (8).

until, respectively, 26, 34, and 34 days, with respective APCs of 7.2, 6.6, and 5.5/g (78). It was suggested that the use of 50% CO_2 + 50% O_2 is technically more feasible than the use of 100% CO_2. While the practical upper limit of CO_2 for red meats is around 20%, higher concentrations can be used with fish since they contain lower levels of myoglobin.

The overall effect of high concentrations of CO_2 in meat packs is to shift the flora from a heterogeneous one consisting of gram-negative bacteria to one consisting primarily of lactobacilli and other lactic acid bacteria. In a study by Blickstad et al. (7), >90% of the flora of pork consisted of *Pseudomonas* spp. after aerobic storage for 8 days at 4°C and 3 days at 14°C, but after 5 atm CO_2 storage, the lactobacilli dominated at both temperatures. This effect can be seen from Table 3-5, where smoked pork loins and frankfurter sausage were held from 48 to 140 days at 4°C in the presence of 100% CO_2, 100% N_2, and in vacuum. While pH did not decrease with the pork loins, decrease did occur with frankfurters in CO_2 and in vacuum, reflective of the domination by the lactobacilli. The shelf life of both products increased in the order, vacuum < N_2 < CO_2 (8). Several investigators have found the shelf life of raw meats to increase in the order N_2 < vacuum < CO_2. In another study, the initial flora of normal, low-pH beef consisted of 13% *Pseudomonas fluorescens* and 87% nonfluorescent *Pseudomonas* spp. After vacuum storage for 21 days at 4°C, 96% of the flora consisted of homofermentative lactic acid bacteria, while after 51 days in 100% CO_2, 100% of the flora consisted of these organisms

(35). In a study employing inoculated, high-pH slices of meat stored in a mixture of 20% CO_2 + 80% air, *Enterobacter* and *Brochothrix thermosphacta* were unaffected at 3°C compared to air-stored controls (40).

With regard to the relative sensitivity of microorganisms to CO_2 inhibition, the fungus *Sclerotium rolfsii* was inhibited by 0.12 atm CO_2, while *Lactococcus cremoris* required ~11 atm (33). *Proteus* and *Micrococcus* spp. are less sensitive than the pseudomonads but not as resistant as *L. cremoris*. From their study of CO_2 atmospheres on the flora of normal, low-pH beef, Erichsen and Molin (35) found resistance to CO_2 as follows: *Pseudomonas* spp. < *B. thermosphacta* < lactic acid bacteria. Storage of meat in 100% CO_2 extended shelf life the greatest, followed by vacuum packaging. To effect a 50% reduction in growth rate compared to controls, 0.5 atm of CO_2 was required for *Pseudomonas fragi*, 1.3 atm for *Bacillus cereus*, and 0.6 atm for *L. cremoris* (33).

With regard to the mechanism of CO_2 inhibition of microorganisms, two explanations have been offered. King and Nagel (51) found that CO_2 blocked the metabolism of *Pseudomonas aeruginosa* and appeared to effect a mass action on enzymatic decarboxylations. Sears and Eisenberg (72) found that CO_2 affected the permeability of cell membranes, and Enfors and Molin (32) found support for the latter hypothesis in their studies on the germination of *Clostridium sporogenes* and *C. perfringens* endospores. At 1 atm CO_2, spore germination of these two species was stimulated, while *B. cereus* spore germination was inhibited. As was shown by others, CO_2 is more stimulatory at low pH than high. With 55 atm CO_2, only 4% germination of *C. sporogenes* spores occurred, while with *C. perfringens* 50 atm reduced germination to 4% (32). These authors suggested that CO_2 inhibition was due to its accumulation in the membrane lipid bilayer such that increased fluidity results. In an earlier study, the generation time for eight strains of *C. perfringens* in 100% CO_2 and 100% N_2 grown in thioglycollate medium at 1 atm was about the same—12.9–17.2 and 12.9–16.9, respectively (59). With *P. fluorescens* in a minimal medium, CO_2 pressure of 100 mm Hg stimulated growth, but at higher concentrations, the growth rate declined (40). In a complex medium at 30°C, maximum inhibition of this organism was achieved at 250 mm Hg. These investigators, along with Enfors and Molin (34), found that CO_2 inhibition increases as temperature is lowered. Although CO_2 has been known to be an inhibitor of certain microorganisms for over 100 years, there is still no clear understanding of how inhibition is achieved (30).

Since extremely high pressures of CO_2 are required to prevent the germination of *C. sporogenes* and *C. perfringens* spores, the storage of meats at temperatures of growth of *C. botulinum* in 1 atm CO_2 may present concerns relative to the possible germination of botulinal spores. In a culture medium flushed with CO_2, germination of types A and B spores of *C. botulinum* was enhanced (37). In the case of salmonellae and staphylococci, the use of up to 60% CO_2 had no stimulatory effect, although *Yersinia enterocolitica* may be of concern (76). High-pressure CO_2 treatment (61.2 atm) of *Listeria monocytogenes* in distilled water for 2 h at 35°C has been shown to be lethal, but no lethal effects were observed with N_2 under the same conditions (84). In several food products, lethality rates from 95 to 100% were observed for *Salmonella* and *Listeria* when treated with CO_2 pressures of 136.1 atm for 2 h at 35°C (84).

The effect of vacuum packaging and gas atmospheres on microorganisms is discussed further in Chapter 4.

References
1. Ahmad, H. A., and J. A. Marchello. 1989. Effect of gas atmosphere packaging on psychrotrophic growth and succession on steak surfaces. *J. Food Sci.* 54:274–276, 310.
2. Anderson, C. B., and L. D. Witter. 1982. Glutamine and proline accumulation by *Staphylococcus aureus* with reduction in water activity. *Appl. Environ. Microbiol.* 43:1501–1503.
3. Bala, K., W. C. Stringer, and H. D. Naumann. 1977. Effect of spray sanitation treatment and gaseous atmospheres on the stability of prepackaged fresh beef. *J. Food Sci.* 42:743–746.
4. Barnes, E. M., and M. Ingram. 1955. Changes in the oxidation-reduction potential of the sterno-cephalicus muscle of the horse after death in relation to the development of bacteria. *J. Sci. Food Agr.* 6:448–455.
5. Barnes, E. M., and M. Ingram. 1956. The effect of redox potential on the growth of *Clostridium welchii* strains isolated from horse muscle. *J. Appl. Bacteriol.* 19:117–128.
6. Bate-Smith, E. C. 1948. The physiology and chemistry of rigor mortis, with special reference to the aging of beef. *Adv. Food Res.* 1:1–38.
7. Blickstad, E., S.-O. Enfors, and G. Molin. 1981. Effect of hyperbaric carbon dioxide pressure on the microbial flora of pork stored at 4° or 14°C. *J. Appl. Bacteriol.* 50:493–504.
8. Blickstad, E., and G. Molin. 1983. The microbial flora of smoked pork loin and frankfurter sausage stored in different gas atmospheres at 4°C. *J. Appl. Bacteriol.* 54:45–56.
9. Briskey, E. J. 1964. Etiological status and associated studies of pale, soft, exudative porcine musculature. *Adv. Food Res.* 13:89–178.
10. Brock, T. D., D. W. Smith, and M. T. Madigan. 1984. *Biology of Microorganisms.* Englewood Cliffs, N.J.: Prentice-Hall.
11. Brown, A. D. 1964. Aspects of bacterial response to the ionic environment. *Bacteriol. Rev.* 28:296–329.
12. Brown, A. D. 1974. Microbial water relations: Features of the intracellular composition of sugar-tolerant yeasts. *J. Bacteriol.* 118:769–777.
13. Brown, A. D. 1976. Microbial water stress. *Bacteriol. Rev.* 40:803–846.
14. Brown, M. H., and O. Emberger. 1980. Oxidation-reduction potential. In *Microbial ecology of foods*, vol. 1, ICMSF, 112–115. New York: Academic Press.
15. Brown, W. D., M. Albright, D. A. Watts, B. Heyer, B. Spruce, and R. J. Price. 1980. Modified atmosphere storage of rockfish (*Sebastes miniatus*) and silver salmon (*Oncorhynchus kisutch*). *J. Food Sci.* 45:93–96.
16. Burleson, G. R., T. M. Murray, and M. Pollard. 1975. Inactivation of viruses and bacteria by ozone, with and without sonication. *Appl. Microbiol.* 29:340–344.
17. Callow, E. H. 1949. Science in the imported meat industry. *J. Roy. Sanitary Inst.* 69:35–39.
18. Charlang, G., and N. H. Horowitz. 1974. Membrane permeability and the loss of germination factor from *Neurospora crassa* at low water activities. *J. Bacteriol.* 117:261–264.
19. Christian, J. H. B. 1955. The water relations of growth and respiration of *Salmonella oranienburg* at 30°C. *Aust. J. Biol. Sci.* 8:490–497.
20. Christian, J. H. B. 1963. Water activity and the growth of microorganisms. In *Recent Advances in Food Science*, vol. 3., ed. J. M. Leitch and D. N. Rhodes, 248–255. London: Butterworths.

21. Christian, J. H. B. 1980. Reduced water activity. In *Microbial Ecology of Foods*, vol. 1, ICMSF, 70–91. New York: Academic Press.
22. Christian, J. H. B., and J. A. Waltho. 1966. Water relations of *Salmonella oranienburg;* stimulation of respiration by amino acids. *J. Gen. Microbiol.* 43:345–355.
23. Chung, K. C., and J. M. Goepfert. 1970. Growth of *Salmonella* at low pH. *J. Food Sci.* 35:326–328.
24. Clark, D. S., and C. P. Lentz. 1973. Use of mixtures of carbon dioxide and oxygen for extending shelf-life of prepackaged fresh beef. *Can. Inst. Food Sci. Technol. J.* 6:194–196.
25. Conway, E. J., and M. Downey. 1950. pH values of the yeast cell. *Biochem. J.* 47:355–360.
26. Corlett, D. A., Jr., and M. H. Brown. 1980. pH and acidity. In *Microbial Ecology of Foods,* vol. 1, ICMSF, 92–111. New York: Academic Press.
27. Csonka, L. N. 1989. Physiological and genetic responses of bacteria to osmotic stress. *Microbiol. Rev.* 51:121–147.
28. Daniels, J. A., R. Krishnamurthi, and S. S. H. Rizvi. 1985. A review of effects of carbon dioxide on microbial growth and food quality. *J. Food Protect.* 48:532–537.
29. Daud, H. B., T. A. McMeekin, and J. Olley. 1978. Temperature function integration and the development and metabolism of poultry spoilage bacteria. *Appl. Environ. Microbiol.* 36:650–654.
30. Dixon, N. M., and D. B. Kell. 1989. The inhibition by CO_2 of the growth and metabolism of micro-organisms. *J. Appl. Bacteriol.* 67:109–136.
31. Edgley, M., and A. D. Brown. 1978. Response of xerotolerant and nontolerant yeasts to water stress. *J. Gen. Microbiol.* 104:343–345.
32. Enfors, S.-O., and G. Molin. 1978. The influence of high concentrations of carbon dioxide on the germination of bacterial spores. *J. Appl. Bacteriol.* 45:279–285.
33. Enfors, S.-O., and G. Molin. 1980. Effect of high concentrations of carbon dioxide on growth rate of *Pseudomonas fragi, Bacillus cereus* and *Streptococcus cremoris. J. Appl. Bacteriol.* 48:409–416.
34. Enfors, S.-O., and G. Molin. 1981. The influence of temperature on the growth inhibitory effect of carbon dioxide on *Pseudomonas fragi* and *Bacillus cereus. Can. J. Microbiol.* 27:15–19.
35. Erichsen, I., and G. Molin. 1981. Microbial flora of normal and high pH beef stored at 4°C in different gas environments. *J. Food Protect.* 44:866–869.
36. Evans, J. B., and C. F. Niven, Jr. 1960. Microbiology of meat: Bacteriology. In *The Science of Meat and Meat Products.* San Francisco: Freeman.
37. Foegeding, P. M., and F. F. Busta. 1983. Effect of carbon dioxide, nitrogen and hydrogen gases on germination of *Clostridium botulinum* spores. *J. Food Protect.* 46:987–989.
38. Frazier, W. C. 1968. *Food Microbiology.* 2d ed., 171. New York: McGraw-Hill.
39. Genigeorgis, C. A. 1985. Microbial and safety implications of the use of modified atmospheres to extend the storage life of fresh meat and fish. *Int. J. Food Microbiol.* 1:237–251.
40. Gill, C. O., and K. N. Tan. 1980. Effect of carbon dioxide on growth of meat spoilage bacteria. *Appl. Environ. Microbiol.* 39:317–319.
41. Grothe, S., R. L. Krogsrud, D. J. McClellan, J. L. Milner, and J. M. Wood. 1986. Proline transport and osmotic stress response in *Escherichia coli* K-12. *J. Bacteriol.* 166:253–259.
42. Hanks, H., R. Nickelson, II, and G. Finne. 1980. Shelf-life studies on carbon dioxide packaged finfish from the Gulf of Mexico. *J. Food Sci.* 45:157–162.
43. Hewitt, L. F. 1950. *Oxidation-Reduction Potentials in Bacteriology and Biochemistry,* 6th ed. Edinburgh: Livingston.

44. Hobbs, B. C. 1968. *Food Poisoning and Food Hygiene,* 2d ed. London: Edward Arnold.

45. Horner, K. J., and G. D. Anagnostopoulos. 1973. Combined effects of water activity, pH and temperature on the growth and spoilage potential of fungi. *J. Appl. Bacteriol.* 36:427–436.

46. Huffman, D. L. 1974. Effect of gas atmospheres on microbial quality of pork. *J. Food Sci.* 39:723–725.

47. Jakobsen, M., and W. G. Murrell. 1977. The effect of water activity and the a_w-controlling solute on germination of bacterial spores. *Spore Res.* 2:819–834.

48. Jakobsen, M., and W. G. Murrell. 1977. The effect of water activity and a_w-controlling solute on sporulation of *Bacillus cereus* T. *J. Appl. Bacteriol.* 43:239–245.

49. Juven, B. J. 1976. Bacterial spoilage of citrus products at pH lower than 3.5. *J. Milk Food Technol.* 39:819–822.

50. Kang, C. K., M. Woodburn, A. Pagenkopf, and R. Cheney. 1969. Growth, sporulation, and germination of *Clostridium perfringens* in media of controlled water activity. *Appl. Microbiol.* 18:798–805.

51. King, A. D., Jr., and C. W. Nagel. 1975. Influence of carbon dioxide upon the metabolism of *Pseudomonas aeruginosa. J. Food Sci.* 40:362–366.

52. Marshall, B. J., F. Ohye, and J. H. B. Christian. 1971. Tolerance of bacteria to high concentrations of NaCl and glycerol in the growth medium. *Appl. Microbiol.* 21:363–364.

53. Matches, J. R., and M. E. Lavrisse. 1985. Controlled atmosphere storage of spotted shrimp (*Pandalus platyceros*). *J. Food Protect.* 48:709–711.

54. Morris, E. O. 1962. Effect of environment on micro-organisms. In *Recent Advances in Food Science,* vol. 1, ed. J. Hawthorn and J. M. Leitch, 24–36. London: Butterworths.

55. McMullen, L., and M. E. Stiles. 1989. Storage life of selected meat sandwiches at 4°C in modified gas atmospheres. *J. Food Protect.* 52:792–798.

56. Mossel, D. A. A., and M. Ingram. 1955. The physiology of the microbial spoilage of foods. *J. Appl. Bacteriol.* 18:232–268.

57. Newton, K. G., J. C. L. Harrison, and K. M. Smith. 1977. The effect of storage in various gaseous atmospheres on the microflora of lamb chops held at −1°C. *J. Appl. Bacteriol.* 43:53–59.

58. Olley, J., and D. A. Ratkowsky. 1973. The role of temperature function integration in monitoring fish spoilage. *Food Technol. New Zealand* 8:13–17.

59. Parekh, K. G., and M. Solberg. 1970. Comparative growth of *Clostridium perfringens* in carbon dioxide and nitrogen atmospheres. *J. Food Sci.* 35:156–159.

60. Parkin, K. L., M. J. Wells, and W. D. Brown. 1981. Modified atmosphere storage of rockfish fillets. *J. Food Sci.* 47:181–184.

61. Peña, A., G. Cinco, A. Gómez-Puyou, and M. Tuena. 1972. Effect of pH of the incubation medium on glycolysis and respiration in *Saccharomyces cerevisiae. Arch. Biochem. Biophys.* 153:413–425.

62. Pitt, J. I. 1975. Xerophilic fungi and the spoilage of foods of plant origin. In *Water Relations of Foods,* ed. R. B. Duckworth, 273–307. London: Academic Press.

63. Prior, B. A. 1978. The effect of water activity on the growth and respiration of *Pseudomonas fluorescens. J. Appl. Bacteriol.* 44:97–106.

64. Ratkowsky, D. A., J. Olley, T. A. McMeekin, and A. Ball. 1982. Relationship between temperature and growth rate of bacterial cultures. *J. Bacteriol.* 149:1–5.

65. Rattray, J. B. M., A. Schibeci, and D. K. Kidby. 1975. Lipids of yeasts. *Bacteriol. Rev.* 39:197–231.

66. Reay, G. A., and J. M. Shewan. 1949. The spoilage of fish and its preservation by chilling. *Adv. Food Res.* 2:343–398.

67. Reed, R. K., J. A. Chudek, R. Foster, and G. M. Gadd. 1987. Osmotic significance of glycerol accumulation in exponentially growing yeasts. *Appl. Environ. Microbiol.* 53:2119–2123.

68. Roberts, R. F., and G. S. Torrey. 1988. Inhibition of psychrotrophic bacterial growth in refrigerated milk by addition of carbon dioxide. *J. Dairy Sci.* 71:52–60.

69. Rose, A. H. 1965. *Chemical Microbiology*, ch. 3. London: Butterworths.

70. Rothstein, A., and G. Demis. 1953. The relationship of the cell surface to metabolism. The stimulation of fermentation by extracellular potassium. *Arch. Biochem. Biophys.* 44:18–29.

71. Salisbury, F. B., and C. Ross. 1969. *Plant Physiology.* 467. Belmont, Cal.: Wadsworth.

72. Sears, D. F., and R. M. Eisenberg. 1961. A model representing a physiological role of CO_2 at the cell membrane. *J. Gen. Physiol.* 44:869–887.

73. Seideman, S. C., and P. R. Durland. 1984. The utilization of modified gas atmosphere packaging for fresh meat: A review. *J. Food Qual.* 6:239–252.

74. Shelef, L. A. 1983. Antimicrobial effects of spices. *J. Food Safety* 6:29–44.

75. Sherman, J. M., and G. E. Holm. 1922. Salt effects in bacterial growth. II. The growth of *Bacterium coli* in relation to H-ion concentration. *J. Bacteriol.* 7:465–470.

76. Silliker, J. H., and S. K. Wolfe. 1980. Microbiological safety considerations in controlled-atmosphere storage of meats. *Food Technol.* 34(3):59–63.

77. Sperber, W. H. 1983. Influence of water activity on foodborne bacteria—A review. *J. Food Protect.* 46:142–150.

78. Stenstrom, I.-M. 1985. Microbial flora of cod fillets (*Gadus morhua*) stored at 2°C in different mixtures of carbon dioxide and nitrogen/oxygen. *J. Food Protect.* 48:585–589.

79. Stier, R. F., L. Bell, K. A. Ito, B. D. Shafer, L. A. Brown, M. L. Seeger, B. H. Allen, M. N. Porcuna, and P. A. Lerke. 1981. Effect of modified atmosphere storage on *C. botulinum* toxigenesis and the spoilage microflora of salmon fillets. *J. Food Sci.* 46:1639–1642.

80. Troller, J. A. 1986. Water relations of foodborne bacterial pathogens—An updated review. *J. Food Protect.* 49:656–670.

81. Troller, J. A. 1984. Effect of low moisture environments on the microbial stability of foods. In *Food Microbiology*, ed. A. H. Rose, 173–198. New York: Academic Press.

82. Wabeck, C. J., C. E. Parmalee, and W. J. Stadelman. 1968. Carbon dioxide preservation of fresh poultry. *Poultry Sci.* 47:468–474.

83. Walden, W. C., and D. J. Hentges. 1975. Differential effects of oxygen and oxidation-reduction potential on the multiplication of three species of anaerobic intestinal bacteria. *Appl. Microbiol.* 30:781–785.

84. Wei, C. I., M. O. Balaban, S. Y. Fernando, and A. J. Peplow. 1991. Bacterial effect of high pressure CO_2 treatment on foods spiked with *Listeria* or *Salmonella*. *J. Food Protect.* 54:189–193.

85. Wodzinski, R. J., and W. C. Frazier. 1961. Moisture requirements of bacteria. II. Influence of temperature, pH, and maleate concentration on requirements of *Aerobacter aerogenes*. *J. Bacteriol.* 81:353–358.

4

Incidence and Types
of Microorganisms in Foods

In general, the numbers and types of microorganisms present in a finished food product are influenced by

1. The general environment from which the food was originally obtained
2. The microbiological quality of the food in its raw or unprocessed state
3. The sanitary conditions under which the product is handled and processed
4. The adequacy of subsequent packaging, handling, and storage conditions in maintaining the flora at a low level

In producing good-quality market foods, it is important to keep microorganisms at a low level for reasons of aesthetics, public health, and product shelf life. Other than those foods that have been made sterile, all foods should be expected to contain a certain number of microorganisms of one type or another. Ideally, the numbers of organisms should be as low as is possible under good conditions of production. Excessively high numbers of microorganisms in fresh foods present cause for alarm. It should be kept in mind that the inner parts of healthy plant and animal tissues are generally sterile and that it is theoretically possible to produce many foods free of microorganisms. This objective becomes impractical, however, when mass production and other economic considerations are realized. The number of microorganisms in a fresh food product, then, may be taken to reflect the overall conditions of raw product quality, processing, handling, storage, and so forth. This is one of the most important uses made of the standard plate count in food microbiology. The question immediately arises as to the attainability of low numbers under the best and most economic production conditions known. With few exceptions, it is difficult to know what is the lowest number of microorganisms attainable under good production conditions because of the many variables that must be considered. With advances in modern technology, it has been possible to reduce the microbial load in a large number of foods over what was possible even fifteen years ago.

The microbiology of various groups of foods is treated below. With each group, the sources of organisms are listed and discussed, along with their incidence as reported by various workers. It should be noted that the numbers reported for a particular product by different investigators do not always reflect the microbiology of that product under ideal conditions of production, handling, and storage. The chapters on food spoilage and food poisoning should be consulted for additional information on the types of organisms associated with the foods discussed.

MEATS

Fresh Red Meats

The reported incidence of microorganisms in red meats is presented in Table 4-1. Comminuted meats such as ground beef invariably have higher numbers

TABLE 4-1.   Relative Percentage of Organisms in Red Meats That Meet Specified Target Numbers

Products	Number of Samples	Microbial Group/Target	% Samples Meeting Target	References
Raw	735	APC: $\log_{10}$ 6.00 or less/g	76	146
beef	735	Coliforms: log 2.00 or less/g	84	146
patties	735	E. coli: log 2.00 or less/g	92	146
	735	S. aureus: 2.00 or less/g	85	146
	735	Presence of salmonellae	0.4	146
Fresh	1,830	APC: 6.70 or less/g	89	13
ground	1,830	S. aureus: 3.00 or less/g	92	13
beef[a]	1,830	E. coli: 1.70 or less/g	84	13
	1,830	Presence of salmonellae	2	13
	1,830	Presence of C. perfringens	20	13
Fresh	1,090	APC: $\geq$7.00 or less/g at 35°C	88	111
ground	1,090	Fecal coliforms: $\leq$2.00/g	76	111
beef	1,090	S. aureus: <2.00/g	91	111
Frozen	604	APC: 6.00 or less/g	67	48
ground beef	604	E. coli: <2.70/g	85	48
patties	604	E. coli: >3.00/g MPN	9	48
Fried	107	APC at 21°C: 72 h, <3.00/g	76	25
hamburger	107	Absence of enterococci, coliforms, S. aureus, salmonellae	100	25
Comminuted	113	Coliforms: 2.00 or less/g	42	131
big game	113	E. coli: 2.00 or less/g	75	131
meats	113	S. aureus: 2.00 or less/g	96	131

TABLE 4-1. (Continued)

Products	Number of Samples	Microbial Group/Target	% Samples Meeting Target	References
Fresh	67	APC: 5.70 or less/g	75	147
pork	67	*E. coli:* 2.00 or less/g	88	147
sausage	67	*S. aureus:* 2.00 or less/g	75	147
	560	Presence of salmonellae	28	147
Pork trimmings for above sausage	528	Presence of salmonellae	28	147
Vacuum-packaged	180	APC: <3.00/g	97	145
sliced imported canned ham samples[b]	180	Absence of *E. coli* or *S. aureus* in 0.1 g; or salmonellae in 25 g	100	145
	180	Coliforms in 0.1 g portions	2	145
Fresh breakfast-	183	APC: $<10^7$	91	32
type sausage	180	Coliforms: $<10^3$	59	32
	149	*E. coli:* $<10^3$	93	32
	175	*S. aureus:* $<10^3$	83	32
Frozen breakfast-	186	APC: $<10^7$	95.2	32
type sausage	180	Coliforms: $<10^3$	76	32
	147	*E. coli:* $<10^3$	97	32
	161	*S. aureus:* $<10^3$	99	32

[a] Under Oregon law that was in effect at the time (see Chap. 17).
[b] The machine-sliced and vacuum-packed samples were obtained from 16- to 21-lb canned, refrigerated, imported hams.

of microorganisms than noncomminuted meats such as steaks. This has been reported rather consistently for over 70 years, and there are several reasons for the generally higher counts on these types of products than one finds on whole meats:

1. Commercial ground meats generally consist of trimmings from various cuts. These pieces have been handled excessively and consequently normally contain more microorganisms than meat cuts such as steaks.

2. Ground meat provides a greater surface area, which itself accounts in part for the increased flora. It should be recalled that as particle size is reduced, the total surface area increases with a consequent increase in surface energy.

3. This greater surface area of ground meat favors the growth of aerobic bacteria, the usual low-temperature spoilage flora.

4. In some commercial establishments, the meat grinders, cutting knives, and storage utensils are rarely cleaned as often and as thoroughly as is necessary to prevent the successive buildup of microbial numbers. This may be illustrated by data I obtained from a study of the bacteriology of several areas in the meat department of a large grocery store. The blade of the meat saw and the cutting block were swabbed immediately after they were cleaned on three different occasions with the following mean results: the saw blade had a total log/in^2 count of 5.28, with 2.3 coliforms, 3.64 enterococci, 1.60 staphylococci, and 3.69 micrococci; the cutting block had a mean log/in^2 count of 5.69, with 2.04 coliforms, 3.77 enterococci, <1.00 staphylococci, and 3.79 micrococci. These are among the sources of the high total bacterial count to comminuted meats.

5. One heavily contaminated piece of meat is sufficient to contaminate others, as well as the entire lot, as they pass through the grinder. This heavily contaminated portion is often in the form of lymph nodes, which are generally embedded in fat. These organs have been shown to contain high numbers of microorganisms and account in part for hamburger meat's having a generally higher total count than ground beef. In some states, the former may contain up to 30% beef fat while the latter should not contain more than 20% fat.

Both bacilli and clostridia are found in meats of all types. In a study of the incidence of putrefactive anaerobe (P.A.) spores in fresh and cured pork trimmings and canned pork luncheon meat, Steinkraus and Ayres (139) found these organisms to occur at very low levels, generally less than 1/g. In a study of the incidence of clostridial spores in meats, Greenberg et al. (49) found a mean P.A. spore count/g of 2.8 from 2,358 meat samples. Of the 19,727 P.A. spores isolated, only 1 was a *Clostridium botulinum* spore, and it was recovered from chicken. The large number of meat samples studied by these investigators consisted of beef, pork, and chicken, obtained from all parts of the United States and Canada. The significance of P.A. spores in meats is due to the problems encountered in the heat destruction of these forms in the canning industry (see Chap. 14).

Erysipelothrix rhusiopathiae was isolated from about 34% of retail pork samples in Japan and from 4 to 54% of pork loins in Sweden. A variety of serovars have been found in pork, and some isolates possessed mouse virulence. (See Chap. 21 for the taxonomic relationship between *E. rhusiopathiae* and *Listeria monocytogenes*.)

The incidence of *Clostridium perfringens* in a variety of American foods was studied by Strong et al. (142). They recovered the organism from 16.4% of raw meats, poultry, and fish tested; from 5% of spices; from 3.8% of fruits and vegetables; from 2.7% of commercially prepared frozen foods; and from 1.8% of home-prepared foods. Others have found low numbers of this organism in both fresh and processed meats. In ground beef, *C. perfringens* at 100 or less/g was found in 87% of 95 samples, while 45 of the 95 (47%) samples contained this organism at levels <1,000/g (81). One group was unable to recover *C. perfringens* from pork carcasses, hearts, and spleens, but 21.4% of livers were positive (7). Commercial pork sausage was found to have a prevalence of 38.9%. The significance of this organism in foods is discussed in Chapter 20.

Some members of the family Enterobacteriaceae have been found to be common in fresh and frozen beef, pork, and related meats. Of 442 meat samples examined by Stiles and Ng (141), 86% yielded enteric bacteria, with all 127 ground beef samples being positive. The most frequently found were *Escherichia coli* biotype I (29%), *Serratia liquefaciens* (17%), and *Pantoea agglomerans*

(12%). A total of 721 isolates (32%) were represented by *Citrobacter freundii*, *Klebsiella pneumoniae*, *Enterobacter cloacae*, and *E. hafniae*. In an examination of 702 foods for fecal coliforms by the most-probable-numbers (MPN) method representing ten food categories, the highest number was found in the 119 ground beef samples, with geometric mean by AOAC procedure being 59/g (3).* Mean number for 94 pork sausage samples was 7.9/g. From 32 samples of minced goat meat, mean coliform count was log 2.88, mean Enterobacteriaceae was 3.07, while log APC mean was 6.57 (97). Information on the incidence of coliforms, enterococci, and other indicator organisms is presented in Chapter 17.

Soy-Extended Ground Meats

The addition of soy protein (soybean flour, soy flakes, texturized soy protein) at levels of 10 to 30% to ground meat patties is fairly widespread in the fast-food industry, at least in the United States, and the microbiology of these soy blends has been investigated. The earliest, most detailed study is that of Craven and Mercuri (20), who found that when ground beef or chicken was extended with 10 or 30% soy, APCs of these products increased over unextended controls when both were stored at 4°C for up to 8–10 days. While coliforms were also higher in beef-soy mixtures than in controls, this was not true for the chicken-soy blends. In general, APCs were higher at the 30% level of soy than at 10%. These findings have been supported by several investigators (8, 58, 73, 151). In one study in which 25% soy was used with ground beef, the mean time to spoilage at 4°C for the beef-soy blend was 5.3 days compared to 7.5 days for the unextended ground beef (8). In another study using 10, 20, and 30% soy, APC increased significantly with both time and concentration of soy in the blend (73).

With regard to the microbiological quality of soy products, the geometric mean APC of 1,226 sample units of seasoned product was found to be 1,500/g, with fungi, coliforms, *E. coli*, and *Staphylococcus aureus* counts of 25, 3, 3, and 10/g, respectively (150).

Why bacteria grow faster in the meat-soy blends than in nonsoy controls is not clear. The soy itself does not alter the initial flora, and the general spoilage pattern of meat-soy blends is not unlike that of all-meat controls. One notable difference is a slightly higher pH (0.3–0.4 unit) in soy-extended products, and this alone could account for the faster growth rate. This was assessed by Harrison et al. (57) by using organic acids to lower the pH of soy blends to that of beef. By adding small amounts of a 5% solution of acetic acid to 20% blends, spoilage was delayed about 2 days over controls, but not all of the inhibitory activity was due to pH depression alone. With 25% fat in the ground meat, bacterial counts did not increase proportionally to those of soy-extended beef (73). It is possible that soy protein increases the surface area of soy-meat mixtures so that aerobic bacteria of the type that predominate on meats at refrigerator temperatures are favored, but data along these lines are wanting. The spoilage of soy-meat blends is discussed further in Chapter 9, and the subject has been reviewed by Draughon (23).

---

* AOAC = Association of Official Analytical Chemists.

Mechanically Deboned Meat, Poultry, and Fish

When meat animals are slaughtered for human consumption, meat from the carcasses is removed by meat cutters. However, the most economical way to salvage the small bits and pieces of lean meat left on carcass bones is by mechanical means (mechanical deboning). Mechanically deboned meat (MDM) is removed from bones by machines. The production of MDM began in the 1970s, preceded by chicken meat in the late 1950s and fish in the late 1940s (33, 40). During the deboning process, small quantities of bone powder become part of the finished product, and the 1978 U.S. Department of Agriculture (USDA) regulation limits the amount of bone (based on calcium content) to no more than 0.75% (the calcium content of meat is 0.01%). MDM must contain a minimum of 14% protein and no more than 30% fat. The most significant parametrical difference between MDM and conventionally processed meat relative to microbial growth is the higher pH of the former, typically 6.0–7.0 (33, 34). The increased pH is due to the incorporation of marrow in MDM.

While most studies on the microbiology of MDM have shown these products to be not unlike those produced by conventional methods, some have found higher counts. The microbiological quality of deboned poultry was compared to other raw poultry products, and while the counts were comparable, MPN coliform counts of the commercial MDM products ranged from 460 to >1,100/g. Six of 54 samples contained salmonellae, four contained *C. perfringens*, but none contained *S. aureus* (107). The APC of hand-boned lamb breasts was found to be 680,000, while for mechanically deboned lamb allowed to age for 1 week, APC was 650,000/g (35). Commercial samples of mechanically deboned fish were found to contain tenfold higher numbers of organisms that conventionally processed fish, but different methods were used to perform the counts on fish frames and the mechanically deboned flesh (117). These investigators did not find *S. aureus* and concluded that the spoilage of MDF was similar to that for the traditionally processed products. In a later study, MDM was found to support the more rapid growth of psychrotrophic bacteria than lean ground beef (119).

Several studies have revealed the absence of *S. aureus* in MDM, reflecting perhaps the fact that these products are less handled by meat cutters. In general, the mesophilic flora count is a bit higher than that for psychrotrophs, and fewer gram negatives tend to be found. Field (33) concluded that with good manufacturing practices, MDM should present no microbiological problems, and a similar conclusion was reached by Froning (40) relative to deboned poultry and fish.

Hot-Boned Meats

In the conventional processing of meats (cold boning), carcasses are chilled after slaughter for 24+ h and processed in the chilled state (postrigor). Hot boning (hot processing) involves the processing of meats generally within 1–2 h after slaughter (prerigor) while the carcass is still "hot."

In general, the microbiology of hot-boned meats is comparable to that of cold-boned meats, but some differences have been reported. One of the earliest studies on hot-boned hams evaluated the microbiological quality of cured hams made from hot-boned meat (hot-processed hams). These hams were found to contain a significantly higher APC (at 37°C) than cold-boned hams, and 67% of the former yielded staphylococci to 47% of the latter (116). Mesophiles counted at 35°C were significantly higher on hot-boned prime cuts than comparable cold-boned cuts, both before and after vacuum-packaged storage at 2°C for 20 days (78). Coliforms, however, were apparently not affected by hot boning. Another early study is that of Barbe et al. (5), who evaluated 19 paired hams (hot and cold boned) and found that the former contained 200 bacteria/g, while 220/g were found in the latter. In a study of hot-boned carcasses held at 16°C and cold-boned bovine carcasses held at 2°C for up to 16 h postmortem, no significant differences in mesophilic and psychro-trophic counts were found (70). Both hot-boned and cold-boned beef initially contained low bacterial counts, but after a 14-day storage period, the hot-boned meats contained higher numbers than the cold boned (41). These investigators found that the temperature control of hot-boned meat during the early hours of chilling is critical and in a later study found that chilling to 21°C within 3–9 h was satisfactory (42).

In a study of sausage made from hot-boned pork, significantly higher counts of mesophiles and lipolytics were found in the product made from hot-boned pork than in the cold-boned product, but no significant differences in psychrotrophs were found (88).

The effect that delayed chilling might have on the flora of hot-boned beef taken about 1 h after slaughter was examined by McMillin et al. (91). Portions were chilled for 1, 2, 4, and 8 h after slaughter and subsequently ground, formed into patties, frozen, and examined. No significant differences were found between this product and a cold-boned product relative to coliforms, staphylococci, psychrotrophs, and mesophiles. A numerical taxonomy study of the flora from hot-boned and cold-boned beef at both the time of processing and after 14 days of vacuum storage at 2°C revealed no statistically significant differences in the flora (83). The predominant organisms, after storage, for both products were streptococci and lactobacilli, while in the freshly prepared hot-boned product (before storage), more staphylococci and bacilli were found. Overall, though, the two products were comparable.

Restructured lamb roast made from 10% and 30% MDM and hot-boned meat was examined for microorganisms, and overall the two uncooked products were of good quality (118). The uncooked products had counts <3.0 × 10⁴/g, with generally higher numbers in products containing the higher amounts of MDM. Coliforms and fecal coliforms especially were higher in products with 30% MDM, and this was thought to be caused by contamination of shanks and pelvic regions during slaughtering and evisceration. Not detected in either uncooked product (in 0.1 g) were *S. aureus* and *C. perfringens*, while no salmonellae, *Yersinia enterocolitica*, or *Campylobacter jejuni* were found in 25 g samples. Cooking reduced cell counts in all products to <30/g.

A summary of the work of ten groups of investigators made by Kotula (77) on the effect of hot boning on the microbiology of meats revealed that six found no effect, three found only limited effects, and only one found higher counts. Kotula concluded that hot boning per se has no effect on microbial counts. Hot boning is often accompanied by prerigor pressurization consisting of the application of around 15,000 psi for 2 min. This process improves muscle color and overall shelf appearance and increases tenderization. It appears not to have any effect on the microbial flora.

Effect of Electrical Stimulation.   Before hot boning, carcasses are electrically stimulated to speed the conversion of glycogen to lactic acid. The resulting rapid drop in carcass pH eliminates the toughening associated with cutting up prerigor meat. By this method, an electric stunner is attached to a carcass, and repeated pulses of 0.5 to 1.0 or more seconds are administered to the product at 400+ volt potential differences between the electrodes. A summary of the findings of ten groups of researchers on what effect, if any, electrical stimulation had on the microbial flora revealed that six found no effect, two found a slight effect, and two found some effect (77). The meats studied included beef, lamb, and pork.

Among investigators who found a reduction of APC by electrical stimulation were Ockerman and Szczawinski (106), who found that the process significantly reduced APC of samples of beef inoculated before electrical stimulation, but when samples were inoculated immediately after the treatment, no significant reductions occurred. The latter finding suggests that the disruption of lysosomal membranes and the consequent release of catheptic enzymes, which has been shown to accompany electrical stimulation (26), should not affect microorganisms. The tenderization associated with electrical stimulation of meats is presumed to be at least in part the result of lysosomal destruction (26). In one study, no significant reduction in surface flora was observed, whereas significant reduction was found to occur on the muscle above the aitch bone of beef carcasses (87). These workers exposed meatborne bacteria to electrical stimulation on culture media and found that gram-positive bacteria were the most sensitive to electrical stimulation, followed by gram negatives and sporeformers. When exposed to a 30-volt, 5-min treatment in saline or phosphate-buffered saline, a 5 log-cycle reduction occurred with *E. coli*, *Shewanella putrefaciens*, and *Pseudomonas fragi*, while in 0.1% peptone or 2.5 M sucrose solutions, essentially no changes occurred.

It appears that electrical stimulation per se does not exert measurable effects on the microbial flora of hot-boned meats.

Organ and Variety Meats

The meats discussed in this section are livers, kidneys, hearts, and tongues of bovine, porcine, and ovine origins. They differ from the skeletal muscle parts of the respective animals in having both higher pH and glycogen levels, especially in the case of liver. The pH of fresh beef and pork liver ranges from 6.1 to 6.5 and that of kidneys from 6.5 to 7.0. Most investigators have found generally low numbers of microorganisms on these products, with surface numbers ranging from $\log_{10} 1.69$ to $4.20/cm^2$ for fresh livers, kidneys, hearts, and tongues (53, 54, 105, 123). The initial flora has been reported to consist largely of gram-positive cocci, coryneforms, aerobic sporeformers, *Moraxella-Acinetobacter*, and *Pseudomonas* spp. (53, 54, 105, 123, 128). In a detailed study by Hanna et al. (54), micrococci, streptococci, and coryneforms were clearly the three most dominant groups on fresh livers, kidneys, and hearts. In one study, coagulase-positive staphylococci, coliforms, and *C. perfringens* counts ranged from $\log_{10} 0.19-1.37/cm^2$, but no salmonellae were found (123).

Vacuum-packaged beef and pork livers and beef kidneys had lower counts after 7- and 14-day storage at 2°C, or 7 days at 5°C, than those stored in air or PVC film (45, 54). As is the case for vacuum-packaged nonorgan meats, pH values of livers decreased during storage for up to 28 days concomitant with increases in homo- and heterofermentative lactic acid bacteria in livers and kidneys (54).

Vacuum-Packaged Meats

It is estimated that at least 80% of fresh beef in the United States leaves the packing plant in vacuum packages, and because of the longer shelf life of this product, its microbiological quality is of great interest. Vacuum packaging is achieved by placing meats into plastic bags or pouches followed by the removal of air using a vacuum packaging machine and the closure of bag or pouch, often by a heat sealer. From the microbiological standpoint, the most dramatic effect of this process is the change in the gaseous environment of the product. While not all O_2 is removed before sealing, some of what remains is consumed by the aerobic flora and the meat itself, resulting in an increased level of CO_2, which is inhibitory to the flora (see Chap. 3). The relative quantities of O_2 and CO_2 in the headspace of stored vacuum packages are controlled largely by the degree to which the plastic package impedes the flow of these gases.

A large number of investigators have shown that when O_2-permeable packaging is used, refrigerated fresh meats undergo gram-negative spoilage accompanied by increased pH and foul odors, with *Pseudomonas* spp. being the predominant organisms (see Chap. 9).

It has been shown that the storage life of vacuum-packaged beef is inversely related to film permeability, with the longest shelf life (>15 weeks) obtained with a "zero" O_2 film and the shortest (2–4 weeks) with a highly permeable film (920 cc $O_2/m^2/24$ h/atm at 25°C, R.H. 100%). Growth of *Pseudomonas* spp. increases with increasing film permeability (101). On the other hand, if O_2-impermeable packaging is used, the growth of lactic acid bacteria and sometimes that of *Brochothrix thermosphacta* is favored because of increased levels of CO_2 and a lowered oxidation-reduction potential (Eh) (124, 125, 154). These organisms typically effect a decrease in pH and create an unfavorable environment for most foodborne pathogens and gram-negative bacteria. With an initial pH of 5.45–5.65, vacuum-packaged beef inoculated with lactic acid bacteria showed pH of 5.30–5.45 after 24–35 days at 5°C (29). With vacuum-packaged cooked luncheon meats held for 16 days at 5°C, plate counts on LBS agar essentially equaled the APC on APT agar, indicating that the flora was composed essentially of lactics (74). After 15 weeks, 0.6–0.8% lactic acid was produced, and the pH of some products decreased to <5.0.

The findings by Henry et al. (61) for vacuum-packaged lamb chops and steaks are typical of those with other vacuum-packaged fresh meats. The initial count averaged $\log_{10} 2.53/cm^2$, and the flora consisted of streptococci, micrococci, coagulase-negative staphylococci, *Moraxella-Acinetobacter* types, and *Pseudomonas*. After a 4-day storage at 2°C in PVC film, the pseudomonads constituted 86% of the flora. However, after 5–15 days at 2°C in high O_2-barrier film, *Lactobacillus* spp. constituted 80–90% of the flora, with *L. cellobiosus* being the most prominent. The $\log_{10}$ APC after the 15-day storage was 6.26/ cm^2. In another study of vacuum-packaged lamb stored up to 21 days at 1–3°C, the initial flora was dominated by groups II and III *Corynebacterium* spp. and *B. thermosphacta*, while after 21 days, lactobacilli and *Moraxella-Acinetobacter* types dominated (55). For vacuum-packaged beef, *B. thermosphacta* constituted about 5% of the flora (148).

In a study by Patterson and Gibbs (110), pH 6.6 beef was vacuum packaged 1 day after slaughter. The initial surface count was log $2.48/cm^2$, and after 0.2°C storage for 8 weeks, the count increased to log $7.69/cm^2$. With regard to the initial flora, 85% consisted of *Aeromonas* spp. and 15% of gram-positive, catalase-positive bacteria. After 8 weeks, *B. thermosphacta* constituted 39% of the flora, homofermentative lactobacilli 22%, and psychrotrophic Enterobacteriaceae 39%. Most of the latter types (93%) were similar to *Serratia liquefaciens*. While vacuum packaging has been shown to extend the shelf life of both whole and cut-up fresh poultry (80), its effectiveness for poultry is highly temperature dependent, with temperatures at or below 0°C showing the greatest effect (92). In a comparison of vacuum and modified-atmosphere packaging (MAP) for the storage of venison at 1°C, MAP did not improve keeping quality over what was achieved by vacuum packaging (126).

Vacuum-packaged cooked meats have been studied in the same way as raw meats. Of 113 samples of sliced bologna, 63% contained $<10^2$ and 93% $<10^4/g$ of *B. thermosphacta*, while *C. perfringens*, *S. aureus*, and salmonellae were generally absent (109). Only 3 of 153 isolates from vacuum-packaged sliced cooked meat products were *B. thermosphacta*, while most of the others were unclassified lactic cocci and rods (94). With meat loaves packaged in vacuum or nitrogen, no significant differences in numbers of organisms were found after 49-day storage at temperatures from −4 to 7°C, but the initial flora of *Pseudomonas* (32–34%), *Brochothrix* (24–38%), micrococci (9–22%), and lactobacilli (7–20%) was changed to predominantly lactobacilli (62–72%) among the psychrotrophic flora by day 49 for both treatments (82). In another study, beef roasts were stored in a combination of atmospheres ranging from 100% O_2 to no O_2, and controls were vacuum packaged, but after 35 days at 1–3°C, counts of psychrotrophic flora and lactobacilli were not statistically significant (18). The possible reason for this was the high initial counts of about $10^6/in^2$.

When nitrites are present in vacuum-packaged meats, domination of the flora by lactic acid bacteria is even more pronounced, since these organisms are insensitive to nitrite and since the microaerophilic environment is more favorable to them than to gram negatives (103). *B. thermosphacta* and the Enterobacteriaceae were more inhibited by nitrites under these conditions. Although lactics outgrow the more aerobic types in the vacuum package environment, their lag phase is increased under these conditions, resulting in increased product shelf life.

The relative dominance of the flora of vacuum-packaged meats by lactobacilli versus *B. thermosphacta* is an interesting one. The latter organism is more sensitive to nitrites than the former so that its relative absence in nitrite-treated products would be expected. Lamb chops with a storage life of 2 weeks at −1°C in air showed a storage life increase of 8 weeks in O_2-free atmospheres, and *B. thermosphacta* was the major organism in each of a combination of gas atmospheres, including nitrogen, O_2, and CO_2 (100). It has been reported that lactobacilli markedly inhibit the growth of *B. thermosphacta* in vacuum-packaged beef when the two organisms are present in approximately equal numbers but not on beef incubated in air (122). In the latter study, up to 75% CO_2 did not inhibit the ultimate growth of *B. thermosphacta*. However, when the initial flora of vacuum-packaged meats contains significant numbers of *B. thermosphacta*, this organism grows faster than the homo- or heterofermentative lactics and becomes dominant (27).

Reports vary on the relative dominance by *B. thermosphacta* and the lactic acid bacteria in vacuum-packaged meats. In one study, vacuum-packaged luncheon meats with lactobacilli as the dominant flora were considered acceptable up to 21 days with counts of $10^8/g$ (27). These investigators found that more homofermentative lactics were required to produce product offness than heterolactics because of the end products produced by the latter. While about 9 days were required for *B. thermosphacta* and heterofermentative lactics to reach a count of $10^8/g$, homofermentative lactics required 12–20 days. With N_2- and CO_2-packaged frankfurters stored at $-4°$ to $7°C$ for 49 days, *B. thermosphacta* decreased from about 48% to about 5%, while lactobacilli increased from around 6% to 94–96% under CO_2 (130). With raw pork stored at $2°C$ for 14 days, about 30% of the flora consisted of lactobacilli, about 18% of *B. thermosphacta*, and the remainder of pseudomonad types (46). The initial counts were approximately $10^3/g$, and the CO_2 content in packages was around 12.3%.

In a study of the populations of lactic acid bacteria on vacuum-packaged beef, 18 of 177 psychrotrophic lactic acid bacteria were *Leuconostoc mesenteroides*, and the remainder were lactobacilli, atypical streptobacteria, or atypical betabacteria (61). From vacuum-packaged beef loins stored at $4°C$, fewer pseudomonads and more lactobacilli were found after 3 weeks than in controls (99). While some of the reported variations may result from product differences, it is not easy to classify correctly the lactic rods to genus and species, and until a common taxonomic scheme is followed by all investigators, variations in the lactic flora of vacuum-packaged meats as reported by different investigators may be expected to continue.

With regard to safety, raw beef was inoculated with types A and B spores of *Clostridium botulinum* and stored for up to 15 days at $25°C$ in one study. Toxin was first detected after 6 days, always accompanied by significant organoleptic changes indicating that the vacuum-packaged toxic samples should be rejected before consumption (60). In another study, the storage of vacuum-packaged, N_2-flushed, CO_2-flushed packages of fresh cod, whiting, or flounder fillets that were inoculated with five strains of type E *C. botulinum* spores resulted in toxin production in all products after 1–3 days when they were stored at $26°C$, at which time the products were generally unacceptable organoleptically (112). Continuous storage of cod fillets at $4°C$ resulted in toxin being detected prior to sensory rejection. The vacuum packaging of cooked bologna-type sausage resulted in the restriction by the normal flora of growth of *Yersinia enterocolitica* and salmonellae but not *S. aureus* (104). *C. perfringens* was completely inhibited by the normal flora, and all pathogens were inhibited by the lactic flora, with greater inhibition occurring as the storage temperature was decreased.

Overall, vacuum-packaged meats of all types may be regarded as being safe and generally free of foodborne pathogens with the possible exception of *Y. enterocolitica* and *S. aureus*. The generally present lactic flora represses the growth of most nonlactics, and this inhibitory effect cannot be explained by the reduced pH alone. The spoilage of vacuum-packaged meats is discussed in Chapter 9.

Sausage, Bacon, Bologna, and Related Meat Products

In addition to the meat components, sausages and frankfurters have additional sources of organisms in the seasoning and formulation ingredients that are usually added in their production. Many spices and condiments have high microbial counts. The lactic acid bacteria and yeasts in some composition products are usually contributed by milk solids. In the case of pork sausage, natural casings have been shown to contain high numbers of bacteria. In their study of salt-packed casings, Riha and Solberg (121) found counts to range from log 4.48 to 7.77 and from 5.26 to 7.36 for wet-packed casings. Over 60% of the isolates from these natural casings consisted of *Bacillus* spp. followed by clostridia and pseudomonads. Of the individual ingredients of fresh pork sausage, casings have been shown to contribute the largest number of bacteria (146).

Processed meats such as bologna and salami may be expected to reflect the sum of their ingredient makeup with regard to microbial numbers and types. The microflora of frankfurters has been shown to consist largely of gram-positive organisms with micrococci, bacilli, lactobacilli, microbacteria, strepto-cocci, and leuconostocs along with yeasts (22). In a study of slime from frank-furters, these investigators found that 275 of 353 isolates were bacteria, and 78 were yeasts. *B. thermosphacta* was the most conspicuous single isolate. With regard to the incidence of *C. botulinum* spores in liver sausage, 3 of 276 heated (75°C for 20 min) and 2 of 276 unheated commercial preparations contained type A botulinal toxin (59). The most probable number of botulinal spores in this product was estimated to be 0.15/kg.

Wiltshire bacon has been reported to have a total count generally in the range of log 5–6/g (65), while high-salt vacuum-packaged bacon has been reported to have a generally lower count—about log 4/g. The flora of vacuum-packaged sliced bacon consists largely of catalase-positive cocci, such as micro-cocci and coagulase-negative staphylococci, as well as catalase-negative bacteria of the lactic acid types, such as lactobacilli, leuconostocs, pediococci, and group D streptococci (2, 15, 76). The flora in cooked salami has been found to consist mostly of lactobacilli.

So-called soul foods may be expected to contain high numbers of organisms since they consist of offal parts that are in direct contact with the intestinal-tract flora, as well as other parts, such as pig feet and pig ears, that do not receive much care during slaughtering and processing. This was confirmed in a study by Stewart (140), who found the geometric mean APC of log 7.92/g for chitterlings (pig intestines), 7.51/g for maws, and 7.32/g for liver pudding. For *S. aureus*, log 5.18, 5.70, and 5.15/g respectively were found for chitterlings, maws, and liver pudding.

POULTRY

Whole poultry tends to have a lower microbial count than cut-up poultry. Most of the organisms on such products are at the surface, so surface counts/cm^2 are generally more valid than counts on surface and deep tissues. May (90) has shown how the surface counts of chickens build up through successive stages of processing. In a study of whole chickens from six commercial processing plants, the initial mean total surface count was log 3.30/cm^2. After the chickens were cut up, the mean total count increased to log 3.81 and further increased to log 4.08 after packaging. The conveyor over which these birds moved showed a count of log 4.76/cm^2. When the procedures were repeated for five retail grocery stores, May found that the mean count before cutting was log 3.18, which increased to log 4.06 after cutting and packaging. The cutting block was shown to have a total count of log 4.68/cm^2.

The changes in enteric bacteria during various stages of poultry chilling were studied by Cox et al. (19). Carcass counts before chilling were log 3.17/cm^2 for APC and log 2.27/cm^2 for Enterobacteriaceae. However, after chilling, the

TABLE 4-2.   General Microbiological Quality of Turkey Meat Products

Products	Number of Samples	Microbial Group/Target	% Samples Meeting Target	References
Precooked	6	APC: log 3.00/g	100	93
turkey	6	Coliforms: log 2.00 or less/g	67	93
rolls	6	Enterococci: log 2.00 or less/g	83	93
	48	Presence of salmonellae	4	93
	48	Presence of C. perfringens	0	93
Precooked	30	APC: <log 2.00/g	20	160
turkey rolls/	29	Presence of coliforms	21	160
sliced turkey meat	29	Presence of E. coli or salmonellae	0	160
Ground fresh	74	APC: log 7.00 or less/g	51	51
turkey meat	75	Presence of coliforms	99	51
	75	Presence of E. coli	41	51
	75	Presence of "fecal streptococci"	95	51
	75	Presence of S. aureus	69	51
	75	Presence of salmonellae	28	51
Frozen ground	50	APC 32°C: <10^6/g	54	52
turkey meat	50	Psychrotrophs: <10^6/g	32	52
	50	MPN E. coli: <10/g	80	53
	50	MPN S. aureus: <10/g	94	52
	50	MPN "fecal streptococci": <10/g	54	52
Processed cheese	234	Aerobic sporeformers, <10^3	86	156
products	234	Anaerobic spores <10^3	91	156
	234	E. coli, 1.8/g	100	156

latter organisms were reduced more than the APC. *Escherichia* spp. constituted 85% of enterics at day 0, but after 10 days at 4°C, they were reduced to 14%, while *Enterobacter* spp. increased from 6% to 88% during the same time.

Poultry represents an important food source of salmonellae to humans. Of 50 frozen comminuted turkey meat samples examined, 38% yielded salmonellae (52). Their incidence in dressed broiler-fryer chickens was investigated by Woodburn (159). It was found that 72 of 264 birds (27%) harbored salmonellae representing 13 serovars. Among the serovars, *S. infantis*, *S. reading*, and *S. blockley* were the most common. Salmonellae were isolated from the surfaces of 24 of 208 (11.5%) turkey carcasses before further processing (11). After processing into uncooked rolls, 90 of 336 (26.8%) yielded salmonellae. From the processing plants, 24% of processing equipment yielded salmonellae. Almost one-third of the workers had the organisms on their hands and gloves. Of 23 serovars recovered, *S. sandiego* and *S. anatum* were recovered most frequently. In fresh-ground turkey meats, salmonellae were found in 28% of 75 samples by another group of workers (51). Almost one-half of the samples had total counts above log 7.00/g. Ninety-nine percent harbored coliforms, 41% *E. coli*, 52% *C. perfringens*, and 69% *S. aureus*. About 14% of 101 chicken samples were positive for fecal coliforms by MPN (3).

Of the various cooked poultry products, precooked turkey rolls have been found to have considerably lower microbial numbers of all types (Table 4-2). In an examination of 118 samples of cooked broiler products, *C. perfringens* was found in 2.6% (86).

The microbial flora of fresh poultry consists largely of pseudomonads and other closely related gram-negative bacteria, as well as coryneforms, yeasts, and other organisms (79). Other microorganisms normally present on poultry products are discussed in Chapter 9, and more on salmonellae in poultry can be found in Chapter 22.

SEAFOOD

The incidence of microorganisms in seafood such as shrimp, oysters, and clams depends greatly on the quality of water from which these animals are harvested. Assuming good-quality waters, most of the organisms are picked up during processing. In the case of breaded raw shrimp, the breading process is expected to add organisms if not properly done or if the ingredients are of poor microbiological quality. In their study of 91 samples of shrimp of various types, Silverman et al. (129) found that all precooked samples except one had total counts of < log 4.00/g. Of the raw samples, 59% had total counts below log 5.88, while 31% were below log 5.69/g. In a study of 204 samples of frozen, cooked, and peeled shrimp, 52% had total counts < log 4.70/g, and 71% had counts of log 5.30 or less/g (85). The general microbiological quality of a variety of seafood is presented in Table 4-3.

In a study of haddock fillets, most microbial contamination was found to occur during filleting and subsequent handling prior to packaging (102). These investigators showed that total count increased from log 5.61 in the morning to log 5.65 at noon and to log 5.94/g in the evening for one particular processor. According to their study, results obtained in other companies were generally similar if the nighttime cleanup was good. In the case of shucked, soft-shell

TABLE 4-3.   General Microbiological Quality of Various Seafood Products

Products	Number of Samples	Microbial Group/Target	% Samples Meeting Target	References
Frozen catfish	41	APC 32°C: 10^5/g or less	100	36
fillet	41	MPN coliforms: <3/g	100	36
	41	MPN *S. aureus:* <3/g	100	36
Frozen salmon	43	APC 32°C: 10^5/g or less	98	36
steaks	43	MPN coliforms: <3/g	93	36
	43	MPN *S. aureus:* <3/g	98	36
Fresh clams	53	APC 32°C: 10^5/g or less	53	36
	53	MPN coliforms: <3/g	51	36
	53	MPN *S. aureus:* <3/g	91	36
Fresh oysters	59	APC 32°C: 10^7/g or less	49	36
	59	MPN coliforms: 1,100 or less/g	22	36
	59	MPN *S. aureus:* <3/g	90	36
Shucked oysters	1,337	APC 30°C: 10^6/g or less	51	157
(retail)	1,337	MPN coliforms: 460 or less/g	94	157
	1,337	MPN fecal coliforms: 460 or less/g	96	157
Blue crabmeat	896	APC 30°C: 10^6/g or less	61	157
(retail)	896	MPN coliforms: 1,100/g or less	93	157
	896	MPN *E. coli:* <3/g	97	157
	896	MPN *S. aureus:* 1,100/g or less	94	157
Hard-shell clams	1,124	APC 30°C: 10^6/g or less	99.8	157
(wholesale)	1,130	MPN coliforms: 460/g or less	96	157
	161	MPN fecal coliforms: <3/g	91	157
Soft-shell clams	351	APC 30°C: 10^6/g or less	96	157
(wholesale)	363	MPN coliforms: 460/g or less	98	157
	75	MPN fecal coliforms: <3/g	72	157
Peeled shrimp	1,468	APC 30°C: 10^7/g or less	94	149
(raw)	1,468	MPN coliforms: 64/g or less	97	149
	1,468	MPN *E. coli:* <3/g	97	149
	1,468	MPN *S. aureus:* 64/g or less	97	149
Peeled shrimp	1,464	APC 30°C: 10^5/g or less	81	149
(cooked)	1,464	MPN coliforms: <3/g	86	149
	1,464	MPN *E. coli:* <3/g	99	149
	1,464	MPN *S. aureus:* <3/g	99	149
Lobster tail	1,315	APC 30°C: 10^6/g or less	74	149
(frozen, raw)	1,315	MPN coliforms: 64 or less/g	91	149
	1,315	MPN *E. coli:* <3/g	95	149
	1,315	MPN *S. aureus:* <3/g	76	149
Retail frozen,	27	APC: 6.00 or less/g	52	154
breaded, raw	27	Coliforms: 3.00 or less/g	100	154
shrimp	27	Presence of *E. coli*	4	154
	27	Presence of *S. aureus*	59	154

TABLE 4-3.   (Continued)

Products	Number of Samples	Microbial Group/Target	% Samples Meeting Target	References
Fresh channel	335	APC: ≤7.00/g	93	4
catfish	335	Fecal coliforms: 2.60/g	70.7	4
	335	Presence of salmonellae	4.5	4
Frozen channel	342	APC: ≤7.00/g	94.5	4
catfish	342	Fecal coliforms: 2.60/g	92.4	4
	342	Presence of salmonellae	1.5	4
Frozen, cooked,	204	APC: <4.70/g	52	85
peeled shrimp	204	APC: 5.30 or less/g	71	85
	204	Coliforms: none or <0.3/g	52.4	85
	204	Coliforms: <3/g	75.2	85

clams, the same general pattern of buildup was demonstrated from morning to evening. The mean clostridial count for both haddock fillets and soft-shell clams was less than 2/g, with clams being slightly higher than haddock fillets for these organisms, although both were low. Total counts on fresh perch fillets produced under commercial conditions were found to average log 5.54/g with yeast and mold counts of about log 2.69/g (72).

Clams may be expected to contain the organisms that inhabit the waters from which they are obtained. Of 60 clam samples from the coast of Florida, 43% contained salmonellae, which were also found in oysters at a level of 2.2/100 g oyster meats (39). Hard-shell clams have been shown to retain *S. typhimurium* more efficiently than *E. coli* (152).

In a study of the flora of raw Pacific shrimp taken from docks, *Moraxella* spp. constituted 30–60%, followed by types I and II pseudomonads (8–22%), *Acinetobacter* (4–24%), *Flavobacterium-Cytophaga* (7–16%), and *Pseudomonas* types III and IV (8–22%); but following blanching and machine peeling, *Acinetobacter* spp. represented 16–35%, *Pseudomonas* types III and IV 2–76%, and *Flavobacterium-Cytophaga* 3–37% (84). The initial flora of herring fillets has been found to be dominated by *S. putrefaciens* and *Pseudomonas* spp., with the latter dominating at 2°C and *S. putrefaciens* more predominant at 2–15°C (95). In a *C. perfringens* survey of 287 retail samples of fresh fish and shellfish, 10% were positive for this organism (1). Freshly harvested shrimp has been shown to contain *Listeria monocytogenes* at low levels (see Chap. 21).

In general, frozen seafood and other frozen products have lower microbial counts than the comparable fresh products. In a study of 597 fresh and frozen seafood from retail stores, APC geometric means for the 240 frozen products ranged from log 3.54–4.97/g and from 4.89–8.43/g for the 357 fresh products (36). For coliforms, geometric mean MPN counts ranged from 1 to 7.7 cells/g for frozen and from 7.8 to 4,800/g for fresh. By MPN, only 4.7% of the 597 were positive for *E. coli*, 7.9% were positive for *S. aureus*, and 2% were positive for *C. perfringens*. All were negative for salmonellae and *Vibrio parahaemolyticus* (Table 4-3).

Plate counts are generally higher on seafood when incubated at 30°C than at 35°C, and this is reflected in results from fresh crabmeat, clams, and oysters evaluated by Wentz et al. (157). APC geometric means for 896 crabmeat samples at 35°C was log 5.15 and 5.72 at 30°C; for 1,337 shucked oysters 5.59 at 35° and 5.95 at 30°; and for 358 soft-shell clams, log APC was 2.83 at 35° and 4.43 at 30°. This was also seen in raw in-shell shrimp and frozen raw lobster tails, where geometric mean APC for shrimp at 35°C was log 5.48 and 5.90/g at 30°C, while for lobster tail, 4.62 at 35° and 5.15/g at 30° (149).

VEGETABLES

The incidence of microorganisms in vegetables may be expected to reflect the sanitary quality of the processing steps and the microbiological condition of the raw product at the time of processing. In a study of green beans before blanching, Splittstoesser et al. (137) showed that the total counts ranged from log 5.60 to over 6.00 in two production plants. After blanching, the total numbers were reduced to log 3.00–3.60/g. After passing through the various processing stages and packaging, the counts ranged from log 4.72 to 5.94/g. In the case of french-style beans, one of the greatest buildups in numbers of organisms occurred immediately after slicing. This same general pattern was shown for peas and corn. Preblanch green peas from three factories showed total counts/g between log 4.94 and 5.95. These numbers were reduced by blanching and again increased successively with each processing step. In the case of whole-kernel corn, the postblanch counts rose both after cutting and at the end of the conveyor belt to the washer. Whereas the immediate postblanch count was about log 3.48, the product had total counts of about log 5.94/g after packaging. Between 40 and 75% of the bacterial flora of peas, snap beans, and corn was shown to consist of leuconostocs and "streptococci," while many of the gram-positive, catalase-positive rods resembled corynebacteria (135, 136).

Lactic acid cocci have been associated with many raw and processed vegetables (96). These cocci have been shown to constitute from 41–75% of the APC flora of frozen peas, snap beans, and corn (132). It has been shown that fresh peas, green beans, and corn all contained coagulase-positive staphylococci after processing (135). Peas were found to have the highest count (log 0.86/g), while 64% of corn samples contained this organism. These authors found that a general buildup of staphylococci occurred as the vegetables underwent successive stages of processing, with the main source of organisms coming from the hands of employees. Although staphylococci may be found on vegetables during processing, they are generally unable to proliferate in the presence of the more normal lactic flora. Both coliforms (but not *E. coli*) and enterococci have been found at most stages during raw vegetable processing, but they appear to present no public health hazard (131).

In a study of the incidence of *C. botulinum* in 100 commercially available frozen vacuum pouch-pack vegetables, the organism was not found in 50 samples of string beans, but types A and B spores were found in 6 of 50 samples of spinach (66). The general microbiological quality of some vegetables is presented in Table 4-4.

TABLE 4-4. General Microbiological Quality of Frozen Vegetables and Cheddar Cheese

Products	Number of Samples	Microbial Group/Target	% Samples Meeting Target	References
Cauliflower	1,556	APC at 35°C: 10^5/g or less	75	6
	1,556	MPN coliforms: <20/g	79	6
	1,556	MPN *E. coli:* <3/g	98	6
Corn	1,542	APC at 35°C: 10^5/g or less	94	6
	1,542	MPN coliforms: <20/g	71	6
	1,542	MPN *E. coli:* <3/g	99	6
Peas	1,564	APC at 35°C: 10^5/g or less	95	6
	1,564	MPN coliforms: <20/g	78	6
	1,564	MPN *E. coli:* <3/g	99	6
Blanched vegetables (17 different)	575	Absence of fecal coliforms	63	136
	575	$n = 5, c = 3, m = 10, M = 10^3$	33	136
	575	$n = 5, c = 3, m = 50, M = 10^3$	70	136
Cut green beans, leaf spinach, peas	144	Mean APC range for group: log 4.73–4.93/g	—	133
Lima beans, corn, broccoli spears, brussels sprouts	170	Mean APC range for group: 5.30–5.36/g	—	133
French-style green beans, chopped greens, squash	135	Mean APC range: log 5.48–5.51/g	—	133
Chopped spinach, cauliflower	80	Mean APC range: log 5.54–5.65/g	—	133
Chopped broccoli	45	Mean APC: 6.26/g	—	133
Freshly formed cheddar cheese	236	<500 coliforms/g	95	10
	237	<1,000 *S. aureus*/g	100	10
	250	Presence of salmonellae	0	10

In a study of 575 packages of frozen vegetables processed by 24 factories in 12 states, Splittstoesser and Corlett (133) found that peas yielded some of the lowest counts (mean of approximately log 1.93/g), while chopped broccoli yielded the highest mean APCs—log 3.26/g. Using the three-class sampling plan of the International Commission on Microbiological Specifications for Foods (ICMSF), the acceptance rate for the 115 lots would have been 74% for the m specification of 10^5/g and 84% for M of 10^6/g. In a study of 17 different frozen blanched vegetables, 63% were negative for fecal coliforms, and 33% of the 575 examined were acceptable when $n = 5$, $c = 3$, $m = 10$, and $M = 10^3$, while 70% were acceptable if $n = 5$, $c = 3$, $m = 50$, and $M = 10^3$ (136). In another study, mean APC at 30°C for 1,556 frozen retail cauliflower samples was log 4.65/g; for 1,542 sample units of frozen corn, log 3.93/g; and for 1,564 units of frozen peas, log 3.83/g with 5/g or less of coliforms and <3/g of $E.$ $coli$ for all samples (6). Based on APC, 97.2–99.6% of the latter foods were acceptable by ICMSF's sampling plan $n = 5$, $c = 3$, $m = 10^5$, and $M = 10^6$.

DAIRY PRODUCTS

The microbial flora of raw milk consists of those organisms that may be present on the cow's udder and hide and on milking utensils or lines. Under proper handling and storage conditions, the predominant flora is gram positive. While yeasts, molds, and gram-negative bacteria may be found along with lactic acid bacteria, most or all of these types are more heat sensitive than gram positives and are more likely to be destroyed during pasteurization. Studies have revealed the presence of psychrotrophic sporeformers and mycobacteria in raw milks. Psychrotrophic *Bacillus* spp. were found in 25–35% of 97 raw milk samples (126). These organisms were shown to grow at or below 7°C. Psychrotrophic clostridia were isolated from 4 of 48 raw milk samples (9). Sporeformers can survive pasteurization temperatures in their spore state and cause problems in the refrigerated products because of their proteolytic abilities. *Mycobacterium* and *Nocardia* spp. have been isolated from about 69% of 51 raw milks (64), but 43 samples of pasteurized milk were negative. The presence of these organisms in raw milk is not surprising since they are quite abundant in soils. Also abundant in soils are psychrotrophic pseudomonads, and they are not uncommon in raw milks.

The incidence of microorganisms in commercial cream cheese was studied by Fanelli et al. (31), who found that the total count/g was about log 2.69. In commercial sour cream, the total count/g ranged from log 4.79–8.58, with most being lactic acid bacteria. Commercial onion dips with sour cream bases had total counts of log 7.76–8.28 and log 4.34/g of yeasts and molds. The majority of bacteria in this product were found to be lactic acid types also. The microbial flora of fermented dairy products is discussed in Chap. 16.

The potential botulism hazard posed by imitation cheeses was investigated by Kautter et al. (71). Spores of *C. botulinum* types A and B were inoculated into 50 samples of 11 imitation cheeses with a_w ranging from 0.942–0.973 and pH from 5.53–6.14, with examination for toxin after 289 days. Only one sample became toxic. From other tests, toxic samples were all organoleptically unacceptable, indicating that these products do not present a botulism hazard.

Milk continues to serve as a vehicle for certain disease, and an excellent review on the subject has been prepared by Bryan (11). Outbreaks generally involve the consumption of raw milk, certified raw milk, homemade ice cream containing fresh eggs, and dried and pasteurized milks contaminated *after* heat processing. For 1980–1981, over 538 cases of salmonellosis in the United States were traced to dairy products with cheddar cheese, raw milk, and certified raw milk as vehicles. During 1980–1982, over 172 cases of campylobacteriosis were traced to raw milk and certified raw milk in the United States, and milkborne outbreaks have been reported in Canada and several European countries. For nine different microbial diseases traced to dairy products for the period 1970–1979, ice cream (often contaminated by eggs) was the leading vehicle, followed by cheese and unspecified types of milk (11).

DELICATESSEN AND RELATED FOODS

Delicatessen foods, such as salads and sandwiches, are sometimes involved in food-poisoning outbreaks. These foods are often prepared by hand, and this direct contact may lead to an increased incidence of food-poisoning agents such as *S. aureus*. Once organisms such as these enter meat salads or sandwiches, they may grow well because of the reduction in numbers of the normal food flora by the prior cooking of salad ingredients.

In a study of retail salads and sandwiches, 36% of 53 salads were found to have total counts > log 6.00/g, but only 16% of the 60 sandwiches had counts as high (17). With respect to coliforms, 57% of sandwiches were found to harbor < log 2.00/g. *S. aureus* was present in 60% of sandwiches and 39% of salads. Yeasts and molds were found in high numbers, with six samples containing > log 6.00/g.

In a study of 517 salads from around 170 establishments, 71–96% were found to have APC < log 5.00/g (109). Ninety-six to 100% of salads contained coagulase-positive *S. aureus* at levels < log 2.00/g. Salads included chicken, egg, macaroni, and shrimp. *S. aureus* was recovered in low numbers from 6 of 64 salads in another study (37). The 12 different salads examined by these investigators had total counts between log 2.08 and 6.76, with egg, shrimp, and some of the macaroni salads having the highest counts. Neither salmonellae nor *C. perfringens* were found in any product. A study of 42 salads by Harris et al. (57) revealed the products to be of generally good microbial quality. Mean APC was log 5.54/g, and mean coliform counts were log 2.66/g for the six different products. Staphylococci were found in some products, especially ham salad.

Fresh green salads (green, mixed green, and coleslaw) were found to contain mean total counts of log 6.67 for coleslaw to log 7.28 for green salad (38). Fecal coliforms were found in 26% of mixed, 28% of green, and 29% of coleslaw, while the respective percentage findings for *S. aureus* were 8, 14, and 3. With respect to parsley, *E. coli* was found on 11 of 64 samples of fresh and unwashed products and on over 50% of frozen samples (68). Mean APC of fresh washed parsley was log 7.28/g. Neither salmonellae nor *S. aureus* were found in any samples.

In a study of the microbiological quality of imitation-cream pies from plants operated under poor sanitary conditions, Surkiewicz (143) found that the microbial load increased successively as the products were carried through the various processing steps. For example, in one instance the final mixture of the synthetic pie base contained fewer than log 2.00 bacteria/g after final heating to 160°F. After overnight storage, however, the count rose to log 4.15. The pie topping ingredients to be mixed with pie base had a rather low count: log 2.78/g. After being deposited on the pies, the pie topping showed a total count of log 7.00/g. In a study of the microbiological quality of french fries,

TABLE 4-5.   General Microbiological Quality of Miscellaneous Food Products

Products	Number of Samples	Microbial Group/Target	% Samples Meeting Target	References
Frozen cream-type pies	465	APC: $\leq 10^4$/g	96	153
	465	Fungi: 10^3/g or less	98	153
	465	Coliforms: <10/g	89	153
	465	E. coli: 10/g or less	99	153
	465	S. aureus: <25/g	99	153
	465	O salmonellae	100	153
Frozen breaded onion rings (pre- or partially cooked)	1,590	APC 30°C: 10^5/g or less	99	158
	1,590	MPN coliforms: <3/g	89	158
	1,590	MPN E. coli: <3/g	99	158
	1,590	MPN S. aureus: <10/g	99.6	158
Frozen tuna pot pies	1,290	APC 30°C: 10^5/g or less	97.6	158
	1,290	MPN coliforms: 64/g or less	93	158
	1,290	MPN E. coli: <3/g	97	158
	1,290	MPN S. aureus: <10/g	98	158
Tofu (commercial)	60	APC: $>10^6$/g	83	120
	60	Psychrotrophs: $<10^4$/g	83	120
	60	Coliforms: $\sim 10^3$/g	67	120
	60	S. aureus: <10/g	100	120
Dry food-grade gelatin	185	APC: 3.00 or less/g	74	85
Delicatessen salads	764	Within Army and Air Force Exchange Service microbial limits	44	36
	764	APC: 5.00 or less/g	84	36
	764	Coliforms: 1.00 or less/g	78	36
	764	Yeasts and molds: 1.30 or less/g	55	36
	764	"Fecal streptococci": 1.00/g	77	36
	764	Presence of S. aureus	9	36
	764	Pres. of C. perfringens; salmonellae	0	36
	517	APC: 5.00 or less/g	26–85	108
	517	Coliforms: 2.00 or less/g	36–79	108
	517	S. aureus: 2.00 or less/g	96–100	108
Retail trade salads	53	APC: >6.00/g	36	17
	53	Coliforms: 2.00 or less/g	57	17
	53	Presence of S. aureus	39	17
Retail trade sandwiches	62	APC: >6.00/g	16	17
	62	Coliforms: >3.00/g	12	17
	62	Presence of S. aureus	60	17

TABLE 4-5.   (Continued)

Products	Number of Samples	Microbial Group/Target	% Samples Meeting Target	References
Imported spices	113	APC: 6.00 or less/g	73	67
and herbs	114	Spores: 6.00 or less/g	75	67
	113	Yeasts and molds: 5.00 or less/g	97	67
	114	TA spores: 3.00 or less/g	70	67
	114	Pres. of *E. coli, S. aureus,* salmonellae	0	67
Processed spices	114	APC: 5.00 or less/g	70	115
	114	APC: 6.00 or less/g	91	115
	114	Coliforms: 2.00 or less/g	97	115
	114	Yeasts and molds: 4.00 or less/g	96	115
	114	*C. perfringens:* <2.00/g	89	115
	110	Presence of *B. cereus*	53	114
Dehydrated	129	APC: <4.00/g	93	113
space foods	129	Coliforms: <1/g	98	113
	129	*E. coli:* negative in 1 g	99	113
	102	"Fecal streptococci": 1.30/g	88	113
	104	*S. aureus:* negative in 5 g	100	113
	104	Salmonellae: negative in 10 g	98	113

Surkiewicz et al. (144) demonstrated the same pattern—that is, the successive buildup of microorganisms as the fries underwent processing. Since these products are cooked late in their processing, the incidence of organisms in the finished state does not properly reflect the actual state of sanitation during processing.

Geometric mean APC of 1,187 sample units of refrigerated biscuit dough was found to be 34,000/g, while for fungi, coliforms, *E. coli*, and *S. aureus*, the mean counts were 46, 11, <3, and <3/g, respectively (150). In the same study, geometric mean APC of 1,396 units of snack cake was 910/g, with <3/g of coliforms, *E. coli*, and *S. aureus* (see Table 4-5).

A bacteriological study of 580 frozen cream-type pies (lemon, cocount, chocolate, and banana) showed them to be of excellent quality, with 98% having APC of log 4.70 or less/g (85). The overall microbiological quality of other related products is presented in Table 4-5.

FROZEN MEAT PIES

The microbiological quality of frozen meat pies has steadily improved since these products were first marketed. Any and all of the ingredients added may increase the total number of organisms, and the total count of the finished product may be taken to reflect the overall quality of ingredients, handling, and storage. Many investigators have suggested that these products should be produced with total counts not to exceed log 5.00/g. In a study of 48 meat pies, 84% had APC < log 5 (88), while in another study of 188 meat pies, 93% had counts less than log 5.00 (75). Accordingly, a microbiological criterion of log 5.00 seems attainable for such products (see Chap. 18 for further information on microbiological standards and criteria).

In a study of 1,290 frozen tuna pot pies, geometric mean APC at 35°C was log 3.20 while at 30°C it was log 3.38/g (158). Coliforms averaged 5/g, *E. coli* <3/g, and *S. aureus* <10/g (Table 4-5).

DEHYDRATED FOODS

In a detailed study of the microbiology of dehydrated soups, Fanelli et al. (30) showed that approximately 17 different kinds of dried soups from nine different processors had total counts of less than log 5.00/g. These soups included chicken noodle, chicken rice, beef noodle, vegetable, mushroom, pea, onion, tomato, and others. Some of these products had total counts as high as log 7.30/g, and some had counts as low as around log 2.00. These investigators further found that reconstituted dehydrated onion soup showed a mean total count of log 5.11/ml, with log 3.00 coliforms, log 4.00 aerobic sporeformers, and log 1.08/ml of yeast and molds. Upon cooking, the total counts were reduced to a mean of log 2.15, while coliforms were reduced to < log 0.26, sporeformers to log 1.64, and yeasts and molds to < log 1.00/ml. In a study of dehydrated sauce and gravy mixes, soup mixes, spaghetti sauce mixes, and cheese sauce mixes, *C. perfringens* was isolated from 10 of 55 samples (98). The facultative anaerobe counts ranged from log 3.00 to > log 6.00/g.

In a study of 185 samples of food-grade dry gelatin, no samples exceeded an APC of log. 3.70/g (85). Of 129 dehydrated space food samples examined, 93% contained total counts < log 4.00/g (113).

Powdered eggs and milk often contain high numbers of microorganisms—on the order of log 6–8/g. One reason for the generally high numbers in dried products is that the organisms have been concentrated on a per gram basis along with product concentration. The same is generally true for fruit juice concentrates, which tend to have higher numbers of microorganisms than the fresh, nonconcentrated products.

I have investigated the incidence and types of organisms on raw squash seeds to be roasted for food use. Some 12 samples of this product showed a mean total count of log 7.99 and the presence of log 4.72 coliforms. Most of the latter were of the nonfecal type. By adding flour batter and salt to these seeds and roasting, the total count was reduced to less than log 2.00/g. The microbiology of desiccated foods is discussed further in Chap. 15.

ENTERAL NUTRIENT SOLUTIONS (ENTERAL FOODS)

Enteral nutrient solutions (ENS) are liquid foods administered by tube. They are available as powdered products requiring reconstitution or as liquids. They are generally administered to certain patients in hospitals or other patient care facilities but may be administered in the home. Administration is by continuous drip from enteral feeding bags, and the process may go on for 8 h or longer, with the ENS at room temperature. Enteral foods are made by several commercial companies as complete diets that only require reconstituting with water before use or as incomplete meals that require supplementation with milk, eggs, or the like prior to use. ENS-use preparations are nutritionally complete, with varying concentrations of proteins, peptides, carbohydrates, and so forth, depending upon patient need.

The microbiology of ENS has been addressed by some hospital researchers, who have found the products to contain varying numbers and types of bacteria and to be the source of patient infections. Numbers as high as 10^8/ml have been found in some ENS at time of infusion (43). In a study of one reconstituted commercial ENS, the initial count of 9×10^3/ml increased to 7×10^4/ml after 8 h at room temperature (63). Numbers as high as 1.2×10^5/ml were found in another sample of the same preparation. The most frequently isolated organism was *Staphylococcus epidermidis*, with *Corynebacterium*, *Citrobacter*, and *Acinetobacter* spp. among the other isolates. From a British study, enteral feeds yielded 10^4–10^6 organisms/ml, with coliforms and *Pseudomonas aeruginosa* as the predominant types (47).

The capacity of five different commercial ENS to support the growth of *Enterobacter cloacae* under use conditions has been demonstrated (29), and the addition of 0.2% potassium sorbate was shown to reduce numbers of this organism by 3 log cycles over controls. Patients are known to have contracted *E. cloacae* and *Salmonella enteritidis* infections from ENS (14, 47). Procedures that should be employed in the preparation/handling of ENS to minimize microbial problems have been noted (50). For more information, see Fagerman et al. (29).

References

1. Abeyta, C., Jr. 1983. Comparison of iron milk and official AOAC methods for enumeration of *Clostridium perfringens* from fresh seafoods. *J. Assoc. Off. Anal. Chem.* 66:1175–1177.
2. Allen, J. R., and E. M. Foster. 1960. Spoilage of vacuum-packed sliced processed meats during refrigerated storage. *Food Res.* 25:1–7.
3. Andrews, W. H., A. P. Duran, F. D. McClure, and D. E. Gentile. 1979. Use of two rapid A-1 methods for the recovery of fecal coliforms and *Escherichia coli* from selected food types. *J. Food Sci.* 44:289–291, 293.
4. Andrews, W. H., C. R. Wilson, P. L. Poelma, and A. Romero. 1977. Bacteriological survey of the channel catfish (*Ictalurus punctalus*) at the retail level. *J. Food Sci.* 42:359–363.
5. Barbe, C. D., R. W. Mandigo, and R. L. Henrickson. 1966. Bacterial flora associated with rapid-processed ham. *J. Food Sci.* 31:988–993.
6. Barnard, R. J., A. P. Duran, A. Swartzentruber, A. H. Schwab, B. A. Wentz, and R. B. Read, Jr. 1982. Microbiological quality of frozen cauliflower, corn, and peas obtained at retail markets. *Appl. Environ. Microbiol.* 44:54–58.
7. Bauer, F. T., J. A. Carpenter, and J. O. Reagan. 1981. Prevalence of *Clostridium perfringens* in pork during processing. *J. Food Protect.* 44:279–283.
8. Bell, W. N., and L. A. Shelef. 1978. Availability and microbial stability of retail beef-soy blends. *J. Food Sci.* 43:315–318, 333.
9. Bhadsavle, C. H., T. E. Shehata, and E. B. Collins. 1972. Isolation and identification of psychrophilic species of *Clostridium* from milk. *Appl. Microbiol.* 24:699–702.
10. Brodsky, M. H. 1984. Bacteriological survey of freshly formed cheddar cheese. *J. Food Protect.* 47:546–548.
11. Bryan, F. L. 1983. Epidemiology of milk-borne diseases. *J. Food. Protect.* 46:637–649.

12. Bryan, F. L., J. C. Ayres, and A. A. Kraft. 1968. Salmonellae associated with further-processed turkey products. *Appl. Microbiol.* 16:1–9.
13. Carl, K. E. 1975. Oregon's experience with microbiological standards for meat. *J. Milk Food Technol.* 38:483–486.
14. Casewell, M. W., J. E. Cooper, and M. Webster. 1981. Enteral feeds contaminated with *Enterobacter cloacae* as a cause of septicaemia. *Brit. Med. J.* 282:973.
15. Cavett, J. J. 1962. The microbiology of vacuum packed sliced bacon. *J. Appl. Bacteriol.* 25:282–289.
16. Chipley, J. R., and E. K. Heaton. 1971. Microbial flora of pecan meat. *Appl. Microbiol.* 22:252–253.
17. Christiansen, L. N., and N. S. King. 1971. The microbial content of some salads and sandwiches at retail outlets. *J. Milk Food Technol.* 34:289–293.
18. Christopher, F. M., S. C. Seideman, Z. L. Carpenter, G. C. Smith, and C. Vanderzant. 1979. Microbiology of beef packaged in various gas atmospheres. *J. Food Protect.* 42:240–244.
19. Cox, N. A., A. J. Mercuri, B. J. Juven, and J. E. Thomson. 1975. *Enterobacteriaceae* at various stages of poultry chilling. *J. Food Sci.* 40:44–46.
20. Craven, S. E., and A. J. Mercuri. 1977. Total aerobic and coliform counts in beef-soy and chicken-soy patties during refrigerated storage. *J. Food Protect.* 40:112–115.
21. DeBoer, E., and E. M. Boot. 1983. Comparison of methods for isolation and confirmation of *Clostridium perfringens* from spices and herbs. *J. Food Protect.* 46:533–536.
22. Drake, S. D., J. B. Evans, and C. F. Niven, Jr. 1958. Microbial flora of packaged frankfurters and their radiation resistance. *Food Res.* 23:291–296.
23. Draughon, F. A. 1980. Effect of plant-derived extenders on microbiological stability of foods. *Food Technol.* 34(10):69–74.
24. Duitschaever, C. L., D. R. Arnott, and D. H. Bullock. 1973. Bacteriological quality of raw refrigerated ground beef. *J. Milk Food Technol.* 36:375–377.
25. Duitschaever, C. L., D. H. Bullock, and D. R. Arnott. 1977. Bacteriological evaluation of retail ground beef, frozen beef patties, and cooked hamburger. *J. Food Protect.* 40:378–381.
26. Dutson, T. R., G. C. Smith, and Z. L. Carpenter. 1980. Lysosomal enzyme distribution in electrically stimulated ovine muscle. *J. Food Sci.* 45:1097–1098.
27. Egan, A. F., A. L. Ford, and B. J. Shay. 1980. A comparison of *Microbacterium thermosphactum* and lactobacilli as spoilage organisms of vacuum-packaged sliced luncheon meats. *J. Food Sci.* 45:1745–1748.
28. Egan, A. F., and B. J. Shay. 1982. Significance of lactobacilli and film permeability in the spoilage of vacuum-packaged beef. *J. Food Sci.* 47:1119–1122, 1126.
29. Fagerman, K. E., J. D. Paauw, M. A. McCamish, and R. E. Dean. 1984. Effects of time, temperature, and preservative on bacterial growth in enteral nutrient solutions. *Amer. J. Hosp. Pharm.* 41:1122–1126.
30. Fanelli, M. J., A. C. Peterson, and M. F. Gunderson. 1965. Microbiology of dehydrated soups. I. A survey. *Food Technol.* 19:83–86.
31. Fanelli, M. J., A. C. Peterson, and M. F. Gunderson, 1965. Microbiology of dehydrated soups. III. Bacteriological examination of rehydrated dry soup mixes. *Food Technol.* 19:90–94.
32. Farber, J. M., S. A. Malcolm, K. F. Weiss, and M. A. Johnston. 1988. Microbiological quality of fresh and frozen breakfast-type sausages sold in Canada. *J. Food Protect.* 51:397–401.
33. Field, R. A. 1976. Mechanically deboned red meat. *Food Technol.* 30(9):38–48.
34. Field, R. A. 1981. Mechanically deboned red meat. *Adv. Food Res.* 27:23–107.
35. Field, R. A., and M. L. Riley. 1974. Characteristics of meat from mechanically

deboned lamb breasts. *J. Food Sci.* 39:851–852.

36. Foster, J. F., J. L. Fowler, and J. Dacey. 1977. A microbial survey of various fresh and frozen seafood products. *J. Food Protect.* 40:300–303.

37. Fowler, J. L. and W. S. Clark, Jr. 1975. Microbiology of delicatessen salads. *J. Milk Food Technol.* 38:146–149.

38. Fowler, J. L., and J. F. Foster. 1976. A microbiological survey of three fresh green salads: Can guidelines be recommended for these foods? *J. Milk Food Technol.* 39:111–113.

39. Fraiser, M. B., and J. A. Koburger. 1984. Incidence of salmonellae in clams, oysters, crabs and mullet. *J. Food Protect.* 47:343–345.

40. Froning, G. W. 1981. Mechanical deboning of poultry and fish. *Adv. Food Res.* 27:109–147.

41. Fung, D. Y. C., C. L. Kastner, M. C. Hunt, M. E. Dikeman, and D. H. Kropf. 1980. Mesophilic and psychrotrophic bacterial populations on hot-boned and conventionally processed beef. *J. Food Protect.* 43:547–550.

42. Fung, D. Y. C., C. L. Kastner, C.-Y. Lee, M. C. Hunt, M. E. Dikeman, and D. H. Kropf. 1981. Initial chilling rate effects of bacterial growth on hot-boned beef. *J. Food Protect.* 44:539–544.

43. Furtado, D., A. Parrish, and P. Beyer. 1980. Enteral nutrient solutions (ENS): In vitro growth supporting properties of ENS for bacteria. *J. Paren. Ent. Nutri.* 4:594.

44. Gabis, D. A., B. E. Langlois, and A. W. Rudnick, 1970. Microbiological examination of cocoa powder. *Appl. Microbiol.* 20:644–645.

45. Gardner, G. A. 1971. A note on the aerobic microflora of fresh and frozen porcine liver stored at 5°C. *J. Food Technol.* 6:225–231.

46. Gardner, G. A., A. W. Carson, and J. Patton. 1967. Bacteriology of prepacked pork with reference to the gas composition within the pack. *J. Appl. Bacteriol.* 30:321–333.

47. Gill, K. J., and P. Gill. 1981. Contaminated enteral feeds. *Brit. Med. J.* 282:1971.

48. Goepfert, J. M. 1977. Aerobic plate count and *Escherichia coli* determination on frozen ground-beef patties. *Appl. Environ. Microbiol.* 34:458–460.

49. Greenberg, R. A., R. B. Tompkin, B. O. Bladel, R. S. Kittaka, and A. Anellis. 1966. Incidence of mesophilic *Clostridium* spores in raw pork, beef, and chicken in processing plants in the United States and Canada. *Appl. Microbiol.* 14:789–793.

50. Gröschel, D. H. M. 1983. Infection control considerations in enteral feeding. *Nutri. Supp. Serv.* 3(6):48–49.

51. Guthertz, L. S., J. T. Fruin, R. L. Okoluk, and J. L. Fowler. 1977. Microbial quality of frozen comminuted turkey meat. *J. Food Sci.* 42:1344–1347.

52. Guthertz, L. S., J. T. Fruin, D. Spicer, and J. L. Fowler. 1976. Microbiology of fresh comminuted turkey meat. *J. Milk Food Technol.* 39:823–829.

53. Hanna, M. O., G. C. Smith, J. W. Savell, F. K. McKeith, and C. Vanderzant. 1982. Microbial flora of livers, kidneys and hearts from beef, pork and lamb: Effects of refrigeration, freezing and thawing. *J. Food Protect.* 45:63–73.

54. Hanna, M. O., G. C. Smith, J. W. Savell, F. K. McKeith, and C. Vanderzant. 1982. Effects of packaging methods on the microbial flora of livers and kidneys from beef or pork. *J. Food Protect.* 45:74–81.

55. Hanna, M. O., C. Vanderzant, Z. L. Carpenter, and G. C. Smith. 1977. Microbial flora of vacuum-packaged lamb with special reference to psychrotrophic, gram-positive, catalase-positive pleomorphic rods. *J. Food Protect.* 40:98–100.

56. Harris, N. D., S. R. Martin, and L. Ellias. 1975. Bacteriological quality of selected delicatessen foods. *J. Milk Food Technol.* 38:759–761.

57. Harrison, M. A., F. A. Draughton, and C. C. Melton. 1983. Inhibition of spoilage bacteria by acidification of soy extended ground beef. *J. Food Sci.* 48:825–828.

58. Harrison, M. A., C. C. Melton, and F. A. Draughon. 1981. Bacterial flora of ground

beef and soy extended ground beef during storage. *J. Food Sci.* 46:1088–1090.

59. Hauschild, A. H. W., and R. Hilsheimer. 1983. Prevalence of *Clostridium botulinum* in commercial liver sausage. *J. Food Protect.* 46:242–244.

60. Hauschild, A. H. W., L. M. Poste, and R. Hilsheimer. 1985. Toxin production by *Clostridium botulinum* and organoleptic changes in vacuum-packaged raw beef. *J. Food Protect.* 48:712–716.

61. Henry, K. G., J. W. Savell, G. C. Smith, J. G. Ehlers, and C. Vanderzant. 1983. Physical, sensory and microbiological characteristics of lamb retail cuts vacuum packaged in high oxygen-barrier film. *J. Food Sci.* 48:1735–1740, 1749.

62. Hitchener, B. J., A. F. Egan, and P. J. Rogers. 1982. Characteristics of lactic acid bacteria isolated from vacuum-packaged beef. *J. Appl. Bacteriol.* 52:31–37.

63. Hostetler, C., T. O. Lipman, M. Geraghty, and R. H. Parker, 1982. Bacterial safety of reconstituted continuous drip tube feeding. *J. Paren. Ent. Nutri.* 6:232–235.

64. Hosty, T. S., and C.I. McDurmont. 1975. Isolation of acid-fast organisms from milk and oysters. *Hlth Lab. Sci.* 12:16–19.

65. Ingram, M. 1960. Bacterial multiplication in packed Wiltshire bacon. *J. Appl. Bacteriol.* 23:206–215.

66. Insalata, N. F., J. S. Witzeman, J. H. Berman, and E. Borker, 1968. A study of the incidence of the spores of *Clostridium botulinum* in frozen vacuum pouch-pack vegetables. *Proc., 96th Ann. Meet., Am. Pub. Health Assoc.*, 124.

67. Julseth, R. M., and R. H. Deibel. 1974. Microbial profile of selected spices and herbs at import. *J. Milk Food Technol.* 37:414–419.

68. Käferstein, F. K. 1976. The microflora of parsley. *J. Milk Food Technol.* 39:837–840.

69. Kajs, T. M., R. Hagenmaier, C. Vanderzant, and K. F. Mattil. 1976. Microbiological evaluation of coconut and coconut products. *J. Food Sci.* 41:352–356.

70. Kastner, C. L., L. O. Leudecke, and T. S. Russell. 1976. A comparison of microbial counts on conventionally and hot-boned bovine carcasses. *J. Milk Food Technol.* 39:684—685.

71. Kautter, D. A., R. K. Lynt, T. Lilly, Jr., and H. M. Solomon. 1981. Evaluation of the botulism hazard from imitation cheeses. *J. Food Sci.* 46:749–750, 764.

72. Kazanas, N., J. A. Emerson, H. L. Seagram, and L. L. Kempe. 1966. Effect of γ-irradiation on the microflora of fresh-water fish. I. Microbial load, lag period, and rate of growth on yellow perch (*Perca flavescens*) fillets. *Appl. Microbiol.* 14:261–266.

73. Keeton, J. T., and C. C. Melton. 1978. Factors associated with microbial growth in ground beef extended with varying levels of textured soy protein. *J. Food Sci.* 43:1125–1129.

74. Kempton, A. G., and S. R. Bobier. 1979. Bacterial growth in refrigerated, vacuum-packed luncheon meats. *Can. J. Microbiol.* 16:287–297.

75. Kereluk, K., and M. F. Gunderson. 1959. Studies on the bacteriological quality of frozen meat pies. I. Bacteriological survey of some commercially frozen meat pies. *Appl. Microbiol.* 7:320–323.

76. Kitchell, A. G. 1962. Micrococci and coagulase negative staphylococci in cured meats and meat products. *J. Appl. Bacteriol.* 25:416–431.

77. Kotula, A. W. 1981. Microbiology of hot-boned and electrostimulated meat. *J. Food Protect.* 44:545–549.

78. Kotula, A. W., and B. S. Emswiler-Rose. 1981. Bacteriological quality of hot-boned primal cuts from electrically stimulated beef carcasses. *J. Food Sci.* 46:471–474.

79. Kraft, A. A., J. C. Ayres, G. S. Torrey, R. H. Salzer, and G. A. N. DaSilva. 1966. Coryneform bacteria in poultry, eggs and meat. *J. Appl. Bacteriol.* 29:161–166.

80. Kraft, A. A., V. Reddy, R. J. Hasiak, K. D. Lind, and D. E. Galloway. 1982. Microbiological quality of vacuum packaged poultry with or without chlorine treatment. *J. Food Sci.* 47:380–385.

81. Ladiges, W. C., J. F. Foster, and W. M. Ganz. 1974. Incidence and viability of *Clostridium perfringens* in ground beef. *J. Milk Food Technol.* 37:622–623.

82. Lee, B. H., R. E. Simard, C. L. Laleye, and R. A. Holley. 1984. Shelf-life of meat loaves packaged in vacuum or nitrogen gas. I. Effect of storage temperature, light and time on the micro-flora change. *J. Food Protect.* 47:128–133.

83. Lee, C. Y., D. Y. C. Fung, and E. L. Kastner. 1982. Computer-assisted identification of bacteria on hot-boned and conventionally processed beef. *J. Food Sci.* 47:363–367, 373.

84. Lee, J. S., and D. K. Pfeifer. 1977. Microbiological characteristics of Pacific shrimp (*Pandalus jordani*). *Appl. Environ. Microbiol.* 33:853–859.

85. Leininger, H. V., L. R. Shelton, and K. H. Lewis. 1971. Microbiology of frozen cream-type pies, frozen cooked-peeled shrimp, and dry food-grade gelatin. *Food Technol.* 25:224–229.

86. Lillard, H. S. 1971. Occurrence of *Clostridium perfringens* in broiler processing and further processing operations. *J. Food Sci.* 36:1008–1010.

87. Lin, C. K., W. H. Kennick, W. E. Sandine, and M. Koohmaraie. 1984. Effect of electrical stimulation on meat microflora: Observations on agar media, in suspensions and on beef carcasses. *J. Food Protect.* 47:279–283.

88. Lin, H.-S., D. G. Topel, and H. W. Walker. 1979. Influence of prerigor and postrigor muscle on the bacteriological and quality characteristics of pork sausage. *J. Food Sci.* 44:1055–1057.

89. Litsky, W., I. S. Fagerson, and C. R. Fellers. 1957. A bacteriological survey of commercially frozen beef, poultry and tuna pies. *J. Milk Food Technol.* 20:216–219.

90. May, K. N. 1962. Bacterial contamination during cutting and packaging chicken in processing plants and retail stores. *Food Technol.* 16:89–91.

91. McMillin, D. J., J. G. Sebranek, and A. A. Kraft. 1981. Microbial quality of hot-processed frozen ground beef patties processed after various holding times. *J. Food Sci.* 46:488–490.

92. Mead, G. C. 1983. Effect of packaging and gaseous environment on the microbiology and shelf life of processed poultry products. In *Food Microbiology: Advances and Prospects*, ed. T. A. Roberts and F. A. Skinner, 203–216. New York and London: Academic Press.

93. Mercuri, A. J., G. J. Banwart, J. A. Kinner, and A. R. Sessoms. 1970. Bacteriological examination of commercial precooked Eastern-type turkey rolls. *Appl. Microbiol.* 19:768–771.

94. Mol, J. H. H., J. E. A. Hietbrink, H. W. M. Mollen, and J. van Tinteren. 1971. Observations on the microflora of vacuum packed sliced cooked meat products. *J. Appl. Bacteriol.* 34:377–397.

95. Molin, G., and I.-M. Stenstrom. 1984. Effect of temperature on the microbial flora of herring fillets stored in air or carbon dioxide. *J. Appl. Bacteriol.* 56:275–282.

96. Mundt, J. O., W. F. Graham, and I. E. McCarty. 1967. Spherical lactic acid-producing bacteria of southern-grown raw and processed vegetables. *Appl. Microbiol.* 15:1303–1308.

97. Murthy, T. R. K. 1984. Relative numbers of coliforms, *Enterobacteriaceae* (by two methods), and total aerobic bacteria counts as determined from minced goat meat. *J. Food Protect.* 47:142–144.

98. Nakamura, M., and K. D. Kelly. 1968. *Clostridium perfringens* in dehydrated soups and sauces. *J. Food Sci.* 33:424–426.

99. Newsome, R. L., B. E. Langlois, W. G. Moody, N. Gay, and J. D. Fox. 1984. Effect of time and method of aging on the composition of the microflora of beef

loins and corresponding steaks. *J. Food Protect.* 47:114–118.

100. Newton, K. G., J. C. L. Harrison, and K. M. Smith. 1977. The effect of storage in various gaseous atmospheres on the microflora of lamb chops held at −1°C. *J. Appl. Bacteriol.* 43:53–59.

101. Newton, K. G., and W. J. Rigg. 1979. The effect of film permeability on the storage life and microbiology of vacuum-packed meat. *J. Appl. Bacteriol.* 47: 433–441.

102. Nickerson, J. T. R., and S. A. Goldblith. 1964. A study of the microbiological quality of haddock fillets and shucked, soft-shelled clams processed and marketed in the greater Boston area. *J. Milk Food Technol.* 27:7–12.

103. Nielsen, H.-J. S. 1983. Influence of nitrite addition and gas permeability of packaging film on the microflora in a sliced vacuum-packed whole meat product under refrigerated storage. *J. Food Technol.* 18:573–585.

104. Nielsen, H.-J. S. and P. Zeuthen. 1985. Influence of lactic acid bacteria and the overall flora on development of pathogenic bacteria in vacuum-packed, cooked emulsion-style sausage. *J. Food Protect.* 48:28–34.

105. Oblinger, J. L., J. E. Kennedy, Jr., C. A. Rothenberg, B. W. Berry, and N. J. Stern. 1982. Identification of bacteria isolated from fresh and temperature abused variety meats. *J. Food Protect.* 45:650–654.

106. Ockerman, H. W., and J. Szczawinski. 1983. Effect of electrical stimulation on the microflora of meat. *J. Food Sci.* 48:1004–1005, 1007.

107. Ostovar, K., J. H. MacNeil, and K. O'Donnell. 1971. Poultry product quality. 5. Microbiological evaluation of mechanically deboned poultry meat. *J. Food Sci.* 36:1005–1007.

108. Pace, P. J. 1975. Bacteriological quality of delicatessen foods: Are standards needed? *J. Milk Food Technol.* 38:347–353.

109. Paradis, D. C., and M. E. Stiles. 1978. A study of microbial quality of vacuum packaged, sliced bologna. *J. Food Protect.* 41:811–815.

110. Patterson, J. T., and P. A. Gibbs. 1977. Incidence and spoilage potential of isolates from vacuum-packaged meat of high pH value. *J. Appl. Bacteriol.* 43:25–38.

111. Pivnick, H., I. E. Erdman, D. Collins-Thompson, G. Roberts, M. A. Johnston, D. R. Conley, G. Lachapelle, U. T. Purvis, R. Foster, and M. Milling. 1976. Proposed microbiological standards for ground beef based on a Canadian survey. *J. Milk Food Technol.* 39:408–412.

112. Post, L. S., D. A. Lee, M. Solberg, D. Furgang, J. Specchio, and C. Graham. 1985. Development of botulinal toxin and sensory deterioration during storage of vacuum and modified atmosphere packaged fish fillets. *J. Food Sci.* 50:990–996.

113. Powers, E. M., C. Ay, H. M. El-Bisi, and D. B. Rowley. 1971. Bacteriology of dehydrated space foods. *Appl. Microbiol.* 22:441–445.

114. Powers, E. M., T. G. Latt, and T. Brown. 1976. Incidence and levels of *Bacillus cereus* in processed spices. *J. Milk Food Technol.* 39:668–670.

115. Powers, E. M., R. Lawyer, and Y. Masuoka. 1975. Microbiology of processed spices. *J. Milk Food Technol.* 38:683–687.

116. Pulliam, J. D., and D. C. Kelley. 1965. Bacteriological comparisons of hot processed and normally processed hams. *J. Milk Food Technol.* 28:285–286.

117. Raccach, M., and R. C. Baker. 1978. Microbial properties of mechanically deboned fish flesh. *J. Food Sci.* 43:1675–1677.

118. Ray, B., and R. A. Field. 1983. Bacteriology of restructured lamb roasts made with mechanically deboned meat. *J. Food Protect.* 46:26–28.

119. Ray, B., C. Johnson, and R. A. Field. 1984. Growth of indicator, pathogenic and psychrotrophic bacteria in mechanically separated beef, lean ground beef and beef bone marrow. *J. Food Protect.* 47:672–677.

120. Rehberger, T. G., L. A. Wilson, and B. A. Glatz. 1984. Microbiological quality of

commercial tofu. *J. Food Protect.* 47:177–181.

121. Riha, W. E., and M. Solberg. 1970. Microflora of fresh pork sausage casings. 2. Natural casings. *J. Food Sci.* 35:860–863.

122. Roth, L. A., and D. S. Clark. 1975. Effect of lactobacilli and carbon dioxide on the growth of *Microbacterium thermosphactum* on fresh beef. *Can. J. Microbiol.* 21:629–632.

123. Rothenberg, C. A., B. W. Berry, and J. L. Oblinger. 1982. Microbiological characteristics of beef tongues and livers as affected by temperature-abuse and packaging systems. *J. Food Protect.* 45:527–532.

124. Seideman, S. C., C. Vanderzant, G. C. Smith, M. O. Hanna, and Z. L. Carpenter. 1976. Effect of degree of vacuum and length of storage on the microflora of vacuum packaged beef wholesale cuts. *J. Food Sci.* 41:738–742.

125. Seideman, S. C., and P. R. Durland. 1983. Vacuum packaging of fresh beef: A review. *J. Food Qual.* 6:29–47.

126. Seman, D. L., K. R. Drew, and R. P. Littlejohn. 1989. Packaging venison for extended chilled storage: Comparison of vacuum and modified atmosphere packaging containing 100% carbon dioxide. *J. Food Protect.* 52:886–893.

127. Shehata, T. E., and E. B. Collins. 1971. Isolation and identification of psychrophilic species of *Bacillus* from milk. *Appl. Microbiol.* 21:466–469.

128. Shelef, L. A. 1975. Microbial spoilage of fresh refrigerated beef liver. *J. Appl. Bacteriol.* 39:273–280.

129. Silverman, G. J., J. T. R. Nickerson, D. W. Duncan, N. S. Davis, J. S. Schachter, and M. M. Joselow. 1961. Microbial analysis of frozen raw and cooked shrimp. I. General results. *Food Technol.* 15:455–458.

130. Simard, R. E., B. H. Lee, C. L. Laleye, and R. A. Holley. 1983. Effects of temperature, light and storage time on the microflora of vacuum- or nitrogen-packed frankfurters. *J. Food Protect.* 46:199–205.

131. Smith, F. C., R. A. Field, and J. C. Adams. 1974. Microbiology of Wyoming big game meat. *J. Milk Food Technol.* 37:129–131.

132. Splittstoesser, D. F. 1973. The microbiology of frozen vegetables. How they get contaminated and which organisms predominate. *Food Technol.* 27:54–56.

133. Splittstoesser, D. F., and D. A. Corlett, Jr. 1980. Aerobic plate counts of frozen blanched vegetables processed in the United States. *J. Food Protect.* 43:717–719.

134. Splittstoesser, D. F., and I. Gadjo. 1966. The groups of micro-organisms composing the "total" count population in frozen vegetables. *J. Food Sci.* 31:234–239.

135. Splittstoesser, D. F., G. E. R. Hervey, II, and W. P. Wettergreen, 1965. Contamination of frozen vegetables by coagulase-positive staphylococci. *J. Milk Food Technol.* 28:149–151.

136. Splittstoesser, D. F., D. T. Queale, J. L. Bowers, and M. Wilkison. 1980. Coliform content of frozen blanched vegetables packed in the United States. *J. Food Safety* 2:1–11.

137. Splittstoesser, D. F., W. P. Wettergreen, and C. S. Pederson. 1961. Control of microorganisms during preparation of vegetables for freezing. I. Green beans. *Food Technol.* 15:329–331.

138. Splittstoesser, D. F., M. Wexler, J. White, and R. R. Colwell. 1967. Numerical taxonomy of gram-positive and catalase-positive rods isolated from frozen vegetables. *Appl. Microbiol.* 15:158–162.

139. Steinkraus, K. H., and J. C. Ayres. 1964. Incidence of putrefactive anaerobic spores in meat. *J. Food Sci.* 29:87–93.

140. Stewart, A. W. 1983. Effect of cooking on bacteriological population of "soul foods." *J. Food Protect.* 46:19–20.

141. Stiles, M. E., and L.-K. Ng. 1981. Biochemical characteristics and identification of *Enterobacteriaceae* isolated from meats. *Appl. Environ. Microbiol.* 41:639–645.

142. Strong, D. H., J. C. Canada, and B. B. Griffiths. 1963. Incidence of *Clostridium perfringens* in American foods. *Appl. Microbiol.* 11:42–44.

143. Surkiewicz, B. F. 1966. Bacteriological survey of the frozen prepared foods industry. *Appl. Microbiol.* 14:21–26.

144. Surkiewicz, B. F., R. J. Groomes, and A. P. Padron. 1967. Bacteriological survey of the frozen prepared foods industry. III. Potato products. *Appl. Microbiol.* 15:1324–1331.

145. Surkiewicz, B. F., M. E. Harris, and J. M. Carosella. 1977. Bacteriological survey and refrigerated storage test of vacuum-packed sliced imported canned ham. *J. Food Protect.* 40:109–111.

146. Surkiewicz, B. F., M. E. Harris, R. P. Elliott, J. F. Macaluso, and M. M. Strand. 1975. Bacteriological survey of raw beef patties produced at establishments under federal inspection. *Appl. Microbiol.* 29:331–334.

147. Surkiewicz, B. F., R. W. Johnston, R. P. Elliott, and E. R. Simmons. 1972. Bacteriological survey of fresh pork sausage produced at establishments under federal inspection. *Appl. Microbiol.* 23:515–520.

148. Sutherland, J. P., J. T. Patterson, and J. G. Murray. 1975. Changes in the microbiology of vacuum-packaged beef. *J. Appl. Bacteriol.* 39:227–237.

149. Swartzentruber, A., A. H. Schwab, A. P. Duran, B. A. Wentz, and R. B. Read, Jr. 1980. Microbiological quality of frozen shrimp and lobster tail in the retail market. *Appl. Environ. Microbiol.* 40:765–769.

150. Swartzentruber, A., A. H. Schwab, B. A. Wentz, A. P. Duran, and R. B. Read, Jr. 1984. Microbiological quality of biscuit dough, snack cakes and soy protein meat extender. *J. Food Protect.* 47:467–470.

151. Thompson, S. G., H. W. Ockerman, V. R. Cahill, and R. F. Plimpton. 1978. Effect of soy protein flakes and added water on microbial growth (total counts, coliforms, proteolytics, staphylococci) and rancidity in fresh ground beef. *J. Food Sci.* 43:289–291.

152. Timoney, J. F., and A. Abston. 1984. Accumulation and elimination of *Escherichia coli* and *Salmonella typhimurium* by hard clams in an in vitro system. *Appl. Environ. Microbiol.* 47:986–988.

153. Todd, E. C. D., G. A. Jarvis, K. F. Weiss, G. W. Riedel, and S. Charbonneau. 1983. Microbiological quality of frozen cream-type pies sold in Canada. *J. Food Protect.* 46:34–40.

154. Vanderzant, C., M. O. Hanna, J. G. Ehlers, J. W. Savell, G. C. Smith, D. B. Griffin, R. N. Terrell, K. D. Lind, and D. E. Galloway. 1982. Centralized packaging of beef loin steaks with different oxygen-barrier films: Microbiological characteristics. *J. Food Sci.* 47:1070–1079.

155. Vanderzant, C., A. W. Matthys, and B. F. Cobb, III. 1973. Microbiological, chemical, and organoleptic characteristics of frozen breaded raw shrimp. *J. Milk Food Technol.* 36:253–261.

156. Warburton, D. W., P. I. Peterkin, and K. F. Weiss. 1986. A survey of the microbiological quality of processed cheese products. *J. Food Protect.* 49:229–231.

157. Wentz, B. A., A. P. Duran, A. Swartzentruber, A. H. Schwab, and R. B. Read, Jr. 1983. Microbiological quality of fresh blue crabmeat, clams and oysters. *J. Food Protect.* 46:978–981.

158. Wentz, B. A., A. P. Duran, A. Swartzentruber, A. S. Schwab, and R. B. Read, Jr. 1984. Microbiological quality of frozen breaded onion rings and tuna pot pies. *J. Food Protect.* 47:58–60.

159. Woodburn, M. 1964. Incidence of Salmonellae in dressed broiler-fryer chickens. *Appl. Microbiol.* 12:492–495.

160. Zottola, E. A., and F. F. Busta. 1971. Microbiological quality of further-processed turkey products. *J. Food Sci.* 36:1001–1004.

PART III

Determining Microorganisms and/or Their Products in Foods

If one assumes that a given science discipline is no better than its methodology, the material presented in these three chapters is critical to food microbiology. Traditional methods are presented in Chapters 5 and 6, along with some newer developments that are designed to be more precise, accurate, and rapid than the former. The areas covered are being pursued actively in research laboratories, and complete treatments go beyond the scope of this work. The references noted below should be consulted for in-depth coverage.

The animal and tissue culture assay methods covered in Chapter 7 are designed to provide the basic principles of these bioassay methods. For most foodborne pathogens, additional information is provided in the chapters in Part VII.

The following references should be consulted for in-depth coverage of the topics included in this part:

A. D. King, Jr., J. I. Pitt, L. R. Beuchat, and J. E. L. Corry. 1986. *Methods for the Mycological Examination of Food*. New York: Plenum. This book provides detailed coverage of the many methods that may be used to examine foods for fungi.

W. H. Nelson. 1991. *Physical Methods for Microorganisms Detection*. Boca Raton, Fla.: CRC Press. The ATP assay to detect microorganisms is detailed.

B. Ray, ed. 1989. *Injured Index and Pathogenic Bacteria: Occurrence and Detection in Foods, Water and Feeds*. Boca Raton, Fla.: CRC Press. This volume covers metabolically injured organisms and methods for their recovery from foods with emphasis on indicators and pathogens.

5

Culture, Microscopic, and Sampling Methods

The examination of foods for the presence, types, and numbers of micro-organisms and/or their products is basic to food microbiology. In spite of the importance of this, none of the methods in common use permits the determination of exact numbers of microorganisms in a food product. Although some methods of analysis are better than others, every method has certain inherent limitations associated with its use.

The four basic methods employed for "total" numbers are

1. standard plate counts (SPC) for viable cells
2. the most-probable numbers (MPN) method as a statistical determination of viable cells
3. dye-reduction techniques to estimate numbers of viable cells that possess reducing capacities
4. direct microscopic counts (DMC) for both viable and nonviable cells.

All of these are discussed in this chapter, along with their uses in determining microorganisms from various sources. Detailed procedures for their use can be obtained from references in Table 5-1. In addition, variations of these basic methods for examining the microbiology of surfaces and of air are presented along with a summary of methods and attempts to improve their overall efficiency.

CONVENTIONAL SPC

By this method, portions of food samples are blended or homogenized, serially diluted in an appropriate diluent, plated in or onto a suitable agar medium, incubated at an appropriate temperature for a given time, after which all visible colonies are counted by use of a Quebec or electronic counter.

TABLE 5-1.   Some Standard References for Methods of Microbiological Analysis of Foods

	Reference						
	17	79	73	120	80	13	11
Direct microscopic counts			X	X	X	X	X
Standard plate counts		X	X	X		X	X
Most probable numbers		X	X	X		X	X
Dye reductions			X				
Coliforms		X	X	X		X	X
Fungi	X			X		X	X
Fluorescent antibodies				X		X	X
Sampling plans		X		X	X	X	

The SPC is by far the most widely used method for determining the numbers of viable cells or colony-forming units (cfu) in a food product. When total viable counts are reported for a product, the counts should be viewed as a function of at least some of the following factors:

1. sampling methods employed
2. distribution of the organisms in the food sample
3. nature of the food flora
4. nature of the food material
5. the preexamination history of the food product
6. nutritional adequacy of the plating medium employed
7. incubation temperature and time used
8. pH, a_w, and oxidation-reduction potential (Eh) of the plating medium
9. type of diluent used
10. relative number of organisms in food sample
11. existence of other competing or antagonistic organisms

In addition to the limitations noted, plating procedures for selected groups are further limited by the degree of inhibition and effectiveness of the selective and/or differential agents employed.

While the SPC is more often determined by pour plating, essentially comparable results can be obtained by surface plating. By the latter method, prepoured and hardened agar plates with dry surfaces are employed. The diluted specimens are planted onto the surface of replicate plates, and with the aid of bent glass rods ("hockey sticks") the 0.1-ml inoculum/plate is carefully and evenly distributed over the entire surface. Surface plating offers advantages in determining the numbers of heat-sensitive psychrotrophs in a food product because the organisms do not come in contact with melted agar. It is the method of choice when the colonial features of a colony are important to its presumptive identification and for most selective media. Strict aerobes are obviously favored by surface plating, but microaerophilic organisms tend to grow slower. Among the disadvantages of surface plating are the problem of spreaders (especially when the agar surface is not adequately dry prior to plating) and the crowding of colonies, which makes enumeration more difficult.

Homogenization of Food Samples

Prior to the mid- to late 1970s, microorganisms were extracted from food specimens for plating almost universally by use of mechanical blenders (Waring type). Around 1971, the Colwell Stomacher was developed in England by Sharpe and Jackson (111), and this device is now the method of choice in many laboratories for homogenizing foods for counts. The Stomacher, a relatively simple device, homogenizes specimens in a special plastic bag by the vigorous pounding of two paddles. The pounding effects the shearing of food specimens, and microorganisms are released into the diluent. Three models of the machine are available, but the model 400 is most widely used in food microbiology laboratories. It can handle samples (diluent and specimen) of 40 to 400 ml.

The Stomacher has been compared to a high-speed blender for food analysis by a large number of investigators. Plate counts from Stomacher-treated samples are similar to those treated by blender (15, 32, 108, 110). The instrument is generally preferred over blending for the following reasons:

1. The need to clean and store blender containers is obviated.
2. Heat buildup does not occur during normal operational times (usually 2 min).
3. The homogenates can be stored in the Stomacher bags in a freezer for further use.
4. The noise level is not as unpleasant as that of mechanical blenders.

In a study by Sharpe and Harshman (110), the Stomacher was shown to be less lethal than a blender to *Staphylococcus aureus, Enterococcus faecalis*, and *Escherichia coli*. One investigator reported that counts by using a Stomacher were significantly higher than when a blender was used (127) while other investigators obtained higher overall counts by blender than by Stomacher (6). The latter investigators showed that the Stomacher is food specific; it is better than high-speed blending for some types of foods but not for others. In another study, SPC determinations made by Stomacher, blender, and shaking were not significantly different, although significantly higher counts of gram-negative bacteria were obtained by Stomacher than by either of the other two methods (62).

Another advantage of the Stomacher over blending is the homogenization of meats for dye-reduction tests. Holley et al. (54) showed that the extraction of bacteria from meat by using a Stomacher does not cause extensive disruption of meat tissue, and consequently fewer reductive compounds were present to interfere with resazurin reduction, while with blending the level of reductive compounds released made resazurin reduction results meaningless.

The Spiral Plater

The spiral plater is a mechanical device that distributes the liquid inoculum on the surface of a rotating plate containing a suitable poured and hardened agar medium. The dispensing arm moves from the near center of the plate toward the outside, depositing the sample in an Archimedes spiral. The attached special syringe dispenses a continuously decreasing volume of sample so that a

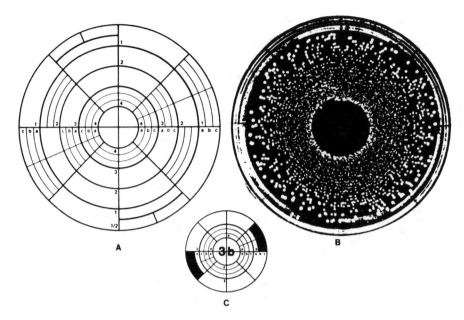

FIGURE 5-1.   Special counting grid for spiral plater (*A*); growth of organisms on an inoculated spiral plate (*B*); and areas of plate enumerated (*C*). In this example, inoculum volume was 0.0018 ml, counts for the two areas shown were 44 and 63, and the averaged count was 6.1 × 10⁴ bacteria/ml.
Courtesy of Spiral System Instruments, Bethesda, Maryland.

concentration range of up to 10,000:1 is effected on a single plate. Following incubation at an appropriate temperature, colony development reveals a higher density of deposited cells near the center of the plate with progressively fewer toward the edge.

The enumeration of colonies on plates prepared with a spiral plater is achieved by use of a special counting grid (Fig. 5-1*A*). Depending on the relative density of colonies, colonies that appear in one or more specific area(s) of the superimposed grid are counted. An agar plate prepared by a spiral plater is shown in Figure 5-1*B*, and the corresponding grid area counted is shown in Figure 5-1*C*. In this example, a total sample volume of 0.0018 ml was deposited, and the two grid areas counted contained 44 and 63 colonies, respectively, resulting in a total count of 6.1 × 10⁴ bacteria/ml.

The spiral plating device here described was devised by Gilchrist et al. (43), although some of its principles were presented by earlier investigators among whom were Reyniers (102) and Trotman (126). The method has been studied by a rather large number of investigators and compared to other methods of enumerating viable organisms. It was compared to the SPC method by using 201 samples of raw and pasteurized milk; overall good agreement was obtained (31). A collaborative study from six analysts on milk samples showed that the spiral plater compared favorably with the SPC. A standard deviation of 0.109 was obtained by using the spiral plater compared to 0.110 for the SPC (91). In another study, the spiral plater was compared with three other methods (pour, surface plating, and drop count), and no difference was found among the methods at the 5% level of significance (61). In yet another study, the spiral plate maker yielded counts as good as those by the droplette method (50). Spiral plating is an official Association of Official Analytical Chemists (AOAC) method (11).

Among the advantages of the spiral plater over standard plating are: less agar is used; fewer plates, dilution blanks, and pipettes are required; and three to four times more samples/h can be examined (68). Also, 50–60 plates/h can be prepared, and little training is required for its operation (61). Among the disadvantages is the problem that food particles may cause in blocking the dispensing stylus. It is more suited for use with liquid foods such as milk. A laser-beam counter has been developed for use with the plater. Because of the expense of the device, it is not likely to be available in laboratories that do not analyze large numbers of plates. The method is further described by AOAC (11) and in the *Bacteriological Analytical Manual* (13).

MEMBRANE FILTERS

Membranes with a pore size that will retain bacteria (generally 0.45 µm) but allow water or diluent to pass are used. Following the collection of bacteria upon filtering a given volume, the membrane is placed on an agar plate or an absorbent pad saturated with the culture medium of choice, and incubated appropriately. Following growth, the colonies are enumerated. Alternatively, a DMC can be made. In this case, the organisms collected on the membrane are viewed and counted microscopically following appropriate staining, washing, and treatment of the membrane to render it transparent. These methods are especially suited for samples that contain low numbers of bacteria. While relatively large volumes of water can be passed through a membrane without clogging it, only small samples of dilute homogenates from certain foods can be used for a single membrane.

The overall efficiency of membrane filter methods for determining microbial numbers by the DMC has been improved by the introduction of fluorescent dyes. The use of fluorescent dyes and epifluorescent microscopes to enumerate bacteria in waters has been employed rather widely since the early 1970s. Cellulose filters were among the earliest used; however, polycarbonate Nucleopore filters offer the advantage of retaining all bacteria on top of the filter. When lake and ocean waters were examined by using the two kinds of membranes, counts were twice as high with Nucleopore membranes as with cellulose membranes (52).

Direct Epifluorescent Filter Technique

This membrane filter technique may be viewed as an improved modification of the basic method. Direct epifluorescent filter technique (DEFT) employs fluorescent dyes and fluorescent microscopy (52), and it has been evaluated by a number of investigators as a rapid method for microorganisms in foods. Typically, a diluted food homogenate is filtered through a 5 µm nylon filter, and the filtrate is collected and treated with 2 ml of Triton X-100 and 0.5 ml trypsin. The latter reagents are used to lyse somatic cells and to prevent clogging of filters. After incubation, the treated filtrate is passed through a 0.6 µm Nucleopore polycarbonate membrane, and the filter is stained with acridine orange. After drying, the stained cells are enumerated by epifluorescence microscopy, and the number of cells/g is calculated by multiplying the average number/field by the microscope factor. Results can be obtained in 25–30 min, and numbers as low as around 6,000 cfu/g can be obtained from meats and milk products.

DEFT has been employed on milk (94) and found to compare favorably with results obtained by APC, and standard Breed DMC on raw milk that contained between 5×10^3 and 5×10^8 bacteria/ml. It has been adapted to the enumeration of viable gram-negative and all gram-positive bacteria in milk in about 10 min (105). As few as 5,700 bacteria/ml could be detected in heat-treated milk and milk products in about 20 min (95). In a collaborative study by six laboratories that compared DEFT and APC, the correlation coefficient was generally above 0.9, but the repeatability of DEFT was 1.5 times worse than APC and reproducibility was only three times that for APC (93). Solid foods can be examined by DEFT after proper filtrations, and <60,000 organisms/g could be detected in one study (96). DEFT has been employed successfully to estimate numbers of microorganisms on meat and poultry (117) and on food contact surfaces (53). For more information, see Pettipher (92).

Microcolony-DEFT

DEFT allows for the direct microscopic determination of cells; microcolony-DEFT is a variation that allows one to determine viable cells only. Typically, food homogenates are filtered through DEFT membranes, and the latter are then placed on the surface of appropriate culture media and incubated for microcolony development. A 3-h incubation can be used for gram-negative bacteria and a 6-h for gram-positives (104). The microcolonies that develop must be viewed with a microscope. For coliforms, pseudomonads, and staphylococci, as few as 10^3/g could be detected within 8 h (104).

In another variation, a microcolony epifluorescence microscopy method that combines DEFT with HGMF was devised (103). By this method, nonenzyme detergent-treated samples are filtered through Nucleopore polycarbonate membranes, which are transferred to the surface of a selective agar medium and incubated for 3 or 6 h for gram-negative or gram-positive bacteria as for microcolony-DEFT. The membranes are then stained with acridine orange, and the microcolonies are enumerated by epifluorescence microscopy. The method allowed results to be obtained in <6 h without a repair step for injured organisms, and in about 12 h when a repair step was employed (103).

Hydrophobic Grid Membrane Filter

The hydrophobic grid membrane filter (HGMF) technique was advanced by Sharpe and Michaud (115, 116) and has since been further developed and used to enumerate microorganisms from a variety of food products. The method employs a specially constructed filter that consists of 1,600 wax grids on a single membrane filter that restricts growth and colony size to individual grids. On one filter, from 10 to 9×10^4 cells can be enumerated by an MPN procedure, and enumeration can be automated (22). The method can detect as few as 10 cells/g, and results can be achieved in 24 h or so (113). It can be used to enumerate all cfu's or specific groups such as indicator organisms (9, 20, 34), fungi (21), salmonellae (33), and pseudomonads (65). It has been given AOAC approval for total coliforms, fecal coliforms, and salmonellae.

In a typical application, 1 ml of a 1:10 homogenate is filtered through a filter membrane, followed by the placing of the membrane on a suitable agar medium for incubation overnight to allow colonies to develop. The grids that contain colonies are enumerated, and MPN is calculated. The method allows the filtering of up to 1 g of food/membrane (114).

When compared to a five-tube MPN for coliforms, the HGMF method, employing a resuscitative step, produced statistically equivalent results for coliforms and fecal coliforms (22). In the latter application, HGMF filters were placed first on trypticase soy agar for 4 to 4.5 h at 35°C (for resuscitation of injured cells) followed by removal to m-FC agar for additional incubation. An HGMF-based enzyme-labeled antibody (ELA) procedure has been developed for the recovery of E. coli 0157:H7 (hemorrhagic colitis, HC) strains from foods (124). The method employs the use of a special plating medium that permits HC strains to grow at 44.5°C. The special medium, HC agar, contains only 0.112% bile salts #3 in contrast to 0.15%. With its use, about 90% of HC strains could be recovered from ground beef (122). The HGMF-ELA method employs the use of HC agar incubated at 43°C for 16 h, washing of colony growth from membranes, exposure of membranes to a blocking solution, and immersion in a horseradish peroxidase-protein A-monoclonal antibody complex. By the method, ELA-positive colonies stain purple, and 95% of HC strains could be recovered within 24 h with a detection limit of 10 HC strains/g of meat.

MICROSCOPE COLONY COUNTS

Microscope colony count methods involve the counting of microcolonies that develop in agar layered over microscope slides. The first was that of Frost, which consisted of spreading 0.1 ml of a milk-agar mixture over a 4-cm^2 area on a glass slide. Following incubation, drying, and staining, microcolonies are counted with the aid of a microscope.

In another method, 2 ml of melted agar are mixed with 2 ml of warmed milk and after mixing, 0.1 ml of the inoculated agar is spread over a 4-cm^2 area. Following staining with thionin blue, the slide is viewed with the 16-mm objective of a wide-field microscope (64).

AGAR DROPLETS

In the agar droplet method of Sharpe and Kilsby (112), the food homogenate is diluted in tubes of melted agar (at 45°C). For each food sample, three tubes of agar are used, the first tube being inoculated with 1 ml of food homogenate. After mixing, a sterile capillary pipette (ideally delivering 0.033 ml/drop) is used to transfer a line of 5 × 0.1-ml droplets to the bottom of an empty petri dish. With the same capillary pipette, three drops (0.1 ml) from the first 9-ml tube are transferred to the second tube, and after mixing, another line of

5×0.1-ml droplets is placed next to the first. This step is repeated for the third tube of agar. Petri plates containing the agar droplets are incubated for 24 h, and colonies are enumerated with the aid of a $10 \times$ viewer. Results using this method from pure cultures, meats, and vegetables compared favorably to those obtained by conventional plate counts; droplet counts from ground meat were slightly higher than plate counts. The method was about three times faster, and 24-h incubations gave counts equal to those obtained after 48 h by the conventional plate count. Dilution blanks are not required, and only one petri dish/sample is needed.

DRY FILMS

A dry film method consisting of two plastic films attached together on one side and coated with culture media ingredients and a cold-water-soluble jelling agent was developed by the 3M Company and designated **Petrifilm** (77). The method can be used with nonselective ingredients to make aerobic plate counts (APCs), and with selective ingredients certain specific groups can be detected. Use of this method to date indicates that it is an acceptable alternative to standard plate count methods that employ petri dishes, and it has been approved by AOAC.

For use, 1 ml of diluent is placed between the two films and spread over the nutrient area by pressing. Following incubation, microcolonies appear red on the nonselective film because of the presence of a tetrazolium dye in the nutrient phase. In addition to its use for APC, Petrifilm methods exist for the detection and enumeration of specific groups, such as coliforms. For APC determination on 108 milk samples, this dry film method correlated highly with the conventional plate count method and was shown to be a suitable alternative (44). When compared to violet red bile agar (VRBA) and MPN for coliform enumeration on 120 samples of raw milk, Petrifilm-VRB compared favorably to VRBA counts, and both were comparable to MPN results (84). A dry medium EC count (*E. coli*) method has been developed; it employs the substrate for β-glucuronidase so that *E. coli* is distinguished from other coliforms by the formation of a blue halo around colonies. When compared to the classical confirmed MPN and VRBA on 319 food samples, the EC dry medium gave comparable results (75). Petrifilm has been shown to be suitable for determining coliforms in frozen ground beef (101) and other foods.

MOST PROBABLE NUMBERS

In this method, dilutions of food samples are prepared as for the SPC. Three serial aliquots or dilutions are then planted into 9 or 15 tubes of appropriate medium for the 3- or 5-tube method, respectively. Numbers of organisms in the original sample are determined by use of standard MPN tables. The method is statistical in nature, and MPN results are generally higher than SPC.

This method was introduced by McCrady in 1915. It is not a precise method of analysis; the 95% confidence intervals for a three-tube test range from 21 to 395. When the three-tube test is used, 20 of the 62 possible test combinations account for 99% of all results, while with the five-tube test, 49 of the possible 214 combinations account for 99% of all results (128). In a collaborative study on coliform densities in foods, a three-tube MPN value of 10 was found to be as high as 34, while in another phase of the study, the upper limit could be as high as 60 (118). Although Woodward (128) concluded that many MPN values are improbable, this method of analysis has gained popularity. Among the advantages it offers are the following:

1. It is relatively simple.
2. Results from one laboratory are more likely than SPC results to agree with those from another laboratory.
3. Specific groups of organisms can be determined by use of appropriate selective and differential media.
4. It is the method of choice for determining fecal coliform densities.

Among the drawbacks to its use is the large volume of glassware required (especially for the five-tube method), the lack of opportunity to observe the colonial morphology of the organisms, and its lack of precision.

DYE REDUCTION

Two dyes are commonly employed in this procedure to estimate the number of viable organisms in suitable products: methylene blue and resazurin. To conduct a dye-reduction test, properly prepared supernatants of foods are added to standard solutions of either dye for reduction from blue to white for methylene blue and from slate blue to pink or white for resazurin. The time for dye reduction to occur is inversely proportional to the number of organisms in the sample.

Methylene blue and resazurin reduction by 100 cultures was studied in milk; with two exceptions, a good agreement was found between numbers of bacteria and time needed for reduction of the two dyes (42). In a study of resazurin reduction as a rapid method for assessing ground beef spoilage, reduction to the colorless state, odor scores, and SPC correlated significantly (107). One of the problems of using dye reduction for some foods is the existence of inherent reductive substances. This is true of raw meats, and Austin and Thomas (12) reported that resazurin reduction was less useful than with cooked meats. For the latter, approximately 600 samples were successfully evaluated by resazurin reduction by adding 20 ml of a 0.0001% resazurin solution to 100 g of sliced meat in a plastic pouch. Another way of getting around the reductive compounds in fresh meats is to homogenize samples by Stomacher rather than by Waring blender. Using Stomacher homogenates, raw meat was successfully evaluated by resazurin reduction when Stomacher homogenates were added to a solution of resazurin in 10% skim milk (54). Stomacher homogenates contained less disrupted tissue and consequently lower concentrations of reductive compounds. The method of Holley et al. (54) was evaluated further by Dodsworth and Kempton (30), who found that raw meat with an SPC > 10^7 bacteria/g could be detected within 2 h. When compared to nitro blue tetrazolium (NT) and indophenyl nitrophenyl tetrazolium (INT), resazurin produced faster results (100). With surface samples from sheep carcasses, resazurin was reduced in 300 min by 18,000 cfu/m^2, NT in 600 min by 21,000 cfu/m^2, and INT in 660 min by 18,000 cfu/m^2 (100). Methylene blue reduction was compared to APC on 389 samples of frozen peas, and the results were linear over the APC range of log 2–6 cfu. Average decolorization times were 8 and 11 h for 10^5 and 10^4 cfu/g, respectively (1).

Dye-reduction tests have a long history of use in the dairy industry for assessing the overall microbial quality of raw milk. Among their advantages are: they are simple, rapid, and inexpensive; and only viable cells actively reduce the dyes. Disadvantages are: not all organisms reduce the dyes equally; and they are not applicable to food specimens that contain reductive enzymes unless special steps are employed. The use of fluorogenic and chromogenic substrates in food microbiology is discussed in Chapter 6.

ROLL TUBES

Screw-capped tubes or bottles of varying sizes are used in this method. Predetermined amounts of the melted and inoculated agar are added to the tube and the agar is made to solidify as a thin layer on the inside of the vessel. Following appropriate incubation, colonies are counted by rotating the vessel. It has been found to be an excellent method for enumerating fastidious anaerobes. For a review of the method, see Anderson and Fung (4).

DIRECT MICROSCOPIC COUNT

In its simplest form, the DMC consists of making smears of food specimens or cultures onto a microscope slide, staining with an appropriate dye, and viewing and counting cells with the aid of a microscope (oil immersion objective). DMCs are most widely used in the dairy industry for assessing the microbial quality of raw milk and other dairy products, and the specific method employed is that originally developed by R. S. Breed (Breed count). Briefly, the method consists of adding 0.01 ml of a sample to a 1-cm^2 area on a microscope slide, and following fixing, defatting of sample, and staining, the organisms or clumps of organisms are enumerated. The latter involves the use of a calibrated microscope (for further details, see 73). The method lends itself to the rapid microbiological examination of other food products, such as dried and frozen foods.

Among the advantages of DMC are: it is rapid and simple; cell morphology can be assessed; and it lends itself to fluorescent probes for improved efficiency. Among its disadvantages are: it is a microscopic method and therefore fatiguing to the analyst; both viable and nonviable cells are enumerated; food particles are not always distinguishable from microorganisms; microbial cells are not uniformly distributed relative to single cells and clumps; some cells do not take the stain well and may not be counted; and DMC counts are invariably higher than counts by SPC. In spite of its drawbacks, it remains the fastest way to make an assessment of microbial cells in a food product.

A slide method to detect and enumerate viable cells has been developed (16). The method employs the use of the tetrazolium salt 2-(p-iodophenyl-3-(p-nitrophenyl)-5-phenyl tetrazolium chloride (INT). Cells are exposed to filter-sterilized INT for 10 min at 37°C in a water bath followed by filtration on 0.45 µm membranes. Following drying of membranes for 10 min at 50°C, the special membranes are mounted in cottonseed oil and viewed with coverslip in place. The method was found to be workable for pure cultures of bacteria and yeasts, but it underestimated APC by 1–1.5 log cycles when compared using milk. By use of fluorescence microscopy and Viablue (modified aniline blue fluochrome), viable yeast cells could be differentiated from nonviable cells (60, 66).

Howard Mold Counts.   This is a microscope slide method developed by B. J. Howard in 1911 primarily for the purpose of monitoring tomato products. The method requires the use of a special chamber (slide) designed to enumerate mold mycelia. It is not valid on tomato products that have been comminuted. Similar to the Howard mold count is a method for quantifying *Geotrichum candidum* in canned beverages and fruits, and this method, as well as the Howard mold count method, is fully described by AOAC (11). The DEFT method has been shown to correlate well with the Howard mold count method on autoclaved and unautoclaved tomato concentrate, and it could be used as an alternative to the Howard mold count (97).

MICROBIOLOGICAL EXAMINATION OF SURFACES

The need to maintain food contact surfaces in a hygienic state is of obvious importance. The primary problem that has to be overcome when examining surfaces or utensils for microorganisms is the removal of a significant percentage of the resident flora. Although a given method may not recover all organisms, its consistent use in specified areas of a food processing plant can still provide valuable information as long as it is realized that not all organisms are being recovered. The most commonly used methods for surface assessment in food operations are presented below.

Swab/Swab-Rinse Methods

Swabbing is the oldest and most widely used method for the microbiological examination of surfaces not only in the food and dairy industries but also in hospitals and restaurants. The swab-rinse method was developed in 1917 by W. A. Manheimer and T. Ybanez. Either cotton or calcium alginate swabs are used. If one wishes to examine given areas of a surface, templates may be prepared with openings corresponding to the size of the area to be swabbed, for example, $1\,in^2$ or cm^2. The sterile template is placed over the surface, and the exposed area is rubbed thoroughly with a moistened swab. The exposed swab is returned to its holder (test tube) containing a suitable diluent and stored at refrigerator temperatures until plated. The diluent should contain a neutralizer, if necessary. When cotton swabs are used, the organisms must be dislodged from the fibers. When calcium alginate swabs are used, the organisms are released into the diluent upon dissolution of the alginate by sodium hexametaphosphate. The organisms in the diluent are enumerated by a suitable method such as SPC, but any of the culture methods in this chapter may be used. Also, a battery of selective and differential culture media may be used to test specifically for given groups of organisms. In an innovation in the swab-rinse method presented by Koller (67), 1.5 ml of fluid is added to a flat surface, swabbed for 15 sec over a 3-cm area, and volumes of 0.1 and 0.5 ml collected in microliter pipettes. The fluid may be surface or pour plated using plate count agar or selective media.

Concerning the relative efficacy of cotton and calcium alginate swabs, most agree that higher numbers of organisms are obtained by use of the latter. Using swabs, some researchers recovered as little as 10% of organisms from bovine carcasses (86); 47% of *Bacillus subtilis* spores from stainless steel surfaces (8); and up to 79% from meat surfaces (24, 90). Swab results from bovine carcasses were on the average 100 times higher than by contact plate method, and the deviation was considerably lower (86). The latter investigators found the swab method to be best suited for flexible, uneven, and heavily contaminated surfaces. The ease of removal of organisms depends on the texture of the surface and the nature and types of flora. Even with its limitations, the swab-rinse method remains a rapid, simple, and inexpensive way to assess the microbiological flora of food surfaces and utensils.

Contact Plate

The replicate organism direct agar contact (RODAC) method employs special petri plates, which are poured with 15.5–16.5 ml of an appropriate plating medium, resulting in a raised agar surface. When the plate is inverted, the hardened agar makes direct contact with the surface. Originated by Gunderson and Gunderson in 1945, it was further developed in 1964 by Hall and Hartnett. When surfaces are examined that have been cleaned with certain detergents, it is necessary to include a neutralizer (lecithin, Tween 80, and so on) in the medium. Once exposed, plates are covered, incubated, and the colonies enumerated.

Perhaps the most serious drawbacks to this method are the covering of the agar surface by spreading colonies and its ineffectiveness for heavily contaminated surfaces. These can be minimized by using plates with dried agar surfaces and by using selective media (29). The RODAC plate has been shown to be the method of choice when the surfaces to be examined are smooth, firm, and nonporous (8, 86). While it is not suitable for heavily contaminated surfaces, it has been estimated that a solution that contaminates a surface needs to contain at least 10 cells/ml before results can be achieved either by contact or by swabs (86). The latter investigators found that the contact plate removed only about 0.1% of surface flora. This suggests that 10 cfu/cm^2 detected by this method are referable to a surface that actually contains about 10^4 cfu/cm^2. When stainless steel surfaces were contaminated by *B. subtilis* endospores, 41% were recovered by the RODAC plate compared to 47% by the swab method (8). In another study, swabs were better than contact plates when the contamination level was 100 or more organisms/21–25 cm^2 (109). On the other hand, contact plates give better results where low numbers exist. In terms of ranking of surface contamination, the two methods correlated well.

Agar Syringe/"Agar Sausage" Methods

The agar syringe method was proposed by W. Litsky in 1955 and subsequently modified (7). By this method, a 100-ml syringe is modified by removing the needle end to create a hollow cylinder that is filled with agar. A layer of agar is pushed beyond the end of the barrel by means of the plunger and pressed against the surface to be examined. The exposed layer is cut off and placed in a petri dish, followed by incubation and colony enumeration. The "agar sausage" method proposed by ten Cate (123) is similar but employs plastic tubing rather than a modified syringe. The latter method has been used largely by European workers for assessing the surfaces of meat carcasses, as well as for food plant surfaces. Both methods can be viewed as variations of the RODAC plate, and both have the same disadvantages: spreading colonies and applicability limited to low levels of surface contaminants. Because clumps or chains of organisms on surfaces may yield single colonies, the counts obtained by these methods are lower than those obtained by methods that allow for the breaking up of chains or clumps.

For the examination of meat carcasses, Nortje et al. (87) compared three methods: a double swab, excision, and agar sausage. While the excision method was found to be the most reliable of the three, the modified agar sausage method correlated more closely with it than double swab, and the investigators recommended the agar sausage method because of its simplicity, speed, and accuracy.

Other Surface Methods

Direct Surface.   A number of workers have employed direct surface agar plating methods, in which melted agar is poured onto the surface or utensil to be assessed. Upon hardening, the agar mold is placed in a petri dish and incubated. Angelotti and Foter (7) proposed this as a reference method for assessing surface contamination, and it is excellent for enumerating particulates containing viable microorganisms (35). It was used successfully to determine the survival of *Clostridium sporogenes* endospores on stainless steel surfaces (83). Although effective as a research tool, the method does not lend itself to routine use for food plant surfaces.

Sticky Film.   The sticky film method of Thomas has been used with some success by Mossel et al. (82). The method consists of pressing sticky film or tape against the surface to be examined and pressing the exposed side on an agar plate. It was shown to be less effective than swabs in recovering bacteria from wooden surfaces (82). An adhesive tape method has been employed successfully to assess microorganisms on meat surfaces (40). In a recent study, the swab, RODAC, and adhesive tape (Mylar) methods were compared for the examination of pork carcasses, and the correlation between adhesive tape and RODAC was better than that between adhesive tape and swab or between RODAC and swab (27). Plastic strips attached to pads containing culture media have been used to monitor microorganisms on bottles (28).

Swab/Agar Slant.   The swab/agar slant method described in 1962 by N.-H. Hansen has been used with success by some European workers. The method involves sampling with cotton swabs that are transferred directly to slants. Following incubation, slants are grouped into one-half $\log_{10}$ units based on estimated numbers of developed colonies. The average number of colonies is determined by plotting the distribution on probability paper. A somewhat similar method, the swab/agar plate, was proposed by Ølgaard (88). It requires a template, a comparator disc, and a reference table, making it a bit more complicated than the other methods noted.

Ultrasonic Devices.   Ultrasonic devices have been used to assess the microbiological contamination of surfaces, but the surfaces to be examined must be small in size and removable so that they can be placed inside a container immersed in diluent. Once the container is placed in an ultrasonic apparatus, the energy generated effects the release of microorganisms into the diluent. A more practical use of ultrasonic energy may be the removal of bacteria from cotton swabs in the swab-rinse method (99).

Spray Gun.   A spray gun method was devised by Clark (24, 25) based on the impingement of a spray of washing solution against a circumscribed area of surface and the subsequent plating of the washing solution. Although the device is portable, a source of air pressure is necessary. It was shown to be much more effective than the swab method in removing bacteria from meat surfaces.

AIR SAMPLING

A variety of methods and devices exists for sampling air in food plants for the presence and relative numbers of microorganisms. They include impingement in liquids, impaction on solid surfaces, filtration, sedimentation, centrifugation, electrostatic precipitation, and thermal precipitation (120). Most commonly used are sedimentation, impaction, and impingement. These and other methods are more fully described in a manual that is periodically updated (72).

One of the simplest methods of air sampling is to open prepoured petri dishes for specific periods of time in the area to be assessed. While this method gives only an approximation of microbial numbers, it is effective in certain situations (51). Results are influenced by size of particles and by the speed and direction of air flow. If the agar surface is exposed for too long, drying of it will affect growth of deposited organisms. The presence and relative numbers of different types of organisms can be assessed by use of appropriate selective media.

The two most popular types of air samplers are the all-glass impinger and the Andersen (3) sieve sampler. They offer the advantage that a specific volume of air can be sampled. With the all-glass impinger, a specific quantity of diluent is placed in the impinger, and air is pulled by vacuum through a capillary orifice with the air impinging on the diluent, where microorganisms are trapped in the liquid. The sample may then be plated as in the SPC or by use of selective media. This method does not work well when numbers of organisms are too small, and some gram-negative bacteria may be destroyed by high impingement velocity.

In the Andersen sampler, air is drawn by vacuum through one or more sieves, and the microorganisms are trapped on the surface of one or more agar plates. Optimum flow rate is 28.3 liters/min. Single- and multiple-stage sieve samplers have been compared (69), and the ranges and coefficients of variation for the all-glass and Andersen samplers have been determined (71). Of somewhat similar design are the Fort Detrick Slit Sampler and the Casella-type single-stage slit sampler, both described by D. A. Gabis et al. in Speck (120).

METABOLICALLY INJURED ORGANISMS

When microorganisms are subjected to environmental stresses such as sublethal heat and freezing, many of the individual cells undergo metabolic injury, resulting in their inability to form colonies on selective media that uninjured cells can tolerate. Whether a culture has suffered metabolic injury can be determined by plating aliquots separately on a nonselective and a selective medium and enumerating the colonies that develop after suitable incubation. The colonies that develop on the nonselective medium represent both injured and uninjured cells, while only the uninjured cells develop on the selective medium. The difference between the number of colonies on the two media is a measure of the number of injured cells in the original culture or population. This principle is illustrated in Figure 5-2 by data from Tomlins et al. (125) on sublethal heat injury of *S. aureus*. These investigators subjected the organism to 52°C for 15 min in a phosphate buffer at pH 7.2 to inflict cell injury. The plating of cells at zero time and up to 15 min of heating on nonselective trypticase soy agar (TSA) and selective TSA + 7.0% NaCl (stress medium; TSAS) reveals only a slight reduction in numbers on TSA, while the numbers on TSAS were reduced considerably, indicating a high degree of injury relative to a level of salt that uninjured *S. aureus* can withstand. To allow the heat-injured cells to repair, the cells were placed in nutrient broth (recovery medium) followed by incubation at 37°C for 4 h. With hourly plating of aliquots from the recovery medium onto TSAS, it can be seen that the injured cells regained their capacity to withstand the 7.0% NaCl in TSAS after the 4-h incubation.

The existence of metabolically injured cells in foods and their recovery during culturing procedures is of great importance not only from the standpoint of pathogenic organisms but for spoilage organisms as well. The data cited suggest that if a high-salt medium had been employed to examine a heat-pasteurized product for *S. aureus*, the number of viable cells found would have been lower than the actual number by a factor of 3 log cycles. Injury of foodborne microorganisms has been shown by a large number of investigators to be induced not only by sublethal heat and freezing but also by freeze drying, drying, irradiation, aerosolization, dyes, sodium azide, salts, heavy metals, antibiotics, essential oils, and other chemicals, such as EDTA and sanitizing compounds.

The recognition of sublethal stresses on foodborne microorganisms and their effect on growth under varying conditions dates back to the turn of the century. However, a full appreciation of this phenomenon did not come until the late 1960s. During the early 1960s, it was observed that an initial rapid decrease in numbers of a metabolically injured organism was followed by only a limited recovery during the resuscitation process ("Phoenix phenomenon"). The increased nutritional requirement of bacteria that had undergone heat treatment was noted by Nelson (85) in 1943 (Nelson also reviewed the work of

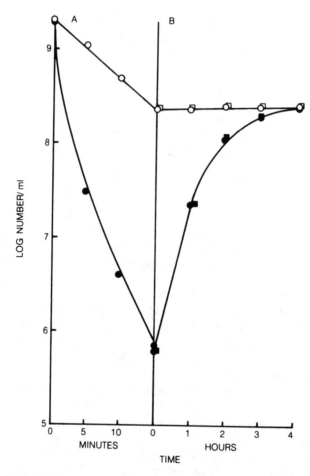

FIGURE 5-2.   Survival and recovery curve for *S. aureus* MF = 31. (*A*) Heat injury at 52°C for 15 min in 100 mM potassium phosphate buffer. (*B*) Recovery from heat injury in NB at 37°C. Symbols: ○, samples plated on TSA to give a total viable count; ●, samples plated on TSAS to give an estimate of the uninjured population—cells recovered in NB containing 100 μg/ml of chloramphenicol; □, samples plated on TSA; ■, samples plated on TSAS.
Tomlins et al., 125; reproduced by permission of Nat'l Res. Counc. of Canada from Can. J. Microbiol. 17:759–765, 1971.

others up to that time). Gunderson and Rose (45) noted the progressive decrease in numbers of coliforms from frozen chicken products that grew on VRBA with increasing storage time of products. Hartsell (49) inoculated foods with salmonellae, froze the inoculated foods, and then studied the fate of the organisms during freezer storage. More organisms could be recovered on highly nutritive nonselective media than on selective media such as MacConkey, deoxycholate, or VRBAs. The importance of the isolation medium in recovering stressed cells was noted also by Postgate and Hunter (98) and by Harris (47). In addition to the more exacting nutritional requirements of foodborne organisms that undergo environmental stresses, these organisms may be expected to manifest their injury via increased lag phases of growth, increased sensitivity to a variety of selective media agents, damage to cell membranes and TCA-cycle enzymes, breakdown of ribosomes, and DNA damage. While damage to ribosomes and cell membranes appears to be a common consequence of sublethal heat injury, not all harmful agents produce identifiable injuries.

Recovery/Repair

Metabolically injured cells can recover, at least in *S. aureus*, in no-growth media (59) and at a temperature of 15°C but not 10°C (41). In some instances at least, the recovery process is not instantaneous, for it has been shown that not all stressed coliforms recover to the same degree but that the process takes place in a step-wise manner (76). Not all cells in a population suffer the same degree of injury. Hurst et al. (56) found dry-injured *S. aureus* cells that failed to develop on the nonselective recovery medium (TSA) but did recover when pyruvate was added to this medium. These cells were said to be severely injured in contrast to injured and uninjured cells. It has been found that sublethally heated *S. aureus* cells may recover their NaCl tolerance before certain membrane functions are restored (58). It is well established that injury repair occurs in the general absence of cell wall and protein synthesis. It can be seen from Figure 5-2 that the presence of chloramphenicol in the recovery medium had no effect on the recovery of *S. aureus* from sublethal heat injury. The repair of cell ribosomes and membrane appears to be essential for recovery at least from sublethal heat, freezing, drying, and irradiation injuries.

The protection of cells from heat and freeze injury is favored by complex media and menstra or certain specific components thereof. Milk provides more protection than saline or mixtures of amino acids (81), and the milk components that are most influential appear to be phosphate, lactose, and casein. Sucrose appears to be protective against heat injury (2, 70), while glucose has been reported to decrease heat protection for *S. aureus* (81). Nonmetabolizable sugars and polyols such as arabinose, xylose, and sorbitol have been found to protect *S. aureus* against sublethal heat injury, but the mechanism of this action is unclear (119).

The consequences of not employing a recovery step have been reviewed by Busta (23). The use of TSB with incubations ranging from 1 to 24 h at temperatures from 20° to 37°C is widely used for various organisms. The enumeration of sublethally heated *S. aureus* strains on various media has been studied (19, 36, 56). In one of these studies, seven staphylococcal media were compared on their capacity to recover 19 strains of sublethally heated *S. aureus*, and the Baird-Parker medium was found to be clearly the best of those studied, including nonselective TSA. Similar findings by others led to the adoption of this medium in the official methods of AOAC for the direct determination of *S. aureus* in foods that contain ≥10 cells/g (11). The greater efficacy of the Baird-Parker medium has been shown to result from its content of pyruvate. The use of this medium following recovery in an antibiotic-containing, nonselective medium has been suggested (56). While this approach may be suitable for *S. aureus* recovery, some problems may be expected to occur with the widespread use of antibiotics in recovery media to prevent cell growth. It has been shown that heat-injured spores of *C. perfringens* are actually sensitized to polymyxin and neomycin (14), and it is well established that the antibiotics that affect cell wall synthesis are known to induce L-phase variations in many bacteria.

Pyruvate is well established as an injury repair agent not only for injured *S. aureus* cells but for other organisms such as *E. coli*. Higher counts are obtained on media containing this compound when injured by a variety of agents. When added to TSB containing 10% NaCl, higher numbers of both stressed and nonstressed *S. aureus* were achieved (19), and the repair-detection of freeze- or heat-injured *E. coli* was significantly improved by pyruvate (78).

Catalase is another agent that increases recovery of injured aerobic organisms. First reported by Martin et al. (74), it has been found effective by many other investigators. It is effective for sublethally heated *S. aureus*, *Pseudomonas fluorescens*, *Salmonella typhimurium*, and *E. coli* (74). It is effective also for *S. aureus* in the presence of 10% NaCl (19) and for water-stressed *S. aureus* (36). Another compound, shown to be as effective as pyruvate for heat-injured *E. coli*, is 3,3'-thiodipropionic acid (78).

Radiation injury of *Clostridium botulinum* type E spores by 4 kGy resulted in the inability to grow at 10°C in the presence of polymyxin and neomycin (106). The injured cells had a damaged postgermination system and formed aseptate filaments during outgrowth, but the germination lytic system was not damaged. The radiation injury was repaired at 30°C in about 15 h on TPEY agar without antibiotics. When *C. botulinum* spores are injured with hypochlorite, the L-alanine germination sites are modified, resulting in the need for higher concentrations of alanine for repair (38). The L-alanine germination sites could be activated by lactate, and hypochlorite-treated spores could be germinated by lysozyme, indicating that the chloride removed spore coat proteins (39). More detailed information on spore injury has been provided by Foegeding and Busta (37).

Sublethally heat-stressed yeasts are inhibited by some essential oils (spices) at concentrations as low as 25 ppm (26). The spice oils affected colony size and pigment production.

Special plating procedures have been found by Speck et al. (121) and Hartman et al. (48) to allow for recovery from injury and subsequent enumeration in essentially one step. The procedures consist of using the agar overlay plating technique with one layer consisting of TSA, onto which are plated the stressed organisms. Following a 1- to 2-h incubation at 25°C for recovery, the TSA layer is overlaid with VRBA and incubated at 35°C for 24 h. The overlay method of Hartman et al. involved the use of a modified VRBA. The principle involved in the overlay technique could be extended to other selective media, of course. An overlay technique has been recommended for the recovery of coliforms. By this method, coliforms are plated with TSA and incubated at 35°C for 2 h followed by an overlay of VRBA (89).

In his comparison of 18 plating media and seven enrichment broths to recover heat-stressed *Vibrio parahaemolyticus*, Beuchat (18) found that the two most efficient plating media were water blue-alizarin yellow agar and arabinose-ammonium-sulphate-cholate agar; arabinose-ethyl violet broth was the most suitable enrichment broth.

Mechanism

Pyruvate and catalase both act to degrade peroxides, suggesting that metabolically injured cells lack this capacity. The inability of heat-damaged *E. coli* cells to grow as well when surface plated as when pour plated with the same medium (46) may be explained by the loss of peroxides.

A large number of investigators have found that metabolic injury is accompanied by damage to cell membranes, ribosomes, DNA, or enzymes. The cell membrane appears to be the most universally affected (55). The lipid components of the membrane are the most likely targets, especially for sublethal heat injuries. Ribosomal damage is believed to result from the loss of Mg^{2+} and not to heat effects per se (57). On the other hand, ribosome-free areas have been observed by electron microscopy in heat-injured *S. aureus* cells (63). Following prolonged heating at 50°C, virtually no ribosomes were detected, and in addition, the cells were characterized by the appearance of surface blebs and exaggerated internal membranes (63). When *S. aureus* was subjected to acid injury by exposure to acetic, hydrochloric, and lactic acids at 37°C, coagulase and thermostable nuclease activities were reduced in injured cells (129). While acid injury did not affect cell membranes, RNA synthesis was affected.

For more information on cell injury and on methods of recovery, see Andrew and Russell (5).

References

1. Alcock, S. J., L. P. Hall, and J. H. Blanchard. 1987. Methylene blue test to assess the microbial contamination of frozen peas. *Food Microbiol.* 4:3–10.
2. Allwood, M. C., and A. D. Russell. 1967. Mechanism of thermal injury in *Staphylococcus aureus*. I. Relationship between viability and leakage. *Appl. Microbiol.* 15:1266–1269.
3. Andersen, A. A. 1958. New sampler for the collection, sizing, and enumeration of viable airborne particles. *J. Bacteriol.* 76:471–484.
4. Anderson, K. L., and D. Y. C. Fung. 1983. Anaerobic methods, techniques and principles for food bacteriology: A review. *J. Food Protect.* 46:811–822.
5. Andrew, M. H. E., and A. D. Russell. 1984. *The Revival of Injured Microbes*. London: Academic Press.
6. Andrews, W. H., C. R. Wilson, P. L. Poelma, A. Romero, R. A. Rude, A. P. Duran, F. D. McClure, and D. E. Gentile. 1978. Usefulness of the Stomacher in a microbiological regulatory laboratory. *Appl. Environ. Microbiol.* 35:89–93.
7. Angelotti, R., and M. J. Foter. 1958. A direct surface agar plate laboratory method for quantitatively detecting bacterial contamination on nonporous surfaces. *Food Res.* 23:170–174.
8. Angelotti, R., J. L. Wilson, W. Litsky, and W. G. Walter. 1964. Comparative evaluation of the cotton swab and rodac methods for the recovery of *Bacillus subtilis* spore contamination from stainless steel surfaces. *Hlth. Lab. Sci.* 1:289–296.
9. Association of Official Analytical Chemists. 1983. Enumeration of coliforms in selected foods. Hydrophobic grid membrane filter method, official first action. *J. Assoc. Off. Anal. Chem.* 66:547–548.
10. Association of Official Analytical Chemists. 1977. Spiral plate method for bacterial count: Official first action. *J. Assoc. Off. Anal. Chem.* 60:493–494.
11. Association of Official Analytical Chemists. 1990. *Official Methods of Analysis*, 15th ed. Washington, D.C.: AOAC.
12. Austin, B. L., and B. Thomas. 1972. Dye reduction tests on meat products. *J. Sci. Food Agric.* 23:542.
13. *Bacteriological Analytical Manual*, 6th ed. 1984. Washington, D.C.: AOAC.

14. Barach, J. T., R. S. Flowers, and D. M. Adams. 1975. Repair of heat-injured *Clostridium perfringens* spores during outgrowth. *Appl. Microbiol.* 30:873–875.

15. Baumgart, J. 1973. Der "Stomacher"—ein neues Zerkleinerungsgerat zur Herstellung von Lebensmittelsuspensionen für die Keimzahlbestimmung. *Fleischwirtschaft* 53:1600.

16. Betts, R. P., P. Bankes, and J. G. Board. 1989. Rapid enumeration of viable micro-organisms by staining and direct microscopy. *Lett. Appl. Microbiol.* 9: 199–202.

17. Beuchat, L. R., ed. 1987. *Food and Beverage Mycology*, 2d ed. New York: Van Nostrand Reinhold.

18. Beuchat, L. R., and R. V. Lechowich. 1968. Effect of salt concentration in the recovery medium on heat-injured *Streptococcus faecalis*. *Appl. Microbiol.* 16: 772–776.

19. Brewer, D. G., S. E. Martin, and Z. J. Ordal. 1977. Beneficial effects of catalase or pyruvate in a most-probable-number technique for the detection of *Staphylococcus aureus*. *Appl. Environ. Microbiol.* 34:797–800.

20. Brodsky, M. H., P. Entis, A. N. Sharpe, and G. A. Jarvis. 1982. Enumeration of indicator organisms in foods using the automated hydrophobic grid membrane filter technique. *J. Food Protect.* 45:292–296.

21. Brodsky, M. H., P. Entis, M. P. Entis, A. N. Sharpe, and G. A. Jarvis. 1982. Determination of aerobic plate and yeast and mold counts in foods using an automated hydrophobic grid membrane filter technique. *J. Food Protect.* 45: 301–304.

22. Brodsky, M. H., P. Boleszczuk, and P. Entis. 1982. Effect of stress and resuscitation on recovery of indicator bacteria from foods using hydrophobic grid-membrane filtration. *J. Food Protect.* 45:1326–1331.

23. Busta, F. F. 1976. Practical implications of injured microorganisms in food. *J. Milk Food Technol.* 39:138–145.

24. Clark, D. S. 1965. Method of estimating the bacterial population of surfaces. *Can. J. Microbiol.* 11:407–413.

25. Clark, D. S. 1965. Improvement of spray gun method of estimating bacterial populations on surfaces. *Can. J. Microbiol.* 11:1021–1022.

26. Conner, D. E., and L. R. Beuchat. 1984. Sensitivity of heat-stressed yeasts to essential oils of plants. *Appl. Environ. Microbiol.* 47:229–233.

27. Cordray, J. C., and D. L. Huffman. 1985. Comparison of three methods for estimating surface bacteria on pork carcasses. *J. Food Protect.* 48:582–584.

28. Cousin, M. A. 1982. Evaluation of a test strip used to monitor food processing sanitation. *J. Food Protect.* 45:615–619, 623.

29. deFigueiredo, M. P., and J. M. Jay. 1976. Coliforms, enterococci, and other microbial indicators. In *Food Microbiology: Public Health and Spoilage Aspects*, ed. M. P. deFigueiredo and D. F. Splittstoesser, 271–297. Westport, Conn.:AVI.

30. Dodsworth, P. J., and A. G. Kempton. 1977. Rapid measurement of meat quality by resazurin reduction. II. Industrial application. *Can. Inst. Food Sci. Technol. J.* 10:158–160.

31. Donnelly, C. B., J. E. Gilchrist, J. T. Peeler, and J. E. Campbell. 1976. Spiral plate count method for the examination of raw and pasteurized milk. *Appl. Environ. Microbiol.* 32:21–27.

32. Emswiler, B. S., C. J. Pierson, and A. W. Kotula. 1977. Stomaching vs. blending. A comparison of two techniques for the homogenization of meat samples for microbiological analysis. *Food Technol.* 31(10):40–42.

33. Entis, P. 1985. Rapid hydrophobic grid membrane filter method for *Salmonella* detection in selected foods. *J. Assoc. Off. Anal. Chem.* 68:555–564.

34. Entis, P. 1983. Enumeration of coliforms in non-fat dry milk and canned custard

by hydrophobic grid membrane filter method: Collaborative study. *J. Assoc. Off. Anal. Chem.* 66:897–904.

35. Favero, M. S., J. J. McDade, J. A. Robertsen, R. K. Hoffman, and R. W. Edwards. 1968. Microbiological sampling of surfaces. *J. Appl. Bacteriol.* 31: 336–343.

36. Flowers, R. S., S. E. Martin, D. G. Brewer, and Z. J. Ordal. 1977. Catalase and enumeration of stressed *Staphylococcus aureus* cells. *Appl. Environ. Microbiol.* 33:1112–1117.

37. Foegeding, P. M., and F. F. Busta. 1981. Bacterial spore injury—an update. *J. Food Protect.* 44:776–786.

38. Foegeding, P. M., and F. F. Busta. 1983. Proposed role of lactate in germination of hypochlorite-treated *Clostridium botulinum* spores. *Appl. Environ. Microbiol.* 45:1369–1373.

39. Foegeding, P. M., and F. F. Busta. 1983. Proposed mechanism for sensitization by hypochlorite treatment of *Clostridium botulinum* spores. *Appl. Environ. Microbiol.* 45:1374–1379.

40. Fung, D. Y. C., C.-Y. Lee, and C. L. Kastner. 1980. Adhesive tape method for estimating microbial load on meat surfaces. *J. Food Protect.* 43:295–297.

41. Fung. D. Y. C., and L. L. VandenBosch. 1975. Repair, growth, and entero-toxigenesis of *Staphylococcus aureus* S-6 injured by freeze-drying. *J. Milk Food Technol.* 38:212–218.

42. Garvie, E. I., and A. Rowlands. 1952. The role of micro-organisms in dye-reduction and keeping-quality tests. II. The effect of micro-organisms when added to milk in pure and mixed culture. *J. Dairy Res.* 19:263–274.

43. Gilchrist, J. E., J. E. Campbell, C. B. Donnelly, J. T. Peeler, and J. M. Delaney. 1973. Spiral plate method for bacterial determination. *Appl. Microbiol.* 25: 244–252.

44. Ginn, R. E., V. S. Packard, and T. L. Fox. 1984. Evaluation of the 3M dry medium culture plate (Petrifilm™ SM) method for determining numbers of bacteria in raw milk. *J. Food Protect.* 47:753–755.

45. Gunderson, M. F., and K. D. Rose. 1948. Survival of bacteria in a precooked, fresh-frozen food. *Food Res.* 13:254–263.

46. Harries, D., and A. D. Russell. 1966. Revival of heat-damaged *Escherichia coli*. *Experientia* 22:803–804.

47. Harris, N. D. 1963. The influence of the recovery medium and the incubation temperature on the survival of damaged bacteria. *J. Appl. Bacteriol.* 26:387–397.

48. Hartman, P. A., P. S. Hartman, and W. W. Lanz. 1975. Violet red bile 2 agar for stressed coliforms. *Appl. Microbiol.* 29:537–539.

49. Hartsell, S. E. 1951. The longevity and behavior of pathogenic bacteria in frozen foods: The influence of plating media. *Amer. J. Pub. Hlth.* 41:1072–1077.

50. Hedges, A. J., R. Shannon, and R. P. Hobbs. 1978. Comparison of the precision obtained in counting viable bacteria by the spiral plate maker, the droplette and the Miles & Misra methods. *J. Appl. Bacteriol.* 45:57–65.

51. Hill, R. A., D. M. Wilson, W. R. Burg, and O. L. Shotwell. 1984. Viable fungi in corn dust. *Appl. Environ. Microbiol.* 47:84–87.

52. Hobbie, J. E., R. J. Daley, and S. Jasper. 1977. Use of nucleopore filters for counting bacteria by fluorescence microscopy. *Appl. Environ. Microbiol.* 33: 1225–1228.

53. Holah, J. T., R. P. Betts, and R. H. Thorpe. 1988. The use of direct epi-fluorescent microscopy (DEM) and the direct epifluorescent filter technique (DEFT) to assess microbial populations on food contact surfaces. *J. Appl. Bacteriol.* 65:215–221.

54. Holley, R. A., S. M. Smith, and A. G. Kempton. 1977. Rapid measurement of

meat quality by resazurin reduction. I. Factors affecting test validity. *Can. Inst. Food Sci. Technol. J.* 10:153–157.

55. Hurst, A. 1977. Bacterial injury: A review. *Can. J. Microbiol.* 23:935–944.
56. Hurst, A., G. S. Hendry, A. Hughes, and B. Paley. 1976. Enumeration of sublethally heated staphylococci in some dried foods. *Can. J. Microbiol.* 22: 677–683.
57. Hurst, A., and A. Hughes. 1978. Stability of ribosomes of *Staphylococcus aureus* S-6 sublethally heated in different buffers. *J. Bacteriol.* 133:564–568.
58. Hurst, A., A. Hughes, J. L. Beare-Rogers, and D. L. Collins-Thompson. 1973. Physiological studies on the recovery of salt tolerance by *Staphylococcus aureus* after sublethal heating. *J. Bacteriol.* 116:901–907.
59. Hurst, A., A. Hughes, D. L. Collins-Thompson, and B. G. Shah. 1974. Relationship between loss of magnesium and loss of salt tolerance after sublethal heating of *Staphylococcus aureus*. *Can. J. Microbiol.* 20:1153–1158.
60. Hutcheson, T. C., T. McKay, L. Farr, and B. Seddon. 1988. Evaluation of the stain Viablue for the rapid estimation of viable yeast cells. *Lett. Appl. Microbiol.* 6:85–88.
61. Jarvis, B., V. H. Lach, and J. M. Wood. 1977. Evaluation of the spiral plate maker for the enumeration of micro-organisms in foods. *J. Appl. Bacteriol.* 43:149–157.
62. Jay, J. M., and S. Margitic. 1979. Comparison of homogenizing, shaking, and blending on the recovery of microorganisms and endotoxins from fresh and frozen ground beef as assessed by plate counts and the *Limulus* amoebocyte lysate test. *Appl. Environ. Microbiol.* 38:879–884.
63. Jones, S. B., S. A. Palumbo, and J. L. Smith. 1983. Electron microscopy of heat-injured and repaired *Staphylococcus aureus*. *J. Food Safety* 5:145–157.
64. Juffs, H. S., and F. J. Babel. 1975. Rapid enumeration of psychrotrophic bacteria in raw milk by the microscopic colony count. *J. Milk Food Technol.* 38:333–336.
65. Knabel, S. J., H. W. Walker, and A. A. Kraft. 1987. Enumeration of fluorescent pseudomonads on poultry by using the hydrophobic-grid membrane filter method. *J. Food Sci.* 52:837–841, 845.
66. Koch, H. A., R. Bandler, and R. R. Gibson. 1986. Fluorescence microscopy procedure for quantification of yeasts in beverages. *Appl. Environ. Microbiol.* 52:599–601.
67. Koller, W. 1984. Recovery of test bacteria from surfaces with a simple new swab-rinse technique: A contribution to methods for evaluation of surface disinfectants. *Zent. Bakteriol. Hyg. I. Orig. B.* 179:112–124.
68. Konuma, H., A. Suzuki, and H. Kurata. 1982. Improved Stomacher 400 bag applicable to the spiral plate system for counting bacteria. *Appl. Environ. Microbiol.* 44:765–769.
69. Kotula, A. W., J. R. Guilfoyle, B. S. Emswiler, and M. D. Pierson. 1978. Comparison of single and multiple stage sieve samplers for airborne micro-organisms. *J. Food Protect.* 41:447–449.
70. Lee, A. C., and J. M. Goepfert. 1975. Influence of selected solutes on thermally induced death and injury of *Salmonella typhimurium*. *J. Milk Food Technol.* 38:195–200.
71. Lembke, L. L., R. N. Kniseley, R. C. Van Nostrand, and M. D. Hale. 1981. Precision of the all-glass impinger and the Andersen microbial impactor for air sampling in solid-waste handling facilities. *Appl. Environ. Microbiol.* 42:222–225.
72. Lioy, P. J., and M. J. Y. Lioy, eds. 1983. *Air Sampling Instruments for Evaluation of Atmospheric Contamination*, 6th ed. Cincinnati: American Conference of Governmental Industrial Hygienists.
73. Marth, E. H., ed. 1978. *Standard Methods for the Examination of Dairy Products.*

14th ed. Washington, D.C.: American Public Health Association.

74. Martin, S. E., R. S. Flowers, and Z. J. Ordal. 1976. Catalase: Its effect on microbial enumeration. *Appl. Environ. Microbiol.* 32:731–734.

75. Matner, R. R., T. L. Fox, D. E. McIver, and M. S. Curiale. 1990. Efficacy of Petrifilm^tm count plates for *E. coli* and coliform enumeration. *J. Food Protect.* 53:145–150.

76. Maxcy, R. B. 1973. Condition of coliform organisms influencing recovery of subcultures on selective media. *J. Milk Food Technol.* 36:414–416.

77. McAllister, J. S., R. L. Nelson, P. E. Hanson, and K. F. McGoldrick. 1984. A dry media film for use as a replacement for the aerobic pour plate for enumeration of bacteria. In *Bacteriological Proceedings*, 14.

78. McDonald, L. C., C. R. Hackney, and B. Ray. 1983. Enhanced recovery of injured *Escherichia coli* by compounds that degrade hydrogen peroxide or block its formation. *Appl. Environ. Microbiol.* 45:360–365.

79. *Microorganisms in Foods.* 1982. Vol. 1, *Their Significance and Methods of Enumeration*, 2d ed. ICMSF. Toronto: University of Toronto Press.

80. *Microorganisms in Foods.* 1986. Vol. 2, *Sampling for Microbiological Analysis: Principles and Specific Applications*, 2d ed. ICMSF. Toronto: University of Toronto Press.

81. Moats, W. A., R. Dabbah, and V. M. Edwards. 1971. Survival of *Salmonella anatum* heated in various media. *Appl. Microbiol.* 21:476–481.

82. Mossel, D. A. A., E. H. Kampelmacher, and L. M. Van Noorle Jansen. 1966. Verification of adequate sanitation of wooden surfaces used in meat and poultry processing. *Zent. Bakteriol. Parasiten., Infek. Hyg. Abt. I.* 201:91–104.

83. Neal, N. D., and H. W. Walker. 1977. Recovery of bacterial endospores from a metal surface after treatment with hydrogen peroxide. *J. Food Sci.* 42:1600–1602.

84. Nelson, C. L., T. L. Fox, and F. F. Busta. 1984. Evaluation of dry medium film (Petrifilm VRB) for coliform enumeration. *J. Food Protect.* 47:520–525.

85. Nelson, F. E. 1943. Factors which influence the growth of heat-treated bacteria. I. A comparison of four agar media. *J. Bacteriol.* 45:395–403.

86. Niskanen, A., and M. S. Pohja. 1977. Comparative studies on the sampling and investigation of microbial contamination of surfaces by the contact plate and swab methods. *J. Appl. Bacteriol.* 42:53–63.

87. Nortje, G. L., E. Swanepoel, R. T. Naude, W. H. Holzapfel, and P. L. Steyn. 1982. Evaluation of three carcass surface microbial sampling techniques. *J. Food Protect.* 45:1016–1017, 1021.

88. Ølgaard, K. 1977. Determination of relative bacterial levels on carcasses and meats—A new quick method. *J. Appl. Bacteriol.* 42:321–329.

89. Ordal, Z. J., J. J. Iandolo, B. Ray, and A. G. Sinskey. 1976. Detection and enumeration of injured microorganisms. In *Compendium of Methods for the Microbiological Examination of Foods*, ed. M. L. Speck, 163–169. Washington, D.C.: American Public Health Association.

90. Patterson, J. T. 1971. Microbiological assessment of surfaces. *J. Food Technol.* 6:63–72.

91. Peeler, J. T., J. E. Gilchrist, C. B. Donnelly, and J. E. Campbell. 1977. A collaborative study of the spiral plate method for examining milk samples. *J. Food Protect.* 40:462–464.

92. Pettipher, G. L. 1983. *The Direct Epifluorescent Filter Technique for the Rapid Enumeration of Microorganisms*. New York: Wiley.

93. Pettipher, G. L., R. J. Fulford, and L. A. Mabbitt. 1983. Collaborative trial of the direct epifluorescent filter technique (DEFT), a rapid method for counting bacteria in milk. *J. Appl. Bacteriol.* 54:177–182.

94. Pettipher, G. L., R. Mansell, C. H. McKinnon, and C. M. Cousins. 1980. Rapid

membrane filtration-epifluorescent microscopy technique for direct enumeration of bacteria in raw milk. *Appl. Environ. Microbiol.* 39:423–429.

95. Pettipher, G. L., and U. M. Rodrigues. 1981. Rapid enumeration of bacteria in heat-treated milk and milk products using a membrane filtration-epifluorescent microscopy technique. *J. Appl. Bacteriol.* 50:157–166.

96. Pettipher, G. L., and U. M. Rodrigues. 1982. Rapid enumeration of micro-organisms in foods by the direct epifluorescent filter technique. *Appl. Environ. Microbiol.* 44:809–813.

97. Pettipher, G. L., R. A. Williams, and C. S. Gutteridge. 1985. An evaluation of possible alternative methods to the Howard mould count. *Lett. Appl. Microbiol.* 1:49–51.

98. Postgate, J. R., and J. R. Hunter. 1963. Metabolic injury in frozen bacteria. *J. Appl. Bacteriol.* 26:405–414.

99. Puleo, J. R., M. S. Favero, and N. J. Petersen. 1967. Use of ultrasonic energy in assessing microbial contamination on surfaces. *Appl. Microbiol.* 15:1345–1351.

100. Rao, D. N., and V. S. Murthy. 1986. Rapid dye reduction tests for the determination of microbiological quality of meat. *J. Food Technol.* 21:151–157.

101. Restaino, L., and R. H. Lyon. 1987. Efficacy of Petrifilm™ VRB for enumerating coliforms and *Escherichia coli* from frozen raw beef. *J. Food Protect.* 50: 1017–1022.

102. Reyniers, J. A. 1935. Mechanising the viable count. *J. Pathol. Bacteriol.* 40:437–454.

103. Rodrigues, U. M., and R. G. Kroll. 1989. Microcolony epifluorescence microscopy for selective enumeration of injured bacteria in frozen and heat-treated foods. *Appl. Environ. Microbiol.* 55:778–787.

104. Rodrigues, U. M., and R. G. Kroll. 1988. Rapid selective enumeration of bacteria in foods using a micro-colony epifluorescence microscopy technique. *J. Appl. Bacteriol.* 64:65–78.

105. Rodrigues, U. M., and R. G. Kroll. 1985. The direct epifluorescent filter technique (DEFT): Increased selectivity, sensitivity and rapidity. *J. Appl. Bacteriol.* 59:493–499.

106. Rowley, D. B., R. Firstenberg-Eden, and G. E. Shattuck. 1983. Radiation-injured *Clostridium botulinum* type E spores: Outgrowth and repair. *J. Food Sci.* 48:1829–1831, 1848.

107. Saffle, R. L., K. N. May, H. A. Hamid, and J. D. Irby. 1961. Comparing three rapid methods of detecting spoilage in meat. *Food Technol.* 15:465–467.

108. Schiemann, D. A. 1977. Evaluation of the Stomacher for preparation of food homogenates. *J. Food Protect.* 40:445–448.

109. Scott, E., S. F. Bloomfield, and C. G. Barlow. 1984. A comparison of contact plate and calcium alginate swab techniques of environmental surfaces. *J. Appl. Bacteriol.* 56:317–320.

110. Sharpe, A. N., and G. C. Harshman. 1976. Recovery of *Clostridium perfringens, Staphylococcus aureus,* and molds from foods by the Stomacher: Effect of fat content, surfactant concentration, and blending time. *Can. Inst. Food Sci. Technol. J.* 9:30–34.

111. Sharpe, A. N., and A. K. Jackson. 1972. Stomaching: A new concept in bacteriological sample preparation. *Appl. Microbiol.* 24:175–178.

112. Sharpe, A. N., and D. C. Kilsby. 1971. A rapid, inexpensive bacterial count technique using agar droplets. *J. Appl. Bacteriol.* 34:435–440.

113. Sharpe, A. N., M. P. Diotte, I. Dudas, S. Malcolm, and P. I. Peterkin. 1983. Colony counting on hydrophobic grid-membrane filters. *Can. J. Microbiol.* 29: 797–802.

114. Sharpe, A. N., P. I. Peterkin, and N. Malik. 1979. Improved detection of

coliforms and *Escherichia coli* in foods by a membrane filter method. *Appl. Environ. Microbiol.* 38:431–435.

115. Sharpe, A. N., and G. L. Michaud. 1974. Hydrophobic grid-membrane filters: New approach to microbiological enumeration. *Appl. Microbiol.* 28:223–225.

116. Sharpe, A. N., and G. L. Michaud. 1975. Enumeration of high numbers of bacteria using hydrophobic grid-membrane filters. *Appl. Microbiol.* 30:519–524.

117. Shaw, B. G., C. D. Harding, W. H. Hudson, and L. Farr. 1987. Rapid estimation of microbial numbers on meat and poultry by direct epifluorescent filter technique. *J. Food Protect.* 50:652–657.

118. Silliker, J. H., D. A. Gabis, and A. May. 1979. ICMSF methods studies. XI. Collaborative/comparative studies on determination of coliforms using the most probable number procedure. *J. Food Protect.* 42:638–644.

119. Smith, J. L., R. C. Benedict, M. Haas, and S. A. Palumbo. 1983. Heat injury in *Staphylococcus aureus* 196E: Protection by metabolizable and non-metabolizable sugars and polyols. *Appl. Environ. Microbiol.* 46:1417–1419.

120. Speck, M. L., ed. 1984. *Compendium of Methods for the Microbiological Examination of Foods*. Washington, D.C.: American Public Health Association.

121. Speck, M. L., B. Ray, and R. B. Read, Jr. 1975. Repair and enumeration of injured coliforms by a plating procedure. *Appl. Microbiol.* 29:549–550.

122. Szabo, R. A., E. C. D. Todd, and A. Jean. 1986. Method to isolate *Escherichia coli* 0157:H7 from food. *J. Food Protect.* 49:768–772.

123. ten Cate, L. 1963. An easy and rapid bacteriological control method in meat processing industries using agar sausage techniques in Rilsan artificial casing. *Fleischwarts.* 15:483–486.

124. Todd, E. C. D., R. A. Szabo, P. Peterkin, A. N. Sharpe, L. Parrington, D. Bundle, M. A. J. Gidney, and M. B. Perry. 1988. Rapid hydrophobic grid membrane filter-enzyme-labeled antibody procedure for identification and enumeration of *Escherichia coli* 0157 in foods. *Appl. Environ. Microbiol.* 54: 2526–2540.

125. Tomlins, R. I., M. D. Pierson, and Z. J. Ordal. 1971. Effect of thermal injury on the TCA cycle enzymes of *Staphylococcus aureus* MF 31 and *Salmonella typhimurium* 7136. *Can. J. Microbiol.* 17:759–765.

126. Trotman, R. E. 1971. The automatic spreading of bacterial culture over a solid agar plate. *J. Appl. Bacteriol.* 34:615–616.

127. Tuttlebee, J. W. 1975. The Stomacher—Its use for homogenization in food microbiology. *J. Food Technol.* 10:113–122.

128. Woodward, R. L. 1957. How probable is the most probable number? *J. Amer. Water Works Assoc.* 49:1060–1068.

129. Zayaitz, A. E. K., and R. A. Ledford. 1985. Characteristics of acid-injury and recovery of *Staphylococcus aureus* in a model system. *J. Food Protect.* 48: 616–620.

6

Physical, Chemical, and Immunologic Methods

The methods of detecting microorganisms and their products covered in this chapter were developed since 1960. Most may be used to estimate numbers of organisms. Unlike direct microscopic counts, most of the enumeration methods that follow are based on either the metabolic activity of the organisms on given substrates, measurements of growth response, or the measurement of some part of cells.

PHYSICAL METHODS

Impedance

Although the concept of electrical impedance measurement of microbial growth was advanced by G. N. Stewart in 1899, it was not until the 1970s that the method was employed for this purpose.

Impedance is the apparent resistance in an electric circuit to the flow of alternating current corresponding to the actual electrical resistance to a direct current. When microorganisms grow in culture media, they metabolize substrates of low conductivity into products of higher conductivity and thereby decrease the impedance of the media. When the impedance of broth cultures is measured, the curves are reproducible for species and strains, and mixed cultures can be identified by use of specific growth inhibitors. The technique has been shown capable of detecting as few as 10 to 100 cells (Table 6-1). Cell populations of 10^5-10^6/ml can be detected in 3–5 h and 10^4-10^5/ml in 5–7 h (243). The times noted are required for the organisms in question to attain a threshold of 10^6-10^7 cells/ml.

Impedance has been evaluated by a large number of investigators as a means of monitoring the overall microbial quality of various foods (24, 226). Two hundred samples of pureed vegetables were assessed, and a 90–95%

agreement was found between impedance measurements and plate count results relative to unacceptable levels of bacteria (94). Impedance analyses required 5 h, and the method was reported to be applicable to cream pies, ground meat, and other foods. The microbiological quality of pasteurized milk was assessed by using the impedance detection time (IDT) of 7 h or less, which was equivalent to an aerobic plate count (APC) of 10,000/ml or more bacteria. Of 380 samples evaluated, 323 (85%) were correctly assessed by impedance (25). Using the same criterion for 27 samples of raw milk, 10 h were required for assessment. From a collaborative study of raw milk involving six laboratories, impedance results varied less than standard plate count (SPC) results between laboratories (67). Impedance results have been compared to plate counts and Moseley test results to predict shelf life of pasteurized whole milk. IDTs at 18° and 21°C correlated better with Moseley results than plate counts as shelf-life predictors, and results were obtained by impedimetry in 25–38 h, compared to 7–9 days for Moseley test results (17). The findings of these investigators suggested that if the IDT was ≤6.1 h, the potential shelf life would be ≤9 days while an IDT of ≥12.4 h indicated a potential shelf life of ≥9 days. From another study with raw milk, impedance was found useful when a 7-h cut-off time (10^5 cfu/ml) was used to screen samples (85). Findings from this study were compared to SPC data. As a monitor of the postpasteurization contamination of cream, an IDT of >20 h was found to denote cream of acceptable microbiological quality (91). A scattergram relating IDT to APC on 132 raw milk samples is presented in Figure 6-1.

By use of impedance, the brewing industry test for detecting spoilage organisms in beer was shortened from 3 weeks or more to only 2–4 days. Yeasts growing in wort caused an increase in impedance, while bacteria caused a decrease (57). For raw beef, IDTs for 48 samples were plotted against log bacterial counts and a regression coefficient of 0.97 was found (66). The IDT for meats was found to be about 9 h.

Impedance has been used to classify frozen orange juice concentrate as acceptable ($<10^4$ cfu/ml) or unacceptable ($>10^4$ cfu/ml). By using cut-off times of 10.2 h for bacteria, and 15.8 h for yeasts, 96% of 468 retail samples could be correctly classified (237). The method has been employed to detect starter culture failure within 2 h. In this application, complete starter failure occurred with approximately 10^5 phages/ml, resulting in the inhibition of normal impedance decrease caused by growth of normal lactic starter organisms (234). The relative level of contamination of meat surfaces by impedance has been assessed. With 10^7 cells and above/cm^2, detection could be made accurately within 2 h (23).

A relatively large number of studies have been conducted with impedance as a rapid method to detect coliforms. Fecal coliforms could be detected in 6.5–7.7 h when 100 log- or exponential-phase cells were inoculated into a selective medium and incubated at 44.5°C (202). Employing a new selective medium for coliforms, impedance was evaluated on 70 samples of ground beef, and 79% of the impedimetric results fell within the 95% confidence limits of the three-tube most-probable-numbers (MPN) procedure for coliforms, and fewer than 100 to 21,000 cells/g could be detected with results obtained within 24 h (143). In another study of coliforms in meats, the investigators found it necessary to develop a new selective medium that yielded impedance signals

TABLE 6-1. Reported Minimum Detectable Levels of Toxins or Organisms by Physical, Chemical, and Immunologic Methods of Analysis

Methods	Toxin or organism	Sensitivity	Reference
Impedance	Coliforms in meats	10^3/g in 6.5 h	69
	Coliforms in culture media	10 in 3.8 h	66
	Fecal coliforms at 44.5°C	100 in 6.5–7.7 h	202
Microcalorimetry	S. aureus cells	2 cells in 12–13 h	133
	S. aureus	Minimum HPR$^a \sim 10^4$ cells/ml	133
ATP measurement	Total flora of meats	10^4 cells/g in 20–25 min.	205
Radiometry	Frozen orange juice flora	10^4 cells/g in 6–10 h	95
	Salmonellae in foods; at 37°C with poly H antiserum	10^4–10^5 cells/g in 7 h	206
Fluorescent antibody	Coliforms in water	1–10 cells in 6 h	6
	Salmonellae	10^6 cells/ml	108
	Staph. enterotoxin B	ca. 50 ng/ml	79
Thermostable nuclease	From S. aureus	10 ng/g	126
	From S. aureus	2.5–5 ng	54
Limulus lysate test	Gram-negative endotoxins	2–6 pg of E. coli LPS	225
Radioimmunoassay	Staph. enterotoxins A, B, C, D, and E in foods	0.5–1.0 ng/g	15,147
	Staph. enterotoxin B in nonfat dry milk	2.2 ng/ml	159
	Staph. enterotoxins A and B	0.1 ng/ml for A; 0.5 ng/ml for B	4
	Staph. enterotoxin C$_2$	100 pg	183
	E. coli ST$_a$ enterotoxin	50–500 pg/tube	83
	Aflatoxin M$_1$ in milk	5–50 ng/assay	173
	Aflatoxin B$_1$	0.5–5.0 ng	53
	Ochratoxin A	20 ppb	1
	Bacterial cells	500–1,000 cells in 8–10 min.	210
Electroimmunodiffusion	C. perfringens enterotoxin	10 ng	51
	Botulinal toxins	3.7–5.6 mouse LD$_{50}$/0.1 ml	231
Micro-Ouchterlony	S. aureus enterotoxins A and B	10–100 ng/ml	12,27
	C. perfringens type A toxin	500 ng/ml	80
Passive immune hemolysis	E. coli LT enterotoxin	<100 ng	56
Aggregate-hemagglutination	B. cereus enterotoxin	4 ng/ml	88

Method	Agent	Detection limit	Ref.
Latex agglutination	E. coli LT enterotoxin	32 ng/ml	65
	S. aureus enterotoxins	0.3 µg/ml	146
Single radial immunodiffusion			
Hemagglutination-inhibition	Staph. enterotoxin B	1.3 ng/ml	118
Reverse passive hemagglutination	Staph. enterotoxin B	1.5 ng/ml	203
ELISA	C. perfringens type A toxin	1 ng/ml	80
	Staph. enterotoxin A in wieners	0.4 ng	192
	Staph. enterotoxins A, B, and C in foods	0.1 ng/ml	207
	Staph. enterotoxins A, B, C, D, and E in foods	$\geqslant 1$ ng/g	76
	Botulinal toxin type A	about 9 mouse LD_{50}/ml with monoclonal antibody	200
	Botulinal toxin type A	50–100 mouse i.p. LD_{50}	161
	Botulinal toxin type E	100 mouse LD_{50}	160
	Aflatoxin B_1	25 pg/assay	172
	Aflatoxin B_1	10 pg/ml	135
	Aflatoxin M_1 in milk	0.25 ng/ml	173
	Aflatoxin B_1	<1 pg/assay	16
	Ochratoxin A	25 pg/assay	174
	Salmonellae	10^4–10^5 cells/ml	127
	AFB_1	0.2 ng/ml (monoclonal)	26
	AFB_1	0.4 ng/ml (polyclonal)	32
	AFM_1	1.0 ng/ml (monoclonal)	45
Polymerase chain reaction	E. coli	1–5 cells/100 ml H_2O	11
	E. coli	1 cell	165
	L. monocytogenes	1–10 cells	239
	V. vulnificus	10^2 cfu/g (oysters)	99

[a] Exothermic heat production rate.

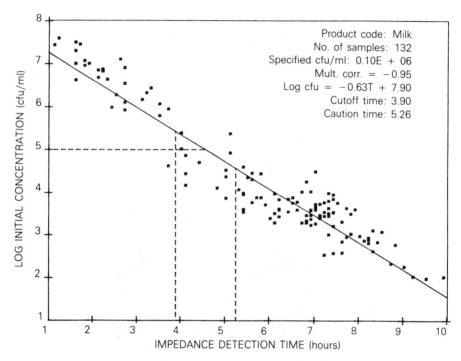

FIGURE 6-1.   Scattergram relating impedance detection time (IDT) to APC on 132 samples of raw milk. Samples containing >10^5 mesophiles/ml were detected within 4 h.
Courtesy of Ruth Firstenberg-Eden, Bactomatic, Inc., Princeton, N.J.

more consistent with colony-forming unit (cfu) results. From an inoculum of ten coliforms into the new medium, the average IDT was 3.8 h. Of 96 meat samples, a correlation coefficient of 0.90 was found between impedance and corrected coliforms counts on violet red bile agar (VRBA) (69). An IDT of 6.5 h was required for meat samples with 10^3 coliforms, and the authors suggested that an impedance signal in 5.5 h or less denoted meat samples with coliforms >10^3/g while the inability to detect by impedance in 7.6 h denoted coliform levels <10^3/g (69). For coliforms in raw and pasteurized milk and two other dairy products, an IDT of <9 h indicated coliforms were >10/ml, while an IDT of >12 h indicated <10 cells/ml (71).

Attention must be paid to culture media for impedimetry since not all media sustain smooth growth of given organisms. For example, it has been found that in brain heart infusion (BHI) broth, *Pseudomonas* spp. exhibited a triphasic-type growth curve, making it difficult to determine the true IDT (70). On the other hand, smooth curves resulted when the organisms were grown in another medium, double-plate-count broth. Similarly, several standard selective media for coliforms were shown to yield less than ideal growth curves relative to impedance changes (69). The latter authors used impedance to determine total, mesophilic, and psychrotrophic numbers in raw milk. Total numbers (samples containing 1×10^5 cfu/ml) could be detected within 16 h, mesophiles within 4 h, and psychrotrophs within 21 h. With each group, correlation with APC was statistically significant (70).

A commercially available conductance method was evaluated by screening raw milk for the presence of Enterobacteriaceae. Using a pass/fail limit of 100 cfu/ml, 90% of samples were correctly classified, and Enterobacteriaceae were detected at levels of <10 cfu/ml in 12 h and at 500 cfu/ml in 6 h (37). The correlation coefficient between detection time and APC was 0.92, and the minimum number of cells needed for earliest detection was 2×10^5 cfu/ml (37).

Impedance has been compared to radiometry for the recovery of heat-stressed fecal coliforms from meat loaf. Radiometry detected in about 17 h and impedance in about 18 h. Either method was found suitable for determining, in cooked food meats, the criterion of 0 fecal coliforms/g (186). For more information on microbiological applications of impedance, see (68).

Microcalorimetry

This is the study of small heat changes: the measurement of the enthalpy change involved in the breakdown of growth substrates. The heat production that is measured is closely related to the cell's catabolic activities (74).

There are two types of calorimeters: batch and flow. Most of the early work was done with batch-type instruments. The thermal events measured by microcalorimetry are those from catabolic activities, as already noted. One of the most widely used microcalorimeters for microbiological work is the Calvet instrument, which is sensitive to a heat flow of 0.01 cal/h from a 10-ml sample (74).

With respect to its use as a rapid method, most attention has been devoted to the identification and characterization of foodborne organisms. Microcalorimetric results vary according to the history of the organism, inoculum size, fermentable substrates, and the like. One group of investigators (169) found the variations such that the identification of microorganisms by this method was questioned, but in a later study in which a synthetic medium was used, Perry et al. (170) successfully characterized commercial yeast strains. The utility of the method to identify yeasts has been questioned (10), but by use of flow microcalorimetry, yeasts could be characterized. The latter method is one in which a microcalorimeter is fitted with a flow-through calorimetric vessel. By use of a chemically defined medium containing seven sugars, thermograms were produced by nine lactic acid bacteria (belonging to the genera *Enterococcus, Leuconostoc,* and *Lactobacillus*) distinctive enough to recommend the method for their identification (77). All cultures were run at 37°C except "*S. cremoris,*" which was run at 30°C, and results were obtained within 24 h.

Microcalorimetry has been applied to the study of spoilage in canned foods (188), to differentiate between the Enterobacteriaceae, to detect the presence of *Staphylococcus aureus*, and to estimate bacteria in ground meat (89). Boling et al. (18) were able to differentiate among 17 species of ten genera of Enterobacteriaceae by inoculating approximately 500 cells into BHI broth and recording the temperature changes over an 8–14-h incubation period. The thermograms produced were distinctive for each organism. In detecting *S. aureus,* Lampi et al. (133) achieved results in 2 h using an initial number of 10^7-10^8 cells/ml and in 12–13 h when only 2 cells/ml were used. Russell et al. (187) examined over 250 cultures representing 24 genera and found that most of the organisms typically produced maximum outputs of $40-60\,\mu$cal/(sec) (ml) and returned to baseline in 5–7 h thereafter. Some profiles developed within 3 h, while others required up to 14 h. A monitoring use of microcalorimetry is suggested by the work of Beezer et al. (9), who used flow microcalorimetry to determine the viability of recovered frozen cells of *Saccharomyces cerevisiae* within 3 h after thawing. When applied to comminuted meat, the peak exothermic heat production rate (HPR) could be recorded within 24 h for meats containing 10^5-10^8 cfu/g, and results by microcalorimetry correlated well with plate count results (89). With 10^2 cfu/ml, a measurable HPR was produced after 6 h, with peak HPR at 10 h.

Flow Cytometry

Flow cytometry is the science of measuring components (cells) and the properties of individual cells in liquid suspension. In essence, suspended cells one by one are brought to a detector by means of a flow channel. Fluidic devices under laminar flow define the trajectories and velocities that cells traverse the detector, and among the cell properties that can be detected are fluorescence, absorbance, and light scatter. By use of flow sorting, individual cells are selected on the basis of their measured properties, and from one to three or more global properties of the cell can be measured (145). Flow cytometers and cell sorters make use of one or more excitation sources such as argon, krypton, or helium-neon ion lasers and one or two fluorescent dyes to measure and characterize several thousand cells/sec. When a dye is used, its excitation spectrum must match the light wavelengths of the excitation source (43). Two dyes may be used in combination to measure—for example, total protein and DNA content. In these instances, both dyes must excite at the same wavelength and emit at different wavelengths so that the light emitted by each dye is measured separately. The early history of flow cytometry has been reviewed by Horan and Wheeless (102).

While most studies have been conducted on mammalian cells, both DNA and protein have been measured in yeast cells. Typically, yeast cells are grown, fixed, and incubated in an RNAse solution for 1 h. Cell protein may be stained with fluorescein isothiocyanate and DNA with propidium iodide. Following necessary washing, the stained cells are suspended in a suitable buffer and are now ready for application to a flow cytometer. The one used by Hutter et al. (105) was equipped with a 50-mW argon laser. Yeast cells were excited at different wavelengths with the aid of special optical filters. By this method, baker's yeast was found to contain 4.6×10^{-14} g of DNA/cell, and the protein content/cell was found to be 1.1×10^{-11} g.

The possible application of flow cytometry to the analysis of foods for their content of microorganisms is suggested by the study of Van Dilla et al. (228), in which bacteria were stained with fluorescent dyes, one of which bound preferentially to DNA rich in guanine-cytosine (G-C) while the other bound to adenine-thymine (A-T)-rich DNA. Thus, the rapid identification of bacteria in food based on their specific A-T/G-C ratios is possible.

In addition to the simultaneous measurement of protein and DNA in cells and the enumeration of cells based on G-C and A-T content, flow cytometry has been used to distinguish between living and dead cells by dual staining; to determine the ploidy of yeast cells (104); to differentiate between spores and vegetative cells in *Bacillus* spp.; and to separate pathogenic and nonpathogenic amoebae (153). It has been used to identify *Listeria monocytogenes* added to milk and to detect this organism in naturally contaminated milk (47). With four bacterial cultures, flow cytometry was compared to plate counts as an enumeration method, and the former compared well with the latter over the range 10^2-10^7 cfu/ml with results obtained within only a few minutes (177). More information on flow cytometry can be obtained from Shapiro (198).

CHEMICAL METHODS

Thermostable Nuclease

The presence of *S. aureus* in significant numbers in a food can be determined by examining the food for the presence of thermostable nuclease (DNAse). This is possible because of the high correlation between the production of coagulase and thermostable nuclease by *S. aureus* strains, especially enterotoxin producers. For example, in one study, 232 of 250 (93%) enterotoxigenic strains produced coagulase, while 242 or 95% produced thermostable nuclease (131). Non–*S. aureus* species that produce DNAse are discussed in Chapter 19.

The examination of foods for this enzyme was first carried out by Chesbro and Auborn (31) employing a spectrophotometric method for nuclease determination. They showed that as the numbers of cells increased in ham sandwiches, there was an increase in the amount of extractable thermostable nuclease of staphylococcal origin. They suggested that the presence of 0.34 unit of nuclease indicated certain staphylococcal growth and that at this level it was unlikely that enough enterotoxin was present to cause food poisoning. The 0.34 unit was shown to correspond to $9.5 \times 10^{-3}\,\mu g$ enterotoxin by *S. aureus* strain 234. The reliability of the thermostable nuclease assay as an indicator of *S. aureus* growth has been shown by many others. It has been found to be as good as coagulase in testing for enterotoxigenic strains (157), and in another study all foods that contained enterotoxin contained thermostable nuclease, which was present in most foods with 1×10^6 *S. aureus* cells/g (167). On the other hand, thermostable nuclease is produced by some enterococci. Of 728 enterococci from milk and milk products, about 30% produced nuclease, with 4.3% of the latter (31 of the 728) being positive for thermostable nuclease (8).

The mean quantity of thermostable nuclease produced by enterotoxigenic strains is less than that for nonenterotoxigenic strains with 19.4 and 25.5 μg/ml, respectively, as determined in one study (157). For detectable levels of nuclease, 10^5-10^6 cells are needed, while for detectable enterotoxin, $>10^6$ cells/ml are needed (158). During the recovery of heat-injured cells in trypticase soy broth (TSB), nuclease was found to increase during recovery but later decreased (245). The reason for the decrease was found to be proteolytic enzymes, and the decrease was reversed by the addition of protease inhibitors.

Several methods have been described for measuring thermostable nuclease in food products. By one, the assay medium containing DNA is flooded with $4N$ HCl after samples have been incubated at 50°C for 1 h. This method produces results in 2.5 h and has been found sensitive to as little as 10 ng nuclease/g of food (126). Another similar method is that described by Lachica et al. (129, 130). It consists of combining DNA and toluidine blue 0 in a buffered salts solution with 1 percent agar. To a microscope slide, 3 ml of the molten DNA-dye preparation are layered. Small wells are cut in the agar layer, and particles of food (about 5 mg) are added to the wells. The inoculated wells are covered, incubated at 37°C for 3 h, and read for the appearance of a bright pink halo around the food particles indicating reaction of DNAse with DNA. Heat-stable nuclease is detected by heating food samples at 97°C for 15 min before adding them to slide wells. These investigators found the technique to be reliable on inoculated beef and pork samples, and Tatini et al. (216) found the technique to be reliable on naturally contaminated products. The latter authors assessed heat-stable nuclease and *S. aureus* growth and enterotoxin production in broth, milk products, ground beef, and bologna. They also followed the production of nuclease in genoa sausage and during the curing and smoking of sausage relative to the efficacy of heat-stable nuclease to assess the safety of these products and found the indirect test to be a reliable product indicator. A high statistical correlation has been found between heat-stable nuclease and *S. aureus* growth in cheddar, colby, and brick cheeses (36). Optimal assay conditions for thermostable nuclease include 50°C incubation at pH 10, which produces better results than 37°C and pH 9 (119).

To test for nuclease in casings of fermented sausage, 0.5-inch disks were removed from casings and placed on assay plates. The presence or absence of nuclease in steamed disks was determined by developing the plates. The method was shown capable of detecting about 2.5–5 ng of nuclease on salami casing disks (54).

The nuclease assay has been combined with a plating method for detecting *S. aureus*. By this method, *S. aureus* is plated on Baird-Parker agar plates and incubated for 24 h followed by a 2-h incubation at 60°C to inactivate heat-labile nuclease. Plates are finally covered with a layer of toluidine blue-DNA agar and observed for color change (128).

The assay of foods containing starter cultures for thermostable nuclease will not necessarily correlate with enterotoxin production. The enzyme was found in dry sausage containing enterotoxin-producing staphylococci, but no toxin was found (158). On the other hand, it has been reported that the enzyme is produced under all conditions that permit cell growth (31).

While *S. epidermidis* and some micrococci produce nuclease, it is not as stable to heating as is that produced by *S. aureus* (131). Thermostable nuclease will withstand boiling for 15 min. It has been found to have a D value (D_{130}) of 16.6 min in BHI broth at pH 8.2, and a z value of 51 (55).

Among the advantages of testing for heat-stable nuclease as an indicator of *S. aureus* growth and activity are the following:

1. Because of its heat-stable nature, the enzyme will persist even if the bacterial cells are destroyed by heat, chemicals, or bacteriophage or if they are induced to L-forms.
2. The heat-stable nuclease can be detected faster than enterotoxin (about 3 h versus several days).
3. The nuclease appears to be produced by enterotoxigenic cells before enterotoxins appear (Fig. 6-2).

4. The nuclease is detectable in unconcentrated cultures of food specimen, while enterotoxin detection requires concentrated samples.
5. The nuclease of concern is stable to heat as are the enterotoxins.

Limulus Lysate for Endotoxins

Gram-negative bacteria are characterized by their production of endotoxins, which consist of a lipopolysaccharide (LPS) layer (outer membrane) of the cell envelope. The LPS is pyrogenic and responsible for some of the symptoms that accompany infections caused by gram-negative bacteria.

The *Limulus* amoebocyte lysate (LAL) test employs a lysate protein obtained from the blood (actually haemolymph) cells (amoebocytes) of the horseshoe crab (*Limulus polyphemous*). The lysate protein is the most sensitive substance known for endotoxins. Of six different LAL preparations tested from five commercial companies, they were found to be from 3–300 times more sensitive to endotoxins than the U.S. Pharmacopeia rabbit pyrogen test (232). The LAL test is performed by adding aliquots of food suspensions or other test materials to small quantities of a lysate preparation, followed by incubation at 37°C for 1 h. The presence of endotoxins causes gel formation of the lysate material. LAL reagent is available that can detect 1.0 pg of LPS. Since the *E. coli* cell

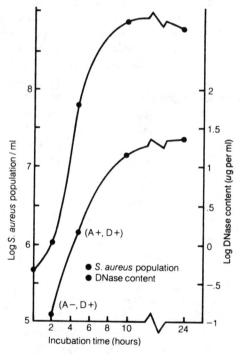

FIGURE 6-2.   Growth of *S. aureus* (196E) and production of DNAse and enterotoxins in BHI broth at 37°C. DNAse and enterotoxin D were detectable within 2 h at a population of 2×10^6, whereas enterotoxin A was detected after 4 h at higher cell populations. DNAse was detectable in unconcentrated cultures, enterotoxins at fifty-fold concentrates.
After Tatini et al. (216); copyright © 1975, Institute of Food Technologists.

contains about 3.0 fg of LPS, it is possible to detect as few as 300 gram-negative cells. From studies with meatborne pseudomonads, as few as 10^2 cfu/ml were detected (58). The various LAL methods employed to detect microorganisms in foods have been reviewed (116).

The first food application was the use of LAL to detect the microbial spoilage of ground beef (111, 112). Endotoxin titers increase in proportion to viable counts of gram-negative bacteria (115). Since the normal spoilage of refrigerated fresh meats is caused by gram-negative bacteria, the LAL test is a good, rapid indicator of the total numbers of gram-negative bacteria. The method has been found to be suitable for the rapid evaluation of the hygienic quality of milk relative to the detection of coliforms before and after pasteurization (217). For raw and pasteurized milk, it represents a method that can be used to determine the history of a milk product relative to its content of gram-negative bacteria. Since both viable and nonviable gram-negative bacteria are detected by LAL, a simultaneous plating is necessary to determine the numbers of cfu's. The method has been applied successfully to monitor milk and milk products (110, 244), microbial quality of raw fish (211), and cooked turkey rolls. In the latter, LAL titers and numbers of Enterobacteriaceae in vacuum-packaged rolls were found to have a statistically significant linear relationship (46).

LAL titers for foods can be determined either by direct serial dilutions or by MPN, with results by the two methods being essentially similar (196). To extract endotoxins from foods, the Stomacher has been found to be generally better than the use of Waring blenders or the shaking of dilution bottles (114).

In this test, the proclotting enzyme of the *Limulus* reagent has been purified. It is a serine protease with a molecular weight of about 150,000 daltons. When activated with Ca^{2+} and endotoxin, gelation of the natural clottable protein occurs. The *Limulus* coagulogen has a molecular weight of 24,500. When it is acted upon by the *Limulus* clotting enzyme, the coagulogen releases a soluble peptide of about 45 amino acid residues and an insoluble coagulin of about 170 amino acids. The latter interacts with itself to form the clot, which involves the cleavage of -arg-lys- or -arg-gly- linkages (215). The process, as summarized from (155), may be viewed as noted below.

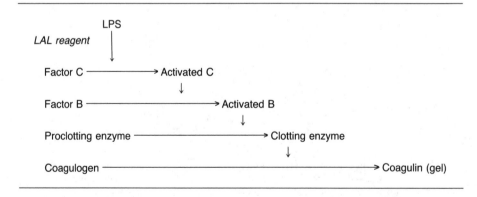

Commercial substrates are available that contain amino acid sequences similar to coagulogen. The chromogenic substrates used for endotoxin consist of these linked to p-nitroaniline. When the endotoxin-activated enzyme attacks the chromogenic substrate, free p-nitroaniline results and can be read at 405 nm. The amount of the chromogenic compound liberated is proportional to the quantity of endotoxin in the sample. Employing a chromogenic substrate, Tsuji et al. (224) devised an automated method for endotoxin assay, and the method was shown sensitive to as little as 30 pg of endotoxin/ml.

Assuming that the quantity of endotoxin/gram-negative bacterial cells is fairly constant, and assuming further that cells of all genera contain the same given quantity, it is possible to calculate the number of cells (viable and nonviable) from which the experimentally determined endotoxin was derived. With a further assumption that the ratio of gram-negative to gram-positive bacteria is more or less constant for given products, one can make a 1-h estimate of the total numbers of bacteria in food products such as fresh ground beef (113). Low values by this procedure are more meaningful than high values, and the latter need to be confirmed by other methods.

Overall, the value of the LAL test lies in the speed at which results can be obtained. Foods that have high LAL titers may be candidates for further testing by other methods; those that have low titers may be placed immediately into categories of lower risk relative to numbers of gram-negative bacteria.

Nucleic Acid (DNA) Probes

A DNA probe consists of the DNA sequence of an organism of interest that can be used to detect homologous DNA or RNA sequences. In effect, the probe DNA must hybridize with that of the strain to be sought. Ideally, the DNA sequence that contains the genes that code for a specific product is used. The probe DNA must be labeled in some way in order to be able to assess whether hybridization has occurred. Radioisotopes are the most widely used labels, and they include ^{32}P, ^{3}H, ^{125}I, and ^{14}C, with ^{32}P being the most widely used. Reporter groups such as alkaline phosphatase, peroxidase, fluorescein, haptens, and biotin have been developed and used (48). One of the drawbacks to the use of some of the reporter methods is the limited number of molecules/ probes. When biotin is used, its presence is detected with avidin-enzyme conjugates, anti-biotin, or photobiotin. Chromosomal DNA is often the source of target nucleic acid, but it contains only one target/cell. Multiple targets are provided by mRNA, rRNA, and plasmid DNA, thus making for a more sensitive detection system. Synthetic oligonucleotide probes may be constructed of 20 to 50 bases, and under proper conditions hybridization times of 30–60 min are possible.

In a typical probe application, DNA fragments of unknown organisms are prepared by use of restriction endonucleases. After separating fragment strands by electrophoresis, they are transferred to cellulose nitrate filters and hybridized to the radiolabeled probe. After gentle washing to remove unreacted probe DNA, the presence of the radiolabel is assessed by autoradiography.

The minimum number of bacterial cells that can be detected with a standard probe is in the 10^{6}–10^{7} range (see Table 6-1) although some investigators report the detection of 10^{4} cells. When probes are used on foods where perhaps only 1 cfu/ml of the target organism exists, enrichment procedures must be employed to allow this cell to attain a level that will provide enough

DNA for detection. With an initial cell number of 10^8 cells, probe results may be obtained in around 10–12 h when radiolabels are employed. Where enrichments are necessary, the time required for results would be enrichment time plus probe assay time, generally 44 h +.

DNA probes have been prepared for the detection of a number of organisms of interest in foods including salmonellae (72, 73), listeriae (42, 163), staphylococcal enterotoxins (62), *Clostridium perfringens* enterotoxin (227), *Entamoeba histolytica* (21), and enterotoxigenic *Escherichia coli* (98, 151, 191). The DNA sequence homology between *E. coli* LT gene and *Vibrio cholerae* DNA was demonstrated by a probe technique (152). Employing a biotin-labeled probe that employed a streptavidin-alkaline phosphatase conjugate to produce an insoluble color precipitate in the presence of a dye, one group of researchers was able to detect 2×10^7 *E. coli* cells/g in about 30 h with the detection system requiring only 3 h for results (48). A single RNA probe has been synthesized that detected simultaneously *E. coli* STa and LT toxins, with either a radioactive or biotin label (189). By employing the *uid* chromosomal region of *E. coli*, a probe was constructed that reacted with both *E. coli* and shigellae species (33). The probe was sensitive to 3×10^4 cells, a polymerase chain reaction (PCR) procedure could detect 8 cells, and the probe could not distinguish between the two groups of organisms. With the commercial availability of digoxigenin-labeled dUTP, a digoxigenin-labeled DNA probe (which also uses an alkaline phosphatase conjugate) was constructed for *E. coli* LT-I, STa, and STB toxins (180). This nonradioactive detection system was more sensitive and showed less nonspecific background contamination than a biotinylated probe for enterotoxigenic strains.

When a commercially available radiolabeled probe for salmonellae was used on 269 poultry carcass and water samples, as few as 0.03 salmonellae/ml could be detected by the use of two enrichments (109). The probe in question had a sensitivity of around 10^4 cells/ml, and it has received Association of Official Analytical Chemists (AOAC) approval. A colorimetric DNA hybridization assay (probe) for salmonellae has been developed and given interim First Action status by the AOAC (29, 241). This method utilizes sequences of 16S and 23S rRNA of salmonellae that are unique to this genus, and the detection system is fluorescein isothiocyanate rather than a radioisotope. It detected 355 of 362 (98%) salmonellae and 206 of 213 (97%) salmonellae serovars and showed no cross-reactivity with 11 other genera of enteric bacteria. When compared with the BAM/AOAC culture method on 20 foods, an overall agreement of 99% was achieved (241).* Pre- and selective enrichments were used for foods, with only about $2\frac{1}{2}$ h required for actual probe assay. By employing a 1600–bp section of *Salmonella enteritidis* DNA, a probe was developed that reacted with all 70 salmonellae serotypes tested and not with any of 101 other enteric bacteria (193).

Among the nucleic acid probes developed for *S. aureus* enterotoxins is one made with DNA that encodes amino acids 207–219 of the SEB and SEC molecules. This probe reacted with SEB and the three SEC enterotoxins (162). A DNA probe for the enterotoxin of *C. perfringens*, consisting of four synthetic oligonucleotides that encode different parts of the enterotoxin gene,

---

*BAM = *Bacteriological Analytical Manual*.

was tested on 245 strains, and 59% were positive with each of the four probes (227). A DNA probe specific for *E. histolytica* was developed and shown capable of detecting <200 trophozoites (21).

DNA probes are used in **colony hybridization** methods where micro- or macrocolonies of the target organism are allowed to develop directly on a membrane following incubation on a suitable agar medium. A replica plate is produced as a duplicate of the master plate or membrane. Colonies that have grown on the duplicate plate are lysed directly on the membrane to release nucleic acid and to convert DNA into single strands. Some of the DNA is transferred to nitrocellulose filters, where hybridization is carried out by applying a labeled DNA or RNA probe. A modification of the traditional DNA colony hybridization technique has been made such that 60 filters with up to 48 organisms/filter can be used (121).

The colony hybridization method developed by Grunstein and Hogness (92) has been employed successfully to detect *Listeria monocytogenes*, entero-toxigenic *E. coli*, and *Yersinia enterocolitica*. In one study, synthetic poly-nucleotide probes were constructed that were homologous to a region of the ST enterotoxin gene of *E. coli* and applied to the detection of producing strains by DNA colony hybridization (100). For the latter, colonies were placed on paper filters to free and denature cellular DNA, hybridized overnight at 40°C, and exposed to autoradiograms. By this procedure, as few as 10^5 ST-producing cells could be detected. In an earlier study from the same laboratory, colony hybridization was used to detect *E. coli* in artifically contaminated food without enrichment, and the method could detect 100 to 1,000 cells/g, or about 1 to 10 cells/filter (98). In another study, the LT enterotoxin of *E. coli* was detected using DNA colony hybridization from a variety of spiked foods (64). In this study, nitrocellulose filters that were inoculated on the surfaces of plate count agar (PCA) and eosin methylene blue (EMB) agar plates were incubated at 35°C for 24 h. Colonies were then lysed and subjected to DNA colony hybridization. The method could detect as few as 100 cells/g and was stated to be nearly equal to PCA and EMB in recovering the organism.

Finally, colony hybridization and **dot blot** analysis (direct application of isolated DNA or RNA to nitrocellulose) were employed to detect Shiga-like toxins I and II of *E. coli* in food and calf fecal samples, and the former could detect as few as 1.3 cfu/g in 3–4, days while the latter could detect the hemorrhagic colitis strain of *E. coli* in 24 h, although the method did not detect the *slt* gene in the latter (191). For more information on nucleic acid probes and illustrations of specific probe methods, see the review by Wolcott (242).

DNA Amplification (Polymerase Chain Reaction)

This technique, first outlined in 1971 by Kleppe et al. (124), is applicable more to the identification of foodborne organisms than to their enumeration. The currently used methodology is that further developed by scientists at the Perkin Elmer–Cetus Corp. (190, 208), among others. In brief, by employing thermostable DNA polymerases and 5′- and 3′-specific oligonucleotide primers, a single molecule of DNA can be amplified to 10^7 molecules after a series of amplification cycles, typically from 20 to 50. The amplified DNA is then detected by use of either agarose gels or Southern hybridization employing a radiolabeled probe. Most reagents for PCR use now exist in kit form from at least two commercial sources.

PCR has been employed to detect enterotoxigenic *E. coli* strains, *L. monocytogenes*, *Vibrio vulnificus*, *Toxoplasma gondii*, and *Shigella* spp. Employing 24-mer oligonucleotide primers, one group of investigators was able to detect 1 to 5 *E. coli* cells/100 ml of water (11). In another application, as few as 20 cells of *E. coli* LT type I were detected within 8 or 12 h, depending on the identification method used. In the latter application, amplification was carried out for 35 cycles, and the amplified product was detected on agarose gels. Employing two synthetic primers of 20 bases each to amplify the LT gene of *E. coli*, DNA from 1 cell could be detected after 30 cycles (165). To detect *L. monocytogenes* and other listeriae, one group of researchers prepared five nucleotide sequences and when used all in combination, three PCR products were possible, including one for all listeriae and one for *L. monocytogenes* (20). Two of the products were specific for the listeriolysin O gene. In another study, PCR was not suitable for the detection of *L. monocytogenes* in cheese, but in pure culture, from 1 to 10 cfu could be detected by this procedure (239). In another application, PCR was used to amplify a 760-base pair fragment with the 220-kbp invasive plasmids of enteroinvasive *E. coli*, *Shigella flexneri*, *S. dysenteriae*, *S. boydii*, and *S. sonnei* as templates (132). Products were detected in agarose gels, and results could be obtained in 6–7 h. For the detection of *V. vulnificus*, a 519-bp portion of the cytolysin-hemolysin gene was targeted, and when 10^2 cfu/g were inoculated into oyster homogenates, the organism was recovered consistently after 24 h in alkaline peptone water (99). The method could detect about 10^2 cfu/g.

Overall, the PCR technique is powerful for detecting the presence of food-borne pathogens that cannot now be detected in faster ways. It could prove to be of great value for foodborne viruses and bacterial L forms, both of which are difficult to recover from foods. It requires a very clean work environment since contaminating DNA can be amplified, along with that from the organisms of interest.

Adenosine-Triphosphate Measurement

Adenosine-triphosphate (ATP) is the primary source of energy in all living organisms. It disappears within 2 h after cell death (154), and the amount per cell is generally constant (222), with values of 10^{-18} to 10^{-17} mole/bacterial cell, which corresponds to around 4×10^4 M ATP/10^5 cfu of bacteria (222). Among procaryotes, ATP in exponentially growing cells is regularly around 2–6 nmole ATP/mg dry weight regardless of mode of nutrition (120). In the case of rumen bacteria, the average cellular content was found to be 0.3 fg/cell, with higher levels found in rumen protozoal cells (164). The complete extraction and accurate measurement of cellular ATP can be equated to individual groups of microorganisms in the same general way as endotoxins for gram-negative bacteria.

One of the simplest ways to measure ATP is by use of the firefly luciferin-luciferase system. In the presence of ATP, luciferase emits light, which is measured with a liquid scintillation spectrometer or a luminometer. The amount of light produced by firefly luciferase is directly proportional to the amount of ATP added (168). Cellular ATP is extracted generally by boiling, and other details of its assay have been presented by Kimmich et al. (123) and Karl (120).

The application of ATP measurement as a rapid method for estimating microbial numbers has been used in clinical microbiology. In the clinical laboratory, it has been employed to screen urine specimens. The numbers of bacteria in urine are considered to be of etiologic importance if they occur at a level of 10^5 cfu/ml urine. In one study of 348 urine specimens, only an 89.4% agreement with culture methods was found (2). Using 10^5 organisms/ml or above to indicate bacteriuria and $<10^5$ as negative, the ATP assay yielded 7% false positive and 27% false negative results. Thore et al. (223) assayed 2,018 urine specimens for ATP by the luciferase method setup to distinguish between samples containing $>10^5$ cfu/ml and those containing $<10^5$ cfu/ml. Ninety-two percent of those with $>10^5$ cfu and 88% of those with $<10^5$ were correctly classified by the ATP assay. The assay was a 15-min test, and samples were set at 13.5 nM ATP to define the limit between negative and positive results. These investigators and others cited by them found ATP analysis for bacteriuria to agree well with plate count results. In a comparative study of plate counts, the LAL test, and ATP measurement to assess contamination of intravenous fluids rapidly, the LAL test was found to be more sensitive than ATP, but the authors suggested that ATP measurements could be used (3).

The successful use of the method for bacteriuria and for assessing biomass in activated sludge (168) suggested that it should be of value for foods. It lends itself to automation and represents an excellent potential method for the rapid estimation of microorganisms in foods. The major problem that has to be overcome for food use is the removal of nonmicrobial ATP. The method was suggested for food use by Sharpe et al. (199) provided that microbial ATP levels are as great as or greater than the intrinsic ATP in the food itself. They found that intrinsic levels decreased with food storage (Table 6-2). The utility of employing ATP analysis for foods was questioned by Williams (240), who called attention to the following:

1. One yeast strain was found to contain 300 times more ATP than the average for bacterial cells (some strains may contain as much as 1,000 times more).
2. Sterile 0.1% peptone diluent gave readout values equivalent to about 200,000 organisms/ml.
3. Heat-killed bacteria contain ATP.
4. ATP is found in food itself, and there are interfering substances in foods such as potato starch and milk.

The utility of the method for measuring rumen biomass has been questioned because of the discrepancy between the bacterial and protozoal content of ATP (164).

The problems noted have been overcome to a large extent. Thore et al. (222) used Triton X-100 and apyrase selectively to destroy nonbacterial ATP in urine specimens and found that the resultant ATP levels were close to values

TABLE 6-2.   Relation between Total Viable Count and ATP during Incubation at 37°C (199)

| | ATP and Mean Viable Count/g after: | | | | | |
| | 0 h | | 6 h | | 24 h | |
Sample	ATP (fg)	Viable Count	ATP (fg)	Viable Count	ATP (fg)	Viable Count
Crinkle-cut chip (frozen)	4.57×10^6 4.21×10^6	4.9×10^2	7.11×10^6 7.69×10^6	1.2×10^3	3.23×10^8 2.62×10^8	5.9×10^8
Frozen peas	5.11×10^7 5.43×10^7	4.1×10^4	3.18×10^8 3.54×10^8	1.3×10^7	3.90×10^8 5.23×10^8	7.7×10^8
Comminuted meat	1.51×10^7 1.60×10^7	1.3×10^3	3.62×10^6 4.94×10^6	6.4×10^3	7.88×10^5 8.34×10^5	2.2×10^3
Beef steaklet (frozen)	2.75×10^7 4.21×10^7 3.30×10^7	4.3×10^3	6.93×10^6 6.61×10^6	4.1×10^3	6.68×10^6 5.71×10^6	1.3×10^6
Beefburger (frozen)	3.92×10^7 4.89×10^7	3.1×10^4	5.65×10^6 4.39×10^6	3.2×10^3	3.76×10^6 3.70×10^6	2.2×10^5
Plaice fillet (frozen)	8.99×10^6 9.33×10^6	1.7×10^4	1.15×10^7 1.28×10^7	4.4×10^5	4.28×10^7 1.89×10^7	3.4×10^8
Fish finger (frozen)	1.96×10^9 1.79×10^9	6.0×10^5	6.08×10^7 4.26×10^7	2.2×10^6	1.38×10^9 1.56×10^9	6.4×10^8

observed in laboratory cultures with detection at 10^5 bacteria/ml. In meats, the problem of nonmicrobial ATP was addressed by Stannard and Wood (205) by use of a three-stage process consisting of centrifugation, use of cation exchange resin, and filtration to get rid of food particles and collect bacteria on 0.22-μm filters. ATP analyses were carried out on bacteria eluted from the filter membranes, and 70–80% of most microorganisms were recovered on the filters. Linear relationship was shown between microbial ATP and bacterial numbers over the range 10^6–10^9 cfu/g. By the methods employed, results on ground beef were obtained in 20–25 min. In another study, 75 samples of ground beef were evaluated, and a high correlation was found between $\log_{10}$ APC and $\log_{10}$ ATP when samples were incubated at 20°C (122). In this study, the amount of ATP/cfu ranged from 0.6 to 17.1 fg, with 51 of the 75 samples containing ≤5.0 fg of ATP. The ATP assay has been employed successfully for seafoods and for the determination of yeasts in beverages.

ATP analysis was combined with benzalkon-crystal violet (BC-ATP) to predict keeping quality of pasteurized milk, and results were compared to those from two commonly used shelf-life tests (233). The BC-ATP results obtained within 24 h were as good as those obtained from the shelf-life tests requiring 5 and 10 days for results. These investigators found that at an ATP content of <1,000 relative light units it was always accompanied by bacterial counts of <10^6/ml.

At the present time, ground beef can be extracted and read in 20 min, and as few as 10^4 organisms/g can be detected by available instrumentation. However, these methods may not be useful when applied to seafoods that may contain luminescent bacteria such as *Photobacterium fisheri* and *Vibrio harveye*, for these organisms produce luciferase.

More recently, high-speed black-and-white film has been used to record emitted light (242). The film, reagents, and specimens are all placed in a special camera. The Spotlight[tm] Dairy Product Test employs this method, and it can detect 10^5 cfu/ml in uncultured dairy products.

Radiometry

The radiometric detection of microorganisms is based upon the incorporation of a ^{14}C-labeled metabolite in a growth medium so that when the organisms utilize this metabolite, $^{14}CO_2$ is released and measured by use of a radioactivity counter. For organisms that utilize glucose, ^{14}C-glucose is usually employed. For those that cannot utilize this compound, others such as ^{14}C-formate or ^{14}C-glutamate are used. The overall procedure consists of using capped 50-ml serum vials to which are added anywhere from 12 to 36 ml of medium containing the labeled metabolite. The vials are made either aerobic or anaerobic by sparging with appropriate gases and are then inoculated. Following incubation, the headspace is tested periodically for the presence of $^{14}CO_2$. The time required to detect the labeled CO_2 is inversely related to the number of organisms in a product.

The use of radiometry to detect the presence of microorganisms was first suggested by Levin et al. (138). It is confined largely to clinical microbiology, but some applications have been made to foods and water. One of the earliest nonclinical uses was the detection of coliforms in water and sewage where it was shown that a direct relationship existed between coliform numbers and the amount of $^{14}CO_2$ produced by the organisms. The experimental detection of *S. aureus, Salmonella typhimurium*, and spores of putrefactive anaerobe 3679 and *Clostridium botulinum* in beef loaf was studied by Previte (179). The inocula employed ranged from about 10^4-10^6/ml of medium, and the detection time ranged from 2 h for *S. typhimurium* to 5–6 h for *C. botulinum* spores. For these studies, 0.0139 μCi of ^{14}C-glucose/ml of tryptic soy broth was employed. In another study, Lampi et al. (133) found that 1 cell/ml of *S. typhimurium* or *S. aureus* could be detected by a radiometric method in 9 h. For 10^4 cells, 3 to 4 h were required. With respect to spores, a level of 90 of putrefactive anaerobe (P.A.) 3679 was detected in 11 h, while 10^4 were detectable within 7 h. These and other investigators have shown that spores required 3–4 h longer for detection than vegetative cells. From the findings of Lampi et al., the radiometric detection procedure could be employed as a screening procedure for foods containing high numbers of organisms, for such foods produced results by this method within 5–6 h, while those with lower numbers required longer times.

The detection of nonfermenters of glucose by this method is possible when metabolites such as labeled formate and/or glutamate are used. It has been shown that a large number of foodborne organisms can be detected by this method in 1 to 6 h. The radiometric detection of 1 to 10 coliforms in water within 6 h was achieved by Bachrach and Bachrach (6) by employing ^{14}C-lactose with incubation at 37°C in a liquid medium. It is conceivable that a differentiation can be made between fecal *E. coli* and total coliforms by employing 45.5°C incubation along with 37°C incubation.

Radiometry has been used to detect organisms in frozen orange juice concentrate (95). The investigators used ^{14}C-glucose, 4 yeasts, and 4 lactic acid bacteria, and at an organism concentration of 10^4 cells, detection was achieved in 6–10 h. Of 600 juice samples examined, 44 with counts of 10^4/ml were detected in 12 h and 41 of these in 8 h. No false negatives occurred, and only two false positives were noted. The method was used for cooked foods to determine if counts were $<10^5$ cfu/ml, and the results were compared to APC. Of 404 samples consisting of seven types of foods, around 75% were correctly classified as acceptable or unacceptable within 6 h (186). No more than five were incorrectly classified. The study employed ^{14}C-glucose, glutamic acid, and sodium formate.

A method for the detection of salmonellae in foods based on salmonellae poly H antibodies preventing salmonellae from producing ^{14}CO$_2$ from (^{14}C) dulcitol has been developed (206). It was shown capable of detecting the organisms in 27 h compared to 4–5 days by culture methods. The agreement with culture methods on 58 samples of food was 91%. The radiometric assay flask contained poly H serum and a control flask without poly H. Positive tests were determined by the difference in radioactivity of the two flasks following incubation at 37°C for 7 h. Results were achieved with an inoculum of 10^4–10^5 cells. The potential use of radiometry for field and surface samples is suggested by a procedure employed by Schrot et al. (194). By this method, membrane filter samples are collected and moistened with a small quantity of labeled medium in a closed container. The evolved ^{14}CO$_2$ is trapped by Ba(OH)$_2$-moistened filter pads, which are assayed later by a radioactivity counter.

Fluoro- and Chromogenic Substrates

Some of the fluorogenic and chromogenic substrates employed in culture media in food microbiology are

- 4-methylumbelliferyl-β-D-glucuronide (MUG)
- 4-methylumbelliferyl-β-D-galactoside (MUGal)
- 4-bromo-4-chloro-3-indolyl-β-D-glucuronide (X-GLUC)
- o-nitrophenyl-β-D-galactopyranoside (ONPG)
- 5-bromo-4-chloro-3-indoxyl-β-D-glucuronide (BCIG)
- L-alanine-p-nitroanilide (LAPN)

Fluorogenic substrates are produced from the first three and chromogenic substrates from the others. These substrates are employed in various ways in plating media, MPN broths, and for membrane filtration methods. MUG is the most widely used of the fluorogenic substrates, and it is hydrolyzed by β-D-glucuronidase (GUD) to release the fluorescent 4-methyl-umbelliferyl moiety, which is detected with long-wave ultraviolet light. The value of MUG is that *E. coli* is the primary producer of GUD, and this substrate has found wide use as a differential agent in media and methods for this organism. A few salmonellae and shigellae are GUD positive, as are some corynebacteria.

First to employ MUG for *E. coli* detection were Feng and Hartman (63), who incorporated it into lauryl tryptose broth (LTB) and other coliform-selective media and found that in LTB-MUG, one *E. coli* cell could be detected in 20 h. Since about 10^7 *E. coli* cells are needed to produce enough GUD to provide detectable MUG results, the time for results depends on the initial number of cells. While most positive reactions occurred in 4 h, some weak GUD-positive strains required up to 16 h for reaction. An important aspect of this method is the occurrence of fluorescence before gas production from lactose. Employing the Feng-Hartman method, another group examined 1,020 specimens by a three-tube MPN and were able to detect more *E. coli*–positive samples than with a conventional MPN (230). The greater effectiveness of LTB-MUG resulted because some *E. coli* strains are anaerogenic. No false-negative results were obtained. In an evaluation of MUG added to lauryl sulfate broth (LSB), a 94.8% agreement was obtained on 270 product samples with 4.8% false positives but no false negatives with (LSB-MUG) (184). Oysters contain endogenous glucuronidase, but an EC broth–MUG method was employed successfully in one study where 102 of 103 fluorescing tubes were positive for *E. coli* (125). A 20-min tube procedure employing MUG was applied to 682 *E. coli* cultures, and 630 (92.4%) were positive (221). Of 188 0157 strains of *E. coli*, 166 were MUG negative, and all were positive for the Vero toxin. By use of this 20-min method, MUG-negative *E. coli* are very likely to be verotoxigenic (221). In a study of molluscan shellfish, EC-MUG broth with MUG employed at 50 ppm, 95% of *E. coli* were positive with 11% being false negative (181). When compared to the AOAC method for *E. coli*, LST-MUG was found to be equivalent for one product and better than AOAC for some others tested (176), while in another, LST-MUG was found to be comparable to the AOAC MPN method (178).

MUG can be employed in filter membrane and plating methods for detecting and enumerating *E. coli*. In one application, filter membranes were incubated 4 h at 37°C on a nonselective medium and then transferred to a tryptone bile agar–MUG agar and incubated at 44.5°C for 18–24 h (62). This method produced better results than a direct plating method based on indole production. A plating method that employed MUG in trypticase soy agar (TSA) spread plates with VRBA overlay produced better results for *E. coli* than the VRBA plating method alone (238). Colonies were fully developed in 24 h, and low false-positive and -negative results were obtained. A modification of the presence-absence (PA) method for coliforms in water has been suggested where Whirl-Pak bags are used with the addition of MUG to effect the detection of *E. coli* by fluorescence (137). Unfortunately, MUG-based media cannot be used to detect the verotoxigenic strain of *E. coli* (50).

MUGal has received limited study as a fluorogenic substrate, but in one study it was used to detect fecal coliforms in water by use of a membrane filter method where as few as 1 cfu/100 ml of water could be detected in 6 h (14). A similar substrate, 4-methylumbelliferone-β-D-galactoside, has been used to differentiate species of enterococci (142). The method employed dyed starch along with the substrate, both of which were added to a medium selective for enterococci. By observing for starch hydrolysis and fluorescence, 86% of enterococci from environmental samples were correctly differentiated.

X-GLUC has been employed in a plating medium for the detection of *E.*

coli (75). With this substrate added at 500 ppm to a peptone-Tergitol agar, *E. coli* produced a blue color in 24 h that did not diffuse from colonies and did not require fluorescent light. When compared with a three-tube standard MPN on 50 ground beef samples, no differences were observed between the two methods (182).

ONPG is a colorimetric substrate that is fairly specific for *E. coli*. The substrate is hydrolyzed by β-galactosidase to produce a yellow color that can be quantitated at 420 nm. To determine *E. coli* in water, the organisms are collected on a 0.45 μm membrane, incubated in EC medium for 1 h, followed by the addition of filter-sterilized ONPG. Incubation is continued at 45.5°C until color develops that can be read at 420 nm (235). The sensitivity of ONPG is similar to that of MUG, with about 10^7 cells required to produce measurable hydrolysis. ONPG is employed in a modification of the classical presence-absence method for coliforms in water (52). By this modified method, tubes that contain coliforms become yellow. To detect *E. coli*, each yellow tube is viewed with a hand-held fluorescent lamp (366 nm); those that contain *E. coli* fluoresce brightly. The Colilert and ColiQuik systems employ both ONPG and MUG as sole nutrient substrates where total coliforms are indicated by a yellow color while *E. coli* is indicated by MUG fluorescence.

BCIG results in *E. coli* colonies' being blue, while other organisms are not blue. When employed in lauryl tryptose agar at a final concentration of 100 ppm, only 1% of 1,025 presumptively positive *E. coli* cultures did not produce the blue color, while 5% of 583 non–*E. coli* colonies were false positive (236). The plating medium was incubated at 35°C for 2 h and then at 44.5°C for 22–24 h.

The LAPN substrate is specific for gram-negative bacteria on the premise that aminopeptidase is restricted to this group. The enzyme cleaves L-alanine-p-nitroanilide to yield p-nitroaniline, a yellow compound that is read spectrophotometrically at 390 nm (28). When used to determine gram-negative bacteria in meats, $10^4–5 \times 10^5$ cfu was the minimum detectable numbers (44). Numbers of 10^6 to 10^7 cfu/cm^2 could be detected in 3 h. The *Limulus* test has received more study for gram-negative bacteria, and since it gives results within 1 h, the LAPN method cannot be considered to be comparable.

IMMUNOLOGIC METHODS

Fluorescent Antibody

This technique has had extensive use in both clinical and food microbiology since its development in 1942. An antibody to a given antigen is made fluorescent by coupling it to a fluorescent compound, and when the antibody reacts with its antigen, the antigen-antibody complex emits fluorescence and can be detected by the use of a fluorescence microscope. The fluorescent markers used are rhodamine B, fluorescein isocyanate, and fluorescein isothiocyanate, with the latter being used most widely. The FA technique can be carried out by use of either of two basic methods. The direct method employs antigen and specific antibody to which is coupled the fluorescent compound (antigen coated by specific antibody with fluorescent label). With the indirect method, the homologous antibody is not coupled with the fluorescent label, but instead an antibody to this antibody is prepared and coupled (antigen coated by homologous antibody, which is in turn coated by antibody to the homologous antibody bearing the fluorescent label). In the indirect method, the labeled compound detects the presence of the homologous antibody, while in the direct method, it detects the presence of the antigen. The use of the indirect method eliminates the need to prepare FA for each organism of interest. The FA technique obviates the necessity of pure culture isolations of salmonellae if H antisera are employed. A commonly employed conjugate is polyvalent salmonellae OH globulin labeled with fluorescein isothiocyanate with somatic groups A to Z represented. Because of the cross-reactivity of salmonellae antisera with other closely related organisms (for example, *Arizona, Citrobacter, E. coli*), false-positive results are to be expected when naturally contaminated foods are examined. The early history and development of the FA technique for clinical microbiology has been reviewed by Cherry and Moody (30) and for food applications by Ayres (5) and Goepfert and Insalata (86).

The first successful use of the FA technique for the detection of foodborne organisms was made by Russian workers, who employed the technique to detect salmonellae in milk. The technique has been employed successfully to detect the presence of salmonellae in a large number of different types of foods and related products (Table 6-3).

A microcolony test was devised (218) and shown to possess several advantages over the direct FA test. Smears are easier to prepare, there is less fluorescence of background material, and sensitivity is greater. A semiautomated method based on the microcolony method was devised and shown

TABLE 6-3.   Application of the Fluorescent Antibody (FA) Technique to the Detection of Salmonellae in Foods

Foods Examined	No. of Samples	FA Method Used	% False Negatives	% False Positives	Reference
Meats	286	Indirect	0	14.7	81
Egg products	20	Indirect	0	0	93
Various foods (656 raw beef)	706	Direct	6.7	7	82
Dried foods and products	420	Indirect	0	0	201
Various inoculated foods	48	Direct	0	0	105
Various foods	3,991	Direct	<0.1	<0.2	195
Animal feed and ingredients	1,013	Direct	2.2	5.7	134
Various foods (39)	894	Direct	0	0	59
Various foods (7)	422	Indirect	0	7.3	87
Variety of human foods and animal feeds	65	Direct slide	0	4.6–14.8	108
Food products	ca. 4,000	Direct	<1	7	60
Meatmeal	100	Auto direct	11.1	1.1	220
Powdered eggs and candy	201	Auto direct	0	5.3–6.6	220
Frog legs and dried products	283	Auto direct	0	4	154

to have an overall agreement of 89.6% with the standard culture techniques employing 144 samples of 67 foods, feeds, and environmental samples (7). However, 13.5% of raw meat products produced false-negative results, which for all products was 3.5%. Only 1.3% false negatives and 5% false-positive samples were found by Gibbs et al. (84) using the buffered peptone preenrichment broth proposed by the International Organization for Standardization. The FA method was compared with standard cultural methods on 546 samples, including over 100 food samples.

Although the FA technique has had its widest food application in the detection of salmonellae, it has been used to trace bacteria associated with plants and to identify clostridia, including *C. perfringens* and *C. botulinum*. It can be applied to any organism to which an antibody can be prepared. Among the many advantages of the procedure for detecting salmonellae in foods are the following:

1. FA is more sensitive than cultural methods.
2. FA is more rapid. Time for results can be reduced from around 5 days to 18–24 hr.
3. Reduction in time for negative tests on food products makes it possible to free foods earlier for distribution.
4. Larger numbers of samples can be analyzed, making for increased sampling of food products.
5. The qualitative determination of specific salmonellae serotypes is possible for FA if desired for epidemiologic reasons.

In spite of its drawbacks, the FA technique is official by AOAC and several other standard reference methods (see Chap. 5). It was designed to screen out salmonellae-negative samples, and all FA-positive samples should be confirmed by cultural and serological methods. It may be improved by use of IgG conjugates and by use of rhodamine-labeled normal rabbit serum as diluent for fluorescein-labeled salmonellae conjugates. Thomason (219) in a status report on FA noted that adequate supplies of reagents for the test are sometimes difficult to obtain.

Enrichment Serology

The use of enrichment serology (ES) is a more rapid method for recovering salmonellae from foods than the conventional culture method (CCM). Originally developed by Sperber and Deibel (204), it is carried out in four steps: preenrichment in a nonselective medium for 18 h; selective enrichment in selenite-cystine and/or tetrathionate broth for 24 h; elective enrichment in M broth for either 6–8 h or 24 h; and agglutination with polyvalent H antisera at 50°C for 1 h. Results can be obtained in 50 h (depending upon elective enrichment time used) compared to 96–120 h by CCM. A modified ES method has been proposed involving a 6-h preenrichment, thus making it possible to obtain results in 32 h (212).

Employing the ES procedure on 105 samples of dried foods and feeds, all positive samples by CCM were positive by ES (204). Similar findings were achieved when 689 samples were examined by ES and CCM (61). The ES method was compared with three FA procedures for salmonellae detection in 347 food samples, and while 52 were positive by CCM, 51 were positive by ES (144). In the latter study, only one false negative was found by ES, and the authors preferred this method with a 6-h elective enrichment over CCM. In yet another comparison of ES with CCM using 2,208 samples, 95% of positive samples by CCM were positive by ES when elective enrichment was carried out for 24 h (19). ES was found to be as good as FA. The 6-h elective enrichment was unsuitable for comminuted meat products. When the ES and FA methods were combined, a greater accuracy was achieved than by use of either alone (97). Employing 126 samples, 66 were salmonellae positive by CCM and FA and 64 by ES. The latter exhibited a relatively low incidence of false positives, while FA showed a higher incidence. The ES method was later shown capable of detecting 97–99% of salmonellae-positive samples (96). Using a modified ES as noted above on 3,486 samples, 96.7% were correctly analyzed (212). Overall, the ES method provides results in 32–50 h compared to 92–120 for CCM; results are comparable to both CCM and FA; and no specialized equipment or training is needed. Possible disadvantages to its use are the need for a minimum of about 10^7 cells/ml and its lack of response to nonmotile salmonellae. The latter can be overcome by use of a slide agglutination test from the elective enrichment broth employing polyvalent O antiserum (204).

The oxoid *Salmonella* rapid test (OSRT) is a variation of ES. It consists of a culture vessel containing two tubes, each of which contains dehydrated enrichment media in their lower compartments and dehydrated selective media in their upper. The media are hydrated with sterile distilled water, and a special salmonellae elective medium is added to the culture vessel along with a novobiocin disc, followed by 1 ml of preenrichment culture of sample. Following incubation at 41°C for 24 h, media in the upper compartment (selective media) of each tube are examined for color change, indicating presence of salmonellae. Positive tubes are further tested with the Oxoid *Salmonella* latex test (2 min). Final confirmation of salmonellae is made by use of traditional biochemical and serologic tests. The OSRT method correctly detected 293 of 296 salmonellae cultures within 42 h, and none of 68 nonsalmonellae gram-negative and 30 gram-positive bacteria gave positive results (101).

Salmonella 1–2 Test

This method is similar to ES and OSRT. ES relies on antibody reaction with flagellated salmonellae strains. Unlike ES, the 1–2 Test employs the use of a semisolid phase. The method is conducted in a specially designed plastic device that has two chambers, one for selective broth and the other for a nonselective motility medium. In addition to selective ingredients, the latter contains the amino acid L-serine, which is elective for salmonellae. Following inoculation of the selective medium chamber, the device is incubated, during which time motile salmonellae move into the nonselective medium chamber. The latter contains flagellar antibodies, and when the motile organisms enter the antibody area, an immunoband develops, indicating antigen-antibody reaction. Following nonselective enrichment, test results can be obtained in 8–14 h.

When compared to a culture method on 196 food and feed samples, the 1–2 Test detected 34 and the culture method detected 26 positive samples (156).

With the addition of a tetrathionate brilliant green broth enrichment step for the 1–2 Test, 84 of 314 samples were salmonellae positive—3 more than the culture method—and results could be obtained 1 day before the culture method (156). That this method produces better results with a preenrichment step was shown by others on 186 foods that contained large numbers of nonsalmonellae (41).

Radioimmunoassay

This technique consists of adding a radioactive label to an antigen, allowing the labeled antigen to react with its specific antibody, and measuring the amount of antigen that combined with the antibody by use of a counter to measure radioactivity. Solid-phase RIA refers to methods that employ solid materials or surfaces onto which a monolayer of antibody molecules binds electrostatically. The solid materials used include polypropylene, polystryrene, and bromacetyl-cellulose. The ability of antibody-coated polymers to bind specifically with radioactive tracer antigens is essential to the basic principle of solid-phase RIA. When the free-labeled antigen is washed out, the radioactivity measurements are quantitative. The label used by many workers is ^{125}I.

Johnson et al. (117) developed a solid-phase RIA procedure for the determination of *S. aureus* enterotoxin B and found the procedure to be 5 to 20 times more sensitive than the immunodiffusion technique. These investigators found the sensitivity of the test to be in the 1–5 ng range employing polystyrene and counting radioactivity with an integral counter. Collins et al. (35) employed RIA for enterotoxin B with the concentrated antibody coupled to bromacetylcellulose. Their findings indicated the procedure to be 100-fold more sensitive than immunodiffusion and to be reliable at an enterotoxin level of 0.01 µg/ml. Staphylococcal enterotoxin A was extracted from a variety of foods, including ham, milk products, and crab meat, by Collins et al. (34) and measured by RIA all within 3–4 h. They agreed with earlier workers that the method was highly sensitive and useful to 0.001 µg/ml and quantitatively reliable to 0.01 µg/ml of enterotoxin A.

By iodination of enterotoxins, solid-phase RIA can be used to detect as little as 1 ng toxin/g (15). When protein A was used as immunoabsorbent to separate antigen-antibody complex from unreacted toxin, a sensitivity of <1.0 ng/g for staphylococcal enterotoxin A (SEA), SEB, SEC, SED, and SEE was achieved within 1 working day (15, 147). In another study, 0.1 ng/ml of SEA and 0.5 ng/ml of SEB could be detected when protein A was used (4). A sensitivity of 100 pg for SEC_2 was achieved by use of a double-antibody RIA (183).

An RIA method was developed for *E. coli* ST_a that detected 50 to 500 pg toxin/tube (83). The method measured ST from both human and porcine strains, and no cross-reactivity with ST_b was noted. Using solid-phase sandwich RIA, colonies of *E. coli* that produced LT and colonies of *V. cholerae* that produced cholera toxin (CT) could be differentiated from nontoxigenic colonies (197). The method detected 5 to 25 pg of pure toxin and was more sensitive than enzyme-linked immunosorbent assay (ELISA). A microtiter solid-phase RIA for LT was developed and found to be comparable in sensitivity to Y-1 adrenal cells (90).

For mycotoxins, an RIA method was developed for ochratoxin A with a sensitivity of 20 ppb (1). A method was developed for aflatoxin M_1 (AFM_1) in milk with a sensitivity of 5–50 ng/assay, but it was subject to interference by whole milk (173). An ELISA method detected as little as 0.25 ng AFM_1 in 3 h. The action level of the U.S. Food and Drug Administration for AFM_1 in milk is 0.5 ng/ml.

The RIA technique lends itself to the examination of foods for other biological hazards such as endotoxins, paralytic shellfish toxins, and the like. The detection and identification of bacterial cells within 8–10 min has been achieved (210) by use of ^{125}I-labeled homologous antibody filtered and washed on a Millipore membrane. Multibacterial species have been detected in one operation when mixtures of homologous antibodies were used (209). Indirect and direct methods may be used, with the former requiring only one labeled globulin, while with the latter a labeled antibody to each organism in a mixed population is needed.

ELISA

The enzyme-linked immunosorbent assay (ELISA, enzyme immunoassay, or EIA) is an immunological method similar to RIA but employing an enzyme coupled to either antigen or antibody rather than a radioactive isotope. Essentially synonymous with ELISA are enzyme-multiplied immunoassay technique (EMIT) and indirect enzyme-linked antibody technique (ELAT). A typical ELISA is performed with a solid-phase (polystyrene) coated with antigen and incubated with antiserum. Following incubation and washing, an enzyme-labeled preparation of anti-immunoglobulin is added. After gentle washing, the enzyme remaining in the tube or microtiter well is assayed to determine the amount of specific antibodies in the initial serum. A commonly used enzyme is horseradish peroxidase, and its presence is measured by the addition of peroxidase substrate. The amount of enzyme present is ascertained by the colorimetric determination of enzyme substrate. Variations of this basic ELISA consist of a "sandwich" ELISA in which the antigen is required to have at least two binding sites. The antigen reacts first with excess solid-phase antibody, and following incubation and washing, the bound antigen is treated with excess labeled antibody. The "double sandwich" ELISA is a variation of the latter method, and it employs a third antibody.

The ELISA technique is used widely to detect and quantitate organisms and/or their products in foods, and synopses of some of these applications are presented below. For more details, the cited references should be consulted.

Salmonellae

1. Employing a polyclonal EIA with the IgG fraction of polyvalent flagellar antibodies and horseradish peroxidase, 92.2% agreement was found with the classical culture method on 142 food samples. False positives by the EIA were 6.4%, and a 95.8% agreement with FA was achieved (214).

2. A polyclonal EIA was used with polystyrene microtiter plates, a capture antibody technique, and a MUG assay. The sensitivity threshold was 10^7 cells/ml, and results could be obtained in 3 working days (148).

3. A monoclonal IgA antibody to flagellar antigens was employed, and 95% of 100 salmonellae were detected with a sensitivity of 10^6 cells/ml with results in 36 h (185).

4. With monoclonal IgA antibodies and polycarbonate-coated metal beads, $<10^6$ cells/ml could be detected in 2 days (144).

5. With an ELAT technique and polyclonal antibodies, salmonellae cells were added to cellulose acetate filter membranes plus enzyme-conjugated antibody and enzyme substrate. The method could detect 10^4–10^5 cells (127).

6. Two EIA methods, Report visual and Salmonella-Tek, were compared to a DNA probe method, Gene-Trak, employing 294 salmonellae and 100 non-salmonellae. Salmonella-Tek detected all salmonellae, but the false-positive rate was 16%. The Report visual EIA detected 98% of the salmonellae and 6% of the nonsalmonellae. The probe method detected 99% of salmonellae and none of the nonsalmonellae. Salmonella-Tek is characterized as a second-generation method that employs microtiter plate wells rather than magnetic beads, as its predecessor, Bio-Enzabead, did (39). The second-generation method required less time than its predecessor. In a collaborative study in which it was compared to the standard culture procedure, no false-negative and only 1.1% false-positive results were found (40). Negative samples could be screened in 48 h (compared to 4 days by the culture procedure), and it has been given AOAC approval (40).

7. Monoclonal antibodies were used in a microtiter plate antibody capture method, and 10 cells/25 g could be detected in 19 h with no cross-reactions with other organisms (136).

8. The Salmonella-TEK micro-ELISA method with monoclonal antibodies could detect 1–5 cfu/25 g with results in 31 h. Sensitivity threshold was 10^4–10^5 cells (229).

9. An enzyme-immunometric assay (EIMA) with polyclonal antibodies was employed with a titanous hydroxide suspension as solid phase substrate. Bound flagellar antibodies were reacted with antirabbit alkaline phosphatase conjugate, followed by addition of enzyme substrate to detect bound enzyme activity. The method was stated to be 100- to 160-fold more sensitive than microtiter plate methods. Results were obtained in 8 h following enrichments, and the sensitivity threshold was 4,000–5,000 cells. With EIMA, 66 of 376 cultures were positive, while for culture method 65 were positive (106, 107).

S. aureus and Its Enterotoxins

1. A double-antibody EIA was developed for staphylococcal enterotoxin A (SEA) that detected 0.4 ng in 20 h in wieners, 3.2 ng/ml in 1–3 h in milk, and 1.6 ng/ml from mayonnaise (192).

2. With polystyrene balls coated individually with respective antibodies, SEA, SEB, and SEC were detected at 0.1 ng or less/ml (207).

3. A standard ELISA for SEA, SEB, SEC, and SEE was used for ground meat, and the method detected $<0.5\,\mu$g/100 g (161).

4. A solid phase "double-antibody sandwich" method with horseradish peroxidase coupled to specific IgG enterotoxin antibodies was employed with polystyrene balls or microtiter plates for solid phase. Levels of >1 ng/g of SEA, SEB, SEC, SED, and SEE were detected in spiked foods in 1 working day (76).

5. A rapid "sandwich" ELISA that employed avid polyclonal antibodies to SEB and a biotin-streptavidin amplification system could detect 0.5–1.0 ng SEB in 1 h (150).

6. Colonies of *S. aureus* were grown on cellulose nitrate filters for 24 h followed by incubation with a fluorescein isothiocyanate-conjugated horseradish peroxidase-protein A conjugate. This membrane-filter ELISA had a sensitivity of 500 pg of SEB, and results could be completed within 27 h (175).

7. A modification of the "sandwich" ELISA for protein A of *S. aureus* cells that employs catalase-labeled antiprotein A antibody was developed for *S. aureus* detection. With H_2O_2 added, catalase releases O_2, which is measured with an amperometric O_2 electrode. Rate of increase in electrode current was proportional to antigen concentration (protein A or *S. aureus* cells). Protein A was detected at 0.1 ng/ml, stated to be 20 times more sensitive than a conventional ELISA. Pure cultures of *S. aureus* were detected at $10^3–10^4$/ml compared to $\sim10^5$ by conventional ELISA. The method could detect 1 cfu/g after 18 h; the overall accuracy is dependent on the protein A content of *S. aureus* cells (149).

8. Another amperometric electrochemical EIA specific for protein A of *S. aureus* employed a "sandwich" plus NAD-NADH enzymatic cycling step. A platinum disk electrode was used to measure final potential, and the method could detect <100 cells/ml within 4 h (22).

Molds and Mycotoxins

1. A standard ELISA was employed to detect three mold species *(Alternaria alternaria, Geotrichum candidum*, and *Rhizopus stolonifer)*. The method could detect ~1 µg of dried mold/g of tomato puree, making it more sensitive than chemical methods (141).

2. Both viable and nonviable molds could be detected with an ELISA method, which produced comparable or better results than the Howard mold count method (140).

3. For the detection of aflatoxin B_1 (AFB$_1$), an ELISA method employing monoclonal antibodies could detect 0.5 ng/ml (45); a commercially available kit can detect 5 ppb (Environmental Diagnostics); a tube ELISA can detect <10 pg/ml (135); a polystyrene microtiter plate method can detect 25 pg/assay (172); a nylon bead or Terasaki plate method can detect 0.1 ng/ml (171); and a monoclonal antibody ELISA can detect 0.2 ng/ml (26).

4. For AFM$_1$ detection, a monoclonal antibody method was sensitive to 0.25 ng/ml (45); a nylon bead or Terasaki plate method to 0.05 ng/ml (171); a direct ELISA to 10–25 pg/ml (103); and another ELISA method to 0.25 ng/ml (173).

5. A commercially available field kit can detect 5 ppb of AFB_2 and AFG_1; T-2 toxin has been detected at a level as low as 0.05 ng/ml (78); and ochratoxin A at a level of 25 pg/assay (174).

Botulinal Toxins

1. For type A toxin, a monoclonal antibody ELISA has been shown to detect ~9 mouse LD_{50}/ml (200); a "double-sandwich" ELISA detected 50–100 mouse LD_{50} of type A and <100 mouse i.p. LD_{50} of type E; and a "double-sandwich" ELISA with alkaline phosphatase and polystyrene plates has been shown capable of detecting 1 mouse i.p. median lethal dose of type G toxin (139).

E. coli Enterotoxins

1. A monoclonal antibody specific for enterohemorrhagic strains of *E. coli* (EHEC) was shown to be highly specific when used in an ELISA to detect EHEC strains (166).

2. Two "sandwich" ELISAs were developed based on toxin-specific murine monoclonal capture antibodies and rabbit polyclonal second antibodies specific for the SLT-I and SLT-II genes of *E. coli*. The SLT-I ELISA could detect 200 pg of purified SLT-I toxin, while the SLT-II could detect 75 pg of SLT-II toxin (49).

3. A competitive ELISA was compared to a DNA probe and the suckling mouse assay for *E. coli* STa enterotoxin. The probe was more specific, but ELISA was more sensitive and the most rapid of the three methods (38).

Gel Diffusion

Gel diffusion methods have been widely used for the detection and quantitation of bacterial toxins and enterotoxins. The four most used are single-diffusion tube (Oudin), microslide double diffusion, micro-Ouchterlony slide, and electroimmunodiffusion. They have been employed to measure enterotoxins of staphylococci and *C. perfringens* and the toxins of *C. botulinum*. The relative sensitivity of the various methods is presented in Table 6-1. Although they should be usable for any soluble protein to which an antibody can be made, they require that the antigen be in precipitable form. Perhaps the most widely used is the Crowle modification of the Ouchterlony slide test as modified by Casman and Bennett (27) and Bennett and McClure (12). The procedure for determining enterotoxins in foods is illustrated in Figure 6-3, and further details on its use are presented in the *Bacteriological Analytical Manual*. The micro-Ouchterlony method can detect 0.1–0.01 μg of staphylococcal enterotoxin, which is the same limit for the Oudin test. The double-diffusion tube test can detect levels as low as 0.1 μg/ml, but the incubation period required for such low levels is 3–6 days. This immunodiffusion method requires that extracts from a 100-g sample be concentrated to 0.2 ml (76). While other methods such as RIA and RPH are more sensitive and rapid than the gel diffusion methods, the latter continue to be widely used. Their

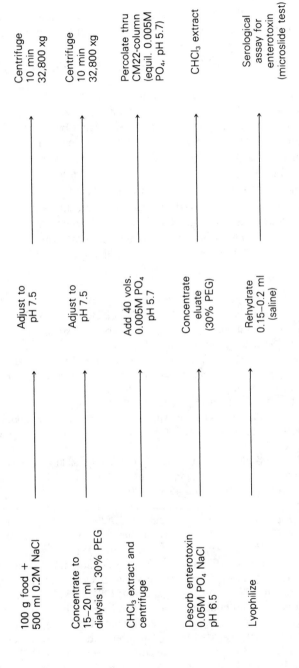

FIGURE 6-3. Schematic diagram for the extraction and serological assay of *S. aureus* enterotoxin in food. *After Bennett and McClure (13); copyright © 1980, Association of Official Analytical Chemists.*

reliability within their range of sensitivity is unquestioned. Recent studies suggest that results can be obtained in <8 h when slides are incubated at 45°C.

Hemagglutination

While gel diffusion methods generally require at least 24 h for results, two comparable serologic methods yield results in 2–4 h: hemagglutination-inhibition (HI) and reverse passive hemagglutination (RPH). Unlike the gel diffusion methods, antigens are not required to be in precipitable form for these two tests.

In the HI test, specific antibody is kept constant and enterotoxin (antigen) is diluted out. Following incubation for about 20 min, treated sheep red blood cells (SRBCs) are added. Hemagglutination (HA) occurs only where antibody is not bound by antigen. HA is prevented (inhibited) where toxin is present in optimal proportions with antibody. The sensitivity of HI in detecting enterotoxins is noted in Table 6-1.

In contrast to HI, antitoxin globulin in RPH is attached directly to SRBCs and used to detect toxin. When diluted toxin preparations are added, the test is read for HA after incubation for 2 h. HA occurs only where optimal antigen-antibody levels occur. No HA occurs if no toxin or enterotoxin is present. The levels of two enterotoxins detected by RPH are indicated in Table 6-1.

References

1. Aalund, G., K. Brunfeldt, B. Hald, P. Krogh, and K. Poulsen. 1975. A radio-immunoassay for ochratoxin A: A preliminary investigation. *Acta Path. Microbiol. Scand. Sect. C.* 83:390–392.
2. Alexander, D. N., G. M. Ederer, and J. M. Jatsen. 1976. Evaluation of an adenosine 5′-triphosphate assay as a screening method to detect significant bacteriuria. *J. Clin. Microbiol.* 3:42–46.
3. Anderson, R. L., A. K. Highsmith, and B. W. Holland. 1981. Comparison of standard pour plate, adenosine triphosphate, and *Limulus* amoebocyte lysate procedures for the detection of microbial contamination in intravenous fluids. *Bacteriol. Proc.* C-272.
4. Areson, P. W. D., S. E. Charm, and B. L. Wong. 1980. Determination of staphylococcal enterotoxins A and B in various food extracts, using staphylococcal cells containing protein. *J. Food Sci.* 45:400–401.
5. Ayres, J. C. 1967. Use of fluorescent antibody for the rapid detection of enteric organisms in egg, poultry and meat products. *Food Technol.* 21:631–640.
6. Bachrach, U., and Z. Bachrach. 1974. Radiometric method for the detection of coliform organisms in water. *Appl. Microbiol.* 28:169–171.
7. Barrell, R. A. E., and A. M. Paton. 1979. A semi-automatic method for the detection of salmonellae in food products. *J. Appl. Bacteriol.* 46:155–159.
8. Bastish, V. K., H. Chander, and B. Ranganathan. 1984. Incidence of enterococcal thermonuclease in milk and milk products. *J. Food Sci.* 49:1610–1611, 1615.
9. Beezer, A. E., D. Newell, and H. J. V. Tyrrell. 1976. Application of flow microcalorimetry to analytical problems: The preparation, storage and assay of frozen inocula of *Saccharomyces cerevisiae*. *J. Appl. Bacteriol.* 41:197–207.
10. Beezer, A. E., D. Newell, and H. J. V. Tyrrell. 1978. Characterisation and metabolic studies of *Saccharomyces cerevisiae* and *Kluyveromyces fragilis* by flow microcalorimetry. *Antonie van Leeuwenhoek* 45:55–63.

11. Bej, A. K., R. J. Steffan, J. DiCesare, L. Haff, and R. M. Atlas. 1990. Detection of coliform bacteria in water by polymerase chain reaction and gene probes. *Appl. Environ. Microbiol.* 56:307–314.

12. Bennett, R. W., and F. McClure. 1976. Collaborative study of the serological identification of staphylococcal enterotoxins by the microslide gel double diffusion test. *J. Assoc. Off. Anal. Chem.* 59:594–600.

13. Bennett, R. W., and F. McClure. 1980. Extraction and separation of staphylococcal enterotoxin in foods: Collaborative study. *J. Assoc. Off. Anal. Chem.* 63:1205–1210.

14. Berg, J. D., and L. Fiksdal. 1988. Rapid detection of total and fecal coliforms in water by enzymatic hydrolysis of 4-methylumbelliferone-β-D-galactoside. *Appl. Environ. Microbiol.* 54:2118–2122.

15. Bergdoll, M. S., and R. Reiser. 1980. Application of radioimmunoassay for detection of staphylococcal enterotoxins in foods. *J. Food Protect.* 43:68–72.

16. Biermann, V. A., and G. Terplan. 1980. Nachweis von Aflatoxin B_1 mittels ELISA. *Arch. Lebensmittelhyg.* 31:51–57.

17. Bishop, J. R., C. H. White, and R. Firstenberg-Eden. 1984. Rapid impedimetric method for determining the potential shelflife of pasteurized whole milk. *J. Food Protect.* 47:471–475.

18. Boling, E. A., G. C. Blanchard, and W. J. Russell. 1973. Bacterial identifications by microcalorimetry. *Nature* 241:472–473.

19. Boothroyd, M., and A. C. Baird-Parker. 1973. The use of enrichment serology for *Salmonella* detection in human foods and animal feeds. *J. Appl. Bacteriol.* 36:165–172.

20. Border, P. M., J. J. Howard, G. S. Plastow, and K. W. Siggens. 1990. Detection of *Listeria* species and *Listeria monocytogenes* using polymerase chain reaction. *Lett. Appl. Microbiol.* 11:158–162.

21. Bracha, R., L. S. Diamond, J. P. Ackers, G. D. Burchard, and D. Mirelman. 1990. Differentiation of clinical isolates of *Entamoeba histolytica* by using specific DNA probes. *J. Clin. Microbiol.* 28:680–684.

22. Brooks, J. L., B. Mirhabibollahi, and R. G. Kroll. 1990. Sensitive enzyme-amplified electrical immunoassay for protein A–bearing *Staphylococcus aureus* in foods. *Appl. Environ. Microbiol.* 56:3278–3284.

23. Bulte, M., and G. Reuter. 1984. Impedance measurement as a rapid method for the determination of the microbial contamination of meat surfaces, testing two different instruments. *Intern. J. Food Microbiol.* 1:113–125.

24. Cady, P. 1975. Rapid automated bacterial identification by impedance measurements. In *New Approaches to the Identification of Microorganisms*, ed. C.-G. Heden and T. Illeni, 73–99. New York: Wiley.

25. Cady, P., D. Hardy, S. Martins, S. W. Dufour, and S. J. Kraeger. 1978. Automated impedance measurements for rapid screening of milk microbial content. *J. Food Protect.* 41:277–283.

26. Candlish, A. A. G., W. H. Stimson, and J. E. Smith. 1985. A monoclonal antibody to aflatoxin B_1: Detection of the mycotoxin by enzyme immunoassay. *Lett. Appl. Microbiol.* 1:57–61.

27. Casman, E. P., and R. W. Bennett. 1965. Detection of staphylococcal enterotoxin in food. *Appl. Microbiol.* 13:181–189.

28. Cernic, G. 1976. Method for the distinction of gram-negative from gram-positive bacteria. *Eur. J. Appl. Microbiol.* 3:223–225.

29. Chan, S. W., S. Wilson, M. Vera-Garcia, K. Whippie, M. Ottaviani, A. Wilby, A. Shah, A. Johnson, M. Mozola, and D. N. Halbert. 1990. Comparative study of a colorimetric DNA hybridization method and the conventional culture procedure for detection of *Salmonella* in foods. *J. Assoc. Off. Anal. Chem.* 73:419–424.

30. Cherry, W. B., and M. D. Moody. 1965. Fluorescent-antibody techniques in diagnostic bacteriol. *Bacteriol. Rev.* 29:222–250.

31. Chesbro, W. R., and K. Auborn. 1967. Enzymatic detection of the growth of *Staphylococcus aureus* in foods. *Appl. Microbiol.* 15:1150–1159.

32. Chu, F. S. 1977. Mycotoxin problems in indigenous fermented foods and new methods for mycotoxin analysis. In *Handbook of Indigenous Fermented Foods*, ed. K. H. Steinkraus, 637–658. New York: Marcel Dekker.

33. Cleuziat, P., and J. Robert-Baudony. 1990. Specific detection of *Escherichia coli* and *Shigella* species using fragments of genes coding for β-glucuronidase. *FEMS Microbiol. Lett.* 72:315–322.

34. Collins, W. S., II, A. D. Johnson, J. F. Metzger, and R. W. Bennett. 1973. Rapid solid-phase radioimmunoassay for staphylococcal enterotoxin A. *Appl. Microbiol.* 25:774–777.

35. Collins, W. S., II, J. F. Metzger, and A. D. Johnson. 1972. A rapid solid phase radioimmunoassay for staphylococcal B enterotoxin. *J. Immunol.* 108:852–856.

36. Cords, B. R., and S. R. Tatini. 1973. Applicability of heat-stable deoxyribonuclease assay for assessment of staphylococcal growth and the likely presence of enterotoxin in cheese. *J. Dairy Sci.* 56:1512–1519.

37. Cousins, D. L., and F. Marlatt. 1990. An evaluation of a conductance method for the enumeration of Enterobacteriaceae in milk. *J. Food Protect.* 53:568–570.

38. Cryan, B. 1990. Comparison of three assay systems for detection of enterotoxigenic *Escherichia coli* heat-stable enterotoxin. *J. Clin. Microbiol.* 28:792–794.

39. Curiale, M. S., D. McIver, S. Weathersby, and C. Planer. 1990. Detection of salmonellae and other Enterobacteriaceae by commercial deoxyribonucleic acid hybridization and enzyme immunoassay kits. *J. Food Protect.* 53:1037–1046.

40. Curiale, M. S., M. J. Klatt, B. J. Robison, and L. T. Beck. 1990. Comparison of colorimetric monoclonal enzyme immunoassay screening methods for detection of *Salmonella* in foods. *J. Assoc. Off. Anal. Chem.* 73:43–50.

41. D'Aoust, J.-Y., and A. M. Sewell. 1988. Reliability of the immunodiffusion 1–2 Test[tm] system for detection of *Salmonella* in foods. *J. Food Protect.* 51:853–856.

42. Datta, A. R., B. A. Wentz, D. Shook, and M. W. Trucksees. 1988. Synthetic oligodeoxyribonucleotide probes for detection of *Listeria monocytogenes*. *Appl. Environ. Microbiol.* 54:2933–2937.

43. Dean, P. N., and D. Pinkel. 1978. High resolution dual laser flow cytometry. *J. Histochem. Cytochem.* 26:622–627.

44. de Castro, B. P., M. A. Asenio, B. Sanz, and J. A. Ordonez. 1988. A method to assess the bacterial content of refrigerated meat. *Appl. Environ. Microbiol.* 54:1462–1465.

45. Dixon-Holland, D. E., J. J. Pestka, B. A. Bidigare, W. L. Casale, R. L. Warner, B. P. Ram, and L. R. Hart. 1988. Production of sensitive monoclonal antibodies to aflatoxin B_1 and aflatoxin M_1 and their application to ELISA of naturally contaminated foods. *J. Food Protect.* 51:201–204.

46. Dodds, K. L., R. A. Holley, and A. G. Kempton. 1983. Evaluation of the catalase and *Limulus* amoebocyte lysate tests for rapid determination of the microbial quality of vacuum-packed cooked turkey. *Can. Inst. Food Sci. Technol. J.* 16:167–172.

47. Donnelly, C. W., and G. J. Baigent. 1986. Method for flow cytometric detection of *Listeria monocytogenes* in milk. *Appl. Environ. Microbiol.* 52:689–695.

48. Dovey, S., and K. J. Towner. 1989. A biotinylated DNA probe to detect bacterial cells in artificially contaminated foodstuffs. *J. Appl. Bacteriol.* 66:43–47.

49. Downes, F. P., J. H. Green, K. Greene, N. Stockbine, J. G. Wells, and I. K. Wachsmuth. 1989. Development and evaluation of enzyme-linked immunosorbent assays for detection of Shiga-like toxin I and Shiga-like toxin II. *J. Clin. Microbiol.* 27:1292–1297.

50. Doyle, M. P., and J. L. Schoeni. 1984. Survival and growth characteristics of *Escherichia coli* associated with hemorrhagic colitis. *Appl. Environ. Microbiol.* 48:855–856.

51. Duncan, C. L., and E. B. Somers. 1972. Quantitation of *Clostridium perfringens* Type A enterotoxin by electroimmunodiffusion. *Appl. Microbiol.* 24:801–804.

52. Edberg, S. C., M. J. Allen, and D. B. Smith, and the National Collaborative Study. 1989. National field evaluation of a defined substrate method for the simultaneous detection of total coliforms and *Escherichia coli* from drinking water: Comparison with presence-absence techniques. *Appl. Environ. Microbiol.* 55:1003–1008.

53. El-Nakib, O., J. J. Pestka, and F. S. Chu. 1981. Determination of aflatoxin B_1 in corn, wheat, and peanut butter by enzyme-linked immunosorbent assay and solid phase radioimmunoassay. *J. Assoc. Off. Anal. Chem.* 64:1077–1082.

54. Emswiler-Rose, B. S., R. W. Johnston, M. E. Harris, and W. L. Lee. 1980. Rapid detection of staphylococcal thermonuclease on casings of naturally contaminated fermented sausages. *Appl. Environ. Microbiol.* 40:13–18.

55. Erickson, A., and R. H. Deibel. 1973. Turbidimetric assay of staphylococcal nuclease. *Appl. Microbiol.* 25:337–341.

56. Evans, D. J., Jr., and D. G. Evans. 1977. Direct serological assay for the heat-labile enterotoxin of *Escherichia coli*, using passive immune hemolysis. *Infect. Immun.* 16:604–609.

57. Evans, H. A. V. 1982. A note on two uses for impedimetry in brewing microbiology. *J. Appl. Bacteriol.* 53:423–426.

58. Fallowfield, H. J., and J. T. Patterson. 1985. Potential value of the *Limulus* lysate assay for the measurement of meat spoilage. *J. Food Technol.* 20:467–479.

59. Fantasia, L. D. 1969. Accelerated immunofluorescence procedure for the detection of *Salmonella* in foods and animal by-products. *Appl. Microbiol.* 18:708–713.

60. Fantasia, L. D., J. P. Schrade, J. F. Yager, and D. Debler. 1975. Fluorescent antibody method for the detection of *Salmonella*: Development, evaluation, and collaborative study. *J. Assoc. Off. Anal. Chem.* 58:828–844.

61. Fantasia, L. D., W. H. Sperber, and R. H. Deibel. 1969. Comparison of two procedures for detection of *Salmonella* in food, feed, and pharmaceutical products. *Appl. Microbiol.* 17:540–541.

62. Farber, J. M. 1986. Potential use of membrane filters and a fluorogenic reagent-based solid medium for the enumeration of *Escherichia coli* in foods. *Can. Inst. Fd. Sci. & Technol. J.* 19:34–37.

63. Feng, P. C. S., and P. A. Hartman. 1982. Fluorogenic assays for immediate confirmation of *Escherichia coli*. *Appl. Environ. Microbiol.* 43:1320–1329.

64. Ferreira, J. L., W. E. Hill, M. K. Hamdy, F. A. Zapatka, and S. G. McCay. 1986. Detection of enterotoxigenic *Escherichia coli* in foods by DNA colony hybridization. *J. Food Sci.* 51:665–667.

65. Finkelstein, R. A., and Z. Yang. 1983. Rapid test for identification of heat-labile enterotoxin-producing *Escherichia coli* colonies. *J. Clin. Microbiol.* 18:23–28.

66. Firstenberg-Eden, R. 1983. Rapid estimation of the number of microorganisms in raw meat by impedance measurement. *Food Technol.* 37(1):64–70.

67. Firstenberg-Eden, R. 1984. Collaborative study of the impedance method for examining raw milk samples. *J. Food Protect.* 47:707–712.

68. Firstenberg-Eden, R., and G. Eden. 1984. *Impedance Microbiology*. New York: Wiley.

69. Firstenberg-Eden, R., and C. S. Klein. 1983. Evaluation of a rapid impedimetric procedure for the quantitative estimation of coliforms. *J. Food Sci.* 48:1307–1311.

70. Firstenberg-Eden, R., and M. K. Tricarico. 1983. Impedimetric determination of total, mesophilic and psychrotrophic counts in raw milk. *J. Food Sci.* 48: 1750–1754.

71. Firstenberg-Eden, R., M. L. Van Sise, J. Zindulis, and P. Kahn. 1984. Impedimetric estimation of coliforms in dairy products. *J. Food Sci.* 49: 1449–1452.

72. Fitts, R. 1985. Development of a DNA-DNA hybridization test for the presence of *Salmonella* in foods. *Food Technol.* 39(3):95–102.

73. Fitts, R., M. L. Diamond, C. Hamilton, and M. Nori. 1983. DNA-DNA hybridization assay for detection of *Salmonella* spp. in foods. *Appl. Environ. Microbiol.* 46:1146–1151.

74. Forrest, W. W. 1972. Microcalorimetry. *Meth. in Microbiol.* 6B:285–318.

75. Frampton, E. W., L. Restaino, and N. Blaszko. 1988. Evaluation of the B-glucuronidase substrate 5-bromo-4-chloro-3-indolyl–β-D-glucuronide (X-GLUC) in a 24-hour direct plating method for *Escherichia coli. J. Food Protect.* 51: 402–404.

76. Freed, R. C., M. L. Evenson. R. F. Reiser, and M. S. Bergdoll. 1982. Enzyme-linked immunosorbent assay for detection of staphylococcal enterotoxins in foods. *Appl. Environ. Microbiol.* 44:1349–1355.

77. Fujita, T., P. R. Monk, and I. Wadso. 1978. Calorimetric identification of several strains of lactic acid bacteria. *J. Dairy Res.* 45:457–463.

78. Gendloff, E. H., J. J. Pestka, S. P. Swanson, and L. P. Hart. 1984. Detection of T-2 toxin in *Fusarium sporotrichioides*–infected corn by enzyme-linked immunosorbent assay. *Appl. Environ. Microbiol.* 47:1161–1163.

79. Genigeorgis, C., and W. W. Sadler. 1966. Immunofluorescent detection of staphylococcal enterotoxin B. II. Detection in foods. *J. Food Sci.* 31:605–609.

80. Genigeorgis, C., G. Sakaguchi, and H. Riemann. 1973. Assay methods for *Clostridium perfringens* Type A enterotoxin. *Appl. Microbiol.* 26:111–115.

81. Georgala, D. L., and M. Boothroyd. 1964. A rapid immunofluorescence technique for detecting salmonellae in raw meat. *J. Hyg.* 62:319–326.

82. Georgala, D. L., M. Boothroyd, and P. R. Hayes. 1965. Further evaluation of rapid immunofluorescence technique for detecting salmonellae in meat and poultry. *J. Appl. Bacteriol.* 28:421–425.

83. Giannella, R. A., K. W. Drake, and M. Luttrell. 1981. Development of a radioimmunoassay for *Escherichia coli* heat-stable enterotoxin: Comparison with the suckling mouse bioassay. *Infect. Immun.* 33:186–192.

84. Gibbs, P. A., J. T. Patterson, and J. Early. 1979. A comparison of the fluorescent antibody method and a standardized cultural method for the detection of salmonellae. *J. Appl. Bacteriol.* 46:501–505.

85. Gnan, S., and L. O. Luedecke. 1982. Impedance measurements in raw milk as an alternative to the standard plate count. *J. Food Protect.* 45:4–7.

86. Goepfert, J. M., and N. F. Insalata. 1969. Salmonellae and the fluorescent anitibody technique: A current evaluation. *J. Milk Food Technol.* 32:465–473.

87. Geopfert, J. M., M. E. Mann, and R. Hicks. 1970. One-day fluorescent-antibody procedure for detecting salmonellae in frozen and dried foods. *Appl. Microbiol.* 20:977–983.

88. Gorina, L. G., F. S. Fluer, A. M. Olovnikov, and Yu. V. Ezepcuk. 1975. Use of the aggregate-agglutination technique for determining exo-enterotoxin of *Bacillus cereus. Appl. Environ. Microbiol.* 29:201–204.

89. Gram, L., and H. Søgaard. 1985. Microcalorimetry as a rapid method for estimation of bacterial levels in ground meat. *J. Food Protect.* 48:341–345.

90. Greenberg, H. B., D. A. Sack, W. Rodriguez, R. B. Sack, R. G. Wyatt, A. R. Kalica, R. L. Horswood, R. M. Chanock, and A. Z. Kapikian. 1977. Microtiter solid-phase radioimmunoassay for detection of *Escherichia coli* heat-labile enterotoxin. *Infect. Immun.* 17:541–545.

91. Griffiths, M. W., and J. D. Phillips. 1984. Detection of post-pasteurization

contamination of cream by impedimetric methods. *J. Appl. Bacteriol.* 57:107–114.

92. Grunstein, M., and D. S. Hogness. 1975. Colony hybridization: A method for the isolation of cloned DNAs that contain a specific gene. *Proc. Nat. Acad. Sci. USA* 72:3961–3965.

93. Haglund, J. R., J. C. Ayres, A. M. Paton, A. A. Kraft, and L. Y. Quinn. 1964. Detection of *Salmonella* in eggs and egg products with fluorescent antibody. *Appl. Microbiol.* 12:447–450.

94. Hardy, D., S. W. Dufour, and S. J. Kraeger. 1975. Rapid detection of frozen food bacteria by automated impedance measurements. *Proceedings, Institute of Food Technologists.*

95. Hatcher, W. S., S. DiBenedetto, L. E. Taylor, and D. I. Murdock. 1977. Radiometric analysis of frozen concentrated orange juice for total viable microorganisms. *J. Food Sci.* 42:636–639.

96. Hilker, J. S. 1975. Enrichment serology and fluorescent antibody procedures to detect salmonellae in foods. *J. Milk Food Technol.* 38:227–231.

97. Hilker, J. S., and M. Solberg. 1973. Evaluation of a fluorescent antibody-enrichment serology combination procedure for the detection of salmonellae in condiments, food products, food by products, and animal feeds. *Appl. Microbiol.* 26:751–756.

98. Hill, W. E., J. M. Madden, B. A. McCardell, D. B. Shah, J. A. Jagow, W. L. Payne, and B. K. Boutin. 1983. Foodborne enterotoxigenic *Escherichia coli*: Detection and enumeration by DNA colony hybridization. *Appl. Environ. Microbiol.* 45:1324–1330.

99. Hill, W. E., S. P. Keasler, M. W. Trucksess, P. Feng, C. A. Kaysner, and K. A. Lampel. 1991. Polymerase chain reaction identification of *Vibrio vulnificus* in artifically contaminated oysters. *Appl. Environ. Microbiol.* 57:707–717.

100. Hill, W. E., W. L. Payne, G. Zon, and S. L. Moseley. 1985. Synthetic oligodeoxyribonucleotide probes for detecting heat-stable enterotoxin-producing *Escherichia coli* by DNA colony hybridization. *Appl. Environ. Microbiol.* 50: 1187–1191.

101. Holbrook, R., J. M. Anderson, A. C. Baird-Parker, L. M. Dodds, D. Sawhney, S. H. Stuchbury, and D. Swaine. 1989. Rapid detection of *Salmonella* in foods—a convenient two-day procedure. *Lett. Appl. Microbiol.* 8:139–142.

102. Horan, P. K., and L. L. Wheeless, Jr. 1977. Quantitative single cell analysis and sorting. *Science* 198:149–157.

103. Hu, W. J., N. Woychik, and F. S. Chu. 1984. ELISA of picogram quantities of aflatoxin M_1 in urine and milk. *J. Food Protect.* 47:126–127.

104. Hutter, K.-J., and H. E. Eipel. 1979. Microbial determinations by flow cytometry. *J. Gen. Microbiol.* 113:369–375.

105. Hutter, K.-J., M. Stöhr, and H. E. Eipel. 1980. Simultaneous DNA and protein measurements of microorganisms. In *Flow Cytometry*, vol. 4, eds. O. D. Laerum, T. Lindmo, and E. Thorud, 100–102. Bergen: Universitetsforlaget.

106. Ibrahim, G. F., and M. J. Lyons. 1987. Detection of salmonellae in foods with an enzyme immunometric assay. *J. Food Protect.* 50:59–61.

107. Ibrahim, G. F., M. J. Lyons, R. A. Walker, and G. H. Fleet. 1985. Rapid detection of salmonellae by immunoassays with titanous hydroxide as the solid phase. *Appl. Environ. Microbiol.* 50:670–675.

108. Insalata, N. F., C. W. Mahnke, and W. G. Dunlap. 1972. Rapid, direct fluorescent-antibody method for the detection of salmonellae in food and feeds. *Appl. Microbiol.* 24:645–649.

109. Izat, A. L., C. D. Driggers, M. Colberg, M. A. Reiber, and M. H. Adams. 1989. Comparison of the DNA probe to culture methods for the detection of *Salmonella* on poultry carcasses and processing waters. *J. Food Protect.* 52:564–570.

110. Jaksch, V. P., K.-J. Zaadhof, and G. Terplan. 1982. Zur Bewertung der hygienischen Qualität von Milchprodukten mit dem *Limulus*-Test. *Molkerei-Zeitung Welt der Milch* 36:5–8.

111. Jay, J. M. 1974. Use of the *Limulus* lysate endotoxin test to assess the microbial quality of ground beef. *Bacteriol. Proc.*, 13.

112. Jay, J. M. 1977. The *Limulus* lysate endotoxin assay as a test of microbial quality of ground beef. *J. Appl. Bacteriol.* 43:99–109.

113. Jay, J. M. 1981. Rapid estimation of microbial numbers in fresh ground beef by use of the *Limulus* test. *J. Food Protect.* 44:275–278.

114. Jay, J. M., and S. Margitic. 1979. Comparison of homogenizing, shaking, and blending on the recovery of microorganisms and endotoxins from fresh and frozen ground beef as assessed by plate counts and the *Limulus* amoebocyte lysate test. *Appl. Environ. Microbiol.* 38:879–884.

115. Jay, J. M., S. Margitic, A. L. Shereda, and H. V. Covington. 1979. Determining endotoxin content of ground beef by the *Limulus* amoebocyte lysate test as a rapid indicator of microbial quality. *Appl. Environ. Microbiol.* 38:885–890.

116. Jay, J. M. 1989. The *Limulus* amoebocyte lysate (LAL) test. In *Progress in Industrial Microbiology: Rapid Methods in Food Microbiology*, ed. M. R. Adams and C. F. A. Hope, 101–119. Amsterdam: Elsevier.

117. Johnson, H. M., J. A. Bukovic, P. E. Kauffman, and J. T. Peeler. 1971. Staphylococcal enterotoxin B: Solid-phase radioimmunoassay. *Appl. Microbiol.* 22:837–841.

118. Johnson, H. M., H. E. Hall, and M. Simon. 1967. Enterotoxin B: Serological assay in cultures by passive hemagglutination. *Appl. Microbiol.* 15:815–818.

119. Kamman, J. F., and S. R. Tatini. 1977. Optimal conditions for assay of staphylococcal nuclease. *J. Food Sci.* 42:421–424.

120. Karl, D. M. 1980. Cellular nucleotide measurements and applications in microbial ecology. *Microbiol. Rev.* 44:739–796.

121. Kaysner, C. A., S. D. Weagant, and W. E. Hill. 1988. Modification of the DNA colony hybridization technique for multiple filter analysis. *Molec. Cell. Probes* 2:255–260.

122. Kennedy, J. E., Jr., and J. L. Oblinger. 1985. Application of bioluminescence to rapid determination of microbial levels in ground beef. *J. Food Protect.* 48: 334–340.

123. Kimmich, G. A., J. Randles, and J. S. Brand. 1975. Assay of picomole amounts of ATP, ADP, and AMP using the luciferase enzyme system. *Anal. Biochem.* 69:187–206.

124. Kleppe, K., E. Ohtsuka, R. Kleppe, I. Molineux, and H. G. Khorana. 1971. Studies on polynucleotides. XCVI. Rapid replication of short synthetic DNA's as catalyzed by DNA polymerases. *J. Mol. Biol.* 56:341–361.

125. Koburger, J. A., and M. L. Miller. 1985. Evaluation of a fluorogenic MPN procedure for determining *Escherichia coli* in oysters. *J. Food Protect.* 48: 244–245.

126. Koupal, A., and R. H. Deibel. 1978. Rapid qualitative method for detecting staphylococcal nuclease in foods. *Appl. Environ. Microbiol.* 35:1193–1197.

127. Krysinski, E. P., and R. C. Heimsch. 1977. Use of enzyme-labeled antibodies to detect *Salmonella* in foods. *Appl. Environ. Microbiol.* 33:947–954.

128. Lachica, R. V. 1980. Accelerated procedure for the enumeration and identification of food-borne *Staphylococcus aureus*. *Appl. Environ. Microbiol.* 39:17–19.

129. Lachica, R. V. F., C. Genigeorgis, and P. D. Hoeprich. 1971. Metachromatic agar-diffusion methods for detecting staphylococcal nuclease activity. *Appl. Microbiol.* 21:585–587.

130. Lachica, R. V. F., P. D. Hoeprich, and C. Genigeorgis. 1972. Metachromatic

agar-diffusion microslide technique for detecting staphylococcal nuclease in foods. *Appl. Microbiol.* 23:168–169.

131. Lachica, R. V., K. F. Weiss, and R. H. Deibel. 1969. Relationships among coagulase, enterotoxin, and heat-stable deoxyribonuclease production by *Staphylococcus aureus. Appl. Microbiol.* 18:126–127.

132. Lampel, K. A., J. A. Jagow, M. Trucksess, and W. E. Hill. 1990. Polymerase chain reaction for detection of invasive *Shigella flexneri* in food. *Appl. Environ. Microbiol.* 56:1536–1340.

133. Lampi, R. A., D. A. Mikelson, D. B. Rowley, J. J. Previte, and R. E. Wells. 1974. Radiometry and microcalorimetry—techniques for the rapid detection of foodborne microorganisms. *Food Technol.* 28(10):52–55.

134. Laramore, C. R., and C. W. Moritz. 1969. Fluorescent-antibody technique in detection of salmonellae in animal feed and feed ingredients. *Appl. Microbiol.* 17:352–354.

135. Lawellin, D. W., D. W. Grant, and B. K. Joyce. 1977. Enzyme-linked immunosorbent analysis for aflatoxin B_1. *Appl. Environ. Microbiol.* 34:94–96.

136. Lee, H. A., G. M. Wyatt, S. Bramham, and M. R. A. Morgan. 1990. Enzyme-linked immunosorbent assay for *Salmonella typhimurium* in food: Feasibility of 1-day *Salmonella* detection. *Appl. Environ. Microbiol.* 56:1541–1546.

137. Lee, R. M., and P. A. Hartman. 1989. Inexpensive, disposable presence-absence test for coliforms and *Escherichia coli* in water. *J. Food Protect.* 52:162–164.

138. Levin, G. V., V. R. Harrison, and W. C. Hess. 1956. Preliminary report on a one-hour presumptive test for coliform organisms. *J. Amer. Water Works Assoc.* 48:75–80.

139. Lewis, G. E., Jr., S. S. Kulinski, D. W. Reichard, and J. F. Metzger. 1981. Detection of *Clostridium botulinum* Type G toxin by enzyme-linked immunosorbent assay. *Appl. Environ. Microbiol.* 42:1018–1022.

140. Lin, H. H., and M. A. Cousin. 1987. Evaluation of enzyme-linked immunosorbent assay for detection of molds in foods. *J. Food Sci.* 52:1089–1094, 1096.

141. Lin, H. H., R. M. Lister, and M. A. Cousin. 1986. Enzyme-linked immunosorbent assay for detection of mold in tomato puree. *J. Food Sci.* 51:180–183, 192.

142. Littel, K. J., and P. A. Hartman. 1983. Fluorogenic selective and differential medium for isolation of fecal streptococci. *Appl. Environ. Microbiol.* 45:622–627.

143. Martins, S B., and M. J. Selby. 1980. Evaluation of a rapid method for the quantitative estimation of coliforms in meat by impedimetric procedures. *Appl. Environ. Microbiol.* 39:518–524.

144. Mattingly, J. A., and W. D. Gehle. 1984. An improved enzyme immunoassay for the detection of *Salmonella. J. Food Sci.* 49:807–809.

145. Mendelsohn, M. L. 1980. The attributes and applications of flow cytometry. In *Flow Cytometry*, vol. 4, ed. O. D. Laerum, T. Lindmo, and E. Thorud, 15–27. Bergen: Universitetsforlaget.

146. Meyer, R. F., and M. J. Palmieri. 1980. Single radial immunodiffusion method for screening staphylococcal isolates for enterotoxin. *Appl. Environ. Microbiol.* 40:1080–1085.

147. Miller, B. A., R. F. Reiser, and M. S. Bergdoll. 1978. Detection of staphylococcal enterotoxins A, B, C, D, and E in foods by radioimmunoassay, using staphylococcal cells containing protein A as immunoadsorbent. *Appl. Environ. Microbiol.* 36:421–426.

148. Minnich, S. A., P. A. Hartman, and R. C. Heimsch. 1982. Enzyme immunoassay for detection of salmonellae in foods. *Appl. Environ. Microbiol.* 43:877–883.

149. Mirhabibollahi, B., J. L. Brooks, and R. G. Kroll. 1990. Development and

performance of an enzyme-linked amperometric immunosensor for the detection of *Staphylococcus aureus* in foods. *J. Appl. Bacteriol.* 68:577–585.

150. Morissette, C., J. Goulet, and G. Lamoureux. 1991. Rapid and sensitive sandwich enzyme-linked immunosorbent assay for detection of staphylococcal enterotoxin B in cheese. *Appl. Environ. Microbiol.* 57:836–842.

151. Moseley, S. L., P. Echeverria, J. Seriwatana, C. Tirapat, W. Chaicumpa, T. Sakuldaipeara, and S. Falkow. 1982. Identification of enterotoxigenic *Escherichia coli* by colony hybridization using three enterotoxin gene probes. *J. Infect. Dis.* 145:863–869.

152. Moseley, S. L., and S. Falkow. 1980. Nucleotide sequence homology between the heat-labile enterotoxin gene of *Escherichia coli* and *Vibrio cholerae* deoxyribonucleic acid. *J. Bacteriol.* 144:444–446.

153. Muldrow, L. L., R. L. Tyndall, and C. B. Fliermans. 1982. Application of flow cytometry to studies of pathogenic free-living amoebas. *Appl. Environ. Microbiol.* 44:1258–1269.

154. Munson, T. E., J. P. Schrade, N. B. Bisciello, Jr., L. D. Fantasia, W. H. Hartung, and J. J. O'Connor. 1976. Evaluation of an automated fluorescent antibody procedure for detection of *Salmonella* in foods and feeds. *Appl. Environ. Microbiol.* 31:514–521.

155. Nakamura, T., T. Morita, and S. Iwanga. 1986. Lipopolysaccharide-sensitive serine-protease zymogen (factor C) found in *Limulus* hemocytes. Isolation and characterization. *Eur. J. Biochem.* 154:511–521.

156. Nath, E. J., E. Neidert, and C. J. Randall. 1989. Evaluation of enrichment protocols for the 1–2 Test^{tm} for *Salmonella* detection in naturally contaminated foods and feeds. *J. Food Protect.* 52:498–499.

157. Niskanen, A., and L. Koiranen. 1977. Correlation of enterotoxin and thermonuclease production with some physiological and biochemical properties of staphylococcal strains isolated from different sources. *J. Food Protect.* 40:543–548.

158. Niskanen, A., and E. Nurmi. 1976. Effect of starter culture on staphylococcal enterotoxin and thermonuclease production in dry sausage. *Appl. Environ. Microbiol.* 31:11–20.

159. Niyomvit, N., K. E. Stevenson, and R. F. McFeeters. 1978. Detection of staphylococcal enterotoxin B by affinity radioimmunoassay. *J. Food Sci.* 43:735–739.

160. Notermans, S., J. Dufrenne, and S. Kozaki. 1979. Enzyme-linked immunosorbent assay for detection of *Clostridium botulinum* type E toxin. *Appl. Environ. Microbiol.* 37:1173–1175.

161. Notermans, S., J. Dufrenne, and M. van Schothorst. 1978. Enzyme-linked immunosorbent assay for detection of *Clostridium botulinum* toxin type A. *Japan. J. Med. Sci. Biol.* 31:81–85.

162. Notermans, S., K. J. Heuvelman, and K. Wernars. 1988. Synthetic enterotoxin B DNA probes for detection of enterotoxigenic *Staphylococcus aureus* strains. *Appl. Environ. Microbiol.* 54:531–533.

163. Notermans, S., T. Chakraborty, M. Leimeister-Wachter, J. Dufrenne, K. J. Heuvelman, H. Maas, W. Janse, K. Wernars, and P. Guinee. 1989. Specific gene probe for detection of biotyped and serotyped *Listeria* strains. *Appl. Environ. Microbiol.* 55:902–906.

164. Nuzback, D. E., E. E. Bartley, S. M. Dennis, T. G. Nagaraja, S. J. Galitzer, and A. D. Dayton. 1983. Relation of rumen ATP concentration to bacterial and protozoal numbers. *Appl. Environ. Microbiol.* 46:533–538.

165. Olive, D. M. 1989. Detection of enterotoxigenic *Escherichia coli* after polymerase chain reaction amplification with a thermostable DNA polymerase. *J. Clin. Microbiol.* 27:261–265.

166. Padhye, N. V., and M. P. Doyle. 1991. Production and characterization of a monoclonal antibody specific for enterohemorrhagic *Escherichia coli* of serotypes 0157:H7 and 026:H11. *J. Clin. Microbiol.* 29:99–103.

167. Park, C. E., H. B. El Derea, and M. K. Rayman. 1978. Evaluation of staphylococcal thermonuclease (TNase) assay as a means of screening foods for growth of staphylococci and possible enterotoxin production. *Can. J. Microbiol.* 24:1135–1139.

168. Patterson, J. W., P. L. Brezonik, and H. D. Putnam. 1970. Measurement and significance of adenosine triphosphate in activated sludge. *Environ. Sci. & Technol.* 4:569–575.

169. Perry, B. F., A. E. Beezer, and R. J. Miles. 1979. Flow microcalorimetric studies of yeast growth: Fundamental aspects. *J. Appl. Bacteriol.* 47:527–537.

170. Perry, B. F., A. E. Beezer, and R. J. Miles. 1983. Characterization of commercial yeast strains by flow microcalorimetry. *J. Appl. Bacteriol.* 54:183–189.

171. Pestka, J. J., and F. S. Chu. 1984. Enzyme-linked immunosorbent assay of mycotoxins using nylon bead and Terasaki plate solid phases. *J. Food Protect.* 47:305–308.

172. Pestka, J. J., P. K. Gaur, and F. S. Chu. 1980. Quantitation of aflatoxin B_1 and aflatoxin B_1 antibody by an enzyme-linked immunosorbent microassay. *Appl. Environ. Microbiol.* 40:1027–1031.

173. Pestka, J. J., V. Li, W. O. Harder, and F. S. Chu. 1981. Comparison of radio-immunoassay and enzyme-linked immunosorbent assay for determining aflatoxin M_1 in milk. *J. Assoc. Off. Anal. Chem.* 64:294–301.

174. Pestka, J. J., B. W. Steinert, and F. S. Chu. 1981. Enzyme-linked immunosorbent assay for detection of ochratoxin A. *Appl. Environ. Microbiol.* 41:1472–1474.

175. Peterkin, P. I., and A. N. Sharpe. 1984. Rapid enumeration of *Staphylococcus aureus* in foods by direct demonstration of enterotoxigenic colonies on membrane filters by enzyme immunoassay. *Appl. Environ. Microbiol.* 47:1047–1053.

176. Peterson, E. H., M. L. Nierman, R. A. Rude, and J. T. Peeler. 1987. Comparison of AOAC method and fluorogenic (MUG) assay for enumerating *Escherichia coli* in foods. *J. Food Sci.* 52: 409–410.

177. Pinder, A. C., P. W. Purdy, S. A. G. Poulter, and D. C. Clark. 1990. Validation of flow cytometry for rapid enumeration of bacterial concentrations in pure cultures. *J. Appl. Bacteriol.* 69:92–100.

178. Poelma, P. L., C. R. Wilson, and W. H. Andrews. 1987. Rapid fluorogenic enumeration of *Escherichia coli* in selected, naturally contaminated high moisture foods. *J. Assoc. Off. Anal. Chem.* 70:991–993.

179. Previte, J. J. 1972. Radiometric detection of some food-borne bacteria. *Appl. Microbiol.* 24:535–539.

180. Riley, L. K., and C. J. Caffrey. 1990. Identification of enterotoxigenic *Escherichia coli* by colony hybridization with nonradioactive digoxigenin-labeled DNA probes. *J. Clin. Microbiol.* 28:1465–1468.

181. Rippey, S. R., L. A. Chandler, and W. D. Watkins. 1987. Fluorometric method for enumeration of *Escherichia coli* in molluscan shellfish. *J. Food Protect.* 50: 685–690.

182. Restaino, L., E. W. Frampton, and R. H. Lyon. 1990. Use of the chromogenic substrate 5-bromo-4-chloro-3-indolyl-β-D-glucuronide (X-GLUC) for enumerating *Escherichia coli* in 24 h from ground beef. *J. Food Protect.* 53:508–510.

183. Robern, H., M. Dighton, Y. Yano, and N. Dickie. 1975. Double-antibody radioimmunoassay for staphylococcal enterotoxin C_2. *Appl. Microbiol.* 30: 525–529.

184. Robison, B. J. 1984. Evaluation of a fluorogenic assay for detection of *Escherichia coli* in foods. *Appl. Environ. Microbiol.* 48:285–288.

185. Robison, B. J., C. I. Pretzman, and J. A. Mattingly. 1983. Enzyme immunoassay

in which a myeloma protein is used for detection of salmonellae. *Appl. Environ. Microbiol.* 45:1816–1821.

186. Rowley, D. B., J. J. Previte, and H. P. Srinivasa. 1978. A radiometric method for rapid screening of cooked foods for microbial acceptability. *J. Food Sci.* 43: 1720–1722.

187. Russell, W. J., J. F. Zettler, G. C. Blanchard, and E. A. Boling. 1975. Bacterial identification by microcalorimetry. In *New Approaches to the Identification of microorganisms*, ed. C.-G. Hedén and T. Illéni, 101–21. New York: Wiley.

188. Sacks, L. E., and E. Menefee. 1972. Thermal detection of spoilage in canned foods. *J. Food Sci.* 37:928–931.

189. Saez-Llorens, X., L. M. Guzman-Verduzco, S. Shelton, J. D. Nelson, and Y. M. Kupersztoch. 1989. Simultaneous detection of *Escherichia coli* heat-stable and heat-labile enterotoxin genes with a single RNA probe. *J. Clin. Microbiol.* 27: 1684–1688.

190. Saiki, R. K., D. H. Gelfand, S. Stoffel, S. J. Scharf, R. Higuchi, G. T. Horn, K. B. Mullis, and H. A. Erlich. 1988. Primer-directed enzymatic amplification of DNA with a thermostable DNA polymerase. *Science* 239:487–491.

191. Samadpour, M., J. Liston, J. E. Ongerth, and P. I. Tarr. 1990. Evaluation of DNA probes for detection of Shiga-like-toxin-producing *Escherichia coli* in food and calf fecal samples. *Appl. Environ. Microbiol.* 56:1212–1215.

192. Saunders, G. C., and M. L. Bartlett. 1977. Double-antibody solid-phase enzyme immunoassay for the detection of staphylococcal enterotoxin A. *Appl. Environ. Microbiol.* 34:518–522.

193. Scholl, D. R., C. Kaufmann, J. D. Jollick, C. K. York, G. R. Goodrum, and P. Charache. 1990. Clinical application of novel sample processing technology for the identification of salmonellae by using DNA probes. *J. Clin. Microbiol.* 28: 237–241.

194. Schrot, J. R., W. C. Hess, and G. V. Levin. 1973. Method for radiorespirometric detection of bacteria in pure culture and in blood. *Appl. Microbiol.* 26:867–873.

195. Schultz, S. J., J. S. Witzeman, and W. M. Hall. 1968. Immunofluorescent screening for *Salmonella* in foods: Comparison with cultural methods. *J. Assoc. Off. Anal. Chem.* 51:1334–1338.

196. Seiter, J. A., and J. M. Jay. 1980. Comparison of direct serial dilution and most-probable-number methods for determining endotoxins in meats by the *Limulus* amoebocyte lysate test. *Appl. Environ. Microbiol.* 40:177–178.

197. Shah, D. B., P. E. Kauffman, B. K. Boutin, and C. H. Johnson. 1982. Detection of heat-labile-enterotoxin-producing colonies of *Escherichia coli* and *Vibrio cholerae* by solid-phase sandwich radioimmunoassays. *J. Clin. Microbiol.* 16: 504–508.

198. Shapiro, H. M. 1988. *Practical Flow Cytometry*, 2d ed. New York: A. R. Liss Pub.

199. Sharpe, A. N., M. N. Woodrow, and A. K. Jackson. 1970. Adenosine-triphosphate (ATP) levels in foods contaminated by bacteria. *J. Appl. Bacteriol.* 33:758–767.

200. Shone, C., P. Wilton-Smith, N. Appleton, P. Hambleton, N. Modi, S. Gatley, and J. Melling. 1985. Monoclonal antibody-based immunoassay for type A *Clostridium botulinum* toxin is comparable to the mouse bioassay. *Appl. Environ. Microbiol.* 50:63–67.

201. Silliker, J. H., A. Schmall, and J. Y. Chiu. 1966. The fluorescent antibody technique as a means of detecting salmonellae in foods. *J. Food Sci.* 31:240–244.

202. Silverman, M. P., and E. F. Munoz. 1979. Automated electrical impedance technique for rapid enumeration of fecal coliforms in effluents from sewage treatment plants. *Appl. Environ. Microbiol.* 37:521–526.

203. Silverman, S.J., A. R. Knott, and M. Howard. 1968. Rapid, sensitive assay for staphylococcal enterotoxin and a comparison of serological methods. *Appl. Microbiol.* 16:1019–1023.

204. Sperber, W. H., and R. H. Deibel. 1969. Accelerated procedure for *Salmonella* detection in dried foods and feeds involving only broth cultures and serological reactions. *Appl. Microbiol.* 17:533–539.

205. Stannard, C. J., and J. M. Wood. 1983. The rapid estimation of microbial contamination of raw meat by measurement of adenosine triphosphate (ATP). *J. Appl. Bacteriol.* 55:429–438.

206. Stewart, B. J., M. J. Eyles, and W. G. Murrell. 1980. Rapid radiometric method for detection of *Salmonella* in foods. *Appl. Environ. Microbiol.* 40:223–230.

207. Stiffler-Rosenberg, G., and H. Fey. 1978. Simple assay for staphylococcal enterotoxins A, B, and D: Modification of enzyme-linked immunosorbent assay. *J. Clin. Microbiol.* 8:473–479.

208. Stoflet, E. S., D. D. Koeberi, G. Sarkar, and S. S. Sommer. 1988. Genomic amplification with transcript sequencing. *Science* 239:491–494.

209. Strange, R. E., and K. L. Martin. 1972. Rapid assays for the detection and determination of sparse populations of bacteria and bacteriophage T7 with radioactively labelled homologous antibodies. *J. Gen. Microbiol.* 72:127–141.

210. Strange, R. E., E. O. Powell, and T. W. Pearce. 1971. The rapid detection and determination of sparse bacterial populations with radioactively labelled homologous antibodies. *J. Gen. Microbiol.* 67:349–357.

211. Sullivan, J. D., Jr., P. C. Ellis, R. G. Lee, W. S. Combs, Jr., and S. W. Watson. 1983. Comparison of the *Limulus* amoebocyte lysate test with plate counts and chemical analyses for assessment of the quality of lean fish. *Appl. Environ. Microbiol.* 45:720–722.

212. Surdy, T. E., and S. G. Haas. 1981. Modified enrichment-serology procedure for detection of salmonellae in soy products. *Appl. Environ. Microbiol.* 42:704–707.

213. Swaminathan, B., J. A. G. Aleixo, and S. A. Minnich. 1985. Enzyme immunoassays for *Salmonella*: One-day testing is now a reality. *Food Technol.* 39(3): 83–89.

214. Swaminathan, B., and J. C. Ayres. 1980. A direct immunoenzyme method for the detection of salmonellae in foods. *J. Food Sci.* 45:352–355, 361.

215. Tai, J. Y., R. C. Seid, Jr., R. D. Hurn, and T.-Y. Liu. 1977. Studies on *Limulus* amoebocyte lysate. II. Purification of the coagulogen and the mechanism of clotting. *J. Biol. Chem.* 252:4773–4776.

216. Tatini, S. R., H. M. Soo, B. R. Cords, and R. W. Bennett. 1975. Heat-stable nuclease for assessment of staphylococcal growth and likely presence of enterotoxins in foods. *J. Food Sci.* 40:352–356.

217. Terplan, V. G., K.-J. Zaadhof, and S. Buchholz-Berchtold. 1975. Zum nachweis von Endotoxinen gramnegativer Keime in Milch mit dem *Limulus*-test. *Arch. Lebensmittelhyg.* 26:217–221.

218. Thomason, B. M. 1971. Rapid detection of *Salmonella* microcolonies by fluorescent antibody. *Appl. Microbiol.* 22:1064–1069.

219. Thomason, B. M. 1981. Current status of immunofluorescent methodology for salmonellae. *J. Food Protect.* 44:381–384.

220. Thomason, B. M., G. A. Hebert, and W. B. Cherry. 1975. Evaluation of a semiautomated system for direct fluorescent antibody detection of salmonellae. *Appl. Microbiol.* 30:557–564.

221. Thompson, J. S., D. S. Hodge, and A. A. Borczyk. 1990. Rapid biochemical test to identify verocytotoxin-positive strains of *Escherichia coli* serotype 0157. *J. Clin. Microbiol.* 28:2165–2168.

222. Thore, A., S. Ånséhn, A. Lundin, and S. Bergman. 1975. Detection of bac-

teriuria by luciferase assay of adenosine triphosphate. *J. Clin. Microbiol.* 1:1–8.

223. Thore, A., A. Lundin, and S. Ånséhn. 1983. Firefly luciferase ATP assay as a screening method for bacteriuria. *J. Clin. Microbiol.* 17:218–224.

224. Tsuji, K., P. A. Martin, and D. M. Bussey. 1984. Automation of chromogenic substrate *Limulus* amebocyte lysate assay method for endotoxin by robotic system. *Appl. Environ. Microbiol.* 48:550–555.

225. Tsuji, K., and K. A. Steindler. 1983. Use of magnesium to increase sensitivity of *Limulus* amoebocyte lysate for detection of endotoxin. *Appl. Environ. Microbiol.* 45:1342–1350.

226. Ur, A., and D. Brown. 1975. Monitoring of bacterial activity by impedance measurements. In *New Approaches to the Identification of Microorganisms*, ed. C.-G. Heden and T. Illeni, 61–71. New York: Wiley.

227. Van Damme-Jongsten, M., J. Rodhouse, R. J. Gilbert, and S. Notermans. 1990. Synthetic DNA probes for detection of enterotoxigenic *Clostridium perfringens* strains isolates from outbreaks of food poisoning. *J. Clin. Microbiol.* 28:131–133.

228. Van Dilla, M. A., R. G. Langlois, D. Pinkel, D. Yajko, and W. K. Hadley. 1983. Bacterial characterization by flow cytometry. *Science* 220:620–622.

229. Van Poucke, L. S. G. 1990. Salmonella-TEK, a rapid screening method for *Salmonella* species in food. *Appl. Environ. Microbiol.* 56:924–927.

230. Van Wart, M., and L. J. Moberg. 1984. Evaluation of a novel fluorogenic-based method for detection of *Escherichia coli*. *Bacteriol. Proc.*, 201.

231. Vermilyea, B. L., H. D. Walker, and J. C. Ayres. 1968. Detection of botulinal toxins by immunodiffusion. *Appl. Microbiol.* 16:21–24.

232. Wachtel, R. E., and K. Tsuji. 1977. Comparison of *Limulus* amebocyte lysates and correlation with the United States Pharmacopeial pyrogen test. *Appl. Environ. Microbiol.* 33:1265–1269.

233. Waes, G. M., and R. G. Bossuyt. 1982. Usefulness of the benzalkon–crystal violet–ATP method for predicting the keeping quality of pasteurized milk. *J. Food Protect.* 45:928–931.

234. Waes, G. M., and R. G. Bossuyt. 1984. Impedance measurements to detect bacteriophage problems in cheddar cheesemaking. *J. Food Protect.* 47:349–351.

235. Warren, L. S., R. E. Benoit, and J. A. Jessee. 1978. Rapid enumeration of fecal coliforms in water by a colorimetric β-galactosidase assay. *Appl. Environ. Microbiol.* 35:136–141.

236. Watkins, W. D., S. R. Rippey, C. S. Clavet, D. J. Kelley-Reitz, and W. Burkhardt III. 1988. Novel compound for identifying *Escherichia coli*. *Appl. Environ. Microbiol.* 54:1874–1875.

237. Weihe, J. L., S. L. Seist, and W. S. Hatcher, Jr. 1984. Estimation of microbial populations in frozen concentrated orange juice using automated impedance measurements. *J. Food Sci.* 49:243–245.

238. Weiss, L. H. and J. Humber. 1988. Evaluation of a 24-hour fluorogenic assay for the enumeration of *Escherichia coli* from foods. *J. Food Protect.* 51:766–769.

239. Wernars, K., C. J. Heuvelman, T. Chakraborty, and S. H. W. Notermans. 1991. Use of the polymerase chain reaction for direct detection of *Listeria monocytogenes* in soft cheese. *J. Appl. Bacteriol.* 70:121–126.

240. Williams, M. L. R. 1971. The limitations of the DuPont luminescence biometer in the microbiological analysis of foods. *Can. Inst. Food Technol. J.* 4:187–189.

241. Wilson, S. G., S. Chan, M. Deroo, M. Vera-Garcia, A. Johnson, D. Lane, and D. N. Halbert. 1990. Development of a colorimetric, second generation nucleic acid hybridization method for detection of *Salmonella* in foods and a comparison with conventional culture procedure. *J. Food Sci.* 55:1394–1398.

242. Wolcott, M. J. 1991. DNA-based rapid methods for the detection of foodborne pathogens. *J. Food Protect.* 54:387–401.

243. Wood, J. M., V. Lach, and B. Jarvis. 1977. Detection of food-associated microbes using electrical impedance measurements. *J. Appl. Bacteriol.* 43:14–15.
244. Zaadhof, K.-J., and G. Terplan. 1981. Der *Limulus*-Test—ein Verfahren zur Beurteilung der mikrobiologischen Qualität von Milch und Milchprodukten. *Deut. Molkereizeitung* 34:1094–1098.
245. Zayaitz, A. E. K., and R. A. Ledford. 1982. Proteolytic inactivation of thermonuclease activity of *Staphylococcus aureus* during recovery from thermal injury. *J. Food Protect.* 45:624–626.

7

Bioassay and Related Methods

After establishing the presence of pathogens or toxins in foods or food products, the next important concern is whether the organisms/toxins are biologically active. For this purpose, experimental animals are employed where feasible. When it is not feasible to use whole animals or animal systems, a variety of tissue culture systems have been developed that, by a variety of responses, provide information on the biological activity of pathogens or their toxic products. These bioassay and related tests are the methods of choice for some foodborne pathogens, and some of the principal ones are described in Table 7-1.

WHOLE-ANIMAL ASSAYS

Mouse Lethality

This method was first employed for foodborne pathogens around 1920 and continues to be an important bioassay method. To test for botulinal toxins in foods, appropriate extracts are made and portions are treated with trypsin (for toxins of nonproteolytic *Clostridium botulinum* strains). Pairs of mice are injected intraperitoneally (i.p.) with 0.5 ml of trypsin-treated and untreated preparations. Untreated preparations that have been heated for 10 min at 100°C are injected into a pair of mice. All injected mice are observed for 72 h for symptoms of botulism or death. Mice injected with the heated preparations should not die since the botulinal toxins are heat labile. Specificity in this test can be achieved by protecting mice with known botulinal antitoxin, and in a similar manner, the specific serologic type of botulinal toxin can be determined (see Chap. 20 for toxin types).

Mouse lethality may be employed for other toxins. Stark and Duncan (100) used the method for *Clostridium perfringens* enterotoxin. Mice were injected

166

TABLE 7-1.   Some Bioassay Models Used to Assess the Biological Activity of Various Foodborne Pathogens and/or Their Products

Organism	Toxin/Product	Bioassay Method	Sensitivity	Reference
A. hydrophila	Cytotoxic enterotoxin	Infant mouse intestines	~30 ng	2, 86
B. cereus	Diarrheagenic toxin	Monkey feeding		104
	Diarrheagenic toxin	Rabbit ileal loop		98
	Diarrheagenic toxin	Rabbit skin		37
	Diarrheagenic toxin	Guinea pig skin		36
	Diarrheagenic toxin	Mouse lethality		70
	Emetic toxin	Rhesus monkey emesis		66
C. jejuni	Viable cells	Adult mice	10^4 cells	6
	Viable cells	Chickens	90 cells	87
	Viable cells	Chickens	$10^3 - 10^6$ cells	91
	Viable cells	Neonatal mice		28
	Culture supernatants	Adult rat jejunal loops		27
	Enterotoxin	Rat ileal loop		51, 88
C. botulinum	A, B, E, F, G, toxins	Mouse lethality		4
C. perfringens A	Enterotoxin	Mouse lethality, LD_{50}	1.8 µg	31
	Enterotoxin	Mouse ileal loop, 90-min test	1.0 µg	114
	Enterotoxin	Rabbit ileal loop, 90-min test	6.25 µg	42
	Enterotoxin	Guinea pig skin (erythemal activity)	0.06–0.125 mg/ml	31, 100
Infant botulism	Endospores	7–12-day-old rats	1,500 spores	69
	Endospores	9-day-old mice	700 spores	101
	Endospores	Adult germ-free mice	10 spores	68
E. coli	LT	Rabbit ileal loop, 18-h test		89
	ST	Suckling mouse (fluid accumulation)		19, 32
	ST	Rabbit ileal loop, 6-h test		25
	ST_a	Suckling mouse		19
	ST_a	1–3-day-old piglets		10
	ST_b	Jejunal loop of pig		71
	ST_b	Weaned piglets, 7–9 weeks old		10
E. coli 0157:H7	ETEC	Mouse colonization		110
		Colonization		38

TABLE 7-1.   (Continued)

Organism	Toxin/Product	Bioassay Method	Sensitivity	Reference
Salmonella spp.	Heat-labile cytotoxin	Rabbit ileal loop (protein synthesis inhibition)		54
S. aureus	SEB	Skin of specially sensitized guinea pigs	0.1–1.0 pg	92
	All enterotoxins	Emesis in rhesus monkeys	5 µg/2–3 kg body wt.	67
	SEA, SEB	Emesis in suckling kittens	0.1, 0.5 µg/kg body wt.	5
V. parahaemolyticus	Broth cultures	Rabbit ileal loop; response in 50% animals	10^2 cells	108
	Viable cells	Adult rabbit ileal loop, invasiveness		8
	Thermostable direct toxin	Mouse lethality, death in 1 min	5 µg/mouse	43, 45
	Thermostable direct toxin	Mouse lethality, LD_{50} by IP route	1.5 µg	115
	Thermostable direct toxin	Rabbit ileal loop	250 µg	115
	Thermostable direct toxin	Guinea pig skin	2.5 µg/g	115
V. vulnificus	Culture filtrates	Rabbit skin permeability		7
V. cholerae (non-01)	Exterotoxin	Suckling mice		74
Y. enterocolitica	Heat-stable toxin	Sereny test		9
	Heat-stable toxin	Suckling mouse (oral)		9
	Heat-stable toxin	Suckling mouse (oral)	110 ng	77
	Enterotoxin	Rabbit ileal loop, 6- and 18-h tests		79
	Viable cells	Mouse diarrhea		93
	Viable cells	Rabbit diarrhea	50% infectious dose = 2.9×10^8	80
	Viable cells	Lethality in suckling mice by IP injection	14 cells	3
	Viable cells	Lethality of gerbils by IP injection	100 cells	94

Note: LT = heat labile toxin; ST = heat-stable toxin; SEA = staphylococcal enterotoxin A.

i.p. with enterotoxin preparations and observed for up to 72 h for lethality. The mouse-lethal dose was expressed as the reciprocal of the highest dilution that was lethal to the mice within 72 h. Genigeorgis et al. (31) employed the method by use of intravenous (i.v.) injections. *C. perfringens* enterotoxin preparations were diluted in phosphate buffer, pH 6.7, to achieve a concentration of 5 to 12 µg/ml. From each dilution prepared, 0.25 ml was injected i.v. into six male mice weighing 12 to 20 g, the number of deaths were recorded, and the LD_{50} was calculated. The mouse is the most widely used animal for virulence assessment of *Listeria* spp. The LD_{50} for *L. monocytogenes* in normal adult mice is 10^5-10^6, and for 15-g infant mice as few as 50 cells may be lethal (Chap. 21).

Suckling (Infant) Mouse

This animal model was introduced by Dean et al. (19) primarily for *Escherichia coli* enterotoxins and is now used for this and some other foodborne pathogens. Typically mice are separated from their mothers and given oral doses of the test material consisting of 0.05 to 0.1 ml with the aid of a blunt 23-gauge hypodermic needle. A drop of 5% Evans blue dye/ml of test material may be used to determine the presence of the test material in the small intestine. The animals are usually held at 25°C for 2 h and then sacrificed. The entire small intestine is removed, and the relative activity of test material is determined by the ratio of gut weight to body weight (GW/BW). Giannella (32) found the following GW/BW ratios for *E. coli* enterotoxins: <0.074 = negative test; 0.075 to 0.082 = intermediate (should be retested); and >0.083 = positive test. The investigator found the day-to-day variability among various *E. coli* strains to range from 10.5 to 15.7% and about 9% for replicate tests with the same strain. A GW/BW of 0.060 was considered negative for *E. coli* ST_a by Mullan et al. (72). In studies with *E. coli* ST, Wood et al. (113) treated as positive GW/BW ratios that were >0.087, while Boyce et al. (9) held mice at room temperature for 4 h for *Yersinia enterocolitica* heat-stable enterotoxin and considered GW/BW of 0.083 or greater to be positive. In studies with *Y. enterocolitica*, Okamoto et al. (78), keeping mice for 3 h at 25°C, considered a GW/BW of 0.083 to be positive.

In using the suckling mouse model, test material may also be injected percutaneously directly into the stomach through the mouse's translucent skin or by administration orogastrically or intraperitoneally. For the screening of large numbers of cultures, the intestines may be examined visually for dilation and fluid accumulation (81). Infant mice along with 1- to 3-day-old piglets are the animals of choice for *E. coli* enterotoxin ST_a, while ST_b is inactive in the suckling mouse but active in piglets and weaned pigs (10, 50). The infant mouse assay does not respond to choleragen or to the LT of *E. coli*. It correlates well with the 6-h rabbit ileal loop assay for the ST_a of *E. coli*.

Suckling mice have been used for lethality studies by employing i.p. injections. Aulisio et al. (3) used 1- to 3-day-old Swiss mice and injected 0.1 ml of diluted culture. The mice were observed for 7 days; deaths that occurred within 24 h were considered nonspecific, while deaths occurring between days 2 and 7 were considered specific for *Y. enterocolitica*. By this method, an LD_{50} can be calculated relative to numbers of cells/inoculum. In the case of *Y. enterocolitica*, Aulisio et al. found the LD_{50} to be 14 cells, and the average time for death of mice to be 3 days.

Rabbit and Mouse Diarrhea

Rabbits and mice have been employed to test for diarrheagenic activity of some foodborne pathogens. Employing young rabbits weighing 500 to 800 g,

Pai et al. (80) inoculated orogastrically with approximately 10^{10} cells of *Y. enterocolitica* suspended in 10% sodium bicarbonate. Diarrhea developed in 87% of 47 rabbits after a mean time of 5.4 days. Bacterial colonization occurred in all animals regardless of dose of cells.

Mice deprived of water for 24 h were employed by Schiemann (93) to test for the diarrheagenic activity of *Y. enterocolitica*. The animals were given inocula of 10^9 cells/ml in peptone water, and fresh drinking water was allowed 24 h later. After 2 days, feces of mice were examined for signs of diarrhea.

Infant rabbits have been used by Smith (95) to assay enterotoxins of *E. coli* and *Vibrio cholerae*. Infant rabbits 6 to 9 days old are administered 1 to 5 ml of culture filtrate via stomach tube. Following return to their mothers, they are observed for diarrhea. Diarrhea after 6 to 8 h is a positive response. If death of animals occurs, a large volume of yellow fluid is found in the small and large intestines. The quantitation of enterotoxin is achieved by ascertaining the ratio of intestinal weight to total body weight. Young pigs have been used in a similar way to assay porcine strains of *E. coli* for enterotoxin activity. Infant rabbits have been employed to detect Shiga-like toxins of *E. coli* (75).

Monkey Feeding

The use of rhesus monkeys (*Macaca mulatta*) to assay staphylococcal enterotoxins was developed in 1931 by Jordan and McBroom (49). Next to humans, this is perhaps the animal most sensitive to staphylococcal enterotoxins. When enterotoxins are to be assayed by this method, young rhesus monkeys weighing 2 to 3 kg are selected. The food homogenate, usually in solution in 50-ml quantities, is administered via stomach tube. The animals are then observed continuously for 5 h. Vomiting in at least two of six animals denotes a positive response. Rhesus monkeys have been shown to respond to levels of enterotoxins A and B as low as approximately 5 µg/2–3 kg body weight (67).

Kitten (Cat) Test

This method was developed by Dolman et al. (22) as an assay for staphylococcal enterotoxins. The original test employed the injection of filtrates into the abdominal cavity of very young kittens (250–500 g). This procedure leads to false positive results. The most commonly used method consists of administering the filtrates i.v. and observing the animals continuously for emesis. When cats weighing 2 to 4 kg are used, positive responses occur in 2 to 6 h (15). Emesis has been reported to occur with 0.1 and 0.5 µg of staphylococcal enterotoxin A (SEA) and SEB/kg body weight (5). The test tends to lack the specificity of the monkey-feeding test since staphylococcal culture filtrates containing other by-products may also induce emesis. Kittens are much easier to obtain and maintain than rhesus monkeys, and in this regard the test has value.

Rabbit and Guinea Pig Skin Tests

The skin of these two animals is used to assay toxins for at least two properties. The **vascular permeability** test is generally done by use of albino rabbits weighing 1.5 to 2.0 kg. Typically, 0.05 to 0.1 ml of culture filtrate is inoculated intradermally (i.d.) in a shaved area of the rabbit's back and sides. From 2 to 18 h later, a solution of Evans blue dye is administered i.v., and 1 to 2 h are allowed for permeation by the dye. The diameters of two zones of blueing are measured and the area approximated by squaring the average of the two values. Areas of $25\,cm^2$ are considered positive. *E. coli* LT gives a positive response in this assay (26). Employing this assay, permeability has been shown to be a function of the *E. coli* diarrheagenic enterotoxin.

Similar to the permeability factor test is a test of **erythemal activity** that employs guinea pigs. The method has been employed by Stark and Duncan (100) to test for erythemal activity of *C. perfringens* enterotoxin. Guinea pigs weighing 300 to 400 g are depilated (back and sides) and marked in 2.5-cm squares, and duplicate 0.05-ml samples of toxic preparations are injected i.d. in the center of the squares. Animals are observed after 18 to 24 h for erythema at the injection site. In the case of *C. perfringens* enterotoxin, a concentric area of erythema is produced without necrosis. A unit of erythemal activity is defined as the amount of enterotoxin producing an area of erythema 0.8 cm in diameter. The enterotoxin preparation used by Stark and Duncan contained 1,000 erythemal units/ml. To enhance readings, 1 ml of 0.5% Evans blue can be injected intracardially (IC) 10 min following the skin injections and the diameters read 80 min later (31). The specificity of the skin reactions can be determined by neutralizing the enterotoxin with specific antisera prior to injections. The erythema test was found to be 1,000 times more sensitive than the rabbit ileal loop technique for assaying the enterotoxin of *C. perfringens* (41).

Sereny and Anton Tests

The Sereny method is used to test for virulence of viable bacterial cultures. It was proposed by Sereny in 1955, and the guinea pig is the animal most often used. The test consists of administering with the aid of a loop a drop of cell suspension, containing 1.5×10^{10} to 2.3×10^{10}/ml in phosphate-buffered saline, into the conjunctivae of guinea pigs weighing about 400 g each. The animal's eyes are examined daily for 5 days for evidence of **keratoconjunctivitis**. When strains of unknown virulence are evaluated, it is important that known positive and negative strains are tested also.

A mouse Sereny test has been developed using Swiss mice and administering half of the dose noted above. A Sereny test for shigellae and enteroinvasive *E. coli* (EIEC) strains has been developed and found useful (73).

The **Anton test** is similar to Sereny; it is used to assess the virulence of *Listeria* spp. Conjunctivitis is produced when about 10^6 cells of *L. monocytogenes* are administered into the eye of a rabbit or guinea pig (1).

ANIMAL MODELS REQUIRING SURGICAL PROCEDURES

Ligated Loop Techniques

These techniques are based on the fact that certain enterotoxins elicit fluid accumulation in the small intestines of susceptible animals. While they may be performed with a variety of animals, rabbits are most often employed. Young rabbits 7 to 20 weeks old and weighing 1.2 to 2.0 kg are kept off food and water for a period of 24 h, or off food for 48 to 72 h with water ad libitum prior to surgery. Under local anesthesia, a midline incision about 2 in. long is made just below the middle of the abdomen through the muscles and peritoneum in order to expose the small intestines (18). A section of the intestine midway between its upper and lower ends or just above the appendix is tied with silk or other suitable ligatures in 8 to 12-cm segments with intervening sections of at least 1 cm or more. Up to six sections may be prepared by single or double ties.

Meanwhile, the specimen or culture to be tested is prepared, suspended in sterile saline, and injected intraluminally into the ligated segments. A common inoculum size is 1 ml, although smaller or larger doses may be used. Different doses of test material may be injected into adjacent loops or into loops separated by a blank loop or by a sham (inoculated with saline). Following injection, the abdomen is closed with surgical thread, and the animal is allowed to recover from anesthesia. The recovered animal may be kept off food and water for an additional 18 to 24 h or water or feed or both may be allowed. With ligatures intact, the animals may not survive beyond 30 to 36 h (11).

To assess the effect of the materials previously injected into ligated loops, the animal is sacrificed, and the loops are examined and measured for fluid accumulation. The fluid may be aspirated and measured. The reaction can be quantitated by measuring loop fluid volume to loop length ratios (11) or by determining the ratio of fluid volume secreted/mg dry weight intestine (65). The appearance of a ligated rabbit ileum 24 h after injection of a *C. perfringens* culture is presented in Figure 7-1. The minimum amount of *C. perfringens* enterotoxin necessary to produce a loop reaction has been reported variously to be 28 to 40 µg and as high as 125 µg of toxin by the standard loop technique. The 90-min loop technique has been found to respond to as little as 6.25 µg and the standard technique to 29 µg of toxin (31).

This technique was developed to study the mode of action of the cholera organism in producing the disease (18). It has been employed widely in studies

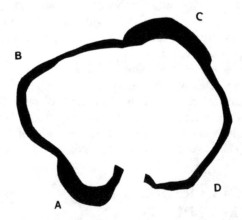

FIGURE 7-1.   Gross appearance of the ligated rabbit ileum 24 h after injection of a 2 ml culture of *Clostridium perfringens* grown 4 h at 37°C in skim milk. Loop A, strain NCTC 8798, 8 ml of fluid in loop; loop B, 2 ml of sterile milk, negative loop; loop C, strain T-65, 10 ml of fluid; loop D, strain 6867, negative loop.

Redrawn from Duncan et al. (24), copyright © 1968, American Society for Microbiology.

on the virulence and pathogenesis of foodborne pathogens, including *Bacillus cereus, C. perfringens, E. coli*, and *Vibrio parahaemolyticus*.

Although the rabbit loop is the most widely used of ligated loop methods, other animal models are used. The **mouse intestinal loop** may be used for *E. coli* enterotoxins. As used by Punyashthiti and Finkelstein (84), Swiss mice 18 to 22 g are deprived of food 8 h before use. The abdomen is opened under light anesthesia, and two 6-cm loops separated by 1-cm interloops are prepared. The loops are inoculated with 0.2 ml of test material followed by closing of the abdomen. Animals are deprived of food and water and killed 8 h later. Fluid is measured and the length of the loops determined. Results are considered positive when the ratio of fluid to length is 50 or more mg/cm. In this study, positive loops generally had ratios between 50 and 100 but occasionally approached 200 or more. Alternatively, the net increase in weight of loops in mg can be used to measure the intensity of a toxic reaction (114). With the mouse loop, 1 μg of enterotoxin can be detected (114). A rat jegunal loop assay has been presented for detecting the ST_b enterotoxin of *E. coli*. A linear dose response was found using 250 to 350 g rats but only after the endogenous protease activity was blocked with soybean trypsin inhibitor (111).

The RITARD Model

The removable intestinal tie-adult rabbit diarrhea (RITARD) method was developed by Spira et al. (99). Rabbits weighing 1.6 to 2.7 kg are kept off food for 24 h but allowed water. Under local anesthesia, the cecum is brought out and ligated close to the ileocecal junction. The small intestine is now brought out and a slip knot tied to close it in the area of the mesoappendix. Test material in 10 ml of phosphate-buffered saline is injected into the lumen of the anterior jejunum. After injection, intestine and cecum are returned to the peritoneal cavity and the incision closed. With the animal kept in a box, the temporary tie is removed 2 to 4 h after administration of the test dose, and the slip knot in the intestine is released. Sutures are applied as needed. The animal is now returned to its cage and provided with food and water. Animals are observed for diarrhea or death at 2-h intervals up to 124 h. At autopsy, small intestine and adjacent sections are tied and removed for fluid measurement. Enterotoxigenic strains of *E. coli* produce severe and watery diarrhea, and the susceptibility of animals to *V. cholerae* infections is similar in this system to that in the infant rabbit model.

The gist of the RITARD model is that the animals are not altered except that the cecum is ligated to prevent it from taking up fluid from the small intestine, and a temporary reversible obstruction is placed on the ileum long enough to allow the inoculated organism to initiate colonization of the small intestine. The method has been successfully used as an animal model for *Campylobacter jejuni* infection (13) and to test virulence of *Aeromonas* strains (83).

CELL CULTURE SYSTEMS

A variety of cell culture systems are employed to assess certain pathogenic properties of viable cells. The properties often assessed are invasiveness,

TABLE 7-2.   Tissue and Cell Culture Systems Employed to Study Biological Activity of Gastroenteritis-causing Organisms or Their Products

Culture system	Pathogen/Toxin	Demonstration/Use	Reference
CHO monolayer	*E. coli* LT; *V. cholerae* toxin	Biological activity	39
	V. parahaemolyticus	Biological activity	44
	Salmonella toxin	Biological activity	90
	C. jejuni enterotoxin	Biological activity	51, 88
CHO floating cell assay	*Salmonella* toxin	Biological activity	46
HeLa cells	*E. coli*	Invasiveness	4
	Y. enterocolitica	Invasiveness	4, 57, 93, 109
	V. parahaemolyticus	Adherence	14, 48
	C. jejuni	Invasiveness	59
Vero	*E. coli* 0157: H7	Shiga-like toxin receptors	(53)
Vero cells	*C. perfringens* enterotoxin	Mode of action	60
	E. coli LT	Biological activity, assay	96, 106
	A. hydrophila toxin	Cytotoxicity	2
	C. perfringens enterotoxin	Binding	61, 62
	C. perfringens enterotoxin	Biological activity	61, 58
	Salmonella cytotoxin	Protein synthesis inhib.	54
	V. vulnificus	Cytotoxicity	7
Y-1 adrenal cells	*E. coli* LT	Biological activity, assay	23, 89
	V. cholerae toxin	Biological activity, assay	23, 89
	V. mimicus	Biological activity	97
Rabbit int. epith. cells	*C. perfringens* enterotoxin	Binding	61
	Salmonella cytotoxin	Protein synthesis inhib.	54
Murine spleen cells	Staph. enterotoxins A, B, and E	Binding	12
Macrophages	*Y. enterocolitica*	Phagocytosis	109
Human peripheral lymphocytes	Staph. enterotoxin A	Biological effects	55
Human laryngeal carcinoma	*E. coli*, *Shigella*	Invasiveness	4
Henle 407 human intestine	*E. coli*, *Shigella*	Invasiveness	4
Henle 407	*L. monocytogenes*	Invasiveness	—
	E. coli 0157: H7	Adherence	—
Caco-2	*V. cholerae* non-01	Adherence	82
	ETEC	Adhesins	16
	L. monocytogenes	Invasion	30
HT29.74	*C. parvum*	Infection model	29
	C. perfringens	Cell lethality	58
Peritoneal macrophages	*L. monocytogenes*	Intracellular survival	—
Murine embryo primary fibroblasts	*L. monocytogenes*	Interleukin production	—
Human fetal intestinal cells	*V. parahaemolyticus*	Adherence	14, 40
	Enteropathogenic *E. coli*	Adherence	64
	B. cereus toxins	Biological activity	107
Human intestinal cells	*V. parahaemolyticus*	Adherence	35, 48
Human ileal cells	Enterotoxigenic *E. coli*	Adherence	20
Human mucosal cells	*E. coli*	Adherence	76, 105
	V. parahaemolyticus	Adherence	85
Human uroepithelial cells	*E. coli*	Adhesion	102
Viable human duodenal biopsies	*E. coli*	Adherence	52
Rat hepatocytes	*C. perfringens* enterotoxin	Amino acid transport	33, 34
	C. perfringens enterotoxin	Membrane permeability	17
Guinea pig intestinal cells	*V. parahaemolyticus*	Adherence	48

permeability, cytotoxicity, adherence/adhesion/binding, and other more general biological activities. Some cell cultures are used to assess various properties of toxins and enterotoxins. Some examples of these models are summarized in Table 7-2, and brief descriptions are presented below.

Human Mucosal Cells

As employed by Ofek and Beachey (76), human buccal mucosa cells (about 2×10^5 in phosphate-buffered saline) are mixed with 0.5 ml of washed E. coli cells—2×10^8/ml. The mixture is rotated for 30 min at room temperature. Epithelial cells are separated from the bacteria by differential centrifugation, followed by drying and staining with gentian violet. Adherence is determined by microscopic counting of bacteria/epithelial cell. As employed by Thorne et al. (105), E. coli cells are labeled with ^{3}H-amino acids (alanine and leucine) or fluorescein isothiocyanate. In their use of this method, Reyes et al. (85) mixed V. parahaemolyticus cells with mucosal epithelial cells and incubated at 37°C for 5 min followed by filtering. The unbound cells are washed off, and the culture is dried, fixed, and stained with Giemsa. Adherence is quantitated by counting the total number of V. parahaemolyticus adhering to 50 buccal cells as compared to controls. Best results are obtained when approximately 10^9 bacterial cells and 10^5 buccal cells were suspended together in phosphate-buffered saline at pH 7.2 for 5 min. All 12 strains tested adhered. Adherence apparently bears no relationship to pathogenicity for V. parahaemolyticus.

Human Fetal Intestine (HFI)

By this adherence model, HFI cells are employed in monolayers. The monolayers are thoroughly washed, inoculated with a suspension of V. parahaemolyticus, and incubated at 37°C for up to 30 min. Adherence is determined by the microscopic examination of stained cells after washing away unattached bacteria. All strains of V. parahaemolyticus tested adhered, but those from food-poisoning cases have a higher adherence ability than those from foods (40). By use of this method, the adherence of an enteropathogenic strain of E. coli of human origin has been found to be plasmid mediated (112).

Human Ileal and Intestinal Cells

To study adherence of enterotoxigenic E. coli (ETEC), Deneke et al. (20) used ileal cells from adult humans in a filtration-binding assay. The cells were mixed with bacteria grown in ^{3}H-alanine and leucine. The amount of binding was determined with a scintillation counter. ETEC strains of human origin bound to a greater extent than controls. Binding to human ileal cells was 10- to 100-fold greater than to human buccal cells.

Monolayers of human intestine cells were employed by Gingras and Howard (35) to study adherence of V. parahaemolyticus. The bacterium was grown in the presence of ^{14}C-labeled valine, and the labeled cells were added to monolayers and incubated for up to 60 min. Following incubation, unattached cells were removed, and those adhering were counted by radioactive counts of monolayers. The adhered cells were also enumerated microscopically. The

Kanagawa-positive and -negative organisms adhered similarly. No correlation was found between hemolysis production and adherence.

Guinea Pig Intestinal Cells

To study adherence of *V. parahaemolyticus*, Iijima et al. (48) employed adult guinea pigs weighing about 300 g and fasted them for 2 days before use. Under anesthesia, the abdomen was opened and the small intestine tied approximately 3 cm distal from the stomach. The intestine was injected with 1.0 ml of a suspension of 2×10^8 cells of adherence-positive and adherence-negative strains, followed by closing of the abdomen. Six h later, the animals were sacrificed and the small intestine removed and cut into four sections. Following homogenization with 3% NaCl, the number of cells in the homogenate was determined by plating. With adherence-positive cells, larger numbers were found in the homogenates, especially in the upper section of the intestine.

Another adhesion model consists of immobilizing soluble mucosal glycoproteins from mouse intestines on polystyrene (56). Using this model, it was shown that two plasmid-bearing strains of *E. coli* (K88 and K99) adhered readily, as do other adhering strains of this organism.

HeLa Cells

This cell line is widely used to test for the invasive potential of intestinal pathogens as well as for adherence. Although HeLa cells seem to be preferred, other cell lines such as **human laryngeal carcinoma** and **Henle 407** human intestine may be employed. In general, monolayers of cells are prepared by standard culture techniques on a chamber slide and inoculated with 0.2 ml of a properly prepared test culture suspension. Following incubation for 3 h at 35°C to allow for bacterial growth, monolayer cells are washed, fixed, and stained for viewing under the light microscope. In the case of invasive *E. coli*, cells will be present in the cytoplasm of monolayer cells but not in the nucleus. In addition, invasive strains are phagocytized to a greater extent than noninvasives and the number of bacteria/cell is >5. According to *BAM* (4), at least 0.5% of the HeLa cells should contain no less than five bacteria. Positive responses to this test are generally confirmed by the Sereny test (see 4).

A modification is used for invasive *Yersinia*. By this method, 0.2 ml of a properly prepared bacterial suspension is inoculated into chamber slides containing the HeLa cell monolayer. Following incubation for 1.5 h at 35°C, the cells are washed, fixed, and stained for microscopic examination. Invasive *Y. enterocolitica* are present in the cytoplasm—usually in the phagolysosome. Infectivity rates are generally greater than 10%. While invasive *E. coli* are confirmed by the Sereny test, this is not done with *Y. enterocolitica*, even though invasive, since this organism may not yield a positive Sereny test.

HeLa cells have been used to test for adherence of *V. parahaemolyticus* and to study the penetration of *Y. enterocolitica*. Strains of the latter that gave an index of 3.7 to 5.0 were considered penetrating (93). The infectivity of HeLa cells by *Y. enterocolitica* has been studied by use of cell monolayers in roller tubes. The number of infecting bacterial cells is counted at random in 100 stained HeLa cells for up to 24 h (21).

Chinese Hamster Ovary (CHO) Cells

The CHO assay was developed by Guerrant et al. (39) for *E. coli* enterotoxins and employs CHO cells grown in a medium containing fetal calf serum. Upon establishment of a culture of cells, enterotoxin is added. Microscopic examinations are made 24 to 30 h later to determine whether cells have become bipolar and elongated at least three times their width and whether their knoblike projections have been lost. The morphological changes in CHO cells caused by both cholera toxin and *E. coli* enterotoxin have been shown to parallel the elevation of cyclic AMP. It has been found to be 100 to 10,000 times more sensitive than skin permeability and ileal loop assays for *E. coli* enterotoxins. For the LT of *E. coli*, CHO has been found to be 5 to 100 times more sensitive than skin permeability and rabbit ileal loop assays (39).

Vero Cells

This monolayer consists of a continuous cell line derived from African green monkey kidneys; it was employed by Speirs et al. (96) to assay for *E. coli* LT. Vero cell results compare favorably with Y-1 adrenal cells (see below), and the test was found by these authors to be the simple and more economical of the two to maintain in the laboratory. Toxigenic strains produce a morphological response to Vero cells similar to Y-1 cells.

A highly sensitive and reproducible biological assay for *C. perfringens* enterotoxin employing Vero cells was developed by McDonel and McClane (63). The assay is based on the observation that the enterotoxin inhibits plating efficiency of Vero cells grown in culture. The inhibition of plating efficiency detected as little as 0.1 ng of enterotoxin, and a linear dose-response curve was obtained with 0.5 to 5 ng (5 to 50 ng/ml). The authors proposed a new unit of biological activity—the plating efficiency unit (PEU)—as that amount of enterotoxin that causes a 25% inhibition of the plating of 200 cells inoculated into 100 μl of medium.

Y-1 Adrenal Cell Assay

In this widely used assay, mouse adrenal cells (Y-1) are grown in a monolayer using standard cell culture techniques. With monolayer cells in microtiter plate wells, test extracts or filtrates are added to the microtiter wells followed by incubation at 37°C. In testing *E. coli* LT, heated and unheated culture filtrates of known positive and negative LT-producing strains are added to monolayers in microtiter plates and results are determined by microscopic examinations. The presence of 50% or more rounded cells in monolayers of unheated filtrates and 10% or less for heated filtrates denotes a positive response. The specificity of the response can be determined by the use of specific antibodies in toxin-containing filtrates. Details of this method for foodborne pathogens are presented in *BAM* (4).

Other Assays

An immunofluorescence method was employed by Boutin et al. (8) using 6-week-old rabbit ileal loops inoculated with *V. parahaemolyticus*. The loops were removed 12 to 18 h after infection and placed in trays, cut into tissue sections, and cleaned by agitation. Tissue sections were fixed and stained with fluorescein isothiocyanate-stained agglutinins to *V. parahaemolyticus*. The reaction of the tagged antibody with *V. parahaemolyticus* cells in the tissue was assessed microscopically. By use of immunofluorescence, it was possible to demonstrate the penetration by this organism into the lamina propria of the ileum and thus the tissue invasiveness of the pathogen. Both Kanagawa-positive and -negative cells penetrated the lamina according to this method.

The chorioallantoic membrane of 10-day-old chick embryos was used to assess the pathogenicity of *Listeria* spp. *L. monocytogenes* causes death within 2 to 5 days with as few as 100 cells, and *L. ivanovii* cells are lethal at levels of 100 to 30,000/egg with death within 72 h (103). Culture filtrates of *L. monocytogenes* and *L. ivanovii* release lactate dehydrogenase (LDH) in rat hepatocyte monolayers following 3-h exposures, but other listerial species had no effect (47).

References

1. Anton, W. 1934. Kritisch-experimenteller Beitrag zur Biologie des *Bacterium monocytogenes*. Mit besonderer Berucksichtigung seiner Beziehung zue infektiosen Mononucleose des Menschen. *Zbt. Bakteriol. Abt. I. Orig.* 131:89–103.
2. Asao, T., Y. Kinoshita, S. Kozaki, T. Uemura, and G. Sakaguchi. 1984. Purification and some properties of *Aeromonas hydrophila* hemolysin. *Infect. Immun.* 46:122–127.
3. Aulisio, C. C. G., W. E. Hill, J. T. Stanfield, and J. A. Morris. 1983. Pathogenicity of *Yersinia enterocolitica* demonstrated in the suckling mouse. *J. Food Protect.* 46:856–860.
4. *Bacteriological Analytical Manual.* 1978. Washington, D.C.: U.S. Food and Drug Administration.
5. Bergdoll, M. S. 1972. The enterotoxins. In *The Staphylococci*, ed. J. O. Cohen, 301–331. New York: Wiley.
6. Blaser, M. J., D. J. Duncan, G. H. Warren, and W.-L. L. Wang. 1983. Experimental *Campylobacter jejuni* infection of adult mice. *Infect. Immun.* 39:908–916.
7. Boutin, B. K., A. L. Reyes, and R. M. Twedt. 1984. Toxicity testing of *Vibrio vulnificus* culture filtrates in the rabbit back permeability model and five cell culture lines. *Bacteriol. Proc.*, 200.
8. Boutin, B. K., S. F. Townsend, P. V. Scarpino, and R. M. Twedt. 1979. Demonstration of invasiveness of *Vibrio parahaemolyticus* in adult rabbits by immunofluorescence. *Appl. Environ. Microbiol.* 37:647–653.
9. Boyce, J. M., D. J. Evans, Jr., D. G. Evans, and H. L. DuPont. 1979. Production of heat-stable, methanol-soluble enterotoxin by *Yersinia enterocolitica*. *Infect. Immun.* 25:532–537.
10. Burgess, M. N., R. J. Bywater, C. M. Cowley, N. A. Mullan, and P. M. Newsome. 1978. Biological evaluation of a methanol-soluble, heat-stable *Escherichia coli* enterotoxin in infant mice, pigs, rabbits, and calves. *Infect. Immun.* 21:526–531.
11. Burrows, W., and G. M. Musteikis. 1966. Cholera infection and toxin in the rabbit ileal loop. *J. Infect. Dis.* 116:183–190.
12. Buxser, S., P. F. Bonventre, and D. L. Archer. 1981. Specific receptor binding of staphylococcal enterotoxins by murine splenic lymphocytes. *Infect. Immun.* 33:827–833.
13. Caldwell, M. B., R. I. Walker, S. D. Stewart, and J. E. Rogers. 1983. Simple

adult rabbit model for *Campylobacter jejuni* enteritis. *Infect. Immun.* 42: 1176–1182.

14. Carruthers, M. M. 1977. In vitro adherence of Kanagawa-positive *Vibrio parahaemolyticus* to epithelial cells. *J. Infect. Dis.* 136:588–592.

15. Clark, W. G., and J. S. Page. 1968. Pyrogenic reponses to staphylococcal enterotoxins A and B in cats. *J. Bacteriol.* 96:1940–1946.

16. Darfeuille-Michaud, A., D. Aubel, G. Chauviere, C. Rich, M. Bourges, A. Servin, and B. Joly. 1990. Adhesion of enterotoxigenic *Escherichia coli* to the human colon carcinoma cell line Caco-2 in culture. *Infect. Immun.* 58:893–902.

17. Dasgupta, B. R., and M. W. Pariza. 1982. Purification of two *Clostridium perfringens* enterotoxin-like proteins and their effects on membrane permeability in primary cultures of adult rat hepatocytes. *Infect. Immun.* 38:592–597.

18. De, S. N., and D. N. Chatterje. 1953. An experimental study of the mechanism of action of *Vibrio cholerae* on the intestinal mucous membrane. *J. Path. Bacteriol.* 66:559–562.

19. Dean, A. G., Y.-C. Ching, R. G. Williams, and L. B. Harden. 1972. Test for *Escherichia coli* enterotoxin using infant mice: Application in a study of diarrhea in children in Honolulu. *J. Infect. Dis.* 125:407–411.

20. Deneke, C. F., K. McGowan, G. M. Thorne, and S. L. Gorbach. 1983. Attachment of enterotoxigenic *Escherichia coli* to human intestinal cells. *Infect. Immun.* 39:1102–1106.

21. Devenish, J. A., and D. A. Schiemann. 1981. HeLa cell infection by *Yersinia enterocolitica*: Evidence for lack of intracellular multiplication and development of a new procedure for quantitative expression of infectivity. *Infect. Immun.* 32: 48–55.

22. Dolman, C. E., R. J. Wilson, and W. H. Cockroft. 1936. A new method of detecting *Staphylococcus* enterotoxin. *Can. J. Pub. Hlth* 27:489–493.

23. Donta, S. T., H. W. Moon, and S. C. Whipp. 1974. Detection of heat-labile *Escherichia coli* enterotoxin with the use of adrenal cells in tissue culture. *Science* 183:334–336.

24. Duncan, C. L., H. Sugiyama, and D. H. Strong. 1968. Rabbit ileal loop response to strains of *Clostridium perfringens*. *J. Bacteriol.* 95:1560–1566.

25. Evans, D. G., D. J. Evans, and N. F. Pierce. 1973. Differences in the response of rabbit small intestine to heat-labile and heat-stable enterotoxins of *Escherichia coli*. *Infect. Immun.* 7:873–880.

26. Evans, D. J., Jr., D. G. Evans, and S. L. Gorbach. 1973. Production of vascular permeability factor by enterotoxigenic *Escherichia coli* isolated from man. *Infect. Immun.* 8:725–730.

27. Fernandez, H., U. F. Neto, F. Fernandes, M. D. A. Pedra, and L. R. Trabulsi. 1983. Culture supernatants of *Campylobacter jejuni* induce a secretory response in jejunal segments of adult rats. *Infect. Immun.* 40:429–431.

28. Field, L. H., J. L. Underwood, L. M. Pope, and L. J. Berry. 1981. Intestinal colonization of neonatal animals by *Campylobacter fetus* subsp. *jejuni*. *Infect. Immun.* 33:884–892.

29. Flanigan, T. P., T. Aji, R. Marshall, R. Soave, M. Aikawa, and C. Kaetzel. 1991. Asexual development of *Cryptosporidium parvum* within a differentiated human enterocyte cell line. *Infect. Immun.* 59:234–239.

30. Gaillard, J.-L., P. Berche, J. Mounier, S. Richard, and P. Sansonetti. 1987. In vitro model of penetration and intracellular growth of *Listeria monocytogenes* in the human enterocyte-like cell line Caco-2. *Infect. Immun.* 55:2822–2829.

31. Genigeorgis, C., G. Sakaguchi, and H. Riemann. 1973. Assay methods for *Clostridium perfringens* type A enterotoxin. *Appl. Microbiol.* 26:111–115.

32. Giannella, R. A. 1976. Suckling mouse model for detection of heat-stable

Escherichia coli enterotoxin: Characteristics of the model. *Infect. Immun.* 14: 95–99.

33. Giger, O., and M. W. Pariza. 1978. Depression of amino acid transport in cultured rat hepatocytes by purified enterotoxin from *Clostridium perfringens*. *Biochem. Biophys. Res. Comm.* 82:378–383.

34. Giger, O., and M. W. Pariza. 1980. Mechanism of action of *Clostridium perfringens* enterotoxin. Effects on membrane permeability and amino acid transport in primary cultures of adult rat hepatocytes. *Biochim. Biophys. Acta* 595:264–276.

35. Gingras. S. P., and L. V. Howard. 1980. Adherence of *Vibrio parahaemolyticus* to human epithelial cell lines. *Appl. Environ. Microbiol.* 39:369–371.

36. Glatz, B. A., and J. M. Goepfert. 1973. Extracellular factor synthesized by *Bacillus cereus* which evokes a dermal reaction in guinea pigs. *Infect. Immun.* 8:25–29.

37. Glatz, B. A., W. M. Spira, and J. M. Goepfert. 1974. Alteration of vascular permeability in rabbits by culture filtrates of *Bacillus cereus* and related species. *Infect. Immun.* 10:299–303.

38. Goldhar, J., A. Zilberberg, and J. Ofek. 1986. Infant mouse model of adherence and colonization of intestinal tissues by enterotoxigenic strains of *Escherichia coli* isolated from humans. *Infect. Immun.* 52:205–208.

39. Guerrant, R. L., L. L. Brunton, T. C. Schaitman, L. L. Rebhun, and A. G. Gilman, 1974. Cyclic adenosine monophosphate and alteration of Chinese hamster ovary cell morphology: A rapid, sensitive in vitro assay for the enterotoxins of *Vibrio cholerae* and *Escherichia coli*. *Infect. Immun.* 10:320–327.

40. Hackney, C. R., E. G. Kleeman, B. Ray, and M. L. Speck. 1980. Adherence as a method for differentiating virulent and avirulent strains of *Vibrio parahaemolyticus*. *Appl. Environ. Microbiol.* 40:652–658.

41. Hauschild, A. H. W. 1970. Erythemal activity of the cellular enteropathogenic factor of *Clostridium perfringens* type A. *Can. J. Microbiol.* 16:651–654.

42. Hauschild, A. H. W., R. Hilsheimer, and C. G. Rogers. 1971. Rapid detection of *Clostridium perfringens* enterotoxin by a modified ligated intestinal loop technique in rabbits. *Can. J. Microbiol.* 17:1475–1476.

43. Honda, T., K. Goshima, Y. Takeda, Y. Sugino, and T. Miwatani. 1976. Demonstration of the cardiotoxicity of the thermostable direct hemolysin (lethal toxin) produced by *Vibrio parahaemolyticus*. *Infect. Immun.* 13:163–171.

44. Honda, T., M. Shimizu, Y. Takeda, and T. Miwatani. 1976. Isolation of a factor causing morphological changes in Chinese hamster ovary cells from the culture filtrate of *Vibrio parahaemolyticus*. *Infect. Immun.* 14:1028–1033.

45. Honda, T., S. Taga, T. Takeda, M. A. Hasibuan, Y. Takeda, and T. Miwatani. 1976. Identification of lethal toxin with the thermostable direct hemolysin produced by *Vibrio parahaemolyticus*, and some physicochemical properties of the purified toxin. *Infect. Immun.* 13:133–139.

46. Houston, C. W., F. C. W. Koo, and J. W. Peterson. 1981. Characterization of *Salmonella* toxin released by mitomycin C-treated cells. *Infect. Immun.* 32: 916–926.

47. Huang, J. C., H. S. Huang, M. Jurima-Romet, F. Ashton, and B. H. Thomas. 1990. Hepatocidal toxicity of *Listeria* species. *FEMS Microbiol. Lett.* 72:249–252.

48. Iijima, Y., H. Yamada, and S. Shinoda. 1981. Adherence of *Vibrio parahaemolyticus* and its relation to pathogenicity. *Can. J. Microbiol.* 27:1252–1259.

49. Jordan, E. O., and J. McBroom. 1931. Results of feeding *Staphylococcus* filtrates to monkeys. *Proc. Soc. Exp. Biol. Med.* 29:161–162.

50. Kennedy, D. J., R. N. Greenberg, J. A. Dunn, R. Abernathy, J. S. Ryerse, and R. L. Guerrant. 1984. Effects of *Escherichia coli* heat-stable enterotoxin ST_b on

intestines of mice, rats, rabbits, and piglets. *Infect. Immun.* 46:639–641.

51. Klipstein, F. A., and R. F. Engert. 1984. Properties of crude *Campylobacter jejuni* heat-labile enterotoxin. *Infect. Immun.* 45:314–319.

52. Knutton, S., D. R. Lloyd, D. C. A. Candy, and A. S. McNeish. 1984. In vitro adhesion of enterotoxigenic *Escherichia coli* to human intestinal epithelial cells from mucosal biopsies. *Infect. Immun.* 44:514–518.

53. Konowalchuk, J., J. I. Speirs, and S. Stavric. 1977. Vero response to a cytotoxin of *Escherichia coli*. *Infect. Immun.* 18:775–779.

54. Koo, F. C. W., J. W. Peterson, C. W. Houston, and N. C. Molina. 1984. Pathogenesis of experimental salmonellosis: Inhibition of protein synthesis by cytotoxin. *Infect. Immun.* 43:93–100.

55. Langford, M. P., G. J. Stanton, and H. M. Johnson. 1978. Biological effects of staphylococcal enterotoxin A on human peripheral lymphocytes. *Infect. Immun.* 22:62–68.

56. Laux, D. C., E. F. McSweegan, and P. S. Cohen. 1984. Adhesion of enterotoxigenic *Escherichia coli* to immobilized intestinal mucosal preparations: A model for adhesion to mucosal surface components. *J. Microbiol. Meth.* 2:27–39.

57. Lee, W. H., P. McGrath, P. H. Carter, and E. L. Eide. 1977. The ability of some *Yersinia enterocolitica* strains to invade HeLa cells. *Can. J. Microbiol.* 23:1714–1722.

58. Mahony, D. E., E. Gilliatt, S. Dawson, E. Stockdale, and S. H. S. Lee. 1989. Vero cell assay for rapid detection of *Clostridium perfringens* enterotoxin. *Appl. Environ. Microbiol.* 55:2141–2143.

59. Manninen, K. I., J. F. Prescott, and I. R. Dohoo. 1982. Pathogenicity of *Campylobacter jejuni* isolates from animals and humans. *Infect. Immun.* 38:46–52.

60. McClane, B. A., and J. L. McDonel. 1980. Characterization of membrane permeability alterations induced in Vero cells by *Clostridium perfringens* enterotoxin. *Biochim. Biophys. Acta* 600:974–985.

61. McDonel, J. L. 1980. Binding of *Clostridium perfringens* (^{125}I)enterotoxin to rabbit intestinal cells. *Biochem.* 19:4801–4807.

62. McDonel, J. L., and B. A. McClane. 1979. Binding versus biological activity of *Clostridium perfringens*. *Biochem. Biophys. Res. Comm.* 87:497–504.

63. McDonel, J. L., and B. A. McClane. 1981. Highly sensitive assay for *Clostridium perfringens* enterotoxin that uses inhibition of plating efficiency of Vero cells grown in culture. *J. Clin. Microbiol.* 13:940–946.

64. McNeish, A. S., P. Turner, J. Fleming, and N. Evans. 1975. Mucosal adherence of human enteropathogenic *Escherichia coli*. *Lancet* 2:946–948.

65. Mehlman, I. J., M. Fishbein, S. L. Gorbach, A. C. Sanders, E. L. Eide, and J. C. Olson, Jr. 1976. Pathogenicity of *Escherichia coli* recovered from food. *J. Assoc. Off. Anal. Chem.* 59:67–80.

66. Melling, J., B. J. Capel, P. C. B. Turnbull, and R. J. Gilbert. 1976. Identification of a novel enterotoxigenic activity associated with *Bacillus cereus*. *J. Clin. Pathol.* 29:938–940.

67. Minor, T. E., and E. H. Marth. 1976. *Staphylococci and Their Significance in Foods*. New York: Elsevier.

68. Moberg, L. J., and H. Sugiyama. 1979. Microbial ecological basis of infant botulism as studied with germfree mice. *Infect. Immun.* 25:653–657.

69. Moberg, L. J., and H. Sugiyama. 1980. The rat as an animal model for infant botulism. *Infect. Immun.* 29:819–821.

70. Molnar, D. M. 1962. Separation of the toxin of *Bacillus cereus* into two components and nonidentity of the toxin with phospholipase. *J. Bacteriol.* 84:147–153.

71. Moon, H. W., E. M. Kohler, R. A. Schneider, and S. C. Whipp. 1980. Prevalence of pilus antigens, enterotoxin types, and enteropathogenicity among K88-

negative enterotoxigenic *Escherichia coli* from neonatal pigs. *Infect. Immun.* 27:222–230.

72. Mullan, N. A., M. N. Burgess, and P. M. Newsome. 1978. Characterization of a partially purified, methanol-soluble heat-stable *Escherichia coli* enterotoxin in infant mice. *Infect. Immun.* 19:779–784.

73. Murayama, S. Y., T. Sakai, S. Makino, T. Kurata, C. Sasakawa, and M. Yoshikawa. 1986. The use of mice in the Sereny test as a virulence assay of shigellae and enteroinvasive *Escherichia coli*. *Infect. Immun.* 51:696–698.

74. Nishibuchi, M., and R. J. Seidler. 1983. Medium-dependent production of extracellular enterotoxins by non-01 *Vibrio cholera, Vibrio mimicus*, and *Vibrio fluvialis*. *Appl. Environ. Microbiol.* 45:228–231.

75. O'Brien, A. D., and R. K. Holmes. 1987. Shiga and Shiga-like toxins. *Microbiol. Rev.* 51:206–220.

76. Ofek, I., and E. H. Beachey. 1978. Mannose binding and epithelial cell adherence of *Escherichia coli*. *Infect. Immun.* 22:247–254.

77. Okamoto, K., T. Inoue, H. Ichikawa, Y. Kawamoto, and A. Miyama. 1981. Partial purification and characterization of heat-stable enterotoxin produced by *Yersinia enterocolitica*. *Infect. Immun.* 31:554–559.

78. Okamoto, K., T. Inoue, K. Shimizu, S. Hara, and A. Miyama. 1982. Further purification and characterization of heat-stable enterotoxin produced by *Yersinia enterocolitica*. *Infect. Immun.* 35:958–964.

79. Pai, C. H., and V. Mors. 1978. Production of enterotoxin by *Yersinia enterocolitica*. *Infect. Immun.* 19:908–911.

80. Pai, C. H., V. Mors, and T. A. Seemayer. 1980. Experimental *Yersinia enterocolitica* enteritis in rabbits. *Infect. Immun.* 28:238–244.

81. Pai, C. H., V. Mors, and S. Toma. 1978. Prevalence of enterotoxigenicity in human and nonhuman isolates of *Yersinia enterocolitica*. *Infect. Immun.* 22: 334–338.

82. Panigrahi, P., B. D. Tall, R. G. Russell, L. J. Detolla, and J. G. Morris, Jr. 1990. Development of an in vitro model for study of non-01 *Vibrio cholerae* virulence using Caco-2 cells. *Infect. Immun.* 58:3415–3424.

83. Pazzaglia, G., R. B. Sack, A. L. Bourgeois, J. Froehlich, and J. Eckstein. 1990. Diarrhea and intestinal invasiveness of *Aeromonas* strains in the removable intestinal tie rabbit model. *Infect. Immun.* 58:1924–1931.

84. Punyashthiti, K., and R. A. Finkelstein. 1971. Enteropathogenicity of *Escherichia coli*. I. Evaluation of mouse intestinal loops. *Infect. Immun.* 4:473–478.

85. Reyes, A. L., R. G. Crawford, P. L. Spaulding, J. T. Peeler, and R. M. Twedt. 1983. Hemagglutination and adhesiveness of epidemiologically distinct strains of *Vibrio parahaemolyticus*. *Infect. Immun.* 39:721–725.

86. Rose, J. M., C. W. Houston, and A. Kurosky. 1989. Bioactivity and immunological characterization of a cholera toxin-cross-reactive cytolytic enterotoxin from *Aeromonas hydrophila*. *Infect. Immun.* 57:1170–1176.

87. Ruiz-Palacios, G., E. Escamilla, and N. Torres. 1981. Experimental *Campylobacter* diarrhea in chickens. *Infect. Immun.* 34:250–255.

88. Ruiz-Palacios, G. M., J. Torres, E. Escamilla, B. R. Ruiz-Palacios, and T. Tamayo. 1983. Cholera-like enterotoxin produced by *Campylobacter jejuni*. *Lancet* 2:250–253.

89. Sack, D. A., and R. B. Sack. 1975. Test for enterotoxigenic *Escherichia coli* using Y-1 adrenal cells in miniculture. *Infect. Immun.* 11:334–336.

90. Sandefur, P. D., and J. W. Peterson. 1977. Neutralization of *Salmonella* toxin-induced elongation of Chinese hamster ovary cells by cholera antitoxin. *Infect. Immun.* 15:988–992.

91. Sanyal, S. C., K. M. N. Islam, P. K. B. Neogy, M. Islam, P. Speelman, and M. I.

Huq. 1984. *Campylobacter jejuni* diarrhea model in infant chickens. *Infect. Immun.* 43:931–936.

92. Scheuber, P. H., H. Mossmann, G. Beck, and D. K. Hammer. 1983. Direct skin test in highly sensitized guinea pigs for rapid and sensitive determination of staphylococcal enterotoxin B. *Appl. Environ. Microbiol.* 46:1351–1356.

93. Schiemann, D. A. 1981. An enterotoxin-negative strain of *Yersinia enterocolitica* serotype 0:3 is capable of producing diarrhea in mice. *Infect. Immun.* 32:571–574.

94. Schiemann, D. A., and J. A. Devenish. 1980. Virulence of *Yersinia enterocolitica* determined by lethality in Mongolian gerbils and by the Sereny test. *Infect. Immun.* 29:500–506.

95. Smith, H. W. 1972. The production of diarrhea in baby rabbits by the oral administration of cell-free preparations of enteropathogenic *Escherichia coli* and *Vibrio cholerae*: The effect of antisera. *J. Med. Microbiol.* 5:299–303.

96. Speirs, J. I., S. Stavric, and J. Konowalchuk. 1977. Assay of *Escherichia coli* heat-labile enterotoxin with Vero cells. *Infect. Immun.* 16:617–622.

97. Spira, W. M., and P. J. Fedorka-Cray. 1983. Production of cholera toxin-like toxin by *Vibrio mimicus* and non-01 *Vibrio cholera*: Batch culture conditions for optimum yields and isolation of hypertoxigenic lincomycin-resistant mutants. *Infect. Immun.* 42:501–509.

98. Spira, W. M., and J. M. Goepfert. 1972. *Bacillus cereus*–induced fluid accumulation in rabbit ileal loops. *Appl. Microbiol.* 24:341–348.

99. Spira, W. M., R. B. Sack, and J. L. Froehlich. 1981. Simple adult rabbit model for *Vibrio cholerae* and enterotoxigenic *Escherichia coli* diarrhea. *Infect. Immun.* 32:739–747.

100. Stark, R. L., and C. L. Duncan. 1971. Biological characteristics of *Clostridium perfringens* type A enterotoxin. *Infect. Immun.* 4:89–96.

101. Sugiyama, H., and D. C. Mills. 1978. Intraintestinal toxin in infant mice challenged intragastrically with *Clostridium botulinum* spores. *Infect. Immun.* 21:59–63.

102. Svanborg Eden, G., R. Eriksson, and L. A. Hanson. 1977. Adhesion of *Escherichia coli* to human uroepithelial cells in vitro. *Infect. Immun.* 18:767–774.

103. Terplan, G., and S. Steinmeyer. 1989. Investigations on the pathogenicity of *Listeria* spp. by experimental infection of the chick embryo. *Int. J. Food Microbiol.* 8:277–280.

104. Terranova, W., and P. A. Blake. 1978. Current concepts: *Bacillus cereus* food poisoning. *New Engl. J. Med.* 298:143–144.

105. Thorne, G. M., C. F. Deneke, and S. L. Gorbach. 1979. Hemagglutination and adhesiveness of toxigenic *Escherichia coli* isolated from humans. *Infect. Immun.* 23:690–699.

106. Tsuji, T., T. Honda, and T. Miwatani. 1984. Comparison of effects of nicked and unnicked *Escherichia coli* heat-labile enterotoxin on Chinese hamster ovary cells. *Infect. Immun.* 46:94–97.

107. Turnbull, P. C. B., J. M. Kramer, K. Jorgensen, R. J. Gilbert, and J. Melling. 1979. Properties and production characteristics of vomiting, diarrheal, and necrotizing toxins of *Bacillus cereus*. *Amer. J. Clin. Nutri.* 32:219–228.

108. Twedt, R. M., J. T. Peeler, and P. L. Spaulding. 1980. Effective ileal loop dose of Kanagawa-positive *Vibrio parahaemolyticus*. *Appl. Environ. Microbiol.* 40:1012–1016.

109. Une, T. 1977. Studies on the pathogenicity of *Yersinia enterocolitica*. II. Interaction with cultured cells *in vitro*. *Microbiol. Immunol.* 21:365–377.

110. Wadolkowski, E. A., J. A. Burris, and A. D. O'Brien. 1990. Mouse model for colonization and disease caused by enterohemorrhagic *Escherichia coli* 0157:H7. *Infect. Immun.* 58:2438–2445.

111. Whipp, S. C. 1990. Assay for enterotoxigenic *Escherichia coli* heat-stable toxin b in rats and mice. *Infect. Immun.* 58:930–934.
112. Williams, P. H., M. I. Sedgwick, N. Evans, P. J. Turner, R. H. George, and A. S. McNeish. 1978. Adherence of an enteropathogenic strain of *Escherichia coli* to human intestinal mucosa is mediated by a colicinogenic conjugative plasmid. *Infect. Immun.* 22:393–402.
113. Wood, L. V., W. H. Wolfe, G. Ruiz-Palacios, W. S. Foshee, L. I. Corman, F. McCleskey, J. A. Wright, and H. L. DuPont. 1983. An outbreak of gastroenteritis due to a heat-labile enterotoxin-producing strain of *Escherichia coli*. *Infect. Immun.* 41:931–934.
114. Yamamoto, K., I. Ohishi, and G. Sakaguchi. 1979. Fluid accumulation in mouse-ligated intestine inoculated with *Clostridium perfringens* enterotoxin. *Appl. Environ. Microbiol.* 37:181–186.
115. Zen-Yoji, H., Y. Kudoh, H. Igarashi, K. Ohta, and K. Fukai. 1975. Further studies on characterization and biological activities of an enteropathogenic toxin of *Vibrio parahaemolyticus*. *Toxicon* 13:134–135.

PART IV
Microbial Spoilage of Foods

Chapters 8 through 10 present the microbial flora of foods, the ecological parameters of the foods that affect the flora, and the outcome of this relationship that leads to products of undesirable microbial quality. Studies on the microbial spoilage of fruits and vegetables are not pursued very actively; the opposite is true for some meat and seafood products, especially those that are processed and packaged under certain conditions.

More detailed information on poultry, red meats, and seafoods can be obtained from the following references:

F. E. Cunningham, and N. A. Cox, eds., *The Microbiology of Poultry Meat Products* (New York: Academic Press, 1987). The microbiology of fresh and spoiled poultry is covered along with poultry as a source of foodborne pathogens.

A. M. Pearson, and T. R. Dutson, *Advances in Meat Research*, vol. 2, *Meat and Poultry Microbiology* (New York: Van Nostrand Reinhold, 1986). The viruses, fungi, parasites, and bacterial pathogens that occur in meats are discussed, as is the spoilage of fresh and processed meats.

D. R. Ward, and C. R. Hackney, *Microbiology of Marine Food Products* (New York: Van Nostrand Reinhold, 1990). The microbiology of seafood processing is discussed along with HACCP and other programs that deal with seafood safety and quality.

8

Spoilage of Fruits and Vegetables

Spoiled food may be defined as food that has been damaged or injured so as to make it undesirable for human consumption. Food spoilage may be caused by insect damage, physical injury of various kinds such as bruising and freezing, enzyme activity, or microorganisms. Only that caused by microorganisms will be treated here.

The microbial spoilage of foods should not be viewed as a sinister plot on the part of microorganisms deliberately to destroy foods but as the normal function of these organisms in the total ecology of all living organisms. While plants and animals have both evolved intrinsic mechanisms that aid them in combating harmful microbes, many microorganisms can and do overcome these forces and bring about the destruction of the plant and animal organic matter by converting it into inorganic compounds. These primary activities of microorganisms on foods bring about spoilage.

It has been estimated that 20% of all fruits and vegetables harvested for human consumption are lost through microbial spoilage by one or more of 250 market diseases (1). The primary causative agents of microbial spoilage are the bacteria, yeasts, and molds. While viruses have the capacity to damage both plant and animal tissues, these agents, along with the *Mycoplasma*, are not generally regarded as being important in food spoilage as it is now recognized. With respect to yeasts, molds, and bacteria, the latter two are by far the most important etiologic agents of food spoilage in general. On the basis of their growth requirements (discussed in Chap. 3), each may be expected to occupy its own niche with respect to the various types of foods and even within the same food.

MICROBIAL SPOILAGE OF VEGETABLES

The general composition of higher plants is presented in Table 8-1, and the composition of 21 common vegetables is presented in Table 8-2. The average

TABLE 8-1.   General Chemical Composition of Higher Plant Material

Carbohydrates and related compounds
 1. Polysaccharides—pentosan (araban), hexosans (cellulose, starch, xylans, fructans, mannans, galactans, levans).
 2. Oligosaccharides—tetrasaccharide (stachyose), trisaccharides (robinose, mannotriose, raffinose), disaccharides (maltose, sucrose, cellobiose, melibiose, trehalose).
 3. Monosaccharides—hexoses (mannose, glucose, galactose, fructose, sorbose), pentoses (arabinose, xylose, ribose, L-rhamnose, L-fucose).
 4. Sugar alcohols—glycerol, ribitol, mannitol, sorbitol, inositols.
 5. Sugar acids—uronic acids, ascorbic acid.
 6. Esters—tannins.
 7. Organic acids—citric, shikimic, D-tartaric, oxalic, lactic, glycolic, malonic, etc.

Proteins—albumins, globulins, glutelins, prolamines, peptides, and amino acids.
Lipids—fatty acids, fatty acid esters, phospholipids, glycolipids, etc.
Nucleic acids and derivatives—purine and pyrimidine bases, nucleotides, etc.
Vitamins—fat-soluble (A, D, E), water-soluble (thiamine, niacin, riboflavin, etc.).
Minerals—Na, K, Ca, Mg, Mn, Fe, etc.
Water
Others—alkaloids, porphyrins, aromatics, etc.

TABLE 8-2.   Vegetable Foods: Approximate Percentage Chemical Composition

Vegetable	Water	Carbohydrates	Proteins	Fat	Ash
Beans, green	89.9	7.7	2.4	0.2	0.8
Beets	87.6	9.6	1.6	0.1	1.1
Broccoli	89.9	5.5	3.3	0.2	1.1
Brussels sprouts	84.9	8.9	4.4	0.5	1.3
Cabbage	92.4	5.3	1.4	0.2	0.8
Cantaloupe	94.0	4.6	0.2	0.2	0.6
Cauliflower	91.7	4.9	2.4	0.2	0.8
Celery	93.7	3.7	1.3	0.2	1.1
Corn	73.9	20.5	3.7	1.2	0.7
Cucumbers	96.1	2.7	0.7	0.1	0.4
Lettuce	94.8	2.9	1.2	0.2	0.9
Onions	87.5	10.3	1.4	0.2	0.6
Peas	74.3	17.7	6.7	0.4	0.9
Potatoes	77.8	19.1	2.0	0.1	1.0
Pumpkin	90.5	7.3	1.2	0.2	0.8
Radishes	93.6	4.2	1.2	0.1	1.0
Spinach	92.7	3.2	2.3	0.3	1.5
Squash, summer	95.0	3.9	0.6	0.1	0.4
Sweet potatoes	68.5	27.9	1.8	0.7	1.1
Tomatoes	94.1	4.0	1.0	0.3	0.6
Watermelon	92.1	6.9	0.5	0.2	0.3
Mean	88.3	8.6	2.0	0.3	0.8

Source: Watt and Merrill (11).

water content of vegetables is about 88%, with an average content of 8.6% carbohydrates, 1.9% proteins, 0.3% fat, and 0.84% ash. The total percentage composition of vitamins, nucleic acids, and other plant constituents is generally less than 1%. From the standpoint of nutrient content, vegetables are capable of supporting the growth of molds, yeasts, and bacteria and, consequently, of being spoiled by any or all of these organisms. The higher water content of vegetables favors the growth of spoilage bacteria, and the relatively low carbohydrate and fat contents suggest that much of this water is in available form. The pH range of most vegetables is within the growth range of a large number of bacteria, and it is not surprising, therefore, that bacteria are common agents of vegetable spoilage. The relatively high oxidation-reduction (O/R) potential of vegetables and their lack of high poising capacity suggest that the aerobic and facultative anaerobic types would be more important than the anaerobes. This is precisely the case; some of the most ubiquitous etiologic agents in the bacterial spoilage of vegetables are species of the genus *Erwinia* and are associated with plants and vegetables in their natural growth environment. The common spoilage pattern displayed by these organisms is referred to as **bacterial soft rot.**

Bacterial Agents

Bacterial Soft Rot.   This type of spoilage is caused by *Erwinia carotovora* and pseudomonads such as *Pseudomonas marginalis*, with the former being the more important. *Bacillus* and *Clostridium* spp. have been implicated, but their roles are probably secondary.

The causative organisms break down pectins, giving rise to a soft, mushy consistency, sometimes a bad odor, and a water-soaked appearance. Some of the vegetables affected by this disease are asparagus, onions, garlic, beans (green, lima, and wax), carrots, parsnips, celery, parsley, beets, endives, globe artichokes, lettuce, rhubarb, spinach, potatoes, cabbage, brussels sprouts, cauliflower, broccoli, radishes, rutabagas, turnips, tomatoes, cucumbers, cantaloupes, peppers, and watermelons.

While the precise manner in which *Erwinia* spp. bring about soft rot is not yet well understood, it is very likely that these organisms, present on the susceptible vegetables at the time of harvest, subsist on vegetable sap until the supply is exhausted. The cementing substance of the vegetable body then induces the formation of pectinases, which act by hydrolyzing pectin, thereby producing the mushy consistency. In potatoes, tissue maceration has been shown to be caused by an endopolygalacturonate transeliminase of *Erwinia* origin (7). Because of the early and relatively rapid growth of these organisms, molds, which tend to be crowded out, are of less consequence in the spoilage of vegetables that are susceptible to bacterial agents. Once the outer plant barrier has been destroyed by these pectinase producers, nonpectinase producers no doubt enter the plant tissues and help bring about fermentation of the simple carbohydrates that are present. The quantities of simple nitrogenous compounds present, the vitamins (especially the B-complex group), and minerals are adequate to sustain the growth of the invading organisms until the vegetables have been essentially consumed or destroyed. The malodors that are produced are probably the direct result of volatile compounds (such as NH_3, volatile acids, and the like) produced by the flora. When growing in acid media, microorganisms tend to decarboxylate amino acids, leaving amines that cause an elevation of pH toward the neutral range and beyond. Complex carbohydrates such as cellulose are generally the last to be degraded, and a varied flora consisting of molds and other soil organisms is usually responsible, since cellulose degradation by *Erwinia* spp. is doubtful. Aromatic constituents and porphyrins are probably not attacked until late in the spoilage process, and again by a varied flora of soil types.

The genes of *E. carotovora* subsp. *carotovora* that are involved in potato tuber maceration have been cloned. Plasmids containing cloned DNA mediated the production of endo-pectate lyases, exo-pectate lyase, endo-polygalacturonase, and cellulases (8). The *Escherichia coli* strains that contained cloned plasmids showed that endo-pectate lyases with endo-polygalacturonase or exo-pectate lyase caused maceration of potato tuber slices. These enzymes, along with phosphatidase C and phospholipase A, are involved in soft rot by this organism. Carrots infected with *Agrobacterium tumefaciens* undergo senescence at a faster rate because of increased ethylene synthesis. In normal uninfected plants, ethylene synthesis is regulated by auxins, but *A. tumefaciens* increases the synthesis of indoleacetic acid, which results in increased levels of ethylene.

The genus *Erwinia* belongs to the family Enterobacteriaceae. Of the 21 species listed in *Bergey's Manual*, all are associated with plants where they are known to cause plant diseases of the rot and wilt types. These are gram-negative rods that are related to the genera *Proteus*, *Serratia*, *Escherichia*, *Salmonella*, and others. *Erwinia* spp. normally do not require organic nitrogen compounds for growth, and the relatively low levels of proteins in vegetables make them suitable for the task of destroying plant materials of this type. The pectinase produced by these organisms is actually a **protopectinase,** since the

TABLE 8-3.   Bacteria That Cause Field and Storage Spoilage of Vegetables

Organisms	Spoilage Condition/Products
Corynebacterium michiganense	Vascular wilt, canker; leaf and fruit spot on tomatoes, others
C. nebraskense	Leaf spot, leaf blight and wilt of corn
C. sepedonicum	Tuber rot of white potatoes
Curtobacterium flaccumfaciens (formerly *Corynebacterium*)	Bacterial wilt of beans
Pseudomonas agarici and *P. tolaasii*	Drippy gill of mushrooms
P. corrupata	Tomato pith necrosis
Pseudomonas cichorii group	Bacterial zonate spot of cabbage and lettuce
Pseudomonas marginalis group	Soft rot of vegetables, side slime of lettuce
P. morsprunorum group (formerly *P. phaseolicola*)	Halo blight of beans
P. syringae group	
Formerly *P. glycinea*	Disease of soybeans
Formerly *P. lachrymans*	Angular leaf spot of cucumbers
Formerly *P. pisi*	Bacterial blight of pears
P. tomato group	Bacterial speck of tomatoes
Xanthomonas campestris . . .	
pv. *campestris*	Black rot of cabbage and cauliflower
pv. *phaseoli*	Common blight of beans
pv. *vesicatoria*	Bacterial spot of tomatoes and peppers
X. oryzae . . .	
pv. *oryzae*	Bacterial blight of rice
pv. *oryzicola*	Bacterial leaf streak of rice

cementing substance of plants as it actually exists in the plant is protopectin. Many *Erwinia* spp. such as *E. carotovora* are capable of fermenting many of the sugars and alcohols that exist in certain vegetables such as rhamnose, cellobiose, arabinose, mannitol, and so forth—compounds that are not utilized by many of the more common bacteria. While most *Erwinia* spp. grow well at about 37°C, most are also capable of good growth at refrigerator temperatures, with some strains reported to grow at 1°C.

Other Bacterial Spoilage Conditions. *E. carotovora* pv. *atroseptica*, *E. carotovora* pv. *carotovora*, and *E. chrysanthemi* cause a rot of potatoes sometimes referred to as "**black leg.**" In temperate regions, *E. carotovora* pv. *atroseptica* is usually involved with *E. chrysanthemi* to a lesser degree. While direct contact with soil may be the source of normal forms of these organisms, L-phase variants may enter healthy tissues and later revert to classical forms (4).

Some of the more important bacteria that cause field and storage spoilage of vegetables are presented in Table 8-3. All genera and species listed are undergoing taxonomic changes. The plant corynebacteria represent a diverse collection, many of which do not belong to this genus. Some have been transferred to the genus *Curtobacter*. The plant pathogenic and field spoilage pseudomonads and xanthomonads are also diverse. *Xanthomonas campestris* alone is represented by six DNA homology groups (10), and reclassification within this group may be expected.

The appearance of some market vegetables undergoing bacterial and fungal spoilage is shown in Figures 8-1 and 8-2.

Fungal Agents

A synopsis of some of the common spoilage conditions of vegetables and fruits is presented in Table 8-4. Some of these spoilage conditions are initiated preharvest and others postharvest. Among the former, *Botrytis* invades the flower of strawberries to cause gray mold rot, *Colletotrichum* invades the epidermis of bananas to initiate banana **anthracnose**, and *Gloeosporium* invades the lenticels of apples to initiate lenticel rot (3). The largest number of market fruit and vegetable spoilage conditions occur after harvesting, and while the fungi most often invade bruised and damaged products, some enter specific areas. For example, *Thielaviopsis* invades the fruit stem of pineapples to cause black rot of this fruit, and *Colletotrichum* invades the crown cushion of bananas to cause **banana crown rot** (3). Black rot of sweet potatoes is caused by *Ceratocystis*, **neck rot** of onions by *Botrytis allii*, and **downey mildew** of lettuce by *Bremia* spp. (2). Some of the spoilage conditions listed in Table 8-4 are discussed further below.

Gray Mold Rot. This condition is caused by *Botrytis cinerea*, which produces a gray mycelium. This type of spoilage is favored by high humidity and warm temperatures. Among the vegetables affected are asparagus, onions, garlic, beans (green, lima, and wax), carrots, parsnips, celery, tomatoes, endives, globe artichokes, lettuce, rhubarb, cabbage, brussels sprouts, cauliflower, broccoli, radishes, rutabagas, turnips, cucumbers, pumpkin, squash, peppers,

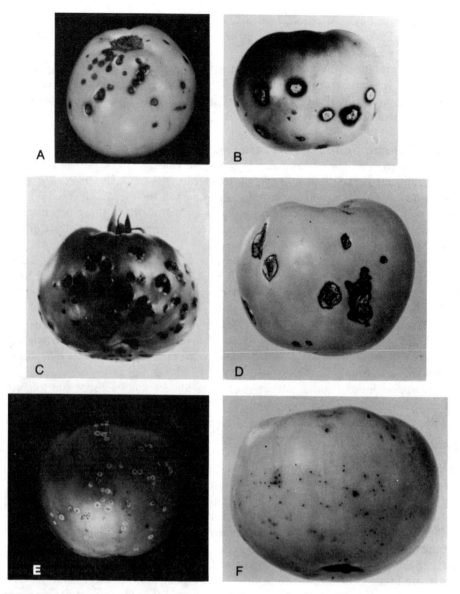

FIGURE 8-1.   Tomato diseases: *A* and *B*, nailhead spot; *C* and *D*, bacterial spot; *E*, bacterial canker; *F*, bacterial spot.
From Agriculture Handbook 28, *USDA, 1968, "Fungus and Bacterial Diseases of Fresh Tomatoes."*

and sweetpotatoes. In this disease, the causal fungus grows on decayed areas in the form of a prominent gray mold. It can enter fruits and vegetables through the unbroken skin or through cuts and cracks.

Sour Rot (Oospora Rot, Watery Soft Rot).   This condition of vegetables is caused by *Geotrichum candidum* and other organisms. Among the vegetables affected are asparagus, onions, garlic, beans (green, lima, and wax), carrots, parsnips,

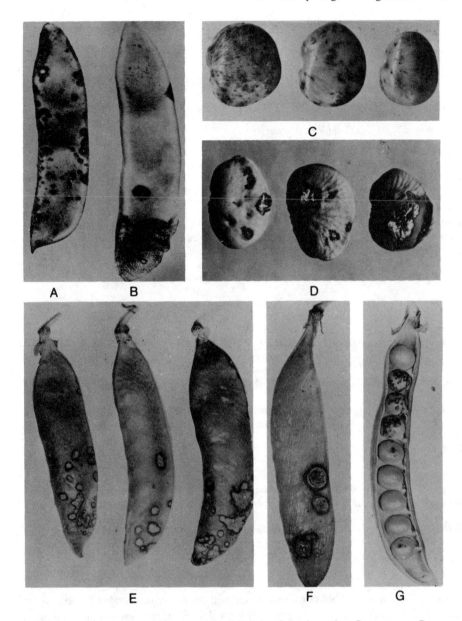

FIGURE 8-2.   Lima bean diseases: *A* and *B*, pod blight; *C*, seed spotting; *D*, yeast spot. Pea diseases: *E*, pod spot; *F*, anthracnose; *G*, scab.
From Agriculture Handbook 303, *USDA, 1966, Chap. 5.*

parsley, endives, globe artichokes, lettuce, cabbage, brussels sprouts, cauliflower, broccoli, radishes, rutabagas, turnips, and tomatoes. The causal fungus is widely distributed in soils and on decaying fruits and vegetables. *Drosophila melanogaster* (fruit fly) carries spores and mycelial fragments on its body from decaying fruits and vegetables to growth cracks and wounds in healthy fruits and vegetables. Since the fungus cannot enter through the unbroken skin, infections usually start in openings of one type or another (6).

TABLE 8-4.   Common Fungal Fruit and Vegetable Spoilage Conditions, Etiologic Agents, and Typical Products Affected

Spoilage Condition	Etiologic Agent	Typical Products Affected
Alternaria rot	A. tenuis	Citrus fruits
Anthracnose (bitter rot)	Colletotrichum musae	Bananas
Anthracnose	C. lindemuthianum	Beans
	C. lagenarium	Watermelons
Black rot	Aspergillus niger	Onions
Black rot	Ceratocystis fimbriata	Sweet potatoes
Blue mold rot	Penicillium digitatum	Citrus fruits
Brown rot	Monilinia fructicola (= Sclerotinia fructicola)	Peaches, cherries
Brown rot	Phytophora spp.	Citrus fruits
Cladosporium rot	C. herbarum	Cherries, peaches
Crown rot	Colletotrichum musae (= Gloeosporium musarum), Fusarium roseum, Verticillium theobromae, Ceratocystis paradoxa	Bananas
Downy mildew	Plasmapara viticole, Phytophora spp., Bremia spp.	Grapes
Dry rot	Fusarium spp.	Potatoes
Gray mold rot	Botrytis cinerea	Grapes, many others
Green mold rot	Penicillium digitatum	Citrus fruits
Lenticel rot	Cryptosporiopsis malicorticis (= Gloeosporium perennans), Phylctaena vagabunda	Apples, pears
Pineapple black rot	Ceratocystis paradoxa (= Thielaviopsis paradoxa)	Pineapples
Phytophora rot	Colletotrichum coccodes	Vegetables
Pink mold rot	Trichothecium roseum	
Rhizopus soft rot	Rhizopus stolonifer	Sweet potatoes, tomatoes
Slimy brown rot	Rhizoctonia spp.	Vegetables
"Smut" (black mold rot)	Aspergillus niger	Peaches, apricots
Sour rot	Geotrichum candidum	Tomatoes, citrus fruits
Stem-end rot	Phomopsis citri, Diplodia natalensis, Alternaria citri	Citrus fruits
Watery soft rot	Sclerotinia sclerotiorum	Carrots

Rhizopus Soft Rot.   This condition is caused by *Rhizopus stolonifer* and other species that make vegetables soft and mushy. Cottony growth of the mold with small black dots of sporangia often covers the vegetables. Among those affected are beans (green, lima, and wax), carrots, sweet potatoes, potatoes, cabbage, brussels sprouts, cauliflower, broccoli, radishes, rutabagas, turnips, cucumbers, cantaloupes, pumpkins, squash, watermelons, and tomatoes. This fungus is spread by *D. melanogaster*, which lays its eggs in the growth cracks on various fruits and vegetables. The fungus is widespread and is disseminated also by other means. Entry usually occurs through wounds and other skin breaks.

Phytophora Rot.   This market condition, caused by *Phytophora* spp., occurs largely in the field as a blight and fruit rot of market vegetables. It appears to be more variable than some other market "diseases" and affects different plants in different ways. Among the vegetables affected are asparagus, onions, garlic, cantaloupes, watermelons, tomatoes, eggplants, and peppers.

Anthracnose.   This plant disease is characterized by spotting of leaves, fruit, or seed pods. It is caused by *Colletotrichum coccodes* and other species. These fungi are considered weak plant pathogens. They live from season to season on plant debris in the soil and on the seed of various plants such as the tomato. Their spread is favored by warm, wet weather. Among the vegetables affected are beans, cucumbers, watermelons, pumpkins, squash, tomatoes, and peppers.

For further information on market diseases of fruits and vegetables, the monographs issued by the Agricultural Research Service of the U.S. Department of Agriculture should be consulted (see Fig. 8-3 for several fungal diseases of onions) and references 5 and 9.

SPOILAGE OF FRUITS

The general composition of 18 common fruits is presented in Table 8-5, which shows that the average water content is about 85% and the average carbohydrate content is about 13%. The fruits differ from vegetables in having somewhat less water but more carbohydrate. The mean protein, fat, and ash content of fruits are, respectively, 0.9, 0.5, and 0.5%—somewhat lower than vegetables except for ash content. Fruits contain vitamins and other organic compounds, just as do vegetables. On the basis of nutrient content, these products would appear to be capable of supporting the growth of bacteria, yeasts, and molds. However, the pH of fruits is below the level that generally favors bacterial growth. This one fact alone would seem to be sufficient to explain the general absence of bacteria in the incipient spoilage of fruits. The wider pH growth range of molds and yeasts suits them as spoilage agents of fruits. With the exception of pears, which sometimes undergo **Erwinia rot,** bacteria are of no known importance in the initiation of fruit spoilage. Just why pears with a reported pH range of 3.8 to 4.6 should undergo bacterial spoilage is not clear. It is conceivable that *Erwinia* initiates its growth on the surface of this fruit where the pH is presumably higher than on the inside.

A variety of yeast genera can usually be found on fruits, and these organisms often bring about the spoilage of fruit products, especially in the field. Many yeasts are capable of attacking the sugars found in fruits and bringing about fermentation with the production of alcohol and carbon dioxide. Due to their generally faster growth rate than molds, they often precede the latter organisms in the spoilage process of fruits in certain circumstances. It is not clear whether some molds are dependent on the initial action of yeasts in the process of fruit and vegetable spoilage. The utilization or destruction of the high-molecular-weight constituents of fruits is brought about more by molds than yeasts. Many molds are capable of utilizing alcohols as sources of energy, and when these and other simple compounds have been depleted, these organisms proceed to destroy the remaining parts of fruits such as the structural polysaccharides and rinds.

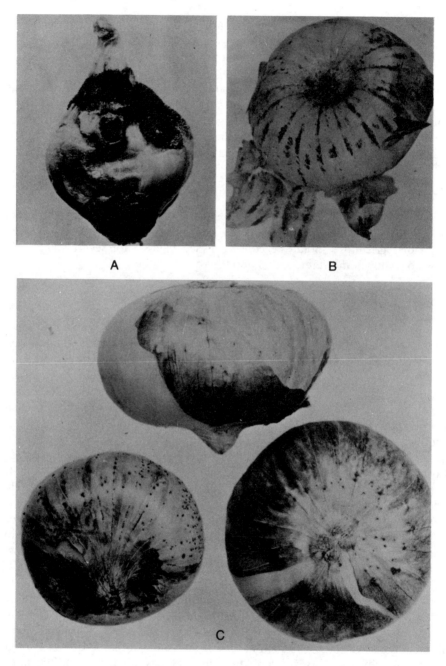

FIGURE 8-3.   Onion diseases: *A*, white rot; *B*, black mold rot; *C*, diplodia stain.
From Agriculture Handbook 303, *USDA, 1966, Chap. 5.*

TABLE 8-5.   Common Fruits: Approximate Percentage Composition

Fruit	Water	Carbohydrate	Protein	Ash	Fat
Apples	84.1	14.9	0.3	0.3	0.4
Apricots	85.4	12.9	1.0	0.6	0.1
Bananas	74.8	23.0	1.2	0.8	0.2
Blackberries	84.8	12.5	1.2	0.5	1.0
Cherries, sweet and sour	83.0	14.8	1.1	0.6	0.5
Figs	78.0	19.6	1.4	0.6	0.4
Grapefruit	88.8	10.1	0.5	0.4	0.2
Grapes, American type	81.9	14.9	1.4	0.4	1.4
Lemons	89.3	8.7	0.9	0.5	0.6
Limes	86.0	12.3	0.8	0.8	0.1
Oranges	87.2	11.2	0.9	0.5	0.2
Peaches	86.9	12.0	0.5	0.5	0.1
Pears	82.7	15.8	0.7	0.4	0.4
Pineapples	85.3	13.7	0.4	0.4	0.2
Plums	85.7	12.9	0.7	0.5	0.2
Raspberries	80.6	15.7	1.5	0.6	1.6
Rhubarb	94.9	3.8	0.5	0.7	0.1
Strawberries	89.9	8.3	0.8	0.5	0.5
Mean	84.9	13.2	0.88	0.53	0.46

Source: Watt and Merrill (11).

References

1. Beraha, L., M. A. Smith, and W. R. Wright. 1961. Control of decay of fruits and vegetables during marketing. *Dev. Indust. Microbiol.* 2:73–77.
2. Brackett, R. E. 1987. [Fungal spoilage of] vegetables and related products. In *Food and Beverage Mycology*, 2d ed., ed. L. R. Beuchat, 129–154. New York: Van Nostrand Reinhold.
3. Eckert, J. W. 1979. Fungicidal and fungistatic agents: Control of pathogenic microorganisms on fresh fruits and vegetables after harvest. In *Food Mycology*, ed. M. E. Rhodes, 164–199. Boston: Hall.
4. Jones, S. M., and A. M. Paton. 1973. The L-phase of *Erwinia carotovora* var. *atroseptica* and its possible association with plant tissue. *J. Appl. Bacteriol.* 36: 729–737.
5. Lund, B. M. 1971. Bacterial spoilage of vegetables and certain fruits. *J. Appl. Bacteriol.* 34:9–20.
6. McColloch, L. P., H. T. Cook, and W. R. Wright. 1968. Market diseases of tomatoes, peppers, and eggplants. *Agricultural Handbook No. 28*. Washington, D.C.: Agricultural Research Service, USDA.
7. Mount, M. S., D. F. Bateman, and H. G. Basham. 1970. Induction of electrolyte loss, tissue maceration, and cellular death of potato tissue by an endopolygalacturonate trans-eliminase. *Phytopathol.* 60:924–1000.
8. Roberts, D. P., P. M. Berman, C. Allen, V. K. Stromberg, G. H. Lacy, and M. S. Mount. 1986. Requirement for two or more *Erwinia carotovora* subsp. *carotovora* pectolytic gene products for maceration of potato tuber tissue by *Escherichia coli*. *J. Bacteriol.* 167:279–284.

9. Splittstoesser, D. F. 1987. [Fungal spoilage of] fruits and fruit products. In *Food and Beverage Mycology*, 2d ed., ed. L. R. Beuchat, 101–128. New York: Van Nostrand Reinhold.

10. Vauterin, L., J. Swings, K. Kersters, M. Gillis, T. W. Mew, M. N. Schroth, N. J. Palleroni, D. C. Hildebrand, D. E. Stead, E. L. Civerolo, A. C. Hayward, H. Maraite, R. E. Stall, A. K. Vidaver, and J. F. Bradbury. 1990. Towards an improved taxonomy of *Xanthomonas*. *Int. J. Syst. Bacteriol.* 40:312–316.

11. Watt, B. K., and A. L. Merrill. 1950. Composition of foods—raw, processed, prepared. *Agricultural Handbook No. 8*, Washington, D.C.: Agricultural Research Service, USDA.

9

Spoilage of Fresh and Processed Meats, Poultry, and Seafood

Meats are the most perishable of all important foods; the reasons are shown in Table 9-1, which lists the chemical composition of a typical adult mammalian muscle postmortem. Meats contain an abundance of all nutrients required for the growth of bacteria, yeasts, and molds, and an adequate quantity of these constituents exist in fresh meats in available form. The general chemical composition of a variety of meats is presented in Table 9–2.

The genera of bacteria most often found on fresh and spoiled meats, poultry, and seafood are listed in Table 9-3. Not all of the genera indicated for a given product are found at all times, of course. Those that are more often found during spoilage are indicated under the various products. In Table 9-4 are listed the genera of molds most often identified from meats and related products; the identified yeasts are listed in Table 9–5. When spoiled meat products are examined, only a few of the many genera of bacteria, molds, or yeasts are found, and in almost all cases one or more genera are found to be characteristic of the spoilage of a given type of meat product. The presence of the more varied flora on nonspoiled meats, then, may be taken to represent the organisms that exist in the original environment of the product in question or contaminants picked up during processing, handling, packaging, and storage.

The question arises, then, as to why only a few types predominate in spoiled meats. It is helpful here to return to the intrinsic and extrinsic parameters that affect the growth of spoilage microorganisms. Fresh meats such as beef, pork, and lamb, as well as fresh poultry, seafood, and processed meats, have pH values within the growth range of most of the organisms listed in Table 9-3. Nutrient and moisture contents are adequate to support the growth of all organisms listed. While the oxidation-reduction (O/R) potential of whole meats is low, O/R conditions at the surfaces tend to be higher so that strict aerobes and facultative anaerobes, as well as strict anaerobes, generally find conditions suitable for growth. Antimicrobial constituents are not known to occur in products of the type in question. Of the extrinsic parameters, temperature of incubation stands out as being of utmost importance in controlling

199

TABLE 9-1.   Chemical Composition of Typical Adult Mammalian Muscle after Rigor Mortis But before Degradative Changes Postmortem (percentage wet weight)

Water	75.5%
Protein	18.0
Myofibrillar	
Myosin, tropomyosin, X protein	7.5
Actin	2.5
Sarcoplasmic	
Myogen, globulins	5.6
Myoglobin	0.36
Haemoglobin	0.04
Mitochondrial—cytochrome C	ca. 0.002
Sarcoplasmic reticulum, collagen, elastin, "reticulin," insoluble enzymes, connective tissue	2.0
Fat	3.0
Soluble nonprotein substances	3.5
Nitrogenous	
Creatine	0.55
Inosine monophosphate	0.30
Di- and tri-phosphopyridine nucleotides	0.07
Amino acids	0.35
Carnosine, anserine	0.30
Carbohydrate	
Lactic acid	0.90
Glucose-6-phosphate	0.17
Glycogen	0.10
Glucose	0.01
Inorganic	
Total soluble phosphorous	0.20
Potassium	0.35
Sodium	0.05
Magnesium	0.02
Calcium	0.007
Zinc	0.005
Traces of glycolytic intermediates, trace metals, vitamins, etc	ca. 0.10

Source: Reprinted with permission from R. A. Lawrie (65), *Meat Science*, copyright 1966, Pergamon Press.

TABLE 9-2.   Meats and Meat Products: Approximate Percentage Chemical Composition

Meats	Water	Carbohydrates	Proteins	Fat	Ash
Beef, hamburger	55.0	0	16.0	28.0	0.8
Beef, round	69.0	0	19.5	11.0	1.0
Bologna	62.4	3.6	14.8	15.9	3.3
Chicken (broiler)	71.2	0	20.2	7.2	1.1
Frankfurters	60.0	2.7	14.2	20.5	2.7
Lamb	66.3	0	17.1	14.8	0.9
Liver (beef)	69.7	6.0	19.7	3.2	1.4
Pork, medium	42.0	0	11.9	45.0	0.6
Turkey, medium fat	58.3	0	20.1	20.2	1.0

Source: Watt and Merrill (114).

TABLE 9-3.   Genera of Bacteria Most Frequently Found on Meats, Poultry, and Seafood

Genus	Gram Reaction	Fresh Meats	Fresh Livers	Processed Meats	Vacuum-Packaged Meats	Bacon[a]	Poultry	Fish and Seafood
Acinetobacter	−	XX	X	X	X	X	XX	X
Aeromonas	−	XX			X	X	X	X
Alcaligenes	−	X	X			X	X	X
Bacillus	+	X		X		X	X	X
Brochothrix	+	X	X	X	XX			
Campylobacter	−						XX	
Carnobacterium	+	X			XX			
Citrobacter	−	X					X	
Clostridium	+	X					X	
Corynebacterium[b]	+	X	X	X	X	X	XX	X
Enterobacter	−	X		X	X		X	X
Enterococcus	+	XX	X	X	XX	X	X	X
Escherichia	−	X					X	X
Flavobacterium	−	X	X				XX	X
Hafnia	−	X			X			
Kurthia[b]	+	X			X			
Lactococcus	+	X		X				
Lactobacillus	+	X		XX	XX			X
Leuconostoc	+	X	X	X	X			
Listeria	+	X		X			XX	X
Microbacterium[b]	+	X		X	X	X	X	X
Micrococcus	+	X	XX	X	X	X	X	
Moraxella	−	XX	X			X	X	X
Pantoea	−	X					X	
Pediococcus	+	X		X	X			
Proteus	−	X					X	
Pseudomonas	−	XX			X		XX	XX
Psychrobacter	−	XX					X	X
Salmonella	−	X					X	
Serratia	−	X		X	X		X	
Shewanella	−	X						XX
Staphylococcus	+	X	X	X	X	X	X	
Vagococcus	+						XX	
Vibrio	−					X		X
Yersinia	−	X			X			

Note: X = known to occur; XX = most frequently reported.

[a] Vacuum packaged not included.

[b] Belong to the coryneform group.

the types of microorganisms that develop on meats, since these products are normally held at refrigerator temperatures. Essentially all studies on the spoilage of meats, poultry, and seafood carried out over the past 40 years or so have dealt with low-temperature-stored products.

SPOILAGE OF FRESH BEEF, PORK, AND RELATED MEATS

Most studies dealing with the spoilage of meats have been done with beef, and most of the discussion in this section is based on beef studies. Pork, lamb, veal, and similar meats are presumed to spoil in a similar way.

TABLE 9-4.   Genera of Molds Most Often Found on Meats, Poultry, and Seafood Products

Genus	Fresh and Refrigerated Meats	Poultry	Fish and Shrimp	Processed and Cured Meats
Alternaria	X	X		X
Aspergillus	X	X	X	XX
Aureobasidium (Pullularia)			XX	
Botrytis				X
Cladosporium	XX	X		X
Fusarium	X			X
Geotrichum	XX	X		X
Monascus	X			
Monilia	X			X
Mucor	XX	X		X
Neurospora	X			
Penicillium	X	X	X	XX
Rhizopus	XX	X		X
Scopulariopsis			X	X
Sporotrichum	XX			
Thamnidium	XX			X

Source: Taken from the literature and from Jay (61).

Note: X = known to occur; XX = most frequently found.

TABLE 9-5.   Yeast Genera Most Often Identified on Meats, Poultry, and Seafood Products

Genus	Fresh and Refrigerated Meats	Poultry	Fish and Shrimp	Processed and Cured Meats
Candida	XX	XX	XX	X
Cryptococcus	X	X	XX	
Debaryomyces	X		X	XX
Hansenula			X	
Pichia			X	
Rhodotorula	X	XX	XX	
Saccharomyces		X		X
Sporobolomyces			X	
Trichosporon	X	X	X	X

Source: Taken from the literature and from Jay (61).

Note: X = known to occur; XX = most frequently found.

Upon the slaughter of a well-rested beef animal, a series of events take place that lead to the production of meat. Lawrie (65) discussed these events in great detail, and they are here presented only in outline form. Following an animal's slaughter,

1. Its circulation ceases; the ability to resynthesize ATP (adenosine triphosphate) is lost; lack of ATP causes actin and myosin to combine to form actomyosin, which leads to a stiffening of muscles.
2. The oxygen supply falls, resulting in a reduction of the O/R potential.
3. The supply of vitamins and antioxidants ceases, resulting in a slow development of rancidity.

4. Nervous and hormonal regulations cease, thereby causing the temperature of the animal to fall and fat to solidify.
5. Respiration ceases, which stops ATP synthesis.
6. Glycolysis begins, resulting in the conversion of most glycogen to lactic acid, which depresses pH from about 7.4 to its ultimate level of about 5.6. This pH depression also initiates protein denaturation, liberates and activates cathepsins, and completes rigor mortis. Protein denaturation is accompanied by an exchange of divalent and monovalent cations on the muscle proteins.
7. The reticuloendothelial system ceases to scavenge, thus allowing microorganisms to grow unchecked.
8. Various metabolites accumulate that also aid protein denaturation.

These events require between 24 and 36 h at the usual temperatures of holding freshly slaughtered beef (2° to 5°C). Meanwhile, part of the normal flora of this meat has come from the animal's own lymph nodes (68), the stick knife used for exsanguination, the hide of the animal, intestinal tract, dust, hands of handlers, cutting knives, storage bins, and the like. Upon prolonged storage at refrigerator temperatures, microbial spoilage begins. In the event that the internal temperatures are not reduced to the refrigerator range, the spoilage that is likely to occur is caused by bacteria of internal sources. Chief among these are *Clostridium perfringens* and genera in the Enterobacteriaceae family (54). On the other hand, bacterial spoilage of refrigerator-stored meats is, by and large, a surface phenomenon reflective of external sources of the spoilage flora (54).

With respect to fungal spoilage of fresh meats, especially beef, the following genera of molds have been recovered from various spoilage conditions of whole beef: *Thamnidium*, *Mucor*, and *Rhizopus*, all of which produce "whiskers" on beef; *Cladosporium*, a common cause of "black spot"; *Penicillium*, which produces green patches; and *Sporotrichum* and *Chrysosporium*, which produce "white spot." Molds apparently do not grow on meats if the storage temperature is below −5°C (71). Among genera of yeasts recovered from refrigerator-spoiled beef with any consistency are *Candida* and *Rhodotorula*, with *C. lipolytica* and *C. zeylanoides* being the two most abundant species in spoiled ground beef (52).

Unlike the spoilage of fresh beef carcasses, ground beef or hamburger meat is spoiled exclusively by bacteria, with the following genera being the most important: *Pseudomonas*, *Alcaligenes*, *Acinetobacter*, *Moraxella*, and *Aeromonas*. Those generally agreed to be the primary cause of spoilage are *Pseudomonas* and *Acinetobacter-Moraxella* spp., with others playing relatively minor roles in the process (36, 60, 77). Findings from two studies suggest that *Acinetobacter* and *Moraxella* spp. may not be as abundant in spoiled beef as once reported (27, 28). *Psychrobacter* is presumably important, but detailed studies are wanting.

A study of the aerobic gram-negative bacteria recovered from beef, lamb, pork, and fresh sausage revealed that all 231 polarly flagellated rods were pseudomonads and that of 110 nonmotile organisms, 61 were *Moraxella* and 49 were *Acinetobacter* (21). The pseudomonads that cause meat spoilage at low temperatures generally do not match the named species in *Bergey's Manual*.

Numerical taxonomic studies by Shaw and Latty (92, 93) led them to group most of their isolates into four clusters based on carbon source utilization tests. Of 787 *Pseudomonas* strains isolated from meats, 89.7% were identified, with 49.6% belonging to their cluster 2, 24.9% to cluster 1, and 11.1% to cluster 3 (93). The organisms in clusters 1 and 2 were nonfluorescent and egg-yolk negative and resembled *P. fragi*; those in cluster 3 were fluorescent and gelatinase positive. *P. fluorescens* biotype I strains were represented by 3.9%, biotype III by 0.9%, and *P. putida* by only one strain. The relative incidence of the clusters on beef, pork, and lamb and on fresh and spoiled meats was similar (93).

Beef rounds and quarters are known to undergo deep spoilage, usually near the bone, especially the "aitch" bone. This type of spoilage is often referred to as "bone taint" or "sours." Only bacteria have been implicated, with the genera *Clostridum* and *Enterococcus* being the primary causative agents (10).

Temperature of incubation is the primary reason that only a few genera of bacteria are found in spoiled meats as opposed to fresh. In one study, only four of the nine genera present in fresh ground beef could be found after the meat underwent frank spoilage at refrigerator temperatures (60). It was noted by Ayres (4) that after processing, more than 80% of the total population of freshly ground beef may be comprised of chromogenic bacteria, molds, yeasts, and sporeforming bacteria, but after spoilage only nonchromogenic, short gram-negative rods are found. While some of the bacteria found in fresh meats can be shown to grow at refrigerator temperatures on culture media, they apparently lack the capacity to compete successfully with the *Pseudomonas* and *Acinetobacter-Moraxella* types.

Beef cuts, such as steaks or roasts, tend to undergo surface spoilage; whether the spoilage organisms are bacteria or molds depends on available moisture. Freshly cut meats stored in a refrigerator with high humidity invariably undergo bacterial spoilage preferential to mold spoilage. The essential feature of this spoilage is surface sliminess in which the causative organisms can nearly always be found. The relatively high Eh, availability of moisture, and low temperature favor the pseudomonads. It is sometimes possible to note discrete bacterial colonies on the surface of beef cuts, especially when the level of contamination is low. The slime layer results from the coalescence of surface colonies and is largely responsible for the tacky consistency of spoiled meats. Ayres (4) presented evidence that odors can be detected when the surface bacterial count is between log 7.0 and $7.5/cm^2$, followed by detectable slime with surface counts usually about log 7.5 to $8.0/cm^2$ (Fig. 9-1). This is further depicted in Figure 9-2, which relates numbers of bacteria not only to surface spoilage of fresh poultry but to red meats and seafoods as well.

Molds tend to predominate in the spoilage of beef cuts when the surface is too dry for bacterial growth or when beef has been treated with antibiotics such as the tetracyclines. Molds virtually never develop on meats when bacteria are allowed to grow freely. The reason appears to be that bacteria grow faster than molds, thus consuming available surface oxygen, which molds also require for their activities.

Unlike the case of beef cuts or beef quarters, mold growth is quite rare on ground beef except when antibacterial agents have been used as preservatives or when the normal bacterial load has been reduced by long-term freezing.

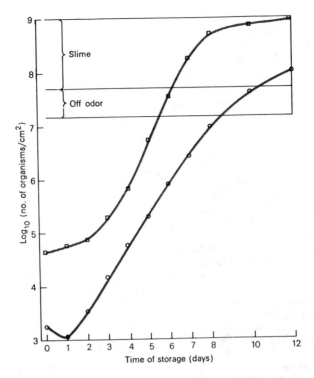

FIGURE 9-1.   The development of off-odor and slime on dressed chicken (squares) and packaged beef (circles) during storage at 5°C. *From Ayres (4).*

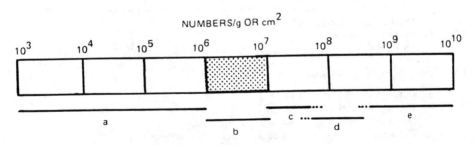

FIGURE 9-2.   Significance of total viable microbial numbers in food products relative to their use as indicators of spoilage. *a*: Microbial spoilage generally not recognized with the possible exception of raw milk, which may sour in the 10^5 to 10^6 range. *b*: some food products show incipiency in this range. Vacuum-packaged meats often display objectionable odors and may be spoiled. *c*: Off-odors generally associated with aerobically stored meats and some vegetables. *d*: Almost all food products display obvious signs of spoilage. Slime is common on aerobically stored meats. *e*: Definite structural changes in product occur at this stage.

Among the early signs of spoilage of ground beef is the development of off-odors followed by tackiness, which indicate the presence of bacterial slime. The sliminess is due to both masses of bacterial growth and the softening or loosening of meat structural proteins.

TABLE 9-6.   Some Methods Proposed for Detecting Microbial Spoilage in Meats, Poultry, and Seafood

Chemical methods

a. Measurement of H_2S production
b. Measurement of mercaptans produced
c. Determination of noncoagulable nitrogen
d. Determination of di- and trimethylamines
e. Determination of tyrosine complexes
f. Determination of indole and skatol
g. Determination of amino acids
h. Determination of volatile reducing substances
i. Determination of amino nitrogen
j. Determination of biochemical oxygen demand (BOD)
k. Determination of nitrate reduction
l. Measurement of total nitrogen
m. Measurement of catalase
n. Determination of creatinine content
o. Determination of dye-reducing capacity
p. Measurement of hypoxanthine
q. ATP measurement
r. Radiometric measurement of CO_2
s. Ethanol production (fish spoilage)
t. Measurement of lactic acid
u. Change in color

Physical methods

a. Measurement of pH changes
b. Measurement of refractive index of muscle juices
c. Determination of alteration in electrical conductivity
d. Measurement of surface tension
e. Measurement of UV illumination (fluorescence)
f. Determination of surface charges
g. Determination of cryoscopic properties
h. Impedance changes
i. Microcalorimetry

Direct bacteriological methods

a. Determination of total aerobes
b. Determination of total anaerobes
c. Determination of ratio of total aerobes to anaerobes
d. Determination of one or more of above at different temperatures
e. Determination of gram-negative endotoxins

Physiochemical methods

a. Determination of extract-release volume (ERV)
b. Determination of water-holding capacity (WHC)
c. Determination of viscosity
d. Determination of meat swelling capacity

In the spoilage of soy-extended ground meats, nothing indicates that the pattern differs from that of unextended ground meats although their rate of spoilage is faster (see Chap. 4).

The precise roles played by spoilage microorganisms that result in the spoilage of meats are not fully understood at this time, but significant progress has been made. Some of the earlier views on the mechanism of meat spoilage are embodied in the many techniques proposed for its detection (Table 9-6).

DETECTION AND MECHANISM OF MEAT SPOILAGE

It is reasonable to assume that reliable methods of determining meat spoilage should be based on the cause and mechanism of spoilage. The chemical methods in Table 9-6 embody the assumption that as meats undergo spoilage, some utilizable substrate is consumed, or some new product or products are created by the spoilage flora. It is well established that the spoilage of meats at low temperatures is accompanied by the production of off-odor compounds such as ammonia, H_2S, indole, and amines. The drawbacks to the use of these methods are that not all spoilage organisms are equally capable of producing them. Inherent in some of these methods is the incorrect belief that low-temperature spoilage is accompanied by a breakdown of primary proteins (58). The physical and direct bacteriological methods all tend to show what is obvious: meat that is clearly spoiled from the standpoint of organoleptic characteristics (odor, touch, appearance, and taste) is, indeed, spoiled. They apparently do not allow one to predict spoilage or shelf life, which a meat freshness test should ideally do.

Among the metabolic by-products of meat spoilage, the diamines cadaverine and putrescine have been studied as spoilage indicators of meats. The production of these diamines occurs in the following manner:

$$Lysine \xrightarrow{\text{decarboxylase}} H_2N(CH_2)_5NH_2$$
$$\text{Cadaverine}$$

$$\text{Ornithine or arginine} \xrightarrow{\text{decarboxylase}} H_2N(CH_2)_4NH_2$$
$$\text{Putrescine}$$

Their use as quality indicators of vacuum-packaged beef that was stored at 1°C for up to 8 weeks has been investigated (22). Cadaverine increased more than putrescine in vacuum-packaged meats, the reverse of findings for aerobically stored samples. Cadaverine levels attained over the incubation period were tenfold higher than the initial levels at total viable counts of $10^6/cm^2$ while there was little change in putrescine at this level. Overall, the findings suggested that these diamines could be of value for vacuum-packaged meats. In fresh beef, pork, and lamb, putrescine occurred at levels from 0.4 to 2.3 ppm and cadaverine from 0.1 to 1.3 ppm (23, 82, 116). Putrescine is the major diamine produced by pseudomonads, while cadaverine is produced more by Enterobacteriaceae (104). It may be noted from Table 9-7 that putrescine increased from 1.2 to 26.1 ppm in one sample of naturally contaminated beef stored at 5°C for 4 days, while cadaverine levels were much lower. In another sample, the two diamines increased to higher levels under the same conditions. Cadaverine was the only amine that correlated with coliforms in ground beef in one study (90). That significant changes in putrescine and cadaverine do not

TABLE 9-7.   Development of Microbial Number and Diamine Concentrations on Naturally Contaminated Minced Beef Stored at 5°C

Sample[a]	Storage Time (d)	Putrescine[b] (µg/g)	Cadaverine[b] (µg/g)	Enterobacteriaceae ($\log_{10}$ no./g)	Aerobic Plate Count ($\log_{10}$ no./g)
E	0	1.2	0.1	3.81	6.29
	1	1.8	0.1	3.56	7.66
	2	4.2	0.5	4.57	8.49
	3	10.0	0.5	5.86	9.48
	4	26.1	0.6	7.54	9.97
F	0	2.3	1.3	6.18	7.49
	1	3.9	4.5	6.23	7.85
	2	12.4	17.9	6.69	8.73
	3	29.9	35.2	7.94	9.69
	4	59.2	40.8	9.00	9.91

Source: Edwards et al. (23), copyright © 1983, Blackwell Scientific Publications, Ltd.

[a] Samples E and F were obtained from two different retail outlets.
[b] Diamine values are the mean of two determinations.

occur in beef until the aerobic plate count (APC) exceeds about 4×10^7 (23) raises questions about their utility to predict meat spoilage. This is a common problem with most, if not all, single metabolites since their production and concentration tend to be related to specific organisms.

The **extract-release volume** (ERV) technique, first described in 1964, has been shown to be of value in determining incipient spoilage in meats as well as in predicting refrigerator shelf life (56, 57, 59). The technique is based on the volume of aqueous extract released by a homogenate of beef when allowed to pass through filter paper for a given period of time. By this method, beef of good organoleptic and microbial quality releases large volumes of extract while beef of poor microbial quality releases smaller volumes or none (Fig. 9-3). One of the more important aspects of this method is the information that it has provided concerning the mechanism of low-temperature beef spoilage.

The ERV method of detecting meat spoilage reveals two aspects of the spoilage mechanism. First, low-temperature meat spoilage occurs in the absence of any significant breakdown of primary proteins—at least not complete breakdown. Although this fact has been verified by total protein analyses on fresh and spoiled meats, it is also implicit in the operation of the method. That is, as meats undergo microbial spoilage, ERV is decreased rather than increased, which would be the case if complete hydrolysis of proteins occurred. The second aspect of meat spoilage revealed by ERV is the increase in hydration capacity of meat proteins by some as yet unknown mechanism, although amino sugar complexes produced by the spoilage flora have been shown to play a role (100). In the absence of complete protein breakdown, the question arises as to how the spoilage flora obtains its nutritional needs for growth.

When fresh meats are placed in storage at refrigerator temperatures, those organisms capable of growth at the particular temperature begin their growth. In the case of fresh meats that have an ultimate pH of around 5.6, enough glucose and other simple carbohydrates are present to support about 10^8 organisms/cm^2 (34). Among the heterogeneous fresh-meat flora, the organisms

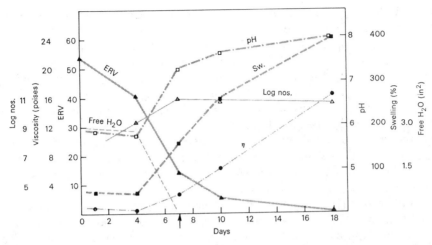

FIGURE 9-3. The response of several physicochemical meat spoilage tests as fresh ground beef was held at 7°C until definite spoilage had occurred. The arrow indicates the first day off-odors were detected. ERV = extract-release volume; free H_2O = measurement of water-holding capacity (inversely related); Sw = meat swelling; η = viscosity; and log Nos. = total aerobic bacteria/g.
From Shelef and Jay (99), copyright © 1969, Institute of Food Technologists.

that grow the fastest and utilize glucose at refrigerator temperatures are the pseudomonads, and available surface O_2 has a definite effect on their ultimate growth (38). *Brochothrix thermosphacta* also utilizes glucose and glutamate but because of its slower growth rate, it is a poor competitor of the pseudomonads. Upon reaching a surface population of about $10^8/cm^2$, the supply of simple carbohydrates is exhausted, and off-odors may or may not be evident at this point, depending on the extent to which free amino acid utilization has occurred. Once simple carbohydrates have been exhausted, pseudomonads along with gram-negative psychrotrophs such as *Moraxella, Alcaligenes, Aeromonas, Serratia*, and *Pantoea* utilize free amino acids and related simple nitrogenous compounds as sources of energy. *Acinetobacter* spp. utilize amino acids first, lactate next, and their growth is reduced at and below pH 5.7 (38).

The foul odors generally associated with spoiling meats owe their origin to free amino acids and related compounds (H_2S from sulfur-containing amino acids, NH_3 from many amino acids, and indole from tryptophane). Off-odors and off-flavors appear only when amino acids begin to be utilized (see below). In the case of dark, firm, and dry meats (DFD), which have ultimate pH >6.0 and a considerably lower supply of simple carbohydrates, spoilage is more rapid and off-odors are detectable with cell numbers of around $10^6/cm^2$ (83). With normal or DFD meats, the primary proteins are not attacked until the supply of the simpler constituents has been exhausted. It has been shown, for example, that the antigenicity of salt-soluble beef proteins is not destroyed under the usual conditions of low-temperature spoilage (72).

In the case of fish spoilage, it has been shown that raw fish press juice displays all the apparent aspects of fish spoilage as may be determined by use of the whole fish (69). This can be taken to indicate a general lack of attack upon insoluble proteins by the fish spoilage flora since these proteins were absent from the filtered press juice.

The same is apparently true for beef and related meats. Incipient spoilage is accompanied by a rise in pH, increase in bacterial numbers, and increase in the hydration capacity of meat proteins, along with other changes. In ground beef, pH may rise as high as 8.5 in putrid meats, although at the time of incipient spoilage, mean pH values of about 6.5 have been found (101). By plotting the growth curve of the spoilage flora, the usual phases of growth can be observed and the phase of decline may be ascribed to the exhaustion of utilizable nutrients by most of the flora and the accumulation of toxic by-products of bacterial metabolism. Precisely how the primary proteins of meat are destroyed at low temperatures is not well understood.

Dainty et al. (19) inoculated beef slime onto slices of raw beef and incubated them at 5°C. Off-odors and slime were noted after 7 days with counts at $2 \times 10^9/cm^2$. Proteolysis was not detected in either sarcoplasmic or myofibrillar fractions of the beef slices. No changes in the sarcoplasmic fractions could be detected even 2 days later when bacterial numbers reached $10^{10}/cm^2$. The first indication of breakdown of myofibrillar proteins occurred at this time with the appearance of a new band and the weakening of another. All myofibrillar bands disappeared after 11 days with weakening of several bands of the sarcoplasmic fraction. With naturally contaminated beef, odors and slime were first noted after 12 days when the numbers were $4 \times 10^8/cm^2$. Changes in myofibrillar proteins were not noted until 18 days of holding. By the use of pure culture studies, these workers showed that Shewan's Group I pseudomonads (see Appendix, Fig. C for the Shewan scheme) were active against myofibrillar proteins while Group II organisms were more active against sarcoplasmics. *Aeromonas* spp. were active on both myofibrillar and sarcoplasmic proteins. With pure cultures, protein changes were not detected until counts were above $3.2 \times 10^9/cm^2$. Borton et al. (9) showed earlier that *P. fragi* (a Group II pseudomonad) effected the loss of protein bands from inoculated pork muscle, but no indication was given as to the minimum numbers that were necessary.

For further information on the mechanism of meat spoilage, the following references should be consulted: 20, 35, 36, and 77.

SPOILAGE OF FRESH LIVERS

The events that occur in the spoilage of beef, pork, and lamb livers are not as well defined as for meats. The mean content of carbohydrate, NH_3, and pH of 10 fresh lamb livers is presented in Table 9-8 (37). Based on the relatively high content of carbohydrates and mean pH of 6.41, these may be expected to undergo a fermentative spoilage, with the pH decreasing below 6.0. This would undoubtedly occur if livers were comminuted or finely diced and stored at refrigerator temperatures, but most studies have been conducted on whole livers, where growth was assessed at the surface, from drip, or from deep tissue. In a study of the spoilage of diced beef livers, the initial pH of 6.3 decreased to about 5.9 after 7 to 10 days at 5°C and the predominant flora at spoilage consisted of lactic acid bacteria (97). In most other studies, the predominant flora at spoilage was found to consist essentially of the same

TABLE 9-8.   pH and Concentrations in Ten Fresh Livers of Glycogen, Glucose, Lactic Acid, and Ammonia

Component	Average Concentration and Range
Glucose	2.73 (0.68–6.33) mg/g
Glycogen	2.98 (0.70–5.43) mg/g
Lactic acid	4.14 (3.42–5.87) mg/g
Ammonia	7.52 (6.44–8.30) μmol/g
pH	6.41 (6.26–6.63)

Source: Gill and DeLacy (37). Copyright © 1982, American Society for Microbiology.

types of organisms that are dominant in the spoilage of muscle meats. In pork livers held at 5°C for 7 days, the predominant organisms found in one study were *Pseudomonas, Alcaligenes, Escherichia*, lactic streptococci, and *B. thermosphacta* (32). In five beef livers stored at 2°C for 14 days, *Pseudomonas* constituted from 75 to 100% of the spoilage flora while the mean initial pH of 6.49 decreased to 5.93 over the 14-day period (44). In another study of beef, pork, and lamb livers, the predominant flora after 5 days at 2°C differed for the three products, with beef livers being dominated by streptococci, yeasts, coryneforms, and pseudomonads; lamb by coryneforms, micrococci, and streptococci; and pork livers by staphylococci, *Moraxella-Acinetobacter*, and streptococci (43). The mean initial pH of each of the three livers declined upon storage, although only slightly. In a study of spoilage of lamb livers by Gill and DeLacy (37), the spoiled surface flora was dominated by *Pseudomonas, Acinetobacter*, and *Enterobacter*; drip from the whole livers was dominated by *Pseudomonas* and *Enterobacter*; while *Enterobacter* and lactobacilli were dominant in the deep tissues. It was shown in this study that the initial pH of around 6.4 decreased to around 5.7 in antibiotic-treated samples, indicating that liver glycolytic events can lead to a decrease in pH in the absence of organisms even though these samples did contain $<10^4$ organisms/cm^2. The high glucose level was sufficient to allow visible surface colony growth before off-odors developed, and herein may lie the explanation for the dominance of the spoilage flora of livers by nonlactic types.

Since most psychrotrophic oxidative gram-negative bacteria grow at a faster rate and are more favored by the higher surface Eh than the lactic fermentative gram positives, their dominance in whole liver spoilage may not be unexpected. The higher concentration of carbohydrates would delay the onset of amino acid utilizers and explain in part why pH does not increase with whole liver spoilage as it does for meats. In this regard, comminuted livers would be expected to support the growth of lactic acid bacteria because of the redistribution of the surface flora throughout the sample where the lactics would be favored by the high carbohydrate content and reduced Eh away from the surface. This would be somewhat analogous to the surface spoilage of meat carcasses, where the slower-growing yeasts and molds develop when conditions are not favorable for bacterial growth. Fungi never dominate the spoilage of fresh comminuted meats unless special steps are taken to inhibit bacteria.

By this analogy, lactic acid bacteria are inconspicuous in the spoilage of whole livers because conditions favor the faster-growing, psychrotrophic gram-negative bacteria.

SPOILAGE OF VACUUM-PACKAGED MEATS

From the research of many groups it is clear that when vacuum-packaged meats undergo long-term refrigerator spoilage, often the predominant organisms are lactobacilli or *B. thermosphacta* or both. Other organisms can be found and, indeed, others may predominate. Among the determining factors are the following:

1. Whether the product is raw or cooked
2. Concentration of nitrites present
3. Relative load of psychrotrophic bacteria
4. The degree to which the vacuum-package film excludes O_2
5. Product pH

Cooked or partially cooked meats, along with DFD and dark-cutting meats, have higher pH than raw and light-cutting meats, and the organisms that dominate these products during vacuum storage are generally different from those found in vacuum-packaged normal meats. In vacuum-packaged DFD meats held at 2°C for 6 weeks, the dominant flora consisted of *Yersinia enterocolitica, Serratia liquefaciens, Shewanella putrefaciens*, and a *Lactobacillus* sp. (39). *S. putrefaciens* caused greening of product, but a pH <6.0 was inhibitory to its growth. When dark-cutting beef of pH 6.6 was vacuum packaged and stored at 0° to 2°C, lactobacilli were dominant after 6 weeks, but after 8 weeks psychrotrophic Enterobacteriaceae became dominant (86). Most of the Enterobacteriaceae resembled *S. liquefaciens* and the remainder resembled *Hafnia alvei*. In vacuum-packaged beef with pH 6.0, *Y. enterocolitica*-like organisms were found at levels of $10^7/g$ after 6 weeks at 0° to 2°C, but on meats with pH <6.0, their numbers did not exceed $10^5/g$ even after 10 weeks (91). The high-pH meat also yielded *S. putrefaciens* with counts as high as log 6.58/g after 10 weeks.

When normal raw beef with an ultimate pH of about 5.6 is vacuum packaged, lactobacilli and other lactic acid bacteria predominate. When the beef was allowed to spoil aerobically, acidic/sour odors were noted when APC was about 10^7–$10^8/cm^2$ with approximately 15% of the flora being *Pseudomonas* spp.; but when vacuum-packaged samples spoiled, the product was accompanied by a slight increase in pH with general increase in ERV (108). After a 9-week storage at 0° to 1°C, Hitchener et al. (49) found that 75% of the flora of vacuum-packaged raw beef consisted of catalase-negative organisms. Upon further characterization of 177 isolates, 18 were found to be *Leuconostoc mesenteroides*, 115 were hetero- and 44 were homofermentative lactobacilli. Using high-barrier oxygen film, the dominant flora of vacuum-packaged beef loin steaks after 12 and 24 days consisted of heterofermentative lactobacilli with *Lactobacillus cellobiosus* being isolated from 92% of the steaks (112). In 59% of the samples, *L. cellobiosus* constituted 50% or more of the flora. The latter investigators found that when medium oxygen-barrier film was used, high percentages of organisms such as *Aeromonas*, *Enterobacter*, *Hafnia*, *B. thermosphacta*, pseudomonads, and *Proteus morganii* were usually found.

When high concentrations of nitrites are present, they generally inhibit *B. thermosphacta* and psychrotrophic Enterobacteriaceae, and the lactic acid bacteria become dominant because they are relatively insensitive to nitrites (84). However, low concentrations of nitrites appear not to affect *B. thermosphacta* growth, especially in cooked, vacuum-packaged products. When Egan et al. (25) inoculated this organism and a homo- and a heterofermentative lactobacillus into corned beef and sliced ham containing 240 ppm nitrate and 20 ppm nitrite, *B. thermosphacta* grew with no detectable lag phase. It had a generation time of 12 to 16 h at 5°C, while the generation time for the heterofermentative lactobacillus was 13 to 16 h and for the homofermentative 18 to 22 h. Time to reach 10^8 cells/g was 9, 9 to 12, and 12 to 20 days, respectively. While off-flavors developed 2 to 3 days after the numbers attained 10^8/g for *B. thermosphacta*, the same did not occur for the homo- and heterofermentative lactobacilli until 11 and 21 days, respectively. The lactic acid bacteria are less significant than *B. thermosphacta* in the spoilage of vacuum-packaged luncheon meats (26). On the other hand, this organism has a longer lag phase and a slower growth rate than the lactobacilli (36). When the two groups are present in equal numbers, the lactobacilli generally dominate.

Findings from a large number of studies make it clear that when red meats are stored for several weeks at 0° to 5°C under packaging conditions that restrict the entry of O_2 or the egress of CO_2, the ultimate flora is dominated by lactic acid bacteria. For example, in a study of vacuum-packaged beef and lamb stored for 28 to 35 days, the initial flora of *Corynebacterium* spp. and *B. thermosphacta* was dominated by lactobacilli (45, 46). With the more widespread use of vacuum and modified atmosphere packaging (MAP) conditions, and with the application of more precise taxonomic methods, it appears that at least two groups of organisms can be regarded as being specific to these product storage conditions.

It appears that at least two *Leuconostoc* spp. are uniquely adapted to vacuum-packaged and MAP meats, *L. carnosum* and *L. gelidum*. Unnamed *Leuconostoc* spp. were found in one study of loins packaged under high O_2 and CO_2 to constitute from 88 to 100% of the flora (89) and to be the dominant members of the flora in another similar study (41). Following an extensive study of lactic acid bacteria isolated from vacuum-packaged meats, *L. carnosum* and *L. gelidum* were established (95). Both species grow at 1°C but not at 37°C, and both produce gas from glucose.

The second group consists of bacteria now classified in the genus *Carnobacterium*. These catalase-negative bacteria are heterofermentative, produce only L(+)-lactic acid, and produce gas from glucose (the typical heterofermentative betabacteria produce both D-and L-lactate). Prior to 1987, the carnobacteria were regarded as being lactobacilli. Of 159 isolates of lactobacilli from vacuum-packaged beef, 115 could not be identified to species level (49). Similar strains were isolated from vacuum-packaged beef, pork, lamb, and bacon (95). Another group of investigators isolated similar organisms from vacuum-packaged beef and upon further study named these unique organisms *Lactobacillus divergens* (51). In the latter study, this organism constituted 6.7% of 120 psychrotrophic isolates, none of which grew either at pH 3.9 or at 4°C in MRS broth. Following DNA-DNA hybridization and other studies, the genus *Carnobacterium* was erected, and *L. divergens* and two other species were placed in the new genus (16). *C. divergens* is associated with vacuum-packaged meats, and *C. piscicola* and *C. mobile* are associated with fish and irradiated chicken, respectively. Since it does not produce H_2S or other foul-odor compounds, *C. divergens* may not be a spoilage bacterium. Indeed, it, along with the two leuconostocs noted, has the potential of being beneficial in gas-impermeable packages where they may produce enough CO_2 to inhibit undesirable organisms.

The behavior of *B. thermosphacta* in vacuum-packaged meats was discussed in Chapter 3. It grows on beef at pH 5.4 when incubated aerobically but does not grow <pH 5.8 anaerobically (11). Under the latter conditions, the apparent minimum growth pH is 6.0. *S. putrefaciens* is also pH sensitive and does not grow on beef of normal pH but grows on DFD meats.

Volatile Components of Vacuum-Packaged Meats, Poultry, and Seafood

The off-odors and off-flavors produced in vacuum-packaged meat products by the spoilage flora are summarized in Table 9-9. In general, short-chain fatty acids are produced by both lactobacilli and *B. thermosphacta*, and spoiled products may be expected to contain these compounds, which confer sharp off-odors. In vacuum-packaged luncheon meats, acetoin and diacetyl have been found to be the most significant relative to spoiled meat odors (105). Using a culture medium (APT) containing glucose and other simple carbohydrates, the formation by *B. thermosphacta* of isobutyric and isovaleric acids was favored by low glucose and near-neutral pH, while acetoin, acetic acid, 2,3-butanediol, 3-methylbutanol, and 3-methylpropanol production were favored by high glucose and low pH (17, 18). According to these authors, acetoin is the major volatile compound produced on raw and cooked meats in O_2-containing atmospheres. This suggests that the volatile compounds produced by *B. thermosphacta* may be expected to vary between products with high and low glucose concentrations. The addition of 2% glucose to raw ground beef has been shown to decrease pH and delay off-odor and slime development without affecting the general spoilage flora (98), and while the studies noted were not conducted with vacuum-packaged meats, it would seem to be a way to shift the volatile components from short-chain fatty acids to acetoin and other compounds that derive from glucose. Since vacuum-packaged, high-pH meats have a much shorter shelf life, the addition of glucose could be of benefit in this regard.

In a study of spoiled vacuum-packaged steaks, a sulfide odor was evident with numbers of $10^7-10^8/cm^2$ (42). The predominant organisms isolated were *H. alvei*, lactobacilli, and *Pseudomonas*. *H. alvei* was the likely cause of the sulfide odor.

From the summary of volatiles in Table 9-9, it is evident that all organisms produced either dimethyl di- or trisulfide, or methyl mercaptan, except *B. thermosphacta*. Dimethyl disulfide was produced in chicken by 8 of 11 cultures evaluated by Freeman et al. (30), ethanol by 7, and methanol and ethyl acetate

TABLE 9-9.   Volatile Compounds Produced by the Spoilage Flora or Spoilage Organisms in Meats, Poultry, Seafood, or Culture Media

Organism/Inoculum	Substrate/Conditions	Principal Volatiles	Reference
Shewanella putrefaciens	Sterile fish muscle, 1–2°C, 15 days	Dimethyl sulfide, dimethyl trisulfide, methyl mercaptan, trimethylamine, propionaldehyde, 1-penten-3-ol, H_2S, etc.	79
"Achromobacter" sp.	As above	Same as above except no dimethyl trisulfide or H_2S	79
P. fluorescens	As above	Methyl sulfide, dimethyl disulfide	79
P. perolens	As above	Dimethyl trisulfide, dimethyl disulfide, methyl mercaptan, 2-methoxy-3-isopropylpyrazine (potatolike odor)	80
Moraxella sp.	TSY agar, 2–4°C, 14 days	16 compounds including dimethyl disulfide, dimethyl trisulfide, methyl isobutyrate, and methyl-2-methyl butyrate	67
P. fluorescens	As above	15 compounds including all above except methyl isobutyrate	67
P. putida	As above	14 compounds including same for *Moraxella* sp. above except methyl isobutyrate and methyl-2-methyl butyrate	67
B. thermosphacta	Inoculated vacuum-packaged corned beef, 5°C	7 compounds including diacetyl, acetoin, nonane, 3-methyl-butanal, and 2-methylbutanol	105
	Aerobically stored, inoculated beef slices, 1°C, 14 days, pH 5.5–5.8	Acetoin, acetic acid, isobutyric/isovaleric acids. Acetic acid increased four fold after 28 days.	17
	As above; pH 6.2–6.6	Acetic acid, isobutyric, isovaleric, and *n*-butyric acids	17
	APT broth, pH 6.5, 0.2% glucose	Acetoin, acetic acid, isobutyric and isovaleric acids	17
	APT broth, pH 6.5, no glucose	Same as above but no acetoin	17
B. thermosphacta (15 strains)	APT broth, pH 6.5, 0.2% glucose	Acetoin, acetic acid, isobutyric and isovaleric acids, traces of 3-methylbutanol	18
S. putrefaciens	Grown in radapper-tized chicken, 5 days, 10°C	H_2S, methyl mercaptan, dimethyl disulfide, methanol, ethanol	30
P. fragi	As above	Methanol, ethanol, methyl and ethyl acetate, dimethyl sulfide, methanol, ethanol	30
B. thermosphacta	As above	Methanol, ethanol	30
Flora	Spoiled chicken	11 compounds including H_2S, methanol, ethanol, methyl mercaptan, dimethyl sulfide, dimethyl disulfide	30

by 6 each. *S. putrefaciens* consistently produces H_2S in vacuum-packaged meats on which it grows. From chicken breast muscle inoculated with *Pseudomonas* Group II strains and held at 2°C for 14 days, odors detected from chromatograph peaks were described by McMeekin (75) as being "sulfide-like," "evaporated milk," and "fruity."

SPOILAGE OF FRANKFURTERS, BOLOGNA, SAUSAGE, AND LUNCHEON MEATS

Unlike other meats covered in this chapter, frankfurters, bologna, sausage, and luncheon meats are prepared from various ingredients, any one or all of which may contribute microorganisms to the final product. Bacteria, yeasts, and molds may be found in and upon processed meats, but the former two groups are by far the most important in the microbial spoilage of these products.

Spoilage of these products is generally of three types: sliminess, souring, and greening. **Slimy spoilage** occurs on the outside of casings, especially of frankfurters, and may be seen in its early stages as discrete colonies, which may later coalesce to form a uniform layer of gray slime. From the slimy material may be isolated yeasts, lactic acid bacteria of the genera *Lactobacillus*, *Enterococcus*, and *B. thermosphacta*. *L. viridescens* produces both sliminess and greening. Slime formation is favored by a moist surface and is usually confined to the outer casing. Removal of this material with hot water leaves the product essentially unchanged.

Souring generally takes place underneath the casing of these meats and results from the growth of lactobacilli, enterococci, and related organisms. The usual sources of these organisms to processed meats are milk solids. The souring results from the utilization of lactose and other sugars by the organisms with the production of acids. Sausage usually contains a more varied flora than most other processed meats due to the different seasoning agents employed, almost all of which contribute their own flora. *B. thermosphacta* has been found to be the most predominant spoilage organism for sausage by many investigators.

Although mold spoilage of these meats is not common, it can and does occur under favorable conditions. When the products are moist and stored under conditions of high humidity, they tend to undergo bacterial and yeast spoilage. Mold spoilage is likely to occur only when the surfaces become dry or when the products are stored under other conditions that do not favor bacteria or yeasts.

Two types of **greening** occur on stored and processed red meats, one caused by H_2O_2 and the other by H_2S. The former occurs commonly on frankfurters as well as on other cured and vacuum-packaged meats. It generally appears after an anaerobically stored meat product is exposed to air. Upon exposure to air, H_2O_2 forms and reacts with nitrosohemochrome to produce a greenish oxidized porphyrin (40). Greening also occurs from growth of causative organisms in the interior core, where the low O/R potential allows H_2O_2 to accumulate. *Lactobacillus viridescens* is the most common organism in this type of greening, but leuconostocs, *Enterococcus faecium*, and *Enterococcus faecalis* are capable of producing greening of products. Greening can be produced also by H_2O_2 producers such as *Lactobacillus fructovorans* and *Lactobacillus jensenii*. *L. viridescens* is resistant to >200 ppm $NaNO_2$, and it can grow in the presence of 2 to 4% NaCl but not in 7% (40). *L. viridescens* has been recovered from anaerobically spoiled frankfurters and from both smoked pork loins and frankfurter sausage stored in atmospheres of CO_2 and N_2 (8). In spite of the discoloration, the green product is not known to be harmful when eaten.

TABLE 9-10. Pigments Found in Fresh, Cured, or Cooked Meat

Pigment	Mode of Formation	State of Iron	State of Haematin Nucleus	State of Globin	Color
1. Myoglobin	Reduction of metmyoglobin; de-oxygenation of oxymyoglobin	Fe^{++}	Intact	Native	Purplish red
2. Oxymyoglobin	*Oxygenation* of myoglobin	Fe^{++}	Intact	Native	Bright red
3. Metmyoglobin	*Oxidation* of myoglobin, oxymyoglobin	Fe^{+++}	Intact	Native	Brown
4. Nitric oxide myoglobin	Combination of myoglobin with nitric oxide	Fe^{++}	Intact	Native	Bright red
5. Metmyoglobin nitrite	Combination of metmyoglobin with excess nitrite	Fe^{+++}	Intact	Native	Red
6. Globin haemochromogen	Effect of heat, denaturing agents on myoglobin, oxymyoglobin; irradiation of globin haemichromogen	Fe^{++}	Intact	Denatured	Dull red
7. Globin haemichromogen	Effect of heat, denaturing agents on myoglobin, oxymyoglobin, metmyoglobin, haemochromogen	Fe^{+++}	Intact	Denatured	Brown
8. Nitric oxide haemochromogen	Effect of heat, salts on nitric oxide myoglobin	Fe^{++}	Intact	Denatured	Bright red
9. Sulphmyoglobin	Effect of H_2S and oxygen on myoglobin	Fe^{+++}	Intact but reduced	Denatured	Green
10. Choleglobin	Effect of hydrogen peroxide on myoglobin or oxymyoglobin; effect of ascorbic or other reducing agent on oxymyoglobin	Fe^{++} or Fe^{+++}	Intact but reduced	Denatured	Green
11. Verdohaem	Effect of reagents as in 9 in excess	Fe^{+++}	Porphyrin ring opened	Denatured	Green
12. Bile pigments	Effect of reagents as in 9 in large excess	Fe absent	Porphyin ring destroyed: chain of porphyrins	Absent	Yellow or colorless

Source: Reprinted with permission from R. A. Lawrie (65), copyright © 1966, Pergamon Press.

**TABLE 9-11.   Effect of Meat pH and Glucose on Hydrogen
Sulphide Production by a Pure Culture of
Lactobacillus sake L13 Growing under
Anaerobic Conditions at 5°C on Beef**

	Hydrogen Sulphide Production		
Days	pH 5.6–5.7	pH 6.4–6.6	pH 6.4–6.6 with 250 µg Glucose per g of Meat
8	$-^a$	–	–
9	–	$+^c$	–
11	–	+	–
15	–	+	–
18	$+^b$	+	$+^c$
21	+	+	+

Source: Egan et al. (24).

[a] Each treatment done in triplicate. – = all three tubes negative: + = all
three positive.
[b] One tube out of three positive.
[c] Two tubes out of three positive.

The second type of greening occurs generally on fresh red meats that are
held at 1° to 5°C and stored in gas-impermeable or vacuum-packaging containers; it is caused by H_2S production. H_2S reacts with myoglobin to form
sulphmyoglobin (Table 9-10). This type of greening usually does not occur
when meat pH is below 6.0. The responsible organism in one study was
thought to be *Pseudomonas mephitica* (85), but in another study of DFD
meats, *S. putrefaciens* was the H_2S producer (39). In the latter, greening
occurred even with glucose present, and it could be prevented by lowering
pH to below 6.0. H_2S-producing lactobacilli were recovered from vacuum-
packaged fresh beef and found to produce H_2S in the pH range 5.4 to 6.5 (96).
Only slight greening was produced, and the H_2S was from cysteine, a system
that was plasmid borne. The organism reached $3 \times 10^7/cm^2$ after 7 days, and
ultimately reached about $10^8/cm^2$ at 50°C. No offness of vacuum-packaged sliced
luncheon meat was observed when another lactobacillus attained $10^8/cm^2$.

At least one strain of *Lactobacillus sake* has been shown to produce H_2S on
vacuum-packaged beef; the effect of pH and glucose on production is presented in Table 9-11 (24). The investigators found that greening by *L. sake* was
not as intense as that caused by *S. putrefaciens* and that it occurred only after
about 6 weeks at 0°C. Further, the lactobacillus produced H_2S only in the
absence of O_2 and utilizable sugars. No greening was observed when films with
an O_2 transmission rate of $1\,ml\ O_2/m^2$ or $300\,ml\ O_2/m^2$ were used, but it did
occur with films that had O_2 transmission rates between 25 and $200\,ml/m^2/24\,h$
(24). Visible greening was seen only on samples packaged in films with O_2
transmission rates of 100 and $200\,ml/m^2/24\,h$ and only after 75 days' storage.
With meat in the pH 6.4 to 6.6 range, H_2S was detected when cell numbers
reached $10^8/g$.

A **yellow** discoloration of vacuum-packaged luncheon-style meat was caused
apparently by *Enterococcus casseliflavus*. The discoloration appeared as small

spots on products stored at 4.4°C, and it was fluorescent under long-wave ultraviolet light (115). Between 3 and 4 weeks were required for the condition to develop, and the responsible organism survived 71.1°C for 20 min but not 30 min. In addition to 4.4°C, it occurred also at 10° but not at 20°C or above. Although tentatively identified as *E. casseliflavus*, the causative organism did not react with Group D antisera. The other yellow-pigmented enterococcal species is *E. mundtii*; both are discussed further in Chapter 17.

SPOILAGE OF BACON AND CURED HAMS

The nature of these products and the procedures employed in preparing certain ones, such as smoking and brining, make them relatively insusceptible to spoilage by most bacteria. The most common form of bacon spoilage is moldiness, which may be due to *Aspergillus*, *Alternaria*, *Fusarium*, *Mucor*, *Rhizopus*, *Botrytis*, *Penicillium*, and other molds (Table 9-4). The high fat content and low a_w make it somewhat ideal for this type of spoilage. Bacteria of the genera *Enterococcus*, *Lactobacillus*, and *Micrococcus* are capable of growing well on certain types of bacon such as Wiltshire, and *E. faecalis* is often present on several types. Vacuum-packed bacon tends to undergo souring due primarily to micrococci and lactobacilli. Vacuum-packed, low-salt bacon stored above 20°C may be spoiled by staphylococci (111).

Cured hams undergo a type of spoilage different from that of fresh or smoked hams. This is due primarily to the fact that curing solutions pumped into the hams contain sugars that are fermented by the natural flora of the ham and also by those organisms pumped into the product in the curing solution, such as lactobacilli. The sugars are fermented to produce conditions referred to as "**sours**" of various types, depending on their location within the ham. A large number of genera of bacteria have been implicated as the cause of ham "sours," among them *Acinetobacter*, *Bacillus*, *Pseudomonas*, *Lactobacillus*, *Proteus*, *Micrococcus*, and *Clostridium*. **Gassiness** is not unknown to occur in cured hams where members of the genus *Clostridium* have been found. The spoilage of canned hams is treated in Chapter 10.

In their study of vacuum-packed sliced bacon, Cavett (12) and Tonge et al. (111) found that when high-salt bacon was held at 20°C for 22 days, the catalase-positive cocci dominated the flora, while at 30°C the coagulase-negative staphylococci became dominant. In the case of low-salt bacon (5–7% NaCl versus 8–12% in high-salt bacon) held at 20°C, the micrococci as well as *E. faecalis* became dominant; at 30°C the coagulase-negative staphylococci as well as *E. faecalis* and micrococci became dominant. Spoilage of this type of product is characterized by a "cheesy," sour-scented, and putrid off-odor.

In a study of lean Wiltshire bacon stored aerobically at 5°C for 35 days or 10°C for 21 days, Gardner (31) found nitrates were reduced to nitrites when the microbial load reached about 10^9/g. The predominant organisms at this stage were micrococci, vibrios, and the yeast genera *Candida* and "*Torulopsis*." Upon longer storage, microbial counts reached about 10^{10}/g with the disappearance of nitrites. At this stage, *Acinetobacter*, *Alcaligenes*, and *Arthrobacter-Corynebacterium* spp. became more important. Micrococci were always found while vibrios were found in all bacons with salt contents that were above 4%.

More information on the spoilage of some of these products can be obtained from the review by Gardner (33).

SPOILAGE OF POULTRY

Studies on the bacterial flora of fresh poultry by many investigators have revealed over 25 genera (Table 9-3). However, when these meats undergo low-temperature spoilage, almost all workers agree that the primary spoilage organisms belong to the genus *Pseudomonas*. In a study of 5,920 isolates from chicken carcasses (64), pseudomonads were found to constitute 30.5%, *Acinetobacter* 22.7%, *Flavobacterium* 13.9%, *Corynebacterium* 12.7%, with yeasts, Enterobacteriaceae, and others in lower numbers. Of the pseudomonads, these investigators found that 61.8% were fluorescent on King's medium and that 95.2% of all pseudomonads oxidized glucose. A previous characterization of pseudomonads on poultry undergoing spoilage was made by Barnes and Impey (6), who showed that the pigmented pseudomonads (Shewan's Group I) decreased from 34% to 16% from initial storage to the development of strong off-odors, while the nonpigmented actually increased from 11% to 58% (see the section below on fish spoilage). *Acinetobacter* and other species of bacteria decreased along with the type I pseudomonads. A similar process occurs in spoiling fish.

Fungi are of considerably less importance in poultry spoilage except when antibiotics are employed to suppress bacterial growth. When antibiotics are employed, however, molds become the primary agents of spoilage. The genera *Candida*, *Rhodotorula*, and "*Torula*" are the most important yeasts found on poultry (Table 9-5). The essential feature of poultry spoilage is **sliminess** at the outer surfaces of the carcass or cuts. The visceral cavity often displays sour odors or what is commonly called visceral taint. This is especially true of the spoilage of **New York–dressed poultry**, where the viscera are left inside. The causative organisms here are also bacteria of the type noted above in addition to enterococci.

The primary reasons that poultry spoilage is mainly restricted to the surfaces are as follows. The inner portions of poultry tissue are generally sterile, or contain relatively few organisms, which generally do not grow at low temperatures. The spoilage flora, therefore, is restricted to the surfaces and hide where it is deposited from water, processing, and handling. The surfaces of fresh poultry stored in an environment of high humidity are susceptible to the growth of aerobic bacteria such as pseudomonads. These organisms grow well on the surfaces, where they form minute colonies that later coalesce to produce the sliminess characteristic of spoiled poultry. May et al. (74) showed that poultry skin supports the growth of the poultry spoilage flora better than even the muscle tissue. In the advanced stages of poultry spoilage, the surfaces will often fluoresce when illuminated with ultraviolet light because of the presence of large numbers of fluorescent pseudomonads. Surface spoilage organisms can be recovered directly from the slime for plating, or one can prepare slides for viewing by smearing with portions of slime. Upon Gram staining, one may note the uniform appearance of organisms indistinguishable from those listed.

Tetrazolium (2,3,5-triphenyltetrazolium chloride) can be used also to assess microbial activity on poultry surfaces. When the eviscerated carcass is sprayed with this compound, a red pigment develops in areas of high microbial activity. These areas generally consist of cut muscle surfaces and other damaged areas such as feather follicles (87).

As poultry undergoes spoilage, off-odors are generally noted before sliminess, with the former being first detected when log numbers/cm^2 are about 7.2 to 8.0. Sliminess generally occurs shortly after the appearance of off-odors, with the log counts/cm^2 about 8 (5). Total aerobic plate counts/cm^2 of slimy surface rarely go higher than log 9.5. With the initial growth first confined to poultry surfaces, the tissue below the skin remains essentially free of bacteria for some time. Gradually, however, bacteria begin to enter the deep tissues, bringing about increased hydration of muscle proteins, much as occurs with beef. Whether autolysis plays an important role in the spoilage of inner poultry tissues is not clear.

Vacuum packaging and CO_2-atmosphere storage are effective in delaying the spoilage of poultry. With raw, cut-up poultry stored at 5°C in O_2-permeable film, vacuum packages, and CO_2-flushed high O_2-barrier film, it was unacceptable by day 9, between days 9 and 11, and after 17 days, respectively (110).

S. putrefaciens grows well at 5°C and produces potent off-odors in 7 days when growing on chicken muscle (78). Among odor producers in general, there is a selection of types that produce strong odors among the varied flora that exists on fresh poultry (75). The study noted was conducted with chicken breast muscle, which spoils differently from leg muscles since the latter have a higher pH. With chicken leg muscle stored at 2°C for 16 days, 47% of the flora consisted of Group I pseudomonads, 32% of Group II, 17% of *Acinetobacter/Moraxella*, and 4% of *S. putrefaciens* (76). All isolates of the latter produced sulfidelike odors and, as may be noted from Table 9-9, this organism produces H_2S, methyl mercaptan, and dimethyl sulfide. It was not of significance in the spoilage of chicken breast muscle. Since Shewan's Group II pseudomonads grow faster than the pigment-producing Group I strains, it appears that the strong odor-producing capacity is a property of these strains. Group II pseudomonads have been shown to be consumers of free amino acids in chicken skin, while Group I types effected increases in the quantities of free amino acids and related nitrogenous compounds (2).

When New York–dressed poultry undergoes microbial spoilage, the organisms make their way through the gut walls and invade inner tissues of the intestinal cavity. The characteristic sharpness associated with the spoilage of this type of poultry is referred to as "visceral taint."

SPOILAGE OF FISH AND SHELLFISH

Fish

Both saltwater and freshwater fish contain comparatively high levels of proteins and other nitrogenous constituents (Table 9-12). The carbohydrate content of these fish is nil, while fat content varies from very low to rather high values depending on species. Of particular importance in fish flesh is the nature of the nitrogenous compounds. The relative percentages of total-N and protein-N are

TABLE 9-12.   Fish and Shellfish: Approximate Percentage Chemical Composition

	Water	Carbo-hydrates	Proteins	Fat	Ash
Bony fish					
Bluefish	74.6	0	20.5	4.0	1.2
Cod	82.6	0	16.5	0.4	1.2
Haddock	80.7	0	18.2	0.1	1.4
Halibut	75.4	0	18.6	5.2	1.0
Herring (Atlantic)	67.2	0	18.3	12.5	2.7
Mackerel (Atlantic)	68.1	0	18.7	12.0	1.2
Salmon (Pacific)	63.4	0	17.4	16.5	1.0
Swordfish	75.8	0	19.2	4.0	1.3
Crustaceans					
Crab	80.0	0.6	16.1	1.6	1.7
Lobster	79.2	0.5	16.2	1.9	2.2
Mollusks					
Clams, meat	80.3	3.4	12.8	1.4	2.1
Oysters	80.5	5.6	9.8	2.1	2.0
Scallops	80.3	3.4	14.8	0.1	1.4

Source: Watt and Merrill (114).

TABLE 9-13.   Distribution of Nitrogen in Fish and Shellfish Flesh

Species	Percentage Total N	Percentage Protein N	Ratio of Protein N/Total N
Cod (Atlantic)	2.83	2.47	0.87
Herring (Atlantic)	2.90	2.53	0.87
Sardine	3.46	2.97	0.86
Haddock	2.85	2.48	0.87
Lobster	2.72	2.04	0.75

Source: Jacquot (55).

presented in Table 9-13, from which it can be seen that not all nitrogenous compounds in fish are in the form of proteins. Among the nonprotein nitrogen compounds are the free amino acids, volatile nitrogen bases such as ammonia and trimethylamine, creatine, taurine, the betaines, uric acid, anserine, carnosine, and histamine.

It is generally recognized that the internal flesh of healthy, live fish is sterile (48, 109), although a few reports to the contrary exist. Bacteria that exist on fresh fish are generally found in three places: the outer slime, the gills, and the intestines of feeding fish.

The microorganisms known to cause fish spoilage are indicated in Tables 9-3, 9-4, and 9-5. Fresh iced fish are invariably spoiled by bacteria, while salted and dried fish are more likely to undergo fungal spoilage. The bacterial flora of spoiling fish is found to consist of asporogenous, gram-negative rods of the *Pseudomonas* and *Acinetobacter-Moraxella* types. Many fish spoilage bacteria are capable of good growth between 0° to 1°C. Shaw and Shewan (94) found that a large number of *Pseudomonas* spp. are capable of causing fish spoilage at −3°C, although at a slow rate.

The spoilage of salt- and freshwater fish appears to occur in essentially the same manner, with the chief differences being the requirement of the saltwater flora for a seawater type of environment and the differences in chemical composition between various fish with respect to nonprotein nitrogenous constituents. The most susceptible part of fish is the gill region, including the gills. The earliest signs of organoleptic spoilage may be noted by examining the gills for the presence of off-odors. If feeding fish are not eviscerated immediately, intestinal bacteria soon make their way through the intestinal walls and into the flesh of the intestinal cavity. This process is believed to be aided by the action of proteolytic enzymes, which are from the intestines and may be natural enzymes inherent in the intestines of the fish, or enzymes of bacterial origin from the inside of the intestinal canal, or both. Fish spoilage bacteria apparently have little difficulty in growing in the slime and on the outer integument of fish. Slime is composed of mucopolysaccharide components, free amino acids, trimethylamine oxide, piperidine derivatives, and other related compounds. As is the case with poultry spoilage, plate counts are best done on the surface of fish, with numbers of organisms expressed per cm^2 of examined surface.

It appears that the spoilage organisms first utilize the simpler compounds and in the process release various volatile off-odor components. According to Shewan (103), trimethylamine oxide, creatine, taurine, anserine, and related compounds along with certain amino acids decrease during fish spoilage with the production of trimethylamine, ammonia, histamine, hydrogen sulfide, indole, and other compounds (see Table 9-9). Fish flesh appears to differ from mammalian flesh in regard to autolysis. Flesh of the former type seems to undergo autolysis at more rapid rates. While the occurrence of this process along with microbial spoilage is presumed by some investigators to aid either the spoilage flora or the spoilage process (47), attempts to separate and isolate the events of the two have proved difficult. In a detailed study of fish isolates with respect to the capacity to cause typical fish spoilage by use of sterile fish muscle press juice, Lerke et al. (70) found that the spoilers belonged to the genera *Pseudomonas* and *Acinetobacter-Moraxella*, with none of the coryneforms, micrococci, or flavobacteria being spoilers. In characterizing the spoilers with respect to their ability to utilize certain compounds, these workers found that most spoilers were unable to degrade gelatin or digest egg albumin. This suggests that fish spoilage proceeds much as does that of beef—in the general absence of complete proteolysis by the spoilage flora. Pure culture inoculations of cod and haddock muscle blocks have failed to effect tissue softening (47). In fish that contain high levels of lipids (herrings, mackerel, salmon, and others), these compounds undergo rancidity as microbial spoilage occurs. It should be noted that the skin of fish is rich in collagen. The scales of most fish are composed of a scleroprotein belonging to the keratin group, and it is quite probable that these are among the last parts of fish to be decomposed.

Earlier studies on the interplay of the bacterial flora of fish undergoing spoilage indicated that *Pseudomonas* spp. of Shewan's Group II became

TABLE 9-14.   Microbial Population Change in Pacific Hake Stored at 5°C

Microorganism	Microbial Population after Incubation (%)			
	0 Day	5 Days	8 Days	14 Days
Pseudomonas				
Type I	14.0	7.3	2.7	15.1
Type II	14.0	52.4	53.4	77.4
Types III or IV	3.5	12.2	31.5	7.5
Acinetobacter-Moraxella				
Acinetobacter	31.6	17.0	8.2	0
Moraxella	19.3	9.8	2.7	0
Flavobacterium	17.6	0	0	0
Coliforms	0	1.2	1.4	0
Microbial count of sample	1.5×10^4	3.4×10^7	9.3×10^8	2.7×10^9
Number of microorganisms identified[a]	57	82	73	53

Source: Lee and Harrison (66), copyright © 1968, American Society for Microbiology.

[a] All isolated colonies on initial isolation plates were picked and identified.

dominant after 14 days at 5°C (Table 9-14). Shewan's Groups I, II, and III organisms appear to represent *P. fluorescens*, *P. fragi*, and *S. putrefaciens*, respectively. In a recent study of 159 gram-negative isolates from spoiled freshwater fish with total aerobic flora of about 10^8 cfu/g, about 46% were pseudomonads and 38% were *Shewanella* spp. (106). Because the latter produce H_2S and reduce trimethylamine-*N*-oxide (TMAO), they are believed to be the most significant fish spoilage bacteria. Domination of late spoilage by Shewan's Group II is seen in poultry spoilage (6), and H_2S producers also increase late.

Studies on the skin flora of four different fish revealed the following as the most common organisms: *Pseudomonas/Alteromonas*, 32–60%, and *Moraxella/Acinetobacter*, 18–37% (50). The initial flora of herring fillets was dominated by *S. putrefaciens* and pseudomonads, and after spoilage in air, these organisms constituted 62 to 95% of the flora (81). When allowed to spoil in 100% CO_2 at 4°C, the herring fillets were dominated almost completely by lactobacilli (81). In the case of rock cod fillets stored in 80% CO_2 + 20% air at 4°C for 21 days, the flora consisted of 71 to 87% lactobacilli and some tan-colored pseudomonads (62).

Some of the volatile compounds produced in spoiling fish are noted in Table 9-9. Phenethyl alcohol has been shown to be produced consistently in fish by a specific organism designated "Achromobacter" by Chen et al. (14) and Chen and Levin (13). The compound, along with phenol, was recovered from a high-boiling fraction of haddock fillets held at 2°C. None of ten known *Acinetobacter* and only one of nine known *Moraxella* produced phenethyl alcohol under similar conditions. Ethanol, propanol, and isopropanol are produced by fish spoilers, and of 244 bacteria isolated from king salmon and trout and tested in fish extracts, all produced ethanol; 241, (98.8%) produced isopropanol, and 227 (93%) produced propanol (3).

In detecting microbial fish spoilage, the reduction of (TMAO) to trimethylamine (TMA) has been used with some success.

$$\text{Trimethylamine-}N\text{-oxide} \rightarrow \begin{array}{c} H_3C \\ \diagdown \\ N—CH_3 \\ \diagup \\ H_3C \end{array}$$

Trimethylamine

TMAO is a normal constituent of seafish, while little or no TMA is found in freshly caught fish. The presence of TMA is generally regarded to be of microbial origin, although some fish contain muscle enzymes that reduce TMAO. Also, some TMAO may be reduced to dimethylamine. Not all bacteria are equal in their capacity to reduce TMAO to TMA, and its reduction is pH dependent. Methods employed to detect TMA include its extraction from fish with toluene and potassium/hydroxide followed by reaction with picric acid or its flushing from extracts and subsequent extraction by use of alkaline permanganate solutions (29, 109). More recently, gas chromatography was used to detect headspace TMA, and sampling and analysis could be completed in <5 min, with results being consistent with sensory tests (63).

Histamine, diamines, and total volatile substances are used also as fish spoilage indicators. Histamine is produced from the amino acid histidine by microbially produced histidine decarboxylase, as noted.

$$\text{Histidine} \xrightarrow{\text{Decarboxylase}} \text{Histamine}$$

Histamine is associated with scombroid poisoning (discussed further in Chap. 25). Cadaverine and putrescine are the two most important diamines evaluated as spoilage indicators, and they have been used for fish as well as for meats and poultry.

Total volatile compounds include total volatile bases (TVB), total volatile acids (TVA), total volatile substances (TVS), and total volatile nitrogen (TVN). TVB includes ammonia, dimethylamine, and trimethylamine; TVN includes TVB and other nitrogen compounds that are obtained by steam distillation of samples; and TVS are those that can be aerated from a product and reduce alkaline permanganate solutions. Because of the reducing capacity of the latter, it is sometimes referred to as the **volatile reducing substances** (VRS) method. TVA includes acetic, propionic, and related organic acids. TVN has been employed in Australia and Japan for shrimp where a maximum level for acceptable quality products is 30 mg TVN/100 g along with a maximum of 5 mg trimethylamine nitrogen/100 g. Clear-cut offness of shrimp has been noted when TVN is more than 30 mg N/100 g (15). TVB values of approximately 45 mg TVB-N/100 g of fish were found to correspond to about 10,000 ng of lipopolysaccharide in one study and to be reflective of lean fish of marginal quality (107). Among the advantages of these methods for fish freshness is their lack of reliance on a single metabolite. Among the drawbacks is their inability to measure spoilage incipiency.

Shellfish

Crustaceans

The most widely consumed shellfish within this group are shrimp, lobsters, crabs, and crayfish. Unless otherwise specified, spoilage of each is presumed or known to be essentially the same. The chief differences in spoilage of these various foods are referable generally to the way in which they are handled and their specific chemical composition.

Crustaceans differ from fish in having about 0.5% carbohydrate as opposed to none for the fish presented (Table 9-12). Shrimp has been reported to have a higher content of free amino acids than fish and to contain cathepticlike enzymes that rapidly break down proteins.

The bacterial flora of freshly caught crustaceans should be expected to reflect the waters from which these foods are caught, contaminants from the deck, handlers, and washing waters. Many of the organisms reported for fresh fish have been reported on these foods, with pseudomonads, *Acinetobacter-Moraxella*, and yeast spp. being predominant on microbially spoiled crustacean meats. When shrimp was allowed to spoil at 0°C for 13 days, *Pseudomonas* spp. were the dominant spoilers, with only 2% of the spoilage flora being gram positives in contrast to 38% for the fresh product (73). *Moraxella* dominated spoilage at 5.6° and 11.1°C, while at 16.7° and 22.2°C *Proteus* was dominant (Table 9-15).

The spoilage of crustacean meats appears to be quite similar to that of fish flesh. Spoilage would be expected to begin at the outer surfaces of these foods due to the anatomy of the organisms. It has been reported that the crustacean muscle contains over 300 mg of nitrogen/100 g of meat, which is considerably higher than that for fish (113). The presence of higher quantities of free amino acids in particular and of higher quantities of nitrogenous extractives in crustacean meats in general makes them quite susceptible to rapid attack by the spoilage flora. Initial spoilage of crustacean meats is accompanied by the production of large amounts of volatile base nitrogen, much as is the case with fish. Some of the volatile base nitrogen arises from the reduction of trimethylamine oxide present in crustacean shellfish (lacking in most mollusks). Creatine is lacking among shellfish, both crustacean and molluscan, and arginine is prevalent. Shrimp microbial spoilage is accompanied by increased hydration capacity in a manner similar to that for meats or poultry (102).

Mollusks

The molluscan shellfish considered in this section are oysters, clams, squid, and scallops. These animals differ in their chemical composition from both teleost fish and crustacean shellfish in having a significant content of carbohydrate material and a lower total quantity of nitrogen in their flesh. The carbohydrate is largely in the form of glycogen, and with levels of the type that exist in molluscan meats, fermentative activities may be expected to occur as part of the microbial spoilage. Molluscan meats contain high levels of nitrogen bases, much as do other shellfish. Of particular interest in molluscan muscle tissue is a higher content of free arginine, aspartic, and glutamic acids than is found in fish. The most important difference in chemical composition between crustacean shellfish and molluscan shellfish is the higher content of carbohydrate in the latter. For example, clam meat and scallops have been reported to contain

TABLE 9-15.   Most Predominant Bacteria in Shrimp Held to Spoilage

Temperature (°C)	Days Held	Organisms
0	13	*Pseudomonas*
5.6	9	*Moraxella*
11.1	7	*Moraxella*
16.7	5	*Proteus*
22.2	3	*Proteus*

Source: Matches (73).

3.4% and oysters 5.6% carbohydrate, mostly as glycogen. The higher content of carbohydrate materials in molluscan shellfish is responsible for the different spoilage pattern of these foods over other seafood.

The microbial flora of molluscan shellfish may be expected to vary considerably, depending on the quality of the water from which these fish are taken and the quality of wash water and other factors. The following genera of bacteria have been recovered from spoiled oysters: *Serratia, Pseudomonas, Proteus, Clostridium, Bacillus, Escherichia, Enterobacter, Shewanella, Lactobacillus, Flavobacterium*, and *Micrococcus*. As spoilage sets in and progresses, *Pseudomonas* and *Acinetobacter-Moraxella* spp. predominate, with enterococci, lactobacilli, and yeasts dominating the late stages of spoilage.

Due to the relatively high level of glycogen, the spoilage of molluscan shellfish is basically fermentative. Several investigators, including Hunter and Linden (53) and Pottinger (88), proposed the following pH scale as a basis for determining microbial quality in oysters:

pH 6.2–5.9   = good

pH 5.8   = "off"

pH. 5.7–5.5   = musty

pH 5.2 and below = sour or putrid

A measure of pH decrease is apparently a better test of spoilage in oysters and other molluscan shellfish than volatile nitrogen bases. A measure of volatile acids was attempted by Beacham (7) and found to be unreliable as a test of oyster freshness. While pH is regarded by most investigators as being the best objective technique for examining the microbial quality of oysters, Abbey et al. (1) found that organoleptic evaluations and microbial counts were more desirable indexes of microbial quality in this product.

Clams and scallops appear to display essentially the same patterns of spoilage as do oysters, but squid meat does not. In squid meat, volatile base nitrogen increases as spoilage occurs much in the same manner as for the crustacean shellfish.

References

1. Abbey, A., R. A. Kohler, and S. D. Upham. 1957. Effect of aureomycin chlortetracycline in the processing and storage of freshly shucked oysters. *Food Technol.* 11:265–271.

2. Adamčič, M., D. S. Clark, and M. Yaguchi. 1970. Effect of psychrotolerant bacteria on the amino acid content of chicken skin. *J. Food Sci.* 35:272–275.
3. Ahmed, A., and J. R. Matches. 1983. Alcohol production by fish spoilage bacteria. *J. Food Protect.* 46:1055–1059.
4. Ayres, J. C. 1960. The relationship of organisms of the genus *Pseudomonas* to the spoilage of meat, poultry and eggs. *J. Appl. Bacteriol.* 23:471–486.
5. Ayres, J. C., W. S. Ogilvy, and G. F. Stewart. 1950. Post mortem changes in stored meats. I. Microorganisms associated with development of slime on eviscerated cut-up poultry. *Food Technol.* 4:199–205.
6. Barnes, E. M., and C. S. Impey. 1968. Psychrophilic spoilage bacteria of poultry. *J. Appl. Bacteriol.* 31:97–107.
7. Beacham, L. M. 1946. A study of decomposition in canned oysters and clams. *J. Assoc. Offic. Agric. Chem.* 29:89–92.
8. Blickstad, E., and G. Molin. 1983. The microbial flora of smoked pork loin and frankfurter sausage stored in different gas atmospheres at 4°C. *J. Appl. Bacteriol.* 54:45–56.
9. Borton, R. J., L. J. Bratzler, and J. F. Price. 1970. Effects of four species of bacteria on porcine muscle. 2. Electrophoretic patterns of extracts of salt-soluble protein. *J. Food Sci.* 35:783–786.
10. Callow, E. H., and M. Ingram. 1955. Bone-taint. *Food* (Feb.).
11. Campbell, R. J., A. F. Egan, F. H. Grau, and B. J. Shay. 1979. The growth of *Microbacterium thermosphactum* on beef. *J. Appl. Bacteriol.* 47:505–509.
12. Cavett, J. J. 1962. The microbiology of vacuum packed sliced bacon. *J. Appl. Bacteriol.* 25:282–289.
13. Chen, T. C., and R. E. Levin. 1974. Taxonomic significance of phenethyl alcohol production by *Achromobacter* isolates from fishery sources. *Appl. Microbiol.* 28:681–687.
14. Chen, T. C., W. W. Nawar, and R. E. Levin. 1974. Identification of major high-boiling volatile compounds produced during refrigerated storage of haddock fillets. *Appl. Microbiol.* 28:679–680.
15. Cobb, B. F., III, and C. Vanderzant. 1985. Development of a chemical test for shrimp quality. *J. Food Sci.* 40:121–124.
16. Collins, M. D., J. A. E. Farrow, B. A. Phillips, S. Ferusu, and D. Jones. 1987. Classification of *Lactobacillus divergens*, *Lactobacillus piscicola*, and some catalase-negative, asporogenous, rod-shaped bacteria from poultry in a new genus, *Carnobacterium. Int. J. Syst. Bacteriol.* 37:310–316.
17. Dainty, R. H., and C. M. Hibbard. 1980. Aerobic metabolism of *Brochothrix thermosphacta* growing on meat surfaces and in laboratory media. *J. Appl. Bacteriol.* 48:387–396.
18. Dainty, R. H., and F. J. K. Hofman. 1983. The influence of glucose concentration and culture incubation time on end-product formation during aerobic growth of *Brochothrix thermosphacta. J. Appl. Bacteriol.* 55:233–239.
19. Dainty, R. H., B. G. Shaw, K. A. DeBoer, and E. S. J. Scheps. 1975. Protein changes caused by bacterial growth on beef. *J. Appl. Bacteriol.* 39:73–81.
20. Dainty, R. H., B. G. Shaw, and T. A. Roberts. 1983. Microbial and chemical changes in chill-stored red meats. In *Food Microbiology: Advances and Prospects*, ed. T. A. Roberts and F. A. Skinner, 151–178. New York and London: Academic Press.
21. Davidson, C. M., M. J. Dowdell, and R. G. Board. 1973. Properties of gram negative aerobes isolated from meats. *J. Food Sci.* 38:303–305.
22. Edwards, R. A., R. H. Dainty, and C. M. Hibbard. 1985. Putrescine and cadaverine formation in vacuum packed beef. *J. Appl. Bacteriol.* 58:13–19.
23. Edwards, R. A., R. H. Dainty, and C. M. Hibbard. 1983. The relationship of

bacterial numbers and types of diamine concentration in fresh and aerobically stored beef, pork and lamb. *J. Food Technol.* 18:777–788.

24. Egan, A. F., B. J. Shaw, and P. J. Rogers. 1989. Factors affecting the production of hydrogen sulphide by *Lactobacillus sake* L13 growing on vacuum-packaged beef. *J. Appl. Bacteriol.* 67: 255–262.

25. Egan, A. F., A. L. Ford, and B. J. Shay. 1980. A comparison of *Microbacterium thermosphactum* and lactobacilli as spoilage organisms of vacuum-packaged sliced luncheon meats. *J. Food Sci.* 45:1745–1748.

26. Egan, A. F., and B. J. Shay. 1982. Significance of lactobacilli and film permeability in the spoilage of vacuum-packaged beef. *J. Food Sci.* 47:1119–1122, 1126.

27. Eribo, B. E., and J. M. Jay. 1985. Incidence of *Acinetobacter* spp. and other gram-negative, oxidase-negative bacteria in fresh and spoiled ground beef. *Appl. Environ. Microbiol.* 49:256–257.

28. Eribo, B. E., S. D. Lall, and J. M. Jay. 1985. Incidence of *Moraxella* and other gram-negative, oxidase-positive bacteria in fresh and spoiled ground beef. *Food Microbiol.* 2:237–240.

29. Fields, M. L., B. S. Richmond, and R. E. Baldwin. 1968. Food quality as determined by metabolic by-products of microorganisms. *Adv. Food Res.* 16: 161–229.

30. Freeman, L. R., G. J. Silverman, P. Angelini, C. Merritt, Jr., and W. B. Esselen. 1976. Volatiles produced by microorganisms isolated from refrigerated chicken at spoilage. *Appl. Environ. Microbiol.* 32:222–231.

31. Gardner, G. A. 1971. Microbiological and chemical changes in lean Wiltshire bacon during aerobic storage. *J. Appl. Bacteriol.* 34:645–654.

32. Gardner, G. A. 1971. A note on the aerobic microflora of fresh and frozen porcine liver stored at 5°C. *J. Food Technol.* 6:225–231.

33. Gardner, G. A. 1983. Microbial spoilage of cured meats. In *Food Microbiology: Advances and Prospects*, ed. T. A. Roberts and F. A. Skinner, 179–202. New York and London: Academic Press.

34. Gill, C. O. 1976. Substrate limitation of bacterial growth at meat surfaces. *J. Appl. Bacteriol.* 41:401–410.

35. Gill, C. O. 1982. Microbial interaction with meats. In *Meat Microbiology*, ed. M. H. Brown, 225–264. London: Applied Science.

36. Gill, C. O. 1983. Meat spoilage and evaluation of the potential storage life of fresh meat. *J. Food Protect.* 46:444–452.

37. Gill, C. O., and K. M. DeLacy. 1982. Microbial spoilage of whole sheep livers. *Appl. Environ. Microbiol.* 43:1262–1266.

38. Gill, C. O., and K. G. Newton. 1977. The development of aerobic spoilage flora on meat stored at chill temperatures. *J. Appl. Bacteriol.* 43:189–195.

39. Gill, C. O., and K. G. Newton. 1979. Spoilage of vacuum-packaged dark, firm, dry meat at chill temperatures. *Appl. Environ. Microbiol.* 37:362–364.

40. Grant, G. F., A. R. McCurdy, and A. D. Osborne. 1988. Bacterial greening in cured meats: A review. *Can. Inst. Food Sci. Technol. J.* 21:50–56.

41. Hanna, M. O., C. Vanderzant, G. C. Smith, and J. W. Savell. 1981. Packaging of beef loin steaks in 75% O_2 plus 25% CO_2. II. Microbiological properties. *J. Food Protect.* 44:928–933.

42. Hanna, M. O., G. C. Smith, L. C. Hall, and C. Vanderzant. 1979. Role of *Hafnia alvei* and a *Lactobacillus* species in the spoilage of vacuum-packaged strip loin steaks. *J. Food Protect.* 42:569–571.

43. Hanna, M. O., G. C. Smith, J. W. Savell, F. K. McKeith, and C. Vanderzant. 1982. Microbial flora of livers, kidneys and hearts from beef, pork, and lamb: Effects of refrigeration, freezing and thawing. *J. Food Protect.* 45:63–73.

44. Hanna, M. O., G. C. Smith, J. W. Savell, F. K. McKeith, and C. Vanderzant. 1982. Effects of packaging methods on the microbial flora of livers and kidneys from beef or pork. *J. Food Protect.* 45:74–81.

45. Hanna, M. O., C. Vanderzant, Z. L. Carpenter, and G. C. Smith. 1977. Characteristics of psychrotrophic, gram-positive, catalase-positive, pleomorphic coccoid rods from vacuum-packaged wholesale cuts of beef. *J. Food Protect.* 40:94–97.

46. Hanna, M. O., C. Vanderzant, Z. L. Carpenter, and G. C. Smith. 1977. Microbial flora of vacuum-packaged lamb with special reference to psychrotrophic, gram-positive, catalase-positive pleomorphic rods. *J. Food Protect.* 40:98–100.

47. Herbert, R. A., M. S. Hendrie, D. M. Gibson, and J. M. Shewan. 1971. Bacteria active in the spoilage of certain sea foods. *J. Appl. Bacteriol.* 34:41–50.

48. Hess, E. 1950. Bacterial fish spoilage and its control. *Food Technol.* 4:477–480.

49. Hitchener, B. J., A. F. Egan, and P. J. Rogers. 1982. Characteristics of lactic acid bacteria isolated from vacuum-packaged beef. *J. Appl. Bacteriol.* 52:31–37.

50. Hobbs, G. 1983. Microbial spoilage of fish. In *Food Microbiology: Advances and Prospects*, ed. T. A. Roberts and F. A. Skinner, 217–229. London: Academic Press.

51. Holzapfel, W. H., and E. S. Gerber. 1983. *Lactobacillus divergens* sp. nov., a new heterofermentative *Lactobacillus* species producing L(+)−lactate. *System. Appl. Microbiol.* 4:522–534.

52. Hsieh, D. Y., and J. M. Jay. 1984. Characterization and identification of yeasts from fresh and spoiled ground beef. *Int. J. Food Microbiol.* 1:141–147.

53. Hunter, A. C., and B. A. Linden. 1923. An investigation of oyster spoilage. *Amer. Food J.* 18:538–540.

54. Ingram, M., and R. H. Dainty. 1971. Changes caused by microbes in spoilage of meats. *J. Appl. Bacteriol.* 34:21–39.

55. Jacquot, R. 1961. Organic constituents of fish and other aquatic animal foods. In *Fish as Food*, vol. 1, ed. G. Borgstrom, 145–209. New York: Academic Press.

56. Jay, J. M. 1964. Release of aqueous extracts by beef homogenates, and factors affecting release volume. *Food Technol.* 18:1633–1636.

57. Jay, J. M. 1964. Beef microbial quality determined by extract-release volume (ERV). *Food Technol.* 18:1637–1641.

58. Jay, J. M. 1966. Influence of postmortem conditions on muscle microbiology. In *The Physiology and Biochemistry of Muscle as a Food*, ed. E. J. Briskey et al., chap. 26. Madison: University of Wisconsin Press.

59. Jay, J. M. 1966. Relationship between the phenomena of extract-release volume and water-holding capacity of meats as simple and rapid methods for determining microbial quality of beef. *Hlth. Lab. Sci.* 3:101–110.

60. Jay, J. M. 1967. Nature, characteristics, and proteolytic properties of beef spoilage bacteria at low and high temperatures. *Appl. Microbiol.* 15:943–944.

61. Jay, J. M. 1987. Meats, poultry, and seafoods. In *Food and Beverage Mycology*, ed. L. R. Beuchat, chap. 5, 2nd ed. New York: Van Nostrand Reinhold.

62. Johnson, A. R., and D. M. Ogrydziak. 1984. Genetic adaptation to elevated carbon dioxide atmospheres by *Pseudomonas*-like bacteria isolated from rock cod (*Sebastes* spp.). *Appl. Environ. Microbiol.* 48:486–490.

63. Krzymien, M. E., and L. Elias. 1990. Feasibility study on the determination of fish freshness by trimethylamine headspace analysis. *J. Food Sci.* 55:1228–1232.

64. Lahellec, C., C. Meurier, and G. Bennejean. 1975. A study of 5,920 strains of psychrotrophic bacteria isolated from chickens. *J. Appl. Bacteriol.* 38:89–97.

65. Lawrie, R. A. 1966. *Meat Science*. New York: Pergamon Press.

66. Lee, J. S., and J. M. Harrison. 1968. Microbial flora of Pacific hake (*Merluccius productus*). *Appl. Microbiol.* 16:1937–1938.

67. Lee, M. L., D. L. Smith, and L. R. Freeman. 1979. High-resolution gas chromatographic profiles of volatile organic compounds produced by microorganisms at refrigerated temperatures. *Appl. Environ. Microbiol.* 37:85–90.

68. Lepovetsky, B. C., H. H. Weiser, and F. E. Deatherage. 1953. A microbiological study of lymph nodes, bone marrow and muscle tissue obtained from slaughtered cattle. *Appl. Microbiol.* 1:57–59.

69. Lerke, P., R. Adams, and L. Farber. 1963. Bacteriology of spoilage of fish muscle. I. Sterile press juice as a suitable experimental medium. *Appl. Microbiol.* 11:458–462.

70. Lerke, P., R. Adams, and L. Farber. 1965. Bacteriology of spoilage of fish muscle. III. Characteristics of spoilers. *Appl. Microbiol.* 13:625–630.

71. Lowry, P. D., and C. O. Gill. 1984. Temperature and water activity minima for growth of spoilage moulds from meat. *J. Appl. Bacteriol.* 56:193–199.

72. Margitic, S., and J. M. Jay. 1970. Antigenicity of salt-soluble beef muscle proteins held from freshness to spoilage at low temperatures. *J. Food Sci.* 35:252–255.

73. Matches, J. R. 1982. Effects of temperature on the decomposition of Pacific coast shrimp (*Pandalus jordani*). *J. Food Sci.* 47:1044–1047, 1069.

74. May, K. N., J. D. Irby, and J. L. Carmon. 1961. Shelf life and bacterial counts of excised poultry tissue. *Food Technol.* 16:66–68.

75. McMeekin, T. A. 1975. Spoilage association of chicken breast muscle. *Appl. Microbiol.* 29:44–47.

76. McMeekin, T. A. 1977. Spoilage association of chicken leg muscle. *Appl. Environ. Microbiol.* 33:1244–1246.

77. McMeekin, T. A. 1981. Microbial spoilage of meats. In *Developments in Food Microbiology*, ed. R. Davies, 1–40. London: Applied Science.

78. McMeekin, T. A., and J. T. Patterson. 1975. Characterization of hydrogen sulfide–producing bacteria isolated from meat and poultry plants. *Appl. Microbiol.* 29:165–169.

79. Miller, A., III, R. A. Scanlan, J. S. Lee, and L. M. Libbey. 1973. Volatile compounds produced in sterile fish muscle (*Sebastes melanops*) by *Pseudomonas putrefaciens, Pseudomonas fluorescens*, and an *Achromobacter* species. *Appl. Microbiol.* 26:18–21.

80. Miller, A., III, R. A. Scanlan, J. S. Lee, L. M. Libbey, and M. E. Morgan. 1973. Volatile compounds produced in sterile fish muscle (*Sebastes melanops*) by *Pseudomonas perolens*. *Appl. Microbiol.* 25:257–261.

81. Molin, G., and I.-M. Stenstrom. 1984. Effect of temperature on the microbial flora of herring fillets stored in air or carbon dioxide. *J. Appl. Bacteriol.* 56:275–282.

82. Nakamura, M., Y. Wada, H. Sawaya, and T. Kawabata. 1979. Polyamine content in fresh and processed pork. *J. Food Sci.* 44:515–517.

83. Newton, K. G., and C. O. Gill. 1978. Storage quality of dark, firm, dry meat. *Appl. Environ. Microbiol.* 36:375–376.

84. Nielsen, H.-J. S. 1983. Influence of nitrite addition and gas permeability of packaging film on the microflora in a sliced vacuum-packed whole meat product under refrigerated storage. *J. Food Technol.* 18:573–585.

85. Nicol, D. J., M. K. Shaw, and D. A. Ledward. 1970. Hydrogen sulfide production by bacteria and sulfmyoglobin formation in prepacked chilled beef. *Appl. Microbiol.* 19:937–939.

86. Patterson, J. T., and P. A. Gibbs. 1977. Incidence and spoilage potential of isolates from vacuum-packaged meat of high pH value. *J. Appl. Bacteriol.* 43:25–38.

87. Peel, J. L., and J. M. Gee. 1976. The role of micro-organisms in poultry taints. In *Microbiology in Agriculture, Fisheries and Food*, ed. F. A. Skinner and J. G.

Carr, 151–160. New York: Academic Press.

88. Pottinger, S. R. 1948. Some data on pH and the freshness of shucked eastern oysters. *Comm. Fisheries Rev.* 10(9):1–3.

89. Savell, J. W., M. O. Hanna, C. Vanderzant, and G. C. Smith. 1981. An incident of predominance of *Leuconostoc* sp. in vacuum-packaged beef strip loins—sensory and microbial profile of streaks stored in O_2-CO_2-N_2 atmospheres. *J. Food Protect.* 44:742–745.

90. Sayem-El-Daher, N., and R. E. Simard. 1985. Putrefactive amine changes in relation to microbial counts of ground beef during storage. *J. Food Protect.* 48:54–58.

91. Seelye, R. J., and B. J. Yearbury. 1979. Isolation of *Yersinia enterocolitica*-resembling organisms and *Alteromonas putrefaciens* from vacuum-packed chilled beef cuts. *J. Appl. Bacteriol.* 46:493–499.

92. Shaw, B. G., and J. B. Latty. 1982. A numerical taxonomic study of *Pseudomonas* strains from spoiled meat. *J. Appl. Bacteriol.* 52:219–228.

93. Shaw, B. G., and J. B. Latty. 1984. A study of the relative incidence of different *Pseudomonas* groups on meat using a computer-assisted identification technique employing only carbon source tests. *J. Appl. Bacteriol.* 57:59–67.

94. Shaw, B. G., and J. M. Shewan. 1968. Psychrophilic spoilage bacteria of fish. *J. Appl. Bacteriol.* 31:89–96.

95. Shaw, B. G., and C. D. Harding. 1984. A numerical taxonomic study of lactic acid bacteria from vacuum-packed beef, pork, lamb and bacon. *J. Appl. Bacteriol.* 56:25–40.

96. Shay, B. J., and A. F. Egan. 1981. Hydrogen sulphide production and spoilage of vacuum-packaged beef by a *Lactobacillus*. In *Phychrotrophic Micro-organisms in Spoilage and Pathogenicity*, ed. T. A. Roberts, G. Hobbs, J. H. B. Christian, and N. Skovgaard, 241–251. London: Academic Press.

97. Shelef, L. A. 1975. Microbial spoilage of fresh refrigerated beef liver. *J. Appl. Bacteriol.* 39:273–380.

98. Shelef, L. A. 1977. Effect of glucose on the bacterial spoilage of beef. *J. Food Sci.* 42:1172–1175.

99. Shelef, L. A., and J. M. Jay. 1969. Relationship between meat-swelling, viscosity, extract-release volume, and water-holding capacity in evaluating beef microbial quality. *J. Food Sci.* 34:532–535.

100. Shelef, L. A., and J. M. Jay. 1969. Relationship between amino sugars and meat microbial quality. *Appl. Microbiol.* 17:931–932.

101. Shelef, L. A., and J. M. Jay. 1970. Use of a titrimetric method to assess the bacterial spoilage of fresh beef. *Appl. Microbiol.* 19:902–905.

102. Shelef, L. A., and J. M. Jay. 1971. Hydration capacity as an index of shrimp microbial quality. *J. Food Sci.* 36:994–997.

103. Shewan, J. M. 1961. The microbiology of sea-water fish. In *Fish as Food*, ed. G. Borgstrom, 1:487–560. New York: Academic Press.

104. Slemr, J. 1981. Biogene Amine als potentieller chemischer Qualitätsindikator für Fleisch. *Fleischwirt.* 61:921–925.

105. Stanley, G., K. J. Shaw, and A. F. Egan. 1981. Volatile compounds associated with spoilage of vacuum-packaged sliced luncheon meat by *Brochothrix thermosphacta*. *Appl. Environ. Microbiol.* 41:816–818.

106. Stenstrom, I.-M., and G. Molin. 1990. Classification of the spoilage flora of fish, with special reference to *Shewanella putrefaciens*. *J. Appl. Bacteriol.* 68:601–618.

107. Sullivan, J. D., Jr., P. C. Ellis, R. G. Lee, W. S. Combs, Jr., and S. W. Watson. 1983. Comparison of the *Limulus amoebocyte* lysate test with plate counts and chemical analyses for assessment of the quality of lean fish. *Appl. Environ. Microbiol.* 45:720–722.

108. Sutherland, J. P., J. T. Patterson, and J. G. Murray. 1975. Changes in the microbiology of vacuum-packaged beef. *J. Appl. Bacteriol.* 39:227–237.
109. Tarr, H. L. A. 1954. Microbiological deterioration of fish post mortem, its detection and control. *Bact. Revs.* 18:1–15.
110. Thomas, V. O., A. A. Kraft, R. E. Rust, and D. K. Hotchkiss. 1984. Effect of carbon dioxide flushing and packaging methods on the microbiology of packaged chicken. *J. Food Sci.* 49:1367–1371.
111. Tonge, R. J., A. C. Baird-Parker, and J. J. Cavett. 1964. Chemical and microbiological changes during storage of vacuum packed sliced bacon. *J. Appl. Bacteriol.* 27:252–264.
112. Vanderzant, C., M. O. Hanna, J. G. Ehlers, J. W. Savell, G. C. Smith, D. B. Griffin, R. N. Terrell, K. D. Lind, and D. E. Galloway. 1982. Centralized packaging of beef loin steaks with different oxygen-barrier films: Microbiological characteristics. *J. Food Sci.* 47:1070–1079.
113. Velankar, N. K., and T. K. Govindan. 1958. A preliminary study of the distribution of nonprotein nitrogen in some marine fishes and invertebrates. *Proc. Indian Acad. Sci.* B47:202–209.
114. Watt, B. K., and A. L. Merrill. 1950. Composition of foods—raw, processed, prepared. *Agricultural Handbook No. 8.* Washington, D.C.: USDA.
115. Whiteley, A. M., and M. D. D'Souza. 1989. A yellow discoloration of cooked cured meat products—isolation and characterization of the causative organism. *J. Food Protect.* 52:392–395.
116. Yamamoto, S., H. Itano, H. Kataoka, and M. Makita. 1982. Gas-liquid chromatographic method for analysis of di- and polyamines in foods. *J. Agric. Food Chem.* 30:435–439.

10

Spoilage of Miscellaneous Foods

This chapter covers the microbiological spoilage of eggs, cereals and flour, bakery products, dairy products, sugar and spices, nutmeats, beverages and fermented foods, salad dressings, and canned foods.

EGGS

The hen's egg is an excellent example of a product that normally is well protected by its intrinsic parameters. Externally, a fresh egg has three structures, each effective to some degree in retarding the entry of microorganisms: the outer, waxy shell membrane; the shell; and the inner shell membrane (Fig. 10-1). Internally, lysozyme is present in egg white. This enzyme has been shown to be quite effective against gram-positive bacteria. Egg white also contains avidin, which forms a complex with biotin, thereby making this vitamin unavailable to microorganisms. In addition, egg white has a high pH (about 9.3) and contains conalbumin, which forms a complex with iron, thus rendering it unavailable to microorganisms. On the other hand, the nutrient content of the yolk material and its pH in fresh eggs (about 6.8) make it an excellent source of growth for most microorganisms.

Freshly laid eggs are generally sterile. However, in a relatively short period of time after laying, numerous microorganisms may be found on the outside and under the proper conditions may enter eggs, grow, and cause spoilage. Among the bacteria found are members of the following genera: *Pseudomonas*, *Acinetobacter*, *Proteus*, *Aeromonas*, *Alcaligenes*, *Escherichia*, *Micrococcus*, *Salmonella*, *Serratia*, *Enterobacter*, *Flavobacterium*, and *Staphylococcus*. Among the molds generally found are members of the genera *Mucor*, *Penicillium*, *Hormodendron*, *Cladosporium*, and others, while "*Torula*" is the only yeast found with any degree of consistency.

The most common form of bacterial spoilage of eggs is a condition known as

234

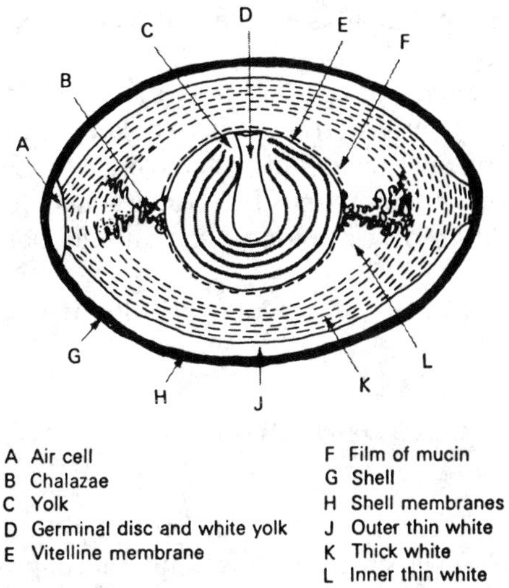

A Air cell
B Chalazae
C Yolk
D Germinal disc and white yolk
E Vitelline membrane

F Film of mucin
G Shell
H Shell membranes
J Outer thin white
K Thick white
L Inner thin white

FIGURE 10-1.   Structure of the hen's egg as shown by a section through the long axis. *From Brooks and Hale (3), reproduced with permission of Elsevier Publishing Co.*

rotting. **Green rots** are caused by *Pseudomonas* spp., especially *P. fluorescens*; **colorless rots** by *Pseudomonas*, *Acinetobacter*, and other species; **black rots** by *Proteus*, *Pseudomonas*, and *Aeromonas*; **pink rots** by *Pseudomonas*; **red rots** by *Serratia* spp., and "**custard**" **rots** by *Proteus vulgaris* and *P. intermedium* (8). Mold spoilage of eggs is generally referred to as **pinspots** from the appearance of mycelial growth on the inside upon candling. *Penicillium* and *Cladosporium* spp. are among the most common causes of pinspots and fungal rotting in eggs. Bacteria also cause a condition in eggs known as **mustiness**. *Pseudomonas graveolens* and *Proteus* spp. have been implicated in this condition, with *P. graveolens* producing the most characteristic spoilage pattern.

The entry of microorganisms into whole eggs is favored by high humidity. Under such conditions, growth of microorganisms on the surface of eggs is favored, followed by penetration through the shell and inner membrane. The latter structure is the most important barrier to the penetration of bacteria into eggs, followed by the shell and the outer membrane (17). More bacteria are found in egg yolk than in egg white, and the reason for a general lack of microorganisms in egg white is quite possibly its content of antimicrobial substances. In addition, upon storage, the thick white loses water to the yolk, resulting in a thinning of yolk and a shrinking of the thick white. This phenomenon makes it possible for the yolk to come into direct contact with the inner membrane, where it may be infected directly by microorganisms. Once inside the yolk, bacteria apparently grow in this nutritious medium, producing by-products of protein and amino acid metabolism such as H_2S and other foul-smelling compounds. The effect of significant growth is to cause the yolk to become "runny" and discolored. Molds generally multiply first in the region of the air sac, where oxygen favors growth of these forms. Under conditions of high humidity, molds may be seen growing over the outer surface of eggs. Under conditions of low humidity and low temperatures, surface growth is not favored, but eggs lose water at a faster rate and thereby become undesirable as products of commerce.

The antimicrobial systems of eggs are noted in Chapter 3. In addition, hen egg albumen contains ovotransferrin, which chelates metal ions, particularly Fe^{3+}, and ovoflavoprotein, which binds riboflavin. At its normal pH of 9.0 to 10.0, egg albumen is cidal to gram-positive bacteria and yeasts at both 30° and 39.5°C (33). The addition of iron reduces the antimicrobial properties of egg albumen.

For a general review of the microbiology of the hen's egg, see Mayes and Takeballi (20).

CEREALS, FLOUR, AND DOUGH PRODUCTS

The microbial flora of wheat, rye, corn, and related products may be expected to be that of soil, storage environments, and those picked up during the processing of these commodities. While these products are high in proteins and carbohydrates, their low a_w is such as to restrict the growth of all micro-organisms if stored properly. The microbial flora of flour is relatively low, since some of the bleaching agents reduce the load. When conditions of a_w favor growth, bacteria of the genus *Bacillus* and molds of several genera are usually the only ones that develop. Many aerobic sporeformers are capable of producing amylase, which enables them to utilize flour and related products as sources of energy, provided that sufficient moisture is present to allow growth to occur. With less moisture, mold growth occurs and may be seen as typical mycelial growth and spore formation. Members of the genus *Rhizopus* are common and may be recognized by their black spores.

The spoilage of fresh refrigerated dough products, including buttermilk biscuits, dinner and sweet rolls, and pizza dough, is caused mainly by lactic acid bacteria. In a study by Hesseltine et al. (11), 92% of isolates were Lactobacillaceae, with more than half belonging to the genus *Lactobacillus*, 36% to the genus *Leuconostoc*, and 3% to "*Streptococcus.*" Molds were found generally in low numbers in spoiled products. The fresh products showed lactic acid bacterial numbers as high as log 8.38/g.

BAKERY PRODUCTS

Commercially produced and properly handled bread generally lacks sufficient amounts of moisture to allow for the growth of any organisms except molds. One of the most common is *Rhizopus stolonifer*, often referred to as the "bread mold." The "red bread mold," *Neurospora sitophila*, may also be seen from time to time. Storage of bread under conditions of low humidity retards mold growth, and this type of spoilage is generally seen only where bread is stored at high humidities or where wrapped while still warm. Homemade breads may undergo a type of spoilage known as **ropiness**, which is caused by the growth of certain strains of *Bacillus subtilis (B. mesentericus)*. The ropiness may be seen as stringiness by carefully breaking a batch of dough into two parts. The source of the organisms is flour, and their growth is favored by holding the dough for sufficient periods of time at suitable temperatures.

Cakes of all types rarely undergo bacterial spoilage due to their unusually high concentrations of sugars, which restrict the availability of water. The most common form of spoilage displayed by these products is moldiness. Common sources of spoilage molds are any and all cake ingredients, especially sugar, nuts, and spices. While the baking process is generally sufficient to destroy these organisms, many are added in icings, meringues, toppings, and so forth. Also, molds may enter baked cakes from handling and from the air. Growth of molds on the surface of cakes is favored by conditions of high humidity. On some fruit cakes, growth often originates underneath nuts and fruits if they are placed on the surface of such products after baking. Continued growth of molds on breads and cakes results in a hardening of the products.

DAIRY PRODUCTS

Dairy products such as milk, butter, cream, and cheese are all susceptible to microbial spoilage because of their chemical composition. Milk is an excellent growth medium for all of the common spoilage organisms, including molds and yeasts. Fresh, nonpasteurized milk generally contains varying numbers of microorganisms, depending on the care employed in milking, cleaning, and handling of milk utensils. Raw milk held at refrigerator temperatures for several days invariably shows the presence of several or all bacteria of the following genera: *Enterococcus*, *Lactococcus*, *Streptococcus*, *Leuconostoc*, *Lactobacillus*, *Microbacterium*, *Propionibacterium*, *Micrococcus*, coliforms, *Proteus*, *Pseudomonas*, *Bacillus*, and others. Those unable to grow at the usual low temperature of holding tend to be present in very low numbers. The pasteurization process eliminates all but thermoduric strains, primarily streptococci and lactobacilli, and sporeformers of the genus *Bacillus* (and clostridia if present in raw milk). The spoilage of pasteurized milk is caused by the growth of heat-resistant streptococci utilizing lactose to produce lactic acid, which depresses the pH to a point (about pH 4.5) where curdling takes place. If present, lactobacilli are able to grow at pH values below that required by *Lactococcus lactis*. These organisms continue the fermentative activities and may bring the pH to 4.0 or below. If mold spores are present, these organisms begin to grow at the surface of the sour milk and raise the pH toward neutrality, thus allowing the more proteolytic bacteria such as *Pseudomonas* spp. to grow and bring about the liquefaction of the milk curd.

The same general pattern may be expected to occur in raw milk, especially if held at refrigerator temperatures. Another condition sometimes seen in raw milk is referred to as **ropiness**. This condition is caused by the growth of *Alcaligenes viscolactis* and is favored by low-temperature holding of raw milk for several days. The rope consists of slime-layer material produced by the bacterial cells, and it gives the product a stringy consistency. This condition is not as common today as it was in years past.

Butter contains around 15% water, 81% fat, and generally less than 0.5% carbohydrate and protein (Table 10-1). Although it is not a highly perishable product, it does undergo spoilage by bacteria and molds. The main source of microorganisms to butter is cream, whether sweet or sour, pasteurized or nonpasteurized. The flora of whole milk may be expected to be found in cream since as the fat droplets rise to the surface of milk, they carry up micro-

TABLE 10-1.   Percentage Composition of Nine Miscellaneous Foods

Food	Water	Carbohydrates	Proteins	Fat	Ash
Beer (4% alcohol)	90.2	4.4	0.6	0.0	0.2
Bread, enriched white	34.5	52.3	8.2	3.3	1.7
Butter	15.5	0.4	0.6	81.0	2.5
Cake (pound)	19.3	49.3	7.1	23.5	0.8
Figbars	13.8	75.8	4.2	4.8	1.4
Jellies	34.5	65.0	0.2	0.0	0.3
Margarine	15.5	0.4	0.6	81.0	2.5
Mayonnaise	1.7	21.0	26.1	47.8	3.4
Peanut butter	16.0	3.0	1.5	78.0	1.5

Source: Watt and Merrill (35).

organisms. The processing of both raw and pasteurized creams to yield butter brings about a reduction in the numbers of all microorganisms, with values for finished cream ranging from several hundred to over 100,000/g having been reported for finished salted butter (19).

Bacteria cause two principal types of spoilage in butter. The first is a condition known as "**surface taint**" or putridity. This condition is caused by "*Pseudomonas putrefaciens*" as a result of its growth on the surface of finished butter. It develops at temperatures within the range 4° to 7°C and may become apparent within 7 to 10 days. The odor of this condition is apparently due to certain organic acids, especially isovaleric acid (5). The second most common bacterial spoilage condition of butter is **rancidity**. This condition is caused by the hydrolysis of butterfat with the liberation of free fatty acids. Lipase from sources other than microorganisms can cause the effect. The causative organism is *Pseudomonas fragi*, although *P. fluorescens* is sometimes found. Bacteria may cause three other less common spoilage conditions in butter. **Malty flavor** is reported to be due to the growth of *Lactococcus lactis* var. "*maltigenes*." **Skunklike** odor is reported to be caused by *Pseudomonas mephitica*, while black discolorations of butter have been reported to be caused by *P. nigrifaciens* (9).

Butter undergoes fungal spoilage rather commonly by species of *Cladosporium*, *Alternaria*, *Aspergillus*, *Mucor*, *Rhizopus*, *Penicillium*, and *Geotrichum*, especially *G. candidum (Oospora lactis)*. These organisms can be seen growing on the surface of butter, where they produce colorations referable to their particular spore colors. Black yeasts of the genus "*Torula*" also have been reported to cause discolorations on butter. The microscopic examination of moldy butter reveals the presence of mold mycelia some distances from the visible growth. The generally high lipid content and low water content make butter more susceptible to spoilage by molds than by bacteria.

Cottage cheese undergoes spoilage by bacteria, yeasts, and molds. The most common spoilage pattern displayed by bacteria is a condition known as **slimy curd**. *Alcaligenes* spp. have been reported to be among the most frequent causative organisms, although *Pseudomonas*, *Proteus*, *Enterobacter*, and *Acinetobacter* spp. have been implicated. *Penicillium*, *Mucor*, *Alternaria*, and *Geotrichum* all grow well on cottage cheese, to which they impart stale, musty, moldy, and yeasty flavors (9). The shelf life of commercially produced cottage cheese in Alberta, Canada, was found to be limited by yeasts and molds (29).

While 48% of fresh samples contained coliforms, these organisms did not increase upon storage in cottage cheese at 40°F for 16 days.

The low moisture content of ripened cheeses makes them insusceptible to spoilage by most organisms, although molds can and do grow on these products as would be expected. Some ripened cheeses have sufficiently low oxidation-reduction (O/R) potentials to support the growth of anaerobes. It is not surprising to find that anaerobic bacteria sometimes cause the spoilage of these products when a_w permits growth to occur. *Clostridium* spp., especially *C. pasteurianum*, *C. butyricum*, and *C. sporogenes*, have been reported to cause **gassiness** of cheeses. One aerobic sporeformer, *Bacillus polymyxa*, has been reported to cause gassiness. All of these organisms utilize lactic acid with the production of CO_2, which is responsible for the gassy condition of these products.

SUGARS, CANDIES, AND SPICES

These products rarely undergo microbial spoilage if properly prepared, processed, and stored, primarily because of the lack of sufficient moisture for growth. Both cane and beet sugars may be expected to contain microorganisms. The important bacterial contaminants are members of the genera *Bacillus* and *Clostridium*, which sometimes cause trouble in the canning industry (see Chap. 14). If sugars are stored under conditions of extremely high humidity, growth of some of these organisms is possible, usually at the exposed surfaces. The successful growth of these organisms depends, of course, upon their getting an adequate supply of moisture and essential nutrients other than carbohydrates. "*Torula*" and osmophilic strains of *Saccharomyces* (*Zygosaccharomyces* spp.) have been reported to cause trouble in high-moisture sugars. These organisms have been reported to cause inversion of sugar. One of the most troublesome organisms in sugar refineries is *Leuconostoc mesenteroides*. This organism hydrolyzes sucrose and synthesizes a glucose polymer referred to as **dextran**. This gummy and slimy polymer sometimes clogs the lines and pipes through which sucrose solutions pass.

Among candies that have been reported to undergo microbial spoilage are chocolate creams, which sometimes undergo explosions. The causative organisms have been reported to be *Clostridium* spp., especially *C. sporogenes*, which finds its way into these products through sugars, starch, and possibly other ingredients.

Although spices do not undergo microbial spoilage in the usual sense of the word, molds and a few bacteria do grow on those that do not contain anti-microbial principals, provided sufficient moisture is available. Prepared mustard has been reported to undergo spoilage by yeasts and by *Proteus* and *Bacillus* spp., usually with a gassy fermentation. The usual treatment of spices with propylene oxide reduces their content of microorganisms, and those that remain are essentially sporeformers and molds. No trouble should be encountered from microorganisms as long as the moisture level is kept low.

NUTMEATS

Due to the extremely high fat and low water content of products such as pecans and walnuts (Table 10-2), these products are quite refractory to spoilage by

TABLE 10-2.   Percentage Composition of Various Nuts

Nut	Water	Carbohydrates	Protein	Fat	Ash
Almonds (dried)	4.7	19.6	18.6	34.1	3.0
Brazil nuts	5.3	11.0	14.4	65.9	3.4
Cashews	3.6	27.0	18.5	48.2	2.7
Peanuts	2.6	23.6	26.9	44.2	2.7
Pecans	3.0	13.0	9.4	73.0	1.6
Mean	3.8	18.8	17.6	57.1	2.7

Source: Watt and Merrill (35).

bacteria. Molds can and do grow on them if they are stored under conditions that permit sufficient moisture to be picked up. Examination of nutmeats will reveal molds of many genera that are picked up by the products during collecting, cracking, sorting, and packaging. (See Chap. 25 for a discussion of aflatoxins as related to nutmeats.)

BEERS, WINES, AND FERMENTED FOODS

The products covered in this section—beers and ales, table wines, sauerkraut, pickles, and olives—are themselves the products of microbial actions.

Beers

The industrial spoilage of beers and ales is commonly referred to as beer infections. This condition is caused by yeasts and bacteria. The spoilage patterns of beers and ales may be classified into four groups: ropiness, sarcinae sickness, sourness, and turbidity. **Ropiness** is a condition in which the liquid becomes characteristically viscous and pours as an "oily" stream. It is caused by *Acetobacter*, *Lactobacillus*, *Pediococcus cerevisiae*, and *Gluconobacter oxydans* (formerly *Acetomonas*) (28, 36). **Sarcinae sickness** is caused by *P. cerevisiae*, which produces a honeylike odor. This characteristic odor is the result of diacetyl production by the spoilage organism in combination with the normal odor of beer. **Sourness** in beers is caused by *Acetobacter* spp. These organisms are capable of oxidizing ethanol to acetic acid, and the sourness that results is referable to increased levels of acetic acid. **Turbidity** and off-odors in beers are caused by *Zymomonas anaerobia* (formerly *Achromobacter anaerobium*) and several yeasts such as *Saccharomyces* spp. Growth of bacteria is possible in beers because of a normal pH range of 4–5 and a good content of utilizable nutrients.

Some gram-negative obligately anaerobic bacteria have been isolated from spoiled beers and pitching yeasts, and the six species are represented by four genera:

Megasphaera cerevisiae   *Selenomonas lacticifex*
Pectinatus cerevisiiphilus   *Zymophilus paucivorans*
P. frisingensis   *Z. raffinosivorans*

All but *M. cerevisiae* produce acetic and propionic acids, and *S. lacticifex* also produces lactate (30). While *M. cerevisiae* produces negligible to minor amounts of acetic and propionic acids, it produces large quantities of isovaleric acid in addition to H_2S (6). *P. cerevisiiphilus* was the first of these to be associated with spoiled beer when they were isolated from turbid and off-flavor beer in 1978 (15). It has since been found in breweries not only in the United States but in several European countries and Japan. Among the unusual features of these organisms as beer spoilers is their gram reaction and obligately anaerobic status. In the past the typical beer spoilers have been regarded as being either lactic acid bacteria or yeasts. *Megasphaera* and *Selenomonas* are best known as members of the rumen flora.

With respect to spoiled packaged beer, one of the major contaminants found is *Saccharomyces diastaticus*, which is able to utilize dextrins that normal brewers' yeasts (*S. "carlsbergensis"* and *S. cerevisiae*) cannot (12). Pediococci, *Flavobacterium proteus* (formerly *Obesumbacterium*), and *Brettanomyces* are sometimes found in spoiled beer.

Wines

Table wines undergo spoilage by bacteria and yeasts, *Candida valida* being the most important yeast. Growth of this organism occurs at the surface of wines, where a thin film is formed. The organisms attack alcohol and other constituents from this layer and create an appearance that is sometimes referred to as **wine flowers**. Among the bacteria that cause wine spoilage are members of the genus *Acetobacter*, which oxidize alcohol to acetic acid (produce vinegar). The most serious and the most common disease of table wines is referred to as **tourne disease** (26). Tourne disease is caused by a facultative anaerobe or anaerobe that utilizes sugars and seems to prefer conditions of low alcohol content. This type of spoilage is characterized by an increased volatile acidity, a silky type of cloudiness, and later in the course of spoilage, a "mousey" odor and taste.

Malo-lactic fermentation is a spoilage condition of great importance in wines. Malic and tartaric acids are two of the predominant organic acids in grape must and wine, and in the malo-lactic fermentation, contaminating bacteria degrade malic acid to lactic acid and CO_2:

$$L(-)\text{-Malic acid} \xrightarrow{\text{"malo-lactic enzyme"}} L(+)\text{-Lactic acid} + CO_2.$$

L-malic acid may be decarboxylated also to yield pyruvic acid (13). The effect of these conversions is to reduce the acid content and affect flavor. The malo-lactic fermentation (which may also occur in cider) can be carried out by many lactic acid bacteria, including leuconostocs, pediococci, and lactobacilli (18, 27). While the function of the malo-lactic fermentation to the fermenting organism is not well understood, it has been shown that *Leuconostoc oenos* is actually stimulated by the process (24). The decomposition in wines of tartaric acid is undesirable also, and this process can be achieved by some strains of *Lactobacillus plantarum* in the following general manner:

$$\text{Tartaric acid} \rightarrow \text{Lactic acid} + \text{acetic acid} + CO_2.$$

The effect is to reduce the acidity of wine. Unlike the malo-lactic fermentation, few lactic acid bacteria break down tartaric acid (27).

Other Products

Root beer undergoes bacterial spoilage on occasion. Lehmann and Byrd (16) investigated spoiled root beer characterized by a musty odor and taste. The causative organism was found to be *"Achromobacter"* sp. By inoculating this organism into the normal product, they found that the characteristic spoilage appeared in 2 weeks.

Sauerkraut is the product of lactic acid fermentation of fresh cabbage, and while the finished product has a pH in the range of 3.1 to 3.7, it is still subject to spoilage by bacteria, yeasts, and molds. The microbial spoilage of sauerkraut generally falls into the following categories: soft kraut, slimy kraut, rotted kraut, and pink kraut. **Soft kraut** results when bacteria that normally do not initiate growth until the late stages of kraut production actually grow earlier. **Slimy kraut** is caused by the rapid growth of *Lactobacillus cucumeris* and *L. plantarum*, especially at elevated temperatures (26). **Rotted** sauerkraut may be caused by bacteria, molds, and/or yeasts, while **pink kraut** is caused by the surface growth of *"Torula"* spp., especially *"T. glutinis."* Due to the high acidity, finished kraut is generally spoiled by molds growing on the surface. The growth of these organisms effects an increase in pH to levels where a large number of bacteria can grow that were previously inhibited by conditions of high acidity.

Pickles result from lactic acid fermentation of cucumbers. The finished product has a pH of around 4.0. These products undergo spoilage by bacteria and molds. **Pickle blackening** may be caused by *Bacillus nigrificans*, which produces a dark water-soluble pigment. *Enterobacter* spp., lactobacilli, and pediococci have been implicated as causes of a condition known as **"bloaters,"** produced by gas formation within the individual pickles. **Pickle softening** is caused by pectolytic organisms of the genera *Bacillus, Fusarium, Penicillium, Phoma, Cladosporium, Alternaria, Mucor, Aspergillus*, and others. The actual softening of pickles may be caused by any one or several of these or related organisms. Pickle softening results from the production of pectinases, which break down the cementlike substance in the wall of the product.

Among the types of microbial spoilage that olives undergo, one of the most characteristic is **zapatera spoilage**. This condition, which sometimes occurs in brined olives, is characterized by a malodorous fermentation. The odor is due apparently to propionic acid, which is produced by certain species of *Propionibacterium* (25).

A **softening** condition of spanish-type green olives has been found to be caused by the yeasts *Rhodotorula glutinis* var. *glutinis, R. minuta* var. *minuta*, and *R. rubra* (34). All of these organisms produce polygalacturonases, which effect olive tissue softening. Under appropriate cultural conditions, the organisms were shown to produce pectin methyl esterase, as well as polygalacturonase. A **sloughing** type of spoilage of California ripe olives was shown by Patel and Vaughn (21) to be caused by *Cellulomonas flavigena*. This organism showed high cellulolytic activity, which was enhanced by the growth of other organisms such as *Xanthomonas, Enterobacter*, and *Escherichia* spp.

MAYONNAISE AND SALAD DRESSINGS

Mayonnaise can be defined as a semisolid emulsion of edible vegetable oil, egg yolk or whole egg, vinegar, and/or lemon juice, and other ingredients such as salt and other seasonings and glucose, in a finished product containing not less than 50% edible oil. The pH of this product ranges from 3.6 to 4.0, with acetic acid as the predominant acid representing 0.29 to 0.5% of total product with an a_w of 0.925. The aqueous phase contains 9 to 11% salt and 7 to 10% sugar (32). Salad dressings are quite similar in composition to mayonnaise, but the finished product contains at least 30% edible vegetable oil and has an a_w of 0.929, a pH of 3.2 to 3.9, with acetic acid usually the predominant acid accounting for 0.9 to 1.2% of total product. The aqueous phase contains 3.0 to 4.0% salt and 20 to 30% sugar (32). While the nutrient content of these products is suitable as food sources for many spoilage organisms, the pH, organic acids, and low a_w restrict spoilers to yeasts, a few bacteria, and molds. The yeast *Zygosaccharomyces bailii* is known to cause the spoilage of salad dressings, tomato catsup, carbonated beverages, and some wines. Yeasts of the genus *Saccharomyces* have been implicated in the spoilage of mayonnaise, salad dressing, and french dressing. One of the few bacteria reported to cause spoilage of products of this type is *Lactobacillus brevis*, which was reported to produce gas in salad dressing. Appleman et al. (1) investigated spoiled mayonnaise and recovered a strain of *B. subtilis* and a yeast that they believed to be the etiologic agents. *Bacillus vulgatus* has been recovered from spoiled thousand islands dressing, where it caused darkening and separation of the emulsion. In one study of the spoilage of thousand island dressing, pepper and paprika were shown to be the sources of *B. vulgatus* (22). Mold spoilage of products of this type occurs only at the surfaces when sufficient oxygen is available. Separation of the emulsion is generally one of the first signs of spoilage of these products, although bubbles of gas and the rancid odor of butyric acid may precede emulsion separation. The spoilage organisms apparently attack the sugars fermentatively. It appears that the pH remains low, thereby preventing the activities of proteolytic and lipolytic organisms. It is not surprising to find yeasts and lactic acid bacteria under these conditions. In a study of 17 samples of spoiled mayonnaise, mayonnaiselike, and blue cheese dressings, Kurtzman et al. (14) found high yeast counts in most samples and high lactobacilli counts in two. The pH of samples ranged from 3.6 to 4.1. Two-thirds of the spoiled samples yielded *Z. bailii*. Common in some samples was *L. fructivorans*, with aerobic sporeformers being found in only two samples. Of 10 unspoiled samples tested, microorganisms were in low numbers or not detectable at all.

In regard to foodborne pathogenic bacteria, the interaction of low pH, acids, and low a_w is such that these products will not support growth of these types of organisms (32).

CANNED FOODS

Although the objective in the canning of foods is the destruction of micro-organisms, these products nevertheless undergo microbial spoilage under certain conditions. The main reasons for this are underprocessing, inadequate cooling, contamination of the can resulting from leakage through seams, and preprocess spoilage. Since some canned foods receive low-heat treatments, it is to be expected that a rather large number of different types of microorganisms may be found upon examining such foods.

As a guide to the type of spoilage that canned foods undergo, the following classification of canned foods based upon acidity is helpful.

Low Acid.   pH > 4.6. Meat and marine products, milk, some vegetables (corn, lima beans), meat and vegetable mixtures, and so on. Spoiled by thermophilic flat-sour group (*Bacillus stearothermophilus, B. coagulans*), sulfide spoilers (*Clostridium nigrificans, C. bifermentans*), and/or gaseous spoilers (*Clostridium thermosaccharolyticum*). Mesophilic spoilers include putrefactive anaerobes (especially P.A. 3679 types). Spoilage and toxin production by proteolytic *C. botulinum* strains may occur if they are present. Medium-acid foods are those with pH range of 5.3 to 4.6, while low-acid foods are those with pH $\geq$ 5.4.

Acid.   pH 3.7–4.0 to 4.6. In this category are fruits such as tomatoes, pears, and figs. Thermophilic spoilers include *B. coagulans* types. Mesophiles include *B. polymyxa, B. macerans (B. betanigrificans), C. pasteurianium, C. butyricum,* lactobacilli, and others.

High Acid.   pH < 4.0–3.7. This category includes fruits and fruit and vegetable products—grapefruit, rhubarb, sauerkraut, pickles, and so forth. Generally spoiled by nonsporeforming mesophiles—yeasts, molds, and/or lactic acid bacteria.

Canned food spoilage organisms may be further characterized as follows:

1. Mesophilic organisms
   a. Putrefactive anaerobes
   b. Butyric anaerobes
   c. Aciduric flat sours
   d. Lactobacilli
   e. Yeasts
   f. Molds
2. Thermophilic organisms
   a. Flat-sour spores
   b. Thermophilic anaerobes producing sulfide
   c. Thermophilic anaerobes not producing sulfide

The canned food spoilage manifestations of these organisms are presented in Table 10-3.

With respect to the spoilage of high-acid and other canned foods by yeasts, molds, and bacteria, several of these organisms have been repeatedly associated with certain specific foods. The yeasts "*Torula lactis-condensi*" and "*T. globosa*" cause blowing or gaseous spoilage of sweetened condensed milk, which is not heat processed. The mold *Aspergillus repens* is associated with the formation of "buttons" on the surface of sweetened condensed milk. *Lactobacillus brevis (L. lycopersici)* causes a vigorous fermentation in tomato

TABLE 10-3.   Spoilage Manifestations in Acid and Low-Acid Canned Foods

Type of Organism	Appearance and Manifestations of Can	Condition of Product
Acid Products		
1. *B. thermoacidurans* (flat sour: tomato juice)	Can flat; little change in vacuum	Slight pH change; off-odor and flavor
2. Butyric anaerobes (tomatoes and tomato juice)	Can swells; may burst	Fermented; butyric odor
3. Nonsporeformers (mostly lactics)	Can swells, usually bursts, but swelling may be arrested	Acid odor
Low-Acid Products		
1. Flat sour	Can flat; possible loss of vacuum on storage	Appearance not usually altered; pH markedly lowered—sour; may have slightly abnormal odor; sometimes cloudy liquor
2. Thermophilic anaerobe	Can swells; may burst	Fermented, sour, cheesey, or butyric odor
3. Sulfide spoilage	Can flat; H_2S gas absorbed by product	Usually blackened; "rotten egg" odor
4. Putrefactive anaerobe	Can swells; may burst	May be partially digested; pH slightly above normal; typical putrid odor
5. Aerobic sporeformers (odd types)	Can flat; usually no swelling, except in cured meats when NO_3 and sugar are present	Coagulated evaporated milk, black beets

Source: Schmitt (31).

catsup, worcestershire sauce, and similar products. *Leuconostoc mesenteroides* has been reported to cause gaseous spoilage of canned pineapples and ropiness in peaches. The mold *Byssochlamys fulva* causes spoilage of bottled and canned fruits. Its actions cause disintegration of fruits as a result of pectin breakdown (2). "*Torula stellata*" has been reported to cause the spoilage of canned bitter lemon, and to grow at a pH of 2.5 (23).

Frozen concentrated orange juice sometimes undergoes spoilage by yeasts and bacteria. Hays and Riester (10) investigated samples of this product spoiled by bacteria. The orange juice was characterized as having a vinegary to buttermilk off-odor with an accompanying off-flavor. From the spoiled product were isolated *L. plantarum* var. *mobilis*, *L. brevis*, *Leuconostoc mesenteroides*, and *Leuconostoc dextranicum*. The spoilage characteristics could be reproduced by inoculating the above isolates into fresh orange juice.

Minimum growth temperatures of spoilage thermophiles are of some importance in diagnosing the cause of spoiled canned foods. *B. coagulans (B. thermoacidurans)* has been reported to grow only slowly at 25°C but grows well between 30° and 55°C. *B. stearothermophilus* does not grow at 37°C, its optimum temperature being around 65°C with smooth variants showing a shorter generation time at this temperature than rough variants (7). *C.*

TABLE 10-4.   Some Features of Canned Food Spoilage Resulting from Understerilization and Seam Leakage

	Understerilization	Leakage
Can	Flat or swelled; seams generally normal	Swelled; may show defects
Product appearance	Sloppy or fermented	Frothy fermentation; viscous
Odor	Normal, sour, or putrid but generally consistent	Sour, fecal, generally varying from can to can
pH	Usually fairly constant	Wide variation
Microscopic and cultural	Pure cultures, sporeformers; growth at 98°F and/or 131°F; may be characteristic on special media, e.g., acid agar for tomato juice	Mixed cultures, generally rods and cocci; growth only at usual temperatures
History	Spoilage usually confined to certain portions of pack. In acid products diagnosis may be less clearly defined. Similar organisms may be involved in understerilization and leakage	Spoilage scattered

Source: Schmitt (31).

thermosaccharolyticum does not grow at 30°C but has been reported to grow at 37°C.

Also of importance in diagnosing the cause of canned food spoilage is the appearance of the unopened can or container. The ends of a can of food are normally flat or slightly concave. When microorganisms grow and produce gases, the can goes through a series of changes visible from the outside. The change is designated a **flipper** when one end can be made convex by striking or heating the can. A **springer** is a can with both ends bulged when one or both remain concave if pushed in or when one end is pushed in and the other pops out. A **soft swell** refers to a can with both ends bulged that may be dented by pressing with the fingers. A **hard swell** has both ends bulged so that neither end can be dented by hand. These events tend to develop successively and become of value in predicting the type of spoilage that might be in effect. Flippers and springers may be incubated under wraps at a temperature appropriate to the pH and type of food in order to allow for further growth of any organisms that might be present. These effects on cans do not always represent microbial spoilage. Soft swells often represent microbial spoilage, as do hard swells. In high-acid foods, however, hard swells are often **hydrogen swells**, which result from the release of hydrogen gas by the action of food acids on the iron of the can. The other two most common gases in cans of spoiled foods are CO_2 and H_2S, both of which are the result of the metabolic activities of microorganisms. Hydrogen sulfide may be noted by its characteristic odor, while CO_2 and hydrogen may be determined by the following test. Construct an apparatus of glass or plastic tubing attached to a hollow punch fitted with a large rubber stopper. Into a test tube filled with dilute KOH, insert the free end of this apparatus and invert it in a beaker filled with dilute KOH. When an opening is made in one end of the can with the hollow punch, the gases will displace the dilute KOH inside the tube. Before removing the open end from the beaker, close the tube by placing the thumb over the end. To test for CO_2, shake the tube and note for a vacuum as evidenced by suction against the finger. To test for hydrogen, repeat the test and apply a match near the top of the tube and then quickly remove the thumb. A "pop" indicates the presence of hydrogen. Both gases may be found in some cans of spoiled foods.

"**Leakage-type**" spoilage of canned foods is characterized by a flora of nonsporeforming organisms that would not survive the heat treatment normally given heat-processed foods. These organisms enter cans at the start of cooling through faulty seams, which generally result from can abuse. The organisms that cause leakage-type spoilage can be found either on the cans or in the cooling water. This problem is minimized if the cannery cooling water contains <100 bacteria/ml. This type of spoilage may be further differentiated from that caused by understerilization (see Table 10-4).

References

1. Appleman, M. D., E. P. Hess, and S. C. Rittenberg. 1949. An investigation of a mayonnaise spoilage. *Food Technol.* 3:201–203.
2. Baumgartner, J. G., and A. C. Hersom. 1957. *Canned foods*. Princeton, N.J.: D. Van Nostrand.
3. Brooks, J., and H. P. Hale. 1959. The mechanical properties of the thick white of the hen's egg. *Biochem. Biophys. Acta* 32:237–250.
4. Cross, T. 1968. Thermophilic *Actinomycetes*. *J. Appl. Bacteriol.* 31:36–53.
5. Dunkley, W. L., G. Hunter, H. R. Thornton, and E. G. Hood. 1942. Studies on surface taint butter. II. An odorous compound in skim milk cultures of *Pseudomonas putrefaciens*. *Scientific Agr.* 22:347–355.
6. Engelmann, U., and N. Weiss. 1985. *Megasphaera cerevisiae* sp. nov.: A new Gram-negative obligately anaerobic coccus isolated from spoiled beer. *System. Appl. Microbiol.* 6:287–290.
7. Fields. M. L. 1970. The flat sour bacteria. *Adv. Food Res.* 18:163–217.
8. Florian, M. L. E., and P. C. Trussell. 1957. Bacterial spoilage of shell eggs. IV. Identification of spoilage organisms. *Food Technol.* 11:56–60.
9. Foster, E. M., F. E. Nelson, M. L. Speck, R. N. Doetsch, and J. C. Olson, Jr. 1957. *Dairy Microbiology*. Englewood Cliffs, N.J.: Prentice-Hall.
10. Hays, G. L., and D. W. Riester. 1952. The control of "off-odor" spoilage in frozen concentrated orange juice. *Food Technol.* 6:386–389.
11. Hesseltine, C. W., R. R. Graves, R. Rogers, and H. R. Burmeister. 1969. Aerobic and facultative microflora of fresh and spoiled refrigerated dough products. *Appl. Microbiol.* 18:848–853.
12. Kleyn, J., and J. Hough. 1971. The microbiology of brewing. *Ann. Rev. Microbiol.* 25:583–608.
13. Kunkee, R. E. 1975. A second enzymatic activity for decomposition of malic acid by malo-lactic bacteria. In *Lactic Acid Bacteria in Beverages and Food*, ed. J. G. Carr et al., 29–42. New York: Academic Press.
14. Kurtzman, C. P., R. Rogers, and C. W. Hesseltine. 1971. Microbiological spoilage of mayonnaise and salad dressings. *Appl. Microbiol.* 21:870–874.
15. Lee, S. Y., M. S. Mabee, and N. O. Jangaard. 1978. *Pectinatus*, a new genus of the family *Bacteroidaceae*. *Int. J. Syst. Bacteriol.* 28:582–594.
16. Lehmann, D. L., and B. E. Byrd. 1953. A bacterium responsible for a musty odor and taste in root beer. *Food Res.* 18:76–78.

17. Lifshitz, A., R. G. Baker, and H. B. Naylor. 1964. The relative importance of chicken egg exterior structures in resisting bacterial penetration. *J. Food Sci.* 29:94–99.
18. London, J. 1976. The ecology and taxonomic status of the lactobacilli. *Ann. Rev. Microbiol.* 30:279–301.
19. Macy, H., S. T. Coulter, and W. B. Combs. 1932. Observations on the quantitative changes in the microflora during the manufacture and storage of butter. *Minn. Agric. Exp. Sta. Techn. Bull.* 82.
20. Mayes, F. J., and M. A. Takeballi. 1983. Microbial contamination of the hen's egg: A review. *J. Food Protect.* 46:1092–1098.
21. Patel, I. B., and R. H. Vaughn. 1973. Cellulolytic bacteria associated with sloughing spoilage of California ripe olives. *Appl. Microbiol.* 25:62–69.
22. Pederson, C. S. 1930. Bacterial spoilage of a thousand island dressing. *J. Bacteriol.* 20:99–106.
23. Perigo, J. A., B. L. Gimbert, and T. E. Bashford. 1964. The effect of carbonation, benzoic acid, and pH on the growth rate of a soft drink spoilage yeast as determined by a turbidostatic continuous culture apparatus. *J. Appl. Bacteriol.* 27:315–332.
24. Pilone, G. J., and R. E. Kunkee. 1976. Stimulatory effect of malo-lactic fermentation on the growth rate of *Leuconostoc oenos*. *Appl. Environ. Microbiol.* 32: 405–408.
25. Plastourgos, S., and R. H. Vaughn. 1957. Species of *Propionibacterium* associated with zapatera spoilage of olives. *Appl. Microbiol.* 5:267–271.
26. Prescott, S. C., and C. G. Dunn. 1959. *Industrial Microbiology*. New York: McGraw-Hill.
27. Radler, F. 1975. The metabolism of organic acids by lactic acid bacteria. In *Lactic Acid Bacteria in Beverages and Food*, ed. J. G. Carr et al., 17–27. New York: Academic Press.
28. Rainbow, C. 1975. Beer spoilage lactic acid bacteria. In *Lactic Acid Bacteria in Beverages and Food*, ed. J. G. Carr et al., 149–158. New York: Academic Press.
29. Roth, L. A., L. F. L. Clegg, and M. E. Stiles. 1971. Coliforms and shelf life of commercially produced cottage cheese. *Can. Inst. Food Technol. J.* 4:107–111.
30. Schleifer, K. H., M. Leuteritz, N. Weiss, W. Ludwig, G. Kirchhof, and H. Seidel-Rufer. 1990. Taxonomic study of anaerobic, Gram-negative, rod-shaped bacteria from breweries: Emended description of *Pectinatus cerevisiiphilus* and description of *Pectinatus frisingensis* sp. nov., *Selenomonas lacticifex* sp. nov., *Zymophilus paucivorans* sp. nov. *Int. J. System. Bacteriol.* 40:19–27.
31. Schmitt, H. P. 1966. Commercial sterility in canned foods, its meaning and determination. *Assoc. Food Drug Off. of U.S., Quart. Bull.* 30:141–151.
32. Smittle, Richard B. 1977. Microbiology of mayonnaise and salad dressing: A review. *J. Food Protect.* 40:415–422.
33. Tranter, H. S., and R. G. Board. 1984. The influence of incubation temperature and pH on the antimicrobial properties of hen egg albumen. *J. Appl. Bacteriol.* 56:53–61.
34. Vaughn, R. H., T. Jakubczyk, J. D. MacMillan, T. E. Higgins, B. A. Dave, and V. M. Crampton. 1969. Some pink yeasts associated with softening of olives. *Appl. Microbiol.* 18:771–775.
35. Watt, B. K., and A. L. Merrill. 1950. Composition of foods—raw, processed, prepared. *Agricultural Handbook No. 8*, Washington, D.C.: USDA.
36. Williamson, D. H. 1959. Studies on lactobacilli causing ropiness in beer. *J. Appl. Bacteriol.* 22:392–402.

PART V

Food Preservation and Some Properties of Psychrotrophs, Thermophiles, Radiation-Resistant, and Lactic Acid Bacteria

The microbiology of a variety of food preservation methods is examined in Chapters 11 through 16. The target organisms for the respective methods are presented along with the mode of action of the respective preservation methods where known. In Chapters 12, 13, and 14, synopses of the respective groups of organisms that these methods are designed to inhibit or kill are presented. Food fermentations as preservation methods are examined in Chapter 16 with emphasis on the specific roles that the fermenters play in effecting product preservation.

More detailed information can be obtained from the following sources:

G. Campbell-Platt, ed., *Fermented Foods of the World* (Stoneham, Mass.: Butterworth-Heinemann, 1987): An excellent treatment of the titled subject.

G. W. Gould, ed., *Mechanisms of Action of Food Preservation Procedures* (N.Y.: Elsevier, 1989): Detailed coverage of the titled subject.

J. N. Sofos, *Sorbate Food Preservatives* (Boca Raton, Fla.: CRC Press, 1989): Thorough coverage of the titled compound in food preservation applications.

W. M. Urbain, *Food Irradiation* (New York: Academic Press, 1986): Excellent coverage of titled subject through the mid-1980s.

B. J. B. Wood, ed., *Microbiology of Fermented Foods*, Vol. 1: *More Developed Food Fermentations* (Amsterdam: Elsevier, 1985): Detailed coverage of many fermented foods.

11

Food Preservation with Chemicals

The use of chemicals to prevent or delay the spoilage of foods derives in part from the fact that such compounds are used with great success in the treatment of diseases of humans, animals, and plants. This is not to imply that any and all chemotherapeutic compounds can or should be used as food preservatives. On the other hand, there are some chemicals of value as food preservatives that would be ineffective or too toxic as chemotherapeutic compounds. With the exception of certain antibiotics, none of the food preservatives now used finds any real use as chemotherapeutic compounds in people and animals. While a large number of chemicals have been described that show potential as food preservatives, only a relatively small number are allowed in food products, due in large part to the strict rules of safety adhered to by the Food and Drug Administration (FDA) and to a lesser extent to the fact that not all compounds that show antimicrobial activity in vitro do so when added to certain foods. Below are described those compounds most widely used, their modes of action where known, and the types of foods in which they are used. Those chemical preservatives generally recognized as safe (GRAS) are summarized in Table 11-1.

BENZOIC ACID AND THE PARABENS

Benzoic acid (C_6H_5COOH) and its sodium salt ($C_7H_5NaO_2$) along with the esters of p-hydroxybenzoic acid (parabens) are considered together in this section. Sodium benzoate was the first chemical preservative permitted in foods by the FDA, and it continues in wide use today in a large number of foods. Its approved derivatives have structural formulas as noted:

TABLE 11-1.   Summary of Some GRAS Chemical Food Preservatives

Preservatives	Maximum Tolerance	Organisms Affected	Foods
Propionic acid/propionates	0.32%	Molds	Bread, cakes, some cheeses, rope inhibitor in bread dough
Sorbic acid/sorbates	0.2%	Molds	Hard cheeses, figs, syrups, salad dressings, jellies, cakes
Benzoic acid/benzoates	0.1%	Yeasts and molds	Margarine, pickle relishes, apple cider, soft drinks, tomato catsup, salad dressings
Parabens[a]	0.1%[b]	Yeasts and molds	Bakery products, soft drinks, pickles, salad dressings
SO_2/sulfites	200–300 ppm	Insects, micro-organisms	Molasses, dried fruits, wine making, lemon juice (not to be used in meats or other foods recognized as sources of thiamine)
Ethylene/propylene oxides[c]	700 ppm	Yeasts, molds, vermin	Fumigant for spices, nuts
Sodium diacetate	0.32%	Molds	Bread
Nisin	1%	Lactics, clostridia	Certain pasteurized cheese spreads
Dehydroacetic acid	65 ppm	Insects	Pesticide on strawberries, squash
Sodium nitrite[c]	120 ppm	Clostridia	Meat-curing preparations
Caprylic acid	—	Molds	Cheese wraps
Ethyl formate	15–200 ppm[d]	Yeasts and molds	Dried fruits, nuts

Note: GRAS (Generally Recognized As Safe) per Section 201 (32) (s) of the U.S. Federal Food, Drug, and Cosmetic Act as amended.

[a] Methyl-, propyl-, and heptyl-esters of *p*-hydroxybenzoic acid.
[b] Heptyl ester—12 ppm in beers; 20 ppm in noncarbonated and fruit-based beverages.
[c] May be involved in mutagenesis and/or carcinogenesis.
[d] As formic acid.

Methylparaben
Methyl *p*-Hydroxybenzoate

HO—⟨benzene ring⟩—$COOCH_3$

Propylparaben
Propyl *p*-Hydroxybenzoate

HO—⟨benzene ring⟩—$COO(CH_2)_2CH_3$

Heptylparaben
n-Heptyl-*p*-hydroxybenzoate

HO—⟨benzene ring⟩—$COO(CH_2)_6CH_3$

The antimicrobial activity of benzoate is related to pH, the greatest activity being at low pH values. The antimicrobial activity resides in the undissociated molecule (see below). These compounds are most active at the lowest pH values of foods and essentially ineffective at neutral values. The pK of benzoate is 4.20 and at a pH of 4.00, 60% of the compound is undissociated, while at a pH of 6.0, only 1.5% is undissociated. This results in the restriction of benzoic acid and its sodium salts to high-acid products such as apple cider, soft drinks, tomato catsup, and salad dressings. High acidity alone is generally sufficient to prevent growth of bacteria in these foods but not that of certain molds and yeasts. As used in acidic foods, benzoate acts essentially as a mold and yeast inhibitor, although it is effective against some bacteria in the 50 to 500 ppm range. Against yeasts and molds at around pH 5.0 to 6.0, from 100 to 500 ppm are effective in inhibiting the former, while for the latter from 30 to 300 ppm are inhibitory.

In foods such as fruit juices, benzoates may impart disagreeable tastes at the maximum level of 0.1%. The taste has been described as being "peppery" or burning.

The three parabens that are permissible in foods in the United States are heptyl-, methyl-, and propyl-, while butyl- and ethylparabens are permitted in food in certain other countries. As esters of *p*-hydroxybenzoic acid, they differ from benzoate in their antimicrobial activity in being less sensitive to pH. Although not as many data have been presented on heptylparaben, it appears to be quite effective against microorganisms, with 10 to 100 ppm effecting complete inhibition of some gram-positive and gram-negative bacteria. Propylparaben is more effective than methylparaben on a ppm basis, with up to 1,000 ppm of the former and 1,000 to 4,000 ppm of the latter needed for bacterial inhibition, with gram-positive bacteria being more susceptible than gram negatives to the parabens in general (27). Heptylparaben has been reported to be effective against the malo-lactic bacteria. In a reduced-broth medium, 100 ppm propylparaben delayed germination and toxin production by *Clostridium botulinum* type A, while 200 ppm effected inhibition up to 120 h at 37°C (119). In the case of methylparaben, 1,200 ppm were required for inhibition similar to that for the propyl- analog.

The parabens appear to be more effective against molds than against yeasts. As in the case of bacteria, the propyl- derivative appears to be the most effective where 100 ppm or less are capable of inhibiting some yeasts and molds, while for heptyl- and methylparabens, 50 to 200 and 500 to 1,000 ppm, respectively, are required.

Like benzoic acid and its sodium salt, the methyl- and propylparabens are permissible in foods up to 0.1% while heptylparaben is permitted in beers to a maximum of 12 ppm and up to 20 ppm in fruit drinks and beverages. The pK for these compounds is around 8.47, and their antimicrobial activity is not increased to the same degree as for benzoate with the lowering of pH as noted. They have been reported to be effective at pH values up to 8.0. For a more thorough review of these preservatives, see Davidson (27).

Similarities between the modes of action of benzoic and salicylic acids have been noted (13). Both compounds, when taken up by respiring microbial cells, were found to block the oxidation of glucose and pyruvate at the acetate level in *Proteus vulgaris*. With *P. vulgaris*, benzoic acid caused an increase in the rate of O_2 consumption during the first part of glucose oxidation (13). The benzoates, like propionate and sorbate, have been shown to act against microorganisms by inhibiting the cellular uptake of substrate molecules (45). The

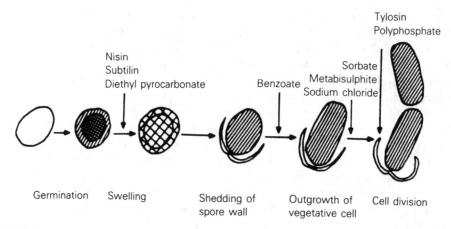

FIGURE 11-1.   Diagrammatic representation of growth of an endospore into vegetative cells showing stages arrested by minimum inhibitory concentrations of some food preservatives. *From Gould (49).*

stage of endospore germination most sensitive to benzoate is noted in Figure 11-1.

The undissociated form is essential to the antimicrobial activity of benzoate as well as for other lipophilics such as sorbate and propionate. In this state, these compounds are soluble in the cell membrane and act apparently as proton ionophores (50). As such, they facilitate proton leakage into cells and thereby increase energy output of cells to maintain their usual internal pH. With the disruption in membrane activity, amino acid transport is adversely affected (50).

SORBIC ACID

Sorbic acid ($CH_3CH{=}CHCH{=}CHCOOH$) is employed as a food preservative, usually as the calcium, sodium, or potassium salt. These compounds are permissible in foods at levels not to exceed 0.2%. Like sodium benzoate, they are more effective in acid foods than in neutral foods and tend to be on par with the benzoates as fungal inhibitors. Sorbic acid works best below pH 6.0 and is generally ineffective above pH 6.5. These compounds are more effective than sodium benzoate between pH 4.0 to 6.0. At pH values of 3.0 and below, the sorbates are slightly more effective than the propionates but about the same as sodium benzoate. The pK of sorbate is 4.80, and at a pH of 4.0, 86% of the compound is undissociated, while at a pH of 6.0, only 6% is undissociated. Sorbic acid can be employed in cakes at higher levels than propionates without imparting flavor to the product.

The sorbates are primarily effective against molds and yeasts, but research has shown them to be effective against a wide range of bacteria. In general, the catalase-positive cocci are more sensitive than the catalase negatives, and aerobes are more sensitive than anaerobes. The resistance of the lactic acid bacteria to sorbate, especially at pH 4.5 or above, permits its use as a fungistat in products that undergo lactic fermentations. Its effectiveness has been shown against *Staphylococcus aureus*, salmonellae, coliforms, psychrotrophic spoilage bacteria (especially the pseudomonads), and *Vibrio parahaemolyticus*. Against the latter organism, concentrations as low as 30 ppm have been shown to be effective. Shelf-life extensions have been obtained by use of sorbates on fresh poultry meat, vacuum-packaged poultry products, fresh fish, and perishable fruits. For further information, see nitrite-sorbate combinations below and the review by Sofos and Busta (136).

The sorbates have been studied by a large number of groups for use in meat products in combination with nitrites. Bacon formulations that contain 120 ppm $NaNO_2$ without sorbate yield products that maintain their desirable organoleptic qualities in addition to being protected from *C. botulinum* growth. When 0.26% (2,600 ppm) potassium sorbate is added along with 40 ppm nitrite, no significant differences are found in the organoleptic qualities or in botulinal protection ((69) and Stevenson and Price, 1976, cited in (105)). The combination of 40 ppm $NaNO_2$ and 0.26% potassium sorbate (along with 550 ppm sodium ascorbate or sodium erythrobate) was proposed by the U.S. Department of Agriculture (USDA) in 1978 but postponed in 1979. The later action was prompted not by the failure of the reduced nitrite level in combination with sorbate but because of taste panel results that characterized finished bacon as having "chemical"-like flavors and producing prickly mouth sensations (8). The combination of sorbate plus reduced nitrite has been shown to be effective in a variety of cured meat products against not only *C. botulinum* but other bacteria such as *S. aureus* and a spoilage *Clostridium* (P.A. 3679). With the latter, a noninhibitory concentration of nitrite and sorbate was bactericidal (124). (For further information see the nitrite section below and references 87, 138, and 149.)

The widest use of sorbates is as fungistats in products such as cheeses, bakery products, fruit juices, beverages, salad dressings, and the like. In the case of molds, inhibition may be due to inhibition of the dehydrogenase enzyme system. Against germinating endospores, sorbate prevents the outgrowth of vegetative cells (Fig. 11-1).

As lipophilic acids, sorbate, benzoate, and propionate appear to inhibit microbial cells by the same general mechanism. The mechanism involves the **proton motive force (PMF)**. Briefly, hydrogen ions (protons) and hydroxyl ions are separated by the cytoplasmic membrane, with the former, outside, giving rise to acidic pH and the latter, inside the cell, giving rise to pH near neutrality. The membrane gradient thus created represents electrochemical potential that the cell employs in the active transport of some compounds such as amino acids. Weak lipophilic acids act as protonophores. After diffusing across the membrane, the undissociated molecule ionizes inside the cell and lowers intracellular pH. This results in a weakening of the transmembrane gradient such that amino acid transport is affected adversely. This hypothesis has been supported by research on P.A. 3679 where sorbate inhibited phenylalanine uptake, decreased protein synthesis, and altered phorphorylated nucleotide accumulation (124, 125). Other possible inhibitory mechanisms have been proposed by various researchers, but the alteration of the PMF appears to be the most feasible at this time.

With respect to safety, sorbic acid is metabolized in the body to CO_2 and H_2O in the same manner as fatty acids normally found in foods (34).

THE PROPIONATES

Propionic acid is a three-carbon organic acid with the structure CH_3CH_2COOH. This acid and its calcium and sodium salts are permitted in breads, cakes, certain cheeses, and other foods primarily as a mold inhibitor. Propionic acid is employed also as a "rope" inhibitor in bread dough. The tendency toward dissociation is low with this compound and its salts, and they are consequently active in low-acid foods. They tend to be highly specific against molds, with the inhibitory action being primarily fungistatic rather than fungicidal.

With respect to the antimicrobial mode of action of propionates, they act in a manner similar to that of benzoate and sorbate. The pK of propionate is 4.87 and at a pH of 4.00, 88% of the compound is undissociated, while at a pH of 6.0, only 6.7% remains undissociated. The undissociated molecule of this lipophilic acid is necessary for its antimicrobial activity. The mode of action of propionic acid is noted above with benzoic acid. See also the section below on medium-chain fatty acids and esters and the review by Doores (37) for further information.

SULFUR DIOXIDE AND SULFITES

Sulfur dioxide (SO_2) and the sodium and potassium salts of sulfite ($=SO_3$), bisulfite ($-HSO_3$), and metabisulfite ($=S_2O_5$) all appear to act similarly and are treated here together. Sulfur dioxide is used in its gaseous or liquid form or in the form of one or more of its neutral or acid salts on dried fruits, in lemon juice, molasses, wines, fruit juices, and others. The parent compound has been used as a food preservative since ancient times. Its use as a meat preservative in the United States dates back to at least 1813; however, it is not permitted in meats or other foods recognizable as sources of thiamine. While SO_2 possesses antimicrobial activity, it is also used in certain foods as an antioxidant.

The predominant ionic species of sulfurous acid depends on pH of milieu with SO_2 being favored by pH < 3.0, HSO_3^- by pH between 3.0 and 5.0, and $SO_3^=$ > pH 6.0 (104). SO_2 has pKs of 1.76 and 7.2. The sulfites react with various food constituents including nucleotides, sugars, disulfide bonds, and others.

With regard to its effect on microorganisms, SO_2 is bacteriostatic against *Acetobacter* spp. and the lactic acid bacteria at low pH, concentrations of 100 to 200 ppm being effective in fruit juices and beverages. It is bactericidal at higher concentrations. When added to temperature-abused comminuted pork, 100 ppm of SO_2 or higher were required to effect significant inhibition of spores of *C. botulinum* at target levels of 100 spores/g (153). The source of SO_2 was sodium metabisulfite. Employing the same salt to achieve an SO_2 concentration of 600 ppm, Banks and Board (6) found that growth of salmonellae and other Enterobacteriaceae were inhibited in British fresh sausage. The most sensitive bacteria were eight salmonellae serovars, which were inhibited by 15 to 109 ppm at pH 7.0, while *Serratia liquefaciens*, *S. marcescens*, and *Hafnia alvei* were the most resistant, requiring 185 to 270 ppm free SO_2 in broth.

Yeasts are intermediate to acetic and lactic acid bacteria and molds in their sensitivity to SO_2, and the more strongly aerobic species are generally more sensitive than the more fermentative species. Sulfurous acid at levels of 0.2 to 20 ppm was effective against some yeasts, including *Saccharomyces*, *Pichia*, and *Candida*, while *Zygosaccharomyces bailii* required up to 230 ppm for inhibition in certain fruit drinks at pH 3.1 (90). Yeasts can actually form SO_2 during juice fermentation; some "*S. carlsbergensis*" and *S. bayanus* strains produce up to 1,000 and 500 ppm, respectively (104). Molds such as *Botrytis* can be controlled on grapes by periodic gassing with SO_2, and bisulfite can be used to destroy aflatoxins (38). Both aflatoxins B_1 and B_2 can be reduced in corn (53, 97). Sodium bisulfite was found to be comparable to propionic acid in its anti-microbial activity in corn containing up to 40% moisture (53). (Aflatoxin degradation is discussed further in Chap. 25.)

Although the actual mechanism of action of SO_2 is not known, several possibilities have been suggested, each supported by some experimental evidence. One suggestion is that the undissociated sulfurous acid or molecular SO_2 is responsible for the antimicrobial activity. Its greater effectiveness at low pH tends to support this. Vas and Ingram (155) suggested the lowering of pH of certain foods by addition of acid as a means of obtaining greater preservation with SO_2. It has been suggested that the antimicrobial action is due to the strong reducing power that allows these compounds to reduce oxygen tension to a point below that at which aerobic organisms can grow or by direct action upon some enzyme system. SO_2 is also thought to be an enzyme poison, inhibiting growth of microorganisms by inhibiting essential enzymes. Its use in the drying of foods to inhibit enzymatic browning is based on this assumption. Since the sulfites are known to act on disulfide bonds, it may be presumed that certain essential enzymes are affected and that inhibition ensues. The sulfites do not inhibit cellular transport. It may be noted from Figure 11-1 that metabisulfite acts on germinating endospores during the outgrowth of vegetative cells.

NITRITES AND NITRATES

Sodium nitrate ($NaNO_3$) and sodium nitrite ($NaNO_2$) are used in curing formulas for meats since they stabilize red meat color, inhibit some spoilage and food poisoning organisms, and contribute to flavor development. The role of NO_2 in cured meat flavor has been reviewed by Gray and Pearson (52). NO_2 has been shown to disappear on both heating and storage. It should be recalled that many bacteria are capable of utilizing nitrate as an electron acceptor and in the process effect its reduction to nitrite. The nitrite ion is by far the more important of the two in preserved meats. This ion is highly reactive and is capable of serving as both a reducing and an oxidizing agent. In an acid environment, it ionizes to yield nitrous acid (3HONO), which further decomposes to yield nitric oxide (NO), the important product from the standpoint of color fixation in cured meats. Ascorbate or erythrobate acts also to reduce NO_2 to NO. Nitric oxide reacts with myoglobin under reducing conditions to produce the desirable red pigment **nitrosomyoglobin** (see also Table 9-11):

When the meat pigment exists in the form of **oxymyoglobin**, as would be the case for comminuted meats, this compound is first oxidized to **metmyoglobin** (brown color). Upon the reduction of the latter, nitric oxide reacts to yield nitrosomyoglobin. Since nitric oxide is known to be capable of reacting with other porphyrin-containing compounds such as catalase, peroxidases, cytochromes, and others, it is conceivable that some of the antibacterial effects of nitrites against aerobes may be due to this action (the mechanism is discussed below). It has been shown that the antibacterial effect of NO_2 increases as pH is lowered within the acid range, and this effect is accompanied by an overall increase in the undissociated HNO_2.

The cooked cured meat pigment is **dinitrosyl ferrohemochrome** (DNFH). It forms when globin in nitrosomyoglobin is replaced with a second NO group (131). A nitrite-free curing formula for wieners has been developed by adding 35 ppm of encapsulated DNFH prepared from bovine erythrocytes, t=butylhydroxyquinoline (TBHQ) as antioxidant (162), and 3,000 ppm sodium hypophosphite as antibotulinal agent (102, 158). Essentially no microbial growth occurred through 4 weeks of storage with this formula, similar to the control nitrite-containing formulation. Sodium hypophosphite was the best of a variety of compounds tested as NO_2 replacements (158).

Organisms Affected

Although the single microorganism of greatest concern relative to nitrite inhibition is *C. botulinum*, the compound has been evaluated as an antimicrobial for other organisms. During the late 1940s it was evaluated as a fish preservative and found to be somewhat effective but generally only at low pH. It is effective against *S. aureus* at high concentrations and, again, the effectiveness increases as pH is lowered. The compound is generally ineffective against Enterobacteriaceae, including the salmonellae, and against the lactic acid bacteria, although some effects are noted in cured and in vacuum-packaged meats and are probably caused by the interaction of nitrite with other environmental parameters rather than to nitrite alone. Nitrite is added to cheeses in some countries to control gassiness caused by *Clostridium butyricum* and *C. tyrobutyricum*. It is effective against other clostridia, including *C. sporogenes* and *C. perfringens*, which are often employed in laboratory studies to assess potential antibotulinal effects not only of nitrites but of other inhibitors that might have value as nitrite adjuncts or sparing agents.

The Perigo Factor

The almost total absence of botulism in cured, canned, and vacuum-packed meats and fish products led some investigators in the mid-1960s to seek reasons as to why meat products that contained viable endospores did not become toxic. Employing culture medium, it was shown in 1967 that about ten times more nitrite was needed to inhibit clostridia if it were added after instead of before the medium was autoclaved. It was concluded that the heating of the medium with nitrite produced a substance or agent about ten times more inhibitory than nitrite alone (106, 107). This agent is referred to as the **Perigo factor**. The existence of this factor or effect has been confirmed by some and questioned by others. While the Perigo factor may be questionable in cured and perishable cured meats, the evidence for an inhibitory factor in culture media involving nitrite, iron, and —SH groups is more conclusive (149).

The factor does not develop in all culture media, and heating to at least 100°C is necessary for its development, although some activity develops in meats when heated to as low as 70°C. The Perigo factor is dialyzable from some culture media and meat suspensions but not from other media (74). It is not found in filter-sterilized solutions of the same medium with nitrite added (123). If meat is added to a medium containing the Perigo factor, the inhibitory activity is lost (75). For this reason, some Canadian workers called the inhibitor formed in meat the "Perigo-type factor" (20).

This inhibitory or antibotulinal effect that results from the heat processing or smoking of certain meat and fish products containing nitrite warrants the continued use of nitrite in such products. The antibotulinal activity of nitrite in cured meats is of greater public health importance than the facts of color and flavor development. For the latter, initial nitrite levels as low as 15 to 50 ppm have been reported to be adequate for various meat products, including Thuringer sausage (33). Nitrite levels of 100 ppm or more have been found to make for maximum flavor and appearance in fermented sausages (84). The antibotulinal effect requires at least 120 ppm for bacon (14, 23), comminuted cured ham (22), and canned, shelf-stable luncheon meat (20). Many of these canned products are given a low heat process (F_o of 0.1–0.6).

Interaction with Cure Ingredients and Other Factors

The interplay of all ingredients and factors involved in heat-processed, cured meats on antibotulinal activity was noted almost 30 years ago by Riemann (117), and several other investigators have pointed out that curing salts in semipreserved meats are more effective in inhibiting heat-injured spores than noninjured (40, 122). With brine and pH alone, higher concentrations of the former are required for inhibition as pH increases, and Chang et al. (20) suggested that the inhibitory effect of salt in shelf-stable canned meats against heat-injured spores may be more important than the Perigo-type factor. With smoked salmon inoculated with 10^2 spores/g of *C. botulinum* types A and E and stored in O_2-impermeable film, 3.8 and 6.1% waterphase NaCl alone inhibited toxin production in 7 days by types E and A, respectively (105). With 100 ppm or more of NO_2, only 2.5% NaCl was required for inhibition of toxin production by type E, and for type A 3.5% NaCl + 150 ppm $NaNO_2$ was inhibitory. With longer incubations or larger spore inocula, more NaCl or $NaNO_2$ is needed.

The interplay of NaCl, $NaNO_2$, $NaNO_3$, isoascorbate, polyphosphate, thermal process temperatures, and temperature/time of storage on spore outgrowth and germination in pork slurries has been studied extensively by Roberts et al. (120), who found that significant reductions in toxin production could be achieved by increasing the individual factors noted. It is well known that low pH is antagonistic to growth and toxin production by *C. botulinum*, whether the acidity results from added acids or the growth of lactic acid bacteria. When 0.9% sucrose was added to bacon along with *Lactobacillus plantarum*, only 1 of 49 samples became toxic after 4 weeks, while with sucrose and no lactobacilli, 50 of 52 samples became toxic in 2 weeks (144). When 40 ppm nitrite was used alone, 47 of 50 samples became toxic after 2 weeks but when 40 ppm nitrite was accompanied by 0.9% sucrose and an inoculum of *L. plantarum*, none of 30 became toxic. While this was most likely a direct pH effect, other factors may have been involved (see the section on lactic antagonism). In later studies, bacon was prepared with 40 or 80 ppm $NaNO_2$ + 0.7% sucrose followed by inoculation with *Pediococcus acidilactici*. When inoculated with *C. botulinum* types A and B spores, vacuum packaged, and incubated up to 56 days at 27°C, the bacon was found to have greater anti-botulinal properties than control bacon prepared with 120 ppm $NaNO_2$ but not sucrose or lactic inoculum (143). Bacon prepared by the above formulation, called the Wisconsin process, was preferred by a sensory panel to that prepared by the conventional method (142). The Wisconsin process employs 550 ppm of sodium ascorbate or sodium erythrobate, as does the conventional process.

Nitrosamines

When nitrite reacts with secondary amines, **nitrosamines** are formed, and many are known to be carcinogenic. The generalized way in which nitrosamines may form is as follows:

$$R_2NH_2 + HONO \xrightarrow{H^+} R_2N-NO + H_2O$$

The amine dimethylamine reacts with nitrite to form N-nitrosodimethylamine:

$$\begin{array}{ccc}
H_3C & & H_3C \\
\diagdown & & \diagdown \\
& N-H + NO_2 \rightarrow & N-N=O \\
\diagup & & \diagup \\
H_3C & & H_3C
\end{array}$$

In addition to secondary amines, tertiary amines and quarternary ammonium compounds also yield nitrosamines with nitrite under acidic conditions. Nitrosamines have been found in cured meat and fish products at low levels. Isoascorbate has an inhibitory effect on nitrosamine formation.

It has been shown that lactobacilli, enterococci, clostridia, and other bacteria will nitrosate secondary amines with nitrite at neutral pH values (60). The fact that nitrosation occurred at near-neutral pH values was taken to indicate that the process was enzymatic, although no cell-free enzyme was obtained (61). Several species of catalase-negative cocci, including *E. faecalis*, *E. faecium*, and *L. lactis*, have been shown to be capable of forming nitrosamines, but the other lactic acid bacteria and pseudomonads tested did not (24). These investigators found no evidence for an enzymatic reaction. *S. aureus* and halobacteria obtained from Chinese salted marine fish (previously shown to contain nitrosamines) produced nitrosamines when inoculated into salted fish homogenates containing 40 ppm of nitrate and 5 ppm nitrite (43).

Nitrite-Sorbate and Other Nitrite Combinations

In an effort to reduce the potential hazard of N-nitrosamine formation in bacon, the USDA in 1978 reduced the input NO_2 level for bacon to 120 ppm and set a 10 ppb maximum level for nitrosamines. While 120 ppm nitrite along with 550 ppm sodium ascorbate or sodium erythrobate is adequate to reduce the botulism hazard, it is desirable to reduce nitrite levels even further if protection against botulinal toxin production can be achieved. To this end, a proposal to allow the use of 40 ppm nitrite in combination with 0.26% potassium sorbate for bacon was made in 1978 but rescinded a year later when taste panel studies revealed undesirable effects. Meanwhile, many groups of researchers have shown that 0.26% sorbate in combination with 40 or 80 ppm nitrite is effective in preventing botulinal toxin production. Extensive reviews of these studies have been provided (91, 138, 149).

In an early study of the efficacy of 40 ppm nitrite + sorbate to prevent or delay botulinal toxin production in commercial-type bacon, Ivey et al. (69) used an inoculum of 1,100 types A and B spores/g and incubated the product at 27°C for up to 110 days. Time for the appearance of toxic samples when neither nitrite nor sorbate was used was 19 days. With 40 ppm nitrite and no sorbate, toxic smaples appeared in 27 days, and for samples containing 40 ppm nitrite plus 0.26% sorbate or no nitrite and 0.26% sorbate, more than 110 days were required for toxic samples. This reduced nitrite level resulted in lower levels of nitrosopyrrolidine in cooked bacon. Somewhat different findings were reported by Sofos et al. (Table 11-2), with 80 ppm nitrite being required for the absence of toxigenic samples after 60 days. In addition to its inhibitory effect on *C. botulinum*, sorbate slows the depletion of nitrite during storage (137).

The effect of isoascorbate is to enhance nitrite inhibition by sequestering iron, although under some conditions it may reduce nitrite efficiency by causing a more rapid depletion of residual nitrite (150, 152). EDTA at 500 ppm appears to be even more effective than erythrobate in potentiating the nitrite effect, but only limited studies have been reported. Another chelate, 8-hydroxyquinoline, has been evaluated as a nitrite-sparing agent. When 200 ppm were combined with 40 ppm nitrite, a *C. botulinum* spore mixture of types A and B strains was inhibited for 60 days at 27°C in comminuted pork (108).

In an evaluation of the interaction of nitrite and sorbate, the relative effectiveness of the combination has been shown to be dependent on other cure ingredients and product parameters. Employing a liver-veal agar medium at pH 5.8 to 6.0, the germination rate of *C. botulinum* type E spores decreased to nearly zero with 1.0, 1.5, or 2.0% sorbate, but with the same concentrations at pH 7.0 to 7.2, germination and outgrowth of abnormally shaped cells occurred

TABLE 11-2.   Effect of Nitrite and Sorbate on Toxin
Production in Bacon Inoculated with *C.
botulinum* Types A and B Spores and
Held up to 60 Days at 27°C

Treatment	Percentage Toxigenic
Control (no NO_2, no sorbate)	90.0
0.26% sorbate, no $NaNO_2$	58.8
0.26% sorbate + 40 ppm $NaNO_2$	22.0
0.26% sorbate + 80 ppm $NaNO_2$	0.0
No sorbate, 120 ppm $NaNO_2$	0.4

Source: Sofos et al. (138).

(130). When 500 ppm nitrite was added to the higher-pH medium along with sorbate, cell lysis was enhanced. These investigators also found that 500 ppm linoleic acid alone at the higher pH prevented emergence and elongation of spores. Potassium sorbate significantly decreased toxin production by types A and B spores in pork slurries when NaCl was increased or pH and storage temperature were reduced (121). For chicken frankfurters, a sorbate-betalains mixture was found to be as effective as a conventional nitrite system for inhibiting *C. perfringens* growth (154).

Mode of Action

It appears that nitrite inhibits *C. botulinum* by interfering with iron-sulfur enzymes such as ferredoxin and thus preventing the synthesis of adenosine-triphosphate (ATP) from pyruvate. The first direct finding in this regard was that of Woods et al. (160), who showed that the phosphoroclastic system of *C. sporogenes* is inhibited by nitric oxide and later that the same occurs in *C. botulinum*, resulting in the accumulation of pyruvic acid in the medium (159).

The phosphoroclastic reaction involves the breakdown of pyruvate with inorganic phosphate and coenzyme A to yield acetyl phosphate. In the presence of adenosine diphosphate (ADP), ATP is synthesized from acetyl phosphate with acetate as the other product. In the breakdown of pyruvate, electrons are transferred first to ferredoxin and from ferredoxin to H^+ to form H_2 in a reaction catalyzed by hydrogenase. Ferredoxin and hydrogenase are iron-sulfur (nonheme) proteins or enzymes (16).

Following the work of Woods and Wood (159), the next most significant finding was that of Reddy et al. (113), who subjected extracts of nitrite-ascorbate-treated *C. botulinum* to electron spin resonance and found that nitric oxide reacted with iron-sulfur complexes to form iron-nitrosyl complexes. The presence of the latter results in the destruction of iron-sulfur enzymes such as ferredoxin.

The resistance of the lactic acid bacteria to nitrite inhibition is well known, but the basis is just now clear: these organisms lack ferredoxin. The clostridia contain both ferredoxin and hydrogenase, which function in electron transport in the anaerobic breakdown of pyruvate to yield ATP, H_2, and CO_2. The ferredoxin in clostridia has a molecular weight of 6,000 and contains 8 Fe atoms/mole and 8-labile sulfide atoms/mole.

Although the first definitive experimental finding was reported in 1981, earlier work pointed to iron-sulfur enzymes as the probable nitrite targets. Among the first were O'Leary and Solberg (103), who showed that a 91% decrease occurred in the concentration of free —SH groups of soluble cellular compounds of *C. perfringens* inhibited by nitrite. Two years later, Tompkin et al. (151) offered the hypothesis that nitric oxide reacted with iron in the vegetative cells of *C. botulinum*, perhaps the iron in ferredoxin. The inhibition by nitrite of active transport and electron transport was noted by several investigators, and these effects are consistent with nitrite inhibition of nonheme enzymes such as ferredoxin and hydrogenase (121, 161). The enhancement of inhibition in the presence of sequestering agent may be due to the reaction of sequestrants to substrate iron: more nitrite becomes available for nitric oxide production and reaction with microorganisms.

Summary of Nitrite Effects

When added to processed meats such as wieners, bacon, smoked fish, and canned cured meats followed by substerilizing heat treatments, nitrite has definite antibotulinal effects. It also forms desirable product color and enhances flavor in cured meat products. The antibotulinal effect consists of inhibition of vegetative cell growth and the prevention of germination and growth of spores that survive heat processing or smoking during postprocessing storage. Clostridia other than *C. botulinum* are affected in a similar manner. While low initial levels of nitrite are adequate for color and flavor development, considerably higher levels are necessary for the antimicrobial effects.

When nitrite is heated in certain laboratory media, an antibotulinal factor or inhibitor is formed, the exact identity of which is not yet known. The inhibitory factor is the Perigo effect/factor or Perigo inhibitor. It does not form in filter-sterilized media. It develops in canned meats only when nitrite is present during heating. The initial level of nitrite is more important to antibotulinal activity than the residual level. Once formed, the Perigo factor is not affected greatly by pH changes.

Measurable preheating levels of nitrite decrease considerably during heating in meats and during postprocessing storage—more at higher storage temperatures than at lower.

The antibotulinal activity of nitrite is interdependent with pH, salt content, temperature of incubation, and numbers of botulinal spores. Heat-injured spores are more susceptible to inhibition than uninjured. Nitrite is more effective under Eh− than under Eh+ conditions.

Nitrite does not decrease the heat resistance of spores. It is not affected by ascorbate in its antibotulinal actions but does act synergistically with ascorbate in pigment formation.

Lactic acid bacteria are relatively resistant to nitrite (see above).

Endospores remain viable in the presence of the antibotulinal effect and will germinate when transferred to nitrite-free media.

Nitrite has a pK of 3.29 and consequently exists as undissociated nitrous acid at low pH values. The maximum undissociated state and consequent greatest antibacterial activity of nitrous acid are between pH 4.5 and 5.5.

With respect to its depletion or disappearance in ham, Nordin (101) found the rate to be proportional to its concentration and to be exponentially related to both temperature and pH. The depletion rate doubled for every 12.2°C increase in temperature or 0.86 pH unit decrease and was not affected by heat denaturation of the ham. These relationships did not apply at room temperature unless the product was first heat treated, suggesting that viable organisms aided in its depletion.

It appears that the antibotulinal activity of nitrite is due to its inhibition of nonheme, iron-sulfur enzymes.

NaCl AND SUGARS

These compounds are grouped together because of the similarity in their modes of action in preserving foods. NaCl has been employed as a food preservative since ancient times. The early food uses of salt were for the purpose of preserving meats. This use is based on the fact that at high concentrations, salt exerts a drying effect on both food and microorganisms. Salt (saline) in water at concentrations of 0.85 to 0.90% produces an **isotonic** condition for nonmarine microorganisms. Since the amounts of NaCl and water are equal on both sides of the cell membrane, water moves across the cell membranes equally in both directions. When microbial cells are suspended in, say, a 5% saline solution, the concentration of water is greater inside the cells than outside (concentration of H_2O is highest where solute concentration is lowest). In diffusion, water moves from its area of high concentration to its area of low concentration. In this case, water passes out of the cells at a greater rate than it enters. The result to the cell is **plasmolysis**, which results in growth inhibition and possibly death. This is essentially what is achieved when high concentrations of salt are added to fresh meats for the purpose of preservation. Both the microbial cells and those of the meat undergo plasmolysis (shrinkage), resulting in the drying of the meat, as well as inhibition or death of microbial cells. Enough salt must be used to effect **hypertonic** conditions. The higher the concentration, the greater are the preservative and drying effects. In the absence of refrigeration, fish and other meats may be effectively preserved by salting. The inhibitory effects of salt are not dependent on pH as are some other chemical preservatives. Most nonmarine bacteria can be inhibited by 20% or less of NaCl, while some molds generally tolerate higher levels. Organisms that can grow in the presence of and require high concentrations of salt are referred to as **halophiles**; those that can withstand but not grow in high concentrations are referred to as **halodurics**. (The interaction of salt with nitrite and other agents in the inhibition of *C. botulinum* is discussed above under nitrites.)

Sugars, such as sucrose, exert their preserving effect in essentially the same manner as salt. One of the main differences is in relative concentrations. It generally requires about six times more sucrose than NaCl to effect the same degree of inhibition. The most common uses of sugars as preserving agents are in the making of fruit preserves, candies, condensed milk, and the like. The shelf stability of certain pies, cakes, and other such products is due in large part to the preserving effect of high concentrations of sugar, which, like salt, makes water unavailable to microorganisms.

Microorganisms differ in their response to hypertonic concentrations of sugars, with yeasts and molds being less susceptible than bacteria. Some yeasts and molds can grow in the presence of as much as 60% sucrose, while most bacteria are inhibited by much lower levels. Organisms that are able to grow in high concentrations of sugars are designated **osmophiles**; **osmoduric** microorganisms are those that are unable to grow but are able to withstand high levels of sugars. Some osmophilic yeasts such as *Zygosaccharomyces rouxii* can grow in the presence of extremely high concentrations of sugars.

INDIRECT ANTIMICROBIALS

The compounds and products in this section are added to foods primarily for effects other than antimicrobial and are thus multifunctional food additives.

TABLE 11-3.   Some GRAS Indirectly Antimicrobial Chemicals Used in Foods

Compound	Primary Use	Most Susceptible Organisms
Butylated hydroxyanisole (BHA)	Antioxidant	Bacteria, some fungi
Butylated hydroxytoluene (BHT)	Antioxidant	Bacteria, viruses, fungi
t-butylhydroxyquinoline (TBHQ)	Antioxidant	Bacteria, fungi
Propyl gallate (PG)	Antioxidant	Bacteria
Nordihydroguaiaretic acid	Antioxident	Bacteria
Ethylenediamine tetraacetic acid (EDTA)	Sequestrant/stabilizer	Bacteria
Sodium citrate	Buffer/sequestrant	Bacteria
Lauric acid	Defoaming agent	Gram-positive bacteria
Monolaurin	Emulsifier	Gram-positive bacteria, yeasts
Diacetyl	Flavoring	Gram-negative bacteria, fungi
d- and l-carvone	Flavoring	Fungi, gram-positive bacteria
Phenylacetaldehyde	Flavoring	Fungi, gram-positive bacteria
Menthol	Flavoring	Bacteria, fungi
Vanillin, ethyl vanillin	Flavoring	Fungi
Spices/spice oils	Flavoring	Bacteria, fungi

Antioxidants

Although used in foods primarily to prevent the auto-oxidation of lipids, the phenolic antioxidants listed in Table 11-3 have been shown to possess antimicrobial activity against a wide range of microorganisms, including some viruses, mycoplasmas, and protozoa. These compounds have been evaluated extensively as nitrite-sparing agents in processed meats and in combination with other inhibitors, and several excellent reviews have been made (15, 46, 77).

Butylated hydroxyanisole (BHA), butylated hydroxytoluene (BHT), and TBHQ are inhibitory to gram-positive and gram-negative bacteria, as well as to yeasts and molds at concentrations ranging from about 10 to 1,000 ppm depending on substrate. In general, higher concentrations are required to inhibit in foods than in culture media, especially in high-fat foods. BHA was about 50 times less effective against *Bacillus* spp. in strained chicken than in nutrient broth (133). BHA, BHT, TBHQ, and propyl gallate (PG) were all less effective in ground pork than in culture media (47). While strains of the same bacterial species may show wide variation in sensitivity to either of these antioxidants, it appears that BHA and TBHQ are more inhibitory than BHT to bacteria and fungi, while the latter is more viristatic. To prevent growth of *C. botulinum* in a prereduced medium, 50 ppm BHA and 200 ppm BHT were required; 200 ppm PG were ineffective (118). Employing 16 gram-negative and 8 gram-positive bacteria in culture media, Gailani and Fung (47) found the gram positives to be more susceptible than gram negatives to BHA, BHT, TBHQ, and PG, with each being more effective in nutrient agar than in brain heart infusion (BHI) broth. In nutrient agar, the relative effectiveness was BHA > PG > TBHQ > BHT, while in BHI, TBHQ > PG > BHA > BHT.

Foodborne pathogens such as *Bacillus cereus*, *V. parahaemolyticus*, salmonellae, and *S. aureus* are effectively inhibited at concentrations <500 ppm, while some are sensitive to as little as 10 ppm. The pseudomonads, especially *P. aeruginosa*, are among the most resistant bacteria. Three toxin-producing penicillia were inhibited significantly in salami by BHA, TBHQ, and a combination of these two at 100 ppm, while BHT and PG were ineffective (88). Combinations of BHA/sorbate and BHT/monolaurin have been shown to be synergistic against *S. aureus* (15, 28) and BHA/sorbate against *S. typhimurium* (28). BHT/TBHQ has been shown to be synergistic against aflatoxin-producing penicillia (88).

Flavoring Agents

Of the many agents used to impart aromas and flavors to foods, some possess definite antimicrobial effects. In general, flavor compounds tend to be more antifungal than antibacterial. The nonlactic, gram-positive bacteria are the most sensitive, and the lactic acid bacteria are rather resistant. The essential oils and spices have received the most attention by food microbiologists, while the aroma compounds have been studied more for their use in cosmetics and soaps.

Of 21 flavoring compounds examined in one study, about half had minimal inhibitory concentrations (MIC) of 1,000 ppm or less against either bacteria or fungi (73). All were pH sensitive, with inhibition increasing as pH and temperature of incubation decreased. Some of these compounds are noted in Table 11-3.

One of the most effective flavoring agents is diacetyl, which imparts the aroma of butter (70). It is somewhat unique in being more effective against gram-negative bacteria and fungi than against gram-positive bacteria. In plate count agar at pH 6.0 and incubation at 30°C, all but 1 of 25 gram-negative bacteria and 15 of 16 yeasts and molds were inhibited by 300 ppm (71). At pH 6.0 and incubation at 5°C in nutrient broth, <10 ppm inhibited *Pseudomonas fluorescens*, *P. geniculata*, and *E. faecalis*, while under the same conditions except with incubation at 30°C, about 240 ppm were required to inhibit these and other organisms (73). It appears that diacetyl antagonizes arginine utilization by reacting with arginine-binding proteins of gram-negative bacteria. The greater resistance of gram-positive bacteria appears to be due to their lack of similar periplasmic binding proteins and their possession of larger amino acid pools. Another flavor compound that imparts the aroma of butter is 2,3-pentanedione, and it has been found to be inhibitory to a limited number of gram-positive bacteria and fungi at 500 ppm or less (73).

The agent *l*-carvone imparts spearmintlike and the agent *d*-carvone imparts carawaylike aromas, and both are antimicrobial, with the *l*-isomer being more effective than the *d*-isomer, while both are more effective against fungi than bacteria at 1,000 ppm or less (73). Phenylacetaldehyde imparts a hyacinthlike aroma and has been shown to be inhibitory to *S. aureus* at 100 ppm and *Candida albicans* at 500 ppm (73, 98). Menthol, which imparts a peppermintlike aroma, was found to inhibit *S. aureus* at 32 ppm, and *E. coli* and *C. albicans* at 500 ppm (73, 98). Vanillin and ethyl vanillin are inhibitory, especially to fungi at levels <1,000 ppm.

Spices and Essential Oils

While used primarily as flavoring and seasoning agents in foods, many spices possess significant antimicrobial activity. In all instances, antimicrobial activity is due to specific chemicals or essential oils (some are noted in Chap. 3). The search for nitrite-sparing agents generated new interest in spices and spice extracts in the late 1970s (132).

It would be difficult to predict what antimicrobial effects, if any, are derived from spices as they are used in foods; the quantities employed differ widely depending on taste, and the relative effectiveness varies depending on product composition. Because of the varying concentrations of the antimicrobial constituents in different spices and because many studies have been conducted employing them on a dry weight basis, it is difficult to ascertain the MIC of given spices against specific organisms. Another reason for conflicting results by different investigators is the assay method employed. In general, higher MIC values are obtained when highly volatile compounds are evaluated on the surface of plating media than when they are tested in pour plates or broth. When eugenol was evaluated by surface plating onto plate count agar (PCA) at pH 6, only 9 of 14 gram-negative and 12 of 20 gram-positive bacteria (including eight lactics) were inhibited by 493 ppm, while in nutrient broth at the same pH, MICs of 32 and 63 were obtained for *Torulopsis candida* and *Aspergillus niger*, and *S. aureus* and *Escherichia coli*, respectively (73). Spice extracts are less inhibitory in media than spices, probably due to a slower release of volatiles by the latter (134). In spite of the difficulties of comparing results from study to study, the antimicrobial activity of spices is unquestioned, and a large number of investigators have shown the effectiveness of at least 20 different spices or their extracts against most food-poisoning organisms, including mycotoxigenic fungi (132).

In general, spices are less effective in foods than in culture media, and gram-positive bacteria are more sensitive than gram negatives, with the lactic acid bacteria being the most resistant among gram positives (163). While results concerning them are debatable, the fungi appear to be in general more sensitive than gram-negative bacteria. Some gram negatives, however, are highly sensitive. Antimicrobial substances vary in content from the allicin of garlic (with a range of 0.3–0.5%) to eugenol in cloves (16–18%) (132). When whole spices are employed, MIC values range from 1 to 5% for sensitive organisms. Sage and rosemary are among the most antimicrobial as reported by various researchers, and it has been reported that 0.3% in culture media inhibited 21 of 24 gram-positive bacteria and were more effective than allspice (134).

With respect to specific inhibitory levels of extracts and essential oils, Huhtanen (65) made ethanol extracts of 33 spices, tested them in broth against *C. botulinum*, and found that achiote and mace extracts produced an MIC of 31 ppm and were the most effective of the 33. Next most effective were nutmeg, bay leaf, and white and black peppers, with MICs of 125 ppm. Employing the essential oils of oregano, thyme, and sassafras, Beuchat (9) found that 100 ppm were cidal to *V. parahaemolyticus* in broth. Growth and aflatoxin production by *Aspergillus parasiticus* in broth were inhibited by 200 to 300 ppm of cinnamon and clove oils, by 150 ppm cinnamic aldehyde, and by 125 ppm eugenol (18).

The mechanisms by which spices inhibit microorganisms are unclear and may be presumed to be different for unrelated groups of spices. That the mechanism for oregano, rosemary, sage, and thyme may be similar is suggested by the finding that resistance development by some lactic acid bacteria to one was accompanied by resistance to the other three (163).

Medium-Chain Fatty Acids and Esters

Acetic, propionic, and sorbic acids are short-chain fatty acids used primarily as preservatives. Medium-chain fatty acids are employed primarily as surface-active or emulsifying agents. The antimicrobial activity of the medium-chain fatty acids is best known from soaps, which are salts of fatty acids. Those most commonly employed are composed of 12 to 16 carbons. For saturated fatty acids, the most antimicrobial chain length is C_{12}; for monounsaturated (containing 1 double bond) $C_{16:1}$; and for polyunsaturated (containing more than one double bond) $C_{18:2}$ is the most antimicrobial (78). In general, fatty acids are effective primarily against gram-positive bacteria and yeasts. While the C_{12} to C_{16} chain lengths are the most active against bacteria, the C_{10} to C_{12} are most active against yeasts (78). Fatty acids and esters and the structure-function relationships among them have been reviewed and discussed by Kabara (77, 78). Saturated aliphatic acids effective against *C. botulinum* have been evaluated by Dymicky and Trenchard (41).

The monoesters of glycerol and the diesters of sucrose are more anti-microbial than the corresponding free fatty acids and compare favorably with sorbic acid and the parabens as antimicrobials (77). Monolaurin is the most effective of the glycerol monoesters, and sucrose dicaprylate is the most effective of the sucrose diesters. Monolaurin (lauricidin) has been evaluated by a large number of investigators and found to be inhibitory to a variety of gram-positive bacteria and some yeasts at 5 to 100 ppm (15, 77). Unlike the short-chain fatty acids, which are most effective at low pH, monolaurin is effective over the range 5.0 to 8.0 (79).

Because the fatty acids and esters have a narrow range of effectiveness and GRAS substances such as EDTA, citrate, and phenolic antioxidants also have limitations as antimicrobial agents when used alone, Karara (77, 78) has stressed the "preservative system" approach to the control of microorganisms in foods by using combinations of chemicals to fit given food systems and preservation needs. By this approach, a preservative system might consist of three compounds—monolaurin/EDTA/BHA, for example. Although EDTA possesses little antimicrobial activity by itself, it renders gram-negative bacteria more susceptible by rupturing the outer membrane and thus potentiating the effect of fatty acids or fatty acid esters. An antioxidant such as BHA would exert effects against bacteria and molds and serve as an antioxidant at the same time. By use of such a system, the development of resistant strains could be minimized and the pH of a food could become less important relative to the effectiveness of the inhibitory system.

ACETIC AND LACTIC ACIDS

These two organic acids are among the most widely employed as preservatives. In most instances, their origin to the subject foods is due to their production within the food by lactic acid bacteria. Products such as pickles, sauerkraut, and fermented milks, among others, are created by the fermentative activities by various lactic acid bacteria, which produce acetic, lactic, and other acids (see Chap. 16 for fermented foods and the review by Doores, 37, for further information).

The antimicrobial effects of organic acids such as propionic and lactic is due to both the depression of pH below the growth range and metabolic inhibition by the undissociated acid molecules. In determining the quantity of organic acids in foods, **titratable acidity** is of more value than pH alone, since the latter is a measure of hydrogen-ion concentration and organic acids do not ionize completely. In measuring titratable acidity, the amount of acid that is capable of reacting with a known amount of base is determined. The titratable acidity of products such as sauerkraut is a better indicator of the amount of acidity present than pH.

The bactericidal effect of acetic acid can be demonstrated by its action on certain pathogens. When two species of *Salmonella* were added to an oil-and-vinegar-based salad dressing, the initial inoculum of 5×10^6 *S. enteritidis* could not be detected after 5 min nor could *S. typhimurium* be detected after 10 min (96).

Organic acids are employed to wash and sanitize animal carcasses after slaughter to reduce their carriage of pathogens and to increase product shelf life. Lactic and acetic acids are the most widely used, but citric, formic, and ascorbic acids have been employed generally in combination with lactic acid. These compounds are employed as sprays at levels of 1 to 3%, with 1% solutions being most often used. They may be applied in cold, warm, or hot water and on carcasses immediately after dehairing or later. Typically, about a 1 log cycle reduction in carcass flora is achieved (135), although differences between treated and control meats are not always noted (1). Mixtures of acids were found to give results similar to single acids (3), and 3% lactic acid at 70°C was found to be more effective than lower concentrations in lower-temperature water (4). Employing irradiated postrigor beef slices inoculated with *S. typhimurium*, *Listeria monocytogenes*, and *E. coli* 0157:H7, up to a 3-log cycle reduction was achieved on fat tissue with a spray of acetic acid ranging from 0.5 to 2.0% (35). Overall, the beneficial effects of carcass spraying appear to result from the lowering of surface pH, in addition perhaps to some direct effect of the acid molecules. The psychrotrophic gram-negative bacteria that cause meat spoilage are generally quite sensitive to acidic conditions, as well as to high temperatures.

ANTIBIOTICS

Antibiotics are secondary metabolites produced by microorganisms that inhibit or kill a wide spectrum of other microorganisms. Most of the useful ones are produced by molds and bacteria of the genus *Streptomyces*. Some antibioticlike substances are produced by *Bacillus* spp., and at least one, nisin, is produced by some strains of *Lactococcus lactis*. While nisin is regarded by many as being an antibiotic, it is more precisely a bacteriocin. Like antibiotics, bacteriocins are chemical compounds produced by microorganisms that inhibit or kill other microorganisms, but, unlike antibiotics, they inhibit or kill only closely related species or different strains of the same species. Bacteriocins and bacteriocinlike substances are discussed further under lactic antagonism. In this section, nisin and subtilin are treated as antibiotics.

Two antibiotics are approved for use in food in a large number of countries (nisin and natamycin), and three others (tetracyclines, subtilin, and tylosin) have been studied and found effective for various food applications. The early history, efficacy, and applications of most were reviewed in 1966 (92), and all have been reviewed and discussed more recently (72). Detailed reviews on nisin have been provided by Hurst (67) and Lipinska (89).

Three antibiotics have been investigated extensively as heat adjuncts for canned foods: subtilin, tylosin, and nisin. Nisin, however, is used most widely in cheeses. Chlortetracycline and oxytetracycline were widely studied for their application to fresh foods, while natamycin is employed as a food fungistat.

In general the use of chemical preservatives in foods is not popular among many consumers; the idea of employing antibiotics is even less popular. Some risks may be anticipated from the use of any food additive, but the risks should not outweigh the benefits overall. The general view in the United States is that the benefits to be gained by using antibiotics in foods do not outweigh the risks, some of which are known and some of which are presumed. Some 15 considerations on the use of antibiotics as food preservatives were noted by Ingram et al., and several of the key ones are summarized below:

- The antibiotic agent should kill, not inhibit, the flora and should ideally decompose into innocuous products or be destroyed on cooking for products that require cooking.
- The antibiotic should not be inactivated by food components or products of microbial metabolism.
- The antibiotic should not readily stimulate the appearance of resistant strains.
- The antibiotic should not be used in foods if used therapeutically or as an animal feed additive.

The tetracyclines are used both clinically and as feed additives, and tylosin is used in animal feeds and only in the treatment of some poultry diseases (Table

TABLE 11-4.   Properties of Some Antibiotics

Property	Tetracyclines	Subtilin	Tylosin	Nisin	Natamycin
Widely used in foods	No	No	No	Yes	Yes
First food use	1950	1950	1961	1951	1956
Chemical nature	Tetracycline	Polypeptide	Macrolide	Polypeptide	Polyene
Used as heat adjunct	No	Yes	Yes	Yes	No
Heat stability	Sensitive	Stable	Stable	Stable	Stable
Microbial spectrum	G^+, G^-	G^+	G^+	G^+	Fungi
Used medically	Yes	No	Yes[a]	No	Yes[b]
Used in feeds	Yes	No	Yes	No	No

Source: Jay (72).

[a] In treating poultry diseases.

[b] Limited.

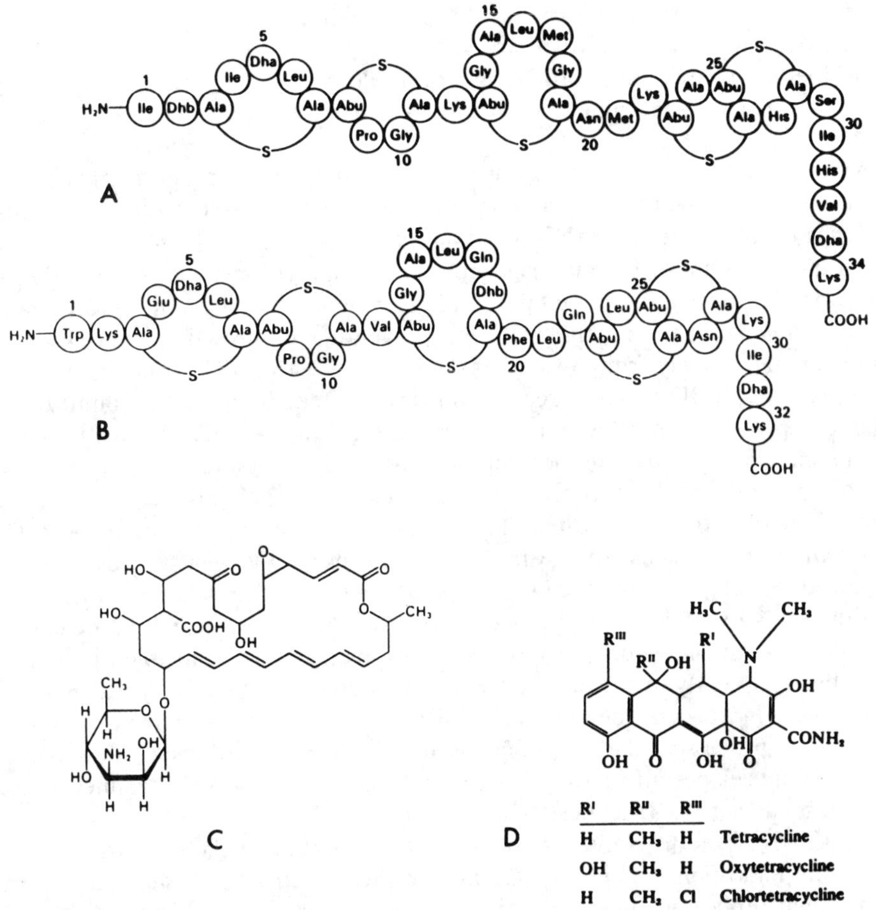

FIGURE 11-2. Structural formulas of nisin (*A*), subtilin (*B*), natamycin (*C*), and the tetracyclines (*D*).

11.4). Neither nisin nor subtilin is used medically or in animal feeds, and while nisin is used in many countries, subtilin is not. The structural similarities of these two antibiotics may be noted from Figure 11-2.

Nisin

This is a polypeptide agent that is structurally related to subtilin, but unlike subtilin it does not contain tryptophane residues (Fig. 11-2). The C-terminal amino acids are similar; the N-terminals are not. The first food use of nisin was by Hirsch et al. (53) to prevent the spoilage of Swiss cheese by *Clostridium butyricum*. It is clearly the most widely used antibiotic for food preservation, with around 46 countries permitting its use in foods to varying degrees (29).

Among some of its desirable properties as a food preservative are the following:

- It is nontoxic.
- It is produced naturally by *Lactococcus lactis* strains.

- It is heat stable and has excellent storage stability.
- It is destroyed by digestive enzymes.
- It does not contribute to off-flavors or off-odors.
- It has a narrow spectrum of antimicrobial activity.

The compound is effective against gram-positive bacteria, primarily spore-formers, and ineffective against fungi and gram-negative bacteria. *Enterococcus faecalis* is one of the most resistant gram positives.

A large amount of research has been carried out with nisin as a heat adjunct in canned foods or as an inhibitor of heat-shocked spores of *Bacillus* and *Clostridium* strains, and the MIC for preventing outgrowth of germinating spores ranges widely from 3 to >5,000 IU/ml or <1 to >125 ppm (1 μg of pure nisin is about 40 IU or RU—reading unit) (66). Depending on the country and the food product, typical usable levels are in the range of about 2.5 to 100 ppm, although some countries do not impose concentration limits.

A conventional heat process for low-acid canned foods requires an F_o treatment of 6 to 8 (see Chap. 14) to inactivate the endospores of both *C. botulinum* and spoilage organisms. By adding nisin, the heat process can be reduced to an F_o of 3 (to inactivate *C. botulinum* spores), resulting in increased product quality of low-acid canned foods. While the low-heat treatment will not destroy the endospores of spoilage organisms, nisin prevents their germination by acting early in the endospore germination cycle (Fig. 11-1). In addition to its use in certain canned foods, nisin is most often employed in dairy products—processed cheeses, condensed milk, pasteurized milk, and so on. Some countries permit its use in processed tomato products and canned fruits and vegetables (66). It is most stable in acidic foods.

Because of the effectiveness of nisin in preventing the outgrowth of germinating endospores of *C. botulinum* and the search to find safe substances that might replace nitrites in processed meats, this antibiotic has been studied as a possible replacement for nitrite. While some studies showed encouraging results employing *C. sporogenes* and other nonpathogenic organisms, a study employing *C. botulinum* types A and B spores in pork slurries indicated the inability of nisin at concentrations up to 550 ppm in combination with 60 ppm nitrite to inhibit spore outgrowth (112). Employed in culture media without added nitrite, the quantity of nisin required for 50% inhibition of *C. botulinum* type E spores was 1 to 2 ppm, 10 to 20 ppm for type B, and 20 to 40 ppm for type A (129). The latter authors found that higher levels were required for inhibition in cooked meat medium than in TPYG medium and suggested that nisin was approximately equivalent to nitrite in preventing the outgrowth of *C. botulinum* spores.

With respect to mode of action, nisin and subtilin may be presumed to act similarly since both are polypeptide antibiotics with highly similar structures. They act at the same site on germinating endospores (Fig. 11-1). Some of the polypeptide antibiotics typically attack cell membranes and act possibly as surfactants or emulsifying agents on membrane lipids. These agents may be presumed to inhibit gram-positive bacteria by inhibiting cell wall murein synthesis since bacitracin (another polypeptide antibiotic that also inhibits gram-positive bacteria) is known to inhibit murein synthesis (58). That nisin affects murein synthesis has been shown by Reisinger et al. (114), and this finding is not inconsistent with its lack of toxicity for humans. A similar lack of toxicity for subtilin may be presumed.

Natamycin

This antibiotic (also known as pimaricin, tennecetin, and myprozine) is a polyene that is quite effective against yeasts and molds but not bacteria. Natamycin is the international nonproprietary name since it was isolated from *Streptomyces natalensis*. Its structural formula is presented in Fig. 11-2.

In granting the acceptance of natamycin as a food preservative, the joint FAO/WHO Expert Committee (44) took the following into consideration: it does not affect bacteria, it stimulates an unusually low level of resistance among fungi, it is rarely involved in cross-resistance among other antifungal polyenes, and DNA transfer between fungi does not occur to the extent that it does with some bacteria. Also, from Table 11–4, it may be noted that its use is limited as a clinical agent, and it is not used as a feed additive. Natamycin has been shown by numerous investigators to be effective against both yeasts and molds, and many of these reports have been summarized (72).

The relative effectiveness of natamycin was compared to sorbic acid and four other antifungal antibiotics by Klis et al. (83) for the inhibition of 16 different fungi (mostly molds), and while from 100 to 1,000 ppm sorbic acid were required for inhibition, from 1 to 25 ppm natamycin were effective against the same strains in the same media. To control fungi on strawberries and raspberries, natamycin was compared with rimocidin and nystatin, and it, along with rimocidin, was effective at levels of 10 to 20 ppm, while 50 ppm nystatin were required for effectiveness. In controlling fungi on salami, the spraying of fresh salami with a 0.25% solution was found to be effective by one group of investigators (62), but another researcher was unsuccessful in his attempts to prevent surface-mold growth on Italian dry sausages when they were dipped in a 2,000-ppm solution (64). Natamycin spray ($2 \times 1,000$ ppm) was as good as or slightly better than 2.5% potassium sorbate.

Natamycin appears to act in the same manner as other polyene antibiotics— by binding to membrane sterols and inducing distortion of selective membrane permeability (54). Since bacteria do not possess membrane sterols, their lack of sensitivity to this agent is thus explained.

Tetracyclines

Chlortetracycline (CTC) and oxytetracycline (OTC) were approved by the FDA in 1955 and 1956, respectively, at a level of 7 ppm to control bacterial spoilage in uncooked refrigerated poultry, but these approvals were subsequently rescinded. The efficacy of this group of antibiotics in extending the shelf life of refrigerated foods was first established by Tarr and associates in Canada working with fish (145). Subsequent research by a large number of workers in many countries established the effectiveness of CTC and OTC in delaying bacterial spoilage of not only fish and seafoods but poultry, red meats, vegetables, raw milk, and other foods (for a review of food applications, see 72, 92). CTC is generally more effective than OTC. The surface treatment of refrigerated meats with 7 to 10 ppm typically results in shelf-life extensions of at least 3 to 5 days and a shift in ultimate spoilage flora from gram-negative bacteria to yeasts and molds. When CTC is combined with sorbate to delay spoilage of fish, the combination has been shown to be effective for up to 14 days. Rockfish fillets dipped in a solution of 5 ppm CTC and 1% sorbate had significantly lower aerobic plate counts (APCs) after vacuum-package storage at 2°C after 14 days than controls (95).

The tetracyclines are both heat sensitive and storage labile in foods, and these factors were important in their initial acceptance for food use. They are used to treat diseases in humans and animals and are used also in feed supplements. The risks associated with their use as food preservatives in developed countries seem clearly to outweigh the benefits.

Subtilin

This antibiotic was discovered and developed by scientists at the Western Regional Laboratory of the USDA, and its properties were described by Dimick et al. (36). It is structurally similar to nisin (Fig. 11-2), although it is produced by some strains of *Bacillus subtilis*. Like nisin, it is effective against gram-positive bacteria, is stable to acid, and possesses enough heat resistance to withstand destruction at 121°C for 30 to 60 min. Subtilin is effective in canned foods at levels of 5 to 20 ppm in preventing the outgrowth of germinating endospores, and its site of action is the same as for nisin (Fig. 11-1). Like nisin, it is used neither in the treatment of human or animal infections nor as a feed additive. This antibiotic may be just as effective as nisin, although it has received little attention since the late 1950s. It mode of action is discussed above along with that of nisin, and its development and evaluation have been reviewed (72).

Tylosin

This antibiotic is a nonpolyene macrolide, as are the clinically useful antibiotics erythromycin, oleandomycin, and others. It is more inhibitory than nisin or subtilin. Denny et al. (31) were apparently the first to study its possible use in canned foods. When 1 ppm was added to cream-style corn containing flat-sour spores and given a "botulinal" cook, no spoilage of product occurred after 30 days with incubation at 54°C (30). Similar findings were made by others in the 1960s, and these have been summarized (72).

Unlike nisin, subtilin, and natamycin, tylosin is used in animal feeds and also to treat some diseases of poultry. As a macrolide, it is most effective against gram-positive bacteria. It inhibits protein synthesis by associating with the 50S ribosomal subunit and shows at least partial cross-resistance with erythromycin.

LACTIC ANTAGONISM

The phenomenon of a lactic acid bacterium inhibiting or killing closely related and food poisoning or spoilage organisms when in mixed culture has been observed for more than 60 years. This antagonism by lactic acid bacteria has received extensive study, and several reviews of the early research have been

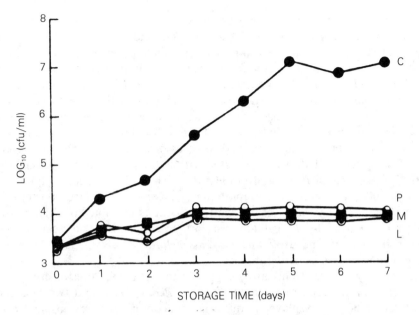

FIGURE 11-3.   Growth of *S. aureus* in pure culture (*C*) and in association with *L. plantarum* (*L*), *P. cerevisiae* (*P*), and the mixture (*M*) in cooked MDPM at 15°C. Lactic acid bacteria were added at a concentration of 10^9 cells/g.
From Raccach and Baker (111), copyright © 1978, International Association of Milk, Food and Environmental Sanitarians.

published (5, 68, 92, 127). More recent reviews are those of Daeschel (25) and Klaenhammer (82). There is renewed interest in this phenomenon as an alternative method to the use of chemicals to control undesirable organisms in foods.

The lactic acid bacteria associated with culture antagonism include the lactococci, enterococci, lactobacilli, carnobacteria, and pediococci. A wide variety of foodborne pathogens are either inhibited or killed, and many spoilage organisms are affected in similar ways, especially gram-negative psychrotrophs. The effect of *Pediococcus cerevisiae* and *Lactobacillus plantarum* on the growth of *Staphylococcus aureus* in cooked, mechanically deboned poultry meat is presented in Figure 11-3.

Among the antagonistic substances produced by lactic acid bacteria are **bacteriocins** or bacteriocinlike agents. They are typically plasmid-borne, heat resistant, and do not affect the producing strains. Many, if not all, are peptides, and they are typically bactericidal in action. The most widely studied are the colicins of *E. coli* strains. It was noted about 30 years ago that at least five distinct types are produced by serologic Group D lactic cocci (17). Nisin is the best known and studied of the bacteriocins produced by lactic acid bacteria, and it is discussed above as an antibiotic. In a study of 280 strains of lactic streptococci, about 5% were found to produce bacteriocinlike factors (48). Lactacin B has been shown to be produced by some *Lactobacillus acidophilus* strains (7); lactacin F by another strain (100); helveticin J by *Lactobacillus helveticus* (76); pediocin by *Pediococcus pentosaceus* (26); and plantacin B by *Lactobacillus plantarum* (156). Lactacin F has been cloned and sequenced (99).

An *Enterococcus faecium* strain was shown to produce an antagonistic substance active against several listerial species (93), and *Carnobacterium piscicola* produced bacteriocins active against closely related lactics (2). *Leuconostoc gelidum* produced a bacteriocinlike substance early in its growth cycle at 1°, 5°, and 25°C (59). The substance was active against *E. faecalis* and *Listeria monocytogenes*. Many other species and strains of lactic acid bacteria have been demonstrated to produce bacteriocins or bacteriocinlike substances.

While many or most lactic bacteriocins are mediated by plasmids, helveticin J was shown to be chromosomal and to be associated with a 37 Mdal protein (76). Bacteriocin activity was associated with a 7.6 Mdal plasmid in one study (59). The reported sizes range from less than 7 kdal for lactacin F (100) up to perhaps 100 kdal for a crude preparation (7). Some of the lactic bacteriocins are heat stable (resist 62°C for up to 30 min); others are not. For example, the inhibitor produced by *L. gelidum* was protease sensitive and heat resistant (59) while that produced by *Lactobacillus delbrueckii* subsp. *lactis* was both protease and heat sensitive (146).

L. monocytogenes strains are susceptible to lactic antagonism, and in one study one species each of *Lactobacillus* and *Pediococcus*; and two lactococci and three pediococci displayed antilisterial antagonism (57). Seven *Lactococcus lactis* strains displayed bactericidal action against *L. monocytogenes*, and the responsible principal was sensitive to proteolytic enzymes (19). A cell-free extract of *L. gelidum* was cidal against three strains of *L. monocytogenes*, in addition to lactobacilli and leuconostocs (56). The active principal withstood heating at 100°C for 60 min, had a molecular weight of more than 10,000 daltons, and was sensitive to protease (56). A substance produced by a strain of *P. acidilactici*, designated PA-1 bacteriocin, was shown to be inhibitory to *L. monocytogenes* in both culture and some refrigerated food systems (109). It was effective over the pH range 5.5 to 7.0 and appeared to be synergistic with lactic acid. Of five lactic acid bacteria tested against *L. monocytogenes*, four that were known to produce bacteriocins were inhibitory (110). In the latter study (110), milk pH was lowered to 4.7 during growth of lactics and the antilisterial effects were more pronounced at 25°C than 37°C. The *C. piscicola* strain produced bacteriocins early in the growth phase in APT broth but not in this broth at pH 5.5 (2). Employing three producing strains, substances (bacteriocins or bacteriocinlike) were produced that inhibited a variety of both gram-positive and gram-negative bacteria (139). Not inhibited were three gram-negative pathogens.

Although many claims have been made, the precise mechanism by which lactic cultures effect microbial inhibition is not yet clear. Among the factors identified are antibiotics, H_2O_2, organic acids, depressed pH, nutrient depletion, and bacteriocins and bacteriocinlike factors. While the activity of lactic bacteriocins seems to be rather straightforward, these substances do not explain all reported instances of lactic antagonism, especially where gram-negative bacteria are inhibited.

LACTOPEROXIDASE SYSTEM

The **lactoperoxidase system** is an inhibitory system that occurs naturally in bovine milk. It consists of three components: lactoperoxidase, thiocyanate, and

H_2O_2. All three components are required for antimicrobial effects, and gram-negative psychrotrophs such as the pseudomonads are quite sensitive. The quantity of lactoperoxidase needed is 0.5 to 1.0 ppm, while bovine milk normally contains about 30 ppm (10). While both thiocyanate and H_2O_2 occur normally in milk, the quantities vary. For H_2O_2 about 100 U/ml are required in the inhibitory system, while only 1 to 2 U/ml normally occurs in milk. An effective level of thiocyanate is around 0.25 mM, while in milk the quantity varies between 0.02 and 0.25 mM (10).

When the lactoperoxidase system in raw milk was activated by adding thiocyanate to 0.25 mM along with an equimolar amount of H_2O_2, the shelf life was extended to 5 days compared to 48 h for controls (10). The system was more effective at 30°C than at 4°C. The bactericidal effect increases with acidity, and the cytoplasmic membrane appears to be the cell target. In addition to the direct addition H_2O_2, an exogenous source can be provided by the addition of glucose and glucose oxidase. To avoid the direct addition of glucose oxidase, this enzyme has been immobilized on glass beads so that glucose is generated only in the amounts needed by the use of immobilized β-galactosidase (11). The lactoperoxidase system can be used to preserve raw milk in countries where refrigeration is uncommon. The addition of about 12 ppm of SCN^- and 8 ppm of H_2O_2 should be harmless to the consumer (116). An interesting aspect of this system is the effect it has on thermal properties. In one study, it was shown to reduce thermal D values at 57.8°C by around 80% for *L. monocytogenes* and by around 86% for *S. aureus* at 55.2°C (80). While the mechanism of this enhanced thermal destruction is unclear, some interesting implications can be envisioned. (For more information on the lactoperoxidase system, see refs. 85 and 116).

ANTIFUNGAL AGENTS FOR FRUITS

Listed in Table 11-5 are some compounds applied to fruits after harvest to control fungi, primarily molds. **Benomyl** is applied uniformly over the entire

TABLE 11-5.   Some Chemical Agents Employed to Control Fungal Spoilage of Fresh Fruits

Compound	Fruits
Thiabendazole	Apples, pears, citrus fruits, pineapples
Benomyl	Apples, pears, bananas, citrus fruits, mangoes, papayas, peaches, cherries, pineapples
Biphenyl	Citrus fruits
SO$_2$ fumigation	Grapes
Sodium-α-phenylphenate	Apples, pears, citrus fruits, pineapples

Source: Eckert (42).

surface of fruits (examples are noted in the table). It is applied at concentrations of 0.5 to 1.0 g/liter. It can penetrate the surface of some vegetables and is used worldwide to control crown rot and anthracnose of bananas, and stem-end rots of citrus fruits. It is more effective than **thiabendazole** and penetrates with greater ease. Both benomyl and thiabendazole are effective in controlling dry rot caused by *Fusarium* spp. To prevent the spread of *Botrytis* from grape to grape, SO_2 is employed for long-term storage. It is applied shortly after harvest and about once a week thereafter. A typical initial treatment consists of a 20-min application of a 1% preparation and about 0.25% in subsequent treatments (the use of SO_2 in other foods is discussed above).

Biphenyl is used to control the decay of citrus fruits by penicillia for long-distance shipments and is generally impregnated into fruit wraps or sheets between fruit layers.

ETHYLENE AND PROPYLENE OXIDES

Ethylene and propylene oxides, along with ethyl and methyl formate ($HCOOC_2H_5$ and $HCOOCH_3$, respectively), are treated together in this section because of their similar actions. The structures of the oxide compounds are as follows:

Ethylene oxide   Propylene oxide

The oxides exist as gases and are employed as fumigants in the food industry. The oxides are applied to dried fruits, nuts, spices, and so forth, primarily as antifungal compounds.

Ethylene oxide is an alkylating agent. Its antimicrobial activity is presumed to be related to this action in the following manner. In the presence of labile H atoms, the unstable three-membered ring of ethylene oxide splits. The H atom attaches itself to the oxygen, forming a hydroxyl ethyl radical, CH_2CH_2OH, which attaches itself to the position in the organic molecule left vacant by the H atom. The hydroxyl ethyl group blocks reactive groups within microbial proteins, thus resulting in inhibition. Among the groups capable of supplying a labile H atom are —COOH, —NH$_2$, —SH, and —OH. Ethylene oxide appears to affect endospores of *C. botulinum* by alkylation of guanine and adenine components of spore DNA (94, 157).

Ethylene oxide is used as a gaseous sterilant for flexible and semirigid containers for packaging aseptically processed foods. All of the gas dissipates from the containers following their removal from treatment chambers. With respect to its action on microorganisms, it is not much more effective against vegetative cells than against endospores, as can be seen from the D values given in Table 11-6.

TABLE 11-6.   D Values for Four Chemical Sterilants of Some Foodborne Microorganisms

Organism	D^a	Concen-tration	Temper-ature[b]	Condition	Reference
Hydrogen Peroxide					
C. botulinum 169B	0.03	35%	88		147
B. coagulans	1.8	26%	25		148
B. stearothermophilus	1.5	26%	25		148
B. subtilis ATCC 95244	1.5	20%	25		141
B. subtilis A	7.3	26%	25		147
Ethylene Oxide					
C. botulinum 62A	11.5	700 mg/L	40	47% R.H.	128
C. botulinum 62A	7.4	700 mg/L	40	23% R.H.	157
C. sporogenes ATCC 7955	3.25	500 mg/L	54.4	40% R.H.	81
B. coagulans	7.0	700 mg/L	40	33% R.H.	12
B. coagulans	3.07	700 mg/L	60	33% R.H.	12
B. stearothermophilus ATCC 7953	2.63	500 mg/L	54.4	40% R.H.	81
L. brevis	5.88	700 mg/L	30	33% R.H.	12
D. radiodurans	3.00	500 mg/L	54.4	40% R.H.	81
Sodium Hypochlorite					
A. niger conidiospores	0.61	20 ppm[c]	20	pH 3.0	21
A. niger conidiospores	1.04	20 ppm[c]	20	pH 5.0	21
A. niger conidiospores	1.31	20 ppm[c]	20	pH 7.0	21
Iodine ($\frac{1}{2}I_2$)					
A. niger conidiospores	0.86	20 ppm[c]	20	pH 3.0	21
A. niger conidiospores	1.15	20 ppm[c]	20	pH 5.0	21
A. niger conidiospores	2.04	20 ppm[c]	20	pH 7.0	21

[a] In minutes.
[b] °C.
[c] As Cl.

MISCELLANEOUS CHEMICAL PRESERVATIVES

Sodium diacetate ($CH_3COONa \cdot CH_3COOH \cdot xH_2O$), a derivative of acetic acid, is used in bread and cakes to prevent moldiness. Organic acids such as **citric**,

$$\begin{array}{c} CH_2COOH \\ | \\ HO-C-COOH, \\ | \\ CH_2COOH \end{array}$$

exert a preserving effect on foods such as soft drinks. **Hydrogen peroxide** (H_2O_2) has received limited use as a food preservative. In combination with heat, it has been used in milk pasteurization and sugar processing, but its widest use is as a sterilant for food-contact surfaces of olefin polymers and polyethylene in aseptic packaging systems (see Chap. 14). The D values of some foodborne microorganisms are presented in Table 11-6. **Ethanol** (C_2H_5OH) is present in flavoring extracts and effects preservation by virtue of its desiccant and denaturant properties. **Dehydroacetic acid,**

$$\underset{\displaystyle O}{\overset{\displaystyle H_3C \diagdown \underset{}{O} \diagup O}{\bigcirc}} COCH_3$$

is used to preserve squash. **Diethylpyrocarbonate** has been used in bottled wines and soft drinks as a yeast inhibitor. It decomposes to form ethanol and CO_2 by either hydrolysis or alcoholysis.

Hydrolysis (reaction with water):

$$\begin{array}{c} C_2H_5O-CO \diagdown \\ \qquad\qquad O \xrightarrow{H_2O} 2C_2H_5OH + 2CO_2; \\ C_2H_5O-CO \diagup \end{array}$$

Alcoholysis (reaction with ethyl alcohol):

$$\begin{array}{c} C_2H_5O-CO \diagdown \qquad\qquad C_2H_5O \diagdown \\ \qquad\qquad O \xrightarrow{C_2H_5OH} \qquad\qquad C = O + CO_2 + C_2H_5OH. \\ C_2H_5O-CO \diagup \qquad\qquad C_2H_5O \diagup \end{array}$$

Saccharomyces cerevisiae and conidia of *A. niger* and *Byssochlamys fulva* have been shown to be destroyed by this compound during the first ½ h of exposure, while the ascospores of *B. fulva* required 4 to 6 h for maximal destruction (140). Cidal concentrations for yeasts range from about 20 to 1,000 ppm depending on species or strain. *L. plantarum* and *Leuconostoc mesenteroides* required 24 h or longer for destruction. Sporeforming bacteria are quite resistant to this compound. Sometimes urethane is formed when this compound is used, and because it is a carcinogen, the use of diethylpyrocarbonate is no longer permissible in the United States.

Wood smoke imparts certain chemicals to smoke products that enable these products to resist microbial spoilage. One of the most important is formaldehyde (CH_2O), which has been known for many years to possess antimicrobial properties. This compound acts as a protein denaturant by virtue of its reaction with amino groups. Also in wood smoke are aliphatic acids, alcohols, ketones, phenols, higher aldehydes, tar, methanol, cresols, and other compounds (39), all of which may contribute to the antibacterial actions of meat smoking. Since a certain amount of heat is necessary to produce smoke, part of the shelf stability of smoked products is due to heat destruction of surface organisms, as well as to the drying that occurs. A study of the antibacterial activity of liquid smoke by Handford and Gibbs (55) revealed that little activity occurred at concentrations of smoke that produced acceptable smoked flavor. Employing an agar medium containing 1:1 dilution of smoked water, they found that micrococci and staphylococci were slightly more inhibited than the lactic acid bacteria. The overall combined effect of smoking and vacuum packaging results in a reduction of numbers of catalase-positive bacteria on the smoked product, while the catalase-negative lactic acid bacteria are better able to withstand the low Eh conditions of vacuum-packaged products.

References

1. Acuff, G. R., C. Vanderzant, J. W. Savell, D. K. Jones, D. B. Griffin, and J. G. Ehlers. 1987. Effect of acid decontamination of beef subprimal cuts on the microbiological and sensory characteristics of steaks. *Meat Sci.* 19:217–226.

2. Ahn, C., and M. E. Stiles. 1990. Plasmid-associated bacteriocin production by a strain of *Carnobacterium piscicola* from meat. *Appl. Environ. Microbiol.* 56: 2503–2510.

3. Anderson, M. E., and R. T. Marshall. 1990. Reducing microbial populations on beef tissues: Concentration and temperature of an acid mixture. *J. Food Sci.* 55:903–905.

4. Anderson, M. E., and R. T. Marshall. 1990. Reducing microbial populations on beef tissues: Concentration and temperature of lactic acid. *J. Food Saf.* 10: 181–190.

5. Babel, F. J. 1977. Antibiosis by lactic culture bacteria. *J. Dairy Sci.* 60:815–821.

6. Banks, J. G., and R. G. Board. 1982. Sulfite inhibition of *Enterobacteriaceae* including *Salmonella* in British fresh sausage and in culture systems. *J. Food Protect.* 45:1292–1297, 1301.

7. Barefoot, S. F., and T. R. Klaenhammer. 1983. Detection and activity of lactacin B, a bacteriocin produced by *Lactobacillus acidophilus*. *Appl. Environ. Microbiol.* 45:1808–1815.

8. Berry, B. W., and T. N. Blumer. 1981. Sensory, physical, and cooking characteristics of bacon processed with varying levels of sodium nitrite and potassium sorbate. *J. Food Sci.* 46:321–327.

9. Beuchat, L. R. 1976. Sensitivity of *Vibrio parahaemolyticus* to spices and organic acids. *J. Food Sci.* 41:899–902.

10. Björck, L. 1978. Antibacterial effect of the lactoperoxidase system on psychrotrophic bacteria in milk. *J. Dairy Res.* 45:109–118.

11. Björck, L., and C.-G. Rosen. 1976. An immobilized two-enzyme system for the activation of the lactoperoxidase antibacterial system in milk. *Biotechnol. Bioengin.* 18:1463–1472.

12. Blake, D. F., and C. R. Stumbo. 1970. Ethylene oxide resistance of microorganisms important in spoilage of acid and high-acid foods. *J. Food Sci.* 35:26–29.

13. Bosund, I. 1962. The action of benzoic and salicylic acids on the metabolism of microorganisms. *Adv. Food Res.* 11:331–353.

14. Bowen, V. G., and R. H. Deibel. 1974. Effects of nitrite and ascorbate on botulinal toxin formation in wieners and bacon. In *Proceedings of the Meat Industry Research Conference*, 63–68. Chicago: American Meat Institute Foundation.

15. Branen, A. L., P. M. Davidson, and B. Katz. 1980. Antimicrobial properties of phenolic antioxidants and lipids. *Food Technol.* 34(5):42–53, 63.

16. Brock, T. D., D. W. Smith, and M. T. Madigan, 1984. *Biology of Micro-organisms*. Englewood Cliffs, N.J.: Prentice-Hall.
17. Brock, T. D., B. Peacher, and D. Pierson. 1963. Survey of the bacteriocines of enterococci. *J. Bacteriol.* 86:702–707.
18. Bullerman, L. B., F. Y. Lieu, and S. A. Seier. 1977. Inhibition of growth and aflatoxin production by cinnamon and clove oils, cinnamic aldehyde and eugenol. *J. Food Sci.* 42:1107–1109, 1116.
19. Carminati, D., G. Giraffa, and M. G. Bossi. 1989. Bacteriocinlike inhibitors of *Streptococcus lactis* against *Listeria monocytogenes*. *J. Food Protect.* 52:614–617.
20. Chang, P.-C., S. M. Akhtar, T. Burke, and H. Pivnick. 1974. Effect of sodium nitrite on *Clostridium botulinum* in canned luncheon meat: Evidence for a Perigo-type factor in the absence of nitrite. *Can. Inst. Food Sci. Technol. J.* 7:209–212.
21. Cheng, M. K. C., and R. E. Levin. 1970. Chemical destruction of *Aspergillus niger conidiospores*. *J. Food Sci.* 35:62–66.
22. Christiansen, L. N., R. W. Johnston, D. A. Kautter, J. W. Howard, and W. J. Aunan. 1973. Effect of nitrite and nitrate on toxin production by *Clostridium botulinum* and on nitrosamine formation in perishable canned comminuted cured meat. *Appl. Microbiol.* 25:357–362.
23. Christiansen, L. N., R. B. Tompkin, A. B. Shaparis, T. V. Kueper, R. W. Johnston, D. A. Kautter, and O. J. Kolari. 1974. Effect of sodium nitrite on toxin production by *Clostridium botulinum* in bacon. *Appl. Microbiol.* 27:733–737.
24. Collins-Thompson, D. L., N. P. Sen, B. Aris, and L. Schwinghamer. 1972. Non-enzymic in vitro formation of nitrosamines by bacteria isolated from meat products. *Can. J. Microbiol.* 18:1968–1971.
25. Daeschel, M. A. 1989. Antimicrobial substances from lactic acid bacteria for use as food preservatives. *Food Technol.* 43(1):164–167.
26. Daeschel, M. A., and T. R. Klaenhammer. 1985. Association of a 13.6-megadalton plasmid in *Pediococcus pentosaceus* with bacteriocin activity. *Appl. Environ. Microbiol.* 50:1538–1541.
27. Davidson, P. M. 1983. Phenolic compounds. In *Antimicrobials in Foods*, ed. A. L. Branen and P. M. Davidson, 37–73. New York: Marcel Dekker.
28. Davidson, P. M., C. J. Brekke, and A. L. Branen. 1981. Antimicrobial activity of butylated hydroxyanisole, tertiary butylhydroquinone, and potassium sorbate in combination. *J. Food Sci.* 46:314–316.
29. Delves-Broughton, J. 1990. Nisin and its uses as a food preservative. *Food Technol.* 44(11):100, 102, 104, 106, 108, 111–112, 117.
30. Denny, C. B., J. M. Reed, and C. W. Bohrer, 1961. Effect of tylosin and heat on spoilage bacteria in canned corn and canned mushrooms. *Food Technol.* 15:338–340.
31. Denny, C. B., L. E. Sharpe, and C. W. Bohrer. 1961. Effects of tylosin and nisin on canned food spoilage bacteria. *Appl. Microbiol.* 9:108–110.
32. Desrosier, N. W. 1963. *The Technology of Food Preservation*. Rev. ed. Westport, Conn.: AVI.
33. Dethmers, A. E., H. Rock, T. Fazio, and R. W. Johnston. 1975. Effect of added sodium nitrite and sodium nitrate on sensory quality and nitrosamine formation in thuringer sausage. *J. Food Sci.* 40:491–495.
34. Deuel, H. J., Jr., C. E. Calbert, L. Anisfeld, H. McKeehan, and H. D. Blunden. 1954. Sorbic acid as a fungistatic agent for foods. II. Metabolism of α,β-unsaturated fatty acids with emphasis on sorbic acid. *Food Res.* 19:13–19.
35. Dickson, J. S. 1991. Control of *Salmonella typhimurium*, *Listeria monocytogenes*, and *Escherichia coli* 0157:H7 on beef in a model spray chilling system. *J. Food Sci.* 56:191–193.
36. Dimick, K. P., G. Alderton, J. C. Lewis, H. D. Lightbody, and H. L. Fevold.

1947. Purification and properties of subtilin. *Arch. Biochem.* 15:1–11.

37. Doores, S. 1983. Organic acids. In *Antimicrobials in Foods*, ed. A. L. Branen and P. M. Davidson, 75–107. New York: Marcel Dekker.

38. Doyle, M. P., and E. H. Marth. 1978. Bisulfite degrades aflatoxins. Effect of temperature and concentration of bisulfite. *J. Food Protect.* 41:774–780.

39. Draudt, H. N. 1963. The meat smoking process: A review. *Food Technol.* 17: 1557–1562.

40. Duncan, C. L., and E. M. Foster. 1968. Role of curing agents in the preservation of shelf-stable canned meat products. *Appl. Microbiol.* 16:401–405.

41. Dymicky, M., and H. Trenchard. 1982. Inhibition of *Clostridium botulinum* 62A by saturated n-aliphatic acids, n-alkyl formates, acetates, propionates and butyrates. *J. Food Protect.* 45:1117–1119.

42. Eckert, J. W. 1979. Fungicidal and fungistatic agents: Control of pathogenic microorganisms on fresh fruits and vegetables after harvest. In *Food Mycology*, ed. M. E. Rhodes, 164–199. Boston: Hall.

43. Fong, Y. Y., and W. C. Chan. 1973. Bacterial production of di-methyl nitrosamine in salted fish. *Nature* 243:421–422.

44. Food and Agriculture Organization/World Health Organization (FAO/WHO). 1976. *Evaluation of Certain Food Additives*. WHO Technical Report Series 599.

45. Freese, E., C. W. Sheu, and E. Galliers. 1973. Function of lipophilic acids as antimicrobial food additives. *Nature* 241:321–325.

46. Fung, D. Y. C., C. C. S. Lin, and M. B. Gailani, 1985. Effect of phenolic antioxidants on microbial growth. *CRC Crit. Rev. Microbiol.* 12:153–183.

47. Gailani, M. B., and D. Y. C. Fung. 1984. Antimicrobial effects of selected antioxidants in laboratory media and in ground pork. *J. Food Protect.* 47:428–433.

48. Geis, A., J. Singh, and M. Teuber. 1983. Potential of lactic streptococci to produce bacteriocin. *Appl. Environ. Microbiol.* 45:205–211.

49. Gould, G. W. 1964. Effect of food preservatives on the growth of bacteria from spores. In *Microbial Inhibitors in Foods*, ed. G. Molin, 17–24. Stockholm: Almquist & Wiksell.

50. Gould, G. W., M. H. Brown, and B. C. Fletcher. 1983. Mechanisms of action of food preservation procedures. In *Food Microbiology: Advances and Prospects*, ed. T. A. Roberts and F. A. Skinner, 67–84. New York and London: Academic Press.

51. Gray, J. I. 1976. N-Nitrosamines and their precursors in bacon. A review. *J. Milk Food Technol.* 39:686–692.

52. Gray, J. I., and A. M. Pearson. 1984. Cured meat flavor. *Adv. Food Res.* 29:1–86.

53. Hagler, W. M., Jr., J. E. Hutchins, and P. B. Hamilton. 1982. Destruction of aflatoxin in corn with sodium bisulfite. *J. Food Protect.* 45:1287–1291.

54. Hamilton-Miller, J. M. T. 1974. Fungal sterols and the mode of action of the polyene antibiotics. *Adv. Appl. Microbiol.* 17:109–134.

55. Handford, P. M., and B. M. Gibbs. 1964. Antibacterial effects of smoke constituents on bacteria isolated from bacon. In *Microbial Inhibitors in Food*, ed. G. Molin, 333–346. Stockholm: Almquist & Wiksell.

56. Harding, C. D., and B. G. Shaw. 1990. Antimicrobial activity of *Leuconostoc gelidum* against closely related species and *Listeria monocytogenes*. *J. Appl. Bacteriol.* 69:648–654.

57. Harris, L. J., M. A. Daeschel, M. E. Stiles, and T. R. Klaenhammer. 1989. Antimicrobial activity of lactic acid bacteria against *Listeria monocytogenes*. *J. Food Protect.* 52:3784–3787.

58. Hash, J. H. 1972. Antibiotic mechanisms. *Ann. Rev. Pharmacol.* 12:35–56.

59. Hastings, J. W., and M. E. Stiles. 1991. Antibiosis of *Leuconostoc gelidum*

isolated from meat. *J. Appl. Bacteriol.* 70:127–134.

60. Hawksworth, G., and M. J. Hill. 1971. The formation of nitrosamines by human intestinal bacteria. *Biochem. J.* 122:28–29P.

61. Hawksworth, G., and M. J. Hill. 1971. Bacteria and the N-nitrosation of secondary amines. *Brit. J. Cancer.* 25:520–526.

62. Hechelman, H., and L. Leistner. 1969. Hemmung von unerwunschtem Schimmelpilzwachstum auf Rohwursten durch Delvocid (Pimaricin). *Fleischw.* 49: 1639–1641.

63. Hirsch, A., E. Grinsted, H. R. Chapman, and A. T. R. Mattick. 1951. Inhibition of an anaerobic sporeformer in Swiss-type cheese by a nisin-producing streptococcus. *J. Dairy Res.* 18:205–206.

64. Holley, R. A. 1981. Prevention of surface mold growth on Italian dry sausage by natamycin and potassium sorbate. *Appl. Environ. Microbiol.* 41:422–429.

65. Huhtanen, C. N. 1980. Inhibition of *Clostridium botulinum* by spice extracts and alphatic alcohols. *J. Food Protect.* 43:195–196, 200.

66. Hurst, A. 1981. Nisin. *Adv. Appl. Microbiol.* 27:85–123.

67. Hurst, A. 1983. Nisin and other inhibitory substances from lactic acid bacteria. In *Antimicrobials in Foods*, ed. A. L. Branen and P. M. Davidson, 327–351. New York: Marcel Dekker.

68. Hurst, A. 1973. Microbial antagonism in foods. *Can. Inst. Food Sci. Technol. J.* 6:80–90.

69. Ivey, F. J., K. J. Shaver, L. N. Christiansen, and R. B. Tompkin. 1978. Effect of potassium sorbate on toxinogenesis by *Clostridium botulinum* in bacon. *J. Food Protect.* 41:621–625.

70. Jay, J. M. 1982. Antimicrobial properties of diacetyl. *Appl. Environ. Microbiol.* 44:525–532.

71. Jay, J. M. 1982. Effect of diacetyl on foodborne microorganisms. *J. Food Sci.* 47:1829–1831.

72. Jay, J. M. 1983. Antibiotics as food preservatives. In *Food Microbiology*, ed. A. H. Rose, 117–143. New York and London: Academic Press.

73. Jay, J. M., and G. M. Rivers. 1984. Antimicrobial activity of some food flavoring compounds. *J. Food Safety* 6:129–139.

74. Johnston, M. A., and R. Loynes. 1971. Inhibition of *Clostridium botulinum* by sodium nitrite as affected by bacteriological media and meat suspensions. *Can. Inst. Food Technol. J.* 4:179–184.

75. Johnston, M. A., H. Pivnick, and J. M. Samson. 1969. Inhibition of *Clostridium botulinum* by sodium nitrite in a bacteriological medium and in meat. *Can. Inst. Food Technol. J.* 2:52–55.

76. Joerger, M. C., and T. R. Klaenhammer. 1986. Characterization and purification of helveticin J and evidence for a chromosomally determined bacteriocin produced by *Lactobacillus helveticus* 481. *J. Bacteriol.* 167:439–446.

77. Kabara, J. J. 1981. Food-grade chemicals for use in designing food preservative systems. *J. Food Protect.* 44:633–647.

78. Kabara, J. J. 1983. Medium-chain fatty acids and esters. In *Antimicrobials in Foods*, ed. A. L. Branen and P. M. Davidson, 109–139. New York: Marcel Dekker.

79. Kabara, J. J., R. Vrable, and M. S. F. Lie Ken Jie. 1977. Antimicrobial lipids: Natural and synthetic fatty acids and monoglycerides. *Lipids* 12:753–759.

80. Kamau, D. N., S. Doores, and K. M. Pruitt. 1990. Enhanced thermal destruction of *Listeria monocytogenes* and *Staphylococcus aureus* by the lactoperoxidase system. *Appl. Environ. Microbiol.* 56:2711–2716.

81. Kereluk, K., R. A. Gammon, and R. S. Lloyd. 1970. Microbiological aspects of ethylene oxide sterilization. II. Microbial resistance to ethylene oxide. *Appl. Microbiol.* 19:152–156.

82. Klaenhammer, T. R. 1988. Bacteriocins of lactic acid bacteria. *Biochimie* 70: 337–349.
83. Klis, J. B., L. D. Witter, and Z. J. Ordal. 1964. The effect of several antifungal antibiotics on the growth of common food spoilage fungi. *Food Technol.* 13: 124–128.
84. Kueper, T. V., and R. D. Trelease. 1974. Variables affecting botulinum toxin development and nitrosamine formation in fermented sausages. In *Proceedings of the Meat Industry Research Conference*, 69–74. Chicago: American Meat Institute Foundation.
85. Law, B. A., and I. A. Mabbitt. 1983. New methods for controlling the spoilage of milk and milk products. In *Food Microbiology: Advances and Prospects*, ed. T. A. Roberts and F. A. Skinner, 131–150. New York and London: Academic Press.
86. Law, B. A., and B. Reiter. 1977. The isolation and bacteriostatic properties of lactoferrin from bovine milk whey. *J. Dairy Res.* 44:595–599.
87. Liewen, M. B., and E. H. Marth. 1985. Growth and inhibition of microorganisms in the presence of sorbic acid: A review. *J. Food Protect.* 48:364–375.
88. Lin, C. C. S., and D. Y. C. Fung. 1983. Effect of BHA, BHT, TBHQ, and PG on growth and toxigenesis of selected aspergilli. *J. Food Sci.* 48:576–580.
89. Lipinska, E. 1977. Nisin and its applications. In *Antibiotics and Antibiosis in Agriculture*, ed. M. Woodbine, 103–130. London: Butterworths.
90. Lloyd, A. C. 1975. Preservation of comminuted orange products. *J. Food Technol.* 10:565–567.
91. Marriott, N. G., R. V. Lechowich, and M. D. Pierson. 1981. Use of nitrite and nitrite-sparing agents in meats: A review. *J. Food Protect.* 44:881–885.
92. Marth, E. H. 1966. Antibiotics in foods—naturally occurring, developed, and added. *Residue Rev.* 12:65–161.
93. McKay, A. M. 1990. Antimicrobial activity of *Enterococcus faecium* against *Listeria* spp. *Lett. Appl. Microbiol.* 11:15–17.
94. Michael, G. T., and C. R. Stumbo. 1970. Ethylene oxide sterilization of *Salmonella senftenberg* and *Escherichia coli*: Death kinetics and mode of action. *J. Food Sci.* 35:631–634.
95. Miller, S. A., and W. D. Brown. 1984. Effectiveness of chlortetracycline in combination with potassium sorbate or tetrasodium ethylene-diaminetetraacetate for preservation of vacuum packed rockfish fillets. *J. Food Sci.* 49:188–191.
96. Miller, M. L., and E. D. Martin. 1990. Fate of *Salmonella enteritidis* and *Salmonella typhimurium* into an Italian salad dressing with added eggs. *Dairy Food Environ. Sanit.* 10(1):12–14.
97. Moerck, K. E., P. McElfresh, A. Wohlman, and B. W. Hilton. 1980. Aflatoxin destruction in corn using sodium bisulfite, sodium hydroxide and aqueous ammonia. *J. Food Protect.* 43:571–574.
98. Morris, J. A., A. Khettry, and E. W. Seitz. 1979. Antimicrobial activity of aroma chemicals and essential oils. *J. Amer. Oil. Chem. Soc.* 56:595–603.
99. Muriana, P. M., and T. R. Klaenhammer. 1991. Cloning, phenotypic expression, and DNA sequence of the gene for lactacin F, an antimicrobial peptide produced by *Lactobacillus* spp. *J. Bacteriol.* 173:1779–1788.
100. Muriana, P. M., and T. R. Klaenhammer. 1991. Purification and partial characterization of lactacin F, a bacteriocin produced by *Lactobacillus acidophilus* 11088. *Appl. Environ. Microbiol.* 57:114–121.
101. Nordin, H. R. 1969. The depletion of added sodium nitrite in ham. *Can. Inst. Food Sci. Technol. J.* 2:79–85.
102. O'Boyle, A. R., L. J. Rubin, L. L. Diosady, N. Aladin-Kassam, F. Comer, and W. Brightwell. 1990. A nitrite-free curing system and its application to the production of wieners. *Food Technol.* 44(5):88, 90–91, 93, 95–96, 98, 100, 102–104.
103. O'Leary, V., and M. Solberg. 1976. Effect of sodium nitrite inhibition on intra-

cellular thiol groups and on the activity of certain glycolytic enzymes in *Clostridium perfringens. Appl. Environ. Microbiol.* 31:208–212.

104. Ough, C. S. 1983. Sulfur dioxide and sulfites. In *Antimicrobials in Foods*, ed. A. L. Branen and P. M. Davidson, 177–203. New York: Marcel Dekker.

105. Paquette, M. W., M. C. Robach, J. N. Sofos, and F. F. Busta. 1980. Effects of various concentrations of sodium nitrite and potassium sorbate on color and sensory qualities of commercially prepared bacon. *J. Food Sci.* 45:1293–1296.

106. Perigo, J. A., and T. A. Roberts. 1968. Inhibition of clostridia by nitrite. *J. Food Technol.* 3:91–94.

107. Perigo, J. A., E. Whiting, and T. E. Bashford. 1967. Observations on the inhibition of vegetative cells of *Clostridium sporogenes* by nitrite which has been autoclaved in a laboratory medium, discussed in the context of sublethally processed meats. *J. Food Technol.* 2:377–397.

108. Pierson, M. D., and N. R. Reddy. 1982. Inhibition of *Clostridium botulinum* by antioxidants and related phenolic compounds in comminuted pork. *J. Food Sci.* 47:1926–1929, 1935.

109. Pucci, M. J., E. R. Vedamuthu, B. S. Kunka, and P. A. Vandenbergh. 1988. Inhibition of *Listeria monocytogenes* by using bacteriocin PA-1 produced by *Pediococcus acidilactici* PAC 1.0. *Appl. Environ. Microbiol.* 54:2349–2353.

110. Raccach, M., R. McGrath, and H. Daftarian. 1989. Antibiosis of some lactic acid bacteria including *Lactobacillus acidophilus* towards *Listeria monocytogenes. Int. J. Food Microbiol.* 9:25–32.

111. Raccach, M., and R. C. Baker. 1978. Lactic acid bacteria as an antispoilage and safety factor in cooked, mechanically deboned poultry meat. *J. Food Protect.* 41:703–705.

112. Rayman, K., N. Malik, and A. Hurst. 1983. Failure of nisin to inhibit outgrowth of *Clostridium botulinum* in a model cured meat system. *Appl. Environ. Microbiol.* 46:1450–1452.

113. Reddy, D., J. R. Lancaster, Jr., and D. P. Cornforth. 1983. Nitrite inhibition of *Clostridium botulinum*: Electron spin resonance detection of iron-nitric oxide complexes. *Science* 221:769–770.

114. Reisinger, P., H. Seidel, H. Tachesche, and W. P. Hammes. 1980. The effect of nisin on murein synthesis. *Arch. Microbiol.* 127:187–193.

115. Reiter, B. 1978. Review of the progress of dairy science: Antimicrobial systems in milk, *J. Dairy Res.* 45:131–147.

116. Reiter, B., and G. Harnulv. 1984. Lactoperoxidase antibacterial system: Natural occurrence, biological functions and practical applications. *J. Food Protect.* 47: 724–732.

117. Riemann, H. 1963. Safe heat processing of canned cured meats with regard to bacterial spores. *Food Technol.* 17:39–49.

118. Robach, M. C., and M. D. Pierson. 1979. Inhibition of *Clostridium botulinum* types A and B by phenolic antioxidants. *J. Food Protect.* 42:858–861.

119. Robach, M. C., and M. D. Pierson. 1978. Influence of para-hydroxybenzoic acid esters on the growth and toxin production of *Clostridium botulinum* 10755A. *J. Food Sci.* 43:787–789, 792.

120. Roberts, T. A., A. M. Gibson, and A. Robinson. 1981. Factors controlling the growth of *Clostridium botulinum* types A and B in pasteurized, cured meats. II. Growth in pork slurries prepared from "high" pH meat (range 6.3–6.8). *J. Food Technol.* 16:267–281.

121. Roberts, T. A., A. M. Gibson, and A. Robinson. 1982. Factors controlling the growth of *Clostridium botulinum* types A and B in pasteurized, cured meats. III. The effect of potassium sorbate. *J. Food Technol.* 17:307–326.

122. Roberts, T. A., and M. Ingram. 1966. The effect of sodium chloride, potassium nitrate and sodium nitrite on the recovery of heated bacterial spores. *J. Food Technol.* 1:147–163.

123. Roberts, T. A., and J. L. Smart. 1974. Inhibition of spores of *Clostridium* spp. by sodium nitrite. *J. Appl. Bacteriol.* 37:261–264.

124. Ronning, I. E., and H. A. Frank. 1988. Growth response of putrefactive anaerobe 3679 to combinations of potassium sorbate and some common curing ingredients (sucrose, salt, and nitrite), and to noninhibitory levels of sorbic acid. *J. Food Protect.* 51:651–654.

125. Ronning, I. E., and H. A. Frank. 1987. Growth inhibition of putrefactive anaerobe 3679 caused by stringent-type response induced by protonophoric activity of sorbic acid. *Appl. Environ. Microbiol.* 53:1020–1027.

126. Rowe, J. J., J. M. Yarbrough, J. B. Rake, and R. G. Egon. 1979. Nitrite inhibition of aerobic bacteria. *Curr. Microbiol.* 2:51–54.

127. Sandine, W. E., K. S. Muralidhara. P. R. Elliker, and D. C. England. 1972. Lactic acid bacteria in food and health: A review with special reference to enteropathogenic *Escherichia coli* as well as certain enteric disease and their treatment with antibiotics and lactobacilli. *J. Milk Food Technol.* 35:691–702.

128. Savage, R. A., and C. R. Stumbo. 1971. Characteristics of progeny of ethylene oxide treated *Clostridium botulinum* type 62A spores. *J. Food Sci.* 36:182–184.

129. Scott, V. N., and S. L. Taylor. 1981. Effect of nisin on the outgrowth of *Clostridium botulinum* spores. *J. Food Sci.* 46:117–120, 126.

130. Seward, R. A., R. H. Deibel, and R. C. Lindsay. 1982. Effects of potassium sorbate and other antibotulinal agents on germination and outgrowth of *Clostidium botulinum* type E spores in microcultures. *Appl. Environ. Microbiol.* 44:1212–1221.

131. Shahidi, F., L. J. Rubin, L. L. Diosady, V. Chew, and D. F. Wood. 1984. Preparation of dinotrosyl ferrohemochrome from hemin and sodium nitrite. *Can. Inst. Food Sci. Technol. J.* 17:33–37.

132. Shelef, L. A. 1983. Antimicrobial effects of spices. *J. Food Safety* 6:29–44.

133. Shelef, L. A., and P. Liang. 1982. Antibacterial effects of butylated hydroxyanisole (BHA) against *Bacillus* species. *J. Food Sci.* 47:796–799.

134. Shelef, L. A., O. A. Naglik, and D. W. Bogen. 1980. Sensitivity of some common food-borne bacteria to the spices sage, rosemary, and allspice. *J. Food Sci.* 45:1042–1044.

135. Smulders, F. J. M., and C. H. J. Woolthuis. 1985. Immediate and delayed microbiological effects of lactic acid decontamination of calf carcasses—influence on conventionally boned versus hot-boned and vacuum-packaged cuts. *J. Food Protect.* 48:838–847.

136. Sofos, J. N. 1989. *Sorbate Food Preservatives*. Boca Raton: CRC Press.

137. Sofos, J. N., F. F. Busta, and C. E. Allen. 1980. Influence of pH on *Clostridium botulinum* control by sodium nitrite and sorbic acid in chicken emulsions. *J. Food Sci.* 45:7–12.

138. Sofos, J. N., F. F. Busta, K. Bhothipaksa, C. E. Allen, M. C. Robach, and M. W. Paquette. 1980. Effects of various concentrations of sodium nitrite and potassium sorbate on *Clostridium botulinum* toxin production in commercially prepared bacon. *J. Food Sci.* 45:1285–1292.

139. Spelhaug, S. R., and S. K. Harlander. 1989. Inhibition of foodborne bacterial pathogens by bacteriocins from *Lactococcus pentosaceus*. *J. Food Protect.* 52: 856–862.

140. Splittstoesser, D. F., and M. Wilkison. 1973. Some factors affecting the activity of diethylpyrocarbonate as a sterilant. *Appl. Microbiol.* 25:853–857.

141. Swartling, P., and B. Lindgren. 1968. The sterilizing effect against *Bacillus subtilis* spores of hydrogen peroxide at different temperatures and concentrations. *J. Dairy Res.* 35:423–428.

142. Tanaka, N., N. M. Gordon, R. C. Lindsay, L. M. Meske, M. P. Doyle, and E. Traisman. 1985. Sensory characteristics of reduced nitrite bacon manufactured by the Wisconsin process. *J. Food Protect.* 48:687–692.

143. Tanaka, N., L. Meske, M. P. Doyle, E. Traisman, D. W. Thayer, and R. W. Johnston. 1985. Plant trials of bacon made with lactic acid bacteria, sucrose and lowered sodium nitrite. *J. Food Protect.* 48:679–686.

144. Tanaka, N., E. Traisman, M. H. Lee, R. G. Cassens, and E. M. Foster. 1980. Inhibition of botulinum toxin formation in bacon by acid development. *J. Food Protect.* 43:450–457.

145. Tarr, H. L. A., B. A. Southcott, and H. M. Bissett. 1952. Experimental preservation of flesh foods with antibiotics. *Food Technol.* 6:363–368.

146. Toba, T., E. Yoshioka, and T. Itoh. 1991. Lacticin, a bacteriocin produced by *Lactobacillus delbrueckii* subsp. *lactis. Lett. Appl. Microbiol.* 12:43–45.

147. Toledo, R. T. 1975. Chemical sterilants for aseptic packaging. *Food Technol.* 29(5):102–107.

148. Toledo, R. T., F. E. Escher, and J. C. Ayres. 1973. Sporicidal properties of hydrogen peroxide against food spoilage organisms. *Appl. Microbiol* 26:592–597.

149. Tompkin, R. B. 1983. Nitrite. In *Antimicrobials in Foods*, ed. A. L. Branen and P. M. Davidson, 205–256. New York: Marcel Dekker.

150. Tompkin, R. B., L. N. Christiansen, and A. B. Shaparis. 1978. Enhancing nitrite inhibition of *Clostridium botulinum* with isoascorbate in perishable canned cured meat. *Appl. Environ. Microbiol.* 35:59–61.

151. Tompkin, B. L., L. N. Christiansen, and A. B. Shaparis. 1978. Causes of variation in botulinal inhibition in perishable canned cured meat. *Appl. Environ. Microbiol.* 35:886–889.

152. Tompkin, B. L., L. N. Christiansen, and A. B. Shaparis. 1979. Iron and the antibotulinal efficacy of nitrite. *Appl. Environ. Microbiol.* 37:351–353.

153. Tompkin, B. L., L. N. Christiansen, and A. B. Shaparis. 1980. Antibotulinal efficacy of sulfur dioxide in meat. *Appl. Environ. Microbiol.* 39:1096–1099.

154. Vareltzis, K., E. M. Buck, and R. G. Labbe. 1984. Effectiveness of a betalains/potassium sorbate system versus sodium nitrite for color development and control of total aerobes, *Clostridium perfringens* and *Clostridium sporogenes* in chicken frankfurters. *J. Food Protect.* 47:532–536.

155. Vas, K., and M. Ingram. 1949. Preservation of fruit juices with less SO_2. *Food Manuf.* 24:414–416.

156. West, C. A., and P. J. Warner. 1988. Plantacin B, a bacteriocin produced by *Lactobacillus plantarum* NCDO 1193. *FEMS Microbiol. Lett.* 49:163–165.

157. Winarno, F. G., and C. R. Stumbo. 1971. Mode of action of ethylene oxide on spores of *Clostridium botulinum* 62A. *J. Food Sci.* 36:892–895.

158. Wood, D. S., D. L. Collins-Thompson, W. R. Usborne, and B. Picard. 1986. An evaluation of antibotulinal activity in nitrite-free curing systems containing dinitrosyl ferrohemochrome. *J. Food Protect.* 49:691–695.

159. Woods, L. F. J., and J. M. Wood. 1982. A note on the effect of nitrite inhibition on the metabolism of *Clostridium botulinum. J. Appl. Bacteriol.* 52:109–110.

160. Woods, L. F. J., J. M. Wood, and P. A. Gibbs. 1981. The involvement of nitric oxide in the inhibition of the phosphoroclastic system in *Clostridium sporogenes* by sodium nitrite. *J. Gen. Microbiol.* 125:399–406.

161. Yarbrough, J. M., J. B. Rake, and R. G. Egon. 1980. Bacterial inhibitory effects of nitrite: Inhibition of active transport, but not of group translocation, and of intracellular enzymes. *Appl. Environ. Microbiol.* 39:831–834.

162. Yun, J., F. Shahidi, L. J. Rubin, and L. L. Diosady. 1987. Oxidative stability and flavour acceptability of nitrite-free meat curing systems. *Can. Inst. Food Sci. Technol. J.* 20:246–251.

163. Zaika, L. L., J. C. Kissinger, and A. E. Wasserman. 1983. Inhibition of lactic acid bacteria by herbs. *J. Food Sci.* 48:1455–1459.

12

Radiation Preservation of Foods and Nature of Microbial Radiation Resistance

Although a patent was issued in 1929 for the use of radiation as a means of preserving foods, it was not until shortly after World War II that this method of food preservation received any serious consideration. While the application of radiation as a food preservation method has been somewhat slow in reaching its maximum potential use, the full application of this method presents some interesting challenges to food microbiologists and other food scientists.

Radiation may be defined as the emission and propagation of energy through space or through a material medium. The type of radiation of primary interest in food preservation is electromagnetic. The electromagnetic spectrum is presented in Figure 12-1. The various radiations are separated on the basis of their wavelengths, with the shorter wavelengths being the most damaging to microorganisms. The electromagnetic spectrum may be further divided as follows with respect to these radiations of interest in food preservation: microwaves, ultraviolet rays, X rays, and gamma rays. The radiations of primary interest in food preservation are **ionizing radiations**, defined as those radiations that have wavelengths of 2,000Å or less—for example, alpha particles, beta rays, gamma rays, X rays, and cosmic rays. Their quanta contain enough energy to ionize molecules in their paths. Since they destroy microorganisms without appreciably raising temperature, the process is termed "cold sterilization."

In considering the application of radiation to foods, there are several useful concepts that should be clarified. A **roentgen** is a unit of measure used for expressing exposure dose of X-ray or gamma radiation. A **milliroentgen** is equal to 1/1,000 of a roentgen. A **curie** is a quantity of radioactive substance in which 3.7×10^{10} radioactive disintegrations occur per second. For practical purposes, 1 g of pure radium possesses the radioactivity of 1 curie of radium. The new unit for a curie is the becquerel (Bq). A **rad** is a unit equivalent to the absorption of 100 ergs/g of matter. A **kilorad** (krad) is equal to 1,000 rads, and a **megarad** (Mrad) is equal to 1 million rads. The newer unit of absorbed dose

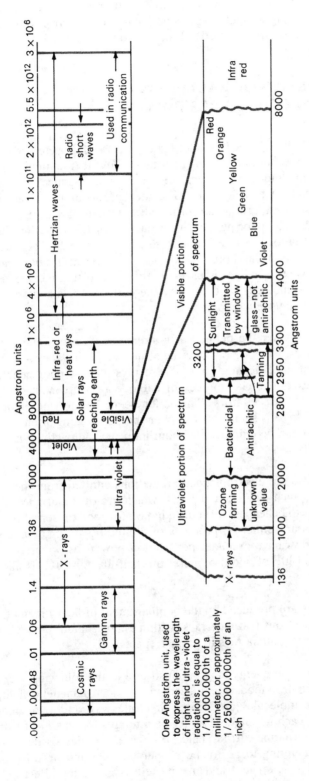

FIGURE 12-1. Spectrum charts.

From the Westinghouse Sterilamp and the Rentschler-James Process of Sterilization, courtesy of the Westinghouse Electric & Manufacturing Co., Inc.

is the gray (1 Gy = 100 rads = 1 joule/kg; 1 kGy = 10^5 rads). The energy gained by an electron in moving through 1 volt is designated **ev** (electron volt). A **mev** is equal to 1 million electron volts. Both the rad and ev are measurements of the intensity of irradiation.

CHARACTERISTICS OF RADIATIONS OF INTEREST IN FOOD PRESERVATION

Ultraviolet Light (UV Light).   Ultraviolet light is a powerful bactericidal agent, with the most effective wavelength being about 2,600Å. It is nonionizing and is absorbed by proteins and nucleic acids, in which photochemical changes are produced that may lead to cell death. The mechanism of UV death in the bacterial cell is due to the production of lethal mutations as a result of action on cell nucleic acids. The poor penetrative capacities of UV light limit its food use to surface applications, where it may catalyze oxidative changes that lead to rancidity, discolorations, and other reactions. Small quantities of ozone may also be produced when UV light is used for the surface treatment of certain foods. UV light is sometimes used to treat the surfaces of baked fruit cakes and related products before wrapping.

Beta Rays.   Beta rays may be defined as a stream of electrons emitted from radioactive substances. Cathode rays are the same except that they are emitted from the cathode of an evacuated tube. These rays possess poor penetration power. Among the commercial sources of cathode rays are Van de Graaff generators and linear accelerators. The latter seem better suited for food preservation uses. There is some concern over the upper limit of energy level of cathode rays that can be employed without inducing radioactivity in certain constituents of foods.

Gamma Rays.   These are electromagnetic radiations emitted from the excited nucleus of elements such as ^{60}Co and ^{137}Cs, which are of importance in food preservation. This is the cheapest form of radiation for food preservation, since the source elements are either by-products of atomic fission or atomic waste products. Gamma rays have excellent penetration power, as opposed to beta rays. ^{60}Co has a half-life of about 5 years; the half-life for ^{137}Cs is about 30 years.

X Rays.   These rays are produced by the bombardment of heavy-metal targets with high-velocity electrons (cathode rays) within an evacuated tube. They are essentially the same as gamma rays in other respects.

Microwaves.   Microwave energy may be illustrated in the following way (19). When electrically neutral foods are placed in an electromagnetic field, the charged asymmetric molecules are driven first one way and then another. During this process, each asymmetric molecule attempts to align itself with the rapidly changing alternating-current field. As the molecules oscillate about their axes while attempting to go to the proper positive and negative poles, intermolecular friction is created and manifested as a heating effect. This is microwave energy. Most food research has been carried out at two frequencies, 915 and 2450 megacycles. At the microwave frequency of 915 megacycles, the molecules oscillate back and forth 915 million times/sec (19). Microwaves lie between the infrared and radio frequency portion of the electromagnetic spectrum (Fig. 12-1). The problem associated with the microwave destruction of trichina larvae in pork products is discussed in Chapter 25.

PRINCIPLES UNDERLYING THE DESTRUCTION OF MICROORGANISMS BY IRRADIATION

Several factors should be considered when the effects of radiation on microorganisms are considered.

Types of Organisms.   Gram-positive bacteria are more resistant to irradiation than gram-negatives. In general, sporeformers are more resistant than nonsporeformers (with the exception of seven species among four genera, which are discussed later in this chapter). Among sporeformers, *Bacillus larvae* seems to possess a higher degree of resistance than most other aerobic sporeformers. Spores of *Clostridium botulinum* type A appear to be the most resistant of all clostridial spores. Apart from the seven extremely resistant species, *Enterococcus faecium* R53, micrococci, and the homofermentative lactobacilli are among the most resistant of nonsporeforming bacteria. Most sensitive to radiations are the pseudomonads and flavobacters, with other gram-negative bacteria being intermediate. A general spectrum of radiation sensitivity from enzymes to higher animals is illustrated in Figure 12-2. Possible mechanisms of radioresistance are discussed below.

With the exception of endospores and the extremely resistant species already noted, radioresistance generally parallels heat resistance among bacteria.

With respect to the radiosensitivity of molds and yeasts, the latter have been reported to be more resistant than the former, with both groups in general being less sensitive than gram-positive bacteria. Some *Candida* strains have been reported to possess resistance comparable to that of some bacterial endospores.

Numbers of Organisms.   The numbers of organisms have the same effect on the efficacy of radiations as in the case of heat, chemical disinfection, and certain other phenomena: the larger the number of cells, the less effective is a given dose.

Composition of Suspending Menstrum (Food).   Microorganisms in general are more sensitive to radiation when suspended in buffer solutions than in protein-containing media. For example, Midura et al. (45) found radiation D values for a strain of *Clostridium perfringens* to be 0.23 in phosphate buffer, while in cooked-meat broth, the D value was 3 kGy. Proteins exert a protective effect against radiations, as well as against certain antimicrobial chemicals and heat. Several investigators have reported that the presence of nitrites tends to make bacterial endospores more sensitive to radiation.

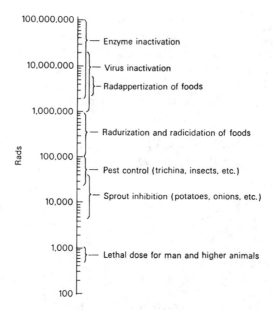

FIGURE 12-2.   Dose ranges of irradiation for various applications.
Adapted from Grünewald (23).

Presence or Absence of Oxygen.   The radiation resistance of microorganisms is greater in the absence of oxygen than in its presence. Complete removal of oxygen from the cell suspension of *Escherichia coli* has been reported to increase its radiation resistance up to threefold (50). The addition of reducing substances such as sulfhydryl compounds generally has the same effect in increasing radiation resistance as an anaerobic environment.

Physical State of Food.   The radiation resistance of dried cells is in general considerably higher than that for moist cells. This is most likely a direct consequence of the radiolysis of water by ionizing radiations, which is discussed later in this chapter. Radiation resistance of frozen cells has been reported to be greater than that of nonfrozen cells (38). Grecz et al. (21) found that the lethal effects of gamma radiation decreased by 47% when ground beef was irradiated at $-196°C$ as compared to $0°C$.

Age of Organisms.   Bacteria tend to be most resistant to radiation in the lag phase just prior to active cell division. The cells become more radiation sensitive as they enter and progress through the log phase and reach their minimum at the end of this phase.

PROCESSING OF FOODS FOR IRRADIATION

Prior to being exposed to ionizing radiations, several processing steps must be carried out in much the same manner as for the freezing or canning of foods.

Selection of Foods.   Foods to be irradiated should be carefully selected for freshness and overall desirable quality. Especially to be avoided are foods that are already in incipient spoilage.

Cleaning of Foods.   All visible debris and dirt should be removed. This will reduce the numbers of microorganisms to be destroyed by the radiation treatment.

Packing.   Foods to be irradiated should be packed in containers that will afford protection against postirradiation contamination. Clear glass containers undergo color changes when exposed to doses of radiation of around 10 kGy, and the subsequent color may be undesirable.

Blanching or Heat Treatment.   Sterilizing doses of radiation are insufficient to destroy the natural enzymes of foods (Fig. 12-2). In order to avoid undesirable postirradiation changes, it is necessary to destroy these enzymes. The best method is a heat treatment—that is, the blanching of vegetables and mild heat treatment of meats prior to irradiation.

APPLICATION OF RADIATION

The two most widely used techniques of irradiating foods are gamma radiation from either ^{60}Co and ^{137}Cs and the use of electron beams from linear accelerators.

Gamma Radiation.   The advantage of gamma radiation is that ^{60}Co and ^{137}Cs are relatively inexpensive by-products of atomic fission. In a common experimental radiation chamber employing these elements, the radioactive material is placed on the top of an elevator that can be moved up for use and down under water when not in use. Materials to be irradiated are placed around the radioactive material (the source) at a suitable distance for the desired dosage. Once the chamber has been vacated by all personnel, the source is raised into position, and the gamma rays irradiate the food. Irradiation at desired temperatures may be achieved either by placing the samples in temperature-controlled containers or by controlling the temperature of the entire concrete- and lead-walled chamber. Among the drawbacks to the use of radioactive material is that the isotope source emits rays in all directions and cannot be turned "on" or "off" as may be desirable (Fig. 12-3). Also, the half-life of ^{60}Co (5.27 yr) requires that the source be changed periodically in order to maintain a given level of radioactive potential. This drawback is overcome by the use of ^{137}Cs, which has a half-life of around 30 years.

Electron Beams.   The use of electron accelerators offers certain advantages over radioactive elements that make this form of radiation somewhat more attractive to potential commercial users. Koch and Eisenhower (33) have listed the following:

- High efficiency for the direct deposition of energy of the primary electron beams means high plant-product capacity.
- The efficient convertibility of electron power to X-ray power means the

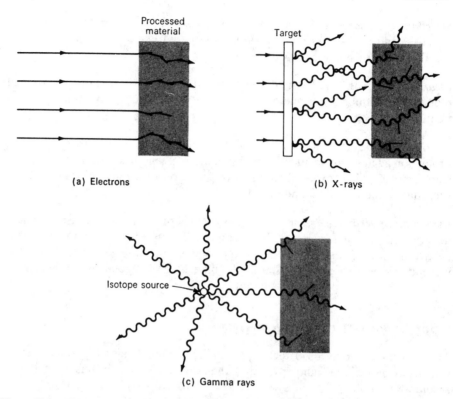

Figure 12-3.   The three basic techniques for radiation processing—interactions of electrons, X rays, and gamma rays in the medium.
From Koch and Eisenhower (33), 1965, Radiation Preservation of Foods, *Publication 1273, Advisory Board on Military Personnel Supplies, National Academy of Sciences, National Research Council.*

capability of handling very thick products that cannot be processed by electron or gamma-ray beams.
- The easy variability of electron-beam current and energy means a flexibility in the choice of surface and depth treatments for a variety of food items, conditions, and seasons.
- The monodirectional characteristic of the primary and secondary electrons and X rays at the higher energies permits a great flexibility in the food package design.
- The ability to program and to regulate automatically from one instant to the next with simple electronic detectors and circuits and various beam parameters means the capability of efficiently processing small, intricate, or nonuniform shapes.
- The ease with which an electron accelerator can be turned off or on means the ability to shut down during off-shifts or off-seasons without a maintenance problem and the ability to transport the radiation source without a massive radiation shield.

There appears to be a definite preference for electron accelerators in countries that lack atomic energy resources. The use of radioactive elements will probably continue in the United States for some time for the reasons cited.

RADAPPERTIZATION, RADICIDATION, AND RADURIZATION OF FOODS

Initially, the destruction of microorganisms in foods by ionizing radiation was referred to by terminology brought over from heat and chemical destruction of microorganisms. While microorganisms can indeed be destroyed by chemicals, heat, and radiation, there is, nevertheless, a lack of precision in the use of this terminology for radiation-treated foods. Consequently, in 1964 an international group of microbiologists suggested the following terminology for radiation treatment of foods (20):

Radappertization: Equivalent to radiation sterilization or "commercial sterility," as it is understood in the canning industry. Typical levels of irradiation are 30 to 40 kGy.

Radicidation: Equivalent to pasteurization—of milk, for example. Specifically, it refers to the reduction of the number of viable specific nonsporeforming *pathogens*, other than viruses, so that none is detectable by any standard method. Typical levels to achieve this process are 2.5 to 10 kGy.

Radurization: May be considered equivalent to pasteurization. It refers to the enhancement of the keeping quality of a food by causing substantial reduction in the numbers of viable specific *spoilage* microbes by radiation. Common dose levels are 0.75 to 2.5 kGy for fresh meats, poultry, seafood, fruits, vegetables, and cereal grains.

Radappertization

Radappertization of any foods may be achieved by application of the proper dose of radiation under the proper conditions. The effect of this treatment on endospores and exotoxins of *C. botulinum* is of obvious interest. Type E spores have been reported to possess radiation D values on the order of 0.12 to 0.17 Mrad (58). Types A and B spores were found by Kempe (31) to have D values of 0.279 and 0.238 Mrad, respectively. Type E spores are the most radiation sensitive of these three types.

The effect of temperature of irradiation on D values of *C. botulinum* spores is presented in Table 12-1: resistance increases at the colder temperatures and decreases at warmer temperatures (22). Different inoculum levels had no significant effect on D values whose calculations were based on a linear destruction rate. D values of four *C. botulinum* strains in three food products are presented in Table 12-2, from which it can be seen that each strain displayed different degrees of radiation resistance in each product. Also, irradiation of cured meat products produced the lowest D values. (The possible significance of this is discussed in Chap. 11 under nitrates and nitrites.) The minimum radiation doses (MRD) in kGy for the radappertization of nine meat and fish products are indicated below (3, 4, 29). With the exception of bacon (irradiated at ambient temperatures), each was treated at $-30°C + 10$:

Bacon	23	Shrimp	37
Beef	47	Codfish cakes	32
Chicken	45	Corned beef	25

TABLE 12-1.   Effect of Irradiation Temperature on D Values of Two Load Levels of *C. botulinum* 33A in Precooked Ground Beef

Temperature (°C)	D (Mrad)	
	Ca. 5×10^6 Spores/Can	Ca. 2×10^8/Can
−196	0.577	0.595
−150	0.532	0.543
−100	0.483	0.486
−50	0.434	0.430
0	0.385	0.373
25	0.360	0.345
65	0.321	0.299

Source: Grecz et al. (22), reproduced by permission of Nat'l. Res. Coun. of Canada from *Can. J. Microbiol.* 17:135–42, 1971.

Note: Data are based on linear spore destroction.

TABLE 12-2.   Variations in Radiation D Values of Strains of *C. botulinum* at −30°C in Three Meat Products

Strain Number	D (Mrad)		
	Codfish Cake	Corned Beef	Pork Sausage
33A	0.203	0.129	0.109
77A	0.238	0.262	0.098
41B	0.245	0.192	0.184
53B	0.331	0.183	0.076

Source: Anellis et al. (3), copyright © 1972, American Society for Microbiology.

Note: Computed by the Schmidt equation.

Ham	37	Pork sausage	24–27
Pork	51		

To achieve 12D treatments of meat products at about 30°C, the following kGy values are necessary (61): beef and chicken, 41.2–42.7; ham and codfish cake, 31.4–31.7; pork, 43.7; and corned beef and pork sausage, 25.5–26.9. Irradiation treatments of the types noted do not make the foods radioactive (61).

The radiation resistance of *C. botulinum* spores in aqueous media was studied by Roberts and Ingram (59), and these values are considerably lower than those obtained in meat products. On three type A strains, D ranged from 0.10 to 0.14; on two strains of type B, 0.10 to 0.11; on two strains of type E, 0.08 to 0.16; and the one type F strain examined by these authors showed a D value of 0.25. All strains were irradiated at 18° to 23°C and an exponential death rate was assumed in the D calculations.

With respect to the effect of radiation on *C. perfringens*, one each of five different strains (types A, B, C, E, and F) was found to have D values between 0.15 and 0.25 in an aqueous environment (59). 12D values for eight strains of this organism were found to range between 30.4 and 41.4 kGy depending upon the strain and method of computing 12D doses (7).

Radiation D_{10} values for *Listeria monocytogenes* in mozzarella cheese and ice cream were found to be 1.4 and 2.0 kGy, respectively, with strain Scott A irradiated at −78°C (24). The respective calculated 12D values were 16.8 and 24.4 kGy. To effect radappertization of ice cream and frozen yogurt, 40 kGy was sufficient but not for mozzarella or cheddar cheeses (25). The radappertization dose for *Bacillus cereus* in cheese and ice cream was 40 to 50 kGy.

As indicated in Fig. 12-2, viruses are considerably more resistant to radiation than bacteria. Radiation D values of 30 viruses were found by Sullivan et al. (64) to range between 3.9 and 5.3 kGy in Eagle's minimal essential medium supplemented with 2% serum. The 30 viruses included coxsackie-, echo-, and poliovirus. Of five selected viruses subjected to ^{60}Co rays in distilled water, the D values ranged from 1.0 to 1.4 kGy. D values of coxsackievirus B-2 in various menstra at −30 and −90°C are presented in Table 12-3. The use of a radiation 12D process for *C. botulinum* in meat products would result in the survival of virus particles unless previously destroyed by other methods such as heating.

Enzymes are also highly resistant to radiation, and a dose of from 20 to 60 kGy has been found to destroy only up to 75% of the proteolytic activity of ground beef (42). When blanching at 65° or 70°C was combined with radiation doses of 45 to 52 kGy, however, at least 95% of the beef proteolytic activity was destroyed. Radiation D values for a variety of organisms are presented in Table 12-4.

The main drawbacks to the application of radiation to some foods are color changes and/or the production of off-odors. Consequently, those food products that undergo relatively minor changes in color and odor have received the greatest amount of attention for commercial radappertization. Bacon is one product that undergoes only slight changes in color and odor development following radappertization. Mean preference scores on radappertized versus control bacon were found to be rather close, with control bacon being scored

TABLE 12-3.   D Values of Coxsackievirus B-2

	D (Mrad)	
Suspending Menstrum	−30°C	−90°C
Eagle's minimal essential medium +2% serum	0.69	0.64
Distilled water	—	0.53
Cooked ground beef	0.68	0.81
Raw ground beef	0.75	0.68

Source: Sullivan et al. (65), copyright © 1973, American Society for Microbiology.

Note: A linear model was assumed in D calculations.

TABLE 12-4.   Radiation D Values Reported

Organism/Substance	D (kGy)	Reference
Bacteria		
Acinetobacter calcoaceticus	0.26	69
Aeromonas hydrophila	0.14	53
Bacillus pumilus spores, ATCC 27142	1.40	69
Clostridium botulinum, type E spores	1.1–1.7	16, 40
C. botulinum, type E Beluga	0.8	42
C. botulinum, 62A spores	1.0	42
C. botulinum, type A spores	2.79	22
C. botulinum, type B Spores	2.38	22
C. botulinum, type F spores	2.5	42
C. botulinum A toxin in meat slurry	36.08	60
C. bifermentans spores	1.4	42
C. butyricum spores	1.5	42
C. perfringens type A spores	1.2	42
C. sporogenes spores (PA 3679/S$_2$)	2.2	42
C. sordellii spores	1.5	42
Enterobacter cloacae	0.18	69
Escherichia coli	0.20	69
Listeria monocytogenes	0.42–0.55	54
L. monocytogenes (mean of 7 strains)	0.35	27
Moraxella phenylpyruvica	0.86	55
Pseudomonas putida	0.08	55
P. aeruginosa	0.13	69
Salmonella typhimurium	0.50	54
Salmonella sp.	0.13	69
Staphylococcus aureus	0.16	69
S. aureus ent. toxin A in meat slurry	61.18; 208.49	60
Yersinia enterocolitica, beef, 25°C	0.195	14
Y. enterocolitica, gr. beef at 30°C	0.388	14
Fungi		
Aspergillus flavus spores (mean)	0.66	57
A. flavus	0.055–0.06	62
A. niger	0.042	62
Penicillium citrinum, NRRL 5452 (mean)	0.88	57
Penicillium sp.	0.42	69
Viruses		
Adenovirus (4 strains)	4.1–4.9	44
Coxsackievirus (7 strains)	4.1–5.0	44
Echovirus (8 strains)	4.4–5.1	44
Herpes simplex	4.3	44
Poliovirus (6 strains)	4.1–5.4	44

just slightly higher (77). Acceptance scores on a larger variety of irradiated products were in the favorable range (29).

Radappertization of bacon is one way to reduce nitrosamines. When bacon containing 20 ppm NaNO$_2$ + 550 ppm sodium ascorbate was irradiated with 30 kGy, the resulting nitrosamine levels were similar to those in nitrite-free bacon (15).

TABLE 12-5.   **Foods and Food Products Approved for Irradiation by Various Countries and by WHO**

Products	Objective	Dose Range (kGy)	Number of Countries[a]
Potatoes	Sprout inhibition	0.1–0.15	17
Onions	Sprout inhibition	0.1–0.15	10
Garlic	Sprout inhibition	0.1–0.15	2
Mushrooms	Growth inhibition	2.5 max.	1
Wheat, wheat flour	Insect disinfestation	0.2–0.75	4
Dried fruits	Insect disinfestation	1.0	2
Cocoa beans	Insect disinfestation	0.7	1
Dry food concentrates	Insect disinfestation	0.7–1.0	1
Poultry, fresh	Radicidation[b]	7.0 max.	2
Cod and redfish	Radicidation	2.0–2.2	1
Spices/condiments	Radicidation	8.0–10.0	1
Semipreserved meats	Radurization	6.0–8.0	1
Fresh fruits[c]	Radurization	2.5	6
Asparagus	Radurization	2.0	1
Raw meats	Radurization	6.0–8.0	1
Cod and haddock fillets	Radurization	1.5 max.	1
Poultry (eviscerated)	Radurization	3.0–6.0	2
Shrimp	Radurization	0.5–1.0	1
Culinary prepared meat products	Radurization	8.0	1
Deep-frozen meals	Radappertization	25.0 min.	2
Fresh, tinned/liquid foodstuffs	Radappertization	25.0 min.	1

Source: Urbain (71).

[a] Including WHO recommendations.
[b] For salmonellae.
[c] Includes tomatoes, peaches, apricots, strawberries, cherries, grapes, etc.

Radicidation

Irradiation at levels of 2 to 5 kGy has been shown by many to be effective in destroying nonsporeforming and nonviral pathogens and to present no health hazard. Kampelmacher (30) notes that raw poultry meats should be given the highest priority since they are often contaminated with salmonellae and since radicidation is effective on prepackaged products, thus eliminating the possibilities of cross-contamination. The treatment of refrigerated and frozen chicken carcasses with 2.5 kGy was highly effective in destroying salmonellae (47). A radiation dosage up to 7 kGy (0.7 Mrad) has been approved by the World Health Organization as being "unconditionally safe for human consumption" (16). When whole cacao beans were treated with 5 kGy, 99.9% of the bacterial flora was destroyed, and *Penicillium citrinum* spores were reduced by about 5 logs/g, and at a level of 4 kGy, *Aspergillus flavus* spores were reduced by about 7 logs/g (57). Fresh poultry, cod and red fish, and spices and condiments have been approved for radicidation in some countries (Table 12-5).

Radurization

Irradiation treatments to extend the shelf life of seafoods, vegetables, and fruits have been verified in many studies. Shelf life of shrimp, crab, haddock, scallops, and clams may be extended from two- to sixfold by radurization with doses from 1 to 4 kGy. Similar results can be achieved for fish and shellfish under various conditions of packaging (51). In one study, scallops stored at 0°C had a shelf life of 13 days, but after irradiation doses of 0.5, 1.5, and 3.0 kGy, shelf life was 18, 23, and 42 days, respectively (56). The gram-negative non-sporeforming rods are among the most radiosensitive of all bacteria, and they are the principal spoilage organisms for these foods. Following the irradiation of vacuum-packaged ground pork at 1.0 kGy and storage at 5°C for 9 days, 97% of the irradiated flora consisted of gram-positive bacteria, with most being coryneforms (13). The gram-negative coccobacillary rods belonging to the genera *Moraxella* and *Acinetobacter* have been found to possess degrees of radiation resistance higher than for all other gram negatives. In studies on ground beef subjected to doses of 272 krad, Tiwari and Maxcy (68) found that 73 to 75% of the surviving flora consisted of these related genera. In un-irradiated meat, they constituted only around 8% of the flora. Of the two genera, the *Moraxella* spp. appeared to be more resistant than *Acinetobacter* spp., with D_{10} values of 273 to 2,039 krad having been found (75). If this finding is correct, some of these organisms are among the most radiation resistant of all bacteria. Among specific species, *M. nonliquefaciens* strains showed D_{10} values of 539 and 583 krad, while the D_{10} for *M. osloensis* strains was 477 up to 1,000 krad.

In comparing the radiosensitivity of some nonsporeforming bacteria in phosphate buffer at −80°C, Anellis et al. (2) found that *Deinococcus radiodurans* survived 18 kGy, *Enterococcus faecium* strains survived 9 to 15, *E. faecalis* survived 6 to 9, and *Lactococcus lactis* did not survive 6 kGy. *Staphylococcus aureus*, *Lactobacillus casei*, and *Lactobacillus arabinosus* did not survive 3-kGy exposures. It was shown that radiation sensitivity decreased as temperature of irradiation was lowered, as is the case for endospores.

The ultimate spoilage of radurized, low-temperature-stored foods is invariably caused by one or more of the *Acinetobacter-Moraxella* or lactic acid types noted above. The application of 2.5 kGy to ground beef destroyed all pseudomonads, Enterobacteriaceae, and *Brochothrix thermosphacta*; reduced aerobic plate counts (APC) from log 6.18/g to 1.78/g; but reduced lactic acid bacteria only by 3.4 log/g (48).

Radurization of fruits with doses of 2 to 3 kGy brings about an extension of shelf life of at least 14 days. Radurization of fresh fruits is permitted by at least six countries, with some meats, poultry, and seafood permitted by several others (Table 12.5). In general, shelf-life extension is not as great for radurized fruits as for meats and seafood because molds are generally more resistant to irradiation than the gram-negative bacteria that cause spoilage of the latter products.

Insect eggs and larvae can be destroyed by 1 kGy, and cysticerci of the pork tapeworm (*Taenia solium*) and the beef tapeworm (*T. saginata*) can be destroyed with even lower doses, with cysticercosis-infested carcasses being rendered free of parasites by exposure to 0.2 to 0.5 kGy (73).

LEGAL STATUS OF FOOD IRRADIATION

At least 36 countries had approved the irradiation of some foods as of mid-1989 (41). At least 20 different food packaging materials have been approved by the U. S. Food and Drug Administration (FDA) at levels of 10 or 60 kGy. In 1983, the FDA permitted spices and vegetable seasonings to be irradiated up to 10 kGy (*U.S. Federal Register*, July 15, 1983). FDA granted permission in 1985 for the irradiation of pork at up to 1 kGy to control *Trichinella spiralis* (*U.S. Federal Register*, July 22, 1985). In 1986, fermented pork sausage (Nham) was irradiated in Thailand at a minimum of 2.0 kGy, and the product was sold in Bangkok (41). Puerto Rican mangoes were irradiated in 1986 at up to 1.0 kGy, flown to Miami, Florida, and sold. Hawaiian papayas were treated at doses of 0.41 to 0.51 kGy to control pests in 1987 and later sold to the public. USDA approval was granted for Hawaiian papayas in 1989 for insect control. Strawberries were irradiated at 2.0 kGy and sold in Lyon, France, in 1987. Overall, the irradiated foods noted were well received by consumers. Sprout inhibition and insect disinfestation continue to be the most widely used direct applications of food irradition.

WHO has given approval for radiation dosages up to 7 kGy (0.7 Mrad) as being unconditionally safe. In the early 1970s, Canada approved for test marketing a maximum dose of 1.5 kGy for fresh cod and haddock fillets. In 1983, the Codex Alimentarius Commission suggested 1.5 or 2.2 kGy for teleost fish and fish products (16). One of the obstacles to getting food irradiation approved on a wider scale in the United States is the way irradiation is defined. It is considered an additive rather than a process, which it is. This means that irradiated foods must be labeled as such. Another area of concern is the fate of *C. botulinum* spores (see below), and yet another is the concern that non-pathogens may become pathogens or that the virulence of pathogens may be increased after exposure to subradappertization doses. There is no evidence that the latter occurs (61).

When low-acid foods are irradiated at doses that do not effect the destruction of *C. botulinum* spores, legitimate questions about the safety of such foods are raised, especially when they are held under conditions that allow for growth and toxin production. Since these organisms would be destroyed by radappertization, only products subjected to radicidation and radurization are of concern here. In regard to the radurization of fish, Giddings (17) has pointed out that the lean whitefish species are the best candidates for irradiation, while high-fat fishes such as herring are not since they are more botulogenic. This investigator notes that when botulinal spores are found on edible lean whitefish, they occur at less than 1/g.

EFFECT OF IRRADIATION ON FOOD QUALITY

The undesirable changes that occur in certain irradiated foods may be caused directly by irradiation or indirectly as a result of postirradiation reactions. Water undergoes radiolysis when irradiated in the following manner:

$$3H_2O \xrightarrow{\text{radiolysis}} H + OH + H_2O_2 + H_2$$

In addition, free radicals are formed along the path of the primary electron and react with each other as diffusion occurs (11). Some of the products formed along the track escape and can then react with solute molecules. By irradiating under anaerobic conditions, off-flavors and -odors are somewhat minimized

TABLE 12-6.   Methods for Reducing Side Effects in Foodstuffs Exposed to Ionizing Radiations

Method	Reasoning
Reducing temperature	Immobilization of free radicals
Reducing oxygen tension	Reduction of numbers of oxidative free radicals to activated molecules
Addition of free radical scavengers	Competition for free radicals by scavengers
Concurrent radiation distillation	Removal of volatile off-flavor, off-odor precursors
Reduction of dose	Obvious

Source: Goldblith (18).

due to the lack of oxygen to form peroxides. One of the best ways to minimize off-flavors is to irradiate at subfreezing temperatures (70). The effect of subfreezing temperatures is to reduce or halt radiolysis and its consequent reactants. Other ways to reduce side effects in foodstuffs are presented in Table 12-6.

Other than water, proteins and other nitrogenous compounds appear to be the most sensitive to irradiation effects in foods. The products of irradiation of amino acids, peptides, and proteins depend on the radiation dose, temperature, amount of oxygen, amount of moisture present, and other factors. The following are among the products reported: NH_3, hydrogen, CO_2, H_2S, amides, and carbonyls. With respect to amino acids, the aromatics tend to be more sensitive than the others and undergo changes in ring structure. Among the most sensitive to irradiation are methionine, cysteine, histidine, arginine, and tyrosine. The amino acid most susceptible to electron beam irradiation is cystine; Johnson and Moser (28) reported that about 50% of this amino acid was lost when ground beef was irradiated. Tryptophan suffered a 10% loss, while little or no destruction of the other amino acids occurred. Amino acids have been reported to be more stable to gamma irradiation than to electron beam irradiation.

Several investigators have reported that the irradiation of lipids and fats results in the production of carbonyls and other oxidation products such as peroxides, especially if irradiation and/or subsequent storage takes place in the presence of oxygen. The most noticeable organoleptic effect of lipid irradiation in air is the development of rancidity.

It has been observed that high levels of irradiation lead to the production of "irradiation odors" in certain foods, especially meats. Wick et al. (76) investigated the volatile components of raw ground beef irradiated with 20 to 60 kGy at room temperature and reported finding a large number of odorous compounds. Of the 45 or more constituents identified by these investigators, there were 17 sulfur-containing, 14 hydrocarbons, and 9 carbonyls, and 5 or more were basic and alcoholic in nature. The higher the level of irradiation, the greater is the quantity of volatile constituents produced. Many of these constituents have been identified in various extracts of nonirradiated, cooked ground beef.

With regard to B vitamins, Liuzzo et al. (40) found that levels of ^{60}Co irradiation between 2 and 6 kGy effected partial destruction of the following B vitamins in oysters: thiamine, niacin, pyridoxine, biotin, and B_{12}. Riboflavin, pantothenic acid, and folic acid were reported to be increased by irradiation, probably owing to release of bound vitamins.

In addition to flavor and odor changes produced in certain foods by irradiation, certain detrimental effects have been reported for irradiated fruits and vegetables. One of the most serious is the softening of these products caused by the irradiation-degradation of pectin and cellulose, the structural polysaccharides of plants. This effect has been shown by Massey and Bourke (43) to be caused by radappertization doses of irradiation. Ethylene synthesis in apples is affected by irradiation so that this product fails to mature as rapidly as nonirradiated controls (43). In green lemons, however, ethylene synthesis is stimulated upon irradiation, resulting in a faster ripening than in controls (44).

Among radiolytic products that develop upon irradiation are some that are antibacterial when exposed in culture media. When 15 kGy were applied to meats, however, no antimicrobial activity was found in the meats (10). The overall wholesomeness and toxicology of irradiated foods have been reviewed (63, 67).

STORAGE STABILITY OF IRRADIATED FOODS

Foods subjected to radappertization doses of ionizing radiation may be expected to be as shelf stable as commercially heat-sterilized foods. There are, however, two differences between foods processed by these two methods that affect storage stability: radappertization does not destroy inherent enzymes, which may continue to act, and some postirradiation changes may be expected to occur. Employing 45 kGy and enzyme-inactivated chicken, bacon, and fresh and barbecued pork, Heiligman (26) found the products to be acceptable after storage for up to 24 months. Those stored at 70°C were more acceptable than those stored at 100°F. The effect of irradiation on beefsteak, ground beef, and pork sausage held at refrigerator temperatures for 12 years was reported by

TABLE 12-7. **Effects of Oxidizing and Reducing Conditions on Resistance to Radiation of *Deinococcus radiodurans* (Table of Means)**

Condition	Log of Surviving Fraction[a]
Buffer, unmodified	−3.11542
Oxygen flushed	−3.89762
Nitrogen flushed	−2.29335
H_2O_2 (100 ppm)	−3.47710
Thioglycolate (0.01 M)	−1.98455
Cysteine (0.1 M)	−0.81880
Ascorbate (0.1 M)	−5.36050

Source: Giddings (17).

Note: Determined by count reduction after exposure to 1 Mrad of gamma radiation in 0.05 M phosphate buffer. LSD: $P = 0.05$ (1.98116); $P = 0.01$ (2.61533).

[a] Averages of four replicates.

Licciardello et al. (39). These foods were packed with flavor preservatives and treated with 10.8 kGy. The authors described the appearance of the meats as excellent after 12 years of storage. A slight irradiation odor was perceptible but was not considered objectionable. The meats were reported to have a sharp, bitter taste, which was presumed to be caused by the crystallization of the amino acid tyrosine. The free amino nitrogen content of the beefsteak was 75 and 175 mg %, respectively, before and after irradiation storage, and 67 and 160 mg % before and after storage for hamburger.

Foods subjected to radurization ultimately undergo spoilage from the surviving flora if stored at temperatures suitable for growth of the organisms in question. The normal spoilage flora of seafoods is so sensitive to ionizing radiations that 99% of the total flora of these products is generally destroyed by doses on the order of 2.5 kGy. Ultimate spoilage of radurized products is the property of the few microorganisms that survive the radiation treatment.

For further information on all aspects of food irradiation, see reviews by Anderson (1), Ley (38), Tsuji (69), and Urbain (71).

NATURE OF RADIATION RESISTANCE OF MICROORGANISMS

The most sensitive bacteria to ionizing radiation are gram-negative rods such as the pseudomonads; the coccobacillary-shaped gram-negative cells of moraxellae and acinetobacters are among the most resistant of gram negatives. Gram-positive cocci are the most resistant of nonsporing bacteria, including micrococci, staphylococci, and enterococci. What makes one organism more sensitive or resistant than another is not only a matter of fundamental biological interest but of interest in the application of irradiation to the preservation of foods. A better understanding of resistance mechanisms can lead to ways of increasing radiation sensitivity and, consequently, to the use of lower doses for food preservation use.

The effect of oxidizing and reducing conditions on the resistance of *Deinococcus radiodurans* in phosphate buffer has been studied; the findings are presented in Table 12-7 (12). The flushing of buffer suspensions with nitrogen or O_2 had no significant effect on radiation sensitivity when compared to the control, nor did the presence of 100 ppm H_2O_2. Treatment with cysteine rendered the cells less sensitive, and ascorbate increased their sensitivity. A study of N-ethylmaleimide (NEM) and indoleacetic acid (IAA) on resistance showed that IAA reduced resistance but NEM did not when tested at nontoxic levels (36). The presence or absence of O_2 had no effect on these two compounds.

Biology of Extremely Resistant Species

The most resistant of all known nonsporeforming bacteria consist of four species of the genus *Deinococcus* and one each of *Deinobacter, Rubrobacter,* and *Acinetobacter*. Some characteristics of these species are presented in Table 12-8. The deinococci were originally assigned to the genus *Micrococcus*, but they, along with *Deinobacter* and the archaebacterial genus *Thermus*, constitute one of the 10 major phyla based on 16S rRNA (72, 74, 78). The deinococci occur in pairs or tetrads, contain red water–insoluble pigments,

TABLE 12-8.   The Extremely Radiation-Resistant Nonsporeforming Bacteria

Organisms	Gram Reaction	Morphology	Pigment	Outer Membrane	Predominant Isoprenoid Quinone	Moles % G + C of DNA	Optimum Growth (°C)	Basic Amino Acid in Peptidoglycan	Radiation D Value (kGy)
Deinococcus									
radiodurans	+	C	R	+	MK-8	67	30	L-orn	
D. radiophilus	+	C	R	+	MK-8	62	30	L-orn	
D. proteolyticus	+	C	R	+	MK-8	65	30	L-orn	
D. radiopugnans	+	C	R	+	MK-8	70	30	L-orn	
Deinobacter									
grandis	−	R	R/P	+	MK-8	69	30–35	Orn	1.0
Acinetobacter									
radioresistens	−	C	R	nd	Q-9	44.1	27–31	nd	1.25–2.2
						44.8			
Rubrobacter									
radiotolerans	+	R	R	nd	MK-8	67.9	46–48	L-lys	1.0

Sources: Brooks and Murray (6), Nishimura et al. (49), Oyaizu et al. (52), and Suzuki et al. (66).

have optimum growth at 30°C, contain L-ornithine as the basic amino acid in their murein (unlike the micrococci, which contain lysine), and are characterized by moles % G + C content between 62 and 70. They do not contain teichoic acids. One of the most unusual features of this genus is the possession of an outer membrane, unlike other gram-positive bacteria. They have been characterized as being gram-negative clones of ancient lineage (9).

Among other unusual features of deinococci is their possesion of palmitoleate (16:1), which makes up about 60% of the fatty acids in their envelope and about 25% of the total cellular fatty acid. The high content of fatty acids is another feature characteristic of gram-negative bacteria. The predominant isoprenoid quinone in their plasma membrane is a menaquinone. The menaquinones represent one of the two groups of naphthoquinones that are involved in electron transport, oxidative phosphorylation, and perhaps active transport (8). The length of the C-3 isoprenyl side chains ranges from 1 to 14 isoprene units (MK), and the deinococci are characterized by the possession of MK-8, as are some micrococci, planococci, staphylococci, and enterococci (8). The deinococci do not contain phosphatidylglycerol or diphophotidylglycerol in their phospholipids but contain instead phosphoglycolipids as the major component.

The genus *Deinobacter* shares many of the deinococcal features except that its members are gram-negative rods. *Rubrobacter radiotolerans* is a gram-positive rod that is highly similar to the deinococci, but the basic amino acid in its murein is L-lysine rather than L-ornithine. *Acinetobacter radioresistens* is a gram-negative coccobacillary rod that differs in several ways from deinococci. Its moles % G + C content of DNA is in the range 44.1 to 44.8, and its predominant isoprenoid quinone is Q-9, not MK-8.

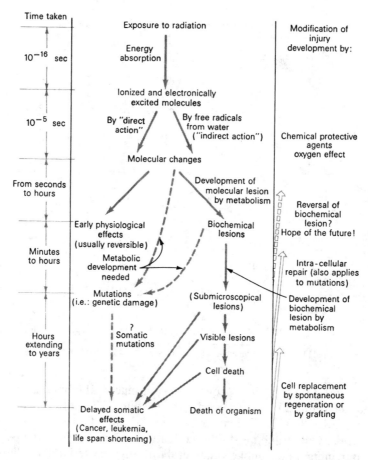

FIGURE 12-4.   Summary of radiation and postirradiation effects in organic matter. *From Bacq and Alexander (5), reprinted with permission of the authors,* Fundamentals of Radiobiology, *copyright © 1961 Pergamon Press.*

Deinococci have been isolated from ground beef, pork sausage, hides of animals, creek water (34) and haddock. It has been reported also to occur in feces, sawdust, and air. *Deinobacter* was isolated from animal feces and freshwater fish, *Rubrobacter* from a radioactive hot spring in Japan, and *A. radioresistens* from cotton and soils.

The seven species noted in Table 12-8 are aerobic, catalase positive, and generally inactive on substrates for biochemical tests. The deinococci possess a variety of carotenoids, and their isolated plasma membrane is bright red.

Radiation D values of the nondeinococcal species are 1.0 to 2.2 kGy, while many strains of the deinococci can survive 15 kGy. *D. radiophilus* is the most radioresistant species.

Apparent Mechanisms of Resistance

Why these organisms are so resistant to radiation is unclear. The extreme resistance of deinococci to desiccation has been observed and presumed to be related in some way to radio-resistance. The complicated cell envelope of these organisms may be a factor, but precise data are wanting. All are highly pigmented and contain various carotenoids, a fact that suggests some relationship to radiation resistance. However, these pigments have been found to play no role in the resistance of *D. radiophilus* (32, 37). Some of the chemical events that occur in organic matter after irradiation are outlined in Figure 12-4. The radiolysis of water leads to the formation of free radicals and peroxides, and radiation-sensitive organisms appear to be unable to overcome their deleterious effects. Chemicals that contain $-SH$ groups tend to be radio-protective, but what role these play, if any, in the extreme resistance of bacteria is still unclear.

Effective nucleic acid repair mechanisms appear to be one reason for extreme radioresistance. Enzymatic repair to the radiation damage of *D. radiodurans* has been demonstrated (46). Also, *D. radiophilus* has been shown to possess an efficient excision repair system (35).

References

1. Anderson, A. W. 1983. Irradiation in the processing of food. In *Food Microbiology*, ed. A. H. Rose, 145–171. New York: Academic Press.
2. Anellis, A., D. Berkowitz, and D. Kemper. 1973. Comparative resistance of nonsporogenic bacteria to low-temperature gamma irradiation. *Appl. Microbiol.* 25:517–523.
3. Anellis, A., D. Berkowitz, W. Swantak, and C. Strojan. 1972. Radiation sterilization of prototype military foods: Low-temperature irradiation of codfish cake, corned beef, and pork sausage. *Appl. Microbiol.* 24:453–462.
4. Anellis, A., E. Shattuck, D. B. Rowley, E. W. Ross, Jr., D. N. Whaley, and V. R. Dowell, Jr. 1975. Low-temperature irradiation of beef and methods for evaluation of a radappertization process. *Appl. Microbiol.* 30:811–820.
5. Bacq, Z. M., and P. Alexander. 1961. *Fundamentals of Radiobiology*. 2d ed. Oxford: Pergamon.
6. Brooks, B. W., and R. G. E. Murray. 1981. Nomenclature for "*Micrococcus radiodurans*" and other radiation-resistant cocci: *Deinococcaceae* fam. nov. and *Deinococcus* gen. nov., including five species. *Int. J. Syst. Bacteriol.* 31:353–360.
7. Clifford, W. J., and A. Anellis. 1975. Radiation resistance of spores of some *Clostridium perfringens* strains. *Appl. Microbiol.* 29:861–863.
8. Collins, M. D., and D. Jones, 1981. Distribution of isoprenoid quinone structural types in bacteria and their taxonomic implications. *Microbiol. Rev.* 45:316–354.
9. Counsell, T. J., and R. G. E. Murray. 1986. Polar lipid profiles of the genus *Deinococcus*. *Int. J. Syst. Bacteriol.* 36:202–206.
10. Dickson, J. S., and R. B. Maxcy. 1984. Effect of radiolytic products on bacteria in a food system. *J. Food Sci.* 49:577–580.
11. Doty, D. M. 1965. Chemical changes in irradiated meats. In *Radiation Preservation of Foods*, pub. no. 1273, 121–25. Washington, D.C.: National Research Council, National Academy of Science.
12. Duggan, D. E., A. W. Anderson, and P. R. Elliker. 1963. Inactivation of the radiation-resistant spoilage bacterium *Micrococcus radiodurans*. II. Radiation inactivation rates as influenced by menstruum temperature, preirradiation heat treatment, and certain reducing agents. *Appl. Microbiol.* 11:413–417.
13. Ehioba, R. M., A. A. Kraft, R. A. Molins, H. W. Walker, D. G. Olson, G. Subbaraman, and R. P. Skowronski. 1988. Identification of microbial isolates from vacuum-packaged ground pork irradiated at 1 kGy. *J. Food Sci.* 53:278–279, 281.

14. El-Zawahry, Y. A., and D. B. Rowley. 1979. Radiation resistance and injury of *Yersinia enterocolitica*. *Appl. Environ. Microbiol.* 37:50–54.
15. Fiddler, W., R. A. Gates, J. W. Pensabene, J. G. Phillips, and E. Wierbicki. 1981. Investigations on nitrosamines in irradiation-sterilized bacon. *J. Agric. Food Chem.* 29:551–554.
16. Food and Agriculture Organization/IAEA/World Health Organization. 1977. *Wholesomeness of Irradiated Food*. Report of joint FAO/IAEA/WHO Expert Committee, WHO Technical Report Series 604.
17. Giddings, G. G. 1984. Radiation processing of fishery products. *Food Technol.* 38(4):61–65, 94–97.
18. Goldblith, S. A. 1963. Radiation preservation of foods—Two decades of research and development. In *Radiation Research*, 155–167. Washington, D.C.: U.S. Department of Commerce, Office of Technical Services.
19. Goldblith, S. A. 1966. Basic principles of microwaves and recent developments. *Adv. in Food Research* 15:277–301.
20. Goresline, H. E., M. Ingram, P. Macuch, G. Mocquot. D. A. A. Mossel, C. F. Niven, and F. S. Thatcher. 1964. Tentative classification of food irradiation processes with microbiological objectives. *Nature* 204:237–238.
21. Grecz, N., O. P. Snyder, A. A. Walker, and A. Anellis. 1965. Effect of temperature of liquid nitrogen on radiation resistance of spores of *Clostridium botulinum*. *Appl. Microbiol.* 13:527–536.
22. Grecz, N., A. A. Walker, A. Anellis, and D. Berkowitz. 1971. Effects of irradiation temperature in the range −196 to 95°C on the resistance of spores of *Clostridium botulinum* 33A in cooked beef. *Can J. Microbiol.* 17:135–142.
23. Grünewald, T. 1961. Behandlung von Lebensmitteln mit energiereichen Strahlen. *Ernährungs-Umschau* 8:239–244.
24. Hashisaka, A. E., S. D. Weagant, and F. M. Dong. 1989. Survival of *Listeria monocytogenes* in mozzarella cheese and ice cream exposed to gamma irradiation. *J. Food Protect.* 52:490–492.
25. Hashisaka, A. E., J. R. Matches, Y. Batters, F. P. Hungate, and F. M. Dong. 1990. Effects of gamma irradiation at −78°C on microbial populations in dairy products. *J. Food Sci.* 55:1284–1289.
26. Heiligman, F. 1965. Storage stability of irradiated meats. *Food Technol.* 19: 114–116.
27. Huhtanen, C. N., R. K. Jenkins, and D. W. Thayer. 1989. Gamma radiation sensitivity of *Listeria monocytogenes*. *J. Food Protect.* 52:610–613.
28. Johnson, B., and K. Moser. 1967. Amino acid destruction in beef by high energy electron beam irradiation. In *Radiation Preservation of Foods*, Advances in Chemistry Series, 171–179. Washington, D.C.: American Chemical Society.
29. Josephson, E. S., A. Brynjolfsson, and E. Wierbicki. 1975. The use of ionizing radiation for preservation of food and feed products. In *Radiation Research—Biomedical, Chemical, and Physical Perspectives*, ed. O. F. Nygaard, H. I. Adler, and W. K. Sinclair, 96–117. New York: Academic Press.
30. Kampelmacher, E. H. 1983. Irradiation for control of *Salmonella* and other pathogens in poultry and fresh meats. *Food Technol.* 37(4):117–119, 169.
31. Kempe, L. L. 1965. The potential problems of type E botulism in radiation-preserved seafoods. In *Radiation Preservation of Foods*, Pub. No. 1273, 211–215. Washington, D.C.: Research Council, National Academy of Science.
32. Kilburn, R. E., W. D. Bellamy, and S. A. Terni. 1958. Studies on a radiation-resistant pigmented Sarcina sp. *Radiat. Res.* 9:207–215.
33. Koch, H. W., and E. H. Eisenhower, 1965. Electron accelerators for food processing. In *Radiation Preservation of Foods*, Pub. No. 1273, 149–180. Washington, D.C.: National Research Council, National Academy of Science.

34. Krabbenhoft, K. L., A. W. Anderson, and P. R. Elliker. 1965. Ecology of *Micrococcus radiodurans*. *Appl. Microbiol.* 13:1030–1037.

35. Lavin, M. F., A. Jenkins, and C. Kidson. 1976. Repair of ultraviolet light-induced damage in *Micrococcus radiophilus*, an extremely resistant microorganism. *J. Bacteriol.* 126:587–592.

36. Lee, J. S., A. W. Anderson, and P. R. Elliker. 1963. The radiation-sensitizing effects of N-ethylmaleimide and iodoacetic acid on a radiation-resistant *Micrococcus*. *Radiat. Res.* 19:593–598.

37. Lewis, N. F., D. A. Madhavesh, and U. S. Kumta. 1974. Role of carotenoid pigments in radio-resistant micrococci. *Can. J. Microbiol.* 20:455–459.

38. Ley, F. J. 1983. New interest in the use of irradiation in the food industry. In *Food microbiology: Advances and Prospects*, ed. T. A. Roberts and F. A. Skinner, 113–129. London: Academic Press.

39. Licciardello, J. J., J. T. R. Nickerson, and S. A. Goldblith. 1966. Observations on radio-pasteurized meats after 12 years of storage at refrigerator temperatures above freezing. *Food Technol.* 20:1232.

40. Liuzzo, J. S., W. B. Barone, and A. F. Novak. 1966. Stability of B-vitamins in Gulf oysters preserved by gamma radiation. *Fed. Proc.* 25:722.

41. Loaharanu, P. 1989. International trade in irradiated foods: Regional status and outlook. *Food Technol.* 43(7):77–80.

42. Losty, T., J. S. Roth, and G. Shults. 1973. Effect of irradiation and heating on proteolytic activity of meat samples. *J. Agr. Food Chem.* 21:275–277.

43. Massey, L. M., Jr., and J. B. Bourke. 1967. Some radiation-induced changes in fresh fruits and vegetables. In *Radiation Preservation of Foods*, Advances in Chemistry Series, 1–11. Washington, D.C.: American Chemical Society.

44. Maxie, E., and N. Sommer. 1965. Irradiation of fruits and vegetables. In *Radiation Preservation of Foods*, Pub. No. 1273, 39–52. Washington, D.C.: National Research Council, National Academy of Science.

45. Midura, T. F., L. L. Kempe, J. T. Graikoski, and N. A. Milone. 1965. Resistance of *Clostridium perfringens* type A spores to gamma-radiation. *Appl. Microbiol.* 13:244–247.

46. Moseley, B. E. B. 1976. Photobiology and radiobiology of *Micrococcus (Deinococcus) radiodurans*. *Photochem. Photobiol. Rev.* 7:223–274.

47. Mulder, R. W. A., S. Notermans, and E. H. Kampelmacher. 1977. Inactivation of salmonellae on chilled and deep frozen broiler carcasses by irradiation. *J. Appl. Bacteriol.* 42:179–185.

48. Niemand, J. G., H. J. van der Linde, and W. H. Holzapfel. 1983. Shelf-life extension of minced beef through combined treatments involving radurization. *J. Food Protect.* 46:791–796.

49. Nishimura, Y., T. Ino, and H. Iizuka. 1988. *Acinetobacter radioresistens* sp. nov. isolated from cotton and soil. *Int. J. Syst. Bacteriol.* 38:290–11.

50. Niven, C. F., Jr. 1958. Microbiological aspects of radiation preservation of food. *Ann. Rev. Microbiol.* 12:507–524.

51. Novak, A. F., R. M. Grodner, and M. R. R. Rao. 1967. Radiation pasteurization of fish and shellfish. In *Radiation Preservation of Foods*, Advances in Chemistry Series, 142–151. Washington, D.C.: American Chemical Society.

52. Oyaizu, H., E. Stackebrandt, K. H. Schleifer, W. Ludwig, H. Pohla, H. Ito, A. Hirata, Y. Oyaizu, and K. Komagata. 1987. A radiation-resistant rod-shaped bacterium, *Deinobacter grandis* gen. nov., sp. nov., with peptidoglycan containing ornithine. *Int. J. Syst. Bacteriol.* 37:62–67.

53. Palumbo, S. A., R. K. Jenkins, R. L. Buchanan, and D. W. Thayer. 1986. Determination of irradiation D-values for *Aeromonas hydrophila*. *J. Food Protect.* 49:189–191.

54. Patterson, M. 1989. Sensitivity of *Listeria monocytogenes* to irradiation on poultry meat and in phosphate-buffered saline. *Lett. Appl. Microbiol.* 8:181–184.

55. Patterson, M. F. 1988. Sensitivity of bacteria to irradiation on poultry meat under various atmospheres. *Lett. Appl. Microbiol.* 7:55–58.

56. Poole, S. E., P. Wilson, G. E. Mitchell, and P. A. Wills. 1990. Storage life of chilled scallops treated with low dose irradiation. *J. Food Protect.* 53:763–766.

57. Restaino, L., J. J. J. Myron, L. M. Lenovich, S. Bills, and K. Tscherneff. 1984. Antimicrobial effects of ionizing radiation on artificially and naturally contaminated cacao beans. *Appl. Environ. Microbiol.* 47:886–887.

58. Roberts, T. A., and M. Ingram. 1965. The resistance of spores of *Clostridium botulinum* Type E to heat and radiation. *J. Appl. Bacteriol.* 28:125–141.

59. Roberts, T. A., and M. Ingram., 1965. Radiation resistance of spores of *Clostridium* species in aqueous suspension. *J. Food Sci.* 30:879–885.

60. Rose, S. A., N. K. Modi, H. S. Tranter, N. E. Bailey, M. F. Stringer, and P. Hambleton. 1988. Studies on the irradiation of toxins of *Clostridium botulinum* and *Staphylococcus aureus*. *J. Appl. Bacteriol.* 65:223–229.

61. Rowley, D. B., and A. Brynjolfsson. 1980. Potential uses of irradiation in the processing of food. *Food Technol.* 34(10):75–77.

62. Saleh, Y. G., M. S. Mayo, and D. G. Ahearn. 1988. Resistance of some common fungi to gamma irradiation. *Appl. Environ. Microbiol.* 54:2134–2135.

63. Skala, J. H., E. L. McGown, and P. P. Waring. 1987. Wholesomeness of irradiated foods. *J. Food Protect.* 50:150–160.

64. Sullivan, R., A. C. Fassolitis, E. P. Larkin, R. B. Read, Jr., and J. T. Peeler. 1971. Inactivation of thirty viruses by gamma radiation. *Appl. Microbiol.* 22:61–65.

65. Sullivan, R., P. V. Scarpino, A. C. Fassolitis, E. P. Larkin, and J. T. Peeler. 1973. Gamma radiation inactivation of coxsackievirus B-2. *Appl. Microbiol.* 26:14–17.

66. Suzuki, K.-I., M. D. Collins, E. Iigima, and K. Komagata. 1988. Chemotaxonomic characterization of a radiotolerant bacterium, *Arthrobacter radiotolerans*: Description of *Rubrobacter radiotolerans* gen. nov., comb. nov. *FEMS Microbiol. Lett.* 52:33–40.

67. Thayer, D. W., J. P. Christopher, L. A. Campbell, D. C. Ronning, R. R. Dahlgren, G. M. Thomson, and E. Wierbicki. 1987. Toxicology studies of irradiation-sterilized chicken. *J. Food Protect.* 50:278–288.

68. Tiwari, N. P., and R. B. Maxcy. 1972. *Moraxella-Acinetobacter* as contaminants of beef and occurrence in radurized product. *J. Food Sci.* 37:901–903.

69. Tsuji, K. 1983. Low-dose cobalt 60 irradiation for reduction of microbial contamination in raw materials for animal health products. *Food Technol.* 37(2):48–54.

70. Urbain, W. M. 1965. Radiation preservation of fresh meat and poultry. In *Radiation Preservation of Foods*, Pub. No. 1273, 87–98. Washington, D.C.: National Research Council, National Academy of Science.

71. Urbain, W. M. 1978. Food irradiation. *Adv. Food Res.* 24:155–227.

72. Van den Eynde, H., Y. Van de Peer, H. Vandenabeele, M. Van Bogaert, and R. de Wachter. 1990. 5S rRNA sequences of myxobacteria and radioresistant bacteria and implications for eubacterial evolution. *Int. J. System. Bacteriol.* 40:399–404.

73. Verster, A., T. A. du Plessis, and L. W. van den Heever. 1977. The eradication of tapeworms in pork and beef carcasses by irradiation. *Radiat. Phys. Chem.* 9:769–771.

74. Weisburg, W. G., S. J. Giovannoni, and C. R. Woese. 1989. The *Deinococcus-Thermus* phylum and the effect of rRNA composition on phylogenetic tree construction. *System. Appl. Microbiol.* 11:128–134.

75. Welch, A. B., and R. B. Maxcy. 1975. Characterization of radiation-resistant vegetative bacteria in beef. *Appl. Microbiol.* 30:242–250.

76. Wick, E., E. Murray. J. Mizutani, and M. Koshika. 1967. Irradiation flavor and the volatile components of beef. In *Radiation Preservation of Foods*, Advances in Chemistry Series, 12–25. Washington, D.C.: American Chemical Society.

77. Wierbicki, E., M. Simon, and E. S. Josephson. 1965. Preservation of meats by sterilizing doses of ionizing radiation. In *Radiation Preservation of Foods*, Pub. No. 1273, 383–409. Washington, D.C.: National Research Council, National Academy of Science.

78. Woese, C. R. 1987. Bacterial evolution. *Microbiol. Rev.* 51:221–271.

13

Low-Temperature Food Preservation and Characteristics of Psychrotrophic Microorganisms

The use of low temperatures to preserve foods is based on the fact that the activities of foodborne microorganisms can be slowed at temperatures above freezing and generally stopped at subfreezing temperatures. The reason is that all metabolic reactions of microorganisms are enzyme catalyzed and that the rate of enzyme-catalyzed reactions is dependent on temperature. With a rise in temperature, there is an increase in reaction rate. The **temperature coefficient** (Q_{10}) may be generally defined as follows:

$$Q_{10} = \frac{(\text{Velocity at a given temp.} + 10°C)}{\text{Velocity at T}}$$

The Q_{10} for most biological systems is 1.5 to 2.5, so that for each 10°C rise in temperature within the suitable range, there is a twofold increase in the rate of reaction. For every 10°C decrease in temperature, the reverse is true. Since the basic feature of low-temperature food preservation consists of its effect on spoilage organisms, most of the discussion that follows will be devoted to the effect of low temperatures on foodborne microorganisms. It should be remembered, however, that temperature is related to relative humidity (R.H.) and that subfreezing temperatures affect R.H. as well as pH and possibly other parameters of microbial growth as well.

DEFINITIONS

The term **psychrophile** was coined by Schmidt-Nielsen in 1902 for micro-organisms that grow at 0°C (38). This term is now applied to organisms that grow over the range of subzero to 20°C, with an optimum range of 10° to 15°C (55). Around 1960, the term **psychrotroph** (*psychros*, "cold," and *trephein*, "to nourish" or "to develop") was suggested for organisms able to grow at 5°C or below (14, 58). It is now widely accepted among food microbiologists that a psychrotroph is an organism that can grow at temperatures between 0° and 7°C and produce visible colonies (or turbidity) within 7 to 10 days. Since some psychrotrophs can grow at temperatures at least as high as 43°C, they are in fact **mesophiles**. By these definitions, psychrophiles would be expected to occur only on products from oceanic waters or from extremely cold climes. The organisms that cause the spoilage of meats, poultry, and vegetables in the 0° to 5°C range would be expected to be psychrotrophs.

Since all psychrotrophs do not grow at the same rate over the 0° to 7°C range, the terms **eurypsychrotroph** (*eurys*, "wide" or "broad") and **stenopsychrotroph** (*stenos*, "narrow," "little," or "close") have been suggested. Eurypsychrotrophs typically do not form visible colonies until sometime between 6 and 10 days; stenopsychrotrophs typically form visible colonies in about 5 days (43). It has been suggested that psychrotrophs can be distinguished from nonpsychrotrophs by their inability to grow on a nonselective medium at 43°C in 24 h while the latter do (58). It has been shown that some bacteria that grow well at 7°C within 10 days also grow well at 43°C, and among these are *Enterobacter cloacae*, *Hafnia alvei*, and *Yersinia enterocolitica* (ATCC 27739) (43). These could be designated eurypsychrotrophs, although there are others that grow well at 43°C but only poorly at 7°C in 10 days. Typical of stenopsychrotrophs are *Pseudomonas fragi* (ATCC 4973) and *Aeromonas hydrophila* (ATCC 7965), which grow well at 7°C in 3 to 5 days and do not grow at 40°C (43).

There are three distinct temperature ranges for low-temperature stored foods. **Chilling temperatures** are those between the usual refrigerator (5–7°C) and ambient temperatures, usually about 10° to 15°C. These temperatures are suitable for the storage of certain vegetables and fruits such as cucumbers, potatoes, and limes. **Refrigerator temperatures** are those between 0° and 7°C (ideally no higher than 40°F). **Freezer temperatures** are those at or below −18°C. Under normal circumstances, growth of all microorganisms is prevented at freezer temperatures; nevertheless, some can and do grow within the freezer range but at an extremely slow rate.

TEMPERATURE GROWTH MINIMA

Bacterial species and strains that can grow at or below 7°C are rather widely distributed among the gram-negative and less so among gram-positive genera (Tables 13-1 and 13-2). The lowest recorded temperature of growth for a microorganism of concern in foods is −34°C, in this case a pink yeast. Growth at temperatures below 0°C is more likely to be that of yeasts and molds than bacteria. This is consistent with the growth of fungi under lower a_w conditions. Bacteria have been reported to grow at −20°C and around −12°C (54). Foods that are likely to support microbial growth at subzero temperatures include fruit juice concentrates, bacon, ice cream, and certain fruits. These products contain cryoprotectants that depress the freezing point of water.

PREPARATION OF FOODS
FOR FREEZING

The preparation of vegetables for freezing includes selecting, sorting, washing, blanching, and packaging prior to actual freezing. Foods in any state of detect-

TABLE 13-1.   Bacterial Genera Containing Species or Strains Known to Grow at or Below 7°C

Gram Negatives	Relative Numbers	Gram Positives	Relative Numbers
Acinetobacter	XX	*Bacillus*	XX
Aeromonas	XX	*Brevibacterium*	X
Alcaligenes	X	*Brochothrix*	XXX
Alteromonas	XX	*Carnobacterium*	XXX
Cedecea	X	*Clostridium*	XX
Chromobacterium	X	*Corynebacterium*	X
Citrobacter	X	*Deinococcus*	X
Enterobacter	XX	*Enterococcus*	XXX
Erwinia	XX	*Kurthia*	X
Escherichia	X	*Lactococcus*	XX
		Lactobacillus	XX
Flavobacterium	XX	*Leuconostoc*	X
Halobacterium	X	*Listeria*	XX
Hafnia	XX	*Micrococcus*	XX
Klebsiella	X	*Pediococcus*	X
Moraxella	XX	*Propionibacterium*	X
Morganella	X	*Vagococcus*	XX
Photobacterium	X		
Pantoea	XX		
Proteus	X		
Providencia	X		
Pseudomonas	XXX		
Psychrobacter	XX		
Salmonella	X		
Serratia	XX		
Shewanella	XXX		
Vibrio	XXX		
Yersinia	XX		

Note: Relative importance and dominance as psychrotrophs: X = minor; XX = intermediate; XXX = very significant.

able spoilage should be rejected for freezing. Meats, poultry, seafoods, eggs, and other foods should be as fresh as possible.

Blanching is achieved either by brief immersion of foods into hot water or the use of steam. Its primary functions are as follows:

- inactivation of enzymes that might cause undesirable changes during freezing storage
- enhancement or fixing of the green color of certain vegetables
- reduction in the numbers of microorganisms on the foods
- facilitating the packing of leafy vegetables by inducing wilting
- displacement of entrapped air in the plant tissues

The method of blanching employed depends on the products in question, size of packs, and other related information. When water is used, it is important that bacterial spores not be allowed to build up sufficiently to contaminate foods. Reductions of initial microbial loads as high as 99% have been claimed upon blanching. Remember that most vegetative bacterial cells can be destroyed

TABLE 13-2.   Minimum Reported Growth Temperatures of Some Foodborne Microbial Species and Strains That Grow at or Below 7°C

Species/Strains	°C	Comments
Pink yeast	−34	
Pink yeasts (2)	−18	
Unspecified molds	−12	
Vibrio spp.	−5	True psychrophiles
Yersinia enterocolitica	−2	
Unspecified coliforms	−2	
Brochothrix thermosphacta	−0.8	Within 7 days; 4°C in 10 days
Aeromonas hydrophila	−0.5	
Enterococcus spp.	0	Various species/strains
Leuconostoc carnosum	1.0	
L. gelidum	1.0	
Listeria monocytogenes	1.0	
Leuconostoc sp.	2.0	Within 12 days
L. sake/curvatus	2.0	Within 12 days; 4°C in 10 days
C. botulinum B, E, F	3.3	
Pantoea agglomerans	4.0	
Salmonella panama	4.0	In 4 weeks
Serratia liquefaciens	4.0	
Vibrio parahaemolyticus	5.0	
Salmonella heidelberg	5.3	
Pediococcus sp.	6.0	Weak growth in 8 days
Lactobacillus brevis	6.0	In 8 days
L. viridescens	6.0	In 8 days
Salmonella typhimurium	6.2	
Staphylococcus aureus	6.7	
Klebsiella pneumoniae	7.0	
Bacillus spp.	7.0	165 of 520 species/strains
Salmonella spp.	7.0	65 of 109, within 4 weeks

Sources: Bonde (7), Mossel et al. (57), Reuter (65).

at milk pasteurization temperatures (145°F for 30 min). This is especially true of most bacteria of importance in the spoilage of vegetables. Although it is not the primary function of blanching to destroy microorganisms, the amount of heat necessary to effect destruction of most food enzymes is also sufficient to reduce vegetative cells significantly.

FREEZING OF FOODS AND FREEZING EFFECTS

The two basic ways to achieve the freezing of foods are quick and slow freezing. **Quick** or **fast freezing** is the process by which the temperature of foods is lowered to about −20°C within 30 min. This treatment may be achieved by direct immersion or indirect contact of foods with the refrigerant and the use of air-blasts of frigid air blown across the foods being frozen.

Slow freezing refers to the process whereby the desired temperature is achieved within 3 to 72 h. This is essentially the type of freezing utilized in the home freezer.

Quick freezing possesses more advantages than slow freezing from the standpoint of overall product quality. The two methods are compared below (8):

Quick Freezing
1. Small ice crystals formed
2. Blocks or suppresses metabolism
3. Brief exposure to concentration of adverse constituents
4. No adaptation to low temperatures
5. Thermal shock (too brutal a transition)
6. No protective effect

7. Microorganisms frozen into crystals?
8. Avoid internal metabolic imbalance

Slow Freezing
1. Large ice crystals formed
2. Breakdown of metabolic rapport
3. Longer exposure to adverse or injurious factors
4. Gradual adaptation

5. No shock effect

6. Accumulation of concentrated solutes with beneficial effects

With respect to crystal formation upon freezing, slow freezing favors large extracellular crystals, and quick freezing favors the formation of small intracellular ice crystals. Crystal growth is one of the factors that limits the freezer life of certain foods, since ice crystals grow in size and cause cell damage by disrupting membranes, cell walls, and internal structures to the point where the thawed product is quite unlike the original in texture and flavor. Upon thawing, foods frozen by the slow freezing method tend to lose more **drip** (drip for meats; **leakage** in the case of vegetables) than quick-frozen foods held for comparable periods of time. The overall advantages of small crystal formation to frozen food quality may be viewed also from the standpoint of what takes place when a food is frozen. During the freezing of foods, water is removed from solution and transformed into ice crystals of a variable but high degree of purity (19). In addition, the freezing of foods is accompanied by changes in properties such as pH, titratable acidity, ionic strength, viscosity, osmotic pressure, vapor pressure, freezing point, surface and interfacial tension, and oxidation-reduction (O/R) potential (see below). Some of the many complexities of this process are discussed by Fennema et al. (20).

STORAGE STABILITY OF FROZEN FOODS

A large number of microorganisms have been reported by many investigators to grow at and below 0°C. In addition to factors inherent within these organisms, their growth at and below freezing temperatures is dependent on nutrient content, pH, and the availability of liquid water. The a_w of foods may be expected to decrease as temperatures fall below the freezing point. The relationship between temperature and a_w of water and ice is presented in Table 13-3. For water at 0°C, a_w is 1.0 but falls to about 0.8 at −20°C and to 0.62 at about −50°C. Organisms that grow at subfreezing temperatures, then, must be able to grow at the reduced a_w levels, unless a_w is favorably affected by food

TABLE 13-3.   Vapor Pressures of Water and Ice at Various Temperatures

°C	Liquid Water (mm. Hg)	Ice (mm. Hg)	$a_w = \dfrac{P_{ice}}{P_{water}}$
0	4.579	4.579	1.00
−5	3.163	3.013	0.953
−10	2.149	1.950	0.907
−15	1.436	1.241	0.864
−20	0.943	0.776	0.823
−25	0.607	0.476	0.784
−30	0.383	0.286	0.75
−40	0.142	0.097	0.68
−50	0.048	0.030	0.62

Source: Scott (71).

constituents with respect to microbial growth. In fruit juice concentrates, which contain comparatively high levels of sugars, these compounds tend to maintain a_w at levels higher than would be expected in pure water, thereby making microbial growth possible even at subfreezing temperatures. The same type of effect can be achieved by the addition of glycerol to culture media. Not all foods freeze at the same initial point (Fig. 13-1). The initial freezing point of a given food is due in large part to the nature of its solute constituents and the relative concentration of those that have freezing-point depressing properties.

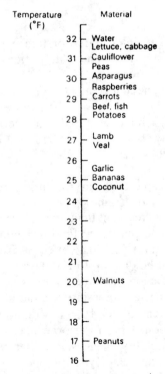

FIGURE 13-1.   Freezing point of selected foods.
From Desrosier (12).

Although the metabolic activities of all microorganisms can be stopped at freezer temperatures, frozen foods may not be kept indefinitely if the thawed product is to retain the original flavor and texture. Most frozen foods are assigned a freezer life. The suggested maximum holding time for frozen foods is not based on the microbiology of such foods but on such factors as texture, flavor, tenderness, color, and overall nutritional quality upon thawing and subsequent cooking.

Some foods that are improperly wrapped during freezer storage undergo **freezer burn**, characterized by a browning of light-colored foods such as the skin of chicken meat. The browning results from the loss of moisture at the surface, leaving the product more porous than the original at the affected site. The condition is irreversible and is known to affect certain fruits, poultry, meats, and fish, both raw and cooked.

EFFECT OF FREEZING ON MICROORGANISMS

In considering the effect of freezing on those microorganisms that are unable to grow at freezing temperatures, it is well known that freezing is one means of preserving microbial cultures, with freeze-drying being perhaps the best method known. However, freezing temperatures have been shown to effect the killing of certain microorganisms of importance in foods. Ingram (39) summarized the salient facts of what happens to certain microorganisms upon freezing:

1. There is a sudden mortality immediately on freezing, varying with species.
2. The proportion of cells surviving immediately after freezing is nearly independent of the rate of freezing.
3. The cells that are still viable immediately after freezing die gradually when stored in the frozen state.
4. This decline in numbers is relatively rapid at temperatures just below the freezing point, especially about $-2°C$, but less so at lower temperatures, and it is usually slow below $-20°C$.

Bacteria differ in their capacity to survive during freezing, with cocci being generally more resistant than gram-negative rods. Of the food-poisoning bacteria, salmonellae are less resistant than *Staphylococcus aureus* or vegetative cells of clostridia, while endospores and food-poisoning toxins are apparently unaffected by low temperatures (24). The effect of freezing several species of *Salmonella* to $-25.5°C$ and holding up to 270 days is presented in Table 13-4. Although a significant reduction in viable numbers occurred over the 270-day storage period with most species, in no instance did all cells die off.

From the strict standpoint of food preservation, freezing should not be regarded as a means of destroying foodborne microorganisms. The type of organisms that lose their viability in this state differ from strain to strain and depend on the type of freezing employed, the nature and composition of the food in question, the length of time of freezer storage, and other factors, such as temperature of freezing. Low freezing temperatures of about $-20°C$ are less

TABLE 13-4.   Survival of Pure Cultures of Enteric Organisms in Chicken Chow Mein at −25.5°C

Organism	Bacterial Count (10^5/g) after Storage for (days)								
	0	2	5	9	14	28	50	92	270
Salmonella newington	7.5	56.0	27.0	21.7	11.1	11.1	3.2	5.0	2.2
S. typhimurium	167.0	245.0	134.0	118.0	11.0	95.5	31.0	90.0	34.0
S. typhi	128.5	45.5	21.8	17.3	10.6	4.5	2.6	2.3	0.86
S. gallinarum	68.5	87.0	45.0	36.5	29.0	17.9	14.9	8.3	4.8
S. anatum	100.0	79.0	55.0	52.5	33.5	29.4	22.6	16.2	4.2
S. paratyphi B	23.0	205.0	118.0	93.0	92.0	42.8	24.3	38.8	19.0

Source: Gunderson and Rose (27); copyright © 1948 by Institute of Food Technologists.

harmful to microorganisms than the median range of temperatures such as −10°C. For example, more microorganisms are destroyed at −4°C than at −15°C or below. Temperatures below −24°C seem to have no additional effect. Food constituents such as egg white, sucrose, corn syrup, fish, glycerol, and undenatured meat extracts have all been reported to increase freezing viability, especially of food-poisoning bacteria, while acid conditions have been reported to decrease cell viability (24).

Consider some of the events that are known to occur when cells freeze:

- The water that freezes is the so-called **free water**. Upon freezing, the free water forms ice crystals. The growth of ice crystals occurs by accretion so that all of the free water of a cell might be represented by a relatively small number of ice crystals. In slow freezing, ice crystals are extracellular; in fast freezing, they are intracellular. Bound water remains unfrozen. The freezing of cells depletes them of usable liquid water and thus dehydrates them.
- Freezing results in an increase in the viscosity of cellular matter, a direct consequence of water being concentrated in the form of ice crystals.
- Freezing results in a loss of cytoplasmic gases such as O_2 and CO_2. A loss of O_2 to aerobic cells suppresses respiratory reactions. Also, the more diffuse state of O_2 may make for greater oxidative activities within the cell.
- Freezing causes changes in pH of cellular matter. Various authors have reported changes ranging from 0.3 to 2.0 pH units. Increases and decreases of pH upon freezing and thawing have been reported.
- Freezing effects a concentration of cellular electrolytes. This effect is also a consequence of the concentration of water in the form of ice crystals.
- Freezing causes a general alteration of the colloidal state of cellular protoplasm. Many of the constituents of cellular protoplasm such as proteins exist in a dynamic colloidal state in living cells. A proper amount of water is necessary to the well-being of this state.
- Freezing causes some denaturation of cellular proteins. Precisely how this effect is achieved is not clear, but it is known that upon freezing, some —SH groups disappear and such groups as lipoproteins break apart from others. The lowered water content, along with the concentration of electrolytes, no doubt affect this change in state of cellular proteins.
- Freezing induces temperature shock in some microorganisms. This is true more for thermophiles and mesophiles than psychrophiles. More cells die

when the temperature decline above freezing is sudden than when it is slow.
- Freezing causes metabolic injury to some microbial cells such as certain *Pseudomonas* spp. Some bacteria have increased nutritional requirements upon thawing from the frozen state and as much as 40% of a culture may be affected in this way (see Chap. 5 for other effects of freeze injury).

Clearly the effects of the freezing process upon living cells such as bacteria and other microorganisms as well as upon foods are complex. According to Mazur (51), the response of microorganisms to subzero temperatures appears to be largely determined by solute concentration and intracellular freezing, although there are only a few cases of clear demonstration of this conclusion.

Why are some bacteria killed by freezing but not all cells? Some small and microscopic organisms are unable to survive freezing as can most bacteria. Examples include the foot-and-mouth disease virus and the causative agent of trichinosis (*Trichinella spiralis*). Protozoa are generally killed when frozen below −5° or −10°C, if protective compounds are not present (51).

Effect of Thawing

Of great importance in the freezing survival of microorganisms is the process of thawing. Repeated freezing and thawing will destroy bacteria by disrupting cell membranes. Also, the faster the thaw, the greater the number of bacterial survivors. Why this is so is not entirely clear. From the changes listed that occur during freezing, it can be seen that the thawing process becomes complicated if it is to lead to the restoration of viable activity. It has been pointed out that thawing is inherently slower than freezing and follows a pattern that is potentially more detrimental. Among the problems attendant on the thawing of speciments and products that transmit heat energy primarily by conduction are the following (20):

- Thawing is inherently slower than freezing when conducted under comparable temperature differentials.
- In practice, the maximum temperature differential permissible during thawing is much less than that which is feasible during freezing.
- The time-temperature pattern characteristic of thawing is potentially more detrimental than that of freezing. During thawing, the temperature rises rapidly to near the melting point and remains there throughout the long course of thawing, thus affording considerable opportunity for chemical reactions, recrystallization, and even microbial growth if thawing is extremely slow.

It has been stated that microorganisms die not upon freezing but, rather, during the thawing process. Whether this is the case remains to be proved. As to why some organisms are able to survive freezing while others are not, Luyet (48) suggested that it is a question of the ability of an organism to survive dehydration and to undergo dehydration when the medium freezes. Luyet further stated that the small size of bacterial cells permits them to undergo dehydration upon freezing. With respect to survival after freeze-drying, Luyet has stated that it might be due to the fact that bacteria do not freeze at all but merely dry. (See Chap. 15 for further discussion of the effect of freeze-drying on microorganisms.)

Most frozen-foods processors advise against the refreezing of foods once they have thawed. Although the reasons are more related to the texture, flavor, and other nutritional qualities of the frozen product, the microbiology of thawed frozen foods is pertinent. Some investigators have pointed out that foods thawed from the frozen state spoil faster than similar fresh products. There are textural changes associated with freezing that would seem to aid the invasion of surface organisms into deeper parts of the product and consequently facilitate the spoilage process. Upon thawing, surface condensation of water is known to occur. There is also, at the surface, a general concentration of water-soluble substances such as amino acids, minerals, B vitamins, and possibly other nutrients. Freezing has the effect of destroying many thermophilic and some mesophilic organisms, making for less competition among the survivors upon thawing. It is conceivable that a greater relative number of psychrotrophs on thawed foods might increase the spoilage rate. Some psychrotrophic bacteria have been reported to have Q_{10} values in excess of 4.0 at refrigerator temperatures. For example *P. fragi* has been reported to possess a Q_{10} of 4.3 at 0°C. Organisms of this type are capable of doubling their growth rate with only a 4 to 5 degree rise in temperature. Whether frozen thawed foods do in fact spoil faster than fresh foods would depend on a large number of factors, such as the type of freezing, the relative numbers and types of organisms on the product prior to freezing, and the temperature at which the product is held to thaw. Although there are no known toxic effects associated with the refreezing of frozen and thawed foods, this act should be minimized in the interest of the overall nutritional quality of the products. One effect of freezing and thawing animal tissues is the release of lysosomal enzymes consisting of cathepsins, nucleases, phosphatases, glycosidases, and others. Once released, these enzymes may act to degrade macromolecules and thus make available simpler compounds that are more readily utilized by the spoilage flora.

More information on the microbiology of frozen foods can be obtained from Robinson (67).

SOME CHARACTERISTICS OF PSYCHROTROPHS AND PSYCHROPHILES

There is an increase in unsaturated fatty acid residues. The usual lipid content of most bacteria is between 2 and 5%, most or all of which is in the cell membrane. Bacterial fats are glycerol esters of two types: neutral lipids, in which all three or only one or two of the −OH groups of glycerol are esterified with long-chain fatty acids, and phospholipids, in which one of the −OH groups is linked through a phosphodiester bond to choline, ethanolamine, glycerol, inositol, or serine. The other two −OH groups are esterified with long-chain fatty acids (68).

Many psychrotrophs synthesize neutral lipids and phospholipids containing an increased proportion of unsaturated fatty acids when grown at low tem-

TABLE 13-5.   Effects of Incubation Temperature on the Fatty Acid Composition of Stationary Cultures of *Candida utilis*

Incubation Temperature (°C)	Cell Concentration (mg/ml)	Fatty Acid Composition[a]				
		16:0	16:1	18:1	18:2	18:3
30	2.0	18.9	4.6	39.1	34.3	2.1
20	2.0	20.3	11.4	31.6	27.7	6.1
10	2.0	27.4	20.6	20.7	17.6	10.7
5	1.7	19.2	15.9	18.2	16.3	27.3

Source: McMurrough and Rose (53); copyright © 1973, American Society for Microbiology.

[a] Values quoted are expressed as percentages of the total fatty acids. Fatty acids are designated *x:y*, where *x* is the number of carbon atoms and *y* is the number of double bonds per molecule.

peratures compared with growth at higher temperatures. As much as a 50% increase in content of unsaturated bonds of fatty acids from mesophilic and psychrotrophic *Candida* spp. was found in cells grown at 10°C compared to 25°C (45). The phospholipid composition of these yeasts was unchanged. The increase in unsaturated fatty acids in *Candida utilis* as growth temperatures were lowered from 30° to 5°C is shown in Table 13-5, Linolenic acid increased at the expense of oleic acid at the lower temperatures.

In a comparative study of four *Vibrio* spp. that grew over the range −5° to 15°C and four *Pseudomonas* spp. that grew over the range of 0° to 25° or 27°C, significant changes were observed in total phospholipids of vibrios as growth temperatures were lowered from 15° to −5°C but not among the pseudomonads over their range of growth (5, 6, 33). A change from saturated to unsaturated lipids would not be expected to occur in pseudomonads as growth temperatures are lowered since the psychrotrophic strains contain between 59 and 72% unsaturated lipids, making them more versatile than many other organisms. In contrast to most other psychrotrophs, *Micrococcus cryophilus* undergoes chain shortening in response to low temperatures, which apparently decreases the melting point of its membrane lipids (69).

The widespread occurrence of low-temperature-induced changes in fatty acid composition suggests that they are associated with physiological mechanisms of the cell. It is known that an increase in the degree of unsaturation of fatty acids in lipids leads to a decrease in lipid melting point. It has been suggested that increased synthesis of unsaturated fatty acids at low temperatures has the function of maintaining the lipid in a liquid and mobile state, thereby allowing membrane activity to continue to function. This concept, referred to as the "**lipid solidification**" theory, was first proposed by Gaughran (23) and Allen (3). It has been shown by Byrne and Chapman (11) that the melting point of fatty-acid side chains in lipids is more important than the entire lipid structure.

Although full support for the lipid solidification idea is wanting, there is circumstantial evidence available such as the phenomenon of "**cold shock**," which is the dying off of many cells of mesophilic bacteria upon the sudden chilling of a suspension of viable cells grown at mesophilic temperatures. It has been shown for a large number of gram-negative bacteria including *Escherichia coli* and is generally a property of gram-negative bacteria and not of gram positives. Cold shock has been shown to be accompanied by the release of certain low-molecular-weight cell constituents, an effect that presumably occurs by virtue of damage to the plasma membrane. According to Rose (68), cold shock seems to result from a sudden release of cell constituents from bacteria following the "freezing" of certain membrane lipids after sudden chilling, with consequent development of "holes" in the membrane. To support this hypothesis, Farrell and Rose (18) grew a mesophilic strain of *Pseudomonas aeruginosa* at 30°C and showed that the cells were susceptible to cold shock while the same strain grown at 10°C was not susceptible. It has been proposed that the growth temperature range of an organism is dependent on the ability of the organism to regulate its lipid fluidity within a given range (21).

Psychrotrophs synthesize high levels of polysaccharides. Well-known examples of this effect include the production of ropy milk and ropy bread dough, both of which are favored by low temperatures. The production of extracellular dextrans by *Leuconostoc* and *Pediococcus* spp. are known to be favored at temperatures below the growth optima of these organisms. The greater production of dextran at lower temperatures is due apparently to the fact that dextransucrase is very rapidly inactivated at temperatures in excess of 30°C (61). A temperature-sensitive dextransucrase synthesizing system has been shown also for a *Lactobacillus* sp. (13).

From a practical standpoint, increased polysaccharide synthesis at low temperatures manifests itself in the characteristic appearance of low-temperature spoiled meats. Slime formation is characteristic of the bacterial spoilage of frankfurters, fresh poultry, and ground beef (Chap. 9). The coalescence of surface colonies leads to the sliminess of such meats and no doubt contributes to the increased hydration capacity that accompanies low-temperature meat spoilage.

Pigment production is favored. This effect appears to be confined to those organisms that synthesize phenazine and carotenoid pigments. The best-documented example of this phenomenon involves pigment production by *Serratia marcescens*. The organism produces an abnormally heat-sensitive enzyme that catalyzes the coupling of a monopyrrole and bipyrrole precursor to give prodigiosin (the red pigment) (86). The increased production of pigments at suboptimum temperatures has been reported by others (80, 86). It is interesting that a very large number of marine psychrotrophs (and perhaps psychrophiles) are pigmented. This is true for bacteria as well as yeasts. On the other hand, none of the more commonly studied thermophiles is pigmented.

Some strains display differential substrate utilization. It has been reported that sugar fermentation at temperatures below 30°C gives rise to both acid and gas, while above 30°C only acid is produced (26). Similarly, others have found psychrotrophs that fermented glucose and other sugars with the formation of acid and gas at 20°C and lower but produced only acid at higher temperatures (82). The latter was ascribed to a temperature-sensitive formic hydrogenase system. These investigators studied a similar effect and attributed the difference to a temperature-sensitive hydrogenase synthesizing system of the cell. Beef spoilage bacteria have been shown to liquefy gelatin and utilize water-soluble beef proteins more at 5°C than at 30°C (42), but whether this effect is due to temperature-sensitive enzymes is not clear.

THE EFFECT OF LOW TEMPERATURES ON MICROBIAL PHYSIOLOGIC MECHANISMS

Of the effects that low incubation temperatures have on the growth and activity of foodborne microorganisms, five have received the most attention and are outlined below.

Psychrotrophs have a slower metabolic rate. The precise reasons that metabolic rates are slowed at low temperatures are not fully understood. Psychrotrophic growth decreases more slowly than that of mesophilic with decreasing temperatures. The temperature coefficients (Q_{10}) for various substrates such as acetate and glucose have been shown by several investigators to be lower for growing psychrotrophs than for mesophiles. The end products of mesophilic and psychrotrophic metabolism of glucose were shown to be the same, with the differences largely disappearing when the cells were broken (36). In other words, the temperature coefficients are about the same for psychrotrophs and mesophiles when cell-free extracts are employed.

As temperature is decreased, the rate of protein synthesis is known to decrease, and this occurs in the absence of changes in the amount of cellular DNA. One reason may be the increase in intramolecular hydrogen bonding that occurs at low temperatures, leading to increased folding of enzymes with losses in catalytic activity (46). On the other hand, the decrease in protein synthesis appears to be related to a decreased synthesis of individual enzymes at low growth temperatures. Although the precise mechanism of reduced protein synthesis is not well understood, it has been suggested that low temperatures affect the synthesis of a repressor protein (50), and that the repressor protein itself is thermolabile (79). Several investigators have suggested that low temperatures may influence the fidelity of the translation of mRNA during protein synthesis. For example, in studies with *E. coli*, it was shown that a leucine-starved auxotroph of this mesophile incorporated radioactive leucine into protein at 0°C (25). It was suggested that at this temperature, all essential steps in protein synthesis apparently go on and involve a wide variety of proteins. The rate of synthesis at 0°C was estimated to be about 350 times slower than at 37°C for this organism. It has been suggested that the cessation of RNA synthesis in general may be the controlling factor in determining low temperature growth (32), and the lack of polysome formation in *E. coli* when shifted to a temperature below its growth minimum has been demonstrated. The formation of polysomes is thus sensitive to low temperatures (at least in some organisms), and protein synthesis would be adversely affected (see 41).

Whatever the specific mechanism of lowered metabolic activity of microorganisms as growth temperature is decreased, psychrotrophs growing at low temperature have been shown to possess good enzymatic activity, since motility, endospore formation, and endospore germination all occur at 0°C (75). *P. fragi*, among other organisms, produces lipases within 2 to 4 days at −7°C, within 7 days at −18°C, and within 3 weeks at −29°C (2). The minimum growth temperature may be determined by the structure of the enzymes and cell membrane, as well as by enzyme synthesis (75). The lack of production of enzymes at high temperatures by psychrotrophs on the other hand is due apparently to the inactive nature of enzyme-synthesizing reactions rather than to enzyme inactivation (75), although the latter is known to occur (see below). With respect to individual groups of enzymes, yields of endocellular proteolytic enzymes are greater in *Pseudomonas fluorescens* grown at 10° than at either 20° or 35°C, (63), while other investigators have shown that *P. fragi* preferentially produces lipase at low temperatures, with none being produced at 30°C or higher (59, 60). *P. fluorescens* has been found to produce just as much lipase at 5°C as at 20°C, but only a slight amount was produced at 30°C (1). On the other hand, a proteolytic enzyme system of *P. fluorescens* showed more activity on egg white and hemoglobin at 25°C than at 15° and 5°C (34).

It has been suggested that there are preformed elements in microbial cells grown at any temperature that are selectively temperature sensitive (44). Microorganisms may cease to grow at a certain low temperature because of excessive sensitivity in one or several control mechanisms, the effectors of which cannot be supplied in the growth medium (37). According to the latter authors, the interaction between effector molecules and the corresponding allosteric proteins may be expected to be a strong function of temperature.

Psychrotroph membranes transport solutes more efficiently. It has been shown in several studies that upon lowering the growth temperature of mesophiles within the psychrotrophic range, solute uptake is decreased. Studies by Baxter and Gibbons (4) indicate that minimum growth temperature of mesophiles is determined by the temperature at which transport permeases are inactivated. Farrell and Rose (17) offered three basic mechanisms by which low temperature could affect solute uptake: (1) inactivation of individual permease-proteins at low temperature as a result of low-temperature-induced conformational changes that have been shown to occur in some proteins, (2) changes in the molecular architecture of the cytoplasmic membrane that prevent permease action, and (3) a shortage of energy required for the active transport of solutes. Although the precise mechanisms of reduced uptake of solutes at low temperatures are not clear at this time, the second mechanism seems the most likely (17).

From studies of four psychrophilic vibrios, maximum uptake of glucose and lactose occurred at 0°C and decreased when temperatures were raised to 15°C while with four psychrotrophic pseudomonads, maximum uptake of these substrates occurred in the 15° to 20°C range and decreased as temperatures were reduced to 0°C (33). The vibrios showed significant changes in total phospholipids at 0°C, while no meaningful changes occurred with pseudomonads as their growth temperature was lowered. In studies with *Listeria monocytogenes* at 10°C, metabolism at low temperatures was believed to be the result of a cold-resistant sugar transport system that provided high concentrations of intracellular substrates (85). The latter authors noted that a cold-resistant sugar transport system is the property most readily identified as a fitness trait for psychrotrophy and that it applies not only to *Listeria monocytogenes* but also to *Erysipelothrix rhusiopathiae* and *Brochothrix thermosphacta* (84). It has been suggested that the minimum growth temperature of an organism may be defined by the inhibition of substrate uptake.

Psychrotrophs tend to possess in their membrane lipids that enable the membrane to be more fluid. The greater mobility of the psychrotrophic membrane may be expected to facilitate membrane transport at low temperatures. In addition, the transport permeases of psychrotrophs are apparently more operative under these conditions than are those of other mesophiles. Whatever the specific mechanism of increased transport might be, it has been demonstrated that psychrotrophs are more efficient than other mesophiles in the uptake of solutes at low temperatures. Baxter and Gibbons (4) showed that a psychrotrophic *Candida* sp. incorporated glucosamine more rapidly than a mesophilic *Candida*. The psychrotroph transported glucosamine at 0°C, while scarcely any was transported by the mesophile at this temperature or even at 10°C.

Some psychrotrophs produce larger cells. Yeasts and molds have been found to produce larger cell sizes when growing under psychrotrophic conditions than when growing under mesophilic. With *C. utilis*, the increased cell size was believed to be due to increases in RNA and protein content of cells (68). Low-temperature synthesis of additional RNA has been reported by others (9, 31, 78), and at least one group found no increase in the amount of RNA at 2°C when *Pseudomonas* strain 92 cells were grown at 2° and 30°C under the same conditions (22). The latter authors found no increase in cell size, protein content, or catalase activity. On the other hand, psychrotrophic organisms are generally regarded as having higher levels of both RNA and proteins (33).

Flagella synthesis is more efficient. Examples of the more efficient production of flagella at low temperatures include *E. coli*, *Bacillus inconstans*, *Salmonella paratyphi* B, and other organisms, including some psychrophiles.

Pyschrotrophs are favorably affected by aeration. The effect of aeration on the generation time of *P. fluorescens* at temperatures from 4° to 32°C, employing three different carbon sources, is presented in Table 13-6. The greatest effect of aeration (shaking) occurred at 4° and 10°C, while at 32°C aerated cultures produced a longer generation time (62). The significance of this effect is not clear. In a study of facultatively anaerobic psychrotrophs under anaerobic conditions, the organisms were shown to grow more slowly, survive longer, die more rapidly at higher temperatures, and produce lower maximal cell yields under anaerobic conditions than under aerobic (81). It has been commonly observed that plate counts on many foods are higher with incubation at low temperatures than at temperatures of 30°C and above. The generally higher counts are due in part to the increased solubility and consequently, availability of O_2 (72). The latter authors found that equally high cell yields can be obtained at both low and high incubation temperatures when O_2 is not limiting. This greater availability of O_2 in refrigerated foods undoubtedly exerts selectivity on the spoilage flora of such foods. The vast majority of psychrotrophic bacteria studied are aerobes or facultative anaerobes, and these are the types associated with the spoilage of foods stored at refrigerator temperatures. Relatively few anaerobic psychrotrophs have been isolated and studied. One of the first was *Clostridium putrefaciens* (52). Other investigators have isolated and studied psychrotrophic clostridia (66, 73).

NATURE OF THE LOW HEAT RESISTANCE OF PSYCHROTROPHS

It has been known for years that psychrotrophic microorganisms are generally unable to grow much above 30° to 35°C. Among the first to suggest reasons for this limitation of growth were Edwards and Rettger (15), who concluded that

TABLE 13-6.   Effect of Growth Temperature, Carbon Source, and Aeration on Generation Times (h) of *Pseudomonas fluorescens*

Growth Medium[a]	Culture	Growth Temperature					
		4°C	10°C	15°C	20°C	25°C	32°C
Glucose	Stationary	8.20	3.52	2.02	1.47	0.97	1.19
	Aerated	5.54	2.61	2.00	1.46	0.93	1.51
Citrate	Stationary	8.20	3.46	2.00	1.43	1.01	1.24
	Aerated	6.68	2.95	2.02	1.26	0.98	1.45
Casamino acids	Stationary	7.55	3.06	1.78	1.36	1.12	0.95
	Aerated	4.17	2.57	1.56	1.12	0.87	1.10

Source: Olsen and Jezeski (62).

[a] Basal salts + 0.02% yeast extract + the carbon source indicated.

TABLE 13-7.   Some Heat-Labile Enzymes of Psychrotrophic Microorganisms

Enzyme	Organism	Temperature of Maximum Growth, (°C)	Temperature of Enzyme Inactivation, (°C)	Reference
Extracellular lipase[a]	*Ps. fragi*		30	60
α-oxoglutarate synthesizing enzymes and others	*Cryptococcus*	~28	30	29
Alcohol dehydrogenase	*Candida* sp.	<30		4
Formic hydrogenlyase	Psychrophile 82	35	45	82
Hydrogenase	Psychrophile 82	35	>20	83
Malic dehydrogenase	Marine *Vibrio*	30	30	10
Pyruvate dehydrogenase	*Candida* sp.	~20	25	16
Isocitrate dehydrogenase	*Arthrobacter* sp.	~35	37	16
Fermentative enzymes	*Candida* sp. P16	~25	35	74
Reduced NAD oxidase	Psychrophile 82	35	46	64
Cytochrome c reductase	Psychrophile 82	35	46	64
Lactic and glycerol dehydrogenase	Psychrophile 82	35	46	64
Pyruvate clastic enzymes	Psychrophile 82	35	46	64
Protein and RNA synthesizing	*Micrococcus cryophilus*	25	30	49

[a] Enzyme-forming system inactivated.

the maximum growth temperatures of bacteria may bear a definite relationship to the minimum temperatures of destruction of respiratory enzymes. Their conclusion has been borne out by results from a large number of investigators. It has been shown that many respiratory enzymes are inactivated at the temperatures of maximum growth of various psychrotrophic types (Table 13-7). Thus, the thermal sensitivity of certain enzymes of psychrotrophs is at least one of the factors that limit the growth of these organisms to low temperatures.

When some psychrotrophs are subjected to temperatures above their growth maxima, cell death is accompanied by the leakage of various intracellular constituents (28, 30, 77). The leakage substances have been shown to consist of proteins, DNA, RNA, free amino acids, and lipid phosphorus. The last was thought by Hagen et al. to represent phosphorus of the cytoplasmic membrane. While the specific reasons for the release of cell constituents are not fully understood, it would appear to involve rupture of the cell membrane. These events appear to follow those of enzyme inactivation.

Whatever is the true mechanism of psychrotroph death at temperatures a few degrees above their growth maxima, their destruction at these relatively low temperatures is characteristic of this group of organisms. This is especially true of those that have optimum growth temperatures at and below 20°C. Reports by several investigators on psychrotrophs isolated and studied over the past two decades reveal that all are capable of growing at 0°C with growth optima at either 15° or between 20° and 25°C and growth maxima between 20° and 35°C (28, 30, 47, 73, 74, 76). Included among these organisms are gram-negative rods, gram-positive aerobic and anaerobic rods, sporeformers and nonsporeformers, gram-positive cocci, vibrios, and yeasts. One of these, *Vibrio fisheri* (*marinus*), was shown by Morita and Albright (56) to have an optimum growth temperature at 15°C and a generation time of 80.7 min at this temperature. In almost all cases, the growth maxima of these organisms were only 5 to 10 degrees above the growth optima.

References

1. Alford, J. A., and L. E. Elliott. 1960. Lipolytic activity of microorganisms at low and intermediate temperatures. I. Action of *Pseudomonas fluorescens* on lard. *Food Res.* 25:296–303.
2. Alford, J. A., and D. A. Pierce. 1961. Lipolytic activity of microorganisms at low and intermediate temperatures. III. Activity of microbial lipases at temperatures below 0°C. *J. Food Sci.* 26:518–524.
3. Allen, M. B. 1953. The thermophilic aerobic sporeforming bacteria. *Bacteriol. Rev.* 17:125–173.
4. Baxter, R. M., and N. E. Gibbons. 1962. Observations on the physiology of psychrophilism in a yeast. *Can. J. Microbiol.* 8:511–517.
5. Bhakoo, M., and R. A. Herbert. 1979. The effect of temperature on psychrophilic *Vibrio* spp. *Arch. Microbiol.* 121:121–127.
6. Bhakoo, M., and R. A. Herbert. 1980. Fatty acid and phospholipid composition of five psychrotrophic *Pseudomonas* spp. grown at different temperatures. *Arch. Microbiol.* 126:51–55.
7. Bonde, G. J. 1981. Phenetic affiliation of psychrotrophic *Bacillus*. In *Psychrotrophic Microorganisms in Spoilage and Pathogenicity*, ed. T. A. Roberts, G. Hobbs, J. H. B. Christian, and N. Skovgaard, 39–54. New York: Academic Press.
8. Borgstrom, G. 1961. Unsolved problems in frozen food microbiology. In *Proceedings, Low Temperature Microbiology Symposium*, 197–251. Camden, N.J.: Campbell Soup Co.
9. Brown, C. M., and A. H. Rose. 1969. Effects of temperature on composition and cell volume of *Candida utilis*. *J. Bacteriol.* 97:261–272.
10. Burton, S. D., and R. Y. Mortia. 1963. Denaturation and renaturation of malic dehydrogenase in a cell-free extract from a marine psychrophile. *J. Bacteriol.* 86:1019–1024.
11. Byrne, P., and D. Chapman. 1964. Liquid crystalline nature of phospholipids. *Nature* 202:987–988.
12. Desrosier, N. W. 1963. *The Technology of Food Preservation*. Westport, Conn.: AVI.

13. Dunican, L. K., and H. W. Seeley. 1963. Temperature-sensitive dextransucrase synthesis by a lactobacillus. *J. Bacteriol.* 86:1079–1083.
14. Eddy, B. P. 1960. The use and meaning of the term "psychrophilic." *J. Appl. Bacteriol.* 23:189–190.
15. Edwards, O. F., and L. F. Rettger. 1937. The relation of certain respiratory enzymes to the maximum growth temperatures of bacteria. *J. Bacteriol.* 34: 489–515.
16. Evison, L. M., and A. H. Rose. 1965. A comparative study on the biochemical bases of the maximum temperatures for growth of three psychrophilic micro-organisms. *J. Gen. Microbiol.* 40:349–364.
17. Farrell, J., and A. Rose. 1967. Temperature effects on micro-organisms. *Ann. Rev. Microbiol.* 21:101–120.
18. Farrell, J., and A. Rose. 1968. Cold shock in mesophilic and psychrophilic pseudomonads. *J. Gen. Microbiol.* 50:429–439.
19. Fennema, O., and W. Powrie. 1964. Fundamentals of low-temperature food preservation. *Adv. Food Res.* 13:219–347.
20. Fennema, O. R., W. D. Powrie, and E. H. Marth. 1973. *Low-Temperature Preservation of foods and Living Matter.* New York: Marcel Dekker.
21. Finne, G., and J. R. Matches. 1976. Spin-labeling studies on the lipids of psychrophilic, psychrotrophic, and mesophilic clostridia. *J. Bacteriol.* 125:211–219.
22. Frank, H. A., A. Reid, L. M. Santo, N. A. Lum, and S. T. Sandler. 1972. Similarity in several properties of psychrophilic bacteria grown at low and moderate temperatures. *Appl. Microbiol.* 24:571–574.
23. Gaughran, E. R. L. 1947. The thermophilic micro-organisms. *Bacteriol. Rev.* 11:189–225.
24. Georgala, D. L., and A. Hurst. 1963. The survival of food poisoning bacteria in frozen foods. *J. Appl. Bacteriol.* 26:346–358.
25. Goldstein, A., D. B. Goldstein, and L. I. Lowney. 1964. Protein synthesis at 0°C in *Escherichia coli*. *J. Mol. Biol.* 9:213–235.
26. Greene, V. W., and J. J. Jezeski. 1954. The influence of temperature on the development of several psychrophilic bacteria of dairy origin. *Appl. Microbiol.* 2:110–117.
27. Gunderson, M. F., and K. D. Rose. 1948. Survival of bacteria in a precooked fresh-frozen food. *Food Res.* 13:254–263.
28. Hagen, P.-O., D. J. Kushner, and N. E. Gibbons. 1964. Temperature-induced death and lysis in a psychrophilic bacterium. *Can. J. Microbiol.* 10:813–822.
29. Hagen, P.-O., and A. H. Rose. 1962. Studies on the biochemical basis of low maximum temperature in a psychrophilic *Cryptococcus*. *J. Gen. Microbiol.* 27: 89–99.
30. Haight, R. D., and R. Y. Morita. 1966. Thermally induced leakage from *Vibrio marinus*, an obligately psychrophilic marine bacterium. *J. Bacteriol.* 92:1388–1393.
31. Harder, W., and H. Veldkamp. 1967. A continuous culture study of an obligately psychrophilic *Pseudomonas* species. *Arch. Mikrobiol.* 59:123–130.
32. Harder, W., and H. Veldkamp. 1968. Physiology of an obligately psychrophilic marine *Pseudomonas* species. *J. Appl. Bacteriol.* 31:12–23.
33. Herbert, R. A. 1981. A comparative study of the physiology of psychrotrophic and psychrophilic bacteria. In *Psychrotrophic Microorganisms in Spoilage and Pathogenicity*, ed. T. A. Roberts, G. Hobbs, J. H. B. Christian, and N. Skovgaard, 3–16. New York: Academic Press.
34. Hurley, W. C., F. A. Gardner, and C. Vanderzant. 1963. Some characteristics of a proteolytic enzyme system of *Pseudomonas fluorescens*. *J. Food Sci.* 28:47–54.
35. Ingraham, J. L. 1962. Newer concepts of psychrophilic bacteria. In *Proceedings, Low Temperature Microbiology Symposium—1961*, 41–62. Camden, N.J.: Campbell Soup Co.

36. Ingraham, J. L., and G. F. Bailey. 1959. Comparative study of effect of temperature on metabolism of psychrophilic and mesophilic bacteria. *J. Bacteriol.* 77:
609–613.

37. Ingraham, J. L., and O. Maaløe. 1967. Cold-sensitive mutants and the minimum
temperature of growth of bacteria. In *Molecular Mechanisms of Temperature
Adaptation,* ed. C. L. Prosser, Pub. No. 84, 297–309. Washington, D.C.:
American Association for the Advancement of Science.

38. Ingraham, J. L., and J. L. Stokes. 1959. Psychrophilic bacteria. *Bacteriol. Rev.*
23:97–108.

39. Ingram, M. 1951. The effect of cold on microorganisms in relation to food. *Proc.
Soc. Appl. Bacteriol.* 14:243.

40. Inniss, W. E. 1975. Interaction of temperature and psychrophilic microorganisms.
Ann. Rev. Microbiol. 29:445–465.

41. Inniss, W. E., and J. L. Ingraham. 1978. Microbial life at low temperatures:
Mechanisms and molecular aspects. In *Microbial Life in Extreme Environments,*
ed. D. J. Kushner, 73–104. New York: Academic Press.

42. Jay, J. M. 1967. Nature, characteristics, and proteolytic properties of beef spoilage
bacteria at low and high temperatures. *Appl. Microbiol.* 15:943–944.

43. Jay, J. M. 1987. The tentative recognition of psychrotrophic gram-negative bacteria
in 48 h by their surface growth at 10°C. *Int. J. Food Microbiol.* 4:25–32.

44. Jezeski, J. J., and R. H. Olsen. 1962. The activity of enzymes at low temperatures.
In *Proceedings, Low Temperature Microbiology Symposium—1961,* 139–155.
Camden, N.J.: Campbell Soup Co.

45. Kates, M., and R. M. Baxter. 1962. Lipid comparison of mesophilic and
psychrophilic yeasts (*Candida* species) as influenced by environmental temperature.
Can. J. Biochem. Physiol. 40:1213–1227.

46. Kavanau, J. L. 1950. Enzyme kinetics and the rate of biological processes. *J. Gen.
Physiol.* 34:193–209.

47. Larkin, J. M., and J. L. Stokes. 1966. Isolation of psychrophilic species of *Bacillus.*
J. Bacteriol. 91:1667–1671.

48. Luyet, B. 1962. Recent developments in cryobiology and their significance in the
study of freezing and freeze-drying of bacteria. In *Proceedings, Low Temperature
Microbiology Symposium,* 63–87. Camden, N.J.: Campbell Soup Co.

49. Malcolm, N. L. 1968. Synthesis of protein and ribonucleic acid in a psychrophile at
normal and restrictive growth temperatures. *J. Bacteriol.* 95:1388–1399.

50. Marr, A. G., J. L. Ingraham, and C. L. Squires. 1964. Effect of the temperature of
growth of *Escherichia coli* on the formation of β-galactosidase. *J. Bacteriol.* 87:
356–362.

51. Mazur, P. 1966. Physical and chemical basis of injury in single-celled microorganisms subjected to freezing and thawing. In *Cryobiology,* ed. H. T. Meryman,
ch. 6. New York: Academic Press.

52. McBryde, C. N. 1911. *A Bacteriological Study of Ham Souring.* Bull. No. 132.
Beltsville, Md.: U.S. Bureau of Animal Industry.

53. McMurrough, I., and A. H. Rose. 1973. Effects of temperature variation on the
fatty acid composition of a psychrophilic *Candida* species. *J. Bacteriol.* 114:
451–452.

54. Michener, H., and R. Elliott. 1964. Minimum growth temperatures for food-
poisoning, fecal-indicator, and psychrophilic microorganisms. *Adv. Food Res.* 13:
349–396.

55. Morita, R. Y. 1975. Psychrophilic bacteria. *Bacteriol. Rev.* 39:144–167.

56. Morita, R. Y., and L. J. Albright. 1965. Cell yields of *Vibrio marinus,* an obligate
psychrophile, at low temperatures. *Can. J. Microbiol.* 11:221–227.

57. Mossel, D. A. A., M. Jansma, and J. De Waart. 1981. Growth potential of 114
strains of epidemiologically most common salmonellae and arizonae between 3 and

17°C. In *Psychrotrophic Microorganisms in Spoilage and Pathogenicity*, ed. T. A. Roberts, G. Hobbs, J. H. B. Christian, and N. Skovgaard, 29–37. New York: Academic Press.

58. Mossel, D. A. A. and H. Zwart. 1960. The rapid tentative recognition of psychrotrophic types among Enterobacteriaceae isolated from foods. *J. Appl. Bacteriol.* 23:183–188.

59. Nashif, S. A., and F. E. Nelson. 1953. The lipase of *Pseudomonas fragi*. I. Characterization of the enzyme. *J. Dairy Sci.* 36:459–470.

60. Nashif, S. A., and F. E. Nelson. 1953. The lipase of *Pseudomonas fragi*. II. Factors affecting lipase production. *J. Dairy Sci.* 36:471–480.

61. Neely, W. B. 1960. Dextran: Structure and synthesis. *Adv. Carbohy. Chem.* 15: 341–369.

62. Olsen, R. H., and J. J. Jezeski. 1963. Some effects of carbon source, aeration, and temperature on growth of a psychrophilic strain of *Pseudomonas fluorescens*. *J. Bacteriol.* 86:429–433.

63. Peterson, A. C., and M. F. Gunderson. 1960. Some characteristics of proteolytic enzymes from *Pseudomonas fluorescens*. *Appl. Microbiol.* 8:98–104.

64. Purohit, K., and J. L. Stokes. 1967. Heat-labile enzymes in a psychrophilic bacterium. *J. Bacteriol.* 93:199–206.

65. Reuter, G. 1981. Psychrotrophic lactobacilli in meat products. In *Psychrotrophic Microorganisms in Spoilage and Pathogenicity*, ed. T. A. Roberts, G. Hobbs, J. H. B. Christian, and N. Skovgaard, 253–258. New York: Academic Press.

66. Roberts, T. A., and G. Hobbs. 1968. Low temperature growth characteristics of Clostridia. *J. Appl. Bacteriol.* 31:75–88.

67. Robinson, R. K. 1985. *Microbiology of Frozen Foods*. New York: Elsevier.

68. Rose, A. H. 1968. Physiology of microorganisms at low temperatures. *J. Appl. Bacteriol.* 31:1–11.

69. Russell, N. J. 1971. Alteration in fatty acid chain length in *Micrococcus cryophilus* grown at different temperatures. *Biochim. Biophy. Acta* 231:254–256.

70. Schmidt, C. F., R. V. Lechowich, and J. F. Folinazzo. 1961. Growth and toxin production by Type E *Clostridium botulinum* below 40°F. *J. Food Sci.* 26:626–630.

71. Scott, W. J. 1962. Available water and microbial growth. In *Proceedings, Low Temperature Microbiology Symposium*, 89–105. Camden, N.J.: Campbell Soup Co.

72. Sinclair, N. A., and J. L. Stokes. 1963. Role of oxygen in the high cell yields of psychrophiles and mesophiles at low temperatures. *J. Bacteriol.* 85:164–167.

73. Sinclair, N. A., and J. L. Stokes. 1964. Isolation of obligately anaerobic psychrophilic bacteria. *J. Bacteriol.* 87:562–565.

74. Sinclair, N. A., and J. L. Stokes. 1965. Obligately psychrophilic yeasts from the polar regions. *Can. J. Microbiol.* 11:259–269.

75. Stokes, J. L. 1967. Heat-sensitive enzymes and enzyme synthesis in psychrophilic microorganisms. In *Molecular Mechanisms of Temperature Adaptation*, ed. C. L. Prosser, 311–323. Pub. No. 84. Washington, D. C.: American Association for the Advancement of Science.

76. Straka, R. P., and J. L. Stokes. 1960. Psychrophilic bacteria from Antarctica. *J. Bacteriol.* 80:622–625.

77. Strange, R. E., and M. Shon. 1964. Effect of thermal stress on the viability and ribonucleic acid of *Aerobacter aerogenes* in aqueous suspensions. *J. Gen. Microbiol.* 34:99–114.

78. Tempest, D. W., and J. R. Hunter. 1965. The influence of temperature and pH value on the macromolecular composition of magnesium-limited and glycerol-limited *Aerobacter aerogenes* growing in a chemostat. *J. Gen. Microbiol.* 41: 267–273.

79. Udaka, S., and T. Horiuchi. 1965. Mutants of *Escherichia coli* having temperature

sensitive regulatory mechanism in the formation of arginine biosynthetic enzymes. *Biochem. Biophys. Res. Commun.* 19:156–160.

80. Uffen, R. L., and E. Canale-Parola. 1966. Temperature-dependent pigment production by *Bacillus cereus* var. *alesti. Can. J. Microbiol.* 12:590–593.

81. Upadhyay, J., and J. L. Stokes. 1962. Anaerobic growth of psychrophilic bacteria. *J. Bacteriol.* 83:270–275.

82. Upadhyay, J., and J. L. Stokes. 1963. Temperature-sensitive formic hydrogenlyase in a psychrophilic bacterium. *J. Bacteriol.* 85:177–185.

83. Upadhyay, J., and J. L. Stokes. 1963. Temperature-sensitive hydrogenase and hydrogenase synthesis in a psychrophilic bacterium. *J. Bacteriol.* 86:992–998.

84. Wilkins, P. O. 1973. Psychrotrophic gram-positive bacteria: Temperature effects on growth and solute uptake. *Can. J. Microbiol.* 19:909–915.

85. Wilkins, P. O., R. Bourgeois, and R. G. E. Murray. 1972. Psychrotrophic properties of *Listeria monocytogenes. Can. J. Microbiol.* 18:543–515.

86. Williams, R. P., M. E. Goldschmidt, and C. L. Gott. 1965. Inhibition by temperature of the terminal step in biosynthesis of prodigiosin. *Biochem. Biophys. Res. Commun.* 19:177–181.

14

High-Temperature Food Preservation and Characteristics of Thermophilic Microorganisms

The use of high temperatures to preserve food is based on their destructive effects on microorganisms. By high temperatures are meant any and all temperatures above ambient. With respect to food preservation, there are two temperature categories in common use: pasteurization and sterilization. Pasteurization by use of heat implies either the destruction of all disease-producing organisms (for example, pasteurization of milk) or the destruction or reduction in number of spoilage organisms in certain foods, as in the pasteurization of vinegar. The pasteurization of milk is achieved by heating as follows:

145°F (63°C) for 30 min (low temperature long time, LTLT)
161°F (72°C) for 15 sec (primary high temperature short time, HTST, method)
191°F (89°C) for 1.0 sec
194°F (90°C) for 0.5 sec
201°F (94°C) for 0.1 sec
212°F (100°C) for 0.01 sec

These treatments are equivalent, and are sufficient to destroy the most heat-resistant of the nonsporeforming pathogenic organisms—*Mycobacterium tuberculosis* and *Coxiella burnetti*. Milk pasteurization temperatures are sufficient to destroy, in addition, all yeasts, molds, gram-negative bacteria, and many gram positives. The two groups of organisms that survive milk pasteurization are placed into one of two groups: thermodurics and thermophiles. **Thermoduric** organisms are those that can survive exposure to relatively high temperatures but do not necessarily grow at these temperatures. The nonsporeforming organisms that survive milk pasteurization generally belong to the genera *Streptococcus* and *Lactobacillus*, and sometimes to other genera. **Thermophilic** organisms are those that not only survive relatively high temperatures but *require* high temperatures for their growth and metabolic activities. The genera *Bacillus* and *Clostridium* contain the thermophiles of greatest importance in foods.

Sterilization means the destruction of all viable organisms as may be measured by an appropriate plating or enumerating technique. Canned foods are sometimes called "commercially sterile" to indicate that no viable organisms can be detected by the usual cultural methods employed or that the number of survivors is so low as to be of no significance under the conditions of canning and storage. Also, microorganisms may be present in canned foods that cannot grow in the product by reason of undesirable pH, Eh, or temperature of storage.

A more recent development in the processing of milk and milk products is the use of **ultrahigh temperatures** (UHT). Milk so produced is a product in its own right and is to be distinguished from pasteurized milk. The primary features of the UHT treatment include its continuous nature, its occurrence outside the package necessitating aseptic storage and aseptic handling of the product downstream from the sterilizer, and the very high temperatures (in the range of 140° to 150°C) and the correspondingly short time (a few seconds) necessary to achieve commercial sterility [28]. UHT-processed milks have higher consumer acceptability than the conventionally heated pasteurized products, and since they are commercially sterile, they may be stored at room temperatures for up to 8 weeks without flavor changes.

FACTORS AFFECTING HEAT RESISTANCE IN MICROORGANISMS

Equal numbers of bacteria placed in physiologic saline and nutrient broth at the same pH are not destroyed with the same ease by heat. Some 11 factors or parameters of microorganisms and their environment have been studied for their effects on heat destruction and are presented below (25).

Water.   The heat resistance of microbial cells increases with decreasing humidity or moisture. Dried microbial cells placed into test tubes and then heated in a water bath are considerably more heat resistant than moist cells of the same type. Since it is well established that protein denaturation occurs at a faster rate when heated in water than in air, it is suggested that protein denaturation is either the mechanism of death by heat or is closely associated with it (see a later section this chapter). The precise manner in which water facilitates heat denaturation of proteins is not entirely clear, but it has been pointed out that the heating of wet proteins causes the formation of free —SH groups with a consequent increase in the water-binding capacity of proteins. The presence of water allows for thermal breaking of peptide bonds, a process that requires more energy in the absence of water and consequently confers a greater refractivity to heat.

Fat.   In the presence of fats, there is a general increase in the heat resistance of some microorganisms (Table 14-1). This is sometimes referred to as fat protection and is presumed to increase heat resistance by directly affecting cell moisture. Sugiyama (48) demonstrated the heat-protective effect of long-chain

TABLE 14-1.   The Effect of the Medium on the Thermal Death Point of *Escherichia coli*

Medium	Thermal Death Point (°C)
Cream	73
Whole milk	69
Skim milk	65
Whey	63
Bouillon (broth)	61

Source: Carpenter (13). Courtesy of W. B. Saunders Co., Philadelphia.

Note: Heating time: 10 min.

fatty acids on *Clostridium botulinum* . It appears that the long-chain fatty acids are better protectors than short-chain acids.

Salts.   The effect of salt on the heat resistance of microorganisms is variable and dependent on the kind of salt, concentration employed, and other factors. Some salts have a protective effect on microorganisms, and others tend to make cells more heat sensitive. It has been suggested that some salts may decrease water activity and thereby increase heat resistance by a mechanism similar to that of drying, while others may increase water activity (for example, Ca^{2+} and Mg^{2+}) and consequently increase sensitivity to heat. It has been shown that supplementation of the growth medium of *Bacillus megaterium* spores with $CaCl_2$ yields spores with increased heat resistance, while the addition of L-glutamate, L-proline, or increased phosphate content decreases heat resistance (33).

Carbohydrates.   The presence of sugars in the suspending menstrum causes an increase in the heat resistance of microorganisms suspended therein. This effect is at least in part due to the decrease in water activity caused by high concentrations of sugars. There is great variation, however, among sugars and alcohols relative to their effect on heat resistance as may be seen in Table 14-2 for D values of *Salmonella senftenberg* 775W. At identical a_w values obtained by use of glycerol and sucrose, wide differences in heat sensitivity occur (5, 24). Corry (15) found that sucrose increased the heat resistance of *S. senftenberg* more than any of four other carbohydrates tested. The following decreasing order was found for the five tested substances: sucrose > glucose > sorbitol > fructose > glycerol.

pH.   Microorganisms are most resistant to heat at their optimum pH of growth, which is generally about 7.0. As pH is lowered or raised from this optimum value, there is a consequent increase in heat sensitivity (Fig. 14-1). Advantage is taken of this fact in the heat processing of high-acid foods, where considerably less heat is applied to achieve sterilization compared to foods at or near neutrality. The heat pasteurization of egg white provides an example of an alkaline food product that is neutralized prior to heat treatment, a practice

TABLE 14-2.   Reported D values of *Salmonella senftenberg* 775W

Temperature (°C)	D Values	Conditions	Reference
61	1.1 min	Liquid whole egg	44
61	1.19 min	Tryptose broth	44
60	9.5 min[a]	Liquid whole egg, pH ca. 5.5	2
60	9.0 min[a]	Liquid whole egg, pH ca. 6.6	2
60	4.6 min[a]	Liquid whole egg, pH ca. 7.4	2
60	0.36 min[a]	Liquid whole egg, pH ca. 8.5	2
65.6	34–35.3 sec	Milk	42
71.7	1.2 sec	Milk	42
70	360–480 min	Milk chocolate	23
55	4.8 min	TSB,[b] log phase, grown 35°C	40
55	12.5 min	TSB,[b] log phase, grown 44°C	40
55	14.6 min	TSB,[b] stationary, grown 35°C	40
55	42.0 min	TSB,[b] stationary, grown 44°C	40
57.2	13.5 min[a]	a_w 0.99 (4.9% glyc.), pH 6.9	24
57.2	31.5 min[a]	a_w 0.90 (33.9% glyc.), pH 6.9	24
57.2	14.5 min[a]	a_w 0.99 (15.4% sucro.), pH 6.9	24
57.2	62.0 min[a]	a_w 0.90 (58.6% sucro.), pH 6.9	24
60	0.2–6.5 min[c]	HIB,[d] pH 7.4	5
60	2.5 min	a_w 0.90, HIB, glycerol	5
60	75.2 min	a_w 0.90, HIB, sucrose	5
65	0.29 min	0.1 M phosphate buf., pH 6.5	15
65	0.8 min	30% sucrose	15
65	43.0 min	70% sucrose	15
65	2.0 min	30% glucose	15
65	17.0 min	70% glucose	15
65	0.95 min	30% glycerol	15
65	0.70 min	70% glycerol	15
55	35 min	a_w 0.997, tryptone soya agar, pH 7.2	26

[a] Mean/average values.   [b] Trypticase soy broth.   [c] Total of 76 cultures.   [d] Heart infusion broth.

not done with other foods. The pH of egg white is about 9.0. When this product is subjected to pasteurization conditions of 60° to 62°C for 3.5 to 4 min, coagulation of proteins occurs along with a marked increase in viscosity. These changes affect the volume and texture of cakes made from such pasteurized egg white. Cunningham and Lineweaver (16) reported that egg white may be pasteurized the same as whole egg if the pH is reduced to about 7.0. This reduction of pH makes both microorganisms and egg white proteins more heat stable. The addition of salts of iron or aluminum increases the stability of the highly heat-labile egg protein conalbumin sufficiently to permit pasteurization at 60° to 62°C. Unlike their resistance to heat in other materials, bacteria are more resistant to heat in liquid whole egg at pH values of 5.4 to 5.6 than at values of 8.0 to 8.5 (Table 14-2). This is true when pH is lowered with an acid such as HCI. When organic acids such as acetic or lactic acid are used to lower pH, a decrease in heat resistance occurs.

Proteins and Other Substances.   Proteins in the heating menstrum have a protective effect on microorganisms. Consequently, high-protein-content foods

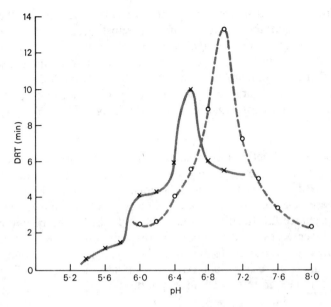

FIGURE 14-1.   Effect of pH on the decimal reduction time (DRT) of *Enterococcus faecalis* (C & G) exposed to 60°C in citrate-phosphate buffer (crosses) and phosphate buffer (circles) solutions at various pH levels.
From White (53).

must be heat processed to a greater degree than low-protein-content foods in order to achieve the same end results. For identical numbers of organisms, the presence of colloidal-size particles in the heating menstrum also offers protection against heat. For example, under identical conditions of pH, numbers of organisms, and so on, it takes longer to sterilize pea purée than nutrient broth.

Numbers of Organisms.   The larger the number of organisms, the higher is the degree of heat resistance (Table 14-3). It has been suggested that the mechanism of heat protection by large microbial populations is due to the pro-

TABLE 14-3.   **Effect of Number of Spores of *Clostridium botulinum* on Thermal Death Time at 100°C**

Number of Spores	Thermal Death Time (minutes)
72,000,000,000	240
1,640,000,000	125
32,000,000	110
650,000	85
16,400	50
328	40

Source: Carpenter (13). Courtesy of W.B. Saunders Co., Philadelphia.

duction of protective substances excreted by the cells, and some authors claim to have demonstrated the existence of such substances. Since proteins are known to offer some protection against heat, many of the extracellular compounds in a culture would be expected to be protein in nature and consequently capable of affording some protection. Of perhaps equal importance in the higher heat resistance of large cell populations over smaller ones is the greater chance for the presence of organisms with differing degrees of natural heat resistance.

Age of Organisms.   Bacterial cells tend to be most resistant to heat while in the stationary phase of growth (old cells) and less resistant during the logarithmic phase. This is true for *S. senftenberg* (see Table 14-2), whose stationary phase cells may be several times more resistant than log phase cells (40). Heat resistance has been reported to be high also at the beginning of the lag phase but decreases to a minimum as the cells enter the log phase. Old bacterial spores are reported to be more heat resistant than young spores. The mechanism of increased heat resistance of less active microbial cells is undoubtedly complex and not well understood.

Growth Temperature.   The heat resistance of microorganisms tends to increase as the temperature of incubation increases. Lechowich and Ordal (28) showed that as sporulation temperature was increased for *Bacillus subtilis* and *Bacillus coagulans*, the thermal resistance of spores of both organisms also increased. Although the precise mechanism of this effect is unclear, it is conceivable that genetic selection favors the growth of the more heat-resistant strains at succeedingly high temperatures. *S. senftenberg* grown at 44°C was found to be approximately three times more resistant than cultures grown at 35°C (Table 14-2).

Inhibitory Compounds.   A decrease in heat resistance of most microorganisms occurs when heating takes place in the presence of heat-resistant antibiotics, SO_2, and other microbial inhibitors. The use of heat plus antibiotics and heat plus nitrite has been found to be more effective in controlling the spoilage of certain foods than either alone. The practical effect of adding inhibitors to foods prior to heat treatment is to reduce the amount of heat that would be necessary if used alone (see Chap. 11).

Time and Temperature.   One would expect that the longer the time of heating, the greater the killing effect of heat. All too often, though, there are exceptions to this basic rule. A more dependable rule is that the higher the temperature, the greater is the killing effect of heat. This is illustrated in Table 14-4 for bacterial spores. As temperature increases, time necessary to achieve the same effect decreases.

These rules assume that heating effects are immediate and not mechanically obstructed or hindered. Also important is the size of the heating vessel or container and its composition (glass, metal, plastic). It takes longer to effect pasteurization or sterilization in large containers than in smaller ones. The same is true of containers with walls that do not conduct heat as readily as others.

TABLE 14-4.   Effect of Temperature on Thermal Death Times of Spores

Temperature	Clostridium botulinum (60 billion spores suspended in buffer at pH 7)	A thermophile (150,000 spores per ml of corn juice at pH 6.1)
100°C	260 min	1140 min
105°C	120	
110°C	36	180
115°C	12	60
120°C	5	17

Source: Carpenter (13). Courtesy of W.B. Saunders Co., Philadelphia.

RELATIVE HEAT RESISTANCE OF MICROORGANISMS

In general, the heat resistance of microorganisms is related to their optimum growth temperatures. Psychrophilic microorganisms are the most heat sensitive, followed by mesophiles and thermophiles. Sporeforming bacteria are more heat resistant than nonsporeformers, and thermophilic sporeformers are in general more heat resistant than mesophilic sporeformers. With respect to gram reaction, gram-positive bacteria tend to be more heat resistant than gram negative, with cocci in general being more resistant than nonsporeforming rods. Yeasts and molds tend to be fairly sensitive to heat, with yeast ascospores being only slightly more resistant than vegetative yeasts. The asexual spores of molds tend to be slightly more heat resistant than mold mycelia. Sclerotia are the most heat resistant of these types and sometimes survive and cause trouble in canned fruits.

Spore Resistance

The extreme heat resistance of bacterial endospores is of great concern in the thermal preservation of foods.

Endospore resistance is believed to be due to three factors: protoplast dehydration, mineralization, and thermal adaptation. Protoplast dehydration appears to be the primary factor, especially for spores of low heat resistance such as those of *B. megaterium*. The spore protoplast is that part bounded by the inner pericytoplasmic membrane. It is rich in Ca^{2+} and dipicolinic acid, which form a calcium-dipicolinate complex with gellike properties. It is generally believed that the water content of the protoplast (and its state) determines the degree of heat resistance of an endospore. That Ca^{2+} and dipicolinic acid are important in the water relations of the protoplast is shown by the increased thermal sensitivity of spores that develop in Ca-deficient media and by mutants that are deficient in dipicolinate. That protoplast dehydration and diminution are major factors of spore thermal resistance has been substantiated (9), but other factors are known to have an additive effect (39).

The endospores of a given species grown at maximum temperature are more heat resistant than those grown at lower temperatures (54). It appears that protoplast water content is lowered by this thermal adaptation, resulting in a more-heat resistant spore (8). Heat resistance is affected extrinsically by changes in mineral content. Although all three factors noted contribute to spore thermal resistance, dehydration appears to be the most important (8).

THERMAL DESTRUCTION OF MICROORGANISMS

In order to understand better the thermal destruction of microorganisms relative to food preservation and canning, it is necessary to understand certain basic concepts associated with this technology. Following are listed some of the more important concepts, but for a more extensive treatment of thermo-bacteriology, the monograph by Stumbo (46) should be consulted.

Thermal Death Time (TDT).   This is the time necessary to kill a given number of organisms at a specified temperature. By this method, the temperature is kept constant and the time necessary to kill all cells is determined. Of less importance is the **thermal death point**, which is the temperature necessary to kill a given number of microorganisms in a fixed time, usually 10 min. Various means have been proposed for determining TDT: the tube, can, "tank," flask, thermo-resistometer, unsealed tube, and capillary tube methods. The general procedure for determining TDT by these methods is to place a known number of cells or spores in a sufficient number of sealed containers in order to get the desired number of survivors for each test period. The organisms are then placed in an oil bath and heated for the required time period. At the end of the heating period, containers are removed and cooled quickly in cold water. The organisms are then placed on a suitable growth medium, or the entire heated containers are incubated if the organisms are suspended in a suitable growth substrate. The suspensions or containers are incubated at a temperature suitable for growth of the specific organisms. Death is defined as the inability of the organisms to form a visible colony.

D *Value*.   This is the decimal reduction time, or the time required to destroy 90% of the organisms. This value is numerically equal to the number of min required for the survivor curve to traverse one log cycle (Fig. 14-2). Mathematically, it is equal to the reciprocal of the slope of the survivor curve and is a measure of the death rate of an organism. When D is determined at 250°F, it is often expressed as D_r. The effect of pH on the D value of *C. botulinum* in various foods is presented in Table 14-5, and D values for *S. senftenberg* 775W under various conditions are presented in Table 14-2. D values of 0.20 to 2.20 min at 150°F have been reported for *S. aureus* strains, D 150°F of 0.50 to 0.60 min for *Coxiella burnetti*, and D 150°F of 0.20 to 0.30 for *Mycobacterium hominis* (46). For pH-elevating strains of *Bacillus licheniformis* spores in tomatoes, a D 95°C of 5.1 min has been reported, while for *B. coagulans* a D 95°C of 13.7 min has been found (38).

z *Value*.   The z value refers to the degrees Fahrenheit required for the thermal destruction curve to traverse one log cycle. Mathematically, this value

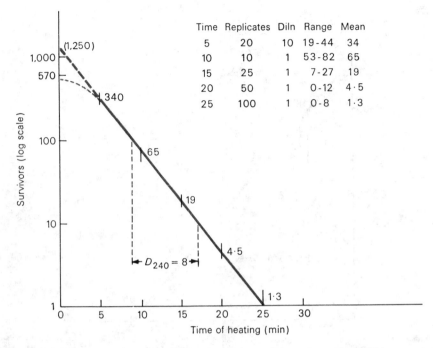

Time	Replicates	Diln	Range	Mean
5	20	10	19-44	34
10	10	1	53-82	65
15	25	1	7-27	19
20	50	1	0-12	4·5
25	100	1	0-8	1·3

FIGURE 14-2.   Rate of destruction curve. Spores of strain F.S. 7 heated at 240°F in canned pea brine pH 6.2.
From Gillespy (22), courtesy of Butterworths Publishers, London.

is equal to the reciprocal of the slope of the TDT curve (Fig. 14-3). While D reflects the resistance of an organism to a specific temperature, z provides information on the relative resistance of an organism to different destructive temperatures; it allows for the calculation of equivalent thermal processes at

TABLE 14-5.   **Effect of pH on D Values for Spores of C.**
botulinum 62A Suspended in Three Food
Products at 240°F

	D Value (min)		
pH	Spaghetti, Tomato Sauce, and Cheese	Macaroni Creole	Spanish Rice
4.0	0.128	0.127	0.117
4.2	0.143	0.148	0.124
4.4	0.163	0.170	0.149
4.6	0.223	0.223	0.210
4.8	0.226	0.261	0.256
5.0	0.260	0.306	0.266
6.0	0.491	0.535	0.469
7.0	0.515	0.568	0.550

Source: Xezones and Hutchings (56); copyright © 1965 by Institute of Food Technologists.

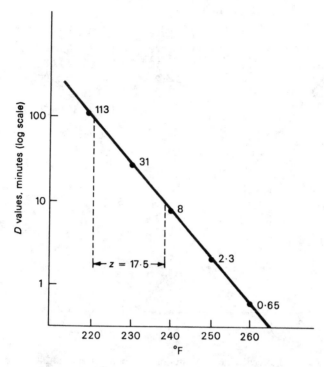

FIGURE 14-3.   Thermal death time curve. Spores of strain F.S. 7 heated in canned pea brine pH 6.2.
From Gillespy (22), courtesy of Butterworths Publishers, London.

different temperatures. If, for example, 3.5 min at 140°F is considered to be an adequate process and $z = 8.0$, either 0.35 min at 148°F or 35 min at 132°F would be considered equivalent processes.

F Value.   This value is the equivalent time, in minutes at 250°F, of all heat considered, with respect to its capacity to destroy spores or vegetative cells of a particular organism. The integrated lethal value of heat received by all points in a container during processing is designated F_s or F_o. This represents a measure of the capacity of a heat process to reduce the number of spores or vegetative cells of a given organism per container. When we assume instant heating and cooling throughout the container of spores, vegetative cells, or food, F_o may be derived as follows:

$$F_o = D_r(\log a - \log b)$$

where a = number of cells in the initial population, and b = number of cells in the final population.

Thermal Death Time Curve.   For the purpose of illustrating a thermal destruction curve and D value, data are employed from Gillespy (22) on the killing of flat sour spores at 240°F in canned pea brine at pH 6.2. Counts were determined at intervals of 5 min with the mean viable numbers indicated as follows:

Time (min.)	Mean viable count
5	340.0
10	65.0
15	19.0
20	4.5
25	1.3

The time of heating in minutes is plotted on semi-log paper along the linear axis, and the number of survivors is plotted along the log scale to produce the TDT curve presented in Fig. 14-2. The curve is essentially linear, indicating that the destruction of bacteria by heat is logarithmic and obeys a first-order reaction. Although difficulty is encountered at times at either end of the TDT curve, process calculations in the canning industry are based upon a logarithmic order of death. From the data presented in Fig. 14-2, the D value is calculated to be 8 min, or $D_{240} = 8.0$.

D values may be used to reflect the relative resistance of spores or vegetative cells to heat. The most heat-resistant strains of *C. botulinum* types A and B spores have a D_r value of 0.21, while the most heat-resistant thermophilic spores have D_r values of around 4.0 to 5.0. Putrefactive anaerobe (P.A.) 3679 was found by Stumbo et al. (47) to have a D_r value of 2.47 in cream-style corn, while flat sour (F.S.) spores 617 were found to have a D_r of 0.84 in whole milk.

The approximate heat resistance of spores of thermophilic and mesophilic spoilage organisms may be compared by use of D_r values.

Bacillus stearothermophilus:	4.0 –5.0
Clostridium thermosaccharolyticum:	3.0 –4.0
Clostridium nigrificans:	2.0 –3.0
C. botulinum (types A and B):	0.10–0.2
Clostridium sporogenes (including P.A. 3679):	0.10–1.5
B. coagulans:	0.01–0.07

The effect of pH and suspending menstrum on D values of *C.botulinum* spores is presented in Table 14-5. As noted above, microorganisms are more resistant at and around neutrality and show different degrees of heat resistance in different foods.

In order to determine the z value, D values are plotted on the log scale, and degrees F are plotted along the linear axis. From the data presented in Fig. 14-3, the z value is 17.5. Values of z for *C.botulinum* range from 14.7 to 16.3, while for P.A. 3679 the range of 16.6 to 20.5 has been reported. Some spores have been reported to have z values as high as 22. Peroxidase has been reported to have a z value of 47, and 50 has been reported for riboflavin and 56 for thiamine.

12-D Concept.   The 12-D concept refers to the process lethality requirement long in effect in the canning industry and implies that the minimum heat process should reduce the probability of survival of the most resistant *C. botulinum* spores to 10^{-12}. Since *C. botulinum* spores do not germinate and produce toxin below pH 4.6, this concept is observed only for foods above this pH value. An example from Stumbo (46) illustrates this concept from the standpoint of canning technology. If it is assumed that each container of food contains only one spore of C. botulinum, F_o may be calculated by use of the general survivor curve equation with the other assumptions noted above in mind:

$$F_o = D_r(\log a - \log b)$$
$$F_o = 0.21 \ (\log 1 - \log 10^{-12})$$
$$F_o = 0.21 \times 12 = 2.52$$

Processing for 2.52 min at 250°F, then, should reduce the C. botulinum spores to one spore in one of 1 billion containers (10^{12}). When it is considered that some flat-sour spores have D_r values of about 4.0 and some canned foods receive F_o treatments of 6.0 to 8.0, the potential number of C. botulinum spores is reduced even more.

ASEPTIC PACKAGING

In traditional canning methods, nonsterile food is placed in nonsterile metal or glass containers followed by container closure and sterilization. In aseptic packaging, sterile food under aseptic conditions is placed in sterile containers, and the packages are sealed under aseptic conditions as well. Although the methodology of aseptic packaging was patented in the early 1960s, the technology was little used until 1981 when the Food and Drug Administration approved the use of hydrogen peroxide for the sterilization of flexible multi-layered packaging materials used in aseptic processing systems (51).

In general, any food that can be pumped through a heat exchanger can be aseptically packaged. The widest application has been to liquids such as fruit juices, and a wide variety of single-serve products of this type has resulted. The technology for foods that contain particulates has been more difficult to develop, with microbiological considerations only one of the many problems to overcome. In determining the sterilization process for foods pumped through heat exchangers, the fastest-moving components (those with the minimum holding time) are used, and where liquids and particulates are mixed, the latter will be the slower moving. Heat penetration rates are not similar for liquids and solids, making it more difficult to establish minimum process requirements that will effectively destroy both organisms and food enzymes.

Some of the advantages of aseptic packaging are these:

- Products such as fruit juices are more flavorful and lack the metallic taste of those processed in metal containers.
- Flexible multilayered cartons can be used instead of glass or metal containers.
- The time a product is subjected to high temperatures is minimized when ultrahigh temperatures are used.
- The technology allows the use of membrane filtration of certain liquids.
- Various container headspace gases such as nitrogen may be used.

Among the disadvantages are that packages may not be equivalent to glass or metal containers in preventing the permeation of oxygen, and the output is lower than that for solid containers.

A wide variety of aseptic packaging techniques now exists, with more under development (12). Sterilization of packages is achieved in various ways, one of which involves the continuous feeding of rolls of packaging material into a machine where hot hydrogen peroxide is used to effect sterilization, followed by the forming, filling with food, and sealing of the containers (27). Sterility of the filling operation may be maintained by a positive pressure of air or gas such as nitrogen. Aseptically packaged fruit juices are shelf stable at ambient temperatures for 6 to 12 months or longer.

The spoilage of aseptically packaged foods differs from foods in metal containers. While hydrogen swells occur in high-acid foods in the latter containers, aseptic packaging materials are nonmetallic. Seam leakage may be expected to be absent in aseptically packaged foods, but the permeation of oxygen by the nonmetal and nonglass containers may allow for other types of spoilage in low-acid foods.

SOME CHARACTERISTICS OF THERMOPHILES

On the basis of growth temperatures, thermophiles may be characterized as organisms with a minimum of around 45°C, optimum between 50° and 60°C, and maximum of 70°C or above. By this definition, thermophilic species/strains are found among the cyanobacteria, archaebacteria, actinomycetes (50), the anaerobic photosynthetic bacteria, thiobacilli, algae, fungi, bacilli, clostridia, lactic acid bacteria, and other groups. Those of greatest importance in foods belong to the genera *Bacillus* and *Clostridium*.

In thermophilic growth, the lag phase is short and sometimes difficult to measure. Spores germinate and grow rapidly. The logarithmic phase of growth is of short duration. Some thermophiles have been reported to have generation times as short as 10 min when growing at high temperatures. The rate of death or "die off" is rapid. Loss of viability or "autosterilization" below the thermophilic growth range is characteristic of organisms of this type. The growth curves of a bacterium at 55°, 37°, and 20°C are compared in Figure 14-4.

Why some organisms require temperatures of growth that are destructive to others is of concern not only from the standpoint of food preservation but also

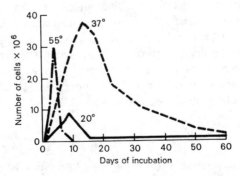

FIGURE 14-4.   Growth curves of bacterial strain incubated at 20°, 37°, and 55°C. *From Tanner and Wallace (49).*

from that of the overall biology of thermophilism. Some of the known features of thermophiles are summarized below.

Enzymes

The enzymes of thermophiles can be divided into three groups:

1. Those that are stable at the temperature of production but require slightly higher temperatures for inactivation—for example, malic dehydrogenase, ATPase, inorganic pyrophosphatase, aldolase, and certain peptidases.
2. Some enzymes are inactivated at the temperature of production in the absence of specific substrates—for example, asparagine deamidase, catalase, pyruvic acid oxidase, isocitrate lyase, and certain membrane-bound enzymes.
3. Some enzymes and proteins are highly heat resistant—for example, alpha-amylase, some proteases, glyceraldehyde-3-phosphate dehydrogenase, certain amino acid activating enzymes, flagellar proteins, esterases, and thermolysin.

In general, the enzymes of thermophiles produced under thermophilic growth conditions are more heat resistant than those of mesophiles (Table 14-6). Of particular note is alpha-amylase produced by a strain of *B. stearothermophilus*, which retained activity after being heated at 70°C for 24 h. In a more recent study, the optimum temperature for the activity of amylase from *B. stearothermophilus* was found to be 82°C with a pH optimum of 6.9 (45). The enzyme required Ca^{2+} for thermostability. The heat stability of cytoplasmic proteins isolated from four thermophiles was greater than that from four mesophiles (30).

Several possibilities exist as to why the enzymes of thermophiles are thermostable. Among these is the existence of higher levels of hydrophobic amino acids than exist in similar enzymes from mesophiles. A more hydrophobic protein presumably would be more heat resistant. Regarding amino acids, it has been shown that lysine in place of glutamine decreased the thermostability of an enzyme while replacements with other amino acids enhanced thermostability (32). Another factor has to do with the binding of metal ions such as Mg^{2+}. The structural integrity of the membrane of *B. stearothermophilus* protoplasts was shown to be affected by divalent cations (55).

Overall, the proteins of thermophiles are similar in molecular weight, amino acid composition, allosteric effectors, subunit composition, and primary sequences to their mesophilic counterparts. Extremely thermophilic and obligately thermophilic organisms synthesize macromolecules that have sufficient intrinsic molecular stability to withstand thermal stress (1). Other aspects of proteins from thermophiles have been reviewed (43).

Ribosomes

In general, the thermal stability of ribosomes corresponds to the maximal growth temperature of a microorganism (Table 14-7). Heat-resistant ribosomes have been reported (18) but not DNA (34). In a study of the ribosomes of *B.*

TABLE 14-6. Comparison of Thermostability and Other Properties of Enzymes from Mesophilic and Thermophilic Bacteria

Species	Enzyme	Heat Stability[a] (%)	Heat Stability[a]	Half-Cystine (mole/mole of protein)	Molecular Weight	Metal Required for Stability
B. subtilis	Subtilisin BPN'	45	(50°C, 30 min)	0	28,000	Yes
B. subtilis	Neutral protease	50	(60°C, 15 min)	0	44,700	Yes
Ps. aeruginosa	Alkaline protease	80	(60°C, 10 min)	0	48,400	Yes
Ps. aeruginosa	Elastase	86	(70°C, 10 min)	4.6	39,500	Yes
		10	(75°C, 10 min)			
Group A streptococci	Streptococcal protease	0	(70°C, 30 min)	1	32,000	
Cl. histolyticum	Collagenase	1.5	(50°C, 20 min)	0	90,000	Yes
St. griseus	Pronase	60	(60°C, 10 min)			Yes
B. thermoproteolyticus	Thermolysin	95	(60°C, 120 min)	0	42,700	Yes
		50	(80°C, 60 min)			
B. subtilis	α-Amylase	55	(65°C, 20 min)	0	50,000	Yes
B. stearothermophilus	α-Amylase	100	(70°C, 24 h)	4	15,500	

Source: Matsubara (36); copyright © 1967 by The American Association for the Advancement of Science.

[a] Activity remaining after heat treatment shown in parentheses.

TABLE 14-7.   Ribosome Melting and Maximal Growth Temperatures

Organism and Strain Number	Maximum Growth Temperature (°C)	Ribosome T_m (°C)
1. *V. marinus* (15381)	18	69
2. 7E-3	20	69
3. 1-1	28	74
4. *V. marinus* (15382)	30	71
5. 2-1	35	70
6. *D. desulfuricans (cholinicus)*[a]	40	73
7. *D. vulgaris* (8303)[a]	40	73
8. *E. coli* (B)	45	72
9. *E. coli* (Q13)	45	72
10. *S. itersonii* (SI–1)[b]	45	73
11. *B. megaterium* (Paris)	45	75
12. *B. subtilis* (SB-19)	50	74
13. *B. coagulans* (43P)	60	74
14. *D. nigrificans* (8351)[c]	60	75
15. Thermophile 194	73	78
16. *B. stearothermophilus* (T-107)	73	78
17. *B. stearothermophilus* (1503R)	73	79
18. Thermophile (Tecce)	73	79
19. *B. stearothermophilus* (10)	73	79

Source: Pace and Campbell (41).

[a] *Desulfovibrio.*
[b] *Spirillum.*
[c] *Desulfotomaculum.*

TABLE 14-8.   Effect of Incubation Temperature on the Nutritional Requirements of Thermophilic Bacteria

Organism and Strain Number	Nutritional Requirements at		
	36°C	45°C	55°C
B. coagulans 2 (F)[a]	his, thi, bio, fol	his, thi, bio, fol	his, thi, bio, fol
B. coagulans 1039 (F)	thi, bio, fol	thi, bio, fol	his, met, thi, nic, bio, fol
B. stearothermophilus 3690 (F)	met, leu, thi, nic, bio, fol	met, thi, bio, fol	met, thi, bio, fol
B. stearothermophilus 4259 (F)	bio, fol	met, his, nic, bio, fol	met, his, nic, bio, fol
B. stearothermophilus 1373b (O)[b]	no growth	glu, his, met, leu, bio	glu, his, met, leu, bio, rib

Source: Campbell and Williams (11).

Note: his = histidine; thi = thiamine; bio = biotin; fol = folic acid; met = methionine; leu = leucine; nic = nicotinic acid; glu = glutamic acid; rib = riboflavin.

[a] F = facultative.
[b] O = obligate thermophile.

stearothermophilus, no unusual chemical features of their proteins could be found that could explain their thermostability (3), and in another study, no significant differences in either the size or the arrangement of surface filaments of *B. stearothermophilus* and *Escherichia coli* ribosomes could be found (6). Base composition of rRNA has been shown to affect thermal stability. In a study of 19 organisms, the G-C content of rRNA molecules increased and the A-U content decreased with increasing maximal growth temperatures (41). The increased G-C content makes for a more stable structure through more extensive hydrogen bonding. On the other hand, the thermal stability of soluble RNA from thermophiles and mesophiles appears to be similar (4, 19, 20).

Flagella

The flagella of thermophiles are more heat stable than those of mesophiles, with the former remaining intact at temperatures as high as 70°C while those of the latter disintegrate at 50°C (29, 31). The thermophilic flagella are more resistant to urea and acetamide than those of mesophiles, suggesting that more effective hydrogen bonding occurs in thermophilic flagella.

OTHER ASPECTS OF THERMOPHILIC MICROORGANISMS

Nutrient Requirements.   Thermophiles generally have a higher nutrient requirement than mesophiles when growing at thermophilic temperatures. This is illustrated for two strains of *B. coagulans* and three of *B. stearothermophilus* in Table 14-8. One strain of *B. coagulans* showed no differences in growth requirements regardless of incubation temperature. The obligately thermophilic strain of *B. stearothermophilus* showed one additional requirement as incubation temperature increased, while one temperature facultative strain of this organism showed additional requirements as the incubation temperature was lowered. Although this aspect of thermophilism has not received much study, changes in nutrient requirements as incubation temperature is raised may be due to a general lack of efficiency on the part of the metabolic complex. Certain enzyme systems might well be affected by the increased temperature of incubation, as well as the overall process of enzyme synthesis.

Oxygen Tension.   Thermophilic growth is affected by oxygen tension. As the temperature of incubation is increased, the growth rate of microorganisms increases, thereby increasing the oxygen demand on the culture medium, while reducing the solubility of oxygen. This is thought by some investigators to be one of the most important limiting factors of thermophilic growth in culture media. Downey (17) has shown that thermophilic growth is optimal at or near the oxygen concentration normally available in the mesophilic range of temperatures—143 to 240 μM. Although it is conceivable that thermophiles are capable of high-temperature growth due to their ability to consume and conserve oxygen at high temperatures, a capacity that mesophiles and psychrophiles lack, further data in support of this notion are wanting.

Cellular Lipids.   The state of cellular lipids affects thermophilic growth. Since an increase in degree of unsaturation of cellular lipids is associated with psychrotrophic growth, it is reasonable to assume that a reverse effect occurs in the case of thermophilic growth. This idea finds support in the investigations of many authors. Gaughran (21) found that mesophiles growing above their maximum range showed decreases in lipid content and more lipid saturation. According to this author, cells cannot grow at temperatures below the solidification point of their lipids. Marr and Ingraham (35) showed a progressive increase in saturated fatty acids and a corresponding decrease in unsaturated fatty acids in *E. coli* as the temperature of growth increased. The general decrease in the proportion of unsaturated fatty acids as growth temperatures increase has been found to occur in a large variety of animals and plants (14). Saturated fatty acids form stronger hydrophobic bonds than do unsaturated. Among the saturated fatty acids are branched-chain acids. The preferential synthesis of branched heptadecanoic acid and the total elimination of unsaturated fatty acids by two thermophilic *Bacillus* spp. has been observed (52).

Cellular Membranes.   The nature of cellular membranes affects thermophilic growth. Brock (10) reported that the molecular mechanism of thermophilism is more likely to be related to the function and stability of cellular membranes than to the properties of specific macromolecules. This investigator pointed out that there is no evidence that organisms are killed by heat because of the inactivation of proteins or other macromolecules, a view that is widely held. According to Brock, an analysis of thermal death curves of various microorganisms shows that this is a first-order process compatible with an effect of heat on some large structure such as the cell membrane, since a single hole in the membrane could result in leakage of cell constituents and subsequent death. Brock has also pointed out that thermal killing due to the inactivation of heat-sensitive enzymes, or heat-sensitive ribosomes, of which there are many copies in the cell, should not result in simple first-order kinetics. The leakage of ultraviolet light–absorbing and other material from cells undergoing "cold shock" would tend to implicate the membrane in high-temperature death. Since most animals die when body temperatures reach between 40° and 45°C and most psychrophilic bacteria are killed at about this temperature range, the suggestion that lethal injury is due to the melting of lipid constituents of the cell or cell membrane is not only plausible; it has been supported by the findings of various investigators. The unit cell membrane consists of layers of lipid surrounded by layers of protein and depends on the lipid layers for its biological functions. The disruption of this structure would be expected to cause cell damage and perhaps death. In view of the changes in cellular lipid saturation noted above, the cell membrane appears to be critical to growth and survival at thermophilic temperatures.

Effect of Temperature.   Brock (10) has called attention to the fact that thermophiles apparently do not grow as fast at their optimum temperatures as one would predict or is commonly believed. Arrhenius plots of thermophile growth compared to *E. coli* over a range of temperatures indicated that, overall, the mesophilic types were more efficient. Brock has noted that thermophile enzymes are inherently less efficient than mesophiles because of thermal stability; that is, the thermophiles have had to discard growth efficiency in order to survive at all.

Genetics.   A significant discovery toward an understanding of the genetic bases of thermophilism was made by McDonald and Matney (37). These investigators effected the transformation of thermophilism in *B. subtilis* by growing cells of a strain that could not grow above 50°C in the presence of DNA extracted from one that could grow at 55°C. The more heat-sensitive strain was transformed at a frequency of 10^{-4}. These authors noted that only 10 to 20% of the transformants retained the high-level streptomycin resistance of the recipient, which indicated that the genetic loci for streptomycin resistance and that for growth at 55°C were closely linked.

Although much has been learned about the basic mechanisms of thermophilism in microorganisms, the precise mechanisms underlying this high-temperature phenomenon remain a mystery. The facultative thermophiles such as some *B. coagulans* strains present a picture as puzzling as the obligate thermophiles. The facultative thermophiles display both mesophilic and thermophilic types of metabolism. In their studies of these types from the genus *Bacillus*, which grew well at both 37° and 55°C, Bausum and Matney (7) reported that the organisms appeared to shift from mesophilism to thermophilism between 44° and 52°C.

References

1. Amelunxen, R. E., and A. L. Murdock. 1978. Microbial life at high temperatures: Mechanisms and molecular aspects. In *Microbial Life in Extreme Environments*, ed. D. J. Kushner, 217–278. New York: Academic Press.
2. Anellis, A., J. Lubas, and M. M. Rayman. 1954. Heat resistance in liquid eggs of some strains of the genus *Salmonella*. *Food Res.* 19:377–395.
3. Ansley, S. B., L. L. Campbell, and P. S. Sypherd. 1969. Isolation and amino acid composition of ribosomal proteins from *Bacillus stearothermophilus*. *J. Bacteriol.* 98:568–572.
4. Arca, M., C. Calvori, L. Frontali, and G. Tecce. 1964. The enzymic synthesis of aminoacyl derivatives of soluble ribonucleic acid from *Bacillus stearothermophilus*. *Biochem. Biophys. Acta* 87:440–448.
5. Baird-Parker, A. C., M. Boothroyd, and E. Jones. 1970. The effect of water activity on the heat resistance of heat sensitive and heat resistant strains of salmonellae. *J. Appl. Bacteriol.* 33:515–522.
6. Bassel, A., and L. L. Campbell. 1969. Surface structure of *Bacillus stearothermophilus* ribosomes. *J. Bacteriol.* 98:811–815.
7. Bausum, H. T., and T. S. Matney. 1965. Boundary between bacterial mesophilism and thermophilism. *J. Bacteriol.* 90:50–53.
8. Beaman, T. C., and P. Gerhardt. 1986. Heat resistance of bacterial spores correlated with protoplast dehydration, mineralization, and thermal adaptation. *Appl. Environ. Microbiol.* 52:1242–1246.
9. Beaman, T. C., J. T. Greenamyre, T. R. Corner, H. S. Pankratz, and P. Gerhardt. 1982. Bacterial spore heat resistance correlated with water content, wet density, and protoplast/sporoplast volume ratio. *J. Bacteriol.* 150:870–877.
10. Brock, T. D. 1967. Life at high temperatures. *Science* 158:1012–1019.
11. Campbell, L. L., and O. B. Williams. 1953. The effect of temperature on the

nutritional requirements of facultative and obligate thermophilic bacteria. *J. Bacteriol.* 65:141–145.

12. Carlson, V. R. 1984. Current aseptic packaging techniques. *Food Technol.* 38(12): 47–50.

13. Carpenter, P. L. 1967. *Microbiology*, 2d ed. Philadelphia: W. B. Saunders.

14. Chapman, D. 1967. The effect of heat on membranes and membrane constituents. In *Thermobacteriology*, ed. A. H. Rose, 123–146. New York: Academic Press.

15. Corry, J. E. L. 1974. The effect of sugars and polyols on the heat resistance of salmonellae. *J. Appl. Bacteriol.* 37:31–43.

16. Cunningham, F. E., and H. Lineweaver. 1965. Stabilization of egg-white proteins to pasteurizing temperatures above 60°C. *Food Technol.* 19:1442–1447.

17. Downey, R. J. 1966. Nitrate reductase and respiratory adaptation in *Bacillus stearothermophilus*. *J. Bacteriol.* 91:634–641.

18. Farrell, J., and A. Rose. 1967. Temperature effects on microorganisms. *Ann. Rev. Microbiol.* 21:101–120.

19. Friedman, S. M., R. Axel, and I. B. Weinstein. 1967. Stability of ribosomes and ribosomal ribonucleic acid from *Bacillus stearothermophilus*. *J. Bacteriol.* 93: 1521–1526.

20. Friedman, S. M., and I. B. Weinstein. 1966. Protein synthesis in a subcellular system from *Bacillus stearothermophilus*. *Biochim. Biophys. Acta* 114:593–605.

21. Gaughran, E. R. L. 1947. The saturation of bacterial lipids as a function of temperature. *J. Bacteriol.* 53:506.

22. Gillespy, T. G. 1962. The principles of heat sterilization. *Recent Adv. in Food Sci.* 2:93–105.

23. Goepfert, J. M., and R. A. Biggie. 1968. Heat resistance of *Salmonella typhimurium* and *Salmonella senftenberg* 775W in milk chocolate. *Appl. Microbiol.* 16:1939–1940.

24. Goepfert, J. M., I. K. Iskander, and C. H. Amundson. 1970. Relation of the heat resistance of salmonellae to the water activity of the environment. *Appl. Microbiol.* 19:429–433.

25. Hansen, N. H., and H. Riemann. 1963. Factors affecting the heat resistance of nonsporing organisms. *J. Appl. Bacteriol.* 26:314–333.

26. Horner, K. J., and G. D. Anagnostopoulos. 1975. Effect of water activity on heat survival of *Staphylococcus aureus*, *Salmonella typhimurium* and *Salm. senftenberg*. *J. Appl. Bacteriol.* 38:9–17.

27. Ito, K. A., and K. E. Stevenson. 1984. Sterilization of packaging materials using aseptic systems. *Food Technol.* 38(3):60–62.

28. Jelen, P. 1982. Experience with direct and indirect UHT processing of milk—A Canadian viewpoint. *J. Food Protect.* 45:878–883.

29. Koffler, H. 1957. Protoplasmic differences between mesophiles and thermophiles. *Bacteriol. Rev.* 21:227–240.

30. Koffler, H., and G. O. Gale. 1957. The relative thermostability of cytoplasmic proteins from thermophilic bacteria. *Arch. Biochem. Biophys.* 67:249–251.

31. Koffler, H., G. E. Mallett, and J. Adye. 1957. Molecular basis of biological stability to high temperatures. *Proc. Nat'l. Acad. Sci. U.S.* 43:464–477.

32. Koizumi, J.-I., M. Zhang, T. Imanaka, and S. Aiba. 1990. Does single-amino-acid replacement work in favor of or against improvement of the thermostability of immobilized enzyme? *Appl. Environ. Microbiol.* 56:3612–3614.

33. Levinson, H. S., and M. T. Hyatt. 1964. Effect of sporulation medium on heat resistance, chemical composition, and germination of *Bacillus megaterium* spores. *J. Bacteriol.* 87:876–886.

34. Marmur, J. 1960. Thermal denaturation of deoxyribonucleic acid isolated from a thermophile. *Biochim. Biophys. Acta* 38:342–343.

35. Marr, A. G., and J. L. Ingraham. 1962. Effect of temperature on the composition of fatty acids in *Escherichia coli*. *J. Bacteriol.* 84:1260–1267.

36. Matsubara, H. 1967. Some properties of thermolysin. In *Molecular Mechanisms of Temperature Adapation*, ed. C. L. Prosser, Pub. No. 84, 283–294. Washington, D.C.: American Association for the Advancement of Science.

37. McDonald, W. C., and T. S. Matney. 1963. Genetic transfer of the ability to grow at 55°C in *Bacillus subtilis*. *J. Bacteriol.* 85:218–220.

38. Montville, T. J., and G. M. Sapers. 1981. Thermal resistance of spores from pH elevating strains of *Bacillus licheniformis*. *J. Food Sci.* 46:1710–1712.

39. Nakashio, S., and P. Gerhardt. 1985. Protoplast dehydration correlated with heat resistance of bacterial spores. *J. Bacteriol.* 162:571–578.

40. Ng, H., H. G. Bayne, and J. A. Garibaldi. 1969. Heat resistance of *Salmonella*: The uniqueness of *Salmonella senftenberg* 775W. *Appl. Microbiol.* 17:78–82.

41. Pace, B., and L. L. Campbell. 1967. Correlation of maximal growth temperature and ribosome heat stability. *Proc. Nat'l. Acad. Sci. U.S.* 57:1110–1116.

42. Read, R. B., Jr., J. G. Bradshaw, R. W. Dickerson, Jr., and J. T. Peeler. 1968. Thermal resistance of salmonellae isolated from dry milk. *Appl. Microbiol.* 16: 998–1001.

43. Singleton, R., Jr., and R. E. Amelunxen. 1973. Proteins from thermophilic microorganisms. *Bacteriol. Rev.* 37:320–342.

44. Solowey, M., R. R. Sutton, and E. J. Calesnick. 1948. Heat resistance of *Salmonella* organisms isolated from spray-dried whole-egg powder. *Food Technol.* 2:9–14.

45. Srivastava, R. A. K., and J. N. Baruah. 1986. Culture conditions for production of thermostable amylase by *Bacillus stearothermophilus*. *Appl. Environ. Microbiol.* 52:179–184.

46. Stumbo, C. R. 1973. *Thermobacteriology in Food Processing*, 2d ed. New York: Academic Press.

47. Stumbo, C. R., J. R. Murphy, and J. Cochran. 1950. Nature of thermal death time curves for P. A. 3679 and *Clostridium botulinum*. *Food Technol.* 4:321–326.

48. Sugiyama, H. 1951. Studies on factors affecting the heat resistance of spores of *Clostridium botulinum*. *J. Bacteriol.* 62:81–96.

49. Tanner, F. W., and G. I. Wallace. 1925. Relation of temperature to the growth of thermophilic bacteria. *J. Bacteriol.* 10:421–437.

50. Tendler, M. D., and P. R. Burkholder. 1961. Studies on the thermophilic *Actinomycetes*. I. Methods of cultivation. *Appl. Microbiol.* 9:394–399.

51. Tillotson, J. E. 1984. Aseptic packaging of fruit juices. *Food Technol.* 38(3):63–66.

52. Weerkamp, A., and W. Heinen. 1972. Effect of temperature on the fatty acid composition of the extreme thermophiles, *Bacillus caldolyticus* and *Bacillus caldotenax*. *J. Bacteriol.* 109:443–446.

53. White, H. R. 1963. The effect of variation in pH on the heat resistance of cultures of *Streptococcus faecalis*. *J. Appl. Bacteriol.* 26:91–99.

54. Williams, O. B., and W. J. Robertson. 1954. Studies on heat resistance. VI. Effect of temperature of incubation at which formed on heat resistance of aerobic thermophilic spores. *J. Bacteriol.* 67:377–378.

55. Wisdom, C., and N. E. Welker. 1973. Membranes of *Bacillus stearothermophilus*. Factors affecting protoplast stability and thermostability of alkaline phosphatase and reduced nicotinamide adenine dinucleotide oxidase. *J. Bacteriol.* 114: 1336–1345.

56. Xezones, H., and I. J. Hutchings. 1965. Thermal resistance of *Clostridium botulinum* (62A) spores as affected by fundamental food constituents. *Food Technol.* 19:1003–1005.

15

Preservation of Foods by Drying

The preservation of foods by drying is based on the fact that microorganisms and enzymes need water in order to be active. In preserving foods by this method, one seeks to lower the moisture content to a point where the activities of food spoilage and food-poisoning microorganisms are inhibited. Dried, desiccated, or low-moisture (LM) foods are those that generally do not contain more than 25% moisture and have an a_w between 0.00 and 0.60. These are the traditional dried foods. Freeze-dried foods are also in this category. Another category of shelf-stable foods are those that contain between 15 and 50% moisture and an a_w between 0.60 and 0.85. These are the intermediate-moisture (IM) foods. Some of the microbiological aspects of IM and LM foods are dealt with in this chapter.

PREPARATION AND DRYING OF LOW-MOISTURE FOODS

The earliest uses of food desiccation consisted of exposing fresh foods to sunlight until drying had been achieved. Through this method of drying, which is referred to as sun drying, certain foods may be successfully preserved if the temperature and relative humidity (R.H.) allow. Fruits such as grapes, prunes, figs, and apricots may be dried by this method, which requires a large amount of space for large quantities of product. The drying methods of greatest commercial importance consist of spray, drum, evaporation, and freeze drying.

Preparatory to drying, foods are handled in much the same manner as for freezing, with a few exceptions. In the drying of fruits such as prunes, alkali dipping is employed by immersing the fruits into hot lye solutions of between 0.1 and 1.5%. This is especially true when sun drying is employed. Light-colored fruits and certain vegetables are treated with SO_2 so that levels of between 1,000 and 3,000 ppm may be absorbed. The latter treatment helps to maintain color, conserve certain vitamins, prevent storage changes, and reduce the microbial load. After drying, fruits are usually heat pasteurized at 150° to 185°F for 30 to 70 min.

Similar to the freezing preparation of vegetable foods, blanching or scalding is a vital step prior to dehydration. This may be achieved by immersion from 1 to 8 min, depending on the product. The primary function of this step is to destroy enzymes that may become active and bring about undesirable changes in the finished products. Leafy vegetables generally require less time than peas, beans, or carrots. For drying, temperatures of 140° to 145°F have been found to be safe for many vegetables. The moisture content of vegetables should be reduced below 4% in order to have satisfactory storage life and quality. Many vegetables may be made more stable if given a treatmen with SO_2 or a sulfite. The drying of vegetables is usually achieved by use of tunnel, belt, or cabinet-type driers.

Meat is usually cooked before being dehydrated. The final moisture content after drying should be approximately 4% for beef and pork.

Milk is dried as either whole milk or nonfat skim milk. The dehydration may be accomplished by either the drum or spray method. The removal of about 60% water from whole milk results in the production of **evaporated** milk, which has about 11.5% lactose in solution. **Sweetened condensed** milk is produced by the addition of sucrose or glucose before evaporation so that the total average content of all sugar is about 54%, or over 64% in solution. The stability of sweetened condensed milk is due in part to the fact that the sugars tie up some of the water and make it unavailable for microbial growth.

Eggs may be dried as whole egg powder, yolks, or egg white. Dehydration stability is increased by reducing the glucose content prior to drying. Spray drying is the method most commonly employed.

In **freeze drying** (lyophilization, cryophilization), actual freezing is preceded by the blanching of vegetables and the precooking of meats. The rate at which a food material freezes or thaws is influenced by the following factors (11):

- the temperature differential between the product and the cooling or heating medium,
- the means of transferring heat energy to, from, and within the product (conduction, convection, radiation),
- the type, size, and shape of the package,
- the size, shape, and thermal properties of the product.

Rapid freezing has been shown to produce products that are more acceptable than slow freezing. Rapid freezing allows for the formation of small ice crystals and consequently less mechanical damage to food structure. Upon thawing, fast-frozen foods take up more water and in general display characteristics more like the fresh product than slow-frozen foods. After freezing, the water in the form of ice is removed by sublimation. This process is achieved by various means of heating plus vacuum. The water content of protein foods can be placed into two groups: freezable and unfreezable. Unfreezable (bound) water has been defined as that which remains unfrozen below −30°C. The removal of freezable water takes place during the first phases of drying, and this phase of drying may account for the removal of anywhere from 40 to 95% of the total moisture. The last water to be removed is generally bound water, some of which may be removed throughout the drying process. Unless heat treatment is given prior to freeze drying, freeze-dried foods retain their enzymes. In studies on freeze-dried meats, it has been shown that 40 to 80% of the enzyme activity is not destroyed and may be retained after 16-months storage at $-20°C$ (24). The final product moisture level in freeze-dried foods may be about 2 to 8%, or have an a_w of 0.10 to 0.25 (38).

Freeze drying is generally preferred to high-temperature vacuum drying. Among the disadvantages of the latter compared to the former are the following (17):

- pronounced shrinkage of solids
- migration of dissolved constituents to the surface when drying solids
- extensive denaturation of proteins
- case hardening: the formation of a relatively hard, impervious layer at the surface of a solid, caused by one or more of the first three changes, that slows rates of both dehydration and reconstitution
- formation of hard, impervious solids when drying liquid solution
- undesirable chemical reactions in heat-sensitive materials
- excessive loss of desirable volatile constituents
- difficulty of rehydration as a result of one or more of the other changes

EFFECT OF DRYING ON MICROORGANISMS

Although some microorganisms are destroyed in the process of drying, this process is not lethal per se to microorganisms, and indeed, many types may be recovered from dried foods, especially if poor-quality foods are used for drying and if proper practices are not followed in the drying steps.

Bacteria require relatively high levels of moisture for their growth, with yeasts requiring less and molds still less. Since most bacteria require a_w values above 0.90 for growth, they play no role in the spoilage of dried foods. With respect to the stability of dried foods, Scott (32) has related a_w levels to the probability of spoilage in the following manner. At a_w values of between 0.80 and 0.85, spoilage occurs readily by a variety of fungi in from 1 to 2 weeks. At a $_w$ values of 0.75, spoilage is delayed, with fewer types of organisms in those products that spoil. At a_w 0.70, spoilage is greatly delayed and may not occur during prolonged holding. At a_w of 0.65, very few organisms are known to grow, and spoilage is most unlikely to occur for even up to 2 years. Some authors have suggested that dried foods to be held for several years should be processed so that the final a_w is between 0.65 and 0.75, with 0.70 suggested by most.

At a_w levels of about 0.90, the organisms most likely to grow are yeasts and molds. This value is near the minimum for most normal yeasts. Even though spoilage is all but prevented at a_w less than 0.65, some molds are known to grow very slowly at a_w 0.60 to 0.62 (37). Osmophilic yeasts such as *Zygosaccharomyces rouxii* strains have been reported to grow at an a_w of 0.65 under certain conditions. The most troublesome group of microorganisms in dried foods are the molds, with the *Aspergillus glaucus* group being the most notorious at low a_w values. The minimum a_w values reported for the germination

TABLE 15-1.   Minimum a_w Reported for the Germination and Growth of Food Spoilage Yeasts and Molds

Organism	Minimum a_w
Candida utilis	0.94
Botrytis cinerea	0.93
Rhizopus stolonifer (nigricans)	0.93
Mucor spinosus	0.93
Candida scottii	0.92
Trichosporon pullulans	0.91
Candida zeylanoides	0.90
Saccharomycopsis vernalis	0.89
Alternaria citri	0.84
Aspergillus glaucus	0.70
Aspergillus echinulatus	0.64
Zygosaccharomyces rouxii	0.62

Note: See Table 3-4, for other organisms.

TABLE 15-2.   "Alarm Water" Content for Miscellaneous Foods

Foods	% Water
Whole milk powder	Ca. 8
Dehydrated whole eggs	10–11
Wheat flour	13–15
Rice	13–15
Milk powder (separated)	15
Fat-free dehydrated meat	15
Pulses	15
Dehydrated vegetables	14–20
Starch	18
Dehydrated fruit	18–25

Source: Mossel and Ingram (27).

Note: R. H. = 70%; temperature = 20°C.

and growth of molds and yeasts are presented in Table 15-1. Pitt and Christian (28) found the predominant spoilage molds of dried and high-moisture prunes to be members of the *A. glaucus* group and *Xeromyces bisporus*. Aleuriospores of *X. bisporus* were able to germinate in 120 days at an a_w of 0.605. Generally higher moisture levels were required for both asexual and sexual sporulation.

As a guide to the storage stability of dried foods, the "**alarm water**" content has been suggested. The alarm water content is the water content that should not be exceeded if mold growth is to be avoided. While these values may be used to advantage, they should be followed with caution since a rise of only 1% may be disastrous in some instances (32). The alarm water content for some miscellaneous foods is presented in Table 15-2. In freeze-dried foods, the rule of thumb has been to reduce the moisture level to 2%. Burke and Decareau (7) pointed out that this low level is probably too severe for some foods that might keep well at higher levels of moisture without the extra expense of removing the last low levels of water.

Although drying destroys some microorganisms, bacterial endospores survive, as do yeasts, molds, and many gram-negative and -positive bacteria. In their study of bacteria from chicken meat after freeze drying and rehydration at room temperature, May and Kelly (25) were able to recover about 32% of the original flora. These workers showed that *Staphylococcus aureus* added prior to freeze drying could survive under certain conditions. Some or all foodborne parasites, such as *Trichinella spiralis*, have been reported to survive the drying process (10). The goal is to produce dried foods with a total count of not more than 100,000/g. It is generally agreed that the coliform count of dried foods should be zero or nearly so, and no food-poisoning organisms should be allowed, with the possible exception of low numbers of *Clostridium perfringens*. With the exception of those that may be destroyed by blanching or precooking, relatively fewer organisms are destroyed during the freeze-drying process. More are destroyed during freezing than during dehydration. During freezing, between 5 and 10% of water remains "bound" to other constituents of the medium. This water is removed by drying. Death or injury from drying may result from denaturation in the still-frozen, undried portions due to concentration resulting from freezing, the act of removing the "bound" water, and/or recrystallization of salts or hydrates formed from eutectic solutions (26). When death occurs during dehydration, the rate is highest during the early stages of drying. Young cultures have been reported to be more sensitive to drying than old cultures (12).

The freeze-drying method is one of the best known ways of preserving microorganisms. Once the process has been completed, the cells may remain viable indefinitely. Upon examining the viability of 277 cultures of bacteria, yeasts, and molds that had been lyophilized for 21 years, Davis (9) found that only 3 failed to survive.

STORAGE STABILITY OF DRIED FOODS

In the absence of fungal growth, desiccated foods are subject to certain chemical changes that may result in the food's becoming undesirable upon holding. In dried foods that contain fats and oxygen, oxidative rancidity is a common form of chemical spoilage. Foods that contain reducing sugars undergo a color change known as **Maillard** reaction or nonenzymic browning. This process is brought about when the carbonyl groups of reducing sugars react with amino groups of proteins and amino acids, followed by a series of other more complicated reactions. Maillard-type browning is quite undesirable in fruits and vegetables not only because of the unnatural color but also because of the bitter taste imparted to susceptible foods. Freeze-dried foods also undergo browning if the moisture content is above 2%. Thus, the moisture content should be held below 2% (13).

With regard to a_w, the maximal browning reaction rates in fruits and vegetable products occur in the 0.65 to 0.75 range, while for nonfat dry milk browning, it seems to occur most readily at about 0.70 (38).

Other chemical changes that take place in dried foods include a loss of vitamin C in vegetables, general discolorations, structural changes leading to the inability of the dried product to rehydrate fully, and toughness in the rehydrated, cooked product.

Conditions that favor one or more of the above changes in dried foods generally tend to favor all, so preventative measures against one are also effective against others to varying degrees. At least four methods of minimizing chemical changes in dried foods have been offered:

1. Keep the moisture content as low as possible. Gooding (14) has pointed out that lowering the moisture content of cabbage from 5 to 3% doubles its storage life at 37° C.
2. Reduce the level of reducing sugars as low as possible. These compounds are directly involved in nonenzymic browning, and their reduction has been shown to increase storage stability.
3. When blanching, use water in which the level of leached soluble solids is kept low. Gooding (14) has shown that the serial blanching of vegetables in the same water increases the chances of browning. The explanation given is that the various extracted solutes (presumably reducing sugars and amino acids) are impregnated on the surface of the treated products at relatively high levels.
4. Use sulfur dioxide. The treatment of vegetables prior to dehydration with this gas protects vitamin C and retards the browning reaction. The precise mechanism of this gas in retarding the browning reaction is not well understood, but it apparently does not block reducing groups of hexoses. It has been suggested that it may act as a free radical acceptor.

One of the most important considerations in preventing fungal spoilage of dried foods is the R.H. of the storage environment. If improperly packed and stored under conditions of high R.H., dried foods will pick up moisture from the atmosphere until some degree of equilibrium has been established. Since the first part of the dried product to gain moisture is the surface, spoilage is inevitable; surface growth tends to be characteristic of molds due to their oxygen requirements.

INTERMEDIATE-MOISTURE FOODS

Intermediate-moisture foods (IMF) are characterized by a moisture content of around 15 to 50% and an a_w between 0.60 and 0.85. These foods are shelf stable at ambient temperatures for varying periods of time. While impetus was given to this class of foods during the early 1960s with the development and marketing of intermediate-moisture dog food, foods for human consumption that meet the basic criteria of this class have been produced for many years. These are referred to as traditional IMFs to distinguish them from the newer IMFs. In Table 15-3 are listed some traditional IMFs along with their a_w values. All of these foods have lowered a_w values, which are achieved by withdrawal of water by desorption, adsorption, and/or the addition of permissible additives such as salts and sugars. The developed IMFs are characterized not only by a_w values of 0.60 to 0.85 but by the use of additives such as glycerol, glycols, sorbitol, sucrose, and so forth, as humectants, and by their content of fungistats such as sorbate and benzoate. The remainder of this chapter is devoted to the newly developed IMFs.

TABLE 15-3.   Traditional Intermediate Moisture
Foods

Food Products	a_w Range
Dried fruits	0.60–0.75
Cake and pastry	0.60–0.90
Frozen foods	0.60–0.90
Sugars, syrups	0.60–0.75
Some candies	0.60–0.65
Commercial pastry fillings	0.65–0.71
Cereals (some)	0.65–0.75
Fruit cake	0.73–0.83
Honey	0.75
Fruit juice concentrates	0.79–0.84
Jams	0.80–0.91
Sweetened condensed milk	0.83
Fermented sausages (some)	0.83–0.87
Maple syrup	0.90
Ripened cheeses (some)	0.96
Liverwurst	0.96

Preparation of IMF

Since *S. aureus* is the only bacterium of public health importance that can grow
at a_w values near 0.86, an IMF can be prepared by formulating the product so
that its moisture content is between 15 and 50 percent, adjusting the a_w to a
value below 0.86 by use of humectants, and adding an antifungal agent to
inhibit the rather large number of yeasts and molds that are known to be
capable of growth at a_w values above 0.70. Additional storage stability is
achieved by reducing pH. While this is essentially all that one needs to produce
an IMF, the actual process and the achievement of storage stability of the
product are considerably more complicated.

The determination of the a_w of a food system is discussed in Chapter 3. One
can use also Raoult's law of mole fractions where the number of moles of water
in a solution is divided by the total number of moles in the solution (3):

$$a_w = \frac{\text{Moles of } H_2O}{\text{Moles of } H_2O + \text{Moles of solute}}$$

For example, a liter of water contains 55.5 moles. Assuming that the water is
pure,

$$a_w = \frac{55.5}{55.5 + 0} = 1.00$$

If, however, 1 mole of sucrose is added,

$$a_w = \frac{55.5}{55.5 + 1} = 0.98$$

This equation can be rearranged to solve for the number of moles of solute required to give a specified a_w value. While the foregoing is not incorrect, it is highly oversimplified, since food systems are complex by virtue of their content of ingredients that interact with water and with each other in ways that are difficult to predict. Sucrose, for example, decreases a_w more than expected so that calculations based on Raoult's law may be meaningless (4). The development of techniques and methods to predict a_w in IMF more accurately has been the concern of several investigators (3, 16, 31), while an extensive evaluation of available a_w-measuring instruments and techniques has been carried out by Labuza et al. (22).

In preparing IMF, water may be removed either by adsorption or desorption. By adsorption, food is first dried (often freeze dried) and then subjected to controlled rehumidification until the desired composition is achieved. By desorption, the food is placed in a solution of higher osmotic pressure so that at equilibrium, the desired a_w is reached (30). While identical a_w values may be achieved by these two methods, IMF produced by adsorption is more inhibitory to microorganisms than that produced by desorption (see below). When sorption isotherms of food materials are determined, adsorption isotherms sometimes reveal that less water is held than for desorption isotherms at the same a_w. The sorption isotherm of a food material is a plot of the amount of water adsorbed as a function of the relative humidity or activity of the vapor space surrounding the material. It is the amount of water that is held after equilibrium has been reached at a constant temperature (21). Sorption isotherms may be either adsorption or desorption, and when the former procedure results in the holding of more water than the latter, the difference is ascribed to a hysteresis effect. This, as well as other physical properties associated with the preparation of IMF, has been discussed by Labuza (21), Sloan et al. (34), and others and will not be dealt with further here. The sorption properties of an IMF recipe, the interaction of each ingredient with water and with other ingredients, and the order of mixing of ingredients add to the complications of the overall IMF preparation procedures, and both direct and indirect effects on the microbiology of these products may result.

The following general techniques are employed to change the water activity in producing an IMF (20):

Moist infusion. Solid food pieces are soaked and/or cooked in an appropriate solution to give the final product the desired water level (desorption).

Dry infusion. Solid food pieces are first dehydrated and then infused by soaking in a solution containing the desired osmotic agents (adsorption).

Component blending. All IMF components are weighed, blended, cooked, and extruded or otherwise combined to give the finished product the desired a_w.

Osmotic drying. Foods are dehydrated by immersion in liquids with a water activity lower than that of the food. When salts and sugars are used, two simultaneous countercurrent flows develop: solute diffuses from solution into food, and water diffuses out of food into solution.

The foods in Table 15-4 were prepared by moist infusion for military use. The 1-cm-thick slices equilibrated following cooking at 95° to 100°C in water and holding overnight in a refrigerator. Equilibration is possible without

TABLE 15-4. Preparation of Representative Intermediate Moisture Foods by Equilibration

Initial Material	% H$_2$O	Processing	Equilibrated Product % H$_2$O	Equilibrated Product a$_w$	Ratio: Solution Weight/Initial Weight	% Components of Solution Glycerol	Water	NaCl	Sucrose	K-sorbate	Na-benzoate
Tuna, canned water pack pieces, 1 cm thick	60.0	cold soak	38.8	0.81	0.59	53.6	38.6	7.1	—	0.7	—
Carrots, diced 0.9 cm, cooked	88.2	cook 95–98°C, refrig.	51.5	0.81	0.48	59.2	34.7	5.5	—	0.6	—
Macaroni, elbow, cooked, drained	63.0	cook 95–98°C, refrig.	46.1	0.83	0.43	42.7	48.8	8.0	—	0.5	—
Pork loin, raw, 1 cm thick	70.0	cook 95–98°C, refrig.	42.5	0.81	0.73	45.6	43.2	10.5	—	0.7	—
Pineapple, canned, chunks	73.0	cold soak	43.0	0.85	0.46	55.0	21.5	—	23.0	0.5	—
Celery, 0.6 cm cross cut, blanch	94.7	cold soak	39.6	0.83	0.52	68.4	25.2	5.9	—	0.5	—
Beef, ribeye, 1 cm thick	70.8	cook 95–98°C, refrig.	—	0.86	2.35	87.9	—	10.1	—	—	2.0

Source: Brockmann (6); copyright © 1970 by Institute of Food Technologists.

cooking over prolonged periods under refrigeration (6). IMF deep-fried catfish, with raw samples of about 2 g each, has been prepared by the moist infusion method (8). Pet foods are more often prepared by component blending. The general composition of one such product is given in Table 15-5. The general way in which a product of this type is made is as follows. The meat and meat products are ground and mixed with liquid ingredients. The resulting slurry is cooked or heat treated and later mixed with the dry ingredient mix (salts, sugars, dry solids, and so on). Once the latter are mixed into the slurry, an additional cook or heat process may be applied prior to extrusion and packaging. The extruded material may be shaped in the form of patties or packaged in loose form. The composition of a model IMF product called Hennican is given in Table 15-6. According to Acott and Labuza (1), this is an adaptation of pemmican, an Indian trail and winter storage food made of buffalo meat and berries. Hennican is the name given to the chicken-based IMF. Both moisture content and a_w of this system can be altered by adjustment of ingredient mix.

The humectants commonly used in pet food formulations are propylene glycol, polyhydric alcohols (sorbitol, for example), polyethylene glycols, glycerol, sugars (sucrose, fructose, lactose, glucose, and corn syrup), and salts (NaCl, KCl, and so on). The commonly used mycostats are propylene glycol, K-sorbate, Na-benzoate, and others. The pH of these products may be as low as 5.4 and as high as 7.0.

Microbial Aspects of IMF

The general a_w range of IMF products makes it unlikely that gram-negative bacteria will proliferate. This is true also for most gram-positive bacteria with the exception of cocci, some sporeformers, and lactobacilli. In addition to the inhibitory effect of lowered a_w, antimicrobial activity results from an interaction of pH, Eh, added preservatives (including some of the humectants), the competitive microflora, generally low storage temperatures, and the pasteurization or other heat processes applied during processing.

The fate of *S. aureus* S-6 in IM pork cubes with glycerol at 25°C is illustrated

TABLE 15-5.   Typical Composition of Soft Moist or Intermediate Moisture Dog Food

Ingredient	%
Meat by-products	32.0
Soy flakes	33.0
Sugar	22.0
Skimmed milk, dry	2.5
Calcium and phosphorus	3.3
Propylene glycol	2.0
Sorbitol	2.0
Animal fat	1.0
Emulsifier	1.0
Salt	0.6
Potassium sorbate	0.3
Minerals, vitamins, and color	0.3

Source: Kaplow (19); copyright © 1970 by Institute of Food Technologists.

TABLE 15-6.   Composition of Hennican

Components	Amount (wt basis, %)
Raisins	30
Water	23
Peanuts	15
Chicken (freeze dried)	15
Nonfat dry milk	11
Peanut butter	4
Honey	2

Source: Acott and Labuza (1); copyright © 1975 by Institute of Food Technologists.

Note: Moisture content = 41 g water/100 g solids, a_w = 0.85.

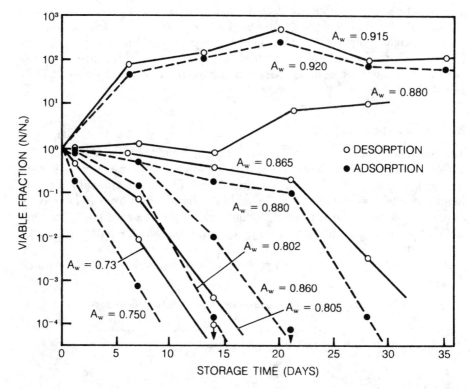

FIGURE 15-1.   Viability of *Staphylococcus aureus* in IMF systems: pork cubes and glycerol at 25°C.
From Plitman et al. (29), copyright © 1973 by Institute of Food Technologists.

in Figure 15-1. In this desorption IM pork at a_w 0.88, the numbers remained stationary for about 15 days and then increased slightly, while in the adsorption IM system at the same a_w the cells died off slowly during the first 3 weeks and thereafter more rapidly. At all a_w values below 0.88, the organisms died off, with the death rate considerably higher at 0.73 than at higher values (29). Findings similar to these have been reported by Haas et al. (15), who found that an inoculum of 10^5 staphylococci in a meat-sugar system at a_w 0.80 decreased to 3×10^3 after 6 days and to 3×10^2 after 1 month. Although growth of *S. aureus* has been reported to occur at an a_w of 0.83, enterotoxin is not produced below a_w 0.86 (35). It appears that enterotoxin A is produced at lower values of a_w than enterotoxin B (36).

Using the model IM Hennican at pH 5.6 and a_w 0.91, Boylan et al. (5) showed that the effectiveness of the IM system against *S. aureus* F265 was a function of both pH and a_w. Adsorption systems are more destructive to microorganisms than desorption systems. Labuza et al. (23) found that the reported minimum a_ws apply in IMF systems when desorption systems are involved but that growth minima are much higher if the food is prepared by an adsorption method. *S. aureus* was inhibited at a_w 0.9 in adsorption while values between 0.75 and 0.84 were required for desorption systems. A similar effect was noted for molds, yeasts, and pseudomonads.

In regard to the effect of IMF systems on the heat destruction of bacteria, heat resistance increases as a_w is lowered and the degree of resistance is dependent on the compounds employed to control a_w (see Table 14-2). In a study of the death rate of salmonellae and staphylococci in the IM range of about 0.8 at pasteurization temperatures (50°–65°C), it has been found that cell death occurs under first-order kinetics (18). These investigators confirmed the findings of many others that the heat destruction of vegetative cells is at a minimum in the IM range, especially when a solid menstrum is employed. Some D values for the thermal destruction of *Salmonella senftenberg* 775W at various a_w values are given in Table 14-2.

With respect to molds in IMF systems, these products would be made quite stable if a_w were reduced to around 0.70, but a dry-type product would then result. A large number of molds are capable of growth in the 0.80 range, and the shelf life of IM pet foods is generally limited by the growth of these

TABLE 15-7.   Time for Growth of Microbes in Inoculated Dog Food with Inhibitors, pH 5.4

Inhibitor	Storage Conditions	
	$a_w = 0.85$ 9-Month Storage	$a_w = 0.88$ 6-Month Storage
No inhibitor added	A. niger—2 wk A. glaucus—1 wk S. epider.—2 wk	A. niger—1 wk A. glaucus—1 wk S. epider.—½ wk
K-sorbate (0.3%)	No mold S. epider.—25 wk	A. niger—5 wk S. epider.—3½ wk
Ca-propionate (0.3%)	A. niger—25 wk A. glaucus—25 wk S. epider.—3½ wk	A. glaucus—2 wk S. epider.—1½ wk

Source: Acott et al. (2); copyright © 1976 by Institute of Food Technologists.

Note: Mold—first visible sign; bacteria—2 log cycle increase.

organisms. The interaction of various IM parameters on the inhibition of molds was shown by Acott et al. (2). In their evaluation of seven chemical inhibitors used alone and in combination to inhibit *Aspergillus niger* and *A. glaucus* inocula, propylene glycol was the only approved agent that was effective alone. None of the agents tested could inhibit alone at a_w 0.88, but in combination the product was made shelf stable. All inhibitors were found to be more effective at pH 5.4 and a_w 0.85 than at pH 6.3. Growth of the two fungi occurred in 2 weeks in the a_w 0.85 formulation without inhibitors but did not occur until 25 weeks when K-sorbate and Ca-propionate were added (Table 15-7). Growth of *Staphylococcus epidermidis* was inhibited by both fungistats, with inhibition being greater at a_w 0.85 than at 0.88. This is probably an example of the combined effects of pH, a_w, and other growth parameters on the growth inhibition of microorganisms in IMF systems.

Storage Stability of IMF

The undesirable chemical changes that occur in dried foods occur also in IM foods. Lipid oxidation and Maillard browning are at their optima in the general IMF ranges of a_w and percentage moisture. However, there are indications that the maximum rate for Maillard browning occurs in the 0.4 to 0.5 a_w range, especially when glycerol is used as humectant (38).

The storage of IMFs under the proper conditions of humidity is imperative in preventing moldiness and for overall shelf stability. The measurement of equilibrium relative humidity (ERH) is of importance in this regard. ERH is an expression of the desorbable water present in a food product and is defined by the following equation:

$$ERH = (P_{equ}/P_{sat})T, P = 1 \text{ atm}$$

where P_{equ} is partial pressure of water vapor in equilibrium with the sample in air at 1 atmosphere total pressure and temperature T; P_{sat} is the saturation partial vapor pressure of water in air at a total pressure of 1 atmosphere and temperature T (16). A food in moist air exchanges water until the equilibrium partial pressure at that temperature is equal to the partial pressure of water in the moist air, so that the ERH value is a direct measure of whether moisture will be sorbed or desorbed. In the case of foods packaged or wrapped in moisture-impermeable materials, the relative humidity of the food-enclosed atmosphere is determined by the ERH of the product, which in turn is controlled by the nature of the dissolved solids present, ratio of solids to moisture, and the like (33). Both traditional and newer IMF products have longer shelf stability under conditions of lower ERH.

In addition to the direct effect of packaging on ERH, gas-impermeable packaging affects the Eh of packaged products with consequent inhibitory effects on the growth of aerobic microorganisms.

References

1. Acott, K. M., and T. P. Labuza. 1975. Inhibition of *Aspergillus niger* in an intermediate moisture food system. *J. Food Sci.* 40:137–139.

2. Acott, K. M., A. E. Sloan, and T. P. Labuza. 1976. Evaluation of antimicrobial agents in a microbial challenge study for an intermediate moisture dog food. *J. Food Sci.* 41:541–546.

3. Bone, D. 1973. Water activity in intermediate moisture foods. *Food Technol.* 27(4):71–76.

4. Bone, D. P. 1969. Water activity—Its chemistry and applications. *Food Prod. Dev.* 3(5):81–94.

5. Boylan, S. L., K. A. Acott, and T. P. Labuza. 1976. *Staphylococcus aureus* challenge study in an intermediate moisture food. *J. Food Sci.* 41:918–921.

6. Brockmann, M. C. 1970. Development of intermediate moisture foods for military use. *Food Technol.* 24:896–900.

7. Burke, R. F., and R. V. Decareau. 1964. Recent advances in the freeze-drying of food products. *Adv. Food Res.* 13:1–88.

8. Collins, J. L., and A. K. Yu. 1975. Stability and acceptance of intermediate moisture, deep-fried catfish. *J. Food Sci.* 40:858–863.

9. Davis, R. J. 1963. Viability and behavior of lyophilized cultures after storage for twenty-one years. *J. Bacteriol.* 85:486–487.

10. Desrosier, N. W. 1963. *The Technology of Food Preservation.* New York: Van Nostrand Reinhold.

11. Fennema, O., and W. D. Powrie. 1964. Fundamentals of low-temperature food preservation. *Adv. Food Res.* 13:219–347.

12. Fry, R. M., and R. I. N. Greaves. 1951. The survival of bacteria during and after drying. *J. Hyg.* 49:220–246.

13. Goldblith, S. A., and M. Karel. 1966. Stability of freeze-dried foods. In *Advances in Freeze-Drying*, ed. L. Rey, 191–210. Paris: Hermann.

14. Gooding, E. G. B. 1962. The storage behaviour of dehydrated foods. In *Recent Advances in Food Science*, ed. J. Hawthorn and J. M. Leitch, 2:22–38. London: Butterworths.

15. Haas, G. J., D. Bennett, E. B. Herman, and D. Collette. 1975. Microbial stability of intermediate moisture foods. *Food Prod. Dev.* 9(4):86–94.

16. Hardman, T. M. 1976. Measurement of water activity. Critical appraisal of methods. In *Intermediate Moisture Foods*, ed. R. Davies, G. G. Birch, and K. J. Parker, 75–88. London: Applied Science.

17. Harper, J. C., and A. L. Tappel. 1957. Freeze-drying of food products. *Adv. Food Res.* 7:171–234.

18. Hsieh, F.-H., K. Acott, and T. P. Labuza. 1976. Death kinetics of pathogens in a pasta product. *J. Food Sci.* 41:516–519.

19. Kaplow, M. 1970. Commercial development of intermediate moisture foods. *Food Technol.* 24:889–893.

20. Karel, M. 1976. Technology and application of new intermediate moisture foods. In *Intermediate Moisture Foods*, ed. R. Davies, G. G. Birch, and K. J. Parker, 4–31. London: Applied Science.

21. Labuza, T. P. 1968. Sorption phenomena in foods. *Food Technol.* 22:263–272.

22. Labuza, T. P., K. Acott, S. R. Tatini, R. Y. Lee, I. Flink, and W. McCall. 1976. Water activity determination: A collaborative study of different methods. *J. Food Sci.* 41:910–917.

23. Labuza, T. P., S. Cassil, and A. J. Sinskey. 1972. Stability of intermediate moisture foods. 2. Microbiology. *J. Food Sci.* 37:160–162.

24. Matheson, N. A. 1962. Enzymes in dehydrated meat. In *Recent Advances in Food Science*, ed. J. Hawthorn and J. M. Leitch, 2:57–64. London: Butterworths.

25. May, K. N., and L. E. Kelly. 1965. Fate of bacteria in chicken meat during freeze-dehydration, rehydration, and storage. *Appl Microbiol.* 13:340–344.

26. Meryman. H. T. 1966. Freeze-drying. In *Cryobiology*, ed. H. T. Meryman, ch. 13. New York: Academic Press.

27. Mossel, D. A. A., and M. Ingram. 1955. The physiology of the microbial spoilage of foods. *J. Appl. Bacteriol.* 18:232–268.
28. Pitt, J. I., and J. H. B. Christian. 1968. Water relations of xerophilic fungi isolated from prunes. *Appl. Microbiol.* 16:1853–1858.
29. Plitman, M., Y. Park, R. Gomez, and A. J. Sinskey. 1973. Viability of *Staphylococcus aureus* in intermediate moisture meats. *J. Food Sci.* 38:1004–1008.
30. Robson, J. N. 1976. Some introductory thoughts on intermediate moisture foods. *In Intermediate Moisture Foods*, ed. R. Davies, G. G. Birch, and K. J. Parker, 32–42. London: Applied Science.
31. Ross, K. D. 1975. Estimation of water activity in intermediate moisture foods. *Food Technol.* 29(3):26–34.
32. Scott, W. J. 1957. Water relations of food spoilage microorganisms. *Adv. Food Res.* 1:83–127.
33. Seiler, D. A. L. 1976. The stability of intermediate moisture foods with respect to mould growth. In *Intermediate Moisture Foods*, ed. R. Davies, G. G. Birch, and K. J. Parker, 166–181. London: Applied Science.
34. Sloan, A. E., P. T. Waletzko, and T. P. Labuza. 1976. Effect of order-of-mixing on a_w-lowering ability of food humectants. *J. Food Sci.* 41:536–540.
35. Tatini, S. R. 1973. Influence of food environments on growth of *Staphylococcus aureus* and production of various enterotoxins. *J. Milk Food Technol.* 36:559–563.
36. Troller, J. A. 1972. Effect of water activity on enterotoxin A production and growth of *Staphylococcus aureus*. *Appl. Microbiol.* 24:440–443.
37. Troller, J. A. 1983. Effect of low moisture environments on the microbial stability of foods. In *Food Microbiology*, ed. A. H. Rose, 173–198. New York: Academic Press.
38. Troller, J. A., and J. H. B. Christian. 1978. *Water Activity and Food*. New York: Academic Press.
39. Warmbier, H. C., R. A. Schnickels, and T. P. Labuza. 1976. Effect of glycerol on nonenzymatic browning in a solid intermediate moisture model food system. *J. Food Sci.* 41:528–531.

16

Fermented Foods and Related Products of Fermentation

Numerous food products owe their production and characteristics to the activities of microorganisms. Many of these, including such foods as ripened cheeses, pickles, sauerkraut, and fermented sausages, are preserved products in that their shelf life is extended considerably over that of the raw materials from which they are made. In addition to being made more shelf stable, all fermented foods have aroma and flavor characteristics that result directly or indirectly from the fermenting organisms. In some instances, the vitamin content of the fermented food is increased along with an increased digestibility of the raw materials. The fermentation process reduces the toxicity of some foods (for example, gari and peujeum), while others may become extremely toxic during fermentation (as in the case of bongkrek). From all indications, no other single group or category of foods or food products is as important as these are and have been relative to nutritional well-being throughout the world. Included in this chapter along with the classical fermented foods are such products as coffee beans, wines, and distilled spirits, for these and similar products either result from or are improved by microbial fermentation activities.

The microbial ecology of food and related fermentations has been studied for many years in the case of ripened cheeses, sauerkraut, wines, and so on, and the activities of the fermenting organisms are dependent on the intrinsic and extrinsic parameters of growth discussed in Chapter 2. For example, when the natural raw materials are acidic and contain free sugars, yeasts develop readily, and the alcohol they produce restricts the activities of most other naturally contaminating organisms (for example, the fermentation of fruits to produce wines). If, on the other hand, the acidity of a plant product permits good bacterial growth and at the same time the product is high in simple sugars, lactic acid bacteria may be expected to develop, and the addition of low levels of NaCl will ensure their growth preferential to yeasts (as in sauerkraut fermentation). Products that contain polysaccharides but no significant levels of simple sugars are normally stable to the activities of yeasts and lactic acid bacteria due to the lack of amylase in most of these organisms. In order to effect their fermentation, an exogenous source of saccharifying enzymes must be supplied. The use of barley malt in the brewing and distilling industries is an example of this. The fermentation of sugars to ethanol that results from malting is then carried out by yeasts. The use of **koji** in the fermentation of soybean products is another example of the way in which alcoholic and lactic acid fermentations may be carried out on products that have low levels of sugars but high levels of starches and proteins. While the saccharifying enzymes of barley malt arise from germinating barley, the enzymes of koji are produced by *Aspergillus oryzae* growing on soaked or steamed rice or other cereals (the commercial product **takadiastase** is prepared by growing *A. oryzae* on wheat bran). The koji hydrolysates may be fermented by lactic acid bacteria and yeasts, as is the case for soy sauce, or the koji enzymes may act directly on soybeans in the production of products such as Japanese miso.

FERMENTATION—DEFINED AND CHARACTERIZED

Fermentation is the metabolic process in which carbohydrates and related compounds are oxidized with the release of energy in the absence of any external electron acceptors. The final electron acceptors are organic compounds produced directly from the breakdown of the carbohydrates. Consequently, only partial oxidation of the parent compound occurs, and only a small amount of energy is released during the process. As fermenting organisms, the lactic acid bacteria lack functional heme-linked electron transport systems or cytochromes, and they obtain their energy by substrate-level phosphorylation while oxidizing carbohydrates; they do not have a functional Krebs cycle. The products of fermentation consist of some that are more reduced than others.

The word **fermentation** has had many shades of meaning in the past. According to one dictionary definition, it is "a process of chemical change with effervescence . . . a state of agitation or unrest . . . any of various transformations of organic substances." The word came into use before Pasteur's studies on wines. Prescott and Dunn (86) and Doelle (25) have discussed the history of the concept of fermentation, and the former authors note that in the broad sense in which the term is commonly used, it is "a process in which chemical changes are brought about in an organic substrate through the action of enzymes elaborated by microorganisms." It is in this broad context that the term is used in this chapter. In the brewing industry, a **top fermentation** refers to the use of a yeast strain that carries out its activity at the upper parts of a large vat such as in the production of ale; a **bottom fermentation** requires the use of a yeast strain that will act in lower parts of the vat—such as in the production of lager beer.

LACTIC ACID BACTERIA

This group is comprised of at least eight genera in contrast to the four genera in the ninth edition of *Bergey's Manual*. The traditional genera of *Lactobacillus*, *Leuconostoc*, *Pediococcus*, and *Streptococcus* have been expanded to include *Carnobacterium*, *Enterococcus*, *Lactococcus*, and *Vagococcus*. The carnobacteria represent strains that were once classified as lactobacilli, and the other three genera are comprised of strains formerly classified as streptococci (see Chaps. 2 and 17 for further details). The species once classified as *Lactobacillus hordniae* and *L. xylosus* have been transferred to the genus *Lactococcus*. With the enterococci and lactococci having been removed from streptococci, the most important member of the latter group of importance in foods is *S. thermophilus*. *S. diacetilactis* is classified as a citrate utilizing strain of *L. Lactis* subsp. *lactis*. Although *Lactococcus cremoris* has been reduced to a subspecies of *L. lactis*, this biovar is important in cheddar cheese production.

The history of our knowledge of the lactic streptococci and their ecology have been reviewed by Sandine et al. (97). These authors believe that plant matter is the natural habitat of this group, but they note the lack of proof of a plant origin for *Lactococcus cremoris*. It has been suggested that plant streptococci may be the ancestral pool from which other species and strains developed (72).

Related to the above four genera of lactic acid bacteria in some respects but generally not considered to fit the group are some *Aerococcus* spp; some *Erysipelothrix* spp; and *Eubacterium*, *Microbacterium*, *Peptostreptococcus*, and *Propionibacterium* (55, 63).

While the lactic acid group is loosely defined with no precise boundaries, all members share the property of producing lactic acid from hexoses. Kluyver divided the lactic acid bacteria into two groups based on end products of glucose metabolism. Those that produce lactic acid as the major or sole product of

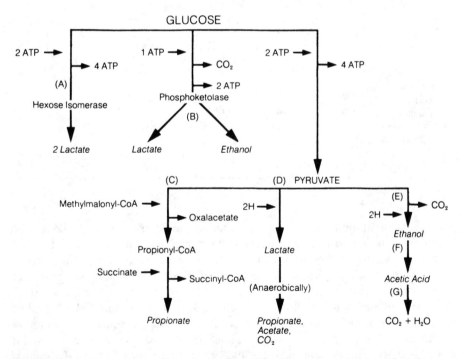

FIGURE 16-1.   Generalized pathways for the production of some fermentation products from glucose by various organisms. *A*: homofermentative lactics; *B*: heterofermentative lactics; *C* and *D*: *Propionibacterium* (see Fig. 16-3); *E*: *Saccharomyces* spp.; *F*: *Acetobacter* spp.; *G*: *Acetobacter* "overoxidizers."

TABLE 16-1.   Homo- and Heterofermentative Lactic Acid Bacteria

	Homofermentative			Heterofermentative	
Organisms	Lactate Config- uration	%G + C	Organisms	Lactate Config- uration	%G + C
Lactobacillus			*Lactobacillus*		
L. acidophilus	DL	36.7	*L. brevis*	DL	42.7–46.4
L. bulgaricus	D(−)	50.3	*L. buchneri*	DL	44.8
L. casei	L(+)	46.4	*L. cellobiosus*	DL	53
L. coryniformis	DL	45	*L. confusus*	DL	44.5–45.0
L. curvatus	DL	43.9	*L. coprophilus*	DL	41.0
L. delbrueckii	D(−)	50	*L. fermentum*	DL	53.4
L. helveticus	DL	39.3	*L. hilgardii*	DL	40.3
L. jugurti	DL	36.5–39.0	*L. sanfrancisco*	DL	38.1–39.7
L. jensenii	D(−)	36.1	*L. trichodes*	DL	42.7
L. lactis	D(−)	50.3	*L. viridescens*	DL	35.7–42.7
L. leichmannii	D(−)	50.8	*Leuconostoc*		
L. plantarum	DL	45	*L. cremoris*	D(−)	39–42
L. salivarius	L(+)	34.7	*L. dextranicum*	D(−)	38–39
Pediococcus			*L. lactis*	D(−)	43–44
P. acidilactici	DL	44.0	*L. mesenteroides*	D(−)	39–42
P. cerevisiae	DL		*L. oenos*	D(−)	39–40
P. pentosaceus	DL	38	*L. paramesenteroides*	D(−)	38–39
Streptococcus			*L. gelidum*	D(−)	37
S. bovis	D(−)	38–42	*L. carnosum*	D(−)	39
S. thermophilus	D(−)	40	*Carnobacterium*		
Lactococcus			*C. divergens*		33.0–36.4
L. lactis subsp. *lactis*	D(−)	38.4–38.6	*C. mobile*		35.5–37.2
L. lactis subsp. cremoris	D(−)	38.0–40.0	*C. gallinarum*		34.3–36.4
L. lactis subsp. hordniae		35.2	*C. piscicola*		33.7–36.4
L. garvieae		38.3–38.7			
L. plantarum		36.9–38.1			
L. raffinolactis		40.0–43			
Vagococcus					
V. fluvialis		33.6			
V. salmoninarum		36.0–36.5			

glucose fermentation are designated **homofermentative** (Fig. 16-1*A*). The homofermentative pattern is observed when glucose is metabolized but not necessarily when pentoses are metabolized, for some homolactics produce acetic and lactic acids when utilizing pentoses. Also, the homofermentative character of homolactics may be shifted for some strains by altering cultural conditions such as glucose concentration, pH, and nutrient limitation (14, 63). The homolactics are able to extract about twice as much energy from a given quantity of glucose as are the heterolactics. Those lactics that produce equal molar amounts of lactate, carbon dioxide, and ethanol from hexoses are designated **heterofermentative** (Fig. 16-1*B*). All members of the genera *Pediococcus*, *Streptococcus*, *Lactococcus*, and *Vagococcus* are homofermenters, along with some of the lactobacilli, while all *Leuconostoc* spp., as well as some lactobacilli,

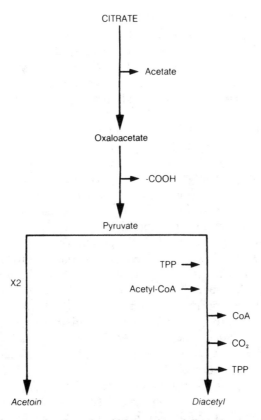

FIGURE 16-2.   The general pathway by which acetoin and diacetyl are produced from citrate by Group N lactococci and *Leuconostoc* spp. Pyruvate may be produced from lactate, and acetylCoA from acetate. (For further details, see refs. 18, 107, and 108.)

are heterofermenters (Table 16-1). The heterolactics are more important than the homolactics in producing flavor and aroma components such as acetylaldehyde and diacetyl (Fig. 16-2).

The end product differences between homo- and heterofermenters when glucose is attacked are a result of basic genetic and physiological differences (Fig. 16-1). The homolactics possess the enzymes aldolase and hexose isomerase but lack phosphoketolase (Fig. 16-1*A*). They use the Embden-Meyerhof-Parnas (EMP) pathway toward their production of 2 lactates/glucose molecule. The heterolactics, on the other hand, have phosphoketolase but do not possess aldolase and hexose isomerase, and instead of the EMP pathway for glucose degradation, these organisms use the hexose monophosphate or pentose pathway (Fig. 16-1*B*).

The genus *Lactobacillus* has been subdivided classically into three subgenera: *Betabacterium*, *Streptobacterium*, and *Thermobacterium*. All of the heterolactic lactobacilli in Table 16-1 are betabacteria. The streptobacteria (for example, *L. casei* and *L. plantarum*) produce up to 1.5% lactic acid with an optimal growth temperature of 30°C, while the thermobacteria (such as *L. acidophilus* and *L. bulgaricus*) can produce up to 3% lactic acid and have an optimal temperature of 40°C (67).

In terms of their growth requirements, the lactic acid bacteria require preformed amino acids, B vitamins, and purine and pyrimidine bases—hence their use in microbiological assays for these compounds. Although they are mesophilic, some can grow below 5°C and some as high as 45°C. With respect to growth pH, some can grow as low as 3.2, some as high as 9.6, and most can grow in the pH range 4.0 to 4.5. The lactic acid bacteria are only weakly proteolytic and lipolytic (109).

In the past, the taxonomy of the lactic acid bacteria has been based largely on the gram reaction, general lack of ability to produce catalase, the production of lactic acid of a given configuration, along with the ability to ferment various carbohydrates. More recently, rRNA sequencing, DNA base composition, DNA homology, cell wall peptidoglycan type, and immunologic specificity of enzymes are used (see Chap. 2). While these organisms may be thought of as having derived from a common ancestor, high degrees of diversity as well as high degrees of relatedness are found within the group. The DNA base composition (expressed as moles % G + C) of those lactics studied varies from a low of around 35 for *L. salivarius* to a high of 53 for *L. fermentum* (127). There is considerable overlap between the homo- and heterolactics (Table 16-1). On the basis of DNA base composition, Miller et al. (68, 69) placed 15 species of lactobacilli into three groups. Group I contained those within 32.4 to 38.3 moles% G + C and includes *L. jugurti*, *L. helveticus*, *L. salivarius*, *L. bulgaricus*, and *L. (jugurti) bulgaricus*. Group II contained 42.7 to 48.0 moles % G + C, and includes *L. buchneri*, *L. brevis*, *L. casei*, *L. viridescens*, and *L. plantarum*. Group III contained 49.0 to 51.9 moles % G + C, and includes *L. lactis*, *L. leichmannii*, *L. delbrueckii*, *L. fermentum*, and *L. cellobiosus*. While the Group I species are all homofermenters, Groups II and III contain both homo- and heterolactic types. The diversity of relatedness of this group has been shown also by both DNA hybridization and immunologic studies. With respect to DNA hybridization, strains of *L. jugurti* and *L. helveticus* have been shown to share between 85 and 100% DNA homology; *L. leichmannii*, *L. delbrueckii*, *L. lactis*, and *L. bulgaricus* to share between 78 and 100%; and *L. fermentum* and *L. cellobiosus* to share from 77 to 100% DNA homology (63). Simonds et al. (103) have shown that *L. bulgaricus* shares 86% DNA homology with *L. lactis*, 4.8% with *L. helveticus*, and none with *L. jugurti*.

DNA-DNA hybridization studies in the genus *Leuconostoc* have been reported (53). Of 45 strains representing six species, six homology groups were distinguished using four reference DNA preparations. Of 19 strains of *L. mesenteroides*, three different hybridization groups were determined.

In regard to immunologic methods, Gasser and Gasser (36) showed that antisera prepared against NAD-dependent D-lactic dehydrogenases of three lactobacilli reacted against crude extracts of almost all other species of lactobacilli containing the enzyme. Extracts of *Leuconostoc* spp. cross reacted with anti-D-lactic dehydrogenases. In a somewhat similar manner, it has been shown that antisera prepared against purified *E. faecalis* fructose diphosphate aldolase reacted to varying degrees with the aldolases of all homofermentative lactics, including some other streptococci, pediococci, and lactobacilli (64).

The cell wall mucopeptides of lactics and other bacteria have been reviewed by Schleifer and Kandler (98) and Williams (127). While there appears to be wide variation within most of the lactic acid genera, the homofermentative lactobacilli of the subgenus *Thermobacterium* appear to be the most homo-geneous in this regard in having L-lysine in the peptidoglycan peptide chain and D-aspartic acid as the interbridge peptide. The Group N streptococci have similar wall mucopeptides.

The measurement of molar growth yields provides information on ferment-ing organisms relative to their fermentation substrates and pathways. By this concept, the μg dry weight of cells produced per μmole substrate fermented is determined as the **molar yield constant**, indicated by Y. It is tacitly assumed that essentially none of the substrate carbon is used for cell biosynthesis, that oxygen does not serve as an electron or hydrogen acceptor, and that all of the energy derived from the metabolism of the substrate is coupled to cell biosynthesis (45). When the substrate is glucose, for example, the molar yield constant for glucose, Y_G, is determined by:

$$Y_G = \frac{\text{g dry weight of cells}}{\text{moles glucose fermented}}$$

If the adenosine-triphospate (ATP) yield or moles ATP produced per mole substrate used is known for a given substrate, the amount of dry weight of cells produced/mole of ATP formed can be determined by:

$$Y_{ATP} = \frac{\text{g dry weight of cells/moles ATP formed}}{\text{moles substrate fermented}}$$

A large number of fermenting organisms have been examined during growth and found to have $Y_{ATP} = 10.5$ or close thereto. This value is assumed to be a constant, so that an organism that ferments glucose by the EMP pathway to produce 2 ATP/mole of glucose fermented should have $Y_G = 21$ (that is, it should produce 21 g of cells dry weight/mole of glucose). This has been verified for *E. faecalis*, *Saccharomyces cerevisiae*, *Saccharomyces rosei*, and *L. plantarum* on glucose (all $Y_G = 21$, $Y_{ATP} = 10.5$, within experimental error). A study by Brown and Collins (14) indicates that Y_G and Y_{ATP} values for *L. diacetilactis* and *L. cremoris* differ when cells are grown aerobically on a partially defined medium with low and higher levels of glucose, and further when grown on a complex medium. On a partially defined medium with low glucose levels (1 to 7 μmol/ml), values for *L. diacetilactis* were $Y_G = 35.3$, $Y_{ATP} = 15.6$, while for *L. cremoris* $Y_G = 31.4$ and $Y_{ATP} = 13.9$. On the same medium with higher glucose levels (1 to 15 μmol/ml), Y_G for *S. diacetilactis* was 21. Y_{ATP} values for these two organisms on the complex medium with 2 μmol glucose/ml were 21.5 and 18.9 for *L. diacetilactis* and *L. cremoris*, respectively. Anaerobic molar growth yields for streptococcal species on low levels of glucose have been studied by Johnson and Collins (56). *Zymomonas mobilis* utilizes the Entner-Doudoroff pathway to produce only 1 ATP/mole of glucose fermented ($Y_G = 8.3$, $Y_{ATP} = 8.3$). If and when the produced lactate is metabolized further, the molar growth yield would be higher. *Bifidobacterium bifidum* produces 2.5 to 3 ATP/mole of glucose fermented with $Y_G = 37$, $Y_{ATP} = 13$ (112).

In addition to the use of molar growth yields to compare organisms on the same energy substrate, this concept can be applied to assess the metabolic routes used by various organisms in attacking a variety of carbohydrates (for further information, see 30, 81, 99, 112).

Molecular genetics have been employed by McKay and co-workers to stabilize lactose fermentation by *L. lactis*. The genes responsible for lactose fermentation by some lactic cocci are plasmid borne, and loss of the plasmid results in the loss of lactose fermentation. In an effort to make lactose fermentation more stable, lac$^+$ genes from *L. lactis* were cloned into a cloning vector, which was incorporated into a *Streptococcus sanguis* strain (48). Thus, the *lac* genes from *L. lactis* were transformed into *S. sanguis* via a vector plasmid, or transformation could be effected by use of appropriate fragments of DNA through which the genes were integrated into the chromosome of the host cells (49). In the latter state, lactose fermentation would be a more stable property than when the *lac* genes are plasmid borne.

PRODUCTS OF FERMENTATION

A selected list of fermented food products is described briefly below. For more detailed information on these and other fermented products, see Beuchat (13), Pederson (83), Rose (95), and Steinkraus (110).

Dairy Products

Some of the many fermented foods and products produced and utilized world-wide are listed in Table 16-2. The commercial and sometimes the home production of many of these is begun by use of appropriate **starter** cultures. A lactic starter is a basic starter culture with widespread use in the dairy industry. For cheese making of all kinds, lactic acid production is essential, and the lactic starter is employed for this purpose. Lactic starters are also used for butter, cultured buttermilk, cottage cheese, and cultured sour cream and are often referred to by product (butter starter, buttermilk starter, and so on). Lactic starters always include bacteria that convert lactose to lactic acid, usually *L. lactis*, *L. cremoris*, or *L. diacetilactis*. Where flavor and aroma compounds such as diacetyl are desired, the lactic starter will include a heterolactic such as *Leuconostoc citrovorum*, "*L. diacetilactis*," or *L. dextranicum* (for biosynthetic pathways, see Fig. 16-2 and reference 18). Starter cultures may consist of single or mixed strains. They may be produced in quantity and preserved by freezing in liquid nitrogen (38), or by freeze drying. The lactococci generally make up around 90% of a mixed dairy starter population, and a good starter culture can convert most of the lactose to lactic acid. The titratable acidity may increase to 0.8 to 1.0%, calculated as lactic acid, and the pH usually drops to 4.3 to 4.5 (31).

Butter, **buttermilk**, and **sour cream** are produced generally by inoculating pasteurized cream or milk with a lactic starter culture and holding until the desired amount of acidity is attained. In the case of butter where cream is inoculated, the acidified cream is then churned to yield butter, which is washed, salted, and packaged (83). Buttermilk, as the name suggests, is the milk that remains after cream is churned for the production of butter. The commercial product is usually prepared by inoculating skim milk with a lactic or buttermilk starter culture and holding until souring occurs. The resulting curd is broken up into fine particles by agitation, and this product is termed **cultured buttermilk**. Cultured sour cream is produced generally by fermenting pasteurized and homogenized light cream with a lactic starter. These products owe their tart flavor to lactic acid and their buttery aroma and taste to diacetyl.

Yogurt (yoghurt) is produced with a yogurt starter, which is a mixed culture of *S. thermophilus* and *L. bulgaricus* in a 1:1 ratio. The coccus grows faster than the rod and is primarily responsible for acid production, while the rod adds flavor and aroma. The associative growth of the two organisms results in lactic acid production at a rate greater than that produced by either when growing alone, and more acetaldehyde (the chief volatile flavor component of yogurt) is produced by *L. bulgaricus* when growing in association with *S. thermophilus* (see 88).

The product is prepared by first reducing the water content of either whole or skim milk by at least one-fourth. This may be done in a vacuum pan following sterilization of the milk. Approximately 5% by weight of milk solids or condensed milk is usually added. The concentrated milk is then heated to 82° to 93°C for 30 to 60 min and cooled to around 45°C (83). The yogurt starter is now added at a level of around 2% by volume and incubated at 45°C for 3 to 5 h, followed by cooling to 5°C. The titratable acidity of a good finished product is around 0.85 to 0.90%, and to get this amount of acidity the fermenting product should be removed from 45°C when the titratable acidity is around 0.65 or 0.70% (20). Good yogurt keeps well at 5°C for 1 to 2 weeks. The coccus grows first during the fermentation followed by the rod, so that after around 3 h, the numbers of the two organisms should be approximately equal. Higher amounts of acidity such as 4% can be achieved by allowing the product to ferment longer, with the effect that the rods will exceed the cocci in number. The streptococci tend to be inhibited at pH values of 4.2 to 4.4, while the lactobacilli can tolerate pHs in the 3.5 to 3.8 range. The lactic acid of yogurt is produced more from the glucose moiety of lactose than the galactose moiety. Goodenough and Kleyn (43) found only a trace of glucose throughout yogurt fermentation, while galactose increased from an initial trace to 1.2%. Samples of commercial yogurts showed only traces of glucose, while galactose varied from around 1.5 to 2.5%.

Freshly produced yogurt typically contains around 10^9 organisms/g but during storage, numbers may decrease to 10^6/g, especially when stored at 5°C for up to 60 days (47). The rod generally decreases more rapidly than the coccus. The addition of fruits to yogurt appears not to affect the numbers of fermenting organisms (47).

The antimicrobial qualities of yogurt, buttermilk, sour cream, and cottage cheese were examined by Goel et al. (40) who inoculated *Enterobacter aerogenes* and *Escherichia coli* separately into commercial products and studied the fate of these organisms when the products were stored at 7.2°C. A sharp decline of both coliforms was noted in yogurt and buttermilk after 24 h. Neither could be found in yogurt generally beyond 3 days. While the numbers of coliforms were reduced also in sour cream, they were not reduced as rapidly as in yogurt. Some cottage cheese samples actually supported an increase in coliform numbers, probably because the products had higher pH values. The initial pH ranges for the products studied by these workers were as follows: 3.65–4.40 for yogurts, 4.1–4.9 for buttermilks, 4.18–4.70 for sour creams,

TABLE 16-2. Fermented Foods and Related Products

Foods and Products	Raw Ingredients	Fermenting Organisms	Commonly Produced
Dairy products			
Acidophilus milk	Milk	*Lactobacillus acidophilus*	Many countries
Bulgarian buttermilk	Milk	*Lactobacillus bulgaricus*	Balkans, other areas
Cheeses (ripened)	Milk curd	Lactic starters; others	Worldwide
Kefir	Milk	*Lactococcus lactis, L. bulgaricus, "Torula"* spp.	Southwestern Asia
Kumiss	Raw mare's milk	*L. bulgaricus, Lactobacillus leichmannii, "Torula"* spp.	Russia
Taette	Milk	*S. lactis* var. *taette*	Scandinavian peninsula
Tarhana[a]	Wheat meal and yogurt	Lactics	Turkey
Yogurt[b]	Milk, milk solids	*Streptococcus thermophilus, L. bulgaricus*	Worldwide
Meat and fishery products			
Country-cured hams	Pork hams	*Aspergillus, Penicillium* spp.	Southern United States
Dry sausages[c]	Pork, beef	*Pediococcus cerevisiae*	Europe, United States
Lebanon bologna	Beef	*P. cerevisiae*	United States
Burong dalag	Dalag fish and rice	*Leuconostoc mesenteroides, P. cerevisiae, L. plantarum*	Philippines
Fish sauces[d]	Small fish	Halophilic *Bacillus* spp.	Southeast Asia
Izushi	Fresh fish, rice, vegetables	*Lactobacillus* spp.	Japan
Katsuobushi	Skipjack tuna	*Aspergillus glaucus*	Japan
Nonbeverage plant products			
Bongkrek	Coconut presscake	*Rhizopus oligosporus*	Indonesia
Cocoa beans	Cacao fruits (pods)	*Candida krusei (Issatchenkia orientalis), Geotrichum* spp.	Africa, South America
Coffee beans	Coffee cherries	*Erwinia dissolvens, Saccharomyces* spp.	Brazil, Congo, Hawaii, India
Gari	Cassava	*"Corynebacterium manihot," Geotrichum* spp.	West Africa
Kenkey	Corn	*Aspergillus* spp., *Penicillium* spp., Lactobacilli, yeasts	Ghana, Nigeria
Kimchi	Cabbage and other veg.	Lactic acid bacteria	Korea
Miso	Soybeans	*Aspergillus oryzae, Zygosaccharomyces rouxii*	Japan
Ogi	Corn	*L. plantarum, L. lactis, Zygosaccharomyces rouxii*	Nigeria
Olives	Green olives	*L. mesenteroides, L. plantarum*	Worldwide
Ontjom[e]	Peanut presscake	*Neurospora sitophila*	Indonesia
Peujeum	Cassava	Molds	Indonesia

Product	Substrate	Microorganisms	Location
Pickles	Cucumbers	P. cerevisiae, L. plantarum	Worldwide
Poi	Taro roots	Lactics	Hawaii
Sauerkraut	Cabbage	L. mesenteroides, L. plantarum	Worldwide
Soy sauce (shoyu)	Soybeans	A. oryzae; or A. soyae; Z. rouxii, L. delbrueckii	Japan
Sufu	Soybeans	Mucor spp.	China and Taiwan
Tao-si	Soybeans	A. oryzae	Philippines
Tempeh	Soybeans	Rhizopus oligosporus; R. oryzae	Indonesia, New Guinea, Surinam

Beverages and related products

Product	Substrate	Microorganisms	Location
Arrack	Rice	Yeasts, bacteria	Far East
Beer and ale	Cereal wort	Saccharomyces cerevisiae	Worldwide
Binuburan	Rice	Yeasts	Philippines
Bourbon whiskey	Corn, rye	S. cerevisiae	United States
Bouza beer	Wheat grains	Yeasts	Egypt
Cider	Apples; others	Saccharomyces spp.	Worldwide
Kaffir beer	Kaffircorn	Yeasts, molds, lactics	Nyasaland
Magon	Corn	Lactobacillus spp.	Bantus of South Africa
Mezcal	Century plant	Yeasts	Mexico
Oo	Rice	Yeasts	Thailand
Pulque[f]	Agave juice	Yeasts and lactics	Mexico, U.S. Southwest
Sake	Rice	Saccharomyces sake (S. cerevisiae)	Japan
Scotch whiskey	Barley	S. cerevisiae	Scotland
Teekwass	Tea leaves	Acetobacter xylinum, Schizosaccharomyces pombe	
Thumba	Millet	Endomycopsis fibuliges	West Bengal
Tibi	Dried figs; raisins	Betabacterium vermiforme, Saccharomyces intermedium	
Vinegar	Cider, wine	Acetobacter spp.	Worldwide
Wines	Grapes, other fruits	Saccharomyces "ellipsoideus" strains	Worldwide
Palm wine	Palm sap	Acetobacter spp., lactics, yeasts	Nigeria

Breads

Product	Substrate	Microorganisms	Location
Idli	Rice and bean flour	Leuconostoc mesenteroides	Southern India
Rolls, cakes, etc.	Wheat flours	S. cerevisiae	Worldwide
San Francisco sourdough bread	Wheat flour	S. exiguus, L. sanfrancisco	Northern California
Sour pumpernickel	Wheat flour	L. mesenteroides	Switzerland, other areas

[a] Similar to Kishk in Syria and Kushuk in Iran.
[b] Also yogurt (matzoon in Armenia; Leben in Egypt; Naja in Bulgaria; Gioddu in Italy; Dadhi in India).
[c] Such as Genoa, Milano, Siciliano.
[d] See text for specific names.
[e] N. sitophila is used to make red ontjom; R. oligosporus for white ontjom.
[f] Distilled to produce tequila.

and 4.80–5.10 for cottage cheese samples. In another study, commercially produced yogurts in Ontario were found to contain the desired 1:1 ratio of coccus to rod in only 15% of 152 products examined (7). Staphylococci were found in 27.6% and coliforms in around 14% of these yogurts. Twenty-six percent of the samples had yeast counts more than 1,000/g and almost 12% had psychrotroph counts more than 1,000/g. In his study of commercial unflavored yogurt in Great Britain, Davis (20) found *S. thermophilus* and *L. bulgaricus* counts to range from a low of around 82 million to a high of over 1 billion/g, and final pH to range from 3.75 to 4.20. The antimicrobial activities of lactic acid bacteria are discussed further in Chapter 11.

Kefir is prepared by the use of kefir grains, which contain *L. lactis*, *L. bulgaricus*, and a lactose-fermenting yeast held together by layers of coagulated protein. Acid production is controlled by the bacteria, while the yeast produces alcohol. The final concentration of lactic acid and alcohol may be as high as 1%. **Kumiss** is similar to kefir except that mare's milk is used, the culture organisms do not form grains, and the alcohol content may reach 2%.

Acidophilus milk is produced by the inoculation into sterile skim milk of an intestinal implantable strain of *L. acidophilus*. The inoculum of 1 to 2% is added, followed by holding of the product at 37°C until a smooth curd develops. **Bulgarian buttermilk** is produced in a similar manner by the use of *L. bulgaricus* as the inoculum or starter, but unlike *L. acidophilus*, *L. bulgaricus* is not implantable in the human intestines.

All **cheeses** result from a lactic fermentation of milk. In general, the process of manufacture consists of the two important steps:

1. Milk is prepared and inoculated with an appropriate lactic starter. The starter produces lactic acid, which, with added rennin, gives rise to curd formation. The starter for cheese production may differ depending on the amount of heat applied to the curds. *S. thermophilus* is employed for acid production in cooked curds since it is more heat tolerant than either of the other more commonly used lactic starters; or a combination of *S. thermophilus* and *L. lactis* is employed for curds that receive an intermediate cook.
2. The curd is shrunk and pressed, followed by salting, and, in the case of ripened cheeses, allowed to ripen under conditions appropriate to the cheese in question.

While most ripened cheeses are the product of metabolic activities of the lactic acid bacteria, several well-known cheeses owe their particular character to other related organisms. In the case of **swiss cheese**, a mixed culture of *L. bulgaricus* and *S. thermophilus* is usually employed along with a culture of *Propionibacterium shermanii*, which is added to function during the ripening process in flavor development and eye formation. (See Fig. 16-1C,D for a summary of propionibacteria pathways and Fig. 16-3 for pathway in detail.) These organisms have been reviewed extensively by Hettinga and Reinbold (51). For blue cheeses such as **roquefort**, the curd is inoculated with spores of *Penicillium roqueforti*, which effect ripening and impart the blue-veined appearance characteristic of this type of cheese. In a similar fashion, either the milk or the surface of **camembert** cheese is inoculated with spores of *Penicillium camemberti*.

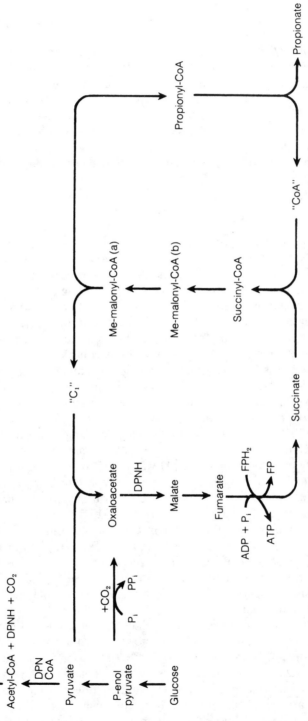

FIGURE 16-3. Reactions of the propionic acid fermentation and the formation of acetate, CO_2, propionate, and ATP. Me-malonyl-CoA is methylmalonyl-CoA and (a) and (b) are the two isomers. FP is flavoprotein, and FPH_2 is reduced flavoprotein. Summary: 1.5 glucose + 6 Pi + 6 ADP → 6 ATP + $2H_2O$ + CO_2 + acetate + 2 propionate.
From Allen et al. (5); copyright © 1964 by American Society for Microbiology.

There are over 400 varieties of cheeses representing fewer than 20 distinct types, and these are grouped or classified according to texture or moisture content, whether ripened or unripened, and if ripened, whether by bacteria or molds. The three textural classes of cheeses are hard, semihard, and soft. Examples of hard cheeses all cheddar, provolone, romano, and edam. All hard cheeses are ripened by bacteria over periods ranging from 2 to 16 months. Semihard cheeses include muenster and gouda and are ripened by bacteria over periods of 1 to 8 months. Blue and roquefort are two examples of semihard cheeses that are mold ripened for 2 to 12 months. Limburger is an example of a soft bacteria-ripened cheese, while brie and camembert are examples of soft mold-ripened cheeses. Among unripened cheeses are cottage, cream, and neufchatel.

Several other less widely produced fermented dairy products are listed in Table 16-2.

Meat and Fishery Products

Fermented sausages are produced generally as dry or semidry products, although some are intermediate. Dry or Italian-type sausages contain 30 to 40% moisture, are generally not smoked or heat processed, and are eaten usually without cooking (83). In their preparation, curing and seasoning agents are added to ground meat, followed by its stuffing into casings and incubation for varying periods of time at 80° to 95°F. Incubation times are shorter when starter cultures are employed. The curing mixtures include glucose as substrate for the fermenters and nitrates and/or nitrites as color stabilizers. When only nitrates are used, it is necessary for the sausage to contain bacteria that reduce nitrates to nitrites, usually micrococci present in the sausage flora or added to the mix. Following incubation, during which fermentation occurs, the products are placed in drying rooms with a relative humidity of 55 to 65% for periods ranging from 10 to 100 days, or, in the case of hungarian salami, up to 6 months (74). Genoa and milano salamis are other examples of dry sausages.

In one study of dry sausages, the pH was found to decrease from 5.8 to 4.8 during the first 15 days of ripening and remained constant thereafter (24). Nine different brands of commercially produced dry sausages were found by these investigators to have pH values ranging from 4.5 to 5.2, with a mean of 4.87. With respect to the changes that occur in the flora of fermenting dry sausage when starters are not used, Urbaniak and Pezacki (118) found the homo-fermenters to predominate overall, with *L. plantarum* being the most commonly isolated species. Heterofermenters such as *L. brevis* and *L. buchneri* increased during the 6-day incubation period as a result of changes in pH and Eh brought about by the homofermenters.

Semidry sausages are prepared in essentially the same way but are subjected to less drying. They contain about 50% moisture and are finished by heating to an internal temperature of 140° to 154°F (60° to 68°C) during smoking. Thuringer, cervelat, and summer sausage and lebanon bologna are some examples of semidry sausages. "Summer sausage" refers to those traditionally of Northern European origin, made during colder months, stored, aged, and then eaten during summer months. They may be dry or semidry.

Lebanon bologna is typical of a semidry sausage. This product, originally produced in the Lebanon, Pennsylvania, area, is an all-beef, heavily smoked and spiced product that may be prepared by use of a *Pediococcus cerevisiae* starter. The product is made by the addition of approximately 3% NaCl along with sugar, seasoning, and either nitrate, nitrite, or both to raw cubed beef. The salted beef is allowed to age at refrigerator temperatures for about 10 days during which time the growth of naturally occurring lactic acid bacteria or the starter organisms is encouraged and gram negatives are inhibited. A higher level of microbial activity along with some drying occurs during the smoking step at higher temperatures. A controlled production process for this product has been studied (78), and it consists of aging salted beef at 5°C for 10 days and smoking at 35°C with high relative humidity (R.H.) for 4 days. Fermentation may be carried out either by the natural flora of the meat or by use of a commercial starter of *P. cerevisiae* or *P. acidilactici*. The amount of acidity produced in lebanon bologna may reach 0.8 to 1.2% (83).

The hazard of eating improperly prepared homemade, fermented sausage is pointed up by an outbreak of trichinosis. Of the 50 persons who actually consumed the raw summer sausage, 23 became ill with trichinosis (85). The sausage was made on two different days in three batches according to a family recipe that called for smoking at cooler smoking temperatures, believed to produce a better-flavored product. All three batches of sausages contained home-raised beef. In addition, two batches eaten by victims contained pork inspected by the U.S. Department of Agriculture (USDA) in one case and home-raised pork in the other, but *Trichinella spiralis* larvae were found only in the USDA-inspected pork. This organism can be destroyed by a heat treatment that results in internal temperatures of at least 140°F (see Chap. 24).

Fermented sausages produced without use of starters have been found to contain large numbers of lactobacilli such as *L. plantarum* (23). The use of a *P. cerevisiae* starter leads to the production of a more desirable product (22, 50). In their study of commercially produced fermented sausages, Smith and Palumbo (104) found total aerobic plate counts to be in the 10^7–10^8/g range, with a predominance of lactic acid types. When starter cultures were used, the final pH of the products ranged from 4.0 to 4.5, while those produced without starters ranged between 4.6 and 5.0. For summer-type sausages, pH values of 4.5 to 4.7 have been reported for a 72-h fermentation (2). These investigators found that fermentation at 30° and 37°C led to a lower final pH than at 22°C and that the final pH was directly related to the amount of lactic acid produced. The pH of fermented sausage may actually increase by 0.1 or 0.2 units during long periods of drying due to uneven buffering produced by increases in amounts of basic compounds (125). The ultimate pH attained following fermentation depends on the type of sugar added. While glucose is most widely used, sucrose has been found to be an equally effective fermentable sugar for low pH production (1). The effect of a commercial frozen concentrate starter (*P. acidilactici*) in fermenting various sugars added to a sausage preparation is illustrated in Figure 16-4.

Prior to the late 1950s, the production of fermented sausages was facilitated by either back inoculations, or a producer took the chance of the desired organisms' being present in the raw materials. The manufacture of these, as well as of many other fermented foods, has been more of an art than a science until recently. With the advent of pure culture starters, not only has production

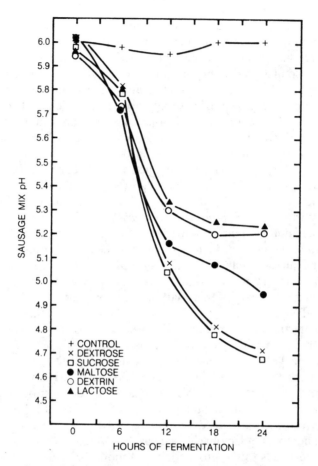

FIGURE 16-4.   Rate of pH reduction in fermenting sausage containing 0 or 1% of various carbohydrates.
From Acton et al. (1); copyright © 1977 by Institute of Food Technologists.

time been shortened but more uniform and safer products can be produced (27). While the use of starter cultures has been in effect for many years in the dairy industry, their use in many nondairy products worldwide is a recent development with great promise. "*Micrococcus aurantiacus*" has been employed along with starters in the production of some European sausages (74).

Molds are known to contribute to the quality of dry European-type sausages such as Italian salami. In an extensive study of the fungi of ripened meat products, Ayres et al. (8) found nine species of penicillia and seven of aspergilli on fermented sausages and concluded that the organisms play a role in the preservation of products of this type. Fewer species of other mold genera were also found.

Country-cured hams are dry-cured hams produced in the southern United States, and during the curing and ripening period of 6 months to 2 years, heavy mold growth occurs on the surfaces. Although Ayres et al. (8) noted that the presence of molds is incidental and that a satisfactory cure does not depend on their presence, it seems quite likely that some aspects of flavor development of these products derive from the heavy growth of such organisms, and to a lesser extent from yeasts. Heavy mold growth obviates the activities of food-poisoning and food-spoilage bacteria, and in this sense the mold flora aids in preservation. Ayres et al. found aspergilli and penicillia to be the predominant types of molds on country-cured hams.

The processing of country-cured hams takes place during the early winter and consists of rubbing sugar cure into the flesh side and onto the hock end. This is followed some time later by rubbing NaCl into all parts of the ham not covered by skin. The hams are then wrapped in paper and individually placed in cotton fabric bags and left lying flat for several days between 32° and 40°C. The hams are hung shank end down in ham houses for 6 weeks or longer and may be given a hickory smoke during this time, although smoking is not essential to a desirable product.

Italian-type country-cured hams are produced with NaCl as the only cure. Curing is carried out for about a month, followed by washing, drying, and ripening for 6 to 12 months or longer (39). While halophilic and halotolerant bacteria increase as Italian hams ripen, the microflora in general is thought to play only a minor role (39). For more detailed information on meat starter cultures and formulations for fermented sausages along with cure ingredients for country-style hams, see Pearson and Tauber (82) and Bacus (9).

Fish sauces are popular products in Southeast Asia, where they are known by various names such as *ngapi* (Burma), *nuoc-mam* (Cambodia and Vietnam), *nam-pla* (Laos and Thailand), *ketjap-ikan* (Indonesia), and so on. The production of some of these sauces begins with the addition of salt to uneviscerated fish at a ratio of approximately 1:3, salt to fish. The salted fish are then transferred to fermentation tanks generally constructed of concrete and built into the ground or placed in earthenware pots and buried in the ground. The tanks or pots are filled and sealed off for at least 6 months to allow the fish to liquefy. The liquid is collected, filtered, and transferred to earthenware containers and ripened in the sun for 1 to 3 months. The finished product is described as being clear dark-brown in color with a distinct aroma and flavor (96). In a study of fermenting Thai fish sauce by the latter authors, pH from start to finish ranged from 6.2 to 6.6 with NaCl content around 30% over the 12-months fermentation period (96). These parameters, along with the relatively high fermentation temperature, result in the growth of halophilic aerobic sporeformers as the predominant microorganisms of these products. Lower numbers of "streptococci," micrococci, and staphylococci were found, and they, along with the *Bacillus* spp., were apparently involved in the development of flavor and aroma. Some part of the liquefaction that occurs is undoubtedly due to the activities of fish proteases. While the temperature and pH of the fermentation are well within the growth range of a large number of undesirable organisms, the safety of products of this type is due to the 30 to 33% NaCl.

Fish pastes are also common in Southern Asia, but the role of fermenting microorganisms in these products appears to be minimal. Among the many other fermented fish, fish-paste, and fish-sauce products are the following: *mam-tom* of China, *mam-ruoc* of Cambodia, *bladchan* of Indonesia, *shiokara* of Japan, *belachan* of Malaya, *bagoong* of the Philippines, *kapi, hoi-dong*, and *pla-mam* of Thailand, *fessik* of Africa, and *nam-pla* of Thailand. Some of these as well as other fish products of Southeast Asia have been reviewed and discussed by van Veen and Steinkraus (123) and Sundhagul et al. (115). See Table 16-2 for other related products.

Nonbeverage Food Products of Plant Origin

Sauerkraut is a fermentation product of fresh cabbage. The starter for sauerkraut production is usually the normal mixed flora of cabbage. The addition of 2.25 to 2.5% salt restricts the activities of gram-negative bacteria, while the lactic acid rods and cocci are favored. *Leuconostoc mesenteroides* and *L. plantarum* are the two most desirable lactic acid bacteria in sauerkraut production, with the former having the shorter generation time and the shorter life span. The activities of the coccus usually cease when the acid content increases to 0.7 to 1.0%. The final stages of kraut production are effected by *L. plantarum* and *L. brevis*. *P. cerevisiae* and *E. faecalis* may also contribute to product development. Final total acidity is generally 1.6 to 1.8%, with lactic acid at 1.0 to 1.3%.

Pickles are fermentation products of fresh cucumbers, and as is the case for sauerkraut production, the starter culture normally consists of the normal mixed flora of cucumbers. In the natural production of pickles, the following lactic acid bacteria are involved in the process in order of increasing prevalence: *L. mesenteroides*, *E. faecalis*, *P. cerevisiae*, *L. brevis*, and *L. plantarum* (26). Of these the pediococci and *L. plantarum* are the most involved, with *L. brevis* being undesirable because of its capacity to produce gas. *L. plantarum* is the most essential species in pickle production as it is for sauerkraut.

In the production of pickles, selected cucumbers are placed in wooden brine tanks with initial brine strengths as low as 5% NaCl (20° salinometer). Brine strength is increased gradually during the course of the 6- to 9-week fermentation until it reaches around 60° salinometer (15.9% NaCl). In addition to exerting an inhibitory effect on the undesirable gram-negative bacteria, the salt extracts from the cucumbers water and water-soluble constituents such as sugars, which are converted by the lactic acid bacteria to lactic acid. The product that results is a salt-stock pickle from which pickles such as sour, mixed sour, chowchow, and so forth, may be made.

The general technique of producing brine-cured pickles briefly outlined has been in use for many years, but it often leads to serious economic loss because of pickle spoilage from such conditions as bloaters, softness, off-colors, and so on. The controlled fermentation of cucumbers brined in bulk has been achieved, and this process not only reduces economic losses of the type noted but leads to a more uniform product over a shorter period of time (26). The controlled fermentation method employs a chlorinated brine of 25° salinometer, acidification with acetic acid, the addition of sodium acetate, and inoculation with *P. cerevisiae* and *L. plantarum*, or the latter alone. The course of the 10- to 12-day fermentation is represented in Figure 16-5 (for more detailed information, see ref. 26).

Olives to be fermented (Spanish, Greek, or Sicilian) are done so by the natural flora of green olives, which consists of a variety of bacteria, yeasts, and molds. The olive fermentation is quite similar to that of sauerkraut except that it is slower, involves a lye treatment, and may require the addition of starters.

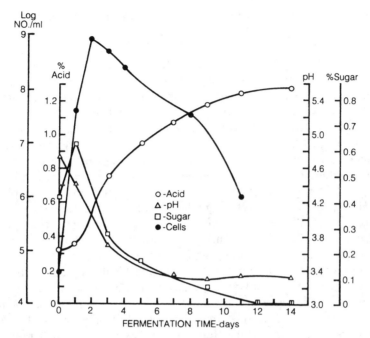

FIGURE 16-5.   Controlled fermentation of cucumbers brined in bulk. Equilibrated brine strength during fermentation, 6.4% NaCl; incubation temperature, 27°C.
From Etchells et al. (26); copyright © 1975 by Academic Press.

The lactic acid bacteria become prominent during the intermediate stage of fermentation. *L. mesenteroides* and *P. cerevisiae* are the first lactics to become prominent, and these are followed by lactobacilli, with *L. plantarum* and *L. brevis* being the most important (124).

The olive fermentation is preceded by a treatment of green olives with from 1.6 to 2.0% lye, depending on type of olive, at 21° to 24°C for 4 to 7 h for the purpose of removing some of the bitter principal. Following the complete removal of lye by soaking and washing, the green olives are placed in oak barrels and brined so as to maintain a constant 28° to 30° salinometer level. Inoculation with *L. plantarum* may be necessary because of destruction of organisms during the lye treatment. The fermentation may take as long as 6 to 10 months, and the final product has a pH of 3.8 to 4.0 following up to 1% lactic acid production.

Soy sauce or shoyu is produced in a two-stage manner. The first stage, the koji (analogous to malting in the brewing industry), consists of inoculating either soybeans or a mixture of beans and wheat flour with *A. oryzae* or *A. soyae* and allowing them to stand for 3 days. This results in the production of large amounts of fermentable sugars, peptides, and amino acids. The second stage, the moromi, consists of adding the fungal-covered product to around 18% NaCl and incubating at room temperatures for at least a year. The liquid obtained at this time is soy sauce. During the incubation of the moromi, lactic acid bacteria, *L. delbrueckii* in particular, and yeasts such as *Zygosaccharomyces rouxii* carry out an anaerobic fermentation of the koji hydrolysate. Pure cultures of *A. oryzae* for the koji and *L. delbrueckii* and *Z.*

rouxii for the moromi stages have been shown to produce good quality soy sauce (129).

Tempeh is a fermented soybean product. Although there are many variations in its production, the general principle of the Indonesian method for tempeh consists of soaking soybeans overnight in order to remove the seedcoats or hulls. Once seedcoats are removed, the beans are cooked in boiling water for about 30 min and spread on a bamboo tray to cool and surface dry. Small pieces of tempeh from a previous fermentation are incorporated as starter followed by wrapping with banana leaves. The wrapped packages are kept at room temperature for 1 or 2 days during which mold growth occurs and binds the beans together as a cake—the tempeh. An excellent product can be made by storing in perforated plastic bags and tubes with fermentations completed in 24 h at 31°C (29). The desirable organism in the fermentation is *Rhizopus oligosporus*, especially for wheat tempeh. Good soybean tempeh can be made with *R. oryzae* or *R. arrhizus*. During the fermentation, the pH of soybeans rises from around 5.0 to values as high as 7.5.

Miso, a fermented soybean product common in Japan, is prepared by mixing or grinding steamed or cooked soybeans with koji and salt and allowing fermentation to take place usually over a 4- to 12-month period. White or sweet miso may be fermented for only a week, while the higher-quality dark brown product (*mame*) may ferment for 2 years (101). In Israel, Ilany-Feigenbaum et al. (54) prepared miso-type products by using defatted soybean flakes instead of whole soybeans and fermenting for around 3 months. The koji for these products was made by growing *A. oryzae* on corn, wheat, barley, millet or oats, potatoes, sugar beets, or bananas, and the authors found that the miso-type products compared favorably to Japanese-prepared miso. Because of the possibility that *A. oryzae* may produce toxic substances, koji was prepared by fermenting rice with *Rhizopus oligosporus* at 25°C for 90 days; the product was found to be an acceptable alternative to *A. oryzae* as a koji fungus (102).

Ogi is a staple cereal of the Yorubas of Nigeria and is the first native food given to babies at weaning. It is produced generally by soaking corn grains in warm water for 2 to 3 days followed by wet-milling and sieving through a screen mesh. The sieved material is allowed to sediment and ferment and is marketed as wet cakes wrapped in leaves. Various food dishes are made from the fermented cakes or the ogi (11). During the steeping of corn, *Corynebacterium* spp. become prominent and appear to be responsible for the diastatic action necessary for the growth of yeasts and lactic acid bacteria (4). Along with the corynebacteria, *S. cerevisiae* and *L. plantarum* have been found to be prominent in the traditional ogi fermentation, as are *Cephalosporium*, *Fusarium*, *Aspergillus*, and *Penicillium*. Most of the acid produced is lactic, which depresses the pH of desirable products to around 3.8. The corynebacteria develop early, and their activities cease after the first day, while those of the lactobacilli and yeasts continue beyond the first day of fermentation. A more recent process for making ogi has been developed, tested, and found to produce a product of better quality than the traditional (10). By the new method, corn is dry-milled into whole corn and dehulled corn flour. Upon the addition of water, the mixture is cooked, cooled, and then inoculated with a mixed culture (starter) of: *L. plantarum*, *L. lactis*, and *Z. rouxii*. The inoculated preparation is incubated at 32°C for 28 h, during which time the pH of the corn drops from 6.1 to 3.8. This process eliminates the need for starch-hydrolyzing bacteria. In addition to the shorter fermentation time, there is also less chance for faulty fermentations.

Gari is a staple food of West Africa prepared from the root of the cassava plant. Cassava roots contain a cyanogenic glucoside that makes them poisonous if eaten fresh or raw. The roots are rendered safe by a fermentation during which the toxic glucoside decomposes with the liberation of gaseous hydrocyanic acid. In the home preparation of gari, the outer peel and the thick cortex of the cassava roots are removed, followed by grinding or grating the remainder. The pulp is pressed to remove the remaining juice, and placed in bags for 3 or 4 days to allow fermentation to occur. The fermented product is cooked by frying. The fermentation has been found to occur in a two-stage manner. In the first stage, *"Corynebacterium manihot"* ferments starch, with the production of acids and the consequent lowering of pH. Under acidic conditions, the cyanogenic glucoside then undergoes spontaneous hydrolysis with the liberation of gaseous hydrocyanic acid (17). The acidic conditions favor the growth of *Geotrichum* spp., and these organisms appear to be responsible for the characteristic taste and aroma of gari.

Bongkrek is an example of a fermented food product that in the past has led to a large number of deaths. Bongkrek or semaji is a coconut presscake product of central Indonesia, and it is the homemade product that may become toxic. The safe products fermented by *R. oligosporus* are finished cakes covered with and penetrated by the white fungus. In order to obtain the desirable fungal growth, it appears to be essential that conditions permit good growth within the first 1 or 2 days of incubation. If, however, bacterial growth is favored during this time and if the bacterium *Pseudomonas cocovenenans* is present, it grows and produces two toxic substances—toxoflavin and bongkrekic acid (121, 122). Both of these compounds show antifungal and antibacterial activity, are toxic for humans and animals, and are heat stable. Production of both is favored by growth of the organisms on coconut (toxoflavin can be produced in complex culture media). The structural formulas of the two antibiotics—toxoflavin, which acts as an electron carrier, and bongkrekic acid, which inhibits oxidative phosphorylation in mitochondria—follow:

Bongkrekic acid

Bongkrekic acid has been shown to be cidal to all 17 molds studied by Subik and Behun (113) by preventing spore germination and mycelial outgrowth. The growth of *P. cocovenenans* in the preparation of bongkrek is not favored if the acidity of starting materials is kept at or below pH 5.5 (120). More recently, it has been shown that 2% NaCl in combination with acetic acid to produce a pH of 4.5 prevent the formation of the bongkrek toxin in tempe (15).

A fermented cornmeal product that is prepared in parts of China has been the cause of food poisoning by strains of *Pseudomonas cocovenenans* (130). The product is prepared by soaking corn in water at room temperature for 2 to 4 weeks, washing in water, and grinding the wet corn into flour for various uses. The toxic organisms apparently grow in the moist product during its storage at room temperature. The responsible organism produced both bongkrekic acid and toxoflavin, as do the strains of *P. cocovenenans* in bongkrek.

Ontjom (Oncom) is a somewhat similar but more popular fermented product of Indonesia made from peanut presscake, the material that remains after oil has been extracted from peanuts. The presscake is soaked in water for about 24 h, steamed, and pressed into molds. The molds are covered with banana leaves and inoculated with *Neurospora sitophila* or *R. oligosporus*. The product is ready for consumption 1 or 2 days later. A more detailed description of ontjom fermentation and the nutritive value of this product has been provided by Beuchat (12).

Beverage and Distilled Products

Beer and **ale** are malt beverages produced by brewing. An essential step in the brewing process is the fermentation of carbohydrates to ethanol. Since most of the carbohydrates in grains used for brewing exist as starches, and since the fermenting yeasts do not produce amylases to degrade the starch, a necessary part of beer brewing includes a step whereby malt or other exogenous sources of amylase are provided for the hydrolysis of starches to sugars. The malt is first prepared by allowing barley grains to germinate. This serves as a source of amylases (fungal amylases may be used also). Both β- and α-amylases are involved, with the latter acting to liquefy starch and the former to increase sugar formation (46). In brief, the brewing process begins with the mixing of malt, malt adjuncts, hops, and water. Malt adjuncts include certain grains, grain products, sugars, and other carbohydrate products to serve as fermentable substances. Hops are added as sources of pyrogallol and catechol tannins, resins, essential oils, and other constituents for the purpose of precipitating unstable proteins during the boiling of wort and to provide for biological stability, bitterness, and aroma. The process by which the malt and malt adjuncts are dissolved and heated and the starches digested is called mashing. The soluble part of the mashed materials is called wort (compare with koji). In some breweries, lactobacilli are introduced into the mash to lower the pH of wort through lactic acid production. The species generally used for this purpose is *L. delbrueckii* (58).

Wort and hops are mixed and boiled for 1.5 to 2.5 h for the purpose of enzyme inactivation, extraction of soluble hop substances, precipitation of coagulable proteins, concentration, and sterilization. Following the boiling of wort and hops, the wort is separated, cooled, and fermented. The fermentation of the sugar-laden wort is carried out by the inoculation of *S. cerevisiae* (Fig. 16-1*E*). Ale results from the activities of top-fermenting yeasts, which depress the pH to around 3.8, while bottom-fermenting yeasts (*S. "carlsbergensis"* strains) give rise to lager and other beers with pH values of 4.1 to 4.2. A top fermentation is complete in 5 to 7 days; a bottom fermentation requires 7 to 12 days (86). The freshly fermented product is aged and finished by the addition of CO_2 to a final content of 0.45 to 0.52% before it is ready for commerce. The pasteurization of beer, at 140°F (60°C) or higher, may be carried out for the purpose of destroying spoilage organisms. When lactic acid bacteria are present in beers, the lactobacilli are found more commonly in top fermentations, while pediococci are found in bottom fermentations (58).

Distilled spirits are alcoholic products that result from the distillation of yeast fermentations of grain, grain products, molasses, or fruit or fruit products. Whiskeys, gin, vodka, rum, cordials, and liqueurs are examples of distilled spirits. While the process for producing most products of these types is quite similar to that for beers, the content of alcohol in the final products is considerably higher than for beers. **Rye** and **bourbon** are examples of whiskeys. In the former, rye and rye malt or rye and barley malt are used in different ratios, but at least 51% rye is required by law. Bourbon is made from corn, barley malt or wheat malt, and usually another grain in different proportions, but at least 51% corn is required by law. A sour wort is maintained to keep down undesirable organisms, the souring occurring naturally or by the addition of acid. The mash is generally soured by inoculating with a homolactic such as *L. delbrueckii*, which is capable of lowering the pH to around 3.8 in 6 to 10 h (83). The malt enzymes (diastases) convert the starches of the cooked grains to dextrins and sugars, and upon completion of diastatic action and lactic acid production, the mash is heated to destroy all microorganisms. It is then cooled to 75° to 80°F (24° to 27°C) and pitched (inoculated) with a suitable strain of *S. cerevisiae* for the production of ethanol. Upon completion of fermentation, the liquid is distilled to recover the alcohol and other volatiles, and these are handled and stored under special conditions relative to the type of product being made. **Scotch whiskey** is made primarily from barley and is produced from barley malt dried in kilns over peat fires. **Rum** is produced from the distillate of fermented sugar cane or molasses. **Brandy** is a product prepared by distilling grape or other fruit wines.

Wines are normal alcoholic fermentations of sound grapes followed by aging. A large number of other fruits such as peaches, pears, and so forth, may be fermented for wines, but in these instances the wine is named by the fruit, such as pear wine, peach wine, and the like. Since fruits already contain fermentable sugars, the use of exogenous sources of amylases is not necessary as it is when grains are used for beers or whiskeys. Wine making begins with the selection of suitable grapes, which are crushed and then treated with a sulfite such as potassium metabisulfite to retard the growth of acetic acid bacteria, wild yeasts, and molds. The pressed juice, called must, is inoculated with a suitable wine strain of *S. "ellipsoideus"*. The fermentation is allowed to continue for 3 to 5 days at temperatures between 70° and 90°F (21° to 32°C), and good yeast strains may produce up to 14 to 18% ethanol (83). Following fermentation, the wine is racked—that is, drawn off from the lees or sediment, which contains potassium bitartrate (cream of tartar). The clearing and development of flavor occur during the storage and aging process. Red wines are made by initially fermenting the crushed grape must "on the skins" during which pigment is extracted into the juice; white wines are prepared generally from the juice of white grapes. **Champagne**, a sparkling wine made by a secondary fermentation of wine, is produced by adding sugar, citric acid, and a champagne yeast starter to bottles of previously prepared, selected table wine. The bottles are corked, clamped, and stored horizontally at suitable temperatures for about 6 months. They are then removed, agitated, and aged for an additional period of up to 4 years. The final sedimentation of yeast cells and tartrates is accelerated by reducing the temperature of the wine to around 25°C and holding for 1 to 2 weeks. Clarification of the champagne is brought about by working the sediment down the bottle onto the cork over a period of 2 to 6 weeks by frequent rotation of the bottle. Finally, the sediment is frozen and disgorged upon removal of the cork. (See references such as Prescott and Dunn [86] for more details of the production and classification of the various types of wine.)

Palm wine or Nigerian palm wine is an alcoholic beverage consumed throughout the tropics and is produced by a natural fermentation of palm sap. The sap is sweet and dirty brown in color, and it contains 10 to 12% sugar, mainly sucrose. The fermentation process results in the sap's becoming milky-white in appearance due to the presence of large numbers of fermenting bacteria and yeasts. This product is unique in that the microorganisms are alive when the wine is consumed. The fermentation has been reviewed and studied by Faparusi and Bassir (28) and Okafor (76), who found the following genera of bacteria to be the most predominant in finished products: *Micrococcus*, *Leuconostoc*, "*Streptococcus*", *Lactobacillus*, and *Acetobacter*. The predominant yeasts found are *Saccharomyces* and *Candida* spp., with the former being the more common (75). The fermentation occurs over a 36 to 48 h period during which the pH of sap falls from 7.0 or 7.2 to less than 4.5. Fermentation products consist of organic acids in addition to ethanol. During the early phases of fermentation, *Serratia* and *Enterobacter* spp. increase in numbers, followed by lactobacilli and leuconostocs. After a 48-h fermentation, *Acetobacter* spp. begin to appear (28, 77).

Sake is an alcoholic beverage commonly produced in Japan. The substrate is the starch from steamed rice, and its hydrolysis to sugars is carried out by *A. oryzae* to yield the koji. Fermentation is carried out by *Saccharomyces sake* over periods of 30 to 40 days, resulting in a product containing 12 to 15% alcohol and around 0.3% lactic acid (83). The latter is produced by hetero- and homolactic lactobacilli.

Cider, in the United States, is a product that represents a mild fermentation of apple juice by naturally occurring yeasts. In making apple cider, the fruits are selected, washed, and ground into a pulp. The pulp "cheeses" are pressed to release the juice. The juice is strained and placed in a storage tank, where sedimentation of particulate matter occurs, usually for 12 to 36 h or several days if the temperature is kept at 40°F or below. The clarified juice is cider. If pasteurization is desired, this is accomplished by heating at 170°F for 10 min. The chemical preservative most often used is sodium sorbate at a level of 0.10%. Preservation may be effected also by chilling or freezing. The finished product contains small amounts of ethanol in addition to acetaldehyde. The holding of nonpasteurized or unpreserved cider at suitable temperatures invariably leads to the development of cider vinegar, which indicates the presence of acetic acid bacteria in these products. The pathway employed by acetic acid bacteria is summarized in Figure 16-1F,G.

In their study of the ecology of the acetic acid bacteria in cider manufacture, Passmore and Carr (79) found six species of *Acetobacter* and noted that those that display a preference for sugars tend to be found early in the cider process, while those that are more acid tolerant and capable of oxidizing alcohols appear after the yeasts have converted most of the sugars to ethanol. *Zymomonas* spp., gram-negative bacteria that ferment glucose to ethanol, have been isolated from ciders, but they are presumed to be present in low numbers. A recently discovered bacterium, *Saccharobacter fermentatus*, is similar to *Zymomonas* in that it ferments glucose to ethanol and CO_2 (128). It was isolated from agave leaf juice, but its presence and possible role in spoiled ciders have yet to be determined.

Coffee beans, which develop as berries or cherries in their natural state, have an outer pulpy and mucilaginous envelope that must be removed before the beans can be dried and roasted. The wet method of removal of this layer seems to produce the most desirable product, and it consists of depulping and demucilaging followed by drying. While depulping is done mechanically, demucilaging is accomplished by natural fermentation. The mucilage layer is composed largely of pectic substances (32), and pectinolytic microorganisms are important in their removal. *Erwinia dissolvens* has been found to be the most important bacterium during the demucilaging fermentation in Hawaiian (33) and Congo coffee cherries (119), although Pederson and Breed (84) indicated that the fermentation of coffee berries from Mexico and Colombia was carried out by typical lactic acid bacteria (leuconostocs and lactobacilli). Agate and Bhat (3) in their study of coffee cherries from the Mysore State of India found that the following pectinolytic yeasts predominated and played important roles in the loosening and removal of the mucilaginous layers: *Saccharomyces marxianus*, *S. bayanus*, *S. "ellipsoideus,"* and *Schizosaccharomyces* spp. Molds are common on green coffee beans, and in one study 99.1% of products from 31 countries contained these organisms, generally on the surface (70). Seven species of aspergilli dominated the flora, with *A. ochraceus* being the most frequently recovered from beans before surface disinfection, followed by *A. niger* and species of the *A. glaucus* group. The toxigenic molds, *A. flavus* and *A. versicolor*, were found, as were *P. cyclopium*, *P. citrinum*, and *P. expansum*, but the penicillia were less frequently found than the aspergilli (70). Microorganisms do not contribute to the development of flavor and aroma in coffee beans as they do in cocoa beans.

Cocoa beans (actually cacao beans—cocoa is the powder and chocolate is the manufactured product), from which chocolate is derived, are obtained from the fruits or pods of the cacao plant in parts of Africa, Asia, and South America. The beans are extracted from the fruits and fermented in piles, boxes, or tanks for 2 to 12 days, depending on the type and size of beans. During the fermentation, high temperatures (45–50°C) and large quantities of liquid develop. Following sun or air drying, during which the water content is reduced to less than 7.5%, the beans are roasted to develop the characteristic flavor and aroma of chocolate. The fermentation occurs in two phases. In the first, sugars from the acidic pulp (about pH 3.6) are converted to alcohol. The second phase consists of the alcohol's being oxidized to acetic acid. In a study of Brazilian cocoa beans by Camargo et al. (16), the flora on the first day of fermentation at 21°C consisted of yeasts. On the third day, the temperature had risen to 49°C, and the yeast count had decreased to no more than 10% of the total flora. Over the 7-day fermentation, the pH increased from 3.9 to 7.1. The cessation of yeast and bacterial activity around the third day is due in part to the unfavorable temperature, lack of fermentable sugars, and increase in alcohol. While some decrease in acetic acid bacteria occurs because of high temperature, not all of these organisms are destroyed. Yeasts and acetic acid bacteria are the most important fermenters of cocoa beans. Of the 142 yeasts isolated by Camargo et al., 105 were asporogenous and 37 were ascosporogenous. *Candida krusei* (*Issatchenkia orientalis*, a thermotolerant yeast) was the most frequently encountered species, and it became predominant after the second day. High numbers of *Geotrichum candidum* and *Candida valida* were found, and both were shown to be pectinolytic. On the other hand, lactic acid bacteria were more abundant than yeasts on fermented cocoa beans in the state of Bahia (Brazil) after 48 h (80). The homolactics were the most abundant of three genera and eight species represented. The only heterolactics were *Leuconostoc mesenteroides* and *L. brevis*. The lactic acid bacteria are thought to be responsible for the acidity of cocoa from this area (80). The gram-negative bacterium, *Zymomonas mobilis*, has been tried, without success, as a starter in cocoa bean fermentation. This organism can convert sugars to ethanol and the latter to acetic acid. Heterofermentative lactics and *Acetobacter* spp. are involved in the fermentation (92).

While yeasts play important roles in producing alcohol in cocoa bean fermentation, their presence appears even more essential to the development of the final, desirable chocolate flavor of roasted beans. Levanon and Rossetini (61) found that the endoenzymes released by autolyzing yeasts are responsible for the development of chocolate precursor compounds. The acetic acid apparently makes the bean tegument permeable to the yeast enzymes. It has been shown that chocolate aroma occurs only after cocoa beans are roasted and that the roasting of unfermented beans does not produce the characteristic aroma (93). Reducing sugars and free amino acids are in some way involved in the final chocolate aroma development (94).

Breads

San Francisco sourdough bread is similar to sourdough breads produced in various countries. Historically, the starter for sourdough breads consists of the natural flora of baker's barm (sour ferment or mother sponge, with a portion of each inoculated dough saved as starter for the next batch). The barm generally contains a mixture of yeasts and lactic acid bacteria. In the case of San Francisco sourdough bread, the yeast has been identified as *Saccharomyces exiguus* (*Candida holmii*, 114) and the responsible bacterium ads *Lactobacillus sanfrancisco* (59). The souring is caused by the acids produced by the bacterium, while the yeast is responsible for the leavening action, although some CO_2 is produced by the bacterium. The pH of these sourdoughs ranges from 3.8 to 4.5. Both acetic and lactic acids are produced, with the former accounting for 20 to 30% of the total acidity (59).

Idli is a fermented bread-type product common in southern India. It is made from rice and black gram mungo (urd beans). These two ingredients are soaked in water separately for 3 to 10 h and then ground in varying proportions, mixed, and allowed to ferment overnight. The fermented and raised product is cooked by steaming and served hot. It is said to resemble a steamed, sourdough bread (111). During the fermentation, the initial pH of around 6.0 falls to values of 4.3 to 5.3. In a particular study, a batter pH of 4.70 after a 20-h fermentation was associated with 2.5% lactic acid, based on dry grain weight (71). In their studies of idli, Steinkraus et al. (111) found total bacterial counts of 10^8–10^9/g after 20 to 22 h of fermentation. Most of the organisms consisted of gram-positive cocci or short rods, with *L. mesenteroides* being the single most abundant species, followed by *E. faecalis*. The leavening action of idli is produced by *L. mesenteroides*. This is the only known instance of a lactic acid bacterium having this role in a naturally fermented bread (71). The latter authors confirmed the work of others in finding the urd beans to be a more important source of lactic acid bacteria than rice. *L. mesenteroides* reaches its peak at around 24 h, with *E. faecalis* becoming active only after about 20 h. Other probable fermenters include *L. delbruecki*, *L. fermenti*, and *Bacillus* spp. (94). Only after idli has fermented for more than 30 h does *P. cerevisiae* become active. The product is not fermented generally beyond 24 h because maximum leavening action occurs at this time and decreases with longer incubations. When idli is allowed to ferment longer, more acidity is produced. It has been found that total acidity (expressed as g lactic acid/g of dry grains) increased from 2.71% after 24 h to 3.70% after 71 h, while pH decreased from 4.55 to 4.10 over the same period (71). (A review of idli fermentation has been made by Reddy et al., [90].)

SINGLE-CELL PROTEIN

The cultivation of unicellular microorganisms as a direct source of human food was suggested in the early 1900s. The expression *single-cell protein* (SCP) was coined at the Massachusetts Institute of Technology around 1966 to depict the idea of microorganisms as food sources (105). Although SCP is a misnomer in that proteins are not the only food constituent represented by microbial cells, it obviates the need to refer to each product generically as in "algal protein," "yeast cell protein," and so on. Although SCP as a potential and real source of food for humans differs from the other products covered in this chapter, with the exception of that from algal cells, it is produced in a similar manner.

Rationale for SCP Production

It is imperative that new food sources be found in order that future generations be adequately fed. A food source that is nutritionally complete and requires a minimum of land, time, and cost to produce is highly desirable. In addition to meeting these criteria, SCP can be produced on a variety of waste materials. Among the overall advantages of SCP over plant and animal sources of proteins are the following (57):

- Microorganisms have a very short generation time and can thus provide a rapid mass increase.
- Microorganisms can be easily modified genetically—to produce cells that bring about desirable results.
- The protein content is high.
- The production of SCP can be based on raw materials readily available in large quantities.
- SCP production can be carried out in continuous culture and thus be independent of climatic changes.

The greater speed and efficiency of microbial protein production compared to plant and animal sources may be illustrated as follows: a 1,000-lb steer produces about 1 lb of new protein/day; soybeans (prorated over a growing season) produce about 80 lb, and yeasts produce about 50 tons.

Organisms and Fermentation Substrates

A large number of algae, yeasts, molds, and bacteria have been studied as SCP sources. Among the most promising genera and species are the following:

Algae: *Chlorella* spp. and *Scenedesmus* spp.
Yeasts: *Candida guilliermondii, C. utilis, C. lipolytica*, and *C. tropicalis; Debaryomyces kloeckeri; Candida famata, C. methanosorbosa; Pichia* spp.; *Kluyveromyces fragilis; Hansenula polymorpha; Rhodotorula* spp.; and *Saccharomyces* spp.
Filamentous fungi: *Agaricus* spp.; *Aspergillus* spp.; *Fusarium* spp.; *Penicillium* spp.; *Saccharomycopsis fibuligera*; and *Trichosporon cutaneum*.
Bacteria: *Bacillus* spp.; *Cellulomonas* spp.; *Acinetobacter calcoaceticus; Nocardia* spp.; *Methylomonas* spp.; *Aeromonas hydrophila; Alcaligenes eutrophus (Hydrogenomonas eutropha), Mycobacterium* sp.; *Spirulina maxima*, and *Rhodopseudomonas* sp.

Of these groups, yeasts have received by far the most attention.

The choice of a given organism is dictated in large part by the type of substrate or waste material in question. The cyanobacterium *Spirulina maxima* grows in shallow waters high in bicarbonate at a temperature of 30°C and pH 8.5 to 11.0. It can be harvested from pond waters and dried for food use. This cell has been eaten by the people of the Chad Republic for many years (105). Other cyanobacteria require sunlight, CO_2, minerals, water, and proper growth temperatures. However, the large-scale use of such cells as SCP sources is said to be practical only in areas below 35° latitude, where sunlight is available most of the year (62).

Bacteria, yeasts, and molds can be grown on a wide variety of materials, including food processing wastes (such as cheese whey and brewery, potato processing, cannery, and coffee wastes), industrial wastes (such as sulfite liquor in the paper industry and combustion gases), and cellulosic wastes (including bagasse, newsprint mill, and barley straw). In the case of cellulosic wastes, it is necessary to use organisms that can utilize cellulose, such as a *Cellulomonas* sp. or *Trichoderma viride*. A mixed culture of *Cellulomonas* and *Alcaligenes* has

Table 16-3.   Substrate Materials That Support the Growth of Microorganisms in the Production of SCP

Substrates	Microorganisms
CO_2 and sunlight	*Chlorella pyrenoidosa*
	Scenedesmus quadricauda
	Spirulina maxima
n-Alkanes, kerosene	*Candida intermedia, C. lipolytica, C. tropicalis*
	Nocardia sp.
Methane	*Methylomonas* sp. (*Methanomonas*)
	Methylococcus capsulatus
	Trichoderma sp.
H_2 and CO_2	*Alcaligenes eutrophus (Hydrogenomonas eutropha)*
Gas oil	*Acinetobacter calcoaceticus (Micrococcus cerificans)*
	Candida lipolytica
Methanol	*Methylomonas methanica (Methanomonas methanica)*
Ethanol	*Candida utilis*
	Acinetobacter calcoaceticus
Sulfite liquor wastes	*Candida utilis*
Cellulose	*Cellulomonas* sp.
	Trichoderma viride
Starches	*Saccharomycopsis fibuligera*
Sugars	*Saccharomyces cerevisiae*
	Candida utilis
	Kluyveromyces fragilis

been employed. For starchy materials, a combination of *Saccharomycopsis fibuligera* and a *Candida* sp. such as *C. utilis* has been employed, in which the former effects hydrolysis of starches and the latter subsists on the hydrolyzed products to produce biomass. Some other representative substrates and organisms are listed in Table 16-3.

SCP Products

The cells may be used directly as a protein source in animal feed formulations, thereby freeing animal feed, such as corn, for human consumption, or they may be used as a protein source or food ingredient for human food. In the case of animal feed or feed supplements, the dried cells may be used without further processing. Whole cells of *Spirulina maxima* are consumed by humans in at least one part of Africa.

For human use, the most likely products are SCP concentrates or isolates that can be further processed into textured or functional SCP products. To produce functional protein fibers, cells are mechanically disrupted, cell walls removed by centrifugation, proteins precipitated from disrupted cells, and the resulting protein extruded from syringelike orifices into suitable menstra such as acetate buffer, $HClO_4$, acetic acid, and the like. The SCP fibers may now be used to form textured protein products. Baker's yeast protein is one product of this type approved for human food ingredient use in the United States.

The Nutrition and Safety of SCP

Chemical analyses of the microorganisms evaluated for SCP reveal that they are comparable in amino acid content and type to plant and animal sources with the possible exception of methionine, which is lower in some SCP sources. All are relatively high in nitrogen. For example, the approximate percentage composition of N on a dry weight basis is as follows: bacteria 12–13, yeast 8–9, algae 8–10, and filamentous fungi 5–8 (57). In addition to proteins, microorganisms contain adequate levels of carbohydrates, lipids, and minerals and are excellent sources of B vitamins. The fat content varies among these sources, with algal cells containing the highest levels and bacteria the lowest. On a dry weight basis, nucleic acids average 3 to 8% for algae, 6 to 12% for yeasts, and 8 to 16% for bacteria (57). B vitamins are high in all SCP sources. The digestibility of SCP in experimental animals has been found to be lower than for animal proteins such as casein. A thorough review of the chemical composition of SCP from a large variety of microorganisms has been made (19, 126).

Success has been achieved in rat-feeding studies with a variety of SCP products, but human-feeding studies have been less successful, except in the case of certain yeast cell products. Gastrointestinal disturbances are common complaints following the consumption of algal and bacterial SCP, and these and other problems associated with the consumption of SCP have been reviewed elsewhere (126). When gram-negative bacteria are used as SCP sources for human use, the endotoxins must be removed or detoxified.

The high nucleic acid content of SCP leads to kidney stone formation and/or gout. The nucleic acid content of bacterial SCP may be as high as 16%, while the recommended daily intake is about 2 g. The problems are caused by an accumulation of uric acid, which is sparingly soluble in plasma. Upon the breakdown of nucleic acids, purine and pyrimidine bases are released. Adenine and guanine (purines) are metabolized to uric acid. Lower animals can degrade uric acid to the soluble compound allantoin (they possess the enzyme uricase), and consequently the consumption of high levels of nucleic acids does not present metabolic problems to these animals as it does to humans. While high nucleic acid contents presented problems in the early development and use of SCP, these compounds can be reduced to levels below 2% by techniques such as acid precipitation, acid or alkaline hydrolysis, or use of endogenous and bovine pancreatic RNAses (62). (For further information on SCP, see refs. 19, 57, 62, 105, 116, 126).

APPARENT HEALTH BENEFITS OF FERMENTED FOODS

The topic of health-promoting effects of certain fermented foods and/or the organisms of fermentation is beset by findings both for and against such effects. While some studies that appear to be well designed support health benefits, other equally well-designed studies do not. The three areas of concern are the possible benefits to lactose-intolerant individuals, the lowering of serum cholesterol, and anticancer activity. More information on these as well as others can be found in the reviews by Deeth and Tamine (21), Friend and Shahani (34), Richardson (91), and Shahani and Ayebo (100).

Lactose Intolerance

Lactose intolerance (lactose malabsorption, intestinal hypolactemia) is the normal state for adult mammals, including most adult humans, and many more groups are intolerant to lactose than are tolerant (60). Among the relatively few groups that have a majority of adults who tolerate lactose are northern Europeans, white Americans, and members of two nomadic pastoral tribes in Africa (60). When lactose malabsorbers consume certain quantities of milk or ice cream, they immediately experience flatulence and diarrhea. The condition is due to the absence or reduced amounts of intestinal lactase, and this allows the bacteria in the colon to utilize lactose with the production of gases. The breath hydrogen test for lactose intolerance is based on the increased levels of H_2 produced by anaerobic and facultatively anaerobic bacteria utilizing the nonabsorbed lactose.

A large number of investigators have found that lactose malabsorbers can consume certain fermented dairy products without harmful effects, while other studies found no beneficial effects. When beneficial effects are found, they are attributed to the reduced level of lactose in the fermented product and to the production of β-galactosidase by the fermenting organisms following ingestion of the products. In one study, the lactose content of yogurt after storage for 11 days decreased about one-half (to about 2.3 g/100 g from 4.8 g/100 g in non-fermented milk). During the same period, galactose increased from traces in milk to 1.3 g/100 g in yogurt, and similar results were found for acidophilus and bifidus milks (6). In a study employing rats, the animals were fed experimental diets containing yogurt, pasteurized yogurt, and simulated yogurt for 7 days. Those that received natural yogurt were able to absorb galactose more efficiently and also had higher levels of intestinal lactase (42). The yogurt bacteria remained viable in the gut for up to 3 h. When eight lactose malabsorbers ingested yogurt or acidophilus milk, they did not experience any of the symptoms that resulted when low-fat milk was ingested (6).

"Sweet" acidophilus milk has been reported by some to prevent symptoms of lactose intolerance, while others have found this product to be ineffective. Developed by M. L. Speck and co-workers, it consists of normal pasteurized milk to which is added large numbers of viable L. acidophilus cells as frozen concentrates. As long as the milk remains under refrigeration, the organisms do not grow, but when it is drunk, the consumer gets the benefit of viable L. acidophilus cells. It is "sweet" because it lacks the tartness of traditional acidophilus milk. When 18 lactase-deficient patients ingested unaltered milk for 1 week followed by "sweet" acidophilus milk for an additional week, they were as intolerant to the latter product as to the unaltered milk (73). In a study with rats, the yogurt bacteria had little effect in preventing the malabsorption of lactose (35). The indigenous lactics in the gut tended to be suppressed by yogurt, and the rat lactobacillus flora changed from one that was predominantly heterofermentative to one that was predominantly homofermentative.

It appears that several factors may be important in the contradictory findings noted: the strains of lactic acid bacteria employed, the basic differences be-tween the digestive tracts of animals and humans, and the degree of lactose intolerance in test subjects. Overall, the amelioration of symptoms of lactose intolerance by lactic acid bacteria is well documented (52).

Cholesterol

Impetus for studies on the effect of fermented milks on cholesterol came from a study of Maasai tribesmen in Africa who, in spite of consuming substantial amounts of meat, have low serum cholesterol and a very low incidence of coronary diseases. This was associated with their common consumption of 4 to 5 liters/day of fermented whole milk (66). Subsequent studies by a large number of groups leave unanswered the true effect of organisms of fermen-tation on serum cholesterol levels in humans, although the weight of evidence tends to support a positive effect. The published findings through 1977 have been reviewed (91).

In a study by Mann (65) using 26 human subjects, large dietary intakes of yogurt were found to lower cholesteremia, and the findings suggested that yogurt contains a factor that inhibits the synthesis of cholesterol from acetate. This factor may be either 3-hydroxy-3-methylglutaric acid and/or orotic acid (91). Rats were employed in a study to evaluate the effect of rat chow plus thermophilus milk and methanol solubles of thermophilus milk on liver cholesterol, and the investigators found that both products significantly reduced liver cholesterol levels compared to controls (89). In another study with rats fed for 4 weeks with a stock diet plus 10% milk fermented by *L. acidophilus*, significantly lower serum cholesterol was found than when those rats were fed two other diets not containing fermented milk (44). While in some studies the lowered cholesterol levels are believed to result from decreased synthesis, in others the bacteria were found to remove cholesterol or its precursors from the gastrointestinal tract. In a study by Gilliland et al. (37), two strains of *L. acidophilus* (recovered from swine) had the ability to grow in the presence of bile. One strain assimilated cholesterol from laboratory culture media in the presence of bile under anaerobic conditions and significantly inhibited increases in serum cholesterol levels in pigs that were fed a high-cholesterol diet. The other strain did not remove cholesterol from laboratory media and did not reduce serum cholesterol when fed to pigs. These investigators thus presented evidence that some strains of *L. acidophilus* reduce serum cholesterol by acting directly on cholesterol in the gastrointestinal tract. More recently, cholesterol was shown to be reduced by 50% in a culture medium after 10 to 14 days of growth at 32°C by *Propionibacterium freudenreichii* (106). The organism did not degrade the compound since up to 70% could be recovered from washed cells.

A total of 68 volunteers (ages 18 to 26) in groups of 10 or 13 were put on a regimen consisting of the following supplements: raw milk, whole milk, skim milk, yogurt, buttermilk, and "sweet" acidophilus milk. The regimen was maintained for 3 weeks, and the findings suggested that cultured buttermilk, yogurt, and acidophilus milk had no noticeable effect on serum cholesterol (117). From a study using rats fed for 4 weeks with chow plus skim milk fermented by *S. thermophilus*, *L. bulgaricus*, and *L. acidophilus* along with appropriate controls, no significant changes in plasma or whole body cholesterol were found (87).

Anticancer Effects

Apparently the first observation of anticancer activity of lactic acid bacteria was that of I. G. Bogdanov and co-workers in the Soviet Union in 1962 (100), who demonstrated an effect against a sarcoma and a carcinoma. Anticancer activities have been demonstrated in animal models by a large number of investigators who variously employed yogurt and yogurt extracts *L. acidophilus*, *L. bulgaricus*, and *L. casei* in addition to extracts of these organisms. The specifics of these findings have been reviewed by Shahani and Ayebo (100) and Friend and Shahani (34).

To study the effect of oral supplements of *L. acidophilus* on fecal bacterial enzyme activity, Goldin and Gorbach (41) used 21 human subjects. The enzymes assayed were β-glucuronidase, nitroreductase, and azoreductase because they can convert indirectly acting carcinogens to proximal carcinogens. The feeding regimen consisted of a 4-week control period followed by 4 weeks of plain milk, 4 weeks of control, 4 weeks of milk containing 2×10^6/ml of viable *L. acidophilus*, and 4 weeks of control. Reductions of two- to fourfold in activities of the three fecal enzymes were observed in all subjects only during the period of lactobacillus feeding; fecal enzyme levels returned to normal during the final 4-week control period. Similar but more limited studies have been reported by others. Findings of the type noted may prove to be significant in colon cancer where the body of evidence supports a role for diet.

References

1. Acton, J. C., R. L. Dick, and E. L. Norris. 1977. Utilization of various carbohydrates in fermented sausage. *J. Food Sci.* 42:174–178.
2. Acton, J. C., J. G. Williams, and M. G. Johnson. 1972. Effect of fermentation temperature on changes in meat properties and flavor of summer sausage. *J. Milk Food Technol.* 35:264–268.
3. Agate, A. D., and J. V. Bhat. 1966. Role of pectinolytic yeasts in the degradation of mucilage layer of *Coffea robusta* cherries. *Appl. Microbiol.* 14:256–260.
4. Akinrele, I. A. 1970. Fermentation studies on maize during the preparation of a traditional African starch-cake food. *J. Sci. Food Agric.* 21:619–625.
5. Allen, S. H. G., R. W. Killermeyer, R. L. Stjernholm, and H. G. Wood. 1964. Purification and properties of enzymes involved in the propionic acid fermentation. *J. Bacteriol.* 87:171–187.
6. Alm, L. 1982. Effect of fermentation of lactose, glucose, and galactose content in milk and suitability of fermented milk products for lactose intolerant individuals. *J. Dairy Sci.* 65:346–352.
7. Arnott, D. R., C. L. Duitschaever, and D. H. Bullock. 1974. Microbiological evaluation of yogurt produced commercially in Ontario. *J. Milk Food Technol.* 37:11–13.
8. Ayres, J. C., D. A. Lillard, and L. Leistner. 1967. Mold ripened meat products. In *Proceedings 20th Annual Reciprocal Meat Conference*, 156–168. Chicago: National Live Stock and Meat Board.
9. Bacus, J. 1984. *Utilization of Microorganisms in Meat Processing: A Handbook for Meat Plant Operators*. New York: Wiley.

10. Banigo, E. O. I., J. M. deMan, and C. L. Duitschaever. 1974. Utilization of high-lysine corn for the manufacture of ogi using a new, improved processing system. *Cereal Chem.* 51:559–572.

11. Banigo, E. O. I., and H. G. Muller. 1972. Manufacture of ogi (a Nigerian fermented cereal porridge): Comparative evaluation of corn, sorghum and millet. *Can. Inst. Food Sci. Technol. J.* 5:217–221.

12. Beuchat, L. R. 1976. Fungal fermentation of peanut press cake. *Econ. Bot.* 30:227–234.

13. Beuchat, L. R. 1987. Traditional fermented food products. In *Food and Beverage Mycology*, 2d ed., L. R. Beuchat, 269–306. New York: Van Nostrand Reinhold.

14. Brown, W. V., and E. B. Collins. 1977. End products and fermentation balances for lactic streptococci grown aerobically on low concentrations of glucose. *Appl. Environ. Microbiol.* 33:38–42.

15. Buckle, K. A., and E. Kartadarma. 1990. Inhibition of bongkrek acid and toxoflavin production in tempe bongkrek containing *Pseudomonas cocovenenans*. *J. Appl. Bacteriol.* 68:571–576.

16. Camargo, R. de, J. Leme, Jr., and A. M. Filho. 1963. General observations on the microflora of fermenting cocoa beans (*Theobroma cacao*) in Bahia (Brazil). *Food Technol.* 17:1328–1330.

17. Collard, P., and S. Levi. 1959. A two-stage fermentation of cassava. *Nature* 183:620–621.

18. Collins, E. B. 1972. Biosynthesis of flavor compounds by microorganisms. *J. Dairy Sci.* 55:1022–1028.

19. Cooney, C. L., C. Rha, and S. R. Tannenbaum. 1980. Single-cell protein: Engineering, economics, and utilization in foods. *Adv. Food Res.* 26:1–52.

20. Davis, J. G. 1975. The microbiology of yoghourt. In *Lactic Acid Bacteria in Beverages and Food*, ed. J. G. Carr et al., 245–263. New York: Academic Press.

21. Deeth, H. C., and A. Y. Tamime. 1981. Yogurt: Nutritive and therapeutic aspects. *J. Food Protect.* 44:78–86.

22. Deibel, R. H., and C. F. Niven, Jr. 1957. *Pediococcus cerevisiae*: A starter culture for summer sausage. *Bacteriol. Proc.*, 14–15.

23. Deibel, R. H., C. F. Niven, Jr., and G. D. Wilson. 1961. Microbiology of meat curing. III. Some microbiological and related technological aspects in the manufacture of fermented sausages. *Appl. Microbiol.* 9:156–161.

24. DeKetelaere, A., D. Demeyer, P. Vandekerckhove, and I. Vervaeke. 1974. Stoichiometry of carbohydrate fermentation during dry sausage ripening. *J. Food Sci.* 39:297–300.

25. Doelle, H. W. 1975. *Bacterial Metabolism*. New York: Academic Press.

26. Etchells, J. L., H. P. Fleming, and T. A. Bell. 1975. Factors influencing the growth of lactic acid bacteria during the fermentation of brined cucumbers. In *Lactic Acid Bacteria in Beverages and Food*, ed. J. G. Carr et al., 281–305. New York: Academic Press.

27. Everson, C. W., W. E. Danner, and P. A. Hammes. 1970. Improved starter culture for semi-dry sausage. *Food Technol.* 24:42–44.

28. Faparusi, S. I., and O. Bassir. 1971. Microflora of fermenting palm-wine. *J. Food Sci. Technol.* 8:206–210.

29. Filho, A. M., and C. W. Hesseltine. 1964. Tempeh fermentation: Package and tray fermentations. *Food Technol.* 18:761–765.

30. Forrest, W. W., and D. J. Walker. 1971. The generation and utilization of energy during growth. *Adv. Microbiol. Physiol.* 5:213–274.

31. Foster, E. M., F. E. Nelson, M. L. Speck, R. N. Doetsch, and J. C. Olson. 1957. *Dairy Microbiology*. Englewood Cliffs, N. J.: Prentice-Hall.

32. Frank, H. A., and A. S. Dela Cruz. 1964. Role of incidental microflora in natural

decomposition of mucilage layer in Kona coffee cherries. *J. Food Sci.* 29:850–853.

33. Frank, H. A., N. A. Lum, and A. S. Dela Cruz. 1965. Bacteria responsible for mucilage-layer decomposition in Kona coffee cherries. *Appl. Microbiol.* 13:201–207.

34. Friend, B. A., and K. M. Shahani. 1984. Antitumor properties of lactobacilli and dairy products fermented by lactobacilli. *J. Food Protect.* 47:717–723.

35. Garvie, E. I., C. B. Cole, R. Fuller, and D. Hewitt. 1984. The effect of yoghurt on some components of the gut microflora and on the metabolism of lactose in the rat. *J. Appl. Bacteriol.* 56:237–245.

36. Gasser, F., and C. Gasser. 1971. Immunological relationships among lactic dehydrogenases in the genera *Lactobacillus* and *Leuconostoc*. *J. Bacteriol.* 106:113–125.

37. Gilliland, S. E., C. R. Nelson, and C. Maxwell. 1985. Assimilation of cholesterol by *Lactobacillus acidophilus*. *Appl. Environ. Microbiol.* 49:377–381.

38. Gilliland, S. E., and M. L. Speck. 1974. Frozen concentrated cultures of lactic starter bacteria: A review. *J. Milk Food Technol.* 37:107–111.

39. Giolitti, G., C. A. Cantoni, M. A. Bianchi, and P. Renon. 1971. Microbiology and chemical changes in raw hams of Italian type. *J. Appl. Bacteriol.* 34:51–61.

40. Goel, M. C., D. C. Kulshrestha, E. H. Marth, D. W. Francis, J. G. Bradshaw, and R. B. Read, Jr. 1971. Fate of coliforms in yogurt, buttermilk, sour cream, and cottage cheese during refrigerated storage. *J. Milk Food Technol.* 34:54–58.

41. Goldin, B. R., and S. L. Gorbach. 1984. The effect of milk and lactobacillus feeding on human intestinal bacterial enzyme activity. *Amer. J. Clin. Nutr.* 39:756–761.

42. Goodenough, E. R., and D. H. Kleyn. 1976. Influence of viable yogurt microflora on digestion of lactose by the rat. *J. Dairy Sci.* 59:601–606.

43. Goodenough, E. R., and D. H. Kleyn. 1976. Qualitative and quantitative changes in carbohydrates during the manufacture of yoghurt. *J. Dairy Sci.* 59:45–47.

44. Grunewald, K. K. 1982. Serum cholesterol levels in rats fed skim milk fermented by *Lactobacillus acidophilus*. *J. Food Sci.* 47:2078–2079.

45. Gunsalus, I. C., and C. W. Shuster. 1961. Energy yielding metabolism in bacteria. In *The Bacteria*, ed. I. C. Gunsalus and R. Y. Stanier, 2:1–58. New York: Academic Press.

46. Haas, G. J. 1976. Alcoholic beverages and fermented food. In *Industrial Microbiology*, ed. B. M. Miller and W. Litsky, 165–191. New York: McGraw-Hill.

47. Hamann, W. T., and E. H. Marth. 1984. Survival of *Streptococcus thermophilus* and *Lactobacillus bulgaricus* in commercial and experimental yogurts. *J. Food Protect.* 47:781–786.

48. Harlander, S. K., and L. L. McKay. 1984. Transformation of *Streptococcus sanguis* Challis with *Streptococcus lactis* plasmid DNA. *Appl. Environ. Microbiol.* 48:342–346.

49. Harlander, S. K., L. L. McKay, and C. F. Schachtels. 1984. Molecular cloning of the lactose-metabolizing genes from *Streptococcus lactis*. *Appl. Environ. Microbiol.* 48:347–351.

50. Harris, D. A., L. Chaiet, R. P. Dudley, and P. Ebert. 1957. The development of a commercial starter culture for summer sausages. *Bacteriol. Proc.*, 15.

51. Hettinga, D. H., and G. W. Reinbold. 1972. The propionic-acid bacteria—A review. *J. Milk Food Technol.* 35:295–301, 358–372, 436–447.

52. Hitchins, A. D., and F. E. McDonough. 1989. Prophylactic and therapeutic aspects of fermented milk. *Am. J. Clin. Nutr.* 49:675–684.

53. Hontebeyrie, M., and F. Gasser. 1977. Deoxyribonucleic acid homologies in the genus *Leuconostoc*. *Intern. J. Syst. Bacteriol.* 27:9–14.

54. Ilany-Feigenbaum, J. Diamant, S. Laxer, and A. Pinsky. 1969. Japanese miso-type products prepared by using defatted soybean flakes and various carbohydrate-containing foods. *Food Technol.* 23:554–556.

55. Ingram, M. 1975. The lactic acid bacteria—A broad view. In *Lactic Acid Bacteria in Beverages and Food*, ed. J. G. Carr et al., 1–13. New York: Academic Press.

56. Johnson, M. G., and E. B. Collins. 1973. Synthesis of lipoic acid by *Streptococcus faecalis* 10C1 and end-products produced anaerobically from low concentrations of glucose. *J. Gen. Microbiol.* 78:47–55.

57. Kihlberg, R. 1972. The microbe as a source of food. *Ann. Rev. Microbiol.* 26:427–466.

58. Kleyn, J., and J. Hough. 1971. The microbiology of brewing. *Ann. Rev. Microbiol.* 25:583–608.

59. Kline, L., and T. F. Sugihara. 1971. Microorganisms of the San Francisco sour dough bread process. II. Isolation and characterization of undescribed bacterial species responsible for the souring activity. *Appl. Microbiol.* 21:459–465.

60. Kretchmer, N. 1972. Lactose and lactase. *Sci. Am.* 227(10):71–78.

61. Levanon, Y., and S. M. O. Rossetini. 1965. A laboratory study of farm processing of cocoa beans for industrial use. *J. Food Sci.* 30:719–722.

62. Litchfield, J. H. 1977. Single-cell proteins. *Food Technol.* 31(5):175–179.

63. London, J. 1976. The ecology and taxonomic status of the lactobacilli. *Ann. Rev. Microbiol.* 30:279–301.

64. London, J., and K. Kline. 1973. Aldolase of lactic acid bacteria: A case history in the use of an enzyme as an evolutionary marker. *Bacteriol. Rev.* 37:453–478.

65. Mann, G. V. 1977. A factor in yogurt which lowers cholesteremia in man. *Atherosclerosis* 26:335–340.

66. Mann, G. V., and A. Spoerry. 1974. Studies of a surfactant and cholesteremia in the Masai. *Amer. J. Clin. Nutr.* 27:464–469.

67. Marth, E. H. 1974. Fermentations. In *Fundamentals of Dairy Chemistry*, ed. B. H. Webb et al., Ch. 13. Westport, Conn.: AVI.

68. Miller, A., III, W. E. Sandine, and P. R. Elliker. 1970. Deoxyribonucleic acid base composition of lactobacilli determined by thermal denaturation. *J. Bacteriol.* 102:278–280.

69. Miller, A., III, W. E. Sandine, and P. R. Elliker. 1971. Deoxyribonucleic acid homology in the genus *Lactobacillus*. *Can. J. Microbiol.* 17:625–634.

70. Mislivec, P. B., V. R. Bruce, and R. Gibson. 1983. Incidence of toxigenic and other molds in green coffee beans. *J. Food Protect.* 46:969–973.

71. Mukherjee, S. K., M. N. Albury, C. S. Pederson, A. G. van Veen, and K. H. Steinkraus. 1965. Role of *Leuconostoc mesenteroides* in leavening the batter of idli, a fermented food of India. *Appl. Microbiol.* 13:227–231.

72. Mundt, J. O. 1975. Unidentified streptococci from plants. *Int. J. Syst. Bacteriol.* 25:281–285.

73. Newcomer, A. D., H. S. Park, P. C. O'Brien, and D. B. McGill. 1983. Response of patients with irritable bowel syndrome and lactase deficiency using unfermented acidophilus milk. *Amer. J. Clin. Nutr.* 38:257–263.

74. Niinivaara, F. P., M. S. Pohja, and S. E. Komulainen. 1964. Some aspects about using bacterial pure cultures in the manufacture of fermented sausages. *Food Technol.* 18:147–153.

75. Okafor, N. 1972. Palm-wine yeasts from parts of Nigeria. *J. Sci. Food Agric.* 23:1399–1407.

76. Okafor, N. 1975. Microbiology of Nigerian palm wine with particular reference to bacteria. *J. Appl. Bacteriol.* 38:81–88.

77. Okafor, N. 1975. Preliminary microbiological studies on the preservation of palm wine. *J. Appl. Bacteriol.* 38:1–7.

78. Palumbo, S. A., J. L. Smith, and S. A. Kerman. 1973. Lebanon bologna. I. Manufacture and processing. *J. Milk Food Technol.* 36:497–503.

79. Passmore, S. M., and J. G. Carr. 1975. The ecology of the acetic acid bacteria with particular reference to cider manufacture. *J. Appl. Bacteriol.* 38:151–158.

80. Passos, F. M. L., D. O. Silva, A. Lopez, C. L. L. F. Ferreira, and W. V. Guimaraes. 1984. Characterization and distribution of lactic acid bacteria from traditional cocoa bean fermentations in Bahia. *J. Food Sci.* 49:205–208.

81. Payne, W. J. 1970. Energy yields and growth of heterotrophs. *Ann. Rev. Microbiol.* 24:17–52.

82. Pearson, A. M., and F. W. Tauber. 1984. *Processed Meats.* 2d ed. Westport, Conn.: AVI.

83. Pederson, C. S. 1979. *Microbiology of Food Fermentations*, 2d ed. Westport, Conn.: AVI.

84. Pederson, C. S., and R. S. Breed. 1946. Fermentation of coffee. *Food Res.* 11:99–106.

85. Potter, M. E., M. B. Kruse, M. A. Matthews, R. O. Hill, and R. J. Martin. 1976. A sausage-associated outbreak of trichinosis in Illinois. *Amer. J. Pub. Hlth* 66:1194–1196.

86. Prescott, S. C., and C. G. Dunn. 1957. *Industrial Microbiology.* New York: McGraw-Hill.

87. Pulusani, S. R., and D. R. Rao. 1983. Whole body, liver and plasma cholesterol levels in rats fed thermophilus, bulgaricus and acidophilus milks. *J. Food Sci.* 48:280–281.

88. Radke-Mitchell, L., and W. E. Sandine. 1984. Associative growth and differential enumeration of *Streptococcus thermophilus* and *Lactobacillus bulgaricus*: A review. *J. Food Protect.* 47:245–248.

89. Rao, D. R., C. B. Chawan, and S. R. Pulusani. 1981. Influence of milk and thermophilus milk on plasma cholesterol levels and hepatic cholesterogenesis in rats. *J. Food Sci.* 46:1339–1341.

90. Reddy, N. R., S. K. Sathe, M. D. Pierson, and D. K. Salunkha. 1981. Idli, an Indian fermented food: A review. *J. Food Qual.* 5:89–101.

91. Richardson, T. 1978. The hypocholesteremic effect of milk—A review. *J. Food Protect.* 41:226–235.

92. Roelofsen, P. A. 1958. Fermentation, drying, and storage of cacao beans. *Adv. Food Res.* 8:225–296.

93. Rohan, T. A. 1964. The precursors of chocolate aroma: A comparative study of fermented and unfermented cocoa beans. *J. Food Sci.* 29:456–459.

94. Rohan, T. A., and T. Stewart. 1966. The precursors of chocolate aroma: Changes in the sugars during the roasting of cocoa beans. *J. Food Sci.* 31:206–209.

95. Rose, A. H. 1982. *Fermented Foods.* Economic Microbiology Series, 7. New York: Academic Press.

96. Saisithi, P., B.-O. Kasemsarn, J. Liston, and A. M. Dollar. 1966. Microbiology and chemistry of fermented fish. *J. Food Sci.* 31:105–110.

97. Sandine, W. E., P. C. Radich, and P. R. Elliker. 1972. Ecology of the lactic streptococci. A review. *J. Milk Food Technol.* 35:176–185.

98. Schleifer, K. H., and O. Kandler. 1972. Peptidoglycan types of bacterial cell walls and their taxonomic implications. *Bacteriol. Rev.* 36:407–477.

99. Senez, J. C. 1962. Some considerations on the energetics of bacterial growth. *Baceriol. Rev.* 26:95–107.

100. Shahani, K. M., and A. D. Ayebo. 1980. Role of dietary lactobacilli in gastrointestinal microecology. *Amer. J. Clin. Nutr.* 33:2448–2457.

101. Shibasaki, K., and C. W. Hesseltine. 1962. Miso fermentation. *Econ. Bot.* 16:180–195.

102. Shieh, Y.-S. G., and L. R. Beuchat. 1982. Microbial changes in fermented peanut and soybean pastes containing kojis prepared using *Aspergillus oryzae* and *Rhizopus oligosporus. J. Food Sci.* 47:518–522.

103. Simonds, J., P. A. Hansen, and S. Lakshmanan. 1971. Deoxyribonucleic acid hybridization among strains of lactobacilli. *J. Bacteriol.* 107:382–384.

104. Smith, J. L., and S. A. Palumbo. 1973. Microbiology of Lebanon bologna. *Appl. Microbiol.* 26:489–496.

105. Snyder, H. E. 1970. Microbial sources of protein. *Adv. Food Res.* 18:85–140.

106. Somkuti, G. A., and T. L. Johnson. 1990. Cholesterol uptake by *Propionibacterium freundenreichii. Curr. Microbiol.* 20:305–309.

107. Speckman, R. A., and E. B. Collins. 1968. Diacetyl biosynthesis in *Streptococcus diacetilactis* and *Leuconostoc citrovorum. J. Bacteriol.* 95:174–180.

108. Speckman, R. A., and E. B. Collins. 1973. Incorporation of radioactive acetate into diacetyl by *Streptococcus diacetilactis. Appl. Microbiol.* 26:744–746.

109. Stamer, J. R. 1976. Lactic acid bacteria. In *Food Microbiology: Public Health and Spoilage Aspects*, ed. M. P. deFigueiredo and D. F. Splittstoesser, 404–426. Westport, Conn.: AVI.

110. Steinkraus, K. H., ed. 1983. *Handbook of Indigenous Fermented Foods.* New York: Marcel Dekker.

111. Steinkraus, K. H., A. G. van Veen, and D. B. Thiebeau. 1967. Studies on idle—As Indian fermented black gram-rice food. *Food Technol.* 21:916–919.

112. Stouthamer, A. H. 1969. Determination and significance of molar growth yields. *Meth. Microbiol.* 1:629–663.

113. Subik, J., and M. Behun. 1974. Effect of bongkrekic acid on growth and metabolism of filamentous fungi. *Arch. Microbiol.* 97:81–88.

114. Sugihara, T. F., L. Kline, and M. W. Miller. 1971. Microorganisms of the San Francisco sour dough bread process. I. Yeasts responsible for the leavening action. *Appl. Microbiol.* 21:456–458.

115. Sundhagul, M., W. Daengsubha, and P. Suyanandana. 1975. Thailand's traditional fermented food products: A brief description. *Thai J. Agr. Sci.* 8:205–219.

116. Tannenbaum, S. R., and D. I. C. Wang, eds. 1975. *Single-Cell Protein*, vol. 2. Cambridge: MIT Press.

117. Thompson, L. U., D. J. A. Jenkins, M. A. Vic Amer, R. Reichert, A. Jenkins, and J. Kamulsky. 1982. The effect of fermented and unfermented milks on serum cholesterol. *Amer. J. Clin. Nutr.* 36:1106–1111.

118. Urbaniak, L., and W. Pezacki. 1975. Die Milchsäure bildende Rohwurst-Mikroflora und ihre technologisch bedingte Veränderung. *Fleischwirtschaft* 55:229–237.

119. Van Pee, W., and J. M. Castelein. 1972. Study of the pectinolytic microflora, particularly the *Enterobacteriaceae*, from fermenting coffee in the Congo. *J. Food Sci.* 37:171–174.

120. van Veen, A. G. 1967. The bongkrek toxins. In *Biochemistry of Some Foodborne Microbial Toxins*, ed. R. I. Mateles and G. N. Wogan, 43–50. Cambridge: MIT Press.

121. van Veen, A. G., and W. K. Mertens. 1934. Die Gifstoffe der sogenannten Bongkrek-vergiftungen auf Java. *Rec. Trav. Chim.* 53:257–268.

122. van Veen, A. G., and W. K. Mertens. 1934. Das Toxoflavin, der gelbe Gifstoff der Bongkrek. *Rec. Trav. Chim.* 53:398–404.

123. van Veen, A. G., and K. H. Steinkraus. 1970. Nutritive value and wholesomeness of fermented foods. *J. Agr. Food Chem.* 18:576–578.

124. Vaughn, R. H. 1975. Lactic acid fermentation of olives with special reference to California conditions. In *Lactic Acid Bacteria in Beverages and Food*, ed. J. G. Carr et al., 307–323. New York: Academic Press.

125. Wardlaw, F. B., G. C. Skelley, M. G. Johnson, and J. C. Acton. 1973. Changes in meat components during fermentation, heat processing and drying of a summer sausage. *J. Food Sci.* 38:1228–1231.
126. Waslien, C. I. 1976. Unusual sources of proteins for man. *CRC Crit. Rev. Food Sci. Nutri.* 6:77–151.
127. Williams, R. A. D. 1975. A review of biochemical techniques in the classification of lactobacilli. In *Lactic Acid Bacteria in Beverages and Food*, ed. J. G. Carr et al., 351–367. New York: Academic Press.
128. Yaping, J., L. Xiaoyang, and Y. Jiaqi. 1990. *Saccharobacter fermentatus* gen. nov., sp. nov., a new ethanol-producing bacterium. *Int. J. Syst. Bacteriol.* 40:412–414.
129. Yong, F. M., and B. J. B. Wood. 1974. Microbiology and biochemistry of the soy sauce fermentation. *Adv. Appl. Microbiol.* 17:157–194.
130. Zhao, N.-X., M.-S. Ma, Y.-P. Zhang, and D.-C. Xu. 1990. Comparative description of *Pseudomonas cocovenenans* (van Damme, Johannes, Cox, and Berends 1960) NCIB 9450[T] and strains isolated from cases of food poisoning caused by consumption of fermented corn flour in China. *Int. J. Syst. Bacteriol.* 40:452–455.

PART VI

Microbial Indicators of Food Safety and Quality, Principles of Quality Control, and Microbiological Criteria

The use of microorganisms and/or their products as quality indicators is presented in Chapter 17, along with the use of coliforms and enterococci as safety indicators. The genus *Enterococcus* now includes at least 16 species, and the utility of this group as sanitary and quality indicators is discussed. The principles of the Hazard Analysis Critical Control Point (HACCP) system are presented in Chapter 18 as the best method to control pathogens in foods. This chapter also contains an introduction to sampling plans and examples of microbiological criteria. The whole area of food quality control has been given much attention; some of the varying approaches and views can be obtained from the following references:

D. A. Shapton and N. F. Shapton, eds., *Principles and Practices for the Safe Processing of Foods* (Stoneham, Mass.: Butterworth Heinemann, 1991). Includes discussions on HACCP and microbiological criteria.

D. D. Bills, Shain-dow Kung, and R. Quatrano, eds., *Biotechnology and Food Quality* (Stoneham, Mass.: Butterworth Heinemann, 1989). Discussions on food quality evaluation and the quality of genetic engineered products.

J. A. Troller, *Sanitation in Food Processing* (New York: Academic Press, 1983). Covers practical approaches to the control of microorganisms.

17

Indicators of Food Microbial Quality and Safety

Indicator organisms may be employed to reflect the microbiological quality of foods relative to product shelf life or their safety from foodborne pathogens. In general, indicators are most often used to assess food sanitation, and most of this chapter treats them in this context; however, quality indicators may be used, and some general aspects of this usage are outlined in the following section.

INDICATORS OF PRODUCT QUALITY

Microbial product quality or shelf-life indicators are organisms and/or their metabolic products whose presence in given foods at certain levels may be used to assess existing quality or, better, to predict product shelf life. When used in this way, the indicator organisms should meet the following criteria:

- They should be present and detectable in all foods whose quality (or lack thereof) is to be assessed.
- Their growth and numbers should have a direct negative correlation with product quality.
- They should be easily detected and enumerated and be clearly distinguishable from other organisms.
- They should be enumerable in a short period of time, ideally within a working day.
- Their growth should not be affected adversely by other components of the food flora.

In general, the most reliable indicators of product quality tend to be product specific; some examples of food products and possible quality indicators are listed in Table 17-1. The products noted have restricted floras, and spoilage is typically the result of the growth of a single organism. When a single organism

413

TABLE 17-1.   Some Organisms That Are Highly Correlated
with Product Quality

Organisms	Products
Acetobacter spp.	Fresh cider
Bacillus spp.	Bread dough
Byssochlamys spp.	Canned fruits
Clostridium spp.	Hard cheeses
Flat-sour spores	Canned vegetables
Lactic acid bacteria	Beers, wines
Lactococcus lactis	Raw milk (refrigerated)
Leuconostoc mesenteroides	Sugar (during refinery)
Pectinatus cerevisiiphilus	Beers
"*Pseudomonas putrefaciens*"	Butter
Yeasts	Fruit juice concentrates
Zygosaccharomyces bailii	Mayonnaise, salad dressing

TABLE 17-2.   Some Microbial Metabolic Products That Correlate
with Food Quality

Metabolites	Applicable Food Product
Cadaverine and putrescine	Vacuum-packaged beef
Diacetyl	Frozen juice concentrate
Ethanol	Apple juice, fishery products
Histamine	Canned tuna
Lactic acid	Canned vegetables
Trimethylamine (TMA)	Fish
Total volatile bases (TVB), total volatile nitrogen (TVN)	Seafoods
Volatile fatty acids	Butter, cream

is the cause of spoilage, its numbers can be monitored by selective culturing or by a method such as impedance with the use of an appropriate selective medium (see Chaps. 5 and 6). Overall microbial quality of the products noted in Table 17-1 is a function of the number of organisms noted, and shelf life can be increased by their control. In effect, microbial quality indicators are spoilage organisms whose increasing numbers result in loss of product quality.

Metabolic products may be used to assess and predict microbial quality in some products; some examples are listed in Table 17-2. The diamines (cadaverine and putrescine), histamine, and polyamines have been found to be of value for several products (discussed further in Chap. 9). Diacetyl was found to be the best negative predictor of quality in frozen orange juice concentrates, where it imparts a buttermilk aroma at levels of 0.8 ppm or above (60). A 30-min method for its detection was developed by Murdock (59). Ethanol has been suggested as a quality index for canned salmon, where 25 to 74 ppm were associated with "offness" and levels higher than 75 ppm indicated spoilage (31). Ethanol was found to be the most predictive of several alcohols in fish extracts stored at 5°C, where 227 of 241 fish spoilage isolates produced this alcohol (2). Lactic acid was the most frequently found organic acid in spoiled canned vegetables, and a rapid (2 h) silica-gel plate method was developed for its detection (1). The production of trimethylamine (TMA) from trimethylamine-*N*-oxide by fish spoilers has been used by a large number of investigators as a quality or spoilage index. Various procedures have been employed to measure total volatile substances as indicators of fish quality, including total volatile bases (TVB)—ammonia, dimethylamine, and TMA— and total volatile nitrogen (TVN)—which include TVB and other nitrogen compounds that are released by steam distillation of fish products. (For more information, see Chap. 9 and ref. 31.)

Total viable count methods have been used to assess product quality. They are of greater value as indicators of the existing state of given products than as predictors of shelf life since the portion of the count represented by the ultimate spoilers is difficult to ascertain.

Overall, microbial quality indicator organisms can be used for food products that have a flora limited by processing parameters and conditions where an undesirable state is associated consistently with a given level of specified organisms. Where product quality is significantly affected by the presence and quantity of certain metabolic products, they may be used as quality indicators. Total viable counts generally are not reliable in this regard but they are better than direct microscopic counts.

INDICATORS OF FOOD SAFETY

Microbial indicators are more often employed to assess food safety and sanitation than quality. Ideally, a food safety indicator should meet certain important criteria. It should

- Be easily and rapidly detectable.
- Be easily distinguishable from other members of the food flora.
- Have a history of constant association with the pathogen whose presence it is to indicate.
- Always be present when the pathogen of concern is present.
- Be an organism whose numbers ideally should correlate with those of the pathogen of concern.
- Possess growth requirements and a growth rate equaling those of the pathogen.
- Have a die-off rate that at least parallels that of the pathogen and ideally persists slightly longer than the pathogen of concern.
- Be absent from foods that are free of the pathogen except perhaps at certain minimum numbers.

These criteria apply to most, if not all, foods that may be vehicles of foodborne pathogens, regardless of their source to the foods. In the historical use of safety indicators, however, the pathogens of concern were assumed to be of intestinal origin, resulting from either direct or indirect fecal contamination. Thus, such sanitary indicators were used historically to detect fecal contamination of waters and thereby the possible presence of intestinal pathogens. The first fecal indicator was *Escherichia coli*. When the concept of fecal indicators was applied to food safety, some additional criteria were stressed, and those suggested by Buttiaux and Mossel (8) are still valid:

- Ideally the bacteria selected should demonstrate specificity, occuring only in intestinal environments.
- They should occur in very high numbers in feces so as to be encountered in high dilutions.
- They should possess a high resistance to the extraenteral environment, the pollution of which is to be assessed.
- They should permit relatively easy and fully reliable detection even when present in very low numbers.

Following the practice of employing *E. coli* as an indicator of fecal pollution of waters, other organisms were suggested for the same purpose. In time, most of these were applied to foods.

Coliforms

While attempting to isolate the etiologic agent of cholera in 1885, Escherich (19) isolated and studied the organism that is now *E. coli*. It was originally named *Bacterium coli commune* because it was present in the stools of each patient he examined. Schardinger (74) was the first to suggest the use of this organism as an index of fecal pollution because it could be isolated and identified more readily than individual waterborne pathogens. A test for this organism as a measure of drinking water potability was suggested in 1895 by T. Smith (81). This marked the beginning of the use of coliforms as indicators of pathogens in water, a practice that has been extended to foods.

Strains

In a practical sense, coliforms are gram-negative asporogeneous rods that ferment lactose within 48 h and produce dark colonies with a metallic sheen on Endo-type agar (3). By and large, coliforms are represented by four genera of the family Enterobacteriaceae: *Citrobacter*, *Enterobacter*, *Escherichia*, and *Klebsiella*. Occasional strains of *Arizona hinshawii* and *Hafnia alvei* ferment lactose but generally not within 48 h, and some *Pantoea agglomerans* strains are lactose positive within 48 h.

Since *E. coli* is more indicative of fecal pollution than the other genera and species noted (especially *E. aerogenes*), it is often desirable to determine its incidence in a coliform population. The IMViC formula is the classical method used, where I = indole production, M = methyl red reaction, V = Voges-Proskauer reaction (production of acetoin), and C = citrate utilization. By this method, the two organisms noted have the following formulas:

	I	M	V	C
E. coli	+	+	−	−
E. aerogenes	−	−	+	+

The IMViC reaction + + − − designates *E. coli* type I; *E. coli* type II strains are − + − −. The MR reaction is the most consistent for *E. coli*. *Citrobacter* spp. have been referred to as intermediate coliforms, and delayed lactose fermentation by some strains is known. All are MR+ and VP−. Most are citrate + while indole production varies. *Klebsiella* isolates are highly variable with respect to IMViC reactions, although *K. pneumoniae* is generally MR−, VP+, and C+, but variations are known to occur in the MR and I reactions. Fluorogenic substrate methods for differentiating between *E. coli* and other coliforms are discussed in Chapter 6.

Fecal coliforms are defined by the production of acid and gas in EC broth between 44° and 46°C, usually 44.5° or 45.5°C (EC broth, for *E. coli.* was developed in 1942 by Perry and Hajna, 66). A test for fecal coliforms is essentially a test for *E. coli* type I, although some *Citrobacter* and *Klebsiella* strains fit the definition. Notable exceptions are the EHEC strains (see below) that do not grow at 44.5°C in the standard EC medium formulation, but will grow when the bile salts content in the medium is reduced from 0.15% to 0.112% (84).

E. coli strains that cause gastroenteritis in infants and adults are referred to as enteropathogenic strains, and several distinct types are known: enterotoxigenic (ETEC), enteroinvasive (EIEC), enterohemorrhagic (EHEC), and facultatively enteropathogenic (FEEC). The EHEC strains produce a Shiga-like toxin, also referred to as verocytotoxin. Some of these strains have the serotype 0157:H7, and they are further characterized in Chapter 22.

Growth

Like most other nonpathogenic gram-negative bacteria, coliforms grow well on a large number of media and in many foods. They have been reported to grow at temperatures as low as −2°C and as high as 50°C. In foods, growth is poor or very slow at 5°C, although several authors have reported the growth of coliforms at 3° to 6°C. Coliforms have been reported to grow over a pH range of 4.4 to 9.0. *E. coli* can be grown in a minimal medium containing only an organic carbon source such as glucose and a source of nitrogen such as $(NH_4)_2SO_4$ and other minerals. Coliforms grow well on nutrient agar and produce visible colonies within 12 to 16 h at 37°C. They can be expected to grow in a large number of foods under the proper conditions.

Coliforms are capable of growth in the presence of bile salts, which inhibit the growth of gram-positive bacteria. Advantage is taken of this fact in their selective isolation from various sources. Unlike most other bacteria, they have the capacity to ferment lactose with the production of gas, and this characteristic alone is sufficient to make presumptive determinations. The general ease with which coliforms can be cultivated and differentiated makes them nearly ideal as indicators except that their identification may be complicated by the presence of atypical strains. The aberrant lactose fermenters, however, appear to be of questionable sanitary significance (26).

One of the attractive properties of *E. coli* as a fecal indicator for water is its period of survival. It generally dies off about the same time as the more common intestinal bacterial pathogens, although some reports indicate that some bacterial pathogens are more resistant in water. It is not, however, as resistant as intestinal viruses (see below). Buttiaux and Mossel (8) concluded that various pathogens may persist after *E. coli* is destroyed in foods that are frozen, refrigerated, or irradiated. Similarly, pathogens may persist in treated waters after *E. coli* destruction. Only in acid food does *E. coli* have particular value as an indicator organism due to its relative resistance to low pH (8).

TABLE 17-3.   Summary of Some Methods Employed to Detect or Enumerate Coliforms and *E. coli*

Methods	Time for Results	Sensitivity	Primary Use	Comments
Direct counts				
VRBA plating				
Presumptive results	18–24 h	~10/g	Total coliforms	
Confirmed results	24–48 h	~10/g	Total coliforms	
Anderson and Baird-Parker	24 h	~10/g	EC type I	
British roll tube	24 h	1/g	EC	Incubated 44.5°C
Dry medium plates				
VRB medium	24 h	~10/g	Total coliforms	
VRB medium	24 h	~10/g	Fecals	Incubated 44.5°C
EC count	24 h	~10/g	EC	MUG substrate used
Broth culture methods				
Classical MPN, presumptive	24–48 h	<1/100 ml	Total coliforms	LST broth
Classical MPN, confirmed	24–48 h	—	Total coliforms	LST to BGLB broth
Classical MPN, for fecals	24 h	<1/100 ml	Fecals	LST to EC medium
Membrane/other filter methods				
Standard membrane method	24 h	<1/g	Total coliforms	LES Endo agar
M-FC method	24 h	<1/g	Fecals	M-FC broth, 44.5°C
M-7h FC method	7 h	<1/g	Fecals	M-7h FC agar; 41.5°C
Coli-count sampler	24 h	≥10/g	Fecals	
HGMF method	24 h	1/g	Coliforms	
HGMF-ELA	24 h	10/g	EC 0157:H7	Employs HC agar
Fluorogenic substrate methods				
LST + MUG	20 h	1 cell	EC	Incubated 35°C
X-GLUC plating	24 h	~10/g	EC	
MUGal plating	6 h	1/100 ml	EC	
Defined substrate method				
Presence-absence (P-A)	24 h	1/100 ml	EC, coliforms	
Impedance	6.5 h	for~10^3/g	Coliforms	Employs special media
Enzyme capture assay (ECA)	24 h	<1/g	EC	ECA from LST tubes
Radiometric assays	6 h	1–10/g	Coliforms	Radiolabel needed
DNA probes	3–4 days	<2/g	EC 0157:H7	Colony hybridization and dot blot assays
DNA amplification (PCR)	8, 12 h	20 cells	EC	Detects DNA sequence
Glutamate decarboxylase assay	10 h	1 cell	EC	
Ethanol assay	9 h	~10/ml	Coliforms	Gas chromatograph
Coliphage detection	4–6 h	~ 5/100 ml	EC	Plaque assay

Notes: BGLB = brilliant green lactose bile; EC = *E. coli*; ELA = enzyme-labeled antibodies; HC = hemorrhagic colitis; HGMF = hydrophobic grid membrane filter; LST = lauryl sulfate tryptose; M-FC = Millipore fecal count; MPN = most probable number; MUG = 4-methylumbelliferyl-β-D-glucuronide; MUGal = 4-methylumbelliferyl-β-D-galactoside; VRB = violet red bile; X-GLUC = 4-bromo-4-chloro-3-indolyl-β-D-glucuronide. (See Chap. 5 and 6 for further details.)

Detection and Enumeration

A synopsis of most of the methods employed to detect and enumerate *E. coli* and the coliforms is presented in Table 17-3. (More details on most of these methods are presented in Chap. 6.) The large number of methods is a reflection in part of strong current interest in methods that are more rapid and accurate and of the value of coliforms as food safety indicators. Although the direct count and most probable numbers (MPN) methods require 24 to 48 h for results, they continue to be of value as reference methods. The fluorogenic,

TABLE 17-4.   Suggested Coliform/*E. coli* Criteria

Indicator/Products		Class Plan	n	c	m	M
1. Coliforms:	Dried milk	3	5	1	10	10^2
2. Coliforms:	Pasteurized liquid, frozen, and dried egg products	3	5	2	10	10^3
3. Coliforms:	Infants, children, and certain dietetic foods; coated or filled, dried shelf-stable biscuits	3	5	2	10	10^2
4. Coliforms:	Dried and instant products requiring reconstitution	3	5	1	10	10^2
5. Coliforms:	Dried products requiring heating to boiling before consumption	3	5	3	10	10^2
6. Coliforms:	Cooked ready-to-eat crabmeat	3	5	2	500	5,000
7. Coliforms:	Cooked ready-to-eat shrimp	3	5	2	100	10^3
8. *E. coli*:	Fresh, frozen, cold-smoked fish; frozen raw crustaceans	3	5	3	11	500
9. *E. coli*:	Precooked breaded fish; frozen cooked crustaceans	3	5	2	11	500
10. *E. coli*:	Cooked, chilled, frozen crabmeat	3	5	1	11	500
11. *E. coli*:	Frozen vegetable/fruits, pH >4.5; dried vegetables	3	5	2	10^2	10^3
12. *E. coli*:	Fresh/frozen bivalve molluscs	2	5	0	16	—
13. *E. coli*:	Bottled water	2	5	0	0	—

Note: Items 6 and 7 are recommendations of the National Advisory Committee on the Microbiological Criteria for Foods, USDA/FDA, January 1990 and the criteria noted are for process integrity. All other items are from ICMSF (32).

defined substrate, and DNA amplification (PCR) methods are among the most recent and actively investigated. (These, along with others in Table 17-4, are further discussed in Chap. 6.)

Distribution

The prevalence and incidence of coliforms and *E. coli* in a variety of foods are presented in Chapter 4. The primary habitat of *E. coli* is the intestinal tract of most warm-blooded animals, although sometimes it is absent from the gut of hogs. The primary habitat of *E. aerogenes* is vegetation and, occasionally, the intestinal tract. It is not difficult to demonstrate the presence of coliforms in air and dust, on the hands, and in and on many foods. The issue is not simply the presence of coliforms but their relative numbers. For example, most market vegetables harbor small numbers of lactose-fermenting, gram-negative rods of the coliform type, but if these products have been harvested and handled properly, the numbers tend to be quite low and of no real significance from the standpoint of public health.

Coliform Criteria and Standards

While the presence of large numbers of coliforms and *E. coli* in foods is highly undesirable, it would be virtually impossible to eliminate all from fresh and frozen foods. The basic questions regarding numbers are these:

1. Under proper conditions of harvesting, handling, storage, and transport of foods by use of a Hazard Analysis Critical Control Point (HACCP) system, what is the lowest possible and feasible number of coliforms to maintain?
2. At what quantitative level do coliforms or *E. coli* indicate that a product has become unsafe?

In the case of water and dairy products, there is a long history of safety related to allowable coliform numbers. Some coliform and *E. coli* criteria and standards for water, dairy products, and other foods covered by some regulatory agencies are as follows:

- Not over 10/ml for Grade A pasteurized milk and milk products, including cultured products.
- Not over 10/ml for certified raw and not over 1 for certified pasteurized milk.
- Not over 10/ml for precooked and partially cooked frozen foods.
- Not over 100/ml for crabmeat.
- Not over 100/ml for custard-filled items.

Low numbers of coliforms are permitted in sensitive foods at numbers ranging from 1 to not over 100/g or 100 ml. These criteria reflect both feasibility and safety parameters.

Some products for which coliform criteria have been recommended by the International Commission on the Microbiological Specifications for Foods (ICMSF, 32) are listed in Table 17-4. The values noted are not meant to be used apart from the total suggested criteria for these products. They are presented here only to show the acceptable and unacceptable ranges of coliforms or *E. coli* for the products noted. Implicit in the recommendations for the first four products is that one or two of five subsamples drawn from a lot may contain up to 10^3 coliforms and yet be safe for human consumption.

Some Limitations for Food Safety Use

Although the coliform index has been applied to foods for many years, there are limitations to the use of these indicators for certain foods. As a means of assessing the adequacy of pasteurization, a committee of the American Public Health Association in 1920 recommended the use of coliforms (51), and this method was well established in the dairy industry around 1930 (65). Coliform tests for dairy products are not intended to indicate fecal contamination, but do reflect overall dairy farm and plant sanitation (70). For frozen blanched vegetables, coliform counts are of no sanitary significance because some, especially *Enterobacter* types, have common associations with vegetation (82). However, the presence of *E. coli* may be viewed as an indication of a processing problem. For poultry products, coliforms are not good sanitary indicators since salmonellae may exist in a flock prior to slaughter, and thus positive fecal coliform tests may be unrelated to postslaughter contamination (86). The standard coliform test is not suitable for meats because of the widespread occurrence of psychrotrophic enterics and *Aeromonas* spp. in meat environments, but fecal coliform tests are of value (61).

Coliform tests are widely used in shellfish sanitation, but they are not always good predictors of sanitary quality. The U.S. National Shellfish Sanitation Program was begun in 1925, and presence of coliforms was used to assess the sanitation of shellfish-growing waters. Generally shellfish from waters that meet the coliform criteria ("open waters") have a good history of sanitary quality, but some human pathogens may still exist in these shellfish. In oysters, there is no correlation between fecal coliforms and *Vibrio cholerae* (14, 35) or between *E. coli* and either *Vibrio parahaemolyticus* or *Yersinia enterocolitica* (50). Coliforms are of no value in predicting scombroid poisoning (50), nor do they always predict the presence of enteric viruses (see coliphage section below).

In spite of the limitations noted, coliforms are of proved value as safety indicators in at least some foods. They are best employed as a component of a safety program such as the HACCP system described in Chapter 18.

Enterococci as Quality and Safety Indicators

The enterococci were once a subgroup of the genus *Streptococcus*, but they were assigned in 1984 to the genus *Enterococcus*. Currently 16 species are recognized (Table 17-5). Prior to 1984 the "fecal streptococci" consisted of two species and three subspecies, and they, along with *S. bovis* and *S. equinus*, were placed together because each contained Lancefield group D antigens. The latter two species are retained in the genus *Streptococcus* while the former "*S. faecalis*" and "*S. faecium*" (hereafter referred to as the classical enterococci) are now *Enterococcus faecalis* and *E. faecium*. (Some of the background on the new classification of enterococci is discussed in Chap. 2; more details can be obtained from refs. 16, 20, 34, 43, 48, 72, 75, and 92.)

Historical Background

Escherich was the first to describe the organism that is now *E. faecalis*, which he named *Micrococcus ovalis* in 1886. *E. faecium* was recognized first in 1899 and further characterized in 1919 by Orla-Jensen (62). Because of their existence in feces, these classical enterococci were suggested as indicators of water quality around 1900. Ostrolenk et al. (63) and Burton (6) were the first to compare the classical enterococci to coliforms as indicators of safety. Pertinent features of the classical enterococci that led to their use as pollution indicators for water are the following:

- They generally do not multiply in water, especially if the organic matter content is low.
- They are generally less numerous in human feces than *E. coli*, with ratios of fecal coliforms to enterococci of 4.0 or higher being indicative of contamination by human waste. Thus, the classical enterococcal tests presumably reflect more closely the numbers of intestinal pathogens than fecal coliforms.
- The enterococci die off at a slower rate than coliforms in waters and thus would normally outlive the pathogens whose presence they are used to indicate.

The simultaneous use of enterococci and coliforms was advocated in the 1950s by Buttiaux (7) since in his opinion the presence of both suggested the occurrence of fecal contamination. In his review of the literature, Buttiaux noted that 100% of human and pig feces samples contained enterococci, while only 86 to 89% contained coliforms (7). (Additional history of enterococci as sanitary indicators can be found in refs. 15, 42, 78, and 85.)

Classification and Growth Requirements

Although the classical enterococci never achieved the status of coliforms as sanitation indicators for water or foods, their current classification in an ex-

TABLE 17-5.   Characteristics of *Enterococcus* spp.

Property	E. faecalis	E. faecium	E. avium	E. casseliflavus	E. durans	E. malodoratus	E. gallinarum	E. hirae	E. mundtii	E. raffinosus	E. solitarius	E. pseudoavium	E. cecorum	E. saccharolyticus	E. columbae	E. dispar
Growth at/in																
10°C	+	+	+	+	+	+	+	+	+	(+)	+	+	−	+		+
45°C	+	+	+	+	+	−	+	+	+	+	+	+	+	+		−
pH 9.6	+	+	+	+	+/−	+	+	+	+				(+)	(+)		
6.5% NaCl	+	+	+	+	+/−	+	+	+	+	+	+		−	+	−	+
40% bile	+	+	+	+	+	+	+	+	+				+		+	
0.1% Methylene blue	+		−	+	+											
0.04% K-tellurite	+	−	−	+	−	−	−	−	−				−			
0.01% Tetrazolium	+	−	+	−		+										
Resist 60°C/30 min.	+	+	+	+	+/−	−	−						−	−	−	
Serologic Group D	+	+	+[a]	+	+	+	+	+	+	+	+		−	−	−	−
Motility	−/+	−	−	−/+	−	−	+	−	−	−	−	−	−	−	−	−
Pigmented	−	−	−	yel	−	−	−		yel	−			−	−	−	−
Esculin hydrolysis	+/−	+	+	+	+	+	+	+	+	+	+	+	+	+	+	+
Hippurate hydrolysis	+/−	+	v	+/−	v	v	+	−	−	−		+	+	−	−	v
Arginine hydrolase	+	+	−	+	+	−	+	+	+						−	+
Produces H₂S	−	−	+	−	−	+	−	−	−	−	−	−	−	−	−	−
Acid from																
Glycerol	+	+	+	−	−	v	−/+	v	v	+	−		−	−	−	+
Mannitol	+	+	+	+	−/+	+	+	−	+	+	+	−	−	+	+	−
Sucrose	+	v	+	+	−	+	+	+	+	+	+	−	+	+	+	+
Salicin	+	+	+/−					+	+	+	+		+	+	+	+
Lactose	+	+	+	+	+		+	+	+	+	−	+	+	+	v	+
Arabinose	−	+	+	+	−	−	+	−	−	+	−	−	−	−	+	−
Raffinose	−	+/−	−	−	−	+	+	+	+	+			+	+	+	+

Sources: Collins et al. (10–13), Devriese et al. (18), Farrow et al. (20–21), Jones et al. (34), Kilpper-Balz et al. (41), Knight and Shlaes (44), Mundt and Graham (56), Rodrigues and Collins (72), Vaughn et al. (88), Williams et al. (91).

Note: + = positive; − = negative; +/− and v = variable reactions.
[a] Also Group Q.

panded genus could, on one hand, make them more attractive as indicators or, on the other hand, make them less attractive and meaningful. *E. faecalis* is found most frequently in the feces of a variety of mammals and *E. faecium* largely in hogs and wild boars (53, 80); the natural distribution of some other members of the new genus is less well understood. Prior to 1984, enterococci and "fecal strep" were essentially synonymous and consisted principally of only *E. faecalis*, *E. faecium*, and *E. durans*. Currently, a test for enterococci is a test for 13 additional species, some of which clearly have less significance as fecal, sanitary, or quality indicators than the classical species. An inspection of the features in Table 17-5 reveals that *E. cecorum* does not grow at 10°C or in 6.5% NaCl. While *E. pseudoavium* grows at 10°C, it does not grow in the presence of 6.5% NaCl (13). With the exception of *E. cecorum*, apparently all grow at 10°C, and some strains of *E. faecalis* and *E. faecium* have been reported to grow between

0° and 6°C. Most of the enterococci grow at 45°C and some, at least *E. faecalis* and *E. faecium*, grow at 50°C. The phylogenetic relationship of enterococci, other lactic acid bacteria, *Listeria*, and *Brochothrix* is presented in Chapter 21 (Fig. 21-1).

At least 14 of the 16 species grow at pH 9.6 and in 40% bile, while 3 do not grow in 6.5% NaCl. *E. cecorum*, *E. columbae*, *E. dispar*, and *E. saccharolyticus* do not react with serologic group D antisera. In addition to reacting with group D antisera, *E. avium* alone reacts with group Q (11). The murein type possessed by *E. faecalis* is Lys-Ala$_{2-3}$ while the other species contain the Lys-D-Asp murein. The moles % G + C content of DNA of the enterococci is 37 to 45. Regarding biochemical characteristics, esculin is hydrolyzed by all 16 species. Two species produce a yellow pigment (*E. casseliflavus* and *E. mundtii*), two are motile (*E. casseliflavus* and *E. gallinarium*), and two produce H$_2$S (*E. casseliflavus* and *E. malodoratus*).

As is typical of other gram-positive bacteria, enterococci are more fastidious in their nutritional requirements than gram negatives but differ from most other gram positives in having requirements for more growth factors, especially B-vitamins and certain amino acids. The requirement for specific amino acids allows some strains to be used in microbiological assays for these compounds. They grow over a much wider range of pH than all other foodborne bacteria (see Chap. 3). Although they are aerobes, they do not produce catalase (except a pseudocatalase by some strains when grown in the presence of O$_2$), and they are microaerophiles that grow well under conditions of low Eh.

Distribution

While the two classical enterococcal species (*E. faecalis* and *E. faecium*) are known to be primarily of fecal origin, the newly created genus awaits further study of natural occurrence, especially regarding fecal occurrence. *E. hirae* and *E. durans* have been found more often in poultry and cattle than in six other animals, while *E. gallinarum* has been found only in poultry (17). *E. durans* and *E. faecium* tend to be associated with the intestinal tract of swine more than does *E. faecalis*. The last appears to be more specific for the human intestinal tract than are other species. *E. cecorum* was isolated from chicken cecae, *E. columbae* from pigeon intestines, and *E. saccharolyticum* from cows. *E. avium* is found in mammalian and chicken feces; *E. casseliflavus* in silage, soils, and on plants; *E. mundtii* on cows, hands of milkers, soils, and plants; *E. hirae* in chicken and pig intestines; *E. dispar* in certain human specimens; and *E. gallinarium* in the intestines of fowls.

It is well established that the classical enterococci exist on plants and insects and in soils (53, 55, 57). The yellow-pigmented species are especially associated with plants, and *E. cecorum* appears to be closely associated with chicken cecae. In general, enterococci on insects and plants may be from animal fecal matter. Such enterococci may be regarded as temporary residents and are disseminated among vegetation by insects and wind, reaching the soil by these means, by rain, and by gravity (55). Although *E. faecalis* is often regarded as being of fecal origin, some strains appear to be common on vegetation and thus have no sanitary significance when found in foods. Mundt (57) studied *E. faecalis* from humans, plants, and other sources and found that the nonfecal indicators could be distinguished from the more fecal types by their reaction in litmus milk and their fermentation reactions in melizitose and melibiose broths. In another study of 2,334 isolates of *E. faecalis* from dried and frozen foods, a high percentage of strains bore a close similarity to the vegetation-resident types and therefore were not of any sanitary significance (54). When used as indicators of sanitary quality in foods, it is necessary to ascertain whether *E. faecalis* isolates are of the vegetation type or whether they represent those of human origin. Enterococci may also be found in dust. They are rather widely distributed, especially in such places as slaughterhouses and curing rooms, where pork products are handled.

With respect to the use of the classical enterococci as indicators of water pollution, some investigators who have studied their persistence in water have found that they die off at a faster rate than coliforms, while others found the opposite. Leininger and McCleskey (46) noted that enterococci do not multiply in water as coliforms sometimes do. Their more exacting growth requirements may be taken to indicate a less competitive role in water environments. In sewage, coliforms and the classical enterococci were found to exist in high numbers, but approximately 13 times more coliforms than enterococci were found (47).

In a study conducted when the genus *Enterococcus* consisted of only eight species, DeVries et al. (17) studied 264 isolates of enterococci obtained from farm animal intestines. Strains were selected solely on the basis of their growth in 40% bile and 6.5% NaCl. Of the 264 isolates, 255 conformed to one of the eight species, with *E. faecalis*, *E. faecium*, and *E. hirae*, representing 37.6%, 29.8%, and 23% of the isolates, respectively. Other species found were *E. durans* (5.1%), *E. gallinarum* (1.6%), *E. avium* (1.2%), *E. mundtii* (1.2%), and *E. casseliflavus* (<1%). These 255 isolates were obtained from eight animal species, with poultry, cattle, and pigs yielding the largest number of isolates. Not known is the relative incidence of the 16 current enterococcal species in human feces or the incidence of the 8 newest species in either human or animal feces.

Relationship to Sanitary Quality of Foods

In this section, the enterococci discussed are those that were defined prior to 1984. A large number of investigators found the classical enterococci to be better than coliforms as indicators of food sanitary quality, especially for frozen foods. In one study, enterococcal numbers were more closely related to aerobic plate counts (APC) than to coliform counts, while coliforms were more closely related to enterococci than to APC (28). Enterococci have been found in greater numbers than coliforms in frozen foods (Table 17-6). In a study of 376 samples of commercially frozen vegetables, Burton (6) found that coliforms were more efficient indicators of sanitation than enterococci prior to freezing, while enterococci were superior indicators after freezing and storage. In samples stored at −20°C for 1 to 3 months, 81% of enterococci and 75% of coliforms survived. After 1 year, 89% of enterococci survived but only 60% of coliforms. In another study, enterococci remained relatively constant for 400 days when stored at freezing temperatures. Enterococci were recovered from 57% of 14 samples of dried foods, while 87% of 13 different frozen vegetables yielded these organisms, many of which were of the vegetation-resident types

TABLE 17-6. Enterococci and Coliform Most Probable Number Counts in Frozen Precooked Fish Sticks

Number	Enterococci MPN Count/100 g	Coliforms MPN Count/100 g
1	86,000	6
2	18,600	19
3	86,000	0
4	46,000	300
5	48,000	150
6	46,000	28
7	46,000	150
8	18,600	7
9	8,600	0
10	4,600	186
11	4,600	186
12	48,000	1,280
13	8,600	46
14	4,600	480
15	48,000	240
16	10,750	1,075
17	10,750	17,000
18	60,000	23,250
19	10,750	2,275
Average	32,339	2,457

Source: Raj et al. (69).

TABLE 17-7. Effect of $-6°F$ Storage on the Longevity of Coliforms and Enterococci in Precooked Frozen Fish Sticks

Days in Storage at $-6°F$	Most Probable Number[a]	
	Coliform	Enterococci
0	5,600,000	15,000,000
7	6,000,000	20,000,000
14	1,400,000	13,000,000
20	760,000	11,300,000
35	440,000	11,200,000
49	600,000	20,000,000
63	88,000	11,000,000
77	395,000	15,000,000
91	125,000	41,000,000
119	50,000	5,400,000
133	136,000	7,400,000
179	130,000	5,600,000
207	55,000	3,500,000
242	14,000	4,000,000
273	21,000	4,000,000
289	42,000	3,200,000
347	20,000	2,300,000
410	8,000	1,600,000
446	260	2,300,000
481	66	5,000,000

Source: Kereluk and Gunderson (40).

[a] Average of four determinations.

TABLE 17-8.   Coliforms and Enterococci as Indicators of Food Sanitary Quality

Characteristic	Coliforms	Enterococci
Morphology	Rods	Cocci
Gram reaction	Negative	Positive
Incidence in intestinal tract	10^7-10^9/g feces	10^5-10^8/g feces
Incidence in fecal matter of various animal species	Absent from some	Present in most
Specificity to intestinal tract	Generally specific	Generally less specific
Occurrence outside of intestinal tract	Common in low nos.	Common in higher nos.
Ease of isolation and identification	Relatively easy	More difficult
Response to adverse environmental conditions	Less resistant	More resistant
Response to freezing	Less resistant	More resistant
Relative survival in frozen foods	Generally low	High
Relative survival in dried foods	Low	High
Incidence in fresh vegetables	Low	Generally high
Incidence in fresh meats	Generally low	Generally low
Incidence in cured meats	Low or absent	Generally high
Relationship to foodborne intestinal pathogens	Generally high	Lower
Relationship to nonintestinal foodborne pathogens	Low	Low

(54). The relative longevity of coliforms and enterococci in frozen fish sticks is presented in Table 17-7.

Overall, the elevation of the once "fecal strep" to the status of a genus and the expansion of the genus from 4 species to 16, including some species that appear to have no natural association with fecal matter, raise questions about the utility of this group as sanitary indicators. During the 1960s and 1970s enterococcal tolerances were suggested for a variety of foods, but they have been disregarded in this context in recent years. Interest in the enterococci as food safety indicators has clearly waned, probably because of the simultaneous interest in faster and more efficient ways to detect and enumerate *E. coli*. As indicators, the enterococci and coliforms are compared in Table 17-8.

Bifidobacteria

In the course of his research on the stools of infants around 1900, Tissier (87) noted an organism that occurred with great frequency and named it *Bacillus bifidus*; it was later named *Lactobacillus bifidus* and currently is *Bifidobacterium bifidum*. The common occurrence of the bifidobactera in stools led Mossel (52) to suggest the use of these gram-positive anaerobic bacteria as indicators of fecal pollution, especially of waters. Interestingly, some bifidobacteria are employed in the production of fermented milks, yogurt, and other food products, and some are believed to provide some health benefits.

The genus *Bifidobacterium* consists of at least 25 species of catalase-negative, nonmotile rods whose minimum and maximum growth temperature ranges are 25° to 28°C and 43° to 45°C, respectively. They grow best in the pH range 5 to 8 and produce lactic and acetic acids as the major end products of their carbohydrate metabolism.

Distribution

The concentration of bifidobacteria has been reported to be 10 to 100 times higher than coliforms and enterococci in human feces deposited in water. At refrigerator temperatures, their respective die-off rates are bifidobacteria, then coliforms, and finally enterococci (27). They are exclusively of fecal origin, and some are abundant and active in sheep and calf rumens (73). They are found most often in the feces of humans and swine (71) and have been shown to be absent from the feces of chickens, cows, dogs, horses, cats, sheep, beavers, goats, and turkeys. *B. adolescentis* and *B. longum* are most often isolated in highest numbers—about $10^6/100$ ml of raw sewage (71). They have been suggested as indicators of recent fecal contamination in tropical fresh waters since they die off faster than either coliforms or enterococci (58).

Isolation Methods

Nutritional exactness and obligate anaerobic requirements prevent bifidobacteria from growing in waters. Research on their utility as pollution indicators was hampered initially by the lack of suitable selective culture media, but progress has been made over the past several years. One of the first selective media was that of Resnick and Levin, designated YN-6 (71), later modified by Mara and Oragui and designated YN-17 (49). The latter contains yeast extract, polypeptone, lactose, Casamino acids, NaCl, bromcresol green, cysteine HCl, nalidixic acid, kanamycin-SO_4, and polymyxin B. Another medium employed with some success is human bifid sorbitol agar (HBSA), which reveals the presence of *B. adolescentis* and *B. breve* because of their sorbitol fermenting capacities. A more recent recovery and enumeration medium is that of Munoa and Pares (58), designated *Bifidobacterium* iodoacetate medium (BIM-25). It contains reinforced clostrial agar, nalidixic acid, polymyxin B SO_4, kanamycin SO_4, iodoacetic acid, and 2,3,5-triphenyltetrazolium-Cl. The use of Columbia agar at pH 5.0 containing 5 ml/L of propionic acid is reported not only to be selective for bifidobacteria but to enhance their growth (4). With the development of these media, more research on the bifidobacteria as alternative pollution indicators may be expected. Their utility for foods remains to be established.

Overall, the close association of bifidobacteria with feces, their absence where fecal matter does not occur, their lack of growth in water, and the specific association of some only with human feces make these bacteria attractive as pollution indicators. On the other hand, since they are strict anaerobes, they tend to grow slowly and require several days for results. Since they are more likely to grow in meat and seafood products than in vegetables (because of the higher natural Eh of the latter), it is possible that they could serve as indicators for meats and seafoods.

Coliphages

Research during the 1920s revealed that bacteriophages occur in waters in association with their host bacteria, and this led Pasricha and de Monte (64) to suggest that phages specific for several intestinal pathogens could be measured as indirect indicators of their host bacterial species. A coliphage assay procedure for water samples that contain five or more phages/100 ml and that can be completed in 4 to 6 h is described in *Standard Methods for the Examination of Waters and Wastewater* (3). Thus, the utility of the coliphage assay for waters using *E. coli* strain C has been established. Of concern in the detection of coliphages is the capacity of the host strains used to allow plaque development by all viable phages. While the American Public Health Association procedure recommends the use of *E. coli* strain C, other hosts may be used simultaneously to increase the plaque counts. There is no way of enumerating all *E. coli* phages or all phages of any other specific bacterium, suggesting the use of mixed indicators for best results (68).

Since coliphage assay by use of *E. coli* hosts may reflect heterogeneous phages with different survival characteristics, the detection of male-specific phages is one method that leads to a more homogeneous phage population. Male-specific phages are single-stranded, homogeneous, and similar in structure and size to enteroviruses (29). Although their standard hosts are F^+ or Hfr strains of *E. coli* K-12, host cells can be constructed by plasmid insertions in *Salmonella typhimurium*. The latter cells contain F-pili, which serve as receptors for male-specific coliphages and are employed essentially in the same way as *E. coli* hosts. (For a review of methods for isolating phages from various environmental sources, see ref. 76.)

Utility for Water

The prediction of fecal coliforms in water by the enumeration of their phages has been shown to be feasible by some investigators (36, 45) and not feasible by others (e.g., 30). In a study of coliphages and fecal and total coliforms in natural waters from ten cities, a linear relationship was found between the two groups (90).

Since bacteria and viruses possess different properties relative to their persistence in the environment, coliphages have attracted interest from those interested in indicators of enteroviruses, especially in water. The inability of the coliform index to predict correctly the presence of enteric viruses in waters has been reported by a number of investigators (22, 23, 24, 25). The survival of coliphages in water has been shown to parallel that of human enteric viruses (79). In a study of approved waters for oyster harvesting along the Gulf Coast, neither *E. coli* nor coliform levels were predictive of the presence of enteroviruses in oysters (22). In recreational waters considered to be acceptable and safe by coliform standards, enteroviruses were detected 43% of the time and 35% of the time in waters acceptable for shellfish harvesting (23). In a study of open waters along the North Carolina coast, enteric viruses were isolated from 3 of 13 100-g clam samples from open beds, and 6 of 15 were positive from closed beds (89). A well-documented outbreak of human hepatitis A was traced to the consumption of oysters taken from open waters (67).

In a study comparing coliphages, total coliforms, fecal coliforms, enterococci, and standard plate counts on water from different treatment processes, coliphages correlated better with enteroviruses than either of the other groups noted (83). When secondary sewage effluents were tested for male-specific coliphages, up to 8,200 plaque-forming units (pfu) were found (29), but how assays for male-specific coliphages compare to the more traditional assay methods is unclear. Since some coliphages have been reported to have their natural habitat in environmental waters, their numbers may not correlate directly with fecal pollution (76). Male-specific coliphages are more indicative of fecal pollution of waters than total coliphages since they do not form F pili at temperatures less than 30°C and thus cannot infect their F^+ host cells (77).

Regarding human enteric viruses, not only can at least some survive better in water than coliforms, but viruses tend to be more resistant to destruction by chlorine. While chlorine destroyed 99.999% of fecal coliforms, total coliforms, and fecal streptococci in primary sewage effluents, only 85 to 99% of viruses present were destroyed in one study (51).

Utility for Foods

The utility of employing coliphage assays for coliforms in foods was first reported by Kennedy et al. in 1984. They employed a 16- to 18-h incubation at 35°C and recovered coliphages from all 18 fresh chicken and pork sausage samples. The highest numbers of coliphages were found on fresh chicken and ranged from log 3.3 to 4.4 pfu/100 g. High coliphage levels in general reflected products that contained high fecal coliform counts (37). In a later study involving 120 samples of 12 products, coliphages at levels of 10 or more pfu/100 g were found in 56% of the samples and 11 of the products (38). The highest numbers were recovered from fresh meats by the 16- to 18-h incubation procedure, and chickens yielded the highest counts (log 2.66–4.04 pfu/100 g). In general, coliphages correlated better with *E. coli* and fecal coliforms than total coliforms. The recovery of coliphages from foods was not affected by pH in the range 6.0 to 9.0 (39). Results could be achieved in 4 to 6 h, but these investigators preferred incubations of 16 to 18 h. On the other hand, male-specific coliphages employing *S. typhimurium* hosts did not correlate with total coliforms, fecal coliforms, or aerobic plate counts in 472 samples of clams from the Chesapeake Bay (9). The low numbers found may have been due to the general absence of sewage contamination in the clam waters.

Overall, the findings from water and sewage and the limited studies with foods suggest that coliphage assays may be suitable either as an alternative for *E. coli* or coliform determinations or as direct indicators for enteroviruses. Since results can be obtained in 4 to 6 h and since coliphages appear to correlate better with enteroviruses than coliforms, further research seems indicated. Host cell systems need to be developed that will yield plaques from all coliphages without allowing plaque development by phages that normally parasitize other closely related enteric bacteria.

References

1. Ackland, M. R., E. R. Trewhella, J. Reeder, and F. G. Bean. 1981. The detection of microbial spoilage in canned foods using thin-layer chromatography. *J. Appl. Bacteriol.* 51:277–281.
2. Ahamed, A., and J. R. Matches. 1983. Alcohol production by fish spoilage bacteria. *J. Food Protect.* 46:1055–1069.
3. American Public Health Association. 1985. *Standard Methods for the Examination of Water and Wastewater*, 16th ed. Washington, D.C.: APHA.
4. Beerens, H. 1990. An elective and selective isolation medium for *Bifidobacterium* spp. *Lett. Appl. Microbiol.* 11:155–157.
5. Berg, G., D. R. Dahling, G. A. Brown, and D. Berman. 1978. Validity of fecal coliforms, total coliforms, and fecal streptococci as indicators of viruses in chlorinated primary sewage effluents. *Appl. Environ. Microbiol.* 36:880–884.
6. Burton, M. C. 1949. Comparison of coliform and enterococcus organisms as indices

of pollution in frozen foods. *Food Res.* 14:434–448.

7. Buttiaux, R. 1959. The value of the association *Escherichiae*-Group D streptococci in the diagnosis of contamination in foods. *J. Appl. Bacteriol.* 22:153–158.

8. Buttiaux, R., and D. A. A. Mossel. 1961. The significance of various organisms of faecal origin in foods and drinking water. *J. Appl. Bacteriol.* 24:353–364.

9. Chai, T.-J., T.-J. Han, R. R. Cockey, and P. C. Henry. 1990. Microbiological studies of Chesapeake Bay soft-shell clams (*Myarenaria*). *J. Food Protect.* 53: 1052–1057.

10. Collins, M. D., U. M. Rodrigues, N. E. Pigott, and R. R. Facklam. 1991. *Enterococcus dispar* sp. nov. a new *Enterococcus* species from human sources. *Lett. Appl. Microbiol.* 12:95–98.

11. Collins, M. D., D. Jones, J. A. E. Farrow, R. Kilpper-Balz, and K. H. Schleifer. 1984. *Enterococcus avium* nom. rev., comb. nov.; *E. casseliflavus* nom. ref., comb. nov.; *E. durans* nom. rev., comb. nov.; *E. gallinarum comb. nov.*; and *E. malodoratus* sp. nov. *Int. J. Syst. Bacteriol.* 34:220–223.

12. Collins, M. D., J. A. E. Farrow, and D. Jones. 1986. *Enterococcus mundtii* sp. nov. *Int. J. Syst. Bacteriol.* 36:8–12.

13. Collins, M. D., R. R. Facklam, J. A. E. Farrow, and R. Williamson. 1989. *Enterococcus raffinosus* sp. nov.; *Enterococcus solitarius* sp. nov. and *Enterococcus pseudoavium* sp. nov. *FEMS Microbiol. Lett.* 57:283–288.

14. Colwell, R. R., R. J. Seidler, J. Kaper, S. W. Joseph, S. Garves, H. Lockman, D. Maneval, H. Bradford, N. Roberts, E. Remmers, I. Huq, and A. Hug. 1981. Occurrence of *Vibrio cholerae* serotype 01 in Maryland and Louisiana estuaries. *Appl. Environ. Microbiol.* 41:555–558.

15. Deibel, R. H. 1964. The group D streptococci. *Bacteriol. Rev.* 37:330–366.

16. Devriese, L. A., K. Ceyssens, U. M. Rodrigues, and M. D. Collins. 1990. *Enterococcus columbae*, a species from pigeon intestines. *FEMS Microbiol. Lett.* 71:247–252.

17. Devriese, L. A., A. van de Kerckhove, R. Kilpper-Balz, and K. H. Schleifer. 1987. Characterization and identification of *Enterococcus* species isolated from the intestines of animals. *Int. J. Syst. Bacteriol.* 37:257–259.

18. Devriese, L. A., G. N. Dutta, J. A. E. Farrow, A. Van de Kerckhove, and B. A. Phillips. 1983. *Streptococcus cecorum*, a new species isolated from chickens. *Int. J. Syst. Bacteriol.* 33:772–776.

19. Escherich, T. 1885. Die Darmbacterien des Neugeborenen und Sauglings. *Fortschr. Med.* 3:515–522, 547–554.

20. Farrow, J. A. E., D. Jones, B. A. Phillips, and M. D. Collins. 1983. Taxonomic studies on some group D streptococci. *J. Gen. Microbiol.* 129:1423–1432.

21. Farrow, J. A. E., and M. D. Collins. 1985. *Enterococcus hirae*, a new species that includes amino acid assay strain NCDO 1258 and strains causing growth depression in young chickens. *Int. J. Syst. Bacteriol.* 35:73–75.

22. Fugate, K. J., D. O. Cliver, and M. T. Hatch. 1975. Enteroviruses and potential bacterial indicators in Gulf coast oysters. *J. Milk Fd. Technol.* 38:100–104.

23. Gerba, C. P., S. M. Goyal, R. L. LaBelle, I. Cech, and G. F. Bodgan. 1979. Failure of indicator bacteria to reflect the occurrence of enteroviruses in marine waters. *Amer. J. Pub. Hlth.* 69:1116–1119.

24. Goyal, S. M. 1983. Indicators of viruses. In *Viral Pollution of the Environment*, ed. G. Berg, 211–230. Boca Raton, Fla.: CRC Press.

25. Goyal, S. M., C. P. Gerba, and J. L. Melnick. 1979. Human enteroviruses in oysters and their overlying waters. *Appl. Environ. Microbiol.* 37:572–581.

26. Griffin, A. M., and C. A. Stuart. 1940. An ecological study of the coliform bacteria. *J. Bacteriol.* 40:83–100.

27. Gyllenberg, H., S. Niemela, and T. Sormunen. 1960. Survival of bifid bacteria in

water as compared with that of coliform bacteria and enterococci. *Appl. Microbiol.* 8:20–22.

28. Hartman, P. A. 1960. *Enterococcus*: Coliform ratios in frozen chicken pies. *Appl. Microbiol.* 8:114–116.
29. Havelaar, A. H., and W. M. Hogeboom. 1984. A method for the enumeration of male-specific bacteriophages in sewage. *J. Appl. Bacteriol.* 56:439–447.
30. Hilton, M. C., and G. Stotzky. 1973. Use of coliphages as indicators of water pollution. *Can. J. Microbiol.* 19:747–751.
31. Hollingworth, T. A., Jr., and H. R. Throm. 1982. Correlation of ethanol concentration with sensory classification of decomposition in canned salmon. *J. Food Sci.* 47:1315–1317.
32. International Commission on the Microbiological Specifications for Foods. 1986. *Microorganisms in Foods*. 2. *Sampling for Microbiological Analysis: Principles and Specific Application*, 2d ed. Toronto: University of Toronto Press.
33. Jay, J. M. 1986. Microbial spoilage indicators and metabolites. In *Foodborne Microorganisms and Their Toxins: Developing Methodology*, ed. M. D. Pierson and N. J. Stern, 219–240. New York: Marcel Dekker.
34. Jones, D., M. J. Sackin, and P. H. A. Sneath. 1972. A numerical taxonomic study of streptococci of serological group D. *J. Gen. Microbiol.* 72:439–450.
35. Kaper, J., H. Lockman, R. R. Colwell, and S. W. Joseph. 1979. Ecology, serology, and enterotoxin production by *Vibrio cholerae* in Chesapeake Bay. *Appl. Environ. Microbiol.* 37:91–103.
36. Kenard, R. P., and R. S. Valentine. 1974. Rapid determination of the presence of enteric bacteria in water. *Appl. Microbiol.* 27:484–487.
37. Kennedy, J. E., Jr., J. L. Oblinger, and G. Bitton. 1984. Recovery of coliphages from chicken, pork sausage and delicatessen meats. *J. Food Protect.* 47:623–626.
38. Kennedy, J. E., Jr., C. I. Wei, and J. L. Oblinger. 1986. Methodology for enumeration of coliphages in foods. *Appl. Environ. Microbiol.* 51:956–962.
39. Kennedy, J. E., Jr., C. I. Wei, and J. L. Oblinger. 1986. Distribution of coliphages in various foods. *J. Food Protect.* 49:944–951.
40. Kereluk, K., and M. F. Gunderson. 1959. Studies on the bacteriological quality of frozen meats. IV. Longevity studies on the coliform bacteria and enterococci at low temperatures. *Appl. Microbiol.* 7:327–328.
41. Kilpper-Balz, R., G. Fischer, and K. H. Schleifer. 1982. Nucleic acid hybridization of group N and group D streptococci. *Curr. Microbiol.* 7:245–250.
42. Kjellander, J. 1960. Enteric streptococci as indicators of fecal contamination of water. *Acta Pathol. Microbiol. Scand. Suppl. 136* 48:1–133.
43. Knight, R. G., D. M. Shlaes, and L. Messineo. 1984. Deoxyribonucleic acid relatedness among major human enterococci. *Int. J. Syst. Bacteriol.* 34:327–331.
44. Knight, R. G., and D. M. Shlaes. 1986. Deoxyribonucleic acid relatedness of *Enterococcus hirae* and "*Streptococcus durans*" homology group II. *Int. J. Syst. Bacteriol.* 36:111–113.
45. Kott, Y., N. Roze, S. Sperber, and N. Betzer. 1974. Bacteriophages as viral pollution indicators. *Water Res.* 8:165–171.
46. Leininger, H. V., and C. S. McCleskey. 1953. Bacterial indicators of pollution in surface waters. *Appl. Microbiol.* 1:119–124.
47. Litsky, W., M. J. Rosenbaum, and R. L. France. 1953. A comparison of the most probable numbers of coliform bacteria and enterococci in raw sewage. *Appl. Microbiol.* 1:247–250.
48. Magrum, C. R. Woese, G. E. Fox, and E. Stackebrandt. 1985. The phylogenetic position of *Streptococcus* and *Enterococcus*. *J. Gen. Microbiol.* 131:543–551.
49. Mara, D.D., and J. I. Oragui. 1983. Sorbitol-fermenting bifidobacteria as specific indicators of human faecal pollution. *J. Appl. Bacteriol.* 55:349–357.

50. Matches, J. R., and C. Abeyta. 1983. Indicator organisms in fish and shellfish. *Food Technol.* 37(6):114–117.
51. McCrady, M. H., and Em. Langevin. 1932. The coli-aerogenes determination in pasteurization control. *J. Dairy Sci.* 15:321–329.
52. Mossel, D. A. A. 1958. The suitability of bifidobacteria as part of a more extended bacterial association, indicating faecal contamination of foods. In *Proceedings, 7th International Congress on Microbiology, Abstracts of Papers*, 440–441. Uppsala: Almquist & Wikesells.
53. Mundt, J. O. 1982. The ecology of the streptococci. *Microbiol. Ecol.* 8:355–369.
54. Mundt, J. O. 1976. Streptococci in dried and frozen foods. *J. Milk Fd. Technol.* 39:413–416.
55. Mundt, J. O. 1961. Occurrence of enterococci: Bud, blossom, and soil studies. *Appl. Microbiol.* 9:541–544.
56. Mundt, J. O., and W. F. Graham. 1968. *Streptococcus faecium* var. *casseliflavus.* nov. var. *J. Bacteriol.* 95:2005–2009.
57. Mundt, J. O. 1973. Litmus milk reaction as a distinguishing feature between *Streptococcus faecalis* of human and nonhuman origins. *J. Milk Fd. Technol.* 36:364–367.
58. Munoa, F. J., and R. Pares. 1988. Selective medium for isolation and enumeration of *Bifidobacterium* spp. *Appl. Environ. Microbiol.* 54:1715–1718.
59. Murdock, D. I. 1968. Diacetyl test as a quality control tool in processing frozen concentrated orange juice. *Food Technol.* 22:90–94.
60. Murdock, D. I. 1967. Methods employed by the citrus concentrate industry for detecting diacetyl and acetylmethylcarbinol. *Food Technol.* 21:643–672.
61. Newton, K. G. 1979. Value of coliform tests for assessing meat quality. *J. Appl. Bacteriol.* 47:303–307.
62. Orla-Jensen, S. H. 1919. The lactic acid bacteria. *Mem. Acad. Royal Soc. Denmark Ser. 8* 5:81–197.
63. Ostrolenk, M., N. Kramer, and R. C. Cleverdon. 1947. Comparative studies of enterococci and *Escherichia coli* as indices of pollution. *J. Bacteriol.* 53:197–203.
64. Pasricha, C. L., and A. J. H. De Monte. 1941. Bacteriophages as an index of water contamination. *Indian Med. Gaz.* 76:492–493.
65. Peabody, F. R. 1963. Microbial indexes of food quality: The coliform group. In *Microbiological Quality of Foods*, ed. L. W. Slanetz, C. O. Chichester, A. R. Gaufin, and Z. J. Ordal, 113–118. New York: Academic Press.
66. Perry, C. A., and A. A. Hajna. 1944. Further evaluation of EC medium for the isolation of coliform bacteria and *Escherichia coli. Amer. J. Pub. Hlth.* 34:735–738.
67. Portnoy, B. L., P. A. Mackowiak, C. T. Caraway, J. A. Walker, T. W. McKinley, and C. A. Klein, Jr. 1975. Oyster-associated hepatitis: Failure of shellfish certification programs to prevent outbreaks. *J. Amer. Med. Assoc.* 233:1065–1068.
68. Primrose, S. B., N. D. Seeley, K. B. Logan, and J. W. Nicolson. 1982. Methods for studying aquatic bacteriophage ecology. *Appl. Environ. Microbiol.* 43:694–701.
69. Raj, H., W. J. Wiebe, and J. Liston. 1961. Detection and enumeration of fecal indicator organisms in frozen sea foods. *Appl. Microbiol.* 9:433–438.
70. Reinbold, G. W. 1983. Indicator organisms in dairy products. *Food Technol.* 37(6):111–113.
71. Resnick, I. G., and M. A. Levin. 1981. Assessment of bifidobacteria as indicators of human fecal pollution. *Appl. Environ. Microbiol.* 42:433–438.
72. Rodrigues, U., and M. D. Collins. 1990. Phylogenetic analysis of *Streptococcus saccharolyticus* based on 16S rRNA sequencing. *FEMS Microbiol. Lett.* 71: 231–234.
73. Scardovi, V., L. D. Trovatelli, B. Biavati, and G. Zani. 1979. *Bifidobacterium cuniculi, Bifidobacterium choerinum, Bifidobacterium boum,* and *Bifidobacterium*

pseudocatenulatum: Four new species and their deoxyribonucleic acid homology relationships. *Int. J. Syst. Bacteriol.* 29:291–311.

74. Schardinger, F. 1892. Über das Vorkommen Gährung erregender Spaltpilze im Trinkwasser und ihre Bedeutung für die hygienische Beurtheilung desselben. *Wien. Klin. Wachr.* 5:403–405, 421–423.

75. Schleifer, K. H., and R. Kilpper-Balz. 1987. Molecular and chemotaxonomic approaches to the classification of streptococci, enterococci and lactococci: A review. *System. Appl. Microbiol.* 10:1–19.

76. Seeley, N. D., and S. B. Primrose. 1982. The isolation of bacteriophages from the environment. *J. Appl. Bacteriol.* 53:1–17.

77. Seeley, N. D., and S. B. Primrose. 1980. The effect of temperature on the ecology of aquatic bacteriophages. *J. Gen. Virol.* 46:87–95.

78. Sherman, J. M. 1937. The streptococci. *Bacteriol. Rev.* 1:3–97.

79. Simkova, A., and J. Cervenka. 1981. Coliphages as ecological indicators of enteroviruses in various water systems. *Bull. WHO* 59:611–618.

80. Slanetz, L. W., and C. H. Bartley. 1964. Detection and sanitary significance of fecal streptococci in water. *Amer. J. Pub. Hlth* 54:609–614.

81. Smith, T. 1895. Notes on *Bacillus coli commune* and related forms, together with some suggestions concerning the bacteriological examination of drinking water. *Amer. J. Med. Sci.* 110:283–302.

82. Splittstoesser, D. F. 1983. Indicator organisms on frozen vegetables. *Food Technol.* 37(6):105–106.

83. Stetler, R. E. 1984. Coliphages as indicators of enteroviruses. *Appl. Environ. Microbiol.* 48:668–670.

84. Szabo, R. A., E. C. D. Todd, and A. Jean. 1986. Method to isolate *Escherichia coli* 0157:H7 from food. *J. Food Protect.* 49:768–772.

85. Tanner, F. W. 1944. Bacteriology of water and sewage. In *Microbiology of Foods*, 2d ed. Champaign, Ill.: Garrard Press.

86. Tompkin, R. B. 1983. Indicator organisms in meat and poultry products. *Food Technol.* 37(6):107–110.

87. Tissier, H. 1908. Recherches sur la flore intestinale normale des enfants agés d'un an à cinq ans. *Ann. Inst. Past.* 22:189–208.

88. Vaughn, D. H., W. S. Riggsby, and J. O. Mundt. 1979. Deoxyribonucleic acid relatedness of strains of yellow pigmented, Group D streptococci. *Int. J. Syst. Bacteriol.* 29:204–212.

89. Wait, D. A., C. R. Hackney, R. J. Carrick, G. Lovelace, and M. D. Sobsey. 1983. Enteric bacterial and viral pathogens and indicator bacteria in hard shell clams. *J. Food Protect.* 46:493–496.

90. Wentsel, R. S., P. E. O'Neill, and J. F. Kitchens. 1982. Evaluation of coliphage detection as a rapid indicator of water quality. *Appl. Environ. Microbiol.* 43:430–434.

91. Williams, A. M., J. A. E. Farrow, and M. D. Collins. 1989. Reverse transcriptase sequencing of 16S ribosomal RNA from *Streptococcus cecorum. Lett. Appl. Microbiol.* 8:185–189.

92. Williamson, R., L. Gutmann, T. Horaud, F. Delbos, and J. F. Acar. 1986. Use of penicillin-binding proteins for the identification of enterococci. *J. Gen. Microbiol.* 132:1929–1937.

18

Microbiological Safety of Foods

Among the desirable qualities that should be associated with foods is freedom from infectious organisms. While it may not be possible to achieve a zero tolerance for all such organisms under good manufacturing practices (GMP), the production of foods with the lowest possible numbers is the desirable goal. With fewer processors producing more products that lead to foods being held longer and shipped farther before they reach consumers, new approaches are needed to ensure safe products. Classical approaches to microbiological quality control have relied heavily on microbiological determinations of both raw materials and end products, but the time required for results is too long for many products. The development and use of certain rapid methods have been of value but these alone have not obviated the need for newer approaches to ensuring safe foods. The Hazard Analysis Critical Control Point (HACCP) system is presented in this chapter as the method of choice for ensuring the safety of foods from farm to home. When deemed necessary, microbiological criteria may be established for some ingredients and foods, and these in connection with sampling plans are presented as components of the HACCP system.

HAZARD ANALYSIS CRITICAL CONTROL POINT SYSTEM

The HACCP concept was advanced in 1971 by H. E. Bauman and other scientists at the Pillsbury Company in collaboration with the National Aeronautics and Space Administration (NASA) and the U.S. Army Research Laboratories. It was presented at the first national Conference on Food Protection (10, 23). Although the conference proceedings included only a synopsis, details of the system were published in 1974 (4). First applied to low-acid canned foods, the HACCP concept has since been applied throughout the

TABLE 18-1.   Leading Factors Contributing to Outbreaks of Foodborne Illness in the United States

Factors	1961–1982	1973–1982
Improper cooling	44%	56%
Lapse of 12 or more hours between preparation/eating	23	31
Contaminated by handlers	18	24
Raw ingredient added without subsequent heating/cooking	16	9
Inadequate cooking/canning/heating	16	20

Sources: Bryan (7, 8).

Note: $N = 1,918$.

food industry to a large variety of products and to the food service industry (5,6,7). The most detailed report on its use is that published by the International Commission on Microbiological Specifications for Foods (ICMSF) (17). More information on the following aspects of HACCP can be obtained from the references noted: its early development (3, 27); its use in meat and poultry inspection (1), seafoods (15), production of meat and poultry products (32), and refrigerated foods (20); the food industry in general (30); the need for flexibility in its operation (2); its use in Canada (13); and to control salmonellosis (25, 26).

HACCP is a system that leads to the production of microbiologically safe foods by analyzing for the hazards of raw materials—those that may appear throughout processing and those that may occur from consumer abuse. While some classical approaches to food safety rely heavily on end product testing, the HACCP system places emphasis on the quality of all ingredients and all process steps on the premise that safe products will result if these are properly controlled. The system is thus designed to control organisms at the point of production and preparation. The five leading factors that contributed to foodborne illness in the United States for the years 1961–1982 are noted in Table 18–1, and it may be noted that events associated with the handling and preparation of foods were significant (8). Mishandling of foods in food service establishments in Canada in 1984 was involved in about 39% of foodborne incidents (31). Proper implementation of HACCP in food service establishments and the home will lead to a decrease in foodborne illness.

A subcommittee of the U.S. National Research Council, National Academy of Sciences made the following recommendation in 1985 (24):

> Because the application of the HACCP system provides for the most specific and critical approach to the control of microbiological hazards presented by foods, use of this system should be required of industry. Accordingly, this subcommittee believes that government agencies responsible for control of microbiological hazards in foods should promulgate appropriate regulations that would require industry to utilize the HACCP system in their food protection programs.

The HACCP system has been endorsed by ICMSF (16), the CFP, the U.S. National Advisory Committee on the Microbiological Criteria for Foods (NACMCF, 21), and other countries.

Definitions

The following terms and concepts are valuable in the development and execution of a HACCP system and are taken from ICMSF (16) and/or NACMCF (21):

Control point: Any point in a specific food system where loss of control does not lead to an unacceptable health risk.

Critical control point (CCP):   Any point or procedure in a food system where control can be exercised and a hazard can be minimized or prevented.

Critical limit: One or more prescribed tolerances that must be met to ensure that a CCP effectively controls a microbiological health hazard.

Deviation: Failure to meet a required critical limit for a CCP.

HACCP: The written document that delineates the formal procedures to be followed in accordance with these general principles.

Hazard: Any biological, chemical, or physical property that may cause an unacceptable consumer health risk (unacceptable contamination, toxin levels, growth, and/or survival of undesirable organisms).

Monitoring: A planned sequence of observations or measurements of critical limits designed to produce an accurate record and intended to ensure that the critical limit maintains product safety.

Risk category: One of six categories prioritizing risk based on food hazards.

Verification: Methods, procedures, and tests used to determine if the HACCP system is in compliance with the HACCP plan.

HACCP Principles

Although interpreted variously, the ICMSF and NACMCF view HACCP as a natural and systematic approach to food safety and as consisting of the following seven principles:

1. Assess the hazards and risks associated with the growing, harvesting, raw materials, ingredients, processing, manufacturing, distribution, marketing, preparation, and consumption of the food in question.
2. Determine the CCP(s) required to control the identified hazards.
3. Establish the critical limits that must be met at each identified CCP.
4. Establish procedures to monitor the CCP(s).
5. Establish corrective actions to be taken when there is a deviation identified by monitoring a given CCP.
6. Establish effective record keeping systems that document the HACCP plan.
7. Establish procedures for verification that the HACCP system is working correctly.

Each of these principles is discussed in more detail below.

Principle 1

Assess hazards and risks.

Hazards and risks may be assessed for individual food ingredients from the flow diagram or by ranking the finished food product by assigning to it a hazard rating from A through F. A plus sign (+) is assigned when a hazard exists. Six hazard categories have been defined by NACMCF (21), representing an expansion of the three proposed by the NRC (25) for salmonellae control:

A. A special class of foods that consist of nonsterile products designated and intended for consumption by individuals at risk, including infants, the aged, infirmed, and immunoincompetents.
B. Product contains "sensitive" ingredients relative to microbiological hazards (e.g., milk, fresh meats).
C. There is no controlled processing step (such as heat pasteurization) that effectively destroys harmful microorganisms.
D. The product is subject to recontamination after processing but before packaging (e.g., pasteurized in bulk and then packaged separately).
E. Substantial potential for abusive handling exists in distribution and/or by consumers that could render the product harmful when consumed (e.g., products to be refrigerated are held above refrigerator temperatures).
F. There is no terminal heat process after packaging or when cooked in the home.

Next, the formulated product should be assigned to one of six hazard categories, expanded by the NACMCF from four suggested by the NRC (24):

VI. A special category that applies to nonsterile products designated and intended for individuals in hazard category A.
V. Food products subject to all five general hazard characteristics (B, C, D, E, and F).
IV. Food products subject to any four general hazard characteristics.
III. Products subject to any three of the general hazard characteristics.
II. Products subject to any two general hazard characteristics.
I. Products subject to any one of the general hazard characteristics.
0. Products subject to no hazards.

Principle 2

Determine CCP(s).

The ICMSF (1988) recognized two types of CCPs: CCP1, to ensure control of a hazard, and CCP2, to minimize a hazard. Typical of CCPs are the following:

- Heat process steps where time-temperature relations must be maintained to destroy given pathogens.
- Freezing and time to freezing before pathogens can multiply.
- The maintenance of pH of a food product at a level that prevents growth of pathogens.
- Employee hygiene.

Principle 3

Establish critical limits.

A critical limit is one or more prescribed tolerances that must be met to ensure that a CCP effectively controls a microbiological hazard. This could mean

TABLE 18-2.   USDA Cooking and Cooling Parameters for Perishable Uncured Meat and Poultry Products

Cooking parameters

USDA/FSIS has established minimal internal temperatures required for cooking perishable uncured meat and poultry products. These temperature requirements are referenced in Title 9 of the CFRs (CFR 301-390) or in policies disseminated through the FSIS Policy Book or Notices.

Cooking requirements[a]

Cooked beef and roast beef (9 CFR 318.17) (121 min. at 130°F to instantaneous at 145°F)	130°–145°F (54.4°–62.7°C)
Baked meatloaf (9 CFR 317.8)	160°F (71.1°C)
Baked pork cut (9 CFR 317.8)	170°C (76.7°C)
Pork (to destroy trichinae) (9 CFR 318.10) (21 hr at 120°F to instantaneous at 144°F)	120°–144°F (48.9°–62.2°C)
Cooked poultry rolls and other uncured poultry products (9 CFR 381.150)	160°F (71.1°C)
Cooked duck, salted (FSIS Policy Book)	155°F (68.3°C)
Jellied chicken loaf (FSIS Policy Book)	160°F (71.1°C)
Partially cooked, comminuted products (FSIS Notice 92–85)	≥151°F for 1 min ≥148°F for 2 min ≥146°F for 3 min ≥145°F for 4 min ≥144°F for 5 min

Cooling parameters

Similarly, parameters for cooling and storing refrigerated products, including temperatures and times, are reflected in agency regulations (9 CFR) and policies.

Cooling requirements

Guidelines for refrigerated storage temperature and internal temperature control point	40°F (4.4°C)
Recommended refrigerated storage temperature for periods exceeding 1 week (FSIS Directive 7110.3)	35°F (1.7°C)

Cooling procedures require that the product's internal temperature not remain between 130°F (54.4°C) and 80°F (26.7°C) for more than 1.5 hr or between 80°F (26.7°C) and 40°F (4.4°C) for more than 5 hr (FSIS Directive 7110.3)

Cooling procedures for products consisting of intact muscle (e.g., roast beef) require that chilling be initiated within 90 min. of the cooking cycle. Product shall be chilled from 120°F (48°C) to 55°F (12.7°C) in not more than 6 hr. Chilling shall continue and the product shall not be packed for shipment until it has reached 40°F (4.4°C). (9 CFR 318.17)

Roast beef for export to the United Kingdom must be chilled to 68°F (20°C) or less within 5 hr after leaving the cooker and to 46°F (7°C) or less within the following 3 hr

[a] Some temperature requirements are based on appearance and labeling characteristics rather than safety.

keeping refrigeration temperatures within a certain specific and narrow range or making sure that a certain minimum destructive temperature is achieved and maintained long enough to effect pathogen destruction. Examples of the latter include adherence to the temperatures noted in Table 18–2 for the control of the respective organisms.

Principle 4

Establish procedures to monitor CCPs.

The monitoring of a CCP involves the scheduled testing or observation of a CCP and its limits; monitoring results must be documented. If, for example, the temperature for a certain process step should not exceed 40°C, a chart recorder may be installed. Microbial counts generally are not satisfactory at this point since too much time is required for results. Physical and chemical parameters such as time, pH, temperature, and a_w can be tested and results obtained immediately. For more details on the monitoring of a HACCP system, see Corlett (11).

Principle 5

Establish corrective actions to be taken when deviations occur in CCP monitoring.

The actions taken must eliminate the hazard that was created by deviation from the plan. If a product is involved that may be unsafe as a result of the deviation, it must be disposed of. While the actions taken may vary widely, in general they must be shown to bring the CCP under control.

Principle 6

Establish effective record keeping to document the HACCP plan.

The HACCP plan must be on file at the food establishment and must be made available to official inspectors upon request. Forms for recording and documenting the system may be developed, or standard forms may be used with necessary modifications. Typically, these may be forms that are completed on a regular basis and filed away. The forms should provide documentation for all ingredients, processing steps, packaging, storage, and distribution.

Principle 7

Establish procedures for verification that the HACCP system is working correctly.

Verification consists of methods, procedures, and tests used to determine that the system is in compliance with the plan. Verification confirms that all hazards were identified in the HACCP plan when it was developed, and verification measures may include compliance with a set of established microbiological criteria when established. Verification activities include the establishment of verification inspection schedules, including review of the HACCP plan, CCP records, deviations, random sample collection and analysis, and written records of verification inspections. Verification inspection reports should include the designation of persons responsible for administering and updating the HACCP plan, direct monitoring of CCP data while in operation, certification that monitoring equipment is properly calibrated, and deviation procedures employed.

Flow Diagrams

The development of a HACCP plan for a food establishment begins with the construction of a flow diagram for the entire process. The diagram should begin with the acquisition of raw materials and include all steps through packaging and subsequent distribution. A diagram typical of frozen pizza production is illustrated in Figure 18-1. A flow diagram should be specific to a production facility and updated as processing steps are altered. Its construction should be the cooperative effort of all individuals involved in engineering, quality assurance, production, microbiology, sanitation, packaging, and the like (32). The flow diagram will serve to identify hazard sites and CCPs.

The manufacture of frozen pizza is used to illustrate the application of HACCP to this process (Fig. 18-1).

Description of product: Frozen pizza that is to be fabricated and frozen for retail sale to consumers. The product must be safe to eat without reconstitution, but directions on the package will state that the product is intended to be heated prior to consumption.

Hazard analysis of ingredients and microorganisms of concern: These are presented in Table 18-3.

Critical control points: These are listed in Table 18-4 and referenced to the numbers on the flow diagram in Figure 18-1. If CCPs are to be classified, sauce and topping preparations (2 and 4) and freezing (8) are CCP1s, and inspection (6) and storage (10) are CCP2s.

A flow diagram for the production of roast beef is presented in Figure 18-2. Cooking is the most important CCP for this product (CCP1), followed by chilling and prevention of recontamination after cooking. The cooking temperature should reach 145°F or otherwise be sufficient to effect a 4-log cycle reduction of *Listeria monocytogenes*. This will not destroy *Clostridium perfringens* spores, and their germination and growth must be controlled by proper chilling and storage. Cooking and cooling parameters for perishable uncured meats are presented in Table 18-2.

Some Limitations of HACCP

Although it is the best system yet devised for controlling microbial hazards in foods from the farm to the table, the uniform application of HACCP in the food manufacturing and service industries will not be without some debate. Among the lingering questions and concerns raised by Tompkin (32) are the following:

• HACCP requires the education of nonprofessional food handlers, especially in the food service industry and in homes; whether this will be achieved

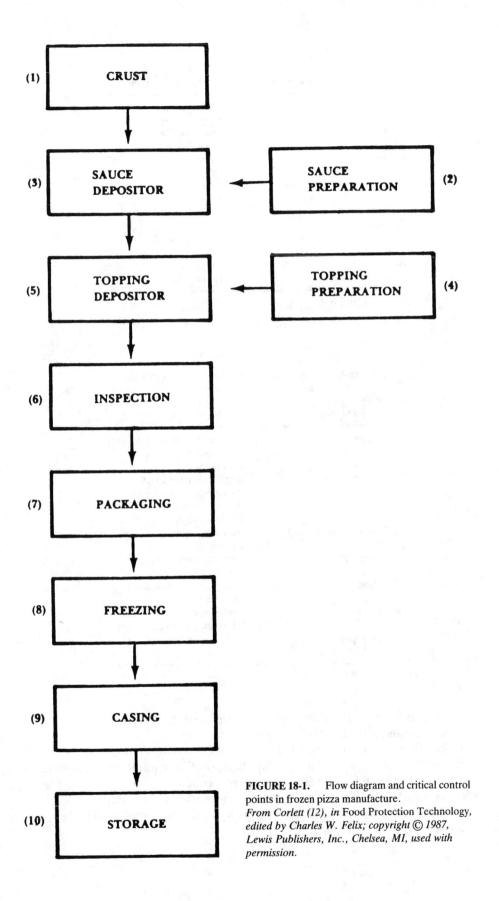

FIGURE 18-1. Flow diagram and critical control points in frozen pizza manufacture.
From Corlett (12), in Food Protection Technology, *edited by Charles W. Felix; copyright © 1987, Lewis Publishers, Inc., Chelsea, MI, used with permission.*

TABLE 18-3.   Microorganisms of Concern in Pizza Ingredients

Ingredients	Microorganisms of Concern
Crust	Staphylococci
Sauce	None, low pH
Cheese	Staphylococci (enterotoxins)
	Salmonellae
	E. coli (EPEC, ETEC, EIEC strains)
Sausage	Salmonellae
	Staphylococci (enterotoxins)
Spices	Salmonellae
Vegetables	Salmonellae

Source: Corlett (12).

TABLE 18-4.   Critical Control Points for the Pizza Process

Flow Diagram Number and Step[a]	Type of Critical Control
1 (crust)	Establish SPC limits on dough; determine proper storage temperature/time.
2,4 (sauce/topping prep.)	Establish microbiological limits on ingredients; sanitation of utensils; storage temperature/time.
3,5 (sauce/topping dep.)	Sanitation of utensils/ equipment
6 (inspection)	Equipment sanitation; employee hygiene
7 (freezing)	Proper internal temperature and time to freezing.
8 (casing)	Establish microbiological limits for finished product.
9 (storage)	Check internal freezing temperature.

[a] From Figure 18-1.

remains to be seen. The failure of these individuals to get a proper understanding of HACCP could lead to its failure.

- To be effective, this concept must be accepted not only by food processors but by food inspectors and the public. Its ineffective application at any level can be detrimental to its overall success for a product.
- It is anticipated that experts will differ as to whether a given step is a CCP and how best to monitor such steps. This has the potential of eroding the confidence of others in HACCP.

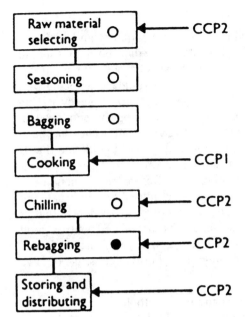

FIGURE 18-2.   Flow diagram for production of roast beef.
From ICMSF (17); copyright © 1988 by Blackwell Scientific Publications. Used with permission.

O indicates a site of minor contamination

● indicates a site of major contamination

CCPI effective CCP. CCP2 not absolute

- The adoption of HACCP by industry has the potential of giving false assurance to consumers that a product is safe, and therefore, there is no need to exercise the usual precautions between the purchase and consumption of a product. Consumers need to be informed that most outbreaks of foodborne illness are caused by errors in food handling in homes and food service establishments and that no matter what steps a processor takes, HACCP principles must be observed after foods are purchased for consumption.

MICROBIOLOGICAL CRITERIA

The concept that microbial limits be assigned to at least some foods to designate their safety or overall quality was suggested as early as 1903 by Marxer (see 14) who suggested an aerobic plate count (APC) limit of 10^6 for hamburger meat. Similarly, APC and indicator organism limits were suggested for many other products through the 1920s and 1930s, with pasteurized milk being notable among those for which limits were widely accepted. The early history of microbiological limits for foods has been reviewed (14). In an effort to eliminate confusion and to agree upon an international language, the Codex Alimentarius Commission (9) has established definitions. The ICMSF has endorsed the Codex definitions with some modifications. The Codex definitions are summarized below with ICMSF modifications noted.

Definitions

Microbiological criteria fall into two main categories: mandatory and advisory. A **mandatory criterion** is a microbiological standard that normally should contain limits only for pathogens of public health significance, but limits for nonpathogens may be set. The ICMSF (16) regards a **standard** as being part of a law or regulation that is enforceable for the regulatory agency having jurisdiction. An **advisory criterion** is either a microbiological end product specification intended to increase assurance that hygienic significance has been met (it may include spoilage organisms), or a microbiological guideline that is applied in a food establishment at a point during or after processing to monitor hygiene (it, too, may include nonpathogens). Before recommending a criterion, the ICMSF (16) believes that each product must be in international trade, must have associated with it good epidemiological evidence that it has been implicated in foodborne disease, and have associated with it good evidence that a criterion will reduce the potential hazard(s) in principle 2.

The Codex definition of a microbiological criterion consists of five components:

1. A statement of the organisms of concern and/or their toxins.
2. The analytical methods for their detection and quantitation.
3. A sampling plan, including when and where samples are to be taken.
4. Microbiological limits considered appropriate to the food.
5. The number of sample units that should conform to these limits.

These five components are embodied in a sampling plan.

Sampling Plans

A sampling plan is a statement of the criteria of acceptance applied to a lot based on appropriate examinations of a required number of sample units by specified methods. It consists of a sampling procedure and decision criteria and may be a two-class or a three-class plan.

A two-class plan consists of the following specifications: n, c, m; a three-class plan requires n, c, m, and M, where

n = the number of sample units (packages, beef patties, and so forth) from a lot that must be examined to satisfy a given sampling plan.

c = the maximum acceptable number, or the maximum allowable number of sample units that may exceed the microbiological criterion m. When this number is exceeded, the lot is rejected.

m = the maximum number or level of relevant bacteria/g; values above this level are either marginally acceptable or unacceptable. It is used to separate acceptable from unacceptable foods in a two-class plan, or, in a three-class plan, to separate good quality from marginally acceptable quality foods. The level of the organism in question that is acceptable and attainable in the food product is m. In the presence/absence situations for two-class plans, it is common to assign $m = 0$. For three-class plans, m is usually some nonzero value.

M = a quantity that is used to separate marginally acceptable quality from unacceptable quality foods. It is used only in three-class plans. Values at or above M in any sample are unacceptable relative to either health hazard, sanitary indicators, or spoilage potential.

A two-class plan is the simpler of the two and in its simplest form may be used to accept or reject a large batch (lot) of food in a presence/absence decision by a plan such as $n = 5$, $c = 0$, where $n = 5$ means that five individual units of the lot will be examined microbiologically for, say, the presence of salmonellae, and $c = 0$ means that all five units must be free of the organisms by the method of examination in order for the lot to be acceptable. If any unit is positive for salmonellae, the entire lot is rejected. If it is desired that two of the five samples may contain coliforms, in a presence/absence test, for example, the sampling plan would be $n = 5$, $c = 2$. By this plan, if three or more of the five-unit samples contained coliforms, the entire lot would be rejected. While presence/absence situations generally obtain for salmonellae, an allowable upper limit for indicator organisms such as coliforms is more often the case. If it is desired to allow up to 100 coliforms/g in two of the five units, the sampling plan would be $n = 5$, $c = 2$, $m = 10^2$. After the five units have been examined for coliforms, the lot is acceptable if no more than two of the five contain as many as 10^2 coliforms/g but is rejected if three or more of the five contain 10^2 coliforms/g. This particular sampling plan may be made more stringent by increasing n (for example, $n = 10$, $c = 2$, $m = 10^2$) or by reducing c (for example, $n = 5$, $c = 1$, $m = 10^2$). On the other hand, it can be made more lenient for a given size n by increasing c.

While a two-class plan may be used to designate acceptable-unacceptable foods, a three-class plan is required to designate acceptable/marginally acceptable/unacceptable foods. To illustrate a typical three-class plan, assume that for a given food product standard plate count (SPC) shall not exceed 10^6/g (M) or be higher than 10^5/g from three or more of five units examined. The specifications are thus $n = 5$, $c = 2$, $m = 10^5$, $M = 10^6$. If any of the five units exceeds 10^6/g, the entire lot its rejected (unacceptable). If not more than c sample units give results above m, the lot is acceptable. Unlike two-class plans, the three-class plan distinguishes values between m and M (marginally acceptable).

With either two- or three-class attributes plans, the numbers n and c may be employed to find the probability of acceptance (P_a) of lots of foods by reference to appropriate tables (16). The decision to employ a two-class or three-class plan may be determined by whether presence/absence tests are desirable, in which case a two-class plan is required, or whether count or concentration tests are desired, in which case a three-class plan is preferred. The latter offers the advantages of being less affected by nonrandom variations between sample units, and of being able to measure the frequency of values in the m to M range. The ICMSF report and recommendations (16) should be consulted for further details on the background, uses, and interpretations of sampling plans. Further information may also be obtained from Kilsby (18).

Microbiological Criteria and Food Safety

The application of criteria to products in the absence of a HACCP program is much less likely to be successful than when the two are combined. Thus, microbiological criteria are best applied as part of a comprehensive program. When criteria are not applied as components of a systematic approach to food safety or quality, the results are known to be less than satisfactory as found by Miskimin et al. (19) and Solberg et al. (29). These investigators studied over 1,000 foods consisting of 853 ready-to-eat and 180 raw products. They applied arbitrary criteria for APC, coliforms, and *E. coli* and tested the efficacy of the criteria to assess safety of the foods with respect to *Staphylococcus aureus, C. perfringens*, and salmonellae. An APC criterion of less than 10^6/g for raw foods resulted in 47% of samples being accepted even though one or more of the three pathogens were present, while 5% were rejected from which pathogens were not isolated, for a total of 52% wrong decisions. An APC of less than 10^5/g for ready-to-eat foods resulted in only 5% being accepted that contained pathogens, while 10% that did not yield pathogens were rejected. In a somewhat similar manner, a coliform criterion of less than 10^2/g resulted in a total of 34% wrong decisions for raw and 15% for ready-to-eat foods. The lowest percentage of wrong decisions for ready-to-eat foods (13%) occurred with an *E. coli* criterion of less than 3/g, while 30% of the decisions were wrong when the same criterion was applied to raw foods. Although the three pathogens were found in both types of foods, no foodborne outbreaks were reported over the four-year period of the study, during which time more than 16 million meals were consumed (28).

These findings represent some initial data from the Rutgers foodservice program. After a seventeen-year experience with modififications in surveillance tests, food audits, laundry evaluations, and more than 30 million meals served, this HACCP-type system has been very effective (28). The microbial guidelines employed by this program for raw and ready-to-eat foods are presented in Table 18-5. Of over 1,600 food samples examined over the period 1983–1989, only 1.24% contained pathogens with protein salads most often contaminated (4.3%). Among the foods that failed microbial surveillance were raw vegetables (they had excessive coliform numbers) (28). The Rutgers Foodservice HACCP-based system is an example of how microbial criteria can be integrated to provide for safe foods; in the 17-year program, no foodborne illness occurred (28).

Microbiological Criteria for Various Products

Prior to the development of the HACCP and sampling plan concepts, microbiological criteria (generally referred to as standards at the time) were applied to a variety of products.

Presented below are foods and food ingredients that are covered under microbiological standards of various organizations along with federal, state, and city standards in effect (after W. C. Frazier, *Food Microbiology*, 1968, courtesy of McGraw-Hill Publishing Company).

1. Standards for Starch and Sugar (National Canners Association).
   A. *Total thermophilic spore count*: Of the five samples from a lot of sugar or starch none shall contain more than 150 spores per 10 g, and the average for all samples shall not exceed 125 spores per 10 g.
   B. *Flat/sour spores*: Of the five samples none shall contain more than 75 spores per 10 g, and the average for all samples shall not exceed 50 spores per 10 g.

    C. *Thermophilic anaerobe spores*: Not more than three (60%) of the five samples shall contain these spores, and in any one sample not more than four (65% + %) of the six tubes shall be positive.

    D. *Sulfide spoilage spores*: Not more than two (40%) of the five samples shall contain these spores, and in any one sample there shall be no more than five colonies per 10 g (equivalent to two colonies in the six tubes).

2. Standard for "Bottlers" Granulated Sugar, Effective July 1, 1953 (American Bottlers of Carbonated Beverages).

    A. *Mesophilic bacteria*: Not more than 200 per 10 g.

    B. *Yeasts*: Not more than 10 per 10 g.

    C. *Molds*: Not more than 10 per 10 g.

3. Standard for "Bottlers" Liquid Sugar, Effective in 1959 (American Bottlers of Carbonated Beverages). All figures based on dry-sugar equivalent (D.S.E.).

    A. *Mesophilic bacteria*: (*a*) Last 20 samples average 100 organisms or less per 10 g D.S.E.; (*b*) 95% of last 20 counts show 200 or less per 10 g; (*c*) 1 of 20 samples may run over 200; other counts as in (*a*) or (*b*).

    B. *Yeasts*: (*a*) Last 20 samples average 10 organisms or less per 10 g D.S.E.; (*b*) 95% of last 20 counts show 18 or less per 10 g; (*c*) 1 of 20 samples may run over 18; other counts as in (*a*) and (*b*).

    C. *Molds*: Standards like those for yeasts.

4. Standards for Dairy Products.

    A. From 1965 recommendations of the U.S. Public Health Service.

        a. *Grade A raw milk for pasteurization*: Not to exceed 100,000 bacteria per milliliter prior to commingling with other producer milk; and not exceeding 300,000 per milliter as commingled milk prior to pasteurization.

        b. *Grade A pasteurized milk and milk products* (except cultured products): Not over 20,000 bacteria per milliliter, and not over 10 coliforms per milliliter.

        c. *Grade A pasteurized cultured products*: Not over 10 coliforms per milliliter.

    NOTE: Enforcement procedures for (*a*), (*b*), and (*c*) require three-out-of-five compliance by samples. Whenever two of four successive samples do not meet the standard, a fifth sample is tested; and if this exceeds any standard, the permit from the health authority may be suspended. It may be reinstated after compliance by four successive samples has been demonstrated.

    B. *Certified milk* (American Association of Medical Milk Commissions, Inc.)

        a. Certified milk (raw): Bacterial plate count not exceeding 10,000 colonies per milliliter; coliform colony count not exceeding 10 per milliliter.

        b. Certified milk (pasteurized): Bacterial plate count not exceeding 10,000 colonies per milliliter before pasteurization and 500 per milliliter in route samples. Milk not exceeding 10 coliforms per milliliter before pasteurization and 1 coliform per milliliter in route samples.

    C. *Milk for manufacturing and processing* (U.S. Dep. Agr. 1955)

        a. Class 1: Direct microscopic clump count (DMC) not over 200,000 per milliliter.

        b. Class 2: DMC not over 3 million per milliliter.

        c. Milk for Grade A dry milk products: must comply with requirements for Grade A raw milk for pasteurization (see above).

    D. *Dry milk*

        a. Grade A dry milk products: at no time a standard plate count over 30,000 per gram, or coliform count over 90 per gram (U.S. Public Health Service).

        b. Standards of Agricultural Marketing Service (U.S. Dep. Agr.):

            (1) Instant nonfat: U.S. Extra Grade, a standard plate count not over 35,000 per gram, and coliform count not over 90 per gram.

(2) Nonfat (roller or spray): U.S. Extra Grade, a standard plate count not over 50,000 per gram; U.S. Standard Grade, not over 100,000 per gram.

(3) Nonfat (roller or spray): Direct microscopic clump count not over 200 million per gram; and must meet the requirements of U.S. Standard Guide. U.S. Extra Grade, such as used for school lunches, has an upper limit of 75 million per gram.

c. Dried milk (International Dairy Federation proposed microbiological specifications, 1982).

Mesophilic count	$n = 5, c = 2, m = 5 \times 10^4, M = 2 \times 10^5$
Coliforms	$n = 5, c = 1, m = 10, M = 100$
Salmonella	$n = 15, c = 0, m = 0$

E. *Frozen desserts*

States and cities that have bacterial standards usually specify a maximal count of 50,000 to 100,000 per milliliter or gram. The U.S. Public Health Ordinance and Code sets the limit at 50,000 and recommends bacteriological standards for cream and milk used as ingredients. Few localities have coliform standards.

5. Standard for Tomato Juice and Tomato Products–Mold-count Tolerances (Food and Drug Administration).

The percentage of positive fields tolerated is 2% for tomato juice and 40% for other comminuted tomato products, such as catsup, purée, paste, etc. A microscopic field is considered positive when aggregate length of not more than three mold filaments present exceeds one-sixth of the diameter of the field (Howard mold count method). This method also has been applied to raw and frozen fruits of various kinds, especially berries.

Other Criteria/Guidelines

1. Sampling plans and microbiological limits for nine products as recommended by ICMSF (15) are presented in Table 18-5 (for an explanation of plan stringency or case, see Table 18-6). The examples presented were selected to reflect different plan stringencies (for two- and three-class plans) and limits for a variety of organisms.

2. Suggested guidelines for further processed deboned poultry products studied in Canada. See Table 18-7.

3. Canadian criteria for cottage cheese and ice cream (24):

Coliforms: $n = 5, c = 1, m = 10, M = 10^3$ (for cottage cheese and ice cream)
Aerobic plate count: $n = 5, c = 2, m = 10^5, M = 10^6$ (for ice cream only)

4. Recommended criteria for cooked ready-to-eat shrimp (22):

S. aureus: $n = 5, c = 2, m = 50, M = 500$
Coliforms: $n = 5, c = 2, m = 10^2, M = 10^3$

5. Recommended criteria for cooked ready-to-eat crabmeat (22):

S. aureus: $n = 5, c = 2, m = 10^2, M = 10^3$
Coliforms: $n = 5, c = 2, m = 500, M = 5,000$

Both products in 4 and 5 should be free of salmonellae and *L. monocytogenes*. The coliform criteria are recommended for process integrity.

TABLE 18-5. ICMSF Sampling Plans and Recommended Microbiological Limits

Products	Tests	Case	Class plan	n	c	m	M	Comments
Precooked breaded fish	APC	2	3	5	2	5×10^5	10^7	
	E. coli	5	3	5	2	11	500	Products likely to be mishandled
	S. aureus	8	3	5	1	10^3	10^4	
Raw chicken (fresh or frozen), during processing	APC	1	3	5	3	5×10^5	10^7	In-plant processing
Frozen vegetables and fruit, pH 4.5	E. coli	5	3	5	2	10^2	10^3	m value is estimate
Comminuted raw meat (frozen) and chilled carcass meat	APC	1	3	5	3	10^6	10^7	In-plant control
Cereals	Molds	5	3	5	2	10^2–10^4	10^5	m values are estimates
Frozen entrées containing rice or corn flour as a main ingredient	S. aureus	8	3	5	1	10^3	10^4	m value is estimated
Noncarbonated natural mineral and bottled noncarbonated waters	Coliforms	5	2	5	0	0	–	Not for use in infant formula or use by highly susceptibles
Roast beef	Salmonella	12	2	20	0	0	–	
Frozen raw crustaceans	S. aureus	7	3	5	2	10^3	10^4	
	V. parahaemolyticus	8	3	5	1	10^2	10^3	
	Salmonella[a]	10	2	5	0	0	–	
	APC[b]	2	3	5	2	5×10^5	10^7	
	E. coli[b]	5	3	5	2	11	500	
	S. aureus[b]	8	2	5	0	10^3	–	

Note: Except where noted for in-plant use, they are intended primarily for foods in international trade and are cited here primarily to illustrate the assignment of produts to case and limits on a variety of organisms. The ICMSF reference (16) should be consulted for methods of analysis and more details in general.

[a] Normal plans and limits.

[b] Additional tests where appropriate.

TABLE 18-6.   Plan Stringency in Relation to Degree of Health Hazard and Conditions of Use

Type of Hazard	Conditions in Which Food Is Expected to Be Handled and Consumed after Sampling		
	Reduce Degree of Hazard	Cause No Change in Hazard	May Increase Hazard
No direct health hazard			
Utility (e.g., general contamination, reduced shelf-life, and spoilage)	Case 1	Case 2	Case 3
Health hazard			
Low, indirect (indicator)	Case 4	Case 5	Case 6
Moderate, direct, limited spread	Case 7	Case 8	Case 9
Moderate, direct, potentially extensive spread	Case 10	Case 11	Case 12
Severe, direct	Case 13	Case 14	Case 15

Source: ICMSF (16); copyright © 1986 by University of Toronto Press, used with permission.

TABLE 18-7.   Suggested Guidelines for Further Processed Deboned Poultry Products

Tests/Conditions	n	c	m	M
APC (heat before serving)	5	3	10^4	10^5
APC (cook before serving)	5	3	10^6	10^7
APC (bring to boil before serving)	5	3	10^5	10^6
S. aureus	5	1	10^2	10^4
E. coli	5	2	10	10^2

Source: Warburton et al. (33).

Note: No salmonellae, yersinae, or campylobacters allowed.

References

1. Adams, C. E. 1990. Use of HACCP in meat and poultry inspection. *Food Technol.* 44(5):169–170.
2. Archer, D. L. 1990. The need for flexibility in HACCP. *Food Technol.* 44(5):174–178.
3. Bauman, H. 1990. HACCP: Concept, development, and application. *Food Technol.* 44(5):156–158.
4. Bauman, H. E. 1974. The HACCP concept and microbiological hazard categories. *Food Technol.* 28(9):30–34, 74.
5. Bryan, F. L. 1981. Hazard analysis of food service operations. *Food Technol.* 35(2):78–87.
6. Bryan, F. L. 1990. Hazard analysis critical control point (HACCP) systems for retail food and restaurant operations. *J. Food Protect.* 53:978–983.
7. Bryan, F. L. 1990. Application of HACCP to ready-to-eat chilled foods. *Food Technol.* 44(7):70–77.
8. Bryan, F. L. 1988. Risks of practices, procedures and processes that lead to outbreaks of foodborne diseases. *J. Food Protect.* 51:663–673.

9. Codex Alimentarius Commission, 14th Session. 1981. *Report of the 17th Session of the Codex Committee on Food Hygiene.* Alinorm 81/13. Rome: Food and Agriculture Organization.

10. Conference on Food Protection. 1971. *Proceedings.* Washington, D.C.: U.S. Government Printing Office.

11. Corlett, D. A., Jr. 1991. Monitoring a HACCP system. *Cereal Foods World* 36(1):33–40.

12. Corlett, D. A., Jr. 1987. Selection of microbiological criteria based on hazard analysis of food. In *Food Protection Technology, Proceedings of the 3rd Conference for Food Protection,* ed. C. W. Felix. Chelsea, Mich.: Lewis Publishers.

13. Dean, K. H. 1990. HACCP and food safety in Canada. *Food Technol.* 44(5): 172–178.

14. Elliott, H. P., and H. D. Michener. 1961. Microbiological standards and handling codes for chilled and frozen foods: A review. *Appl. Microbiol.* 9:452–468.

15. Garrett, E. S., III, and M. Hudak-Roos. 1990. Use of HACCP for seafood surveillance and certification. *Food Technol.* 44(5):159–165.

16. ICMSF. 1986. *Microorganisms in Foods 2. Sampling for Microbiological Analysis: Principles and Specific Applications.* 2d ed. Toronto: University of Toronto Press.

17. ICMSF. 1988. *Microorganisms in Foods 4. Application of the Hazard Analysis Critical Control Point (HACCP) System to Ensure Microbiological Safety and Quality.* London: Blackwell.

18. Kilsby, D. C. 1982. Sampling schemes and limits. In *Meat Microbiology,* ed. M. H. Brown, 387–421. London: Applied Science Publishers.

19. Miskimin, D. K., K. A. Berkowitz, M. Solberg, W. E. Riha, Jr., W. C. Franke, R. L. Buchanan, and V. O'Leary. 1976. Relationships between indicator organisms and specific pathogens in potentially hazardous foods. *J. Food Sci.* 41:1001–1006.

20. Moberg, L. 1989. Good manufacturing practices for refrigerated foods. *J. Food Protect.* 52:363–367.

21. National Advisory Committee on Microbiological Criteria for Foods. 1990. Hazard analysis and critical control point system. Washington, D.C.: U.S. Department of Agriculture.

22. National Advisory Committee on Microbiological Criteria for Foods. 1990. Recommendations of the seafood working group for cooked ready-to-eat shrimp and cooked ready-to-eat crabmeat. Washington, D.C.: U.S. Department of Agriculture.

23. National Conference on Food Protection. 1971. *Proceedings.* Publ. 1712–0134. Washington, D.C.: U.S. Government Printing Office.

24. National Research Council (U.S.A.). 1985. *An Evaluation of the Role of Microbiological Criteria for Foods and Food Ingredients.* Washington, D.C.: National Academic Press.

25. National Research Council (U.S.A.). 1969. *An Evaluation of the Salmonella Problem.* Washington, D.C.: National Academy of Sciences.

26. Simonsen, B., F. L. Bryan, J. H. B. Christian, T. A. Roberts, R. B. Tompkin, and J. H. Silliker. 1987. Prevention and control of food-borne salmonellosis through application of Hazard Analysis Critical Control Point (HACCP). *Int. J. Food Microbiol.* 4:227–247.

27. Sperber, W. H. 1991. The modern HACCP system. *Food Technol.* 45(6):116–120.

28. Solberg, M., J. J. Buckalew, C. M. Chen, D. W. Schaffner, K. O'Neill, J. McDowell, L. S. Post, and M. Boderck. 1990. Microbiological safety assurance system for foodservice facilities. *Food Technol.* 44(12):68–73.

29. Solberg, M., D. K. Miskimin, B. A. Martin, G. Page, S. Goldner, and M. Libfeld. 1977. Indicator organisms, foodborne pathogens and food safety. *Assoc. Food Drug. Off. Quart. Bull.* 41(1):9–21.

30. Stevenson, K. E. 1990. Implementing HACCP in the food industry. *Food Technol.* 44(5):179–180.
31. Todd, E. C. D. 1989. Foodborne and waterborne disease in Canada 1984: Annual summary. *J. Food Protect.* 52:503–511.
32. Tompkin, R. B. 1990. The use of HACCP in the production of meat and poultry products. *J. Food Protect.* 53:795–803.
33. Warburton, D. W., K. F. Weiss, G. Lachapelle, and D. Dragon. 1988. The microbiological quality of further processed deboned poultry products sold in Canada. *Can. Inst. Food Sci. Technol. J.* 21:84–89.

PART VII

Foodborne Diseases

Most of the human diseases that are contracted from foods are covered in Chapters 19 through 25. Each chapter deals with one or more foodborne pathogens, with emphasis on the biology of the respective organisms, vehicle foods, symptoms, mode of action of pathogenesis, and prevention. Some of the animal parasites that are known and presumed to be transmitted by foods are covered in Chapter 24. Most of these pathogens have not been addressed by food microbiologists in the past; this is an area that will demand more attention in the years to come. Much of what is known about each of the syndromes addressed goes beyond the scope of this text; readers are referred to the following references for more extensive information:

D. O. Cliver, ed., *Foodborne Diseases* (New York: Academic Press, 1990). An excellent but less detailed coverage of foodborne diseases than the work edited by Doyle.

M. P. Doyle, ed., *Foodborne Bacterial Pathogens* (New York: Marcel Dekker, 1989). All of the known and most of the suspected foodborne bacterial pathogens are covered extensively in this work.

P. Krogh, ed., *Mycotoxins in Food* (New York: Academic Press, 1988). A more concise work than that by Sharma and Salunkhe.

E. T. Ryser and E. H. Marth, *Listeria, Listeriosis, and Food Safety* (New York: Marcel Dekker, 1991). The most thorough treatment of the titled subject now available. Includes detailed methods for the recovery of listeriae from foods.

R. P. Sharma and D. K. Salunkhe, *Mycotoxins and Phytoalexins* (Boca Raton, Fla.: CRC Press, 1991). Detailed coverage of aflatoxins and other mycotoxins.

19

Staphylococcal Gastroenteritis

The staphylococcal food-poisoning or food-intoxication syndrome was first studied in 1894 by J. Denys and later in 1914 by Barber, who produced in himself the signs and symptoms of the disease by consuming milk that had been contaminated with a culture of *Staphylococcus aureus*. The capacity of some strains of *S. aureus* to produce food poisoning was proved conclusively in 1930 by G. M. Dack et al. (31), who showed that the symptoms could be produced by feeding culture filtrates of *S. aureus*. While some authors refer to food-associated illness of this type as food intoxication rather than food poisoning, the designation **gastroenteritis** obviates the need to indicate whether the illness is an intoxication or an infection.

Staphylococcal gastroenteritis is caused by the ingestion of food that contains one or more **enterotoxins**, which are produced only by some staphylococcal species and strains. Although enterotoxin production is believed generally to be associated with *S. aureus* strains that produce coagulase and thermonuclease (TNase), many species of *Staphylococcus* that produce neither coagulase nor TNase are known to produce enterotoxins (see below).

An extensive literature exists on staphylococci and the food-poisoning syndrome, much of which goes beyond the scope of this chapter. For more extensive information, other references should be consulted (15, 50, 64, 65, 74, 98).

SPECIES OF CONCERN IN FOODS

The genus *Staphylococcus* included at least 27 species at the start of 1991 and those of potential interest in foods are listed in Table 19-1. Of the 18 species and subspecies noted in the table, only 6 are coagulase positive, and they generally produce thermostable nuclease (TNase). Ten of the coagulase negative species have been shown to produce enterotoxins, and they do not produce

TABLE 19-1.   Staphylococcal Species and Subspecies Known to Produce Coagulase, Nuclease, and/or Enterotoxins

Organisms	Coagulase	Nuclease	Enterotoxin	Hemolysis	Mannitol	G + C of DNA
S. aureus subsp.						
anaerobius	+	TS	−	+	−	31.7
aureus	+	TS	+	+	+	32–36
S. intermedius	+	TS	+	+	(+)	32–36
S. hyicus	(+)	TS	+	−	−	33–34
S. delphini	+	−		+	+	39
S. schleiferi subsp.						
coaqulans	+	TS		+	(+)	35–37
schleiferi	−	TS		+	−	37
S. caprae	−	TL	+	(+)	−	36.1
S. chromogens	−	−w	+	−	v	33–34
S. cohnii	−	−	+	−	v	36–38
S. epidermidis	−	−	+	v	−	30–37
S. haemolyticus	−	TL	+	+	v	34–36
S. lentus	−		+	−	+	30–36
S. saprophyticus	−	−	+	−	+	31–36
S. sciuri	−		+	−	+	30–36
S. simulans	−	v		v	+	34–38
S. warneri	−	TL	+	−w	+	34–35
S. xylosus	−	−	+	+	v	30–36

Notes: + = positive; − = negative; −w = negative to weakly positive; (+) = weak reaction; v = variable; TS = thermostable; TL = thermolabile.

nuclease, or those that do, produce a thermolabile form. The coagulase-negative enterotoxigenic strains are not consistent in their production of hemolysins or their fermentation of mannitol. The long-standing practice of examining foods for coagulase positive staphylococci as the strains of importance has undoubtedly led to underestimations of the prevalence of enterotoxin producers.

The relationship between TNase and coagulase production in staphylococci is discussed in Chapter 6. It is common to assume that TNase and coagulase-positive strains are the only staphylococci that warrant further investigations when found in foods, but the existence of both TNase and coagulase-negative enterotoxin-producing strains has been known for some time. In a study reported in 1969 of 31 coagulase-negative strains, 18 were enterotoxigenic (110). Eight of 15 enterotoxigenic strains were TNase and coagulase negative. In another study, 5 enterotoxigenic strains were both TNase and coagulase negative (54). In neither case was the staphylococcal species identified.

Among coagulase-positive species, *S. intermedius* is well known as an enterotoxin producer. This species is found in the nasals and on the skin of carnivores and horses but rarely in humans. They are well known as pathogens in dogs. From pyrodermatitis in dogs in Brazil, 73 staphylococci were recovered, of which 52 were *S. intermedius* (52). Of the 52, all were coagulase positive in rabbit plasma but negative in human plasma, and 13 (25%) were enterotoxigenic. Four produced staphylococcal enterotoxin (SE) D (SED), 5 produced SEE, and one each produced SEB, SEC, SED/E, and SEA/C. All 13 were TNase positive, and three produced the toxic shock syndrome toxin

(TSST). A large number of *S. hyicus* strains are coagulase positive, and it appears that some produce enterotoxins. In one study, *S. hyicus* strains elicited positive enterotoxin responses in cynomologus monkeys, but the enterotoxin was not one of the known types—SEA through SEE (2, 54). In another study of sheep isolates, 2 of 6 coagulase-positive *S. hyicus* produced SEC (107). Enterotoxin production by *S. delphini*, *S. simulans*, and *S. schleiferi* subsp. *coagulans* has not been reported.

At least 10 of the coagulase-negative staphylococcal species listed in Table 19-1 produce enterotoxins. *S. cohnii*, *S. epidermidis*, *S. haemolyticus*, and *S. xylosus* were recovered from sheep milk along with *S. aureus* (6). The one isolate of *S. cohnii* produced SEC; three isolates of *S. epidermidis* produced SEC and SEB/C/D (2 strains); five isolates of *S. haemolyticus* produced SEA, SED, SEB/C/D, and SEC/D (2 strains); while the four isolates of *S. xylosus* all produced SED (6). These investigators noted that mannitol fermentation was best to distinguish between enterotoxin-positive and enterotoxin-negative strains. In another study, 1 of 20 coagulase-negative food isolates was found to be an enterotoxigenic strain of *S. haemolyticus* that produced both SEC and SED (39). In a study of staphylococcal isolates from healthy goats, 74.3% of 70 coagulase positives produced enterotoxins and 22% of 272 coagulase negatives were enterotoxin positive (107). SEC was the most frequently found enterotoxin among the goat isolates. Seven species of the goat isolates produced more than one enterotoxin (*S. caprae*, *S. epidermidis*, *S. haemolyticus*, *S. saprophyticus*, *S. sciuri*, *S. warneri*, and *S. xylosus*) and two species produced only one—SEC by *S. chromogens* and SEE by *S. lentus* (107).

HABITAT AND DISTRIBUTION

In humans, the main reservoir of *S. aureus* is the nasal cavity. From this source, the organisms find their way to the skin and into wounds either directly or indirectly. While the nasal carriage rate varies, it is generally about 50% for adults and somewhat higher among children. The most common skin sources are the arms, hands, and face, where the carriage rate runs between 5 and 30% (35). In addition to skin and nasal cavities, *S. aureus* may be found in the eyes, throat, and intestinal tract. From these sources, the organism finds its way into air and dust, onto clothing, and in other places from which it may contaminate foods. The two most important sources to foods are nasal carriers and individuals whose hands and arms are inflicted with boils and carbuncles, who are permitted to handle foods.

Most domesticated animals harbor *S. aureus*. Staphylococcal mastitis is not unknown among dairy herds, and if milk from infected cows is consumed or used for cheese making, the chances of contracting food intoxication are excellent. There is little doubt that many strains of this organism that cause bovine mastitis are of human origin. However, some are designated as "animal strains." In one study, staphylococcal strains isolated from parts of raw pork products were essentially all of the animal strain type. However, during the manufacture of pickled pork products, these animal strains were gradually replaced by human strains during the production process, to a point where none of the original animal strains could be detected in finished products (95).

In regard to some of the non–*S. aureus* species, *S. cohnii* is found on the skin of humans and occasionally in urinary tract and wound infections. Human skin is the habitat of both *S. epidermidis* and *S. haemolyticus*, and the latter is associated with human infections. *S. hyicus* is found on the skin of pigs, where it sometimes causes lesions, and it has been found in milk and on poultry. The skin of lower primates and other mammals is the habitat of *S. xylosus*, and the skin of humans and other primates is the habitat of *S. simulans*. *S. schleiferi* subsp. *schleiferi* was found in clinical specimens from human patients with decreased resistance to infection (41), and subsp. *coagulans* was isolated from ear infections in dogs (58). *S. aureus* subsp. *anaerobius* causes disease in sheep, and *S. delphini* was recovered from dolphins (109). *S. sciuri* is found on the skin of rodents and *S. lentus* and *S. caprae* are associated with goats, especially goat milk.

While many of the coagulase-negative species noted are adapted primarily to nonhuman hosts, their entry into human foods is not precluded. Once in susceptible foods, their growth may be expected to lead to the production of enterotoxins. All of these species grow in the presence of 10% NaCl.

Since among the staphylococci *S. aureus* has been studied most as a cause of foodborne gastroenteritis, most of the information that follows is about this species.

INCIDENCE IN FOODS

In general, staphylococci may be expected to exist, at least in low numbers, in any or all food products that are of animal origin or in those that are handled directly by humans, unless heat processing steps are applied to effect their destruction. They have been found in a large number of commercial foods by many investigators (see Chap. 4 for foods and relative numbers of *S. aureus* found).

NUTRITIONAL REQUIREMENTS FOR GROWTH

Staphylococci are typical of other gram-positive bacteria in having a requirement for certain organic compounds in their nutrition. Amino acids are required as nitrogen sources, and thiamine and nicotinic acid are required among the B vitamins. When grown anaerobically, they appear to require uracil. In one minimal medium for aerobic growth and enterotoxin production, monosodium glutamate serves as the C, N, and energy sources. This medium contains only three amino acids (arginine, cystine, and phenylalanine) and four vitamins (pantothenate, biotin, niacin, and thiamine), in addition to inorganic salts (73). Arginine appears to be essential for enterotoxin B production (112).

TEMPERATURE GROWTH RANGE

Although it is a mesophile, some strains of *S. aureus* can grow at a temperature as low as 6.7°C (4). The latter authors found three food-poisoning strains that grew in custard at 114°F but decreased at 116° to 120°F, with time of incubation. They grew in chicken à la king at 112°F but failed to grow in ham salad at the same temperature. In general, growth occurs over the range of 7° to 47.8°C, and enterotoxins are produced between 10° and 46°C, with the optimum between 40° and 45°C (98). These minimum and maximum temperatures of growth and toxin production assume optimal conditions relative to the other parameters, and the ways in which they interact to raise minimum growth or lower maximum growth temperatures are noted below.

EFFECT OF SALTS AND OTHER CHEMICALS

While *S. aureus* grows well in culture media without NaCl, it can grow well in 7 to 10% concentrations, and some strains can grow in 20%. The maximum concentrations that permit growth depend on other parameters such as temperature, pH, a_w, and Eh (see below).

S. aureus has a high degree of tolerance to compounds such as tellurite, mercuric chloride, neomycin, polymyxin, and sodium azide, all of which have been used as selective agents in culture media. *S. aureus* can be differentiated from other staphylococcal species by its greater resistance to acriflavine. In the case of borate, *S. aureus* is sensitive, while *S. epidermidis* is resistant (60). With novobiocin, *S. saprophyticus* is resistant, whereas *S. aureus* and *S. epidermidis* are not. The capacity to tolerate high levels of NaCl and certain other compounds is shared by members of the genus *Micrococcus*, which are widely distributed in nature and occur in foods generally in greater numbers than staphylococci, thus making the recovery of the latter more difficult. The effect of other chemicals on *S. aureus* is presented in Chapter 11.

EFFECT OF pH, a_w, AND OTHER PARAMETERS

Regarding pH, *S. aureus* can grow over the range of 4.0 to 9.8, but its optimum is in the range of 6 to 7. As is the case with the other growth parameters, the precise minimum growth pH is dependent on the degree to which all other parameters are at optimal levels. In homemade mayonnaise, enterotoxins were produced when initial pH was as low as 5.15 and when final growth pH was not below 4.7 (49). SEB was produced at a level of 158 ng/100 g with an inoculum of approximately 10^5/g. In general, SEA production is less sensitive to pH than SEB. The buffering of a culture medium at pH 7.0 leads to more SEB than when the medium is unbuffered or buffered in the acid range (72). A similar result was noted at a controlled pH of 6.5 rather than 7.0 (59).

With respect to a_w, the staphylococci are unique in being able to grow at values lower than for any other nonhalophilic bacteria. Growth has been demonstrated as low as 0.83 under otherwise ideal conditions, although 0.86 is the generally recognized minimum a_w.

NaCl and pH.   Using a protein hydrolysate medium incubated at 37°C for 8 days, growth and enterotoxin C production occurred over the pH range 4.00 to 9.83 with no NaCl. With 4% NaCl, the pH range was restricted to 4.4 to 9.43 (Table 19-2). Toxin was produced at 10% NaCl with a pH of 5.45 or higher, but none was produced at 12% NaCl (43).

It has been shown that *S. aureus* growth is inhibited in broth at pH 4.8 and

TABLE 19-2.   The Effect of pH and NaCl on the Production of Enterotoxin C by an Inoculum of 10^8 cells/ml of *S. aureus* 137 in a Protein Hydrolysate Medium Incubated at 37°C for 8 Days

pH range	4.00–9.83	4.4–9.43	4.50–8.55	5.45–7.30	4.50–8.55
NaCl content (%)	0	4	8	10^a	12
Enterotoxin production	+	+	+	+	−

Source: Genigeorgis et al. (43); copyright © 1971 by American Society for Microbiology.
[a] Enterotoxin was detected also with an inoculum of 3.6×10^6 at pH 6.38–7.30.

5% NaCl. Growth and enterotoxin B production by strain S-6 occurred in 10% NaCl at pH 6.9 but not with 4% at pH 5.1 (45). The general effect of increasing NaCl concentration is to raise the minimum pH of growth. At pH 7.0 and 37°C, enterotoxin B was inhibited by 6% or more NaCl (see Fig. 19-1).

pH, a_w, and Temperature.   No growth of a mixture of *S. aureus* strains occurred in brain heart infusion (BHI) broth containing NaCl and sucrose as humectants either at pH 4.3, a_w of 0.85, or 8°C. No growth occurred with a combination of pH < 5.5, 12°C, and a_w of 0.90 or 0.93; and no growth occurred at pH < 4.9, 12°C, and a_w of 0.96 (79).

NaNO$_2$, Eh, pH, and Temperature of Growth.   *S. aureus* strain S-6 grew and produced enterotoxin B in cured ham under anaerobic conditions with a brine content up to 9.2%, but not below pH 5.30 and 30°C, or below pH 5.58 at 10°C. Under aerobic conditions, enterotoxin production occurred sooner than under anaerobic conditions. As the concentration of HNO_2 increased, enterotoxin production decreased (44).

STAPHYLOCOCCAL ENTEROTOXINS: TYPES AND INCIDENCE

By use of serologic methods, seven different enterotoxins are recognized: A, B, C_1, C_2, C_3, D, and E. Enterotoxin C_3 (SEC$_3$) is chemically and serologically related to but not identical to SEC$_1$ and SEC$_2$ (89). Antibodies to each of the SECs cross-react with each other, although they differ slightly from each other antigenically. SEC$_3$ shares 98% nucleotide sequence with SEC$_1$ (27), and SEC$_1$ and SEB share 68% amino acid homology. Also, there is cross-reaction between SEA and SEE, and some antibodies against SEB cross-react with the SECs (15). The TSST is not an enterotoxin. Some enterotoxin-producing strains also produce TSST, and some of the symptoms of TSS appear to be caused by SEA (66) and by SEB and SEC$_1$.

Five of the seven enterotoxins have been cloned and sequenced: SEA, SEB, SEC$_1$, SED, and SEE. SEA and SEE are closely related by nucleotide sequence analysis. In regard to the cellular location of their genes, some are chromosomal, and some are plasmid borne. A chromosomal location has been reported for SEA (84, 94), but more recent findings indicate that it is encoded by lysogenic phages in *S. aureus* (28), which integrate with the bacterial chromosome (17). Some SEE producers have genes that are associated with ultraviolet light–inducible DNA that appears to be defective phages (28). In at least one strain, SEB and SEC$_1$ production is associated with the same plasmid (3), whereas in another strain, the SEB gene was associated with the chromosome (61). SED is located on a penicillinase plasmid (7). The staphylococcal enterotoxins belong to a family of pyrogenic toxins (PT) that include TSST and the streptococcal toxins A–E. All PTs are pyrogenic and immunosuppressive through their capacity to induce nonspecific T-cell mitogenicity. However, these properties are not known to be involved in a typical gastroenteritis case, where the actions of staphylococcal entero-toxins are confined to the gastrointestinal tract. (For more on the genetics of staphylococcal enterotoxins, see ref. 16.)

The relative incidence of the enterotoxins is presented in Table 19-3. In general, SEA is recovered from food-poisoning outbreaks more often than any of the others, with SED being second most frequent. The fewest number of outbreaks are associated with SEE. The incidence of SEA among 3,109 and SED among 1,055 strains from different sources, and by a large number of investigators, was 23 and 14%, respectively (100). For SEB, SEC, and SEE, 11, 10, and 3%, respectively, were found among 3,367, 1,581, and 1,072 strains.

The relative incidence of specific enterotoxins among strains recovered from various sources varies widely. While from human specimens in the United States over 50% of isolates secrete SEA alone or in combination (22), from human isolates in Sri Lanka SEA producers constituted only 7.8% (82). Unlike other reports, the latter study found more SEB producers than any other types. Wide variations are found among *S. aureus* strains isolated from foods. While in one study Harvey et al. (51) found SED to be associated more with poultry isolates than human strains, in another study these investigators found no SED producers among 55 poultry isolates (47). In yet another study, two of three atypical *S. aureus* isolates that produced a slow, weak positive or negative coagulase reaction, and were negative for the anaerobic fermentation of man-nitol, produced SED (37). The isolates were from poultry. From Nigerian ready-to-eat foods, about 39% of 248 isolates were enterotoxigenic, with 44% of these producing SED (1). Among 449 coagulase-positive *S. aureus* isolates from a variety of Nigerian foods, 57%, 15%, 6%, and 5% were SEA, SEB, SED, and SEC respectively (99). From sheep milk, SEA and SED constituted 35% each of 124 strains, including four coagulase-negative strains (6). Of 48 isolates from dairy and 134 from meat products, 46% and 49%, respectively, were enterotoxigenic (85), and of 80 strains from food-poisoning outbreaks, 96% produced SEA (22). SEC was produced by 67.9% of 342 isolates of both coagulase-positive and coagulase-negative species from healthy goats (107). SEA, SEB, and SEC were detected in the milk of 17 of 133 healthy goats (107).

Regarding the percentage of strains that are enterotoxigenic, widely dif-ferent percentages have been found depending on the source of isolates. Only 10% of 236 raw milk isolates were enterotoxigenic (22), while 62.5% of 200 food isolates were positive (85). In a study of *S. aureus* from chicken livers, 40% were enterotoxigenic (46). In another study, 33% of 36 food isolates were enterotoxigenic (97).

Attempts to associate enterotoxigenicity with other biochemical properties

TABLE 19-3. Incidence of Staphylococcal Enterotoxins Alone and in Combination

Source	Number of cultures	Percentage enterotoxic	Enterotoxins					Reference
			A	B	C	D	E	
Human specimens	582	—	54.5	28.1	8.4	41.0	—	21
Raw milk	236	10	1.8	0.8	1.2	6.8	—	21
Frozen foods	260	30	3.4	3.0	7.4	10.4	—	21
Food-poisoning outbreaks	80	96.2	77.8	10.0	7.4	37.5	—	21
Foods	200	62.5	47.5	3.5	12.0	18.5	6.5	85
Poultry	139	25.2	1.4	0	0.7	23.7	0	51
Humans	293	39	7.8	17.7	7.2	6.8	0.7	82
Poultry	55	62	60.0	1.8	3.6	0	0	47

TABLE 19-4.   Some Properties of the Enterotoxins of S. *aureus*

	Enterotoxin						
	A	B	C_1	C_2	C_3	D	E
Emetic dose (ED_{50}) (monkey) (μg/animal)	5	5	5	5–10	$<10^a$	20	10–20
Nitrogen content (%)	16.5	16.10	16.2	16.0	—	—	—
Sedimentation coefficient (s°_{20}, w), S	3.04	2.89 2.78	3.00	2.90	—	—	2.60
Diffusion coefficient (D°_{20}, w), $\times 10^{-7}\,cm^2\,sec^{-1}$	7.94	7.72 8.22	8.10	8.10	—	—	—
Reduced viscosity (ml/g)	4.07	3.92 3.81	3.4	3.7	—	—	—
Molecular weight	27,800	28,366	34,100	34,000	26,900	27,300	29,600
Partial specific volume	0.726	0.743 0.726	0.732	0.742	—	—	—
Isoelectric point	6.8	8.60	8.6	7.0	8.15	7.4	7.0
Maximum absorption (mμ)	277	277	277	277	—	278	277
Extinction ($E^{1\%}_{1cm}$)	14.3	14.00 14.40	12.1	12.1	—	10.8	12.5

Sources: Bergdoll (12), Huang and Bergdoll (55), Reiser et al. (89), Schantz et al. (93).
[a] Per os. 0.05 μg/kg by IV route (89).

of staphylococci such as gelatinase, phosphatase, lysozyme, lecithinase, lipase, and DNAse production or the fermentation of various carbohydrates have been unsuccessful. Enterotoxigenic strains appear to be about the same as other coagulase-positive strains in these respects. Attempts to relate enterotoxigenesis with specific bacteriophage types have been unsuccessful also. Most enterotoxigenic strains belong to phage Group III, but all phage groups are known to contain toxigenic strains. Of 54 strains from clinical specimens that produced SEA, 5.5, 1.9, and 27.8% belonged, respectively, to phage Groups I, II, and III, with 20.4% being untypable (21). Among poultry isolates, 49% were found to be phage untypable (47). In a study of bruised poultry tissue, essentially the same phage types were found on the hands and in lesions of handlers as in the bruised tissues, suggesting that the bruised poultry tissue served as the source of staphylococci to the handlers (91).

Chemical and Physical Properties of Enterotoxins

These properties are summarized in Table 19-4. All are simple proteins that, upon hydrolysis, yield 18 amino acids, with aspartic, glutamic, lysine, and tyrosine being the most abundant. The amino acid sequence of SEB was determined first (55). Its N-terminal is glutamic acid, and lysine is the C-terminal amino acid. SEA, SEB, and SEE are composed of 239 to 296 amino acid residues. SEC_3 contains 236 amino acid residues and the N-terminal is serine, while the N-terminal of SEC_1 is glutamic acid (89). The disulfide bond in enterotoxin B is not essential for biological activity and conformation (32). Biological activity of SEA is destroyed when the abnormal tyrosyl residues are modified (26). The effect of acetylation, succinylation, guanidination, and carbamylation on the biological activity of SEB from strain S-6 has been studied (25). It was found that guanidination of 90% of the lysine residues had no effect on emetic activity or on the combining power of antigen-antibody reactions of SEB. The toxin was reduced, however, when acetylated, succinylated, and carbamylated. The investigators concluded that acetylation and succinylation decreased the net positive charge of the enterotoxin that is contributed by amino groups. The normal positive charge of the enterotoxin is thought to play an important role in both its emetic activity and in its combining with specific antibody. In their activate states, the enterotoxins are resistant to proteolytic enzymes such as trypsin, chymotrypsin, rennin, and papain but sensitive to pepsin at a pH of about 2 (12). While the various enterotoxins differ in certain physiochemical properties, each has about the same potency. Although biological activity and serologic reactivity are generally associated, it has been shown that serologically negative exterotoxin may be biologically active (see below).

The enterotoxins are quite heat resistant. The biological activity of SEB was retained after heating for 16 h at 60°C and pH 7.3 (93). Heating of one preparation of SEC for 30 min at 60°C resulted in no change in serologic reactions (18). The heating of SEA at 80°C for 3 min or at 100°C for 1 min caused it to lose its capacity to react serologically (12). In phosphate-buffered saline, SEC has been found to be more heat resistant than SEA or SEB. The relative thermal resistance of these three enterotoxins was SEC > SEB > SEA (104).

The thermal inactivation of SEA based on cat emetic response was shown by Denny et al. (34) to be 11 min at 250°F ($F_{250}^{48} = 11$ min). When monkeys were employed, thermal inactivation was $F_{250}^{46} = 8$ min. These enterotoxin preparations consisted of a 13.5-fold concentration of casamino acid culture filtrate employing strains 196-E. Using double-gel-diffusion assay, Read and Bradshaw (88) found the heat inactivation of 99+% pure SEB in veronal buffer to be $F_{250}^{58} = 16.4$ min. The end point for enterotoxin inactivation by gel diffusion was identical to that by intravenous injection of cats. The slope of the thermal inactivation curve for SEA in beef bouillon at pH 6.2 was found to be around 27.8°C (50°F) using three different toxin concentrations (5, 20, and 60 μg/ml, (33)). Some D values for the thermal destruction of SEB are presented in Table 19-5. Crude toxin preparations have been found to be more resistant than purified toxins (88). It may be noted from Table 19-5 that

TABLE 19-5.   *D* values for the Heat Destruction of Staphylococcal Enterotoxin B and Staphylococcal Heat-Stable Nuclease

Conditions	D(C)	Reference
Veronal buffer	$D_{110} = 29.7^a$	88
Veronal buffer	$D_{110} = 23.5^b$	88
Veronal buffer	$D_{121} = 11.4^a$	88
Veronal buffer	$D_{121} = 9.9^b$	88
Veronal buffer, pH 7.4	$D_{110} = 18$	67
Beef broth, pH 7.4	$D_{110} = 60$	67
Staph. nuclease	$D_{130} = 16.5$	36

[a] Crude toxin.
[b] 99 + percent purified.

TABLE 19-6.   *D* **and** *z* **Values for the Thermal Destruction of** *S. aureus* **196E in Various Heating Menstra at 140°F**

Products	D(F)	z	Reference
Chicken à la king	5.37	10.5	5
Custard	7.82	10.5	5
Green pea soup	6.7–6.9	8.1	102
Skim milk	3.1–3.4	9.2	102
0.5 percent NaCl	2.2–2.5	10.3	102
Beef bouillon	2.2–2.6	10.5	102
Skim milk alone	5.34	—	62
Raw skim milk + 10% sugar	4.11	—	62
Raw skim milk + 25% sugar	6.71	—	62
Raw skim milk + 45% sugar	15.08	—	62
Raw skim milk + 6% fat	4.27	—	62
Raw skim milk + 10% fat	4.20	—	62
Tris buffer, pH 7.2	2.0	—	56
Tris buffer, pH 7.2, 5.8% NaCl or 5% MSG	7.0	—	56
Tris buffer, pH 7.2 + 5.8% NaCl + 5% MSG	15.5	—	56

staphylococcal thermonuclease displays heat resistance similar to that of SEB (see Chap. 6 for more information on this enzyme). In one study, SEB was found to be more heat sensitive at 80°C than at 100° or 110°C (92). The thermal destruction was more pronounced at 80°C than at either 60° or 100°C when heating was carried out in the presence of meat proteins. SEA and SED in canned infant formula were immunologically nonreactive after thermal processing but were biologically active when injected in kittens (10).

S. aureus cells are considerably more sensitive to heat than the enterotoxins, as may be noted from *D* values presented in Table 19-6 from various heating menstra. The cells are quite sensitive in Ringer's solution at pH 7.2 ($D_{140°F}$ = 0.11) and much more resistant in milk at pH 6.9 ($D_{140°F}$ = 10.0). In frankfurters, heating to 71.1°C was found to be destructive to several strains of *S. aureus* (83), and microwave heating for 2 min was destructive to over 2 million cells/g (111).

The maximum growth temperature and heat resistance of *S. aureus* strain MF 31 were shown to be affected when the cells were grown in heart infusion broth containing soy sauce and monosodium glutamate (MSG). Without these ingredients in the broth, maximum growth temperature was 44°C, but with them the maximum was above 46 (56). The most interesting effect of MSG was on $D_{60°C}$ values determined in Tris buffer at pH 7.2. With cells grown at 37°C, the mean *D* 60°C value in buffer was 2.0 min, but when 5% MSG and 5% NaCl were added to the buffer, *D* 60°C was 15.5 min. Employing cells grown at 46°C, the respective *D* 60°C values were 7.75 and 53.0 min in buffer and buffer-MSG-NaCl. It is well known that heat resistance increases along with increasing growth temperature, but changes of this magnitude in vegetative cells are unusual.

Production

In general, enterotoxin production tends to be favored by optimum growth conditions of pH, temperature, Eh, and so on. It is well established that staphylococci can grow under conditions that do not favor enterotoxin production.

With respect to a_w, enterotoxin production (except for SEA) occurs over a slightly narrower range than growth. In precooked bacon incubated aerobically at 37°C, *S. aureus* A100 grew rapidly at a_w as low as 0.84 and produced SEA (68). The production of the individual enterotoxins is more inherent to the toxin than to the strain that produces them (86). SEA but not SEB has been shown to be produced by L-phase cells (29). In pork, SEA production occurred at a_w 0.86 but not at 0.83 and in beef at 0.88 but not at 0.86 (101). SEA can be produced under conditions of a_w that do not favor SEB (105). SED has been produced at a_w 0.86 in 6 days at 37°C in BHI (40). In general, SEB production is sensitive to a_w while SEC is sensitive to both a_w and temperature. Regarding NaCl and pH, enterotoxin production has been recorded at pH 4.0 in the absence of NaCl (see Table 19-2). The effect of NaCl on SEB synthesis by strain S-6 at pH 7.0 at 37°C is presented in Figure 19-1.

With respect to growth temperature, SEB production in ham at 10°C has

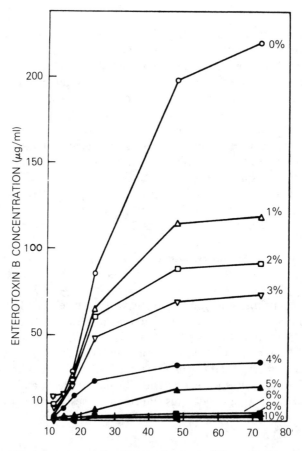

FIGURE 19-1.   Staphylococcal enterotoxin B production in different NaCl concentrations in 4% NZ-Amine NAK medium at pH 7.0 and 37°C.
From Pereira et al. (86); copyright © 1982 by International Association of Milk, Food and Environmental Sanitarians.

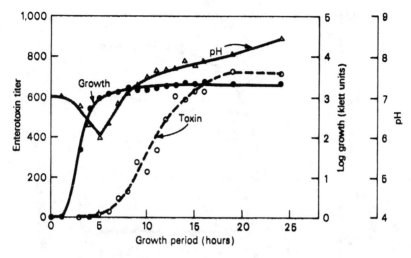

FIGURE 19-2. Enterotoxin B production, growth, and pH changes in *Staphylococcus aureus* at 37°C.
From McLean et al. (71); copyright © 1968 by American Society for Microbiology.

been recorded (44), as well as small amounts of SEA, SEB, SEC, and SED in cooked ground beef, ham, and bologna at 10°C (101). Production has been observed at 46°C, but the optimum temperature for SEB and SEC is 40°C in a protein hydrolysate medium (108), and for SEE, 40°C at pH 6.0 (103). The growth of *S. aureus* on cooked beef at 45.5°C for 24 h has been demonstrated, but at 46.6°C the initial inoculum decreased by 2 log cycles over the same period (19). The optimum for SEB in a culture medium at pH 7.0 was 39.4°C (86). Thus, the optimum temperature for enterotoxin production is in the 40° to 45°C range.

Staphylococcal enterotoxins have been reported to appear in cultures as early as 4 to 6 h (Fig. 19-2) and to increase proportionately through the stationary phase (69) and into the transitional phase (Fig. 19-3). Enterotoxin production has been shown to occur during all phases of growth (30), although earlier studies revealed that with strain S-6, 95% of SEB was released during the latter part of the log phase of growth. Chloramphenicol inhibited the appearance of enterotoxin, suggesting that the presence of toxin was dependent on de novo protein synthesis (75). In ice cream pies, 3.9 ng/g of SEA was produced in 18 h at 25°C and 4.8 ng/g in 14 h at 30°C (53). In the same study, TNase was detectable before SEA, with 72 ng/g being found after 12 h at 37°C. With a 3% pancreatic digest of casein as substrate and incubation at 37°C, SEC_1 and SEC_2 were produced during the exponential growth phase and at the beginning of the stationary phase (81). SEC_1 was detected after 10 h (2 ng/ml) at an *S. aureus* population of 8.3×10^7 cfu/ml, while TNase was detected after 5 h with a cell count of 1.3×10^4 cfu/ml. SEC_2 and TNase first appeared after 7 h with a cell population of 10^7 cfu/ml (81). With both enterotoxins, TNase production ceased before enterotoxin production.

In regard to quantities of enterotoxins produced, levels of 375 and 60 μg/ml or more of SEB and SEC, respectively, have been recorded (90). In a pro-

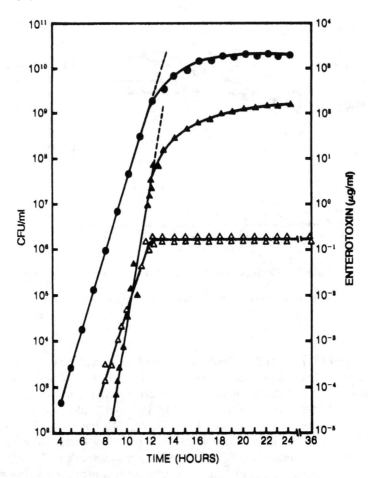

FIGURE 19-3.   Rates of growth and enterotoxins A and B synthesis by *Staphylococcus aureus* S-6.
Symbols: ●, CFU/ml; △, enterotoxin A; ▲, enterotoxin B.
Czop and Bergdoll (30); copyright © 1974 by American Society for Microbiology.

tein hydrolysate medium, up to 500 µg/ml of SEB may be produced (13).
Employing a sac culture assay method, 289 ng/ml of SEA were produced by *S. haemolyticus*, 213 ng/ml of SEC by *S. aureus*, and 779 ng/ml of SED, also by *S. aureus* (6).

SEB production in unbuffered media has been found to be repressed by excess glucose in the medium (75). Streptomycin, actinomycin D, acriflavine, Tween 80, and other compounds have been found to inhibit SEB synthesis in broth (42). SEB production is inhibited by 2-deoxyglucose, and the inhibition is not restored by glucose, indicating that this toxin, at least, is not under catabolite control (57). While actinomycin D has been shown to inhibit SEB synthesis in strain S-6, the inhibition occurred about 1 h after cellular synthesis ceased. The latter was immediately and completely inhibited. A possible conclusion from this finding is that the mRNA responsible for enterotoxin synthesis is more stable than that for cellular synthesis (63).

The minimum number of cells of *S. aureus* required to produce the minimum level of enterotoxin considered necessary to cause the gastroenteritis syndrome in humans (1 ng/g) appears to differ for substrates and for the particular enterotoxin. Detectable SEA has been found with as few as $\sim 10^4$ cfu/g (53). In milk, SEA and SED were detected with counts of 10^7 but not below this level (77). Employing a strain of *S. aureus* that produces SEA, SEB, and SED, SEB and SED were detected when the count reached 6×10^6/ml and the enterotoxin level was 1 ng/ml, while SEA at a level of 4 ng/ml was detected with a count of 3×10^7 cfu/ml (78). In imitation cheese with pH of 5.56 to 5.90 and a_w of 0.94 to 0.97, enterotoxins were first detected at the following counts: SEA at 4×10^6/g; SEC at 1×10^8; SED at 3×10^6; SEE at 5×10^6; and SEC and SEE at 3×10^6/g (11). In precooked bacon, SEA was produced by strain A100 with cells $> 10^6$/g (96). In meat products and vanilla custard, SEA was produced with $\geq \log_{10} 7.2$ cells/g, but in certain vegetable products no toxin was detected with counts up to $\log_{10} 10.00$/g (80).

With respect to enterotoxin synthesis, evidence has been presented for SEB that the kinetic precursor of the extracellular product is a larger membrane-bound form designated pSEB (106). The latter investigators believe that the temporary sequestering on the membrane may be critical to the mechanism that facilitates the transfer of SEB through the cell wall. After its release from the membrane, the enterotoxin appears to be transiently sequestered by the cell wall before its ultimate release into the extracellular environment. A similar pattern of synthesis, sequestering, and release has been shown for SEA (24). Regarding intracellular synthesis, evidence has been presented that SEB and SEC are synthesized in one way while SEA, SED, and SEE are assembled in another (14).

Detection in Foods

Serologic and in vitro methods for detecting enterotoxins are presented in Chapter 6, and in vivo and related methods in Chapter 7. For the extraction of foods for enterotoxins, an appropriate reference in Table 5-1 should be consulted.

THE GASTROENTERITIS SYNDROME

The symptoms of staphylococcal food poisoning usually develop within 4 h of the ingestion of contaminated food, although a range of from 1 to 6 h has been reported. The symptoms—nausea, vomiting, abdominal cramps (which are usually quite severe), diarrhea, sweating, headache, prostration, and sometimes a fall in body temperature—generally last from 24 to 48 h, and the mortality rate is very low or nil. The usual treatment for healthy persons consists of bed rest and maintenance of fluid balance. Upon cessation of symptoms, the victim possesses no demonstrable immunity to recurring attacks, although animals become resistant to enterotoxin after repeated oral doses (13). Since the symptoms are referable to the ingestion of preformed enterotoxin, it is conceivable that stool cultures might be negative for the organisms, although this is rare. Proof of staphylococcal food poisoning is established by recovering enterotoxigenic staphylococci from leftover food and from the stool cultures of victims. Attempts should be made to extract enterotoxin from suspect foods, especially when the number of recoverable viable cells is low.

The minimum quantity of enterotoxin needed to cause illness in humans is about 200 ng. This value is derived from an outbreak of staphylococcal gastroenteritis traced to 2% chocolate milk. From 12 cartons of milk, SEA was found at levels from 94 to 184 ng/carton, with a mean of 144 ng (38). The attack rate was associated with the quantity of milk consumed and somewhat with age; those aged 5 to 9 years were more sensitive than those aged 10 to 19 years. Earlier findings indicated a dose of 20 to 35 µg of pure SEB for adults (87). From 16 incidents of staphylococcal gastroenteritis, SE levels of less than 0.01 to 0.25 µg/g of food were found (48).

The pathogenesis of enterotoxins in humans is not clear. They act on the intestines to induce vomiting and diarrhea, and the same effects can be achieved by intravenous (i.v.) injections. When SEA was administered i.v. to monkeys, an initial state of lymphopenia was induced; it lasted for 1 to 2 days and was followed by the release of new immature cells that had greater DNA synthesis activity (113).

INCIDENCE AND VEHICLE FOODS

The incidence of staphylococci in a variety of foods is presented in Chapter 4 (see Tables 4-1, 4-2, 4-3, 4-4, and 4-5). They may be expected to occur in a wide variety of foods not given heat treatments for their destruction.

With regard to vehicle foods for staphylococcal enteritis, a large number has been incriminated in outbreaks, usually products made by hand and improperly refrigerated after being prepared. Outbreaks and cases of food-borne gastroenteritis reported to the Centers for Disease Control for the years 1973 through 1987 are presented in Table 19-7. From a high of around 16% in 1983, this syndrome accounted for only 1.0% of cases in 1987. The reported cases constitute only a small part of the actual number, however; estimates place the number of cases of staphylococcal foodborne gastroenteritis at between 1 million and 2 million per year in the United States. The six leading vehicle foods for 1973–1987 are listed in Table 19-8, with pork and pork products accounting for more outbreaks than the other five combined.

The problem is one of reporting; all too often the small outbreaks that occur

TABLE 19-7.   **Staphylococcal Foodborne Gastroenteritis Outbreaks and Cases in the United States, 1973–1987**

Years	Outbreaks	Cases	Percentage of All Cases
1973–1987	367	17,248	14.0
1983	14	1,257	15.9
1984	11	1,153	14.1
1985	14	421	1.8
1986	7	250	4.3
1987	1	100	1.0

Sources: Bean and Griffin (8), Bean et al. (9).

TABLE 19-8.   Leading Food Sources for Staphylococcal Gastroenteritis Outbreaks in the United States, 1973–1987

Food Sources	Number of Outbreaks
Pork	96
Bakery products	26
Beef	22
Turkey	20
Chicken	14
Eggs	9

Source: Bean and Griffin (8).

in homes are not reported to public health officials. A large percentage of the reported cases of all types are those that result from banquets, generally involving large numbers of persons.

An unusual outbreak was caused by SEA and SED and traced to wild mushrooms in vinegar (70). The food contained 10 ng SEA and 1 ng of SED/g.

ECOLOGY OF *S. AUREUS* GROWTH

In general, the staphylococci do not compete well with the normal flora of most foods, and this is especially true for those that contain large numbers of lactic acid bacteria where conditions permit the growth of the latter organisms (see Chap. 16). A large number of investigators have shown the inability of *S. aureus* to compete in both fresh and frozen foods. At temperatures that favor staphylococcal growth, the normal food saprophytic flora offers protection against staphylococcal growth through antagonism, competition for nutrients, and modification of the environment to conditions less favorable to *S. aureus*. Bacteria known to be antagonistic to *S. aureus* growth include *Acinetobacter*, *Aeromonas*, *Bacillus*, *Pseudomonas*, *S. epidermidis*, the Enterobacteriaceae, the Lactobacillaceae, enterococci, and others (76). SEA has been shown to be resistant to a variety of environmental stresses, but growth of several lactic acid bacteria did lead to its reduction and to a suggestion that toxin reduction might have resulted from specific enzymes or other metabolites of the lactic acid bacteria (23).

PREVENTION OF STAPHYLOCOCCAL AND OTHER FOOD-POISONING SYNDROMES

When susceptible foods are produced with low numbers of staphylococci, they will remain free of enterotoxins and other food-poisoning hazards if kept either *below* 40° or *above* 140°F until consumed. For the years 1961 through 1972, over 700 foodborne disease outbreaks were investigated by Bryan (20) relative to the factors that contributed to the outbreaks, and of the 16 factors identified, the 5 most frequently involved were:

TABLE 19-9.   Leading Factors That Led to the Outbreaks of Staphylococcal Foodborne Gastroenteritis in the United States, 1973–1987

Causes	Number of Outbreaks
Improper holding temperatures	98
Poor personal hygiene	71
Contaminated equipment	43
Inadequate cooking	22
Food from unsafe source	12
Others	24

Source: Bean and Griffin (8).

1. inadequate refrigeration
2. preparing foods far in advance of planned service
3. infected persons' practicing poor personal hygiene
4. inadequate cooking or heat processing
5. holding food in warming devices at bacterial growth temperatures

Inadequate refrigeration alone comprised 25.5% of the contributing factors. The 5 listed contributed to 68% of outbreaks. For the period 1973 through 1987, the five leading causes are listed in Table 19-9; notice that the leading factors for 1961–1972 continued to be among the leading factors for the later years. Susceptible foods should not be held within the staphylococcal growth range for more than 3 to 4 h.

References

1. Adesiyun, A. A. 1984. Enterotoxigenicity of *Staphylococcus aureus* strains isolated from Nigerian ready-to-eat foods. *J. Food Protect.* 47:438–440.
2. Adesiyun A. A., S. R. Tatini, and D. G. Hoover. 1984. Production of enterotoxin(s) by *Staphylococcus hyicus*. *Vet. Microbiol.* 9:487–495.
3. Altboum, Z., I. Hertman, and S. Sarid. 1985. Penicillinase plasmid-linked genetic determinants for enterotoxins B and C_1 production in *Staphylococcus aureus*. *Infect. Immun.* 47:514–521.
4. Angelotti, R., M. J. Foter, and K. H. Lewis. 1961. Time-temperature effects on salmonellae and staphylococci in foods. *Amer. J. Pub. Hlth* 51:76–88.
5. Angelotti, R., M. J. Foter, and K. H. Lewis. 1960. Time-temperature effects on salmonellae and staphylococci in foods. II. Behavior at warm holding temperatures. Thermal-Death-Time Studies. Cincinnati: Public Health Service, U.S. Department of Health, Education and Welfare.
6. Bautista, L., P. Gaya, M. Medina, and M. Nunez. 1988. A quantitative study of enterotoxin production by sheep milk staphylococci. *Appl. Environ. Microbiol.* 54:566–569.
7. Bayles, K. W., and J. J. Iandolo. 1989. Genetic and molecular analyses of the gene encoding staphylococcal enterotoxin D. *J. Bacteriol.* 171:4799–4806.
8. Bean, N. H., and P. M. Griffin. 1990. Foodborne disease outbreaks in the United States, 1973–1987: Pathogens, vehicles, and trends. *J. Food Protect.* 53:804–817.
9. Bean, N. H., P. M. Griffin, J. S. Goulding, and C. B. Ivey. 1990. Foodborne disease outbreaks, 5-year summary, 1983–1987. *J. Food Protect.* 53:711–728.
10. Bennett, R. W., and M. R. Berry, Jr. 1987. Serological reactivity and in vivo

toxicity of *Staphylococcus aureus* enterotoxins A and D in selected canned foods. *J. Food Sci.* 52:416–418.

11. Bennett, R. W., and W. T. Amos. 1983. *Staphylococcus aureus* growth and toxin production in imitation cheeses. *J. Food Sci.* 48:1670–1673.

12. Bergdoll, M. S. 1967. The staphylococcal enterotoxins. In *Biochemistry of Some Foodborne Microbial Toxins*, ed. R. I. Mateles and G. N. Wogan, 1–25. Cambridge: MIT Press.

13. Bergdoll, M. S. 1972. The enterotoxins. In *The Staphylococci*, ed. J. O. Cohen, 301–331. New York: Wiley-Interscience.

14. Bergdoll, M. S. 1979. Staphylococcal intoxications. In *Food-Borne Infections and Intoxications*, ed. H. Riemann and F. L. Bryan, 443–494. New York: Academic Press.

15. Bergdoll, M. S. 1990. Staphylococcal food poisoning. In *Foodborne Diseases*, ed. D. O. Cliver, 85–106. New York: Academic Press.

16. Betley, M. J., V. L. Miller, and J. J. Mekalanos. 1986. Genetics of bacterial enterotoxins. *Ann. Rev. Microbiol.* 40:577–605.

17. Betley, M. J., and J. J. Mekalanos. 1985. Staphylococcal enterotoxin A is encoded by phage. *Science* 229:185–187.

18. Borja, C. R., and M. S. Bergdoll. 1967. Purification and partial characterization of enterotoxin C produced by *Staphylococcus aureus* strain 137. *J. Biochem.* 6:1457–1473.

19. Brown, D. F., and R. M. Twedt. 1972. Assessment of the sanitary effectiveness of holding temperatures on beef cooked at low temperature. *Appl. Microbiol.* 24:599–603.

20. Bryan, F. L. 1974. Microbiological food hazards today—based on epidemiological information. *Food Technol.* 28(9):52–59.

21. Casman, E. P. 1965. Staphylococcal enterotoxin. In *The Staphylococci: Ecologic Perspectives*, Annals of the New York Academy of Science, vol. 28, 128:124–133.

22. Casman, E. P., R. W. Bennett, A. E. Dorsey, and J. A. Issa. 1967. Identification of a fourth staphylococcal enterotoxin, enterotoxin D. *J. Bacteriol.* 94:1875–1882.

23. Chordash, R. A., and N. N. Potter. 1976. Stability of staphylococcal enterotoxin A to selected conditions encountered in foods. *J. Food Sci.* 41:906–909.

24. Christianson, K. K., R. K. Tweten, and J. J. Iandolo. 1985. Transport and processing of staphylococcal enterotoxin A. *Appl. Environ. Microbiol.* 50:696–697.

25. Chu, F. S., E. Crary, and M. S. Bergdoll. 1969. Chemical modification of amino groups in staphylococcal enterotoxin B. *Biochem.* 8:2890–2896.

26. Chu, F. S., K. Thadhani, E. J. Schantz, and M. S. Bergdoll. 1966. Purification and characterization of staphylococcal enterotoxin A. *Biochem.* 5:3281–3289.

27. Couch, J. L., and M. J. Betley. 1989. Nucleotide sequence of the type C_3 staphylococcal enterotoxin gene suggests that intergenic recombination causes antigenic variation. *J. Bacteriol.* 171:4507–4510.

28. Couch, J. L., M. T. Solis, and M. J. Betley. 1988. Cloning and nucleotide sequence of the type E staphylococcal enterotoxin gene. *J. Bacteriol.* 170:2954–2960.

29. Czop, J. K., and M. S. Bergdoll. 1970. Synthesis of enterotoxins by L-forms of *Staphylococcus aureus*. *Infect. Immun.* 1:169–173.

30. Czop, J. K., and M. S. Bergdoll. 1974. Staphylococcal enterotoxin synthesis during the exponential, transitional, and stationary growth phases. *Infect. Immun.* 9:229–235.

31. Dack, G. M., W. E. Cary, O. Woolpert, and H. Wiggers. 1930. An outbreak of food poisoning proved to be due to a yellow hemolytic staphylococcus. *J. Prev. Med.* 4:167–175.

32. Dalidowicz, J. E., S. J. Silverman, E. J. Schantz, D. Stefanye, and L. Spero. 1966. Chemical and biological properties of reduced and alkylated staphylococcal enterotoxin B. *Biochem.* 5:2375–2381.

33. Denny, C. B., J. Y. Humber, and C. W. Bohrer. 1971. Effect of toxin concentration on the heat inactivation of staphylococcal enterotoxin A in beef bouillon and in phosphate buffer. *Appl. Microbiol.* 21:1064–1066.

34. Denny, C. B., P. L. Tan, and C. W. Bohrer. 1966. Heat inactivation of staphylococcal enterotoxin. *J. Food Sci.* 31:762–767.

35. Elek, S. D. 1959. *Staphylococcus Pyogenes and Its Relation to Disease.* Edinburgh and London: Livingstone.

36. Erickson, A., and R. H. Deibel. 1973. Production and heat stability of staphylococcal nuclease. *Appl. Microbiol.* 25:332–336.

37. Evans, J. B., G. A. Ananaba, C. A. Pate, and M. S. Bergdoll. 1983. Enterotoxin production by atypical *Staphylococcus aureus* from poultry. *J. Appl. Bacteriol.* 54:257–261.

38. Evenson, M. L., M. W. Hinds, R. S. Bernstein, and M. S. Bergdoll. 1988. Estimation of human dose of staphylococcal enterotoxin A from a large outbreak of staphylococcal food poisoning involving chocolate milk. *Int. J. Food Microbiol.* 7:311–316.

39. Ewald, S. 1987. Enterotoxin production by *Staphylococcus aureus* strains isolated from Danish foods. *Int. J. Food Microbiol.* 4:207–214.

40. Ewald, S., and S. Notermans. 1988. Effect of water activity on growth and enterotoxin D production of *Staphylococcus aureus*. *Int. J. Food Microbiol.* 6: 25–30.

41. Freney, J., Y. Brun, M. Bes, H. Meugnier, F. Grimont, P. A. D. Grimont, C. Nervi, and J. Fleurette. 1988. *Staphylococcus lugdunensis* sp. nov. and *Staphylococcus schleiferi* sp. nov., two species from human clinical specimens. *Int. J. Syst. Bacteriol.* 38:168–172.

42. Friedman, M. E. 1966. Inhibition of staphylococcal enterotoxin B formation in broth cultures. *J. Bacteriol.* 92:277–278.

43. Genigeorgis, C., M. S. Foda, A. Mantis, and W. W. Sadler. 1971. Effect of sodium chloride and pH on enterotoxin C production. *Appl. Microbiol.* 21: 862–866.

44. Genigeorgis, C., H. Riemann, and W. W. Sadler. 1969. Production of enterotoxin B in cured meats. *J. Food Sci.* 34:62–68.

45. Genigeorgis, C., and W. W. Sadler. 1966. Effect of sodium chloride and pH on enterotoxin B production. *J. Bacteriol.* 92:1383–1387.

46. Genigeorgis, C., and W. W. Sadler. 1966. Characterization of strains of *Staphylococcus aureus* isolated from livers of commercially slaughtered poultry. *Poultry Sci.* 45:973–980.

47. Gibbs, P. A., J. T. Patterson, and J. Harvey. 1978. Biochemical characteristics and enterotoxigenicity of *Staphylococcus aureus* strains isolated from poultry. *J. Appl. Bacteriol.* 44:57–74.

48. Gilbert, R. J., and A. A. Wieneke. 1973. Staphylococcal food poisoning with special reference to the detection of enterotoxin in food. In *The Microbiological Safety of Food*, ed. B. C. Hobbs and J. H. B. Christian, 273–285. New York: Academic Press.

49. Gomez-Lucia, E., J. Goyache, J. L. Blanco, J. F. F. Garayzabal, J. A. Orden, and G. Suarez. 1987. Growth of *Staphylococcus aureus* and enterotoxin production in homemade mayonnaise prepared with different pH values. *J. Food Protect.* 50:872–875.

50. Halpin-Dohnalek, M. I., and E. H. Marth. 1989. *Staphylococcus aureus*: Production of extracellular compounds and behavior in foods—A review. *J. Food Protect.* 52:267–282.

51. Harvey, J., J. T. Patterson, and P. A. Gibbs. 1982. Enterotoxigenicity of *Staphylococcus aureus* strains isolated from poultry: Raw poultry carcasses as a potential food-poisoning hazard. *J. Appl. Bacteriol.* 52:251–258.

52. Hirooka, E. Y., E. E. Muller, J. C. Freitas, E. Vicente, Y. Yashimoto, and M. S. Bergdoll. 1988. Enterotoxigenicity of *Staphylococcus intermedius* of canine origin. *Int. J. Food Microbiol.* 7:185–191.

53. Hirooka, E. Y., S. P. C. DeSalzberg, and M. S. Bergdoll. 1987. Production of staphylococcal enterotoxin A and thermonuclease in cream pies. *J. Food Protect.* 50:952–955.

54. Hoover, D. G., S. R. Tatini, and J. B. Maltais. 1983. Characterization of staphylococci. *Appl. Environ. Microbiol.* 46:649–660.

55. Huang, I.-Y., and M. S. Bergdoll. 1970. The primary structure of staphylococcal enterotoxin B. III. The cyanogen bromide peptides of reduced and aminoethylated enterotoxin B and the complete amino acid sequence. *J. Biol. Chem.* 245: 3518–3525.

56. Hurst, A., and A. Hughes. 1983. The protective effect of some food ingredients on *Staphylococcus aureus* MF 31. *J. Appl. Bacteriol.* 55:81–88.

57. Iandolo, J. J., and W. M. Shafer. 1977. Regulation of staphylococcal enterotoxin B. *Infect. Immun.* 16:610–616.

58. Igimi, S., E. Takahashi, and T. Mitsuoka. 1990. *Staphylococcus schleiferi* subsp. *coagulans* subsp. nov., isolated from the external auditory meatus of dogs with external ear otitis. *Int. J. Syst. Bacteriol.* 40:409–411.

59. Jarvis, A. W., R. C. Lawrence, and G. G. Pritchard. 1973. Production of staphylococcal enterotoxins A, B, and C under conditions of controlled pH and aeration. *Infect. Immun.* 7:847–854.

60. Jay, J. M. 1970. Effect of borate on the growth of coagulase-positive and coagulase-negative staphylococci. *Infect. Immun.* 1:78–79.

61. Johns, M. B., Jr., and S. A. Khan. 1988. Staphylococcal enterotoxin B gene is associated with a discrete genetic element. *J. Bacteriol.* 170:4033–4039.

62. Kadan, R. S., W. H. Martin, and R. Mickelsen. 1963. Effects of ingredients used in condensed and frozen dairy products on thermal resistance of potentially pathogenic staphylococci. *Appl. Microbiol.* 11:45–49.

63. Katsuno, S., and M. Kondo. 1973. Regulation of staphylococcal enterotoxin B synthesis and its relation to other extracellular proteins. *Jap. J. Med. Sci. Biol.* 26:26–29.

64. Kloos, W. E. 1980. Natural populations of the genus *Staphylococcus*. *Ann. Rev. Microbiol.* 34:559–592.

65. Kloos, W. E. 1990. Systematics and the natural history of staphylococci. 1. *J. Appl. Bacteriol.* 69S:25–37.

66. McCollister, B. D., B. N. Kreiswirth, R. P. Novick, and P. M. Schlievert. 1990. Production of toxic shock syndrome–like illness in rabbits by *Staphylococcus aureus* D4508: Association with enterotoxin A. *Infect. Immun.* 58:2067–2070.

67. Lee, I. C., K. E. Stevenson, and L. G. Harmon. 1977. Effect of beef broth protein on the thermal inactivation of staphylococcal enterotoxin B. *Appl. Environ. Microbiol.* 33:341–344.

68. Lee, R. Y., G. J. Silverman, and D. T. Munsey. 1981. Growth and enterotoxin A production by *Staphylococcus aureus* in precooked bacon in the intermediate moisture range. *J. Food Sci.* 46:1687–1692.

69. Lilly, H. D., R. A. McLean, and J. A. Alford. 1967. Effects of curing salts and temperature on production of staphylococcal enterotoxin. *Bacteriol. Proc.*, 12.

70. Lindroth, S., E. Strandberg, A. Pessa, and M. J. Pellinen. 1983. A study of the growth potential of *Staphylococcus aureus* in *Boletus edulis*, a wild edible mushroom, prompted by a food poisoning outbreak. *J. Food Sci.* 48:282–283.

71. McLean, R. A., H. D. Lilly, and J. A. Alford. 1968. Effects of meat-curing salts

and temperature on production of staphylococcal enterotoxin B. *J. Bacteriol.* 95:1207–1211.

72. Metzger, J. F., A. D. Johnson, W. S. Collins, II, and V. McGann. 1973. *Staphylococcus aureus* enterotoxin B release (excretion) under controlled conditions of fermentation. *Appl. Microbiol.* 25:770–773.

73. Miller, R. D., and D. Y. C. Fung. 1973. Amino acid requirements for the production of enterotoxin B by *Staphylococcus aureus* S-6 in a chemically defined medium. *Appl. Microbiol.* 25:800–806.

74. Minor, T. E., and E. H. Marth. 1976. *Staphylococci and Their Significance in Foods.* New York: Elsevier.

75. Morse, S. A., R. A. Mah, and W. J. Dobrogosz. 1969. Regulation of staphylococcal enterotoxin B. *J. Bacteriol.* 98:4–9.

76. Mossel, D. A. A. 1975. Occurrence, prevention, and monitoring of microbial quality loss of foods and dairy products. *CRC Crit. Rev. Environ. Control* 5: 1–140.

77. Noleto, A. L., and M. S. Bergdoll. 1980. Staphylococcal enterotoxin production in the presence of nonenterotoxigenic staphylococci. *Appl. Environ. Microbiol.* 39:1167–1171.

78. Noleto, A. L., and M. S. Bergdoll. 1982. Production of enterotoxin by a *Staphylococcus aureus* strain that produces three identifiable enterotoxins. *J. Food Protect.* 45:1096–1097.

79. Notermans, S., and C. J. Heuvelman. 1983. Combined effect of water activity, pH and sub-optimal temperature on growth and enterotoxin production of *Staphylococcus aureus*. *J. Food Sci.* 48:1832–1835, 1840.

80. Notermans, S., and R. L. M. van Otterdijk. 1985. Production of enterotoxin A by *Staphylococcus aureus* in food. *Int. J. Food Microbiol.* 2:145–149.

81. Otero, A., M. L. Garcia, M. C. Garcia, B. Moreno, and M. S. Bergdoll. 1990. Production of staphylococcal enterotoxins C_1 and C_2 and thermonuclease throughout the growth cycle. *Appl. Environ. Microbiol.* 56:555–559.

82. Palasuntheram, C., and M. S. Beauchamp. 1982. Enterotoxigenic staphylococci in Sri Lanka. *J. Appl. Bacteriol.* 52:39–41.

83. Palumbo, S. A., J. L. Smith, and J. C. Kissinger. 1977. Destruction of *Staphylococcus aureus* during frankfurter processing. *Appl. Environ. Microbiol.* 34:740–744.

84. Pattee, P. A., and B. A. Glatz. 1980. Identification of a chromosomal determinant of enterotoxin A production in *Staphylococcus aureus*. *Appl. Environ. Microbiol.* 39:186–193.

85. Payne, D. N., and J. M. Wood. 1974. The incidence of enterotoxin production in strains of *Staphylococcus aureus* isolated from foods. *J. Appl. Bacteriol.* 37: 319–325.

86. Pereira, J. L., S. P. Salzberg, and M. S. Bergdoll. 1982. Effect of temperature, pH and sodium chloride concentrations on production of staphylococcal enterotoxins A and B. *J. Food Protect.* 45:1306–1309.

87. Raj, H. D., and M. S. Bergdoll. 1969. Effect of enterotoxin B on human volunteers. *J. Bacteriol.* 98:833–834.

88. Read, R. B., and J. G. Bradshaw. 1966. Thermal inactivation of staphylococcal enterotoxin B in veronal buffer. *Appl. Microbiol.* 14:130–132.

89. Reiser, R. F., R. N. Robbins, A. L. Noleto, G. P. Khoe, and M. S. Bergdoll. 1984. Identification, purification, and some physiochemical properties of staphylococcal enterotoxin C_3. *Infect. Immun.* 45:625–630.

90. Reiser, R. F., and K. F. Weiss. 1969. Production of staphylococcal enterotoxins A, B, and C in various media. *Appl. Microbiol.* 18:1041–1043.

91. Roskey, C. T., and M. K. Hamdy. 1972. Bruised poultry tissue as a possible

source of staphylococcal infection. *Appl. Microbiol.* 23:683–687.

92. Satterlee, L. D., and A. A. Kraft. 1969. Effect of meat and isolated meat proteins on the thermal inactivation of staphylococcal enterotoxin B. *Appl Microbiol.* 17:906–909.

93. Schantz, E. J., W. G. Roessler, J. Wagman, L. Spero, D. A. Dunnery, and M. S. Bergdoll. 1965. Purification of staphylococcal enterotoxin B. *J. Biochem.* 4: 1011–1016.

94. Shafer, W. M., and J. J. Iandolo. 1978. Staphylococcal enterotoxin A: A chromosomal gene product. *Appl. Environ. Microbiol.* 36:389–391.

95. Siems, H., D. Kusch, H.-J. Sinell, and F. Untermann. 1971. Vorkommen und Eigenschaften von Staphylokokken in verschiedenen Produktionsstufen bei der Fleischverarbeitung. *Fleischwirts.* 51:1529–1533.

96. Silverman, G. J., D. T. Munsey, C. Lee, and E. Ebert. 1983. Interrelationship between water activity, temperature and 5.5 percent oxygen on growth and enterotoxin A secretion by *Staphylococcus aureus* in precooked bacon. *J. Food Sci.* 48:1783–1786, 1795.

97. Šimkovičová, M., and R. J. Gilbert. 1971. Serological detection of enterotoxin from food-poisoning strains of *Staphylococcus aureus. J. Med. Microbiol.* 4: 19–30.

98. Smith, J. L., R. L. Buchanan, and S. A. Palumbo. 1983. Effect of food environment on staphylococcal enterotoxin synthesis: A review. *J. Food Protect.* 46: 545–555.

99. Sokari, T. G., and S. O. Anozie. 1990. Occurrence of enterotoxin producing strains of *Staphylococcus aureus* in meat and related samples from traditional markets in Nigeria. *J. Food Protect.* 53:1069–1070.

100. Sperber, W. H. 1977. The identification of staphylococci in clinical and food microbiology laboratories. *CRC Crit. Rev. Clin. Lab. Sci.* 7:121–184.

101. Tatini, S. R. 1973. Influence of food environments on growth of *Staphylococcus aureus* and production of various enterotoxins. *J. Milk Food Technol.* 36:559–563.

102. Thomas, C. T., J. C. White, and K. Longree. 1966. Thermal resistance of salmonellae and staphylococci in foods. *Appl. Microbiol.* 14:815–820.

103. Thota, F. H., S. R. Tatini, and R. W. Bennett. 1973. Effects of temperature, pH and NaCl on production of staphylococcal enterotoxins E and F. *Bacteriol. Proc.*, 1.

104. Tibana, A., K. Rayman, M. Akhtar, and R. Szabo. 1987. Thermal stability of staphylococcal enterotoxins A, B and C in a buffered system. *J. Food Protect.* 50:239–242.

105. Troller, J. A. 1972. Effect of water activity on enterotoxin A production and growth of *Staphylococcus aureus. Appl. Microbiol.* 24:440–443.

106. Tweten, R. K., and J. J. Iandolo. 1983. Transport and processing of staphylococcal enterotoxin B. *J. Bacteriol.* 153:297–303.

107. Valle, J., E. Gomez-Lucia, S. Piriz, J. Goyache, J. A. Orden, and S. Vadillo. 1990. Enterotoxin production by staphylococci isolated from healthy goats. *Appl. Environ. Microbiol.* 56:1323–1326.

108. Vandenbosch, L. L., D. Y. C. Fung, and M. Widomski. 1973. Optimum temperature for enterotoxin production by *Staphylococcus aureus* S-6 and 137 in liquid medium. *Appl. Microbiol.* 25:498–500.

109. Veraldo, P. E., R. Kilpper-Balz, F. Biavasco, G. Satta, and K. H. Schleifer. 1988. *Staphylococcus delphini* sp. nov., a coagulase-positive species isolated from dolphins. *Int. J. System. Bacteriol.* 38:436–439.

110. Victor, R., F. Lachica, K. F. Weiss, and R. H. Deibel. 1969. Relationships among coagulase, enterotoxin, and heat-stable deoxyribonuclease production by *Staphylococcus aureus. Appl. Microbiol.* 18:126–127.

111. Woodburn, M., M. Bennion, and G. E. Vail. 1962. Destruction of salmonellae and staphylococci in precooked poultry products by heat treatment before freezing *Food Technol.* 16:98–100.

112. Wu, C.-H., and M. S. Bergdoll. 1971. Stimulation of enterotoxin B production. *Infect. Immun.* 3:784–792.

113. Zehavi-Willner, T., E. Shenberg, and A. Barnea. 1984. In vivo effect of staphylococcal enterotoxin A on peripheral blood lymphocytes. *Infect. Immun.* 44:401–405.

20

Food Poisoning Caused by Gram-Positive Sporeforming Bacteria

At least three gram-positive sporeforming rods are known to cause bacterial food poisoning: *Clostridium perfringens (welchii)*, *C. botulinum*, and *Bacillus cereus*. The incidence of food poisoning caused by each of these organisms is related to certain specific foods, as is food poisoning in general.

CLOSTRIDIUM PERFRINGENS FOOD POISONING

The causative organism of this syndrome is a gram-positive, anaerobic spore-forming rod widely distributed in nature. Based on their ability to produce certain exotoxins, five types are recognized: types A, B, C, D, and E. The food-poisoning strains belong to type A, as do the classical gas gangrene strains, but unlike the latter, the food-poisoning strains are generally heat resistant and produce only traces of alpha toxin. Some type C strains produce enterotoxin and may cause a food-poisoning syndrome. The classical food-poisoning strains differ from type C strains in not producing beta toxin. The latter, which have been recovered from enteritis necroticans, are compared to type A heat-sensitive and heat-resistant strains in Table 20-1. Type A heat-resistant strains produce theta toxin, which is perfringolysin O (PLO), a thiol-activated hemolysin similar to listeriolysin O (LLO), produced by *Listeria monocytogenes* (discussed in Chap. 21). Like LLO, PLO has a molecular weight of 60 kda, and it has been cloned and sequenced (100).

While *C. perfringens* has been associated with gastroenteritis since 1895, the first clear-cut demonstration of its etiological status in food poisoning was made by McClung (70), who investigated four outbreaks in which chicken was incriminated. The first detailed report of the characteristics of this food-poisoning syndrome was that of Hobbs et al. (41) in Great Britain. While the British workers were more aware of this organism as a cause of food poisoning

TABLE 20-1.   Toxins of *Clostridium welchii* Types A and C

Cl. welchii	Toxins										
	α	β	γ	δ	ε	θ	ι	κ	λ	μ	ν
Heat-sensitive type A	+ + +	−	−	−	−	+ +	−	+ +	−	+ or −	+
Heat-resistant type A	± or tr	−	−	−	−	−	−	+ or −	−	+ + + or −	−
Heat-resistant type C	+	+	+	−	−	−	−	−	−	−	+

Source: B. C. Hobbs, 1962. *Bacterial Food Poisoning*. London: Royal Society of Health.

during the 1940s and 1950s, few incidents were recorded in the United States prior to 1960. It is clear now that *C. perfringens* food poisoning is widespread in the United States and many other countries.

Distribution of *C. perfringens*

The food-poisoning strains of *C. perfringens* exist in soils, water, foods, dust, spices, and the intestinal tract of humans and other animals. Various authors have reported the incidence of the heat-resistant, nonhemolytic strains to range from 2 to 6% in the general population. Between 20 and 30% of healthy hospital personnel and their families have been found to carry these organisms in their feces, and the carrier rate of victims after 2 weeks may be 50% or as high as 88% (20). The heat-sensitive types are common to the intestinal tract of all humans. *C. perfringens* gets into meats directly from slaughter animals or by subsequent contamination of slaughtered meat from containers, handlers, or dust. Since it is a sporeformer, it can withstand the adverse environmental conditions of drying, heating, and certain toxic compounds.

Characteristics of the Organism

Food poisoning as well as most other strains of *C. perfringens* grow well on a variety of media if incubated under anaerobic conditions or if provided with sufficient reducing capacity. Strains of *C. perfringens* isolated from horse muscle grew without increased lag phase at an Eh of −45 or lower, while more positive Eh values had the effect of increasing the lag phase (7). Although it is not difficult to obtain growth of these organisms on various media, sporulation occurs with difficulty and requires the use of special media, such as those described by Duncan and Strong (21), or the employment of special techniques such as dialysis sacs (99).

   C. perfringens is mesophilic, with an optimum between 37° and 45°C. The lowest temperature for growth is around 20°C, and the highest is around 50°C. Optimum growth in thioglycollate medium for six strains was found to occur between 30° and 40°C, and the optimum for sporulation in Ellner's medium was 37° to 40°C (93). Growth at 45°C under otherwise optimal conditions leads to generation times as short as 7 min. Regarding pH, many strains grow over the range 5.5 to 8.0 but generally not below 5.0 or above 8.5. The lowest reported a_w values for growth and germination of spores lie between 0.97 and 0.95 with sucrose or NaCl, or about 0.93 with glycerol employing a fluid thioglycollate base (49). Spore production appears to require higher a_w values than the above minima. While growth of type A was demonstrated at pH 5.5 by Labbe and Duncan (54), no sporulation or toxin production occurred. A pH of 8.5 appears to be the highest for growth. *C. perfringens* is not as strict an anaerobe as are some other clostridia. Its growth at an initial Eh of +320 mv has been observed (89). At least 13 amino acids are required for growth, along with biotin, pantothenate, pyridoxal, adenine, and other related compounds. It is heterofermentative, and a large number of carbohydrates are attacked. Growth is inhibited by around 5% NaCl.

The endospores of food-poisoning strains differ in their resistance to heat, with some being typical of other mesophilic sporeformers and some being highly resistant. A *D* 100°C value of 0.31 for *C. perfringens* (ATCC 3624) and a value of 17.6 for strain NCTC 8238 have been reported (130). For eight strains that produced reactions in rabbits, *D* 100°C values ranged from 0.70 to 38.37; strains that did not produce rabbit reactions were more heat sensitive (118).

In view of the practice of cooking roasts in water baths for long times at low temperatures (LTLT), the heat destruction of vegetative cells of *C. perfringens* has been studied by several groups. For strain ATCC 13124 in autoclaved ground beef, *D* 56.8°C was 48.3 min, essentially similar to the *D* 56.8°C or *D* 47.9°C for phospholipase C (27). Employing strain NCTC 8798, *D* values for cells were found to increase with increasing growth temperatures in autoclaved ground beef. For cells grown at 37°C, *D* 59°C was 3.1 min; cells grown at 45°C had *D* 59°C of 7.2; and cells grown at 49°C had *D* 59°C of 10.6 min (94). While the wide differences in heat resistance between the two strains noted may in part be due to strain differences, the effect of fat in the heating menstrum may also have played a role. With beef roasts cooked in plastic bags in a water bath at 60° to 61°C, holding the product to an internal temperature of 60°C for at least 12 min eliminated salmonellae and reduced the *C. perfringens* population by about 3 log cycles. To effect a 12-log reduction of numbers for roasts weighing 1.5 kg, holding at 60°C for 2.3 h or longer was necessary (104). The thermal destruction of *C. perfringens* enterotoxin in buffer and gravy at 61°C required 25.4 and 23.8 min, respectively (9).

While the wide variations in heat resistance recorded for *C. perfringens* spores may be due to many factors, similar variations have not been recorded for *C. botulinum*, especially types A and B. The latter organisms are less common in the human intestinal tract than *C. perfringens* strains. An organism inhabiting environments as diverse as these may be expected to show wide variations among its strains. Another factor that is important in heat resistance of bacterial spores is that of the chemical environment. Alderton and Snell (2) have pointed out that spore heat resistance is largely an inducible property, chemically reversible between a sensitive and resistant state. Using this hypothesis, it has been shown that spores can be made more heat resistant by treating them in Ca-acetate solutions—for example, 0.1 or 0.5 M at pH 8.5 for 140 h at 50°C. The heat resistance of endospores may be increased five- to tenfold by this method (1). On the other hand, heat resistance may be decreased by holding spores in 0.1 N HCl at 25°C for 16 h or as a result of the exposure of endospores to the natural acid conditions of some foods. It is not inconceivable that the high variability of heat resistance of *C. perfringens* spores may be a more or less direct result of immediate environmental history.

The freezing survival of *C. perfringens* in chicken gravy was studied by Strong and Canada (116), who found that only around 4% of cells survived when frozen to −17.7°C for 180 days. Dried spores, on the other hand, displayed a survival rate of about 40% after 90 days but only about 11% after 180 days.

For epidemiologic studies, serotyping has been employed, but because of the many serovars, there appears to be no consistent relationship between outbreaks and given serovars (see Hatheway et al., 37). The bacteriocin typing of type A has been achieved, and of 90 strains involved in food outbreaks, all were typable by a set of eight bacteriocins and 85.6% consisted of bacteriocin types 1–6 (97).

The Enterotoxin

The causative factor of *C. perfringens* food poisoning is an enterotoxin. It is unusual in that it is a spore-specific protein; its production occurs together with that of sporulation. All known food-poisoning cases by this organism are caused by type A strains. An unrelated disease, necrotic enteritis, is caused by beta toxin produced by type C strains and is only rarely reported outside New Guinea. While necrotic enteritis due to type C has been associated with a mortality rate of 35 to 40%, food poisoning due to type A strains has been fatal only in elderly or otherwise debilitated persons. Some type C strains have been shown to produce enterotoxin, but its role in disease is unclear (101).

The enterotoxin of type A strains was demonstrated by Duncan and Strong (22). The purified enterotoxin has a molecular weight of 35,000 daltons and an isoelectric point of 4.3 (39). It is heat sensitive (biological activity destroyed at 60°C for 10 min) and pronase sensitive but resistant to trypsin, chymotrypsin, and papain (115). L-forms of *C. perfringens* produce the toxin, and in one study they were shown to produce as much enterotoxin as classical forms (68).

The enterotoxin is synthesized by sporulating cells in association with late stages of sporulation. The peak for toxin production is just before lysis of the cell's sporangium, and the enterotoxin is released along with spores. Conditions that favor sporulation also favor enterotoxin production, and this was demonstrated with raffinose (57), caffeine, and theobromine (56). The latter two compounds increased enterotoxin from undetectable levels to 450 μg/ml of cell extract protein. It has been shown to be similar to spore structural proteins covalently associated with the spore coat. Cells sporulate freely in the intestinal tract and in a wide variety of foods (14). In culture media, the enterotoxin is normally produced only when endospore formation is permitted (Fig. 20-1), but vegetative cells are known to produce enterotoxin at low levels (34, 35). A single gene has been shown to be responsible for the enterotoxin trait (20, 21), and enterotoxin and a spore-coat protein have been shown to be controlled by a stable mRNA (55).

The enterotoxin may appear in a growth and sporulation medium about 3 h after inoculation with vegetative cells (19), and from 1 to 100 μg/ml of

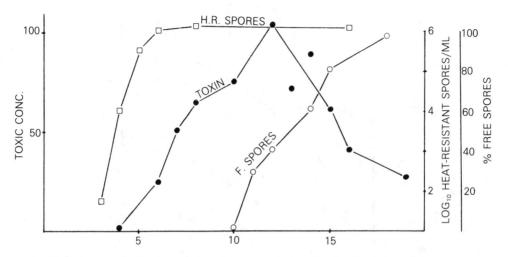

FIGURE 20-1.   Kinetics of sporulation and enterotoxin formation by *Clostridium perfringens* type A.
Redrawn from Labbe (53); copyright © 1980 by Institute of Food Technologists.

enterotoxin production has been shown for three strains of *C. perfringens* in Duncan-Strong (DS) medium after 24 to 36 h (28). It has been suggested that preformed enterotoxin may exist in some foods and in infrequent cases contribute to the early onset of symptoms (84). Purified enterotoxin has been shown to contain up to 3,500 mouse LD/mg N.

The enterotoxin may be detected in the feces of victims. From one case, 13 to 16 µg/g feces were found, and from another victim with a milder case, 3 to 4 µg/g were detected (102).

Binding of enterotoxin is necessary for the expression of biological activity (75). The toxin binds to the brush border of intestinal epithelial cells. Binding is specific, time and temperature dependent (69), and irreversible (Fig. 20-2). Once bound, the enterotoxin is rapidly inserted into membranes. It apparently does not enter cytoplasm. It appears that two eucaryotic membrane proteins serve as toxin receptors, 50 kda and 70 kda proteins (133). Once attached, the perfringens enterotoxin is nicked, and this leads to changes in permeability of epithelial cells, resulting in the reversal of water, Na^+, and Cl^- uptake. As a cytotoxic enterotoxin, cell death results from membrane damage. In rabbits the activity of enterotoxin was shown to affect the intestinal tract in the following order: ileum > jejunum > duodenum (74). According to McDonel (71), in the normal ileum, there is a net absorption of water, Na^+, Cl^-, and glucose, while at the same time there is a net efflux of K^+ and bicarbonate ions. Under the influence of enterotoxin, the net movement of water, Na^+, and Cl^- is reversed while net movement of K^+ and bicarbonate is unaffected (Fig. 20-3). By use of cell cultures and isolated tissues, it has been shown that enterotoxin decreases O_2 consumption, causes rounding of cells, induces holes in outer cell membranes, inhibits macromolecular synthesis, and causes cell death (73). The toxin also causes erythema in test animals and increased capillary permeability, and it exhibits parasympathomimetric properties. Its effect on rabbit ileum is

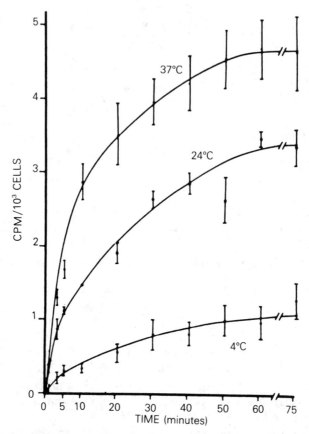

FIGURE 20-2.   Specific binding of (^{125}I) enterotoxin to intestinal cells as a function of temperature and time. Error bars show standard of error of the mean, which if not shown was smaller than the symbol.
McDonel (72); copyright © 1980 by American Chemical Society.

presented in Fig. 20-4. (These and other effects on tissue culture systems and bioassays are treated further in Chap. 7.)

Vehicle Foods and Symptoms

Symptoms appear between 6 and 24 h, especially between 8 and 12 h, after the ingestion of contaminated foods. The symptoms are characterized by acute abdominal pain and diarrhea; nausea, fever, and vomiting are rare. Except in the elderly or in debilitated persons, the illness is of short duration—a day or less. The fatality rate is quite low, and no immunity seems to occur, although circulating antibodies to the enterotoxin may be found in some persons with a history of the syndrome.

The true incidence of *C. perfringens* food poisoning is unknown. Because of the relative mildness of the disease, it is quite likely that only those outbreaks and cases that affect groups of people are ever reported and recorded. The confirmed outbreaks reported to the U.S. Centers for Disease Control for the

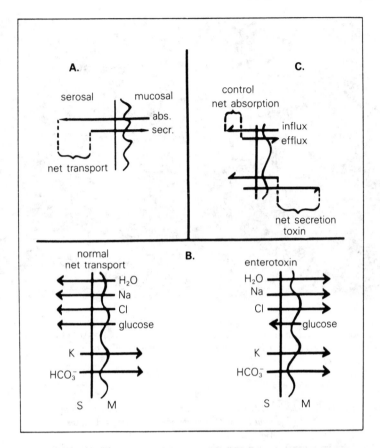

FIGURE 20-3.   Summary of intestinal transport alterations in rate due to the effects of *Clostridium perfringens* enterotoxin. *A*. Most substances are simultaneously absorbed from the luminal (mucosal) side of the intestine to the blood (serosal) side, and in the opposite direction. Net absorption occurs when the absorptive flux (M-S) exceeds the secretory (S-M) flux. *B*. In the normal ileum, net absorption of water, sodium, chloride, and glucose occurs, while there is a net efflux of potassium and bicarbonate ions. Under the influence of enterotoxin, the movement of water, sodium, and chloride is reversed while net movement of potassium and bicarbonate is unaffected. Glucose continues to be absorbed but at a significantly reduced rate. *C*. When net sodium transport is resolved into its component fluxes, it is found that uptake of sodium is the same under control and enterotoxin-treated conditions. Net secretion due to enterotoxin is the result of a significant increase in the secretory flux.

From McDonel (71); used with permission of Amer. J. Clin. Nutrition.

years 1983–1987 are noted in Table 20-2, along with cases. The average number of cases was about 100 for each outbreak.

The foods involved in *C. perfringens* outbreaks are often meat dishes prepared one day and eaten the next. The heat preparation of such foods is presumably inadequate to destroy the heat-resistant endospores, and when the food is cooled and rewarmed, the endospores germinate and grow. Meat dishes are most often the cause of this syndrome, although nonmeat dishes may be contaminated by meat gravy. The greater involvement of meat dishes may be due in part to the slower cooling rate of these foods and also to the higher incidence of food-poisoning strains in meats. Strong et al. (117) found the

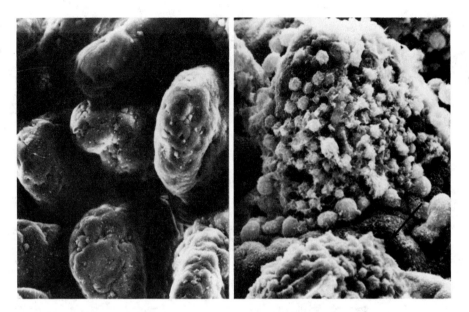

FIGURE 20-4.   Scanning electron micrographs (×370) of control rabbit ileum (*left*) showing typical morphology of villi, and micrograph (×900) of rabbit ileum treated with *Clostridium perfringens* enterotoxin (*right*) for 90 min showing toxin-damaged villus tips. *Courtesy of J. L. McDonel.*

TABLE 20-2.   Outbreaks, Cases, and Deaths from *C. perfringens* Foodborne Gastroenteritis in the United States, 1983–1987

Years	Outbreaks/Cases/Deaths
1983	5/353/0
1984	8/882/2
1985	6/1,016/0
1986	3/202/0
1987	2/290/0

Source: N. H. Bean, P. M. Griffin, J. S. Goulding, and C. B. Ivey. *J. Food Protect.* 53:711–728, 1990.

overall incidence of the organism to be about 6% in 510 American foods. The incidence for various foods was 2.7% for commercially prepared frozen foods, 3.8% for fruits and vegetables, 5% for spices, 1.8% for home-prepared foods, and 16.4% for raw meat, poultry, and fish. Hobbs et al. (41) found that 14 to 24% of veal, pork, and beef samples examined contained heat-resistant endospores, but all 17 samples of lamb were negative. In Japan, enterotoxigenic strains were recovered from food handlers (6% of 80), oysters (12% of 41), and water (10% of 20 samples) (95).

An outbreak of food poisoning involving 375 persons where 140 became ill was shown to be caused by both *C. perfringens* and *Salmonella typhimurium* (91). *C. perfringens* has been demonstrated to grow in a large number of foods. A study of retail, frozen precooked foods revealed that half were positive for vegetative cells, and 15% contained endospores (125). The latter investigators inoculated meat products with the organism and stored them at −29°C for up to 42 days. While spore survival was high, vegetative cells were virtually eliminated during the holding period. The survival of inoculated cells in raw ground beef was studied by Goepfert and Kim (32), who found decreased numbers upon storage at temperatures between 1° and 12.5°C. The raw beef contained a natural flora, and the finding suggests that *C. perfringens* is unable to compete under these conditions. For the recovery of *C. perfringens* from foods, an appropriate reference in Table 5-1 or the review by Walker (129), should be consulted.

Prevention

The *C. perfringens* gastroenteritis syndrome may be prevented by proper attention to the leading causes of food poisoning of all types noted in the previous chapter. Since this syndrome often occurs in institutional cafeterias, some special precautions should be taken. Upon investigating a *C. perfringens* food-poisoning outbreak in a school lunchroom in which 80% of students and teachers became ill, Bryan et al. (10) constructed a time-temperature chart in an effort to determine when, where, and how the turkey became the vehicle (Fig. 20-5). It was concluded that meat and gravy but not dressing were responsible for the illness. As a means of preventing recurrences of such episodes, these authors suggested nine points for the preparation of turkey and dressing:

1. Cook turkeys until the internal breast temperature reaches at least 165°F (74°C), preferably higher.
2. Thoroughly wash and sanitize all containers and equipment that previously had contact with raw turkeys.
3. Wash hands and use disposable plastic gloves when deboning, deicing, or otherwise handling cooked turkey.
4. Separate turkey meat and stock before chilling.
5. Chill the turkey and stock as rapidly as possible after cooking.
6. Use shallow pans for storing stock and deboned turkey in refrigerators.
7. Bring stock to a rolling boil before making gravy or dressing.
8. Bake dressing until all portions reach 165°F or higher.
9. Just prior to serving, heat turkey pieces submerged in gravy until largest portions of meat reach 165°F.

BOTULISM

Unlike *C. perfringens* food poisoning, in which large numbers of viable cells must be ingested, the symptoms of botulism are caused by the ingestion of a highly toxic, soluble exotoxin produced by the organism while growing in foods.

Among the earliest references to what in all probability was human botulism was the order by Emperor Leo VI, one of the Macedonian-era rulers of the Byzantium, during the period 886–912 A.D. forbidding the eating of blood sausage because of its harmful health effects. An outbreak of "sausage poison-

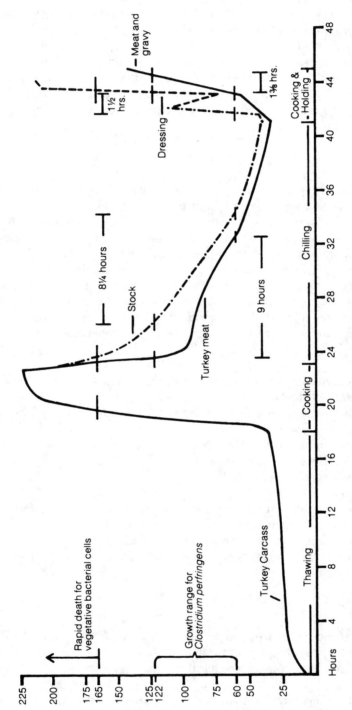

FIGURE 20-5. Illustration of possible time-temperature relationships during turkey preparation in a school lunch kitchen. *From Bryan et al. (10); copyright © 1971 by International Association of Milk, Food and Environmental Sanitarians.*

ing" occurred in 1793 in Wildbad Württemberg, Germany, with 13 cases and 6 deaths. It was traced to blood sausage (pig gut filled with blood and other ingredients). The filled gut was tied, boiled briefly, smoked, and stored at room temperature. Between 1820 and 1822, Justinius Kerner studied 230 cases of "sausage poisoning" in Württemberg and noted that the product did not become toxic if air pockets were left in casings and that toxic sausage had always been boiled. In 1896, 24 music club members in Ellezelles ate raw salted ham; 23 became ill, and 3 died. E. P. M. Van Ermengen of the University of Ghent, studied the outbreak. He found that the ham was neither cooked nor smoked, and the same organism was recovered from ham and the spleen of a victim. Van Ermengen named the causative organism *Bacillus botulinus* (*botulus*, L "sausage"). This strain was later determined to be a type B. (See ref. 106 for more information on the early history of botulism.)

Botulism is caused by certain strains of *C. botulinum*, a gram-positive, anaerobic sporeforming rod with oval to cylindrical, terminal to subterminal spores. On the basis of the serological specificity of their toxins, seven types are recognized: A, B, C, D, E, F, and G. Types A, B, E, F, and G cause disease in humans; type C causes botulism in fowls, cattle, mink, and other animals; and type D is associated with forage poisoning of cattle, especially in South Africa. The types are also differentiated on the basis of their proteolytic activity. Types A and G are proteolytic, as are some types B and F strains. Type E is nonproteolytic, as are some B and F strains (see Table 20-3). The proteolytic activity of type G is slower than that for type A, and its toxin requires trypsin potentiation (109).

A vast literature exists on botulism, and the information that follows is in no way meant to be complete. For more information, see Smith (106) Sperber (112), and Sakaguchi (96).

Distribution of *C. botulinum*

This organism is indigenous to soils and waters. In the United States, type A occurs more frequently in soils in the western states, and type B is found more frequently in the eastern states and in Europe. Soils and manure from various countries have been reported to contain 18% type A and 7% type B spores. Cultivated soil samples examined showed 7% to contain type A and 6% type B endospores. Type E spores tend to be confined more to waters, especially marine waters. In a study of mud samples from the harbor of Copenhagen, Pederson (90) found 84% to contain type E spores, while 26% of soil samples taken from a city park contained the organism. From a study of 684 environmental samples from Denmark, the Faroe Islands, Iceland, Greenland, and Bangladesh, 90% of aquatic samples from Denmark and 86% of marine samples from Greenland contained type E (43). This strain was not found in Danish soils and woodlands, while type B was. Based on these results, Huss (43) suggested that type E is a truly aquatic organism that proliferates in dead aquatic animals and sediments and is disseminated by water currents and migrating fish. Type E spores have been known for some time to exist in waters off the shores of northern Japan. Prior to 1960, the existence of the organisms in Great Lakes and Gulf Coast waters was not known, but their presence in these waters as well as in the Gulf of Maine and the gulfs of Venezuela and

TABLE 20-3. Summary Comparison of *C. botulinum* Strains and Their Toxins

Property		Serologic Types					
	A	B	B	E	F	F	G
Year discovered	1904	1896	1960	1936	1960	1965	1969
Proteolytic (+), nonproteolytic (−)	+	+	−	−	+	−	+ (weak)
Primary habitat	Terrestrial	Terrestrial	Aquatic	Aquatic	Aquatic	Aquatic	Terrestrial
Minimum growth temp. (°C)	~10	~10	3.3	3.3	~10	3.3	~12
Maximum growth temp. (°C)	~50	~50	~45	~45	~50	~45	n.d.
Minimum pH for growth (see text)	4.7	4.7	4.7	4.8	4.8	4.8	4.8
Minimum a_w for growth	0.94	0.94	~0.97	~0.97	0.94?	~0.97	n.d.
Thermal D values for endospores (°C)	$D_{110} = 2.72\text{–}2.89^a$	$D_{110} = 1.34\text{–}1.37^a$	n.d.	$D_{80} = 0.80^b$	$D_{110} = 1.45\text{–}1.82^c$	$D_{82.2} = 0.25\text{–}0.84^d$	$D_{110} = 0.45\text{–}0.54^e$
Radiation D values of spores (kGy)	1.2–1.5	1.1–1.3	n.d.	1.2	1.1; 2.5	1.5	n.d.
Maximum NaCl for growth (%)	~10	~10	5–6	5–6	8–10	5–6	n.d.
Relative frequency of food outbreaks	High	High	n.d.	Highest for seafoods	1 outbreak	1 outbreak	None
H_2S production	+	+	−	−	+	−	+ +
Casein hydrolysis	+	+	−	−	+	−	+
Lipase production	+	+	+	+	+	+	−
Glucose fermentation	+	+	+	+	+	+	−
Mannose fermentation	−	−	+	+	−	+	−
Propionic acid produced	+	+	n.d.	n.d.	+	n.d.	n.d.

Notes:

+ = positive

+ + = strongly positive

− = negative

n.d. = no data

[a] In phosphate buffer (74).

[b] Strain 8-E (54).

[c] (54).

[d] (53).

[e] (55).

Darien is well established. Ten percent of soil samples tested in Russia were found positive for *C. botulinum*, with type E strains being predominant.

A study of sediment samples from the upper Chesapeake Bay area of the U.S. Atlantic Coast revealed the presence of types A and E spores in 4 of 33 samples (98). The investigators believed these organisms to be randomly distributed in sediment and to be autochthonous. From Lake Michigan, 9% of fish caught contained type E spores, and from Green Bay 57% of fish contained the organisms (8). In another study (26), 6.2% of 500 commercially dressed fish taken from Lake Michigan near the Two Rivers, Wisconsin, area yielded type E, while only 0.4% of 427 laboratory-dressed fish were positive. The authors found the type E organisms to exist in relatively low numbers on freshly caught fish but to increase after evisceration.

In soils from 52 locations around Rome, Italy, 7 of 520 (1.3%) contained proteolytic A and B strains (15). In a study of the incidence of *C. botulinum* on whitefish chubs in smoking plants, Pace et al. (88) found the highest incidence (21%) at the brine tank processing stage.

As to the overall incidence of *C. botulinum* in soils, it has been suggested that the numbers/g are probably less than one. The nonproteolytic types are associated more with waters than soils, and it may be noted from Table 20-3 that the discovery of these types occurred between 1960 and 1969. The late recognition is probably a consequence of the low heat resistance of the nonproteolytics, which would be destroyed if specimens were given their usual heat treatment for spore recovery.

The first type F strains were isolated by Moller and Scheibel (82) from a homemade liver paste incriminated in an outbreak of botulism, involving one death, on the Danish island Langeland. Since that time, Craig and Pilcher (13) have isolated type F spores from salmon caught in the Columbia River; Eklund and Poysky (25) found type F spores in marine sediments taken off the coasts of Oregon and California; Williams-Walls (132) isolated two proteolytic strains from crabs collected from the York River in Virginia; and Midura et al. (79) isolated the organism from venison jerky in California.

The type G strain was isolated first in 1969 from soil samples in Argentina (31), and it was isolated later from five human corpses in Switzerland (111). These deaths were not food associated. It has not been incriminated in food-poisoning outbreaks to date, and the reason might be due to the fact that this strain produces considerably less neurotoxin than type A. It has been shown that type G produced 40 LD_{50}/ml of toxin in media in which type A normally produces 10,000 to 1,000,000 LD_{50}/ml, but that under certain conditions the organism could be induced to produce up to 90,000 LD_{50}/ml of medium (12).

Growth of *C. botulinum* Strains

Some of the growth and other characteristics of the strains that cause botulism in humans are summarized in Table 20-3. The discussion that follows emphasizes the differences between the proteolytic and nonproteolytic strains irrespective of serologic type. The proteolytic strains, unlike the nonproteolytics, digest casein and produce H_2S. The latter, on the other hand, ferment mannose, while the proteolytics do not. The proteolytics and nonproteolytics have been shown to form single groups relative to somatic antigens as evaluated by agglutination (110). The absorption of antiserum by any one of a group removes antibodies from all three of that group. Proteolytic strains are placed in Group I, nonproteolytic E, B, and F are placed in Group III, and type G is placed in Group IV (106).

The nutritional requirements of these organisms are complex, with amino acids, B vitamins, and minerals being required. Synthetic media have been devised that support growth and toxin production of most types. The proteolytic strains tend not to be favored in their growth by carbohydrates, while the nonproteolytics are. At the same time, the nonproteolytics tend to be more fermentative than the proteolytic types.

The proteolytics generally do not grow below 12.5°C, although a few reports exist in which growth was detected at 10°C. The upper range for types A and proteolytic B, and presumably for the other proteolytic types, is about 50°C. On the other hand, the nonproteolytic strains can grow as low as 3.3°C with the maximum about 5 degrees below that for proteolytics. Minimum and maximum temperatures of growth of these organisms are dependent on the state of other growth parameters, and the minima and maxima noted may be presumed to be at totally optimal conditions relative to pH, a_w, and the like. In a study of the minimum temperature for growth and toxin production by nonproteolytic types B and F in broth and crabmeat, both grew and produced toxin at 4°C in broth, but in crabmeat growth and toxin production occurred only at 26°C and not at 12°C or lower (108). A type G strain grew and produced toxin in broth and crabmeat at 12°C but not at 8°C (108).

The minimum pH that permits growth and toxin production of *C. botulinum* strains has been the subject of many studies. It is generally recognized that growth does not occur at or below pH 4.5, and it is this fact that determines the degree of heat treatment given to foods with pH values below this level (see Chap. 14). Because of the existence of botulinal toxins in some high-acid, home-canned foods, this area has been the subject of recent studies. In one study, no growth of types A and B occurred in tomato juice at pH around 4.8, but when the product was inoculated with *Aspergillus gracilis*, toxin was produced at pH 4.2 in association with the mycelial mat (86). In another study with the starting pH of tomato juice at 5.8, the pH on the underside of the mold mat increased to 7.0 after 9 days and to 7.8 after 19 days (42). The tomato juice was inoculated with type A botulinal spores, a *Cladosporium* sp., and a *Penicillium* sp. The topmost 0.5 ml of product showed pH increases from 5.3 to 6.4 or 7.5 after 9 and 19 days, respectively. One type B strain was shown to produce gas in tomato juice at pH 5.24 after 30 days and at pH 5.37 after 6 days. In food systems consisting of whole shrimp, shrimp purée, tomato purée, and tomato and shrimp purée acidified to pH 4.2 and 4.6 with acetic or citric acid, none of three type E strains grew or produced toxin at 26°C after 8 weeks (92). Growth and toxin production of a type E strain at pH 4.20 and 26°C in 8 weeks was demonstrated when citric acid but not acetic acid was used to control the pH of a culture medium (126). In general, the pH minima are similar for proteolytic and nonproteolytic strains.

Employing aqueous suspensions of soy proteins inoculated with four type A and two type B strains with incubation at 30°C, growth occurred at pH 4.2, 4.3, and 4.4 (103). The inoculum was 5×10^6 spores/ml, and 4 weeks were required for detectable toxin at pH 4.4 when pH was adjusted with either HCl or citric acid. When lactic or acetic acid was used, 12 and 14 weeks, respectively, were required for toxin at pH 4.4. Inocula of $10^3 - 10^4$/g of botulinal spores represent considerably higher numbers of these organisms than would be found on foods naturally (see below). That growth may occur at pH lower than 4.5 with large inocula does not render invalid the widely held view that this organism does not grow at or below pH 4.5 in raw foods with considerably smaller numbers of spores.

Regarding the interaction of pH, NaCl, and growth temperature, a study with Japanese noodle soup (*tsuyu*) revealed that with types A and B spores, no toxin developed at (1) pH less than 6.5, 4% NaCl, and 20°C; (2) pH less than 5.0, 1% NaCl, and 30°C; (3) pH less than 5.5, 3% NaCl, and 30°C; and (4) pH less than 6.0, 4% NaCl, and 30°C (46).

The minimum a_w that permits growth and toxin production of types A and proteolytic B strains is 0.94, and this value seems to be well established. The minimum for type E is around 0.97. While all strains have not been studied equally, it is possible that the other nonproteolytic strains have a minimum similar to that of type E. The way in which a_w is achieved in culture media affects the minimum values obtained. When glycerol is used as humectant, a_w values tend to be a bit lower than when NaCl or glucose is used (113). Salt at a level of about 10%, or 50% sucrose, will inhibit growth of types A and B, and 3 to 5% salt has been found to inhibit toxin production in smoked fish chubs (11). Lower levels of salt are required when nitrites are present (see Chap. 11).

With respect to heat resistance, the proteolytic strains are much more resistant than nonproteolytics (Table 20-3). Although the values noted in the table suggest that type A is the most heat resistant, followed by proteolytic F and then proteolytic B, these data should be taken only as representative since heating menstra, previous history of strains, and other factors are known to affect heat resistance (Chap. 14). Of those noted, all were determined in phosphate buffer. Among type E, the Alaska and Beluga strains appear to be more heat resistant than others, and in ground whitefish chubs, D 80°C of 2.1 and 4.3 have been reported (16), while in crabmeat, D 82.2°C of 0.51 and 0.74 have been reported, respectively, for Alaska and Beluga (67). With regard to smoked whitefish chubs, it was determined in one study that heating to an internal temperature of 180°C for 30 min produced a nontoxigenic product (26), while in another study, 10 or 1.2% of 858 freshly smoked chubs given the same heat treatment were contaminated, mostly with type E strains (88). (The heat destruction of bacterial endospores is dealt with further in Chap. 14.)

With regard to type G, the Argentine and Swiss strains both produce two kinds of spores: heat labile and heat resistant. The former, which are destroyed at around 80°C after 10 min, represent about 99% of the spores in a culture of the Swiss strains; in the Argentine strain only about 1 in 10,000 endospores is heat resistant (66). The D 230°F of two heat-resistant strains in phosphate buffer was 0.45 to 0.54 min, while for two heat-labile strains, D 180°F was 1.8 to 5.9 min (66). The more heat-resistant spores of type G have not yet been propagated.

Unlike heat, radiation seems to affect the endospores of proteolytic and nonproteolytic strains similarly, with D values of 1.1 to 2.5 kGy having been reported (see Chap. 12). However, the D value of one nonproteolytic type F strain was found to be 1.5 kGy, which was similar to the D value for a type A strain, but a proteolytic type F strain produced a D of 1.16 kGy (5).

Ecology of *C. botulinum* Growth

It appears that this organism cannot grow and produce its toxins in competition with large numbers of other microorganisms. Toxin-containing foods are generally devoid of other types of organisms because of heat treatments. In the presence of yeasts, however, *C. botulinum* has been reported to grow and produce toxin at a pH as low as 4.0. While a synergistic effect between clostridia and lactic acid bacteria has been reported on the one hand, lactobacilli will antagonize growth and toxin production; indirect evidence for this is the absence of botulinal toxins in milk. Yeasts are presumed to produce growth factors needed by the clostridia to grow at low pH, while the lactic acid bacteria may aid growth by reducing the oxidation-reduction (O/R) potential or inhibit growth by "lactic antagonism" (see Chap. 11). In one study, type A was inhibited by soil isolates of *C. sporogenes*, *C. perfringens*, and *B. cereus* (105). Some *C. perfringens* strains produced an inhibitor that was effective on 11 type A strains, on 7 type B proteolytic and one nonproteolytic strains, and on 5 type E and 7 type F strains (50). It is possible for *C. botulinum* spores to germinate and grow in certain canned foods whose pH is less than 4.5 when *Bacillus coagulans* is present. In a study with tomato juice of pH 4.5 inoculated with *B. coagulans*, the pH increased after 6 days at 35°C to 5.07 and to 5.40 after 21 days, thus making it possible for *C. botulinum* to grow (4). Kautter et al. (50) found that type E strains are inhibited by other nontoxic organisms whose biochemical properties and morphological characteristics were similar to type E. These organisms were shown to effect inhibition of type E strains by producing a bacteriocin-like substance designated "boticin E." In a more detailed study, proteolytic A, B, and F strains were found to be resistant to boticin E elaborated by a nontoxic type E, but toxic E cells were susceptible (3). The boticin was found to be sporostatic for nonproteolytic types B, E, and F and nontoxigenic type E.

Vacuum-packaged foods such as bacon are capable of supporting growth and toxin production by *C. botulinum* strains without causing noticeable off-odors, and types A and E spores have been shown to germinate and produce toxin in smoked fish (123). Fish inoculated with type A was offensively spoiled, while type E strains caused a less drastic type of spoilage. The proteolytic strains generally produce more offensive by-products than the nonproteolytics. In a study of type E toxin formation and cell growth in turkey rolls incubated at 30°C, type E spores germinated and produced toxin within 24 h (77). The appearance of toxin coincided with cell growth for 2 weeks, after which time toxins outlasted viable cells. These findings suggest that it is possible to find type E toxin in foods in the absence of type E cells. Toxin could not be demonstrated after 56 days of incubation.

A report on the ecology of type F showed that the absence of this strain in mud samples during certain times of the year was associated with the presence of *Bacillus licheniformis* in the samples during these periods, when the bacillus was apparently inhibiting type F strains (131).

Concerns for Sous Vide and Related Food Products

Special concerns for the growth and toxin production by *C. botulinum* strains are presented by **sous vide** processed foods. By this method, developed in France around 1980, raw food is placed in high barrier bags and cooked under vacuum (*sous vide*, "under vacuum"). Most, if not all, vegetative cells are destroyed, but bacterial spores survive. Thus, the sous vide product is one that contains bacterial spores in an O_2-depressed environment with no microbial competitors. In low-acid foods such as meats, poultry, and seafoods, spores of *C. botulinum* can germinate, grow, and produce toxin. Holding temperature and time are the two parameters that must be carefully monitored to avoid toxic products.

While the proteolytic strains do not grow in the refrigerator temperature range, the nonproteolytic strains can. A summary of published data on the incidence of botulinal spores in meat and poultry reveals that the numbers are extremely low—well below 1 spore/g (Table 20-4). Assuming a mean spore load of 1 spore/g and constant storage at 3° to 5°C, low-acid sous vide meat products should be safe for at least 21 days. With raw rockfish fillets inoculated at a level of 1 spore/sample with a mixture of 13 strains of types E, B, and F, no toxin could be detected in 21 days when stored at 4°C (45) or in red snapper homogenates after 21 days at 4°C (61). In inoculated modified atmosphere packaged (MAP) pork stored at 5°C, no toxin could be detected in 44 days (59). Whatever storage time is possible under constant low-temperature storage is shortened by temperature abuse. Products that have secondary barriers such as a_w less than 0.93 or pH less than 4.6 may be held safely for longer periods of time even with some temperature abuse. Since *Bacillus* spp. spores may be more abundant than botulinal and since some can grow at pH less than 4.6, it is not inconceivable that these forms can germinate, grow, and elevate pH during

TABLE 20-4.   Incidence of *Clostridium botulinum* in Meat and Poultry for a 14–Year Period

Country	Product	Number Positive/ Number Tested	C. botulinum Type (No.)	No./g
United Kingdom	Bacon	36/397	A (23) B (13)	0.00217
United States	Cooked ham	5/100	A (5)	0.00166
	Smoked turkey	1/41	B	0.0081
	Other meats	0/231	—	—
	Frankfurters	1/10	B	0.0066
	Other meats	0/80	—	—
	Luncheon meats	1/73	B	0.0057
	"Sausages"	0/17	—	—
United States and Canada	Raw chicken	1/1,078	C	0.0031
	Raw beef, pork	0/1,279	—	—
Canada	Sliced meats	0/436	—	—

Source: Tompkin (124).

temperature abuse. Botulinal toxin was detectable in anaerobically stored noodles with initial pH less than 4.5 when pH was increased by microbial growth (44). Although it is widely assumed that fish contains more botulinal spores than land animals and consequently should be of more concern, a recent study of 1,074 test samples of commercial vacuum-packaged fresh fish that were held at 12°C for 12 days failed to develop botulinal toxins (60). Inoculated type E strains grew in controls, suggesting that either the samples contained no botulinal spores or they were overgrown by other members of the flora.

The development of mathematical models designed to predict the probability of growth and toxin production in sous vide and MAP foods has been undertaken by several groups of investigators. By use of factorial designs, these models are designed to integrate the individual and combined effects of the parameters of temperature, a_w, pH, inoculum size, and storage time. Equations were developed in one series of studies that could predict the length of time to toxin production and the probability of toxigenesis by a single spore under defined conditions using cooked, vacuum-packed potatoes (18). From the latter study, the response by mixtures of five each of types A and B spores was shown to be linear, while to a_w the response was curvilinear. In another series of studies employing MAP-stored fresh fish inoculated with nonproteolytic strains, 74.6% of experimental variation in the final multiple linear regression model was accounted for by temperature of storage, with spore load accounting for only 7.4% (6). The earliest time to toxicity at 20°C was 1 day, but at 4°C the time increased to 18 days. With type E spores, no growth was observed in chopped meat medium at 3°C in 170 days, but in vacuum-packaged herring inoculated with 10^4 type E spores/g, toxigenesis was detected after 21 days at 3.3°C (6). Among other models reported is one that includes sorbic acid up to 2,270 ppm in combination with some of the other parameters noted (62).

Nature of the Botulinal Toxins

The neurotoxins are formed within the organism and released upon autolysis. They are produced by cells growing under optimal conditions, though resting cells have been reported to form toxin as well. The botulinal toxins are the most toxic substances known, with purified type A reported to contain about 30 million mouse LD_{50}/mg. The minimum lethal dose for mice has been reported also to be 0.4 to 2.5 ng/kg by intravenous or intraperitoneal injection, and a 50% human lethal dose of about 1 ng/kg of body weight has been reported. The first of these toxins to be purified was type A, which was achieved by C. Lamanna et al. and by A. Abrams et al. both in 1946. The purification of B, E, and F has been achieved.

It appears that all botulinal strains synthesize the neurotoxin as a single polypeptide chain (unnicked) with a molecular weight of about 150,000 daltons. The proteolytic strains A, B, and F produce endogenous proteases that cleave the 150,000-dalton unit to form a heavy chain of about 100,000 daltons and a light chain of about 50,000, the two being held together by at least one disulfide bond (24). The nonproteolytic types B, E, and F release into the environment the 150,000-dalton chain, which is referred to as a **progenitor** toxin. It is made toxic by treatment with exogenous proteases (nicking) such as trypsin (17). If the double-chained molecule is reduced to individual chains, each lacks toxicity, and neither is known to possess enzymatic activity. In culture fluids, botulinal toxins actually exist as complexes with nontoxic culture proteins, which may be considered the natural state of botulinal toxins (120). The molecular weight of these complexes may be as high as 900,000 daltons.

As a prerequisite to its neurotoxic activity, botulinal toxin must attach to neural tissue, specifically to ganglioside receptors. It appears that for toxin binding, two sialic acid residues on the inner galactose of the gangliosides are essential, and an additional sialic acid at the nonreducing end of the ganglioside also aids toxin binding (24). Once attached, the toxin exerts its effects by presynaptically blocking the release of acetylcholine from cholinergic nerve endings.

Type A toxin has been reported to be more lethal than B or E. Type B has been reported to have associated with it a much lower case mortality than type A, and case recoveries from type B have occurred even when appreciable amounts of toxin could be demonstrated in the blood.

Symptoms of botulism can be produced by either parenteral or oral administration of the toxins. They may be absorbed into the bloodstream through the respiratory mucous membranes, as well as through the walls of the stomach and intestines. The toxins are not completely inactivated by the proteolytic enzymes of the stomach, and, indeed, those produced by nonproteolytics may be activated. The high molecular weight complexes or the progenitor possess higher resistance to acid and pepsin (119). While the derivative toxin was rapidly inactivated, the progenitor was shown to be resistant to rat intestinal juice in vitro. The progenitor was more stable in the stomach of rats. Similar findings were made by Ohishi et al. (87), who showed that progenitor toxins of nonproteolytics were more toxic orally in mice than the dissociated toxic components of the derived toxins. It appears that the nontoxic component of the progenitor provides protection to the toxin activity. After botulinal toxins are absorbed into the bloodstream, they enter the peripheral nervous system where they affect nerves.

Unlike the staphylococcal enterotoxins and heat-stable toxins of other food-borne pathogens, the botulinal toxins are heat-sensitive and may be destroyed by heating at 80°C (176°F) for 10 min, or boiling temperatures for a few min.

The Adult Botulism Syndrome: Incidence and Vehicle Foods

Symptoms of botulism may develop anywhere between 12 and 72 h after the ingestion of toxin-containing foods. Even longer incubation periods are not unknown. Symptoms consist of nausea, vomiting, fatigue, dizziness, and headache; dryness of skin, mouth, and throat; constipation, lack of fever, paralysis of muscles, double vision, and finally, respiratory failure and death. The duration of the illness is from 1 to 10 or more days, depending upon host resistance and other factors. The mortality rate varies between 30 and 65%, with the rate being generally lower in European countries than in the United States. All symptoms are caused by the exotoxin, and treatment consists of administering specific antisera as early as possible. Although it is assumed that the tasting of toxin-containing foods allows for absorption from the oral cavity, Lamanna et al. (58) found that mice and monkeys are more susceptible to the

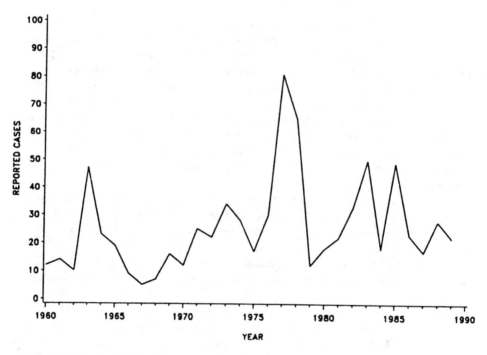

FIGURE 20-6.   Cases of foodborne botulism in the United States, 1960–1989.
From Morb. Mort. Weekly Report, *38:20, 1989.*

toxins when administered by stomach tube than by exposure to the mouth. The botulinal toxins are neurotoxins and attach irreversibly to nerves. Early treatment by use of antisera brightens the prognosis.

Prior to 1963, most cases of botulism in the United States in which the vehicle foods were identified were traced to home-canned vegetables and were caused by types A and B toxins. In almost 70% of the 640 cases reported for the period 1899–1967, the vehicle food was not identified. Among the 640 cases, 17.8% were associated with vegetables, 4.1% fruits, 3.6% fish, 2.2% condiments, 1.4% meats and poultry, and 1.1% for all others. Reported foodborne cases in the United States for the years 1960–1989 are shown in Figure 20-6. The large number of cases in 1977 occurred in a restaurant in Pontiac, Michigan, following consumption of a hot sauce prepared from home-canned jalapeño peppers. No deaths occurred, and type B toxin was identified. Total cases from all sources in the United States rarely exceed 50 per year, with the highest ten-year period being 1930–1939, when 384 cases were reported from noncommercial foods. Between 1899 and 1963, 1,561 cases were reported from noncommercial foods, while 219 were reported from commercial foods between 1906 and 1963, with 24 in 1963 alone.

Of 404 verified cases of type E botulism through 1963, 304 or 75% occurred in Japan. No outbreak of botulism was recorded in Japan prior to 1951. For the period May 1951 through January 1960, 166 cases were recorded, with 58 deaths for a mortality rate of 35%. Most of these outbreaks were traced to a home-prepared food called *izushi*, a preserved food consisting of raw fish, vegetables, cooked rice, malted rice (*koji*), and a small amount of salt and vinegar. This preparation is packed tightly in a wooden tub equipped with a lid and held for 3 weeks or longer to permit lactic acid fermentation. During this time, the O/R potential is lowered, thus allowing for the growth of anaerobes.

Sixty-two outbreaks of botulism resulting from commercially canned foods were recorded for the period 1899–1973 (63), with 41 prior to 1930. Between 1941 and 1982, 7 outbreaks occurred in the United States involving commercially canned foods in metal containers, with 17 cases and 8 deaths (85). Three of these outbreaks were caused by type A and the remainder by type E. In 5 of the outbreaks, can leakage or underprocessing occurred (85). Canned mushrooms have been incriminated in several botulism outbreaks. A study in 1973 and 1974 turned up 30 cans of mushrooms containing botulinal toxin (29 were type B). An additional 11 cans contained viable spores of *C. botulinum* without preformed toxin (63). The capacity of the commercial mushroom (*Agaricus bisporus*) to support the growth of inoculated spores of *C. botulinum* was studied by Sugiyama and Yang (122). Following inoculation of various parts of mushrooms, they were sealed with plastic film and incubated. Toxin was detected as early as 3 to 4 days later, when products were incubated at 20°C. While the plastic film used to wrap the inoculated mushrooms allowed for gas exchange, the respiration of the fresh mushrooms apparently consumed oxygen at a faster rate than it entered the film. No toxin was detected in products stored at refrigerator temperatures.

An unusual outbreak of 36 cases occurred in 1985 with victims in three countries: Canada, the Netherlands, and the United States. The vehicle food was chopped garlic in soybean oil that was packaged in glass bottles. Although labeled with instructions to refrigerate, unopened bottles were stored unrefrigerated for 8 months. The product was used to make garlic-buttered bread, which in turn was used to prepare beef dip sandwiches. Proteolytic type B spores were found, and toxin was produced within 2 weeks when proteolytic and nonproteolytic B strains were inoculated into bottles of chopped garlic and held at 25°C (114). Type A toxin was produced in bottled chopped garlic in 20 days at 35°C when inoculated with 1 spore/g and by type B toxin in 20 days (107). In the latter study, highly toxic bottles looked and smelled acceptable.

One of the recorded outbreaks of botulism (five cases with one death) due to type F involved homemade liver paste. The U.S. outbreak occurred in 1966 from home-prepared venison jerky, with three clinical cases (79).

The greatest hazards of botulism come from home-prepared and home-canned foods that are improperly handled or given insufficient heat treatments to destroy botulinal spores. Such foods are often consumed without heating. The best preventative measure is the heating of suspect foods to boiling temperatures for a few minutes, which is sufficient to destroy the neurotoxins.

Infant Botulism

First recognized as such in California in 1976 (78), infant botulism has since been confirmed in most states in the United States and in many other countries. In the adult form of botulism, preformed toxins are ingested; in infant botulism, viable botulinal spores are ingested, and upon germination in the intestinal tract, toxin is synthesized. While it is possible that in some adults under special conditions botulinal endospores may germinate and produce

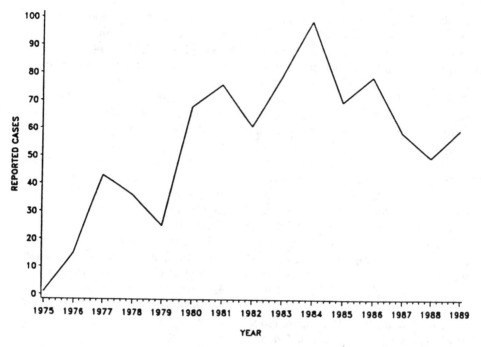

FIGURE 20-7.   Cases of infant botulism in the United States, 1975–1989.
From Morb. Mort. Weekly Report, *38:20, 1989.*

small quantities of toxin, the colonized intestinal tract does not favor spore germination. Infants over 1 year of age tend not to be affected by this syndrome because of the establishment of a more normal intestinal flora. The disease is mild in some infants; in others it can be severe. High numbers of spores are found in the feces of infants during the acute phase of the disease, and as recovery progresses, numbers of organisms abate.

This syndrome is diagnosed by demonstrating botulinal toxins in infant stools and by use of the mouse lethality test. Since *Clostridium difficle* produces mouse-lethal toxins in the intestinal tract of infants, it is necessary to differentiate between these toxins and that of *C. botulinum* (30).

Infants get viable spores from infant foods and possibly from their environment. Vehicle foods are those that do not undergo heat processing to destroy endospores; the two most common products are syrup and honey. Of 90 samples of honey examined, 9 contained viable spores. Six of these had been fed to babies who developed infant botulism (80). Of the 9, 7 were type B and 2 were type A. Of 910 infant foods from 10 product classes, only 2 classes were positive for spores: honey and corn syrup (51). Of 100 honey samples, two contained type A, and 8 of 40 corn syrup samples yielded type B. In Canada, only 1 of 150 samples of honey contained viable botulinal spores (type A), 1 of 40 dry cereal samples (type B), while 43 syrup samples were negative (48). Reported cases in the United States through 1989 are shown in Figure 20-7. The 62 cases for 1982, which occurred among infants aged 2 to 48 weeks, involved type A and B toxins equally. The first two cases reported in

Rome, Italy, were caused by type E toxin, which was produced by *Clostridium butyricum* (15).

Animal models for the study of this syndrome consist of 8–11-day-old mice (121), and 7–13-day-old rats (81). In the mouse model, botulinal toxin was found in the lumen of the large intestine, and it was not associated with the ileum. (The sensitivity of these animal models is noted in Chap. 7.)

BACILLUS CEREUS GASTROENTERITIS

Bacillus cereus is an aerobic, sporeforming rod normally present in soil, dust, and water. It has been associated with food poisoning in Europe since at least 1906. Among the first to report this syndrome with precision was U. Plazikowski. His findings were confirmed by several other European workers in the early 1950s. The first documented outbreak in the United States occurred in 1969, and the first in Great Britain occurred in 1971.

This bacterium has a minimum growth temperature around 4° to 5°C, with a maximum around 48° to 50°C. Growth has been demonstrated over the pH range 4.9 to 9.3 (33). Its spores possess a resistance to heat typical of other mesophiles.

B. cereus Toxins

Food-poisoning strains produce the following toxins and extracellular products: lecithinase, proteases, beta-lactamase, cereolysin (mouse lethal toxin, hemolysin I), emetic enterotoxin, and diarrheagenic enterotoxin. Cereolysin is a thiol-activated toxin analogous to perfringolysin O and listeriolysin O. It has a molecular weight of 55 kda and apparently plays no role in the food-borne gastroenteritis syndromes. The latter are caused by the emetic and the diarrheal enterotoxins.

The **diarrheagenic** toxin induces vascular permeability in the skin of rabbits, elicits fluid accumulation in the rabbit ileal loop, and causes diarrhea in Rhesus monkeys (see Chap. 7). It has been shown to be a protein with a molecular weight variously reported to be 38, 39.5, or 43 kda. It is produced during the exponential phase, but maximum toxin is found early in the stationary phase (36). Toxin has been reported to be detectable at more than $10^{6.6}$/g (128) and approximately 10^7 cells/ml (36). It is heat labile and sensitive to trypsin and pronase. Its production is favored over the pH range 6.0 to 8.5. In one study, no toxin was found at pH 5.0 and a_w 0.94 (128). Some strains produce enterotoxin at 4°C after 24 days, at 7°C in about 12 days, and at 17°C in about 48 h with pH at 7.0 (128). Growth and toxin production have been demonstrated at 4°C (128), and several milk strains were shown to produce toxin between 6° and 21°C (36).

Unlike the *C. perfringens* enterotoxin, the *B. cereus* toxin is a vegetative growth metabolite and can be separated from phospholipase and cereolysin (127). The purified toxin displays mouse lethality. Although its mode of action is not well understood, it apparently induces diarrhea by stimulating the adenylate cyclase-cAMP system.

Enterotoxigenic strains of *B. cereus* have been recovered from a variety of foods, with 85% of 83 strains from raw milk being positive for the diarrheagenic toxin (36). Twenty-seven percent of 534 raw meat samples, 42% of 820 meat products, and 31% of 609 spices and meat additives were found to be positive for *B. cereus*, with levels of 10^2 to 10^4 (52).

In addition to *B. cereus*, *B. mycoides* strains from milk have been shown to produce diarrheagenic enterotoxin in 9 days at temperatures between 6° and 21°C (36). Varying numbers of isolates of the following species were found also to be enterotoxin producers: *B. circulans*, *B. lentus*, *B. thuringiensis*, *B. pumilus*, *B. polymyxa*, *B. carotarum*, and *B. pasteurii* (36).

The **emetic** (vomiting-type) toxin is distinctly different from the diarrheagenic in having a molecular weight of 5,000 daltons (127) and in being heat and pH stable. It is insensitive to trypsin and pepsin (47). The emetic toxin strains grow over the range 15° to 50°C, with an optimum between 35° to 40°C (48). While the emetic syndrome is most often associated with rice dishes, growth of the emetic toxin strains in rice is not favored in general over other *B. cereus* strains, although higher populations and more extensive germination have been noted in this product (48).

Diarrheal Syndrome

This syndrome is rather mild, with symptoms developing within 8 to 16 h, more commonly within 12 to 13 h, and lasting for 6 to 12 (38). Symptoms consist of nausea (with vomiting being rare), cramplike abdominal pains, tenesmus, and watery stools. Fever is generally absent. The similarity between this syndrome and that of *C. perfringens* food poisoning has been noted (29).

Vehicle foods consist of cereal dishes that contain corn and cornstarch, mashed potatoes, vegetables, minced meat, liver sausage, meat loaf, milk, cooked meat, Indonesian rice dishes, puddings, soups, and others (29). Reported outbreaks between 1950 and 1978 have been summarized by Gilbert (29), and when plate counts on leftover foods were recorded, they ranged from 10^5 to 9.5×10^8/g, with many in the 10^7–10^8/g range. The first well-studied outbreaks were those investigated by Hauge (38), which were traced to vanilla sauce; the counts ranged from 2.5×10^7 to 1×10^8/g. From meat loaf involved in a U.S. outbreak in 1969, 7×10^7/g were found (76). Serovars found in diarrheal outbreaks include types 1, 6, 8, 9, 10, and 12. Serovars 1, 8, and 12 have been associated with this as well as with the emetic syndrome (29).

Emetic Syndrome

This form of *B. cereus* food poisoning is more severe and acute than the diarrheal syndrome. The incubation period ranges from 1 to 6 h, with 2 to 5 being most common (83). Its similarity to the staphylococcal food-poisoning syndrome has been noted (29). Unlike staphylococcal enterotoxins, the emetic toxin has a molecular weight of less than 10 kda, and it may be spore associated similar to the perfringens enterotoxin. It is heat stable and is produced almost exclusively by serotype 1 strains. It is often associated with fried or boiled rice dishes. In addition to these, pasteurized cream, spaghetti, mashed potatoes, and vegetable sprouts have been incriminated (29). Outbreaks have been reported from Great Britain, Canada, Australia, the Netherlands, Finland,

Japan, and the United States. The first U.S. outbreak was reported in 1975, with mashed potatoes as the vehicle food.

The numbers of organisms necessary to cause this syndrome seem to be higher than for the diarrheal syndrome, with numbers as high as $2 \times 10^9/g$ having been found (see 29). *B. cereus* serovars associated with the emetic syndrome include 1, 3, 4, 5, 8, 12, and 19 (29).

For the recovery and quantitation of *B. cereus* enterotoxins, see Chapters 6 and 7. The incidence of this organism in certain foods is noted in Chapter 4; for more information, see Gilbert (29) and Johnson (47).

References

1. Alderton, G., K. A. Ito, and J. K. Chen. 1976. Chemical manipulation of the heat resistance of *Clostridium botulinum* spores. *Appl. Environ. Microbiol.* 31:492–498.
2. Alderton, G., and N. Snell. 1969. Bacterial spores: Chemical sensitization to heat. *Science* 163:1212–1213.
3. Anastasio, K. L., J. A. Soucheck, and H. Sugiyama. 1971. Boticinogeny and actions of the bacteriocin. *J. Bacteriol.* 107:143–149.
4. Anderson, R. E. 1984. Growth and corresponding elevation of tomato juice pH by *Bacillus coagulans*. *J. Food Sci.* 49:647, 649.
5. Anellis, A., and D. Berkowitz. 1977. Comparative dose-survival curves of representative *Clostridium botulinum* type F spores with type A and B spores. *Appl. Environ. Microbiol.* 34:600–601.
6. Baker, D. A., and C. Genigeorgis. 1990. Predicting the safe storage of fresh fish under modified atmospheres with respect to *Clostridium botulinum* toxigenesis by modeling length of the lag phase of growth. *J. Food Protect.* 53:131–140.
7. Barnes, E., and M. Ingram. 1956. The effect of redox potential on the growth of *Clostridium welchii* strains isolated from horse muscle. *J. Appl. Bacteriol.* 19:117–128.
8. Bott, T. L., J. S. Deffner, E. McCoy, and E. M. Foster. 1966. *Clostridium botulinum* type E in fish from the Great Lakes. *J. Bacteriol.* 91:919–924.
9. Bradshaw, J. G., G. N. Stelma, V. I. Jones, J. T. Peeler, J. G. Wimsatt, J. J. Corwin, and R. M. Twedt. 1982. Thermal inactivation of *Clostridium perfringens* enterotoxin in buffer and in chicken gravy. *J. Food Sci.* 47:914–916.
10. Bryan, F. L., T. W. McKinely, and B. Mixon. 1971. Use of time-temperature evaluations in detecting the responsible vehicle and contributing factors of food-borne disease outbreaks. *J. Milk Food Technol.* 34:576–582.
11. Christiansen, L. N., J. Deffner, E. M. Foster, and H. Sugiyama. 1968. Survival and outgrowth of *Clostridium botulinum* type E spores in smoked fish. *Appl. Microbiol.* 16:133–137.
12. Ciccarelli, A. S., D. N. Whaley, L. M. McCroskey, D. F. Gimenez, V. R. Dowell, Jr., and C. L. Hatheway. 1977. Cultural and physiological characteristics of *Clostridium botulinum* type G and the susceptibility of certain animals to its toxin. *Appl. Environ. Microbiol.* 34:843–848.
13. Craig, J., and K. Pilcher. 1966. *Clostridium botulinum* type F: Isolation from salmon from the Columbia River. *Science* 153:311–312.
14. Craven, S. E. 1980. Growth and sporulation of *Clostridium perfringens* in foods. *Food Technol.* 34(4):80–87, 95.
15. Creti, R., L. Fenicia, and P. Aureli. 1990. Occurrence of *Clostridium botulinum* in the soil of the vicinity of Rome. *Curr. Microbiol.* 20:317–321.
16. Crisley, F. D., J. T. Peeler, R. Angelotti, and H. E. Hall. 1968. Thermal

resistance of spores of five strains of *Clostridium botulinum* type E in ground whitefish chubs. *J. Food Sci.* 33:411–416.

17. DasGupta, B. R., and H. Sugiyama. 1976. Molecular forms of neurotoxins in proteolytic *Clostridium botulinum* type B cultures. *Infect. Immun.* 14:680–686.

18. Dodds, K. L. 1989. Combined effect of water activity and pH on inhibition of toxin production by *Clostridium botulinum* in cooked, vacuum-packed potatoes. *Appl. Environ. Microbiol.* 55:656–660.

19. Duncan, C. L. 1973. Time of enterotoxin formation and release during sporulation of *Clostridium perfringens* type A. *J. Bacteriol.* 113:932–936.

20. Duncan, C. L. 1976. *Clostridium perfringens.* In *Food Microbiology: Public Health and Spoilage Aspects,* ed. M. P. deFigueiredo and D. F. Splittstoesser, 170–197. Westport, Conn.: AVI.

21. Duncan, C. L., and D. H. Strong. 1968. Improved medium for sporulation of *Clostridium perfringens. Appl. Microbiol.* 16:82–89.

22. Duncan, C. L., and D. H. Strong. 1969. Ileal loop fluid accumulation and production of diarrhea in rabbits by cell-free products of *Clostridium perfringens. J. Bacteriol.* 100:86–94.

23. Duncan, C. L., D. H. Strong, and M. Sebald. 1972. Sporulation and enterotoxin production by mutants of *Clostridium perfringens. J. Bacteriol.* 110:378–391.

24. Eidels, L., R. L. Proia, and D. A. Hart. 1983. Membrane receptors for bacterial toxins. *Microbiol. Rev.* 47:596–620.

25. Eklund, M., and F. Poysky. 1965. *Clostridium botulinum* type E from marine sediments. *Science* 149:306.

26. Fantasia, L. D., and A. P. Duran. 1969. Incidence of *Clostridium botulinum* Type E in commercially and laboratory dressed white fish chubs. *Food Technol.* 23:793–974.

27. Foegeding, P. M., and F. F. Busta. 1980. *Clostridium perfringens* cells and phospholipase C activity at constant and linearly rising temperatures. *J. Food Sci.* 45:918–924.

28. Genigeorgis, C., G. Sakaguchi, and H. Riemann. 1973. Assay methods for *Clostridium perfringens* type A enterotoxin. *Appl. Microbiol.* 26:111–115.

29. Gilbert, R. J. 1979. *Bacillus cereus* gastroenteritis. In *Food-borne Infections and Intoxications,* ed. H. Riemann and F. L. Bryan, 495–518. New York: Academic Press.

30. Gilligan, P. H., L. Brown, and R. E. Berman. 1983. Differentiation of *Clostridium difficile* toxin from *Clostridium botulinum* toxin by the mouse lethality test. *Appl. Environ. Microbiol.* 45:347–349.

31. Gimenez, D. F., and A. S. Ciccarelli. 1970. Another type of *Clostridium botulinum. Zentral. Bakteriol. Orig. A.* 215:221–224.

32. Goepfert, J. M., and H. U. Kim. 1975. Behavior of selected foodborne pathogens in raw ground beef. *J. Milk Food Technol.* 38:449–452.

33. Goepfert, J. M., W. M. Spira, and H. U. Kim. 1972. *Bacillus cereus*: Food poisoning organism. A review. *J. Milk Food Technol.* 35:213–227.

34. Goldner, S. B., M. Solbert, S. Jones, and L. S. Post. 1986. Enterotoxin synthesis by nonsporulating cultures of *Clostridium botulinum. Appl. Environ. Microbiol.* 52:407–412.

35. Granum, P. E., W. Telle, Ø. Olsvik, and A. Stavn. 1984. Enterotoxin formation by *Clostridium perfringens* during sporulation and vegetative growth. *Intern. J. Food Microbiol.* 1:43–49.

36. Griffiths, M. W. 1990. Toxin production by psychrotrophic *Bacillus* spp. present in milk. *J. Food Protect.* 53:790–792.

37. Hatheway, C. L., D. N. Whaley, and V. R. Dowell, Jr. 1980. Epidemiological

aspects of *Clostridium perfringens* foodborne illness. *Food Technol.* 34(4):77–79, 90.

38. Hauge, S. 1955. Food poisoning caused by aerobic spore-forming bacilli. *J. Appl. Bacteriol.* 18:591–595.

39. Hauschild, A. H. W., and R. Hilsheimer. 1971. Purification and characteristics of the enterotoxin of *Clostridium perfringens* type A. *Can. J. Microbiol.* 17:1425–1433.

40. Hauschild, A. H. W., R. Hilsheimer, K. F. Weiss, and R. B. Burke. 1988. *Clostridium botulinum* in honey, syrups and dry infant cereals. *J. Food Protect.* 51:892–894.

41. Hobbs, B., M. Smith, C. Oakley, G. Warrack, and J. Cruickshank. 1953. *Clostridium welchii* food poisoning. *J. Hyg.* 51:75–101.

42. Huhtanen, C. N., J. Naghski, C. S. Custer, and R. W. Russell. 1976. Growth and toxin production by *Clostridium botulinum* in moldy tomato juice. *Appl. Environ. Microbiol.* 32:711–715.

43. Huss, H. H. 1980. Distribution of *Clostridium botulinum*. *Appl. Environ. Microbiol.* 39:764–769.

44. Ikawa, J. Y. 1991. *Clostridium botulinum* growth and toxigenesis in shelf-stable noodles. *J. Food Sci.* 56:264–265.

45. Ikawa, J. Y., and C. Genigeorgis. 1987. Probability of growth and toxin production by nonproteolytic *Clostridium botulinum* in rockfish fillets stored under modified atmospheres. *Int. J. Food Microbiol.* 4:167–181.

46. Imai, H., K. Oshita, H. Hashimoto, and D. Fukushima. 1990. Factors inhibiting the growth and toxin formation of *Clostridium botulinum* types A and B in "tsuyu" (Japanese noodle soup). *J. Food Protect.* 53:1025–1032.

47. Johnson, K. M. 1984. *Bacillus cereus* foodborne illness—An update. *J. Food Protect.* 47:145–153.

48. Johnson, K. M., C. L. Nelson, and F. F. Busta. 1983. Influence of temperature on germination and growth of spores of emetic and diarrheal strains of *Bacillus cereus* in a broth medium and in rice. *J. Food Sci.* 48:286–287.

49. Kang, C. K., M. Woodburn, A. Pagenkopf, and R. Cheney. 1969. Growth, sporulation, and germination of *Clostridium perfringens* in media of controlled water activity. *Appl. Microbiol.* 18:798–805.

50. Kautter, D. A., S. M. Harmon, R. K. Lynt, Jr., and T. Lilly, Jr. 1966. Antagonistic effect on *Clostridium botulinum* type E by organisms resembling it. *Appl. Microbiol.* 14:616–622.

51. Kautter, D. A., T. Lilly, Jr., H. M. Solomon, and R. K. Lynt. 1982. *Clostridium botulinum* spores in infant foods: A survey. *J. Food Protect.* 45:1028–1029.

52. Konuma, H., K. Shinagawa, M. Tokumaru, Y. Onoue, S. Konno, N. Fujino, T. Shigehisa, H. Kurata, Y. Kuwabara, and C. A. M. Lopes. 1988. Occurrence of *Bacillus cereus* in meat products, raw meat and meat product additives. *J. Food Protect.* 51:324–326.

53. Labbe, R. G. 1980. Relationship between sporulation and enterotoxin production in *Clostridium perfringens* type A. *Food Technol.* 34(4):88–90.

54. Labbe, R. G., and C. L. Duncan. 1974. Sporulation and enterotoxin production by *Clostridium perfringens* type A under conditions of controlled pH and temperature. *Can. J. Microbiol.* 20:1493–1501.

55. Labbe, R. G., and C. L. Duncan. 1977. Evidence for stable messenger ribonucleic acid during sporulation and enterotoxin synthesis by *Clostridium perfringens* type A. *J. Bacteriol.* 129:843–849.

56. Labbe, R. G., and L. L. Nolan. 1981. Stimulation of *Clostridium perfringens* enterotoxin formation by caffeine and theobromine. *Infect. Immun.* 34:50–54.

57. Labbe, R. G., and D. K. Rey. 1979. Raffinose increases sporulation and entero-toxin production by *Clostridium perfringens* type A. *Appl. Environ. Microbiol.* 37:1196–1200.

58. Lamanna, C., R. A. Hillowalla, and C. C. Alling. 1967. Buccal exposure to botulinal toxin. *J. Infect. Dis.* 117:327–331.

59. Lambert, A. D., J. P. Smith, and K. L. Dodds. 1991. Combined effect of modified atmosphere packaging and low-dose irradiation on toxin production by *Clostridium botulinum* in fresh pork. *J. Food Protect.* 54:94–101.

60. Lilly, T., Jr., and D. A. Kautter. 1990. Outgrowth of naturally occurring *Clostridium botulinum* in vacuum-packaged fresh fish. *J. Assoc. Off. Anal. Chem.* 73:211–212.

61. Lindroth, S., and C. Genigeorgis. 1986. Probability of growth and toxin produc-tion by non-proteolytic *Clostridium botulinum* in rock fish stored under modified atmospheres. *Int. J. Food Microbiol.* 3:167–181.

62. Lund, B. M., A. F. Graham, S. M. George, and D. Brown. 1990. The combined effect of incubation temperature, pH and sorbic acid on the probability of growth of non-proteolytic type B *Clostridium botulinum*. *J. Appl. Bacteriol.* 69:481–492.

63. Lynt, R. K., D. A. Kautter, and R. B. Read, Jr. 1975. Botulism in commercially canned foods. *J. Milk Food Technol.* 38:546–550.

64. Lynt, R. K., D. A. Kautter, and H. M. Solomon. 1979. Heat resistance of non-proteolytic *Clostridium botulinum* type F in phosphate buffer and crabmeat. *J. Food Sci.* 44:108–111.

65. Lynt, R. K., D. A. Kautter, and H. M. Solomon. 1982. Heat resistance of pro-teolytic *Clostridium botulinum* type F in phosphate buffer and crabmeat. *J. Food Sci.* 47:204–206, 230.

66. Lynt, R. K., H. M. Solomon, and D. A. Kautter. 1984. Heat resistance of *Clostridium botulinum* type G in phosphate buffer. *J. Food Protect.* 47:463–466.

67. Lynt, R. K., H. M. Solomon, T. Lilly, Jr., and D. A. Kautter. 1977. Thermal death time of *Clostridium botulinum* type E in meat of the blue crab. *J. Food Sci.* 42:1022–1025, 1037.

68. Mahony, D. E. 1977. Stable L-forms of *Clostridium perfringens*: Growth, toxin production, and pathogenicity. *Infect. Immun.* 15:19–25.

69. McClane, B. A., and A. P. Wnek. 1990. Studies of *Clostridium perfringens* en-terotoxin action at different temperatures demonstrate a correlation between complex formation and cytotoxicity. *Infect. Immun.* 58:3109–3115.

70. McClung, L. 1945. Human food poisoning due to growth of *Clostridium per-fringens (C. welchii)* in freshly cooked chicken: Preliminary note. *J. Bacteriol.* 50:229–231.

71. McDonel, J. L. 1979. The molecular mode of action of *Clostridium perfringens* enterotoxin. *Amer. J. Clin. Nutri.* 32:210–218.

72. McDonel, J. L. 1980. Binding of *Clostridium perfringens* (^{125}I) enterotoxin to rabbit intestinal cells. *Biochem.* 19:4801–4807.

73. McDonel, J. L. 1980. Mechanism of action of *Clostridium perfringens* enterotoxin. *Food Technol.* 34(4):91–95.

74. McDonel, J. L., and C. L. Duncan. 1977. Regional localization of activity of *Clostridium perfringens* type A enterotoxin in the rabbit ileum, jejunum, and duodenum. *J. Infect. Dis.* 136:661–666.

75. McDonel, J. L., and B. A. McClane. 1979. Binding versus biological activity of *Clostridium perfringens* enterotoxin in Vero cells. *Biochem. Biophys. Res. Comm.* 87:497–504.

76. Midura, T., M. Gerber, R. Wood, and A. R. Leonard. 1970. Outbreak of food poisoning caused by *Bacillus cereus*. *Publ. Hlth Rept.* 85:45–47.

77. Midura, T., C. Taclindo, Jr., G. S. Mygaard, H. L. Bodily, and R. M. Wood. 1968. Use of immunofluorescence and animal tests to detect growth and toxin production by *Clostridium botulinum* type E in food. *Appl. Microbiol.* 16:102–105.

78. Midura, T. F., and S. S. Arnon. 1976. Infant botulism: Identification of *Clostridium botulinum* and its toxins in faeces. *Lancet* 2:934–936.

79. Midura, T. F., G. S. Nygaard, R. M. Wood, and H. L. Bodily. 1972. *Clostridium botulinum* type F: Isolation from venison jerky. *Appl. Microbiol.* 24:165–167.

80. Midura, T. F., S. Snowden, R. M. Wood, and S. S. Arnon. 1979. Isolation of *Clostridium botulinum* from honey. *J. Clin. Microbiol.* 9:282–283.

81. Moberg, L. J., and H. Sugiyama. 1980. The rat as an animal model for infant botulism. *Infect. Immun.* 29:819–821.

82. Moller, V., and I. Scheibel. 1960. Preliminary report on the isolation of an apparently new type of *Cl. botulinum*. *Acta Path. Microbiol. Scan.* 48:80.

83. Mortimer, P. R., and G. McCann. 1974. Food-poisoning episodes associated with *Bacillus cereus* in fried rice. *Lancet* 1:1043–1045.

84. Naik, H. S., and C. L. Duncan. 1977. Enterotoxin formation in foods by *Clostridium perfringens* type A. *J. Food Safety* 1:7–18.

85. NFPA/CMI Task Force. 1984. Botulism risk from post-processing contamination of commercially canned foods in metal containers. *J. Food Protect.* 47:801–816.

86. Odlaug, T. E., and I. J. Pflug. 1979. *Clostridium botulinum* growth and toxin production in tomato juice containing *Aspergillus gracilis*. *Appl. Environ. Microbiol.* 37:496–504.

87. Ohishi, I., S. Sugii, and G. Sakaguchi. 1977. Oral toxicities of *Clostridium botulinum* toxins in response to molecular size. *Infect. Immun.* 16:107–109.

88. Pace, P. J., E. R. Krumbiegel, R. Angelotti, and H. J. Wisniewski. 1967. Demonstration and isolation of *Clostridium botulinum* types from whitefish chubs collected at fish smoking plants of the Milwaukee area. *Appl. Microbiol.* 15:877–884.

89. Pearson, C. B., and H. W. Walker. 1976. Effect of oxidation-reduction potential upon growth and sporulation of *Clostridium perfringens*. *J. Milk Food Technol.* 39:421–425.

90. Pederson, H. O. 1955. On type E botulism. *J. Appl. Bacteriol.* 18:619–629.

91. Peterson, D., H. Anderson, and R. Detels. 1966. Three outbreaks of foodborne disease with dual etiology. *Pub. Hlth Repts.* 81:899–904.

92. Post, L. S., T. L. Amoroso, and M. Solberg. 1985. Inhibition of *Clostridium botulinum* type E in model acidified food systems. *J. Food Sci.* 50:966–968.

93. Rey, C. R., H. W. Walker, and P. L. Rohrbaugh. 1975. The influence of temperature on growth, sporulation, and heat resistance of spores of six strains of *Clostridium perfringens*. *J. Milk Food Technol.* 38:461–465.

94. Roy, R. J., F. F. Busta, and D. R. Thompson. 1981. Thermal inactivation of *Clostridium perfringens* after growth at several constant and linearly rising temperatures. *J. Food Sci.* 46:1586–1591.

95. Saito, M. 1990. Production of enterotoxin by *Clostridium perfringens* derived from humans, animals, foods, and the natural environment in Japan. *J. Food Protect.* 53:115–118.

96. Sakaguchi, G. 1983. *Clostridium botulinum* toxins. *Pharmac. Ther.* 19:165–194.

97. Satija, K. C., and K. G. Narayan. 1980. Passive bacteriocin typing of strains of *Clostridium perfringens* type A causing food poisoning for epidemiologic studies. *J. Infect. Dis.* 142:899–902.

98. Saylor, G. S., J. D. Nelson, Jr., A. Justice, and R. R. Colwell. 1976. Incidence of *Salmonella* spp., *Clostridium botulinum*, and *Vibrio parahaemolyticus* in an estuary. *Appl. Environ. Microbiol.* 31:723–730.

99. Schneider, M., N. Grecz, and A. Anellis. 1963. Sporulation of *Clostridium botulinum* types A, B, and E, *Clostridium perfringens*, and putrefactive anaerobe 3679 in dialysis sacs. *J. Bacteriol.* 85:126–133.

100. Shimizu, T., A. Okabe, J. Minami, and H. Hayashi. 1991. An upstream regulatory sequence stimulates expression of the perfringolysin O gene of *Clostridium perfringens*. *Infect. Immun.* 59:137–142.

101. Skjelkvåle, R., and C. L. Duncan. 1975. Enterotoxin formation by different toxigenic types of *Clostridium perfringens*. *Infect. Immun.* 11:563–575.

102. Skjelkvåle, R., and T. Uemura. 1977. Detection of enterotoxin in faeces and anti-enterotoxin in serum after *Clostridium perfringens* food-poisoning. *J. Appl. Bacteriol.* 42:355–363.

103. Smelt, J. P. P. M., G. J. M. Raatjes, J. S. Crowther, and C. T. Verrips. 1982. Growth and toxin formation by *Clostridium botulinum* at low pH values. *J. Appl. Bacteriol.* 52:75–82.

104. Smith, A. M., D. A. Evans, and E. M. Buck. 1981. Growth and survival of *Clostridium perfringens* in rare beef prepared in a water bath. *J. Food Protect.* 44:9–14.

105. Smith, L. DS. 1975. Inhibition of *Clostridium botulinum* by strains of *Clostridium perfringens* isolated from soil. *Appl. Microbiol.* 30:319–323.

106. Smith, L. DS. 1977. *Botulism—The Organism, Its Toxins, the Disease.* Springfield, Ill.: Thomas.

107. Solomon, H. M., and D. A. Kautter. 1988. Outgrowth and toxin production by *Clostridium botulinum* in bottled chopped garlic. *J. Food Protect.* 51:862–865.

108. Solomon, H. M., D. A. Kautter, and R. K. Lynt. 1982. Effect of low temperatures on growth of nonproteolytic *Clostridium botulinum* types B and F and proteolytic type G in crabmeat and broth. *J. Food Protect.* 45:516–518.

109. Solomon, H. M., D. A. Kautter, and R. K. Lynt. 1985. Common characteristics of the Swiss and Argentine strains of *Clostridium botulinum* type G. *J. Food Protect.* 48:7–10.

110. Solomon, H. M., R. K. Lynt, Jr., D. A. Kautter, and T. Lilly, Jr. 1971. Antigenic relationships among the proteolytic and nonproteolytic strains of *Clostridium botulinum*. *Appl. Microbiol.* 21:295–299.

111. Sonnabend, O., W. Sonnabend, R. Heinzle, T. Sigrist, R. Dirnhofer, and U. Krech. 1981. Isolation of *Clostridium botulinum* type G and identification of type G botulinal toxin in humans: Report of five sudden unexpected deaths. *J. Infect. Dis.* 143:22–27.

112. Sperber, W. H. 1982. Requirements of *Clostridium botulinum* for growth and toxin production. *Food Technol.* 36(12):89–94.

113. Sperber, W. H. 1983. Influence of water activity on foodborne bacteria—A review. *J. Food Protect.* 46:142–150.

114. St. Louis, M. E., S. H. S. Peck, D. Bowering, G. B. Morgan, J. Blatherwick, S. Banerjee, G. D. M. Kettyls, W. A. Black, M. E. Milling, A. H. W. Hauschild, R. V. Tauxe, and P. A. Blake. 1988. Botulism from chopped garlic: Delayed recognition of a major outbreak. *Ann. Int. Med.* 108:363–368.

115. Stark, R. L., and C. L. Duncan. 1971. Biological characteristics of *Clostridium perfringens* type A enterotoxin. *Infect. Immun.* 4:89–96.

116. Strong, D. H., and J. C. Canada. 1964. Survival of *Clostridium perfringens* in frozen chicken gravy. *J. Food Sci.* 29:479–482.

117. Strong, D. H., J. C. Canada, and B. Griffiths. 1963. Incidence of *Clostridium perfringens* in American foods. *Appl. Microbiol.* 11:42–44.

118. Strong, D. H., C. L. Duncan, and G. Perna. 1971. *Clostridium perfringens* type A food poisoning. II. Response of the rabbit ileum as an indication of enteropath-

ogenicity of strains of *Clostridium perfringens* in human beings. *Infect. Immun.* 3:171–178.

119. Sugii, S., I. Ohishi, and G. Sakaguchi. 1977. Correlation between oral toxicity and in vitro stability of *Clostridium botulinum* types A and B toxins of different molecular sizes. *Infect. Immun.* 16:910–914.

120. Sugiyama, H. 1980. *Clostridium botulinum* neurotoxin. *Microbiol. Rev.* 44:419–448.

121. Sugiyama, H., and D. C. Mills. 1978. Intraintestinal toxin in infant mice challenged intragastrically with *Clostridium botulinum* spores. *Infect. Immun.* 21:59–63.

122. Sugiyama, H., and K. H. Yang. 1975. Growth potential of *Clostridium botulinum* in fresh mushrooms packaged in semipermeable plastic film. *Appl. Microbiol.* 30:964–969.

123. Thatcher, F. S., J. Robinson, and I. Erdman. 1962. The "vacuum pack" method of packaging foods in relation to the formation of the botulinum and staphylococcal toxins. *J. Appl. Bacteriol.* 25:120–124.

124. Tompkin, R. B. 1980. Botulism from meat and poultry products—a historical perspective. *Food Technol.* 34(5):229–236, 257.

125. Trakulchang, S. P., and A. A. Kraft. 1977. Survival of *Clostridium perfringens* in refrigerated and frozen meat and poultry items. *J. Food Sci.* 42:518–521.

126. Tsang, N., L. S. Post, and M. Solberg. 1985. Growth and toxin production by *Clostridium botulinum* in model acidified systems. *J. Food Sci.* 50:961–965.

127. Turnbull, P. C. B. 1986. *Bacillus cereus* toxins. In *Pharmacology of Bacterial Toxins*, ed. F. Dorner and J. Drews, 397–448. New York: Pergamon.

128. Van Netten, P., A. van De Moosdijk, P. Van Hoensel, D. A. A. Mossel, and I. Perales. 1990. Psychrotrophic strains of *Bacillus cereus* producing enterotoxin. *J. Appl. Bacteriol.* 69:73–79.

129. Walker, H. W. 1975. Food-borne illness from *Clostridium perfringens*. *CRC Crit. Rev. Food Sci. Nutri.* 7:71–104.

130. Weiss, K. F., and D. H. Strong. 1967. Some properties of heat-resistant and heat-sensitive strains of *Clostridium perfringens*. I. Heat resistance and toxigenicity. *J. Bacteriol.* 93:21–26.

131. Wentz, M., R. Scott, and J. Vennes. 1967. *Clostridium botulinum* type F: Seasonal inhibition by *Bacillus licheniformis*. *Science* 155:89–90.

132. Williams-Walls, N. J. 1968. *Clostridium botulinum* type F: Isolation from crabs. *Science* 162:375–376.

133. Wnek, A. P., and B. A. McClane. 1989. Preliminary evidence that *Clostridium perfringens* type A enterotoxin is present in a 160,000-M_r complex in mammalian membranes. *Infect. Immun.* 57:574–581.

21

Foodborne Listeriosis

The suddenness with which *Listeria monocytogenes* emerged as the etiologic agent of a foodborne disease is unparalled. The acquired immunodeficiency syndrome (AIDS) and legionellosis are examples of two other human diseases that appeared suddenly, but unlike foodborne listeriosis, the etiologic agents of these syndromes were previously unknown as human pathogens, and they proved to be difficult to culture. Not only is *L. monocytogenes* rather easy to culture, but listeriosis was well documented as a disease of many animal species, and human cases were not unknown. Although only a few foodborne outbreaks have been recorded, *L. monocytogenes* in foods has attracted worldwide attention. This organism has been the subject of well over 1,000 research papers and review articles since the early 1980s, in addition to conference proceedings and monographs. For early information on the listeriae, see the 1961 monograph by Seeliger (191), the 1963 monograph by Gray (79), and the 1966 review by Gray and Killinger (80). More recent monographs are those authored or edited by Woodbine (224), Ivanov (95), Courtieu et al. (38), Ralovich (173), and Miller et al. (151). A number of review articles have been published (36, 39, 65, 146, 165, 195).

TAXONOMY OF LISTERIA

The listeriae are gram-positive, nonsporeforming, and nonacid-fast rods that were once classified as "*Listerella*." The generic name was changed in 1940 to *Listeria*. In many ways they are similar to the genus *Brochothrix*. Both genera are catalase positive and tend to be associated with each other in nature, along with *Lactobacillus*. All three genera produce lactic acid from glucose and other fermentable sugars, but unlike *Listeria* and *Brochothrix*, the lactobacilli are catalase negative. At one time the listeriae were believed to be related to coryneform bacteria and in fact were placed in the family Corynebacteriaceae, but it is clear now that they are more closely related to *Bacillus*, *Lactobacillus*, and *Streptococcus*. By 16S rRNA sequence data, *Listeria* places closest to *Brochothrix*, and these two genera, together with *Staphylococcus* and *Kurthia*, occupy a position between the *Bacillus* group and the *Lactobacillus/Streptococcus* group within the *Clostridium-Lactobacillus-Bacillus* branch where the moles % G + C of all members is less than 50 (101). Genetic transfers occur among *Listeria*, *Bacillus*, and *Streptococcus*, and immunological cross-reactions occur among *Listeria*, *Streptococcus*, *Staphylococcus*, and *Lactobacillus*. *Brochothrix* shares 338 common purine and pyrimidine bases with *Listeria* (137). Although *Erysipelothrix* is in the *Mycoplasma* line, it shares at least 23 oligonucleotides in common with *Listeria* and *Brochothrix* (137). *Listeria* spp. contain teichoic and lipoteichoic acids (55), as do the bacilli, staphylococci, streptococci, and lactobacilli, but unlike these groups, their colonies form a blue-green sheen when viewed by obliquely transmitted light.

Seven species of *Listeria* are recognized currently, and they, with some differentiating characteristics, are listed in Table 21-1. The first five species are more closely related to each other than to *L. grayi* and *L. murrayi*. Note from Table 21-1 that the moles % G + C of DNA of the first five species ranges from 36 to 38%, while the latter two range from 41 to 42.5%. Also, the latter two species are serologically distinct from the others (177). Based on degree of DNA-DNA hybridization, the first five species form a rather tight cluster and clearly belong to the same genus, and they are referred to here as the true listerial species. It has been suggested that *L. grayi* and *L. murrayi* be placed in the genus *Murraya* as *M. grayi* subsp. *grayi* and *M. grayi* subsp. *murrayi*, but this proposal has not been validated (206). On the other hand, the retention of these two species in the genus *Listeria* has been proposed (175). By numerical taxonomic studies, the two species share 81 to 87% similarity with the other five species, but their H-antigens are different from the other five species, and no antibody cross-reactions occur (190). Employing reverse transcriptase sequencing of 16S rRNA, a phylogenetic analysis of the genus *Listeria* has been conducted, and it can be seen from Figure 21-1 that *L. grayi* and *L. murrayi* occupy a position between *Brochothrix* and the other five listerial species.

Poly(ribitolphosphate)-type teichoic acids are the prevalent or only accessory cell wall polymer in *Listeria* spp., including *L. grayi* and *L. murrayi* (55). While these two species contain lipoteichoic acids, they are of the modified type, further separating them from the other five species (180). Also, the modified lipoteichoic acids may account for their insensitivity to bacteriophages that lyse the five true species (132). What was once *Listeria denitrificans* is now *Jonesia denitrificans*.

Members of the genus *Erysipelothrix* are often associated with *Listeria*, and some differences between the two genera are noted in Table 21-2. Unlike *Listeria*, *Erysipelothrix* is nonmotile, catalase negative, and H_2S positive and contains L-lysine as the major diamino acid in its murein. Like *L. monocytogenes*, *E. rhusiopathiae* causes disease in animals—in this case, swine erysipelas. The latter organism is also infectious for humans, in whom it causes erysipeloid.

TABLE 21-1. Some Differentiating Characteristics of the Species of *Listeria*

Species	Xylose	Lactose	Galactose	Rhamnose	Mannitol	Hippurate Hydrolysis	CAMP test S. aureus	CAMP test R. equi	Beta Hemolysis	Moles % G + C	Serovars
L. monocytogenes	–	v	v	+	–	+	+	–	+	37–39	[a]
L. innocua	–	+	–	(+)	–	+	–	–	–	36–38	4ab,6a,6b [b]
L. seeligeri	+			–	–		+	–	w	36	
L. welshimeri	+			v	–		–	–	–	36	6a,6b
L. ivanovii	+	+	v	–	–	+	–	+	++	37–38	5
L. grayi	–	+	+	–	+	–	–	–	–	41–42	
L. murrayi	–	+	+	v	+	–	–	–	–	41–42.5	

Notes: v = variable; w = weak; + = most strains positive

[a] 1/2a, b, c; 3a, b, c; 4a, ab, b, c, d, e; "7."

[b] Same as for *L. monocytogenes* and *L. innocua* but no 5 or "7."

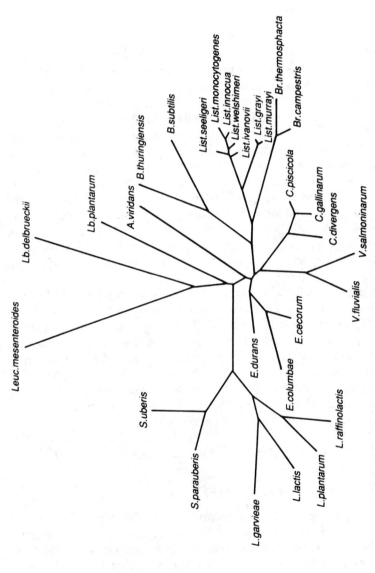

FIGURE 21-1. Unrooted tree or network showing the phylogenetic interrelationships of listeriae and other low-G + C–content gram-positive taxa. The tree is based on a comparison of a continuous stretch of 1,340 nucleotides: the first and last bases in the sequence used to calculate K_{nuc} values correspond to positions 107 (G) and 1433 (A), respectively, in the *E. coli* sequence. Abbreviations: *A, Aerococcus; B, Bacillus; Br, Brochothrix; C, Carnobacterium; E, Enterococcus; L, Lactococcus; Lb, Lactobacillus; Leuc, Leuconostoc; List, Listeria; S, Streptococcus; V, Vagococcus.* From M. D. Collins, S. Wallbanks, D. J. Lane, J. Shah, R. Nietupski, J. Smida, M. Dorsch, and E. Stackebrandt. Int. J. System. Bacteriol. *41:240–246, 1991; copyright © 1991 by American Society for Microbiology. Used with permission.*

TABLE 21-2.   A Comparison of the Genera *Listeria* and *Erysipelothrix*

Genera	Motility	Catalase	H_2S Production	Major Diamino Acid	Moles % G + C
Listeria	+	+	−	meso-DAP	36–38
Erysipelothrix	−	−	+	L-lysine	36–40

Serotypes

The five true species of listeriae are characterized by the possession of antigens that give rise to 17 serovars. The somatic O antigens (1–4) give rise to 15 serovars, and there are 5 flagellar or H antigens (a–e). In order to prepare antibodies to H, cells must be cultured in the temperature range of 18° to 22°C; culturing at 37°C is done for O antigens since flagellae are not produced at this temperature. The O antigenic determinants are teichoic acids and perhaps lipoteichoic acids of the cell envelope (55), and antibodies to listerial O antigens cross-react with many other gram-positive bacteria, especially *Coryne-bacterium* spp., *Staphylococcus aureus*, and enterococci (192). The primary pathogenic species, *L. monocytogenes*, is represented by 13 serovars, some of which are shared by *L. innocua* and *L. seeligeri*. While *L. innocua* is represented by only 3 serovars, it is sometimes regarded as the nonpathogenic variant of *L. monocytogenes*. The greater antigenic heterogeneity of the outer envelope of the latter species may be related to the wide number of animal hosts in which it can proliferate.

The most commonly isolated of these serotypes are types 1 and 4. Prior to the 1960s, it appeared that type 1 existed predominantly in Europe and Africa and type 4 in North America (80), but this pattern appears to have changed. Gray and Killinger noted in 1966 that serotypes of listeriae in no way are related to host, disease process, or geographical origin, and this is generally confirmed by food isolations (see below), although serovars 1/2a and 4b do show some geographical differences (192). In the United States and Canada, serovar 4b has accounted for 65 to 80% of all strains. The most frequently reported in Eastern Europe, West Africa, Central Germany, Finland, and Sweden is serovar 1/2a, while serovars 1/2a and 4b are more often reported in France and the Netherlands in about equal proportions (192).

Multilocus Enzyme Electrophoresis (MEE) Typing

This technique may be employed to estimate the overall genomic relationships among strains of organisms within species by determining the relative electro-phoretic mobilities of a set of water-soluble cellular enzymes. Variation in electrophoretic mobility can be related to allelic variation and to levels of genetic variation within populations of a species. Typically, from 15 to 25 enzymes are tested for on starch gels. Since some of the enzymes may have different mobilities (be polymorphic) among strains of the same species, MEE typing can be used to characterize strains for epidemiological purposes in the same general way as serotyping or phage typing. The basic technique has been described in detail and reviewed by Selander et al. (193).

With 175 isolates of *L. monocytogenes* tested for allelic variation among 16 enzymes, Pifferetti et al. (167) found that 45 allelic profiles or MEE types could be distinguished and further divided into two primary phylogenetic divisions separated at a genetic distance of 0.54. All 4a, 4b, and 1/2b serovars were in one division, and all 1/2a and 1/2c serovars were in the other one. Human and animal isolates were distributed among the 45 MEE types throughout the two divisions, while isolates in one division came from various parts of the world. Phage types were determined on 109 of the 175 isolates, and those with the same MEE type were not of the same phage type (167). On the other hand, the dominant phage types in two outbreaks of human listeriosis were similar, as were the MEE types. All strains from four outbreaks tested by these investigators placed into two MEE typing groups.

The MEE typing of 390 isolates of *L. monocytogenes* by Bibb et al. (16) resulted in 82 MEE types that could be defined into two distinct clusters similar to the findings of Pifferetti et al. (167). One cluster contained all serovars of *L. monocytogenes* that contained H antigen a, and all b H antigens fell into another cluster, with all of the latter being associated with perinatal cases of human listeriosis. Among 328 isolates from patients, 55 MEE types were defined, and from 62 food isolates, 34 were defined. Overall, clinical isolates had an average of 5.2 strains per MEE type, while food isolates averaged 1.67, suggesting a greater diversity among food isolates. Of 100 isolates of serovar 4b, the mean number/MEE type was 10.0 (17). In this and another study (167), MEE typing was effective in confirming the identity of strains from common source outbreaks where one MEE type was represented by all strains while many different MEE types were found when a common source did not exist. For example, all 7 clinical isolates from the 1983 Massachusetts milk outbreak (see below) had one MEE type, while the 22 isolates from the Philadelphia outbreak whose source was questionable had 11 different MEE types (16). Unlike phage typing, where some strains of *L. monocytogenes* are untypable and where no consistent relationship exists between serotypes, MEE typing can be obtained for all strains, and a high correlation exists with serotypes.

More recently, MEE has been employed as a taxonomic procedure on all 7 listerial species. The mean number of alleles/locus was 9.5, and all loci were polymorphic (19). By this method, 6 principal clusters were established at the species level, with the 5 classical species each belonging to separate clusters. Both *L. grayi* and *L. murrayi* fell in the same cluster, reinforcing the view that these two species are in fact biovars of a single species (19).

Phage Types

Since they were first described in 1945, bacteriophages specific for *Listeria* have been studied by a number of investigators relative to their uses for species and strain differentiation, and their epidemiologic value. *Listeria* phages contain dsDNA and belong to two groups: Siphoviridae (noncontractile tails) and Myoviridae (contractile tails). In a study of 823 strains of *L. monocytogenes* collected in France over the period 1958–1978, 69.4% were serotype 4, and a phage typing system was defined using 12 principal and 3 secondary phages (5). Six phages could be used to differentiate serotype 1, 9 phages for serotype 4, and only 2 phages for serotype 2 strains. By employing a set of 20 phages, these investigators were able to type 78.4% of the 823 strains, with 88% of serotype 4 and 57% of serotype 1 being typable. For 552 of the 645 typable strains, eight principal phage patterns could be established (5). When a set of 29 phages was employed in a multicenter study, 77% and 54% of serotypes 4 and 1/2, respectively, were typable (176). The more recent typing set of Audurier and co-workers is divided into three groups: 12 phages for 1/2 strains, 16 for 4b strains, and 7 for other strains (4). Typability of 826 serovar 4b strains isolated in France in the period 1985–1987 was 84% compared with 49% of 1,644 serovar 1 strains employing the 35 phages. By use of this scheme, *L. monocytogenes* isolates involved in three outbreaks of human listeriosis were shown to be of the same phage type, whether recovered from victims or foods.

In a study that employed 127 isolates of *L. monocytogenes* (208), lytic patterns allowed eight phage groups to be established, but more recent findings suggest that the lytic agents were monocins—defective phages with a tail that lack a head region (161). While monocin susceptibility appeared to be associated with serotypes, no relationship was found relative to animal source or geographic origin of *L. monocytogenes* strains (208).

In a study of 807 *L. monocytogenes* cultures collected in Britain from human cases over the period 1967–1984, phage typing was shown to be an effective tool for common source cases of listeriosis involving more than one patient or for recurrent episodes in the same patient (147). The 807 cultures belonged to serotypes 1/2, 3, and 4. In another study that employed a set of 16 phages recovered from lysogenic and environmental sources, 464 strains representing five species were placed in four groups (132). While the results were highly reproducible, species and serovar specificities did not conform to any lytic patterns. Phage susceptibility of *L. monocytogenes* was highest for serotype 4 (98%) followed by serotype 1 (90%) and serotype 3 (10%). No phages were restricted to either one species or serovar in lytic patterns. *L. grayi* and *L. murrayi* were not lysed by any of the selected phages (132). A reversed typing procedure that employs ready-to-use plates containing phage suspensions on tryptose agar plates was developed by Loessner (130). With a set of 21 genus-specific phages, the overall typability rate was 89.5% on 1,087 listerial strains.

The broad lytic spectrum of most listerial phages suggests fairly common receptor sites among these organisms. Since there appears to be no concurrence of serotypes and lytic pattern, this suggests that the O antigenic determinants and phage receptor site substances are not identical.

Restriction Enzyme Analysis (REA)

By this method, chromosomal DNA of test strains is digested by use of an appropriate restriction endonuclease. The latter class of enzymes makes double-stranded breaks in DNA at specific nucleotide sequences. One of the most widely used restriction endonucleases is *EcoRI* (obtained from *Escherichia coli*), which recognizes the DNA base sequence GAATTC and cleaves between GA. Another endonuclease is *HhaI* (obtained from *Haemophilus influenzae*), and it recognizes the sequence GTPyPuAC (Py = any pyrimidine, and Pu = any purine base). The cleavage site for *HhaI* is between PyPu, and it has been found to be of value in studying the epidemiology of *L. monocytogenes* (221).

After some *L. monocytogenes* serovar 4b strains associated with three food-associated outbreaks were subjected to restriction enzyme analysis (REA) using *HhaI*, the method was found to be valuable as both a taxonomic tool and an epidemiologic tracer (220). Of 32 isolates associated with the 1981 outbreak in Nova Scotia, Canada, 29 showed restriction enzyme patterns identical to the strain recovered from coleslaw. Also, the patterns of 9 clinical isolates from the 1983 Boston cases were identical to each other. Some of the isolates from the 1985 California outbreak were subjected to REA, and those examined from patients, suspect cheese samples, and cheese factory environmental samples were found to be identical (220).

It appears that the combined application of REA, serotyping, MEE, and phage typing make it possible to trace the source of *L. monocytogenes* involved in foodborne listeriosis and to exclude incidental strains.

GROWTH

The nutritional requirements of listeriae are typical of those for many other gram-positive bacteria. They grow well in many common media such as brain heart infusion, trypticase soy, and tryptose broths. Although most nutritional requirements have been described for *L. monocytogenes*, the other species are believed to be similar. At least four B-vitamins are required—biotin, riboflavin, thiamin, and thioctic acid (alpha-lipoic acid; a growth factor for some bacteria and protozoa)—and the amino acids cysteine, glutamine, isoleucine, leucine, and valine are required (13). Glucose enhances growth of all species, and L(+)-lactic acid is produced. While all species utilize glucose by the Embden-Meyerhof pathway, various other simple and complex carbohydrates are utilized by some. *Listeria* spp. resemble most enterococci in being able to hydrolyze esculin, grow in the presence of 10% or 40% (w/v) bile, in about 10% NaCl, 0.025% thallous acetate, and 0.04% potassium tellurite, but unlike the enterococci, they do not grow in the presence of 0.02% sodium azide. Unlike most other gram-positive bacteria, they grow on MacConkey agar. Although iron is important in its in vivo growth, *L. monocytogenes* apparently does not possess specific iron-binding compounds, and it obtains its needs through the reductive mobilization of free iron, which binds to surface receptors (1).

Effect of pH.   Although the listeriae grow best in the pH range 6 to 8, the minimum pH that allows growth and survival has been the subject of a large number of studies. Most research has been conducted with *L. monocytogenes* strains, and whether the findings for this species are similar for other listerial species can only be assumed. In general, some species/strains will grow over the pH range of 4.1 to around 9.6 and over the temperature range of 1° to around 45°C. (Details of these parameters follow.)

In general, the minimum growth pH of a bacterium is a function of temperature of incubation, general nutrient composition of growth substrate, a_w, and presence and quantity of NaCl and other salts or inhibitors. Growth of *L. monocytogenes* in culture media has been observed at pH 4.4 in less than 7 days at 30°C (69), at pH 4.5 in tryptose broth at 19°C (24), and at pH 4.66 in

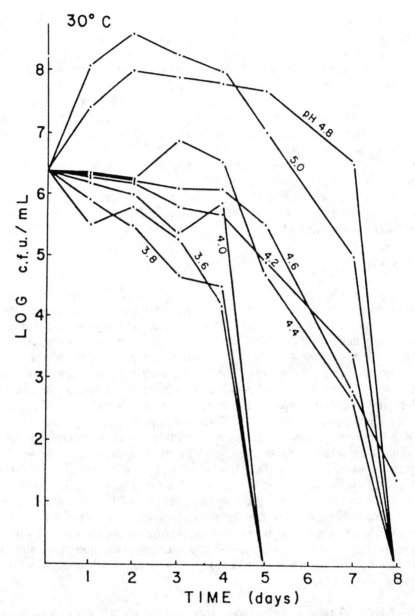

FIGURE 21-2.   Change in cell populations of strain F5069/(4b) in pH-adjusted orange serum incubated at 30°C. Initial cell concentration was 2.2×10^6 cfu/ml.
From Parish and Higgins (163); copyright © 1989 by International Association of Milk, Food and Environmental Sanitarians, used with permission.

60 days at 30°C (33). In the first study, growth at pH 4.4 occurred at 20°C in 14 days and at pH 5.23 at 4°C in 21 days (69). In the second, growth at 4.5 was enhanced by a restriction of oxygen. In the third study, growth of *L. mono-cytogenes* was observed at pH 4.66 in 60 days at 30°C, the minimum at 10°C was 4.83, while at 5°C no growth occurred at pH 5.13. In yet another study, four strains of *L. monocytogenes* grew at pH 4.5 after 30 days in a culture

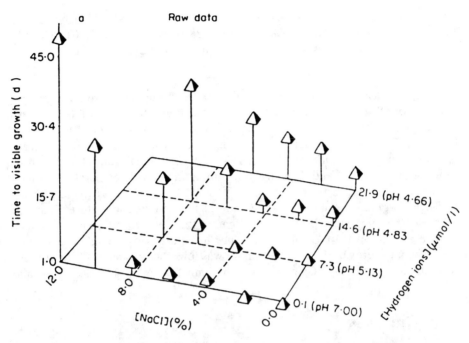

FIGURE 21-3. Effect of salt and hydrogen-ion concentration on the time to reach visible growth of *Listeria monocytogenes*. Three-dimensional scatter plot for the effect of salt (%, x axis) and hydrogen-ion concentration (μmol/l, = axis) on the time to reach visible growth (days, y axis) representing at least a 100-fold increase in numbers of *Listeria monocytogenes* (a) at 30°C and (b) at 10°C. The mean actual values are compared with predicted values determined from polynomial equations 1 and 2.
From Cole et al. (34).

medium incubated at 30°C (163), but no growth occurred at pH 4.0 or lower. pH values of 3.8 to 4.0 were more destructive to one strain than pH 4.2 to 5.0 when held in orange serum at 30°C for 5 days (Fig. 21-2).

When pH of tryptic soy broth was adjusted with various acids, minimum pH for growth of four strains of *L. monocytogenes* was shown to be a function of the acid employed. At the same pH, the antimicrobial activity was acetic acid > lactic acid > citric acid > malic acid > HCl (203). Growth occurred at pH 4.6 at 35°C between 1 and 3 days, and some strains grew at pH 4.4. The growth of two strains of this species in cabbage juice containing no added NaCl has been observed at pH 4.1 within 8 days when incubated at 30°C, but death occurred at 30°C when the organism was inoculated into sterile cabbage juice adjusted to pH less than 4.6 with lactic acid (35).

Combined Effect of pH and NaCl. The interaction of pH with NaCl and incubation temperature has been the subject of several studies (34, 35). The latter authors used factorially designed experiments to determine the interaction of these parameters on the growth and survival of a human isolate (serovar 4b); some of their findings are illustrated in Figure 21-3. At pH 4.66, time to visible growth was 5 days at 30°C with no NaCl added, 8 days at 30°C with 4.0% NaCl, and 13 days at 30°C with 6.0% NaCl, all at the same pH (34).

Growth at 5°C occurred only at pH 7.0 in 9 days with no added NaCl, but 15 days were required for 4.0% NaCl and 28 days for 6.0% NaCl. The pH and NaCl effects were determined to be purely additive and not synergistic in any way.

Effect of Temperature.   The mean minimum growth temperature on trypticase soy agar of 78 strains of *L. monocytogenes* was found to be 1.1° ± 0.3°C, with a range of 0.5° to 3.0°C (103). Two strains grew at 0.5°C, and 8 grew at or below 0.8°C in 10 days as determined with a plate-type continuous temperature gradient incubator. With 22 other strains (19 *L. innocua* and 1 each of *L. welshimeri*, *L. grayi*, and *L. murrayi*), minimum growth temperature ranged from 1.7° to 3.0°C, with a mean of 1.7° + 0.5°C (103). That the *L. monocytogenes* strains had about a 0.6°C lower minimum temperature than the other species suggested to these investigators that the hemolysin may enhance growth and survival of *L. monocytogenes* in cold environments even though the growth of serovars 1/2a, 1/2b, and 4b was lower at around 3.0°C than those with 0I antigens. The maximum growth temperature for listeriae is around 45°C.

DISTRIBUTION

The Environment

The listeriae are widely distributed in nature and can be found on decaying vegetation and in soils, animal feces, sewage, silage, and water (78). The finding of *L. monocytogenes* on 11 of 12 farms only during the spring months suggested to one group that dead vegetation is a better source than green or recently dead vegetation (218). This species was found in each of over 50 samples of sewage, sewage sludge, and river water examined in the United Kingdom (214). The numbers found were often higher than salmonellae, and on two occasions *L. monocytogenes* was found when no salmonellae could be detected. In general, listeriae may be expected to exist where the lactic acid bacteria, *Brochothrix*, and some coryneform bacteria occur. Their association with certain dairy products and silage is well known, as is the association with these products of some other lactic acid producers. In a study of gull feces, rooks, and silage in Scotland, gulls feeding at sewage works had a higher rate of carriage than those elsewhere, and fecal samples from rooks generally had low numbers of listeriae (54). *L. monocytogenes* and *L. innocua* were most often found with only one sample containing *L. seeligeri*. In the same study, *L. monocytogenes* and *L. innocua* were found in 44% of moldy silage samples and in 22.2% of big bale silage. In Denmark, 15% of 75 silage samples were positive for *L. monocytogenes*, as was 52% of 75 fecal samples from cows (199). The organism was found in silage with pH above and below 4.5. *L. monocytogenes* was isolated from 8.4 to 44% of samples taken from grain fields, pastures, mud, animal feces, wildlife feeding grounds, and related sources (216). Its survival in moist soils for 295 days and beyond has been demonstrated (217). From California coastal waters, 62% of 37 samples of fresh or low-salinity water and 17.4% of 46 sediment samples were positive for *L. monocytogenes*, but none could be recovered from 35 oyster samples (33).

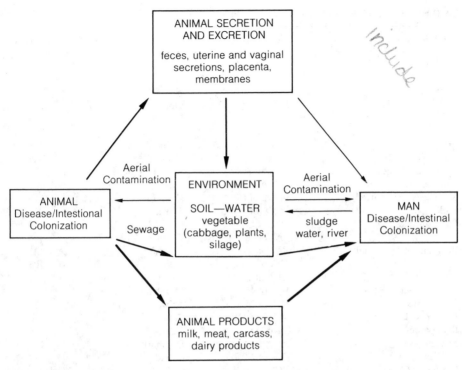

FIGURE 21-4.   Ways in which *L. monocytogenes* is disseminated in the environment, animals, foods, and humans.
From Audurier and Martin (4).

Some of the ways in which *L. monocytogenes* is disseminated throughout the environment, along with the many sources of the organism to humans, are illustrated in Figure 21-4.

Foods and Humans

How *L. monocytogenes* is transmitted among susceptible hosts continues to be the subject of much debate. As far as human cases are concerned, it is clear that this organism exists in many foods; synopses of findings of its occurrence in a variety of products from different countries are presented in Table 21-3. The incidence is as high as 45% in raw milk, 95% in pork, 60% in raw poultry, 79% in ground beef, and 30% in some vegetables. In 650 samples of raw milk, no *L. monocytogenes* was found in 100 from California but 3 and 12% of the respective samples from Ohio and Massachusetts were positive (135). In an extensive review of this organism in meats and poultry, Johnson et al. (99) noted that it has been found in beef, pork, lamb, sausage, and ready-to-eat meats, as well as poultry. Over a 39-month period in the United States, 7.1% of 1,727 raw beef samples collected throughout the country were positive for *L. monocytogenes*, and over a 21-month period 19.3% of 3,700 raw broiler necks and backs were positive (81). From the same survey, 2.8% of a variety of ready-to-eat meats from 4,105 processing plants throughout the United States were positive for the organism. While 15% of 180 samples of turkey

TABLE 21-3.   Prevalence of *L. monocytogenes* in a Variety of Foods

Food Products	Incidence	Source of Products	Reference
Raw milk	1% of 1,004	Germany	187
	45.3% of 95	Spain	42
	4.4% of 137	Netherlands	10
	5.4% of 315	Ontario	200
	12% of 121	United States	87
	4% of 200	Nebraska	128
	4.2% of 350	United States	136
Ice cream, mixes	0.4% of 530	Canada	49
Soft/semisoft cheeses	0.5% of 374	1 of 14 countries	52
Soft cheeses	10% of 222	Britain plus 6 others	169
Soft cheeses (imported)	14.5% of 69	From France	10
Pork	58.5% of 34	Taiwan	223
Pork, ground	95% of 19	Canada	49
	80% of 30	Germany	185
Beef and pork, ground	36% of 100	Austria	23
	44% of 117	Germany	105
Raw beef	7.1% of 1,727	United States	81
Frozen ground beef	49% of 41	Maryland	145
Beef steaks	24% of 25	Taiwan	223
Frozen beef	26.2% of 149	France	156
Beef patties	8% of 62	France	125
Minced beef	28% of 67	Denmark	199
	77% of 22	Canada	49
Frozen seafoods	2.6% of 57	9 of 12 countries	215
Seafoods	10.5% of 57	Taiwan	223
Chicken carcasses	50% of 16	Taiwan	222
Broiler chickens	23% of 90	S.E. United States	6
Broiler necks/backs	19.3% of 3,700	United States	81
Poultry neck skin	47% of 17	Denmark	199
Raw chicken	57% of 35	United Kingdom	122
	33% of 22	Maryland	145
Raw chicken, duck	14.7% of 68	United Kingdom	72
Chicken legs	56% of 16	Canada	49
Fresh/frozen chicken	60% of 100	United Kingdom	169
Turkey parts	38% of 50	Taiwan	223
	15% of 180	California	66
Liquid whole egg	5% of 42	United States	123
Precooked refrigerated chick.	26.5% of 102	United Kingdom	113
Meat roasts, retail	4.5% of 110	Wisconsin	98
Dry raw sausage	21.6% of 37	France	156
Fresh raw sausage	10% of 120	France	156
Pork sausage	52% of 23	Maryland	145
Raw dry-cured sausage	15.6% of 96 lots	Canada	51
Mettwurst	59% of 30	Germany	185
Mettwurst	23% of 100	Austria	23
Frozen semiready foods	34% of 68	Taiwan	223
Fresh salads	7% of 60	Britain	197
Uncooked deli products	6.8% of 480	France	125
Vegetables	12.2% of 49	Taiwan	223
Potatoes	25.8% of 132	Minnesota	88
Radishes	30.3% of 132	Minnesota	88
Cucumbers	10.9% of 92	Minnesota	88
Cabbage	2.2% of 92	Minnesota	88
Mushrooms	12% of 92	Minnesota	88
Lettuce	1.1% of 92	Minnesota	88

parts contained *L. monocytogenes*, none of 60 turkey livers was positive (66). The latter authors showed that the prevalence of listeriae increased during processing, where from chilling to packaging to retail the percentage of samples that contained listeriae were 4, 13, and 23%, respectively. Of the six vegetables listed in Table 21-3, potatoes were most often positive, with radishes being next. No *L. monocytogenes* was found in 92 tests of broccoli, carrots, cauliflower, and tomatoes (88). *L. monocytogenes* could not be recovered from 10 vegetables in another study following cold enrichment for up to 8 weeks (166) or from 110 samples of four vegetables (49).

The most common serovar in meat products from six countries was 1/2 (99). Serotype 4 was isolated from meat products in five countries, and serotype 3 was recovered from products in only two countries. Serotypes 1/2 and 4 are represented more than others in the nonmeat products in Table 21-3. Serotype 1/2 was found more often than serotype 4 in raw milk (135, 168) and cheese (169), and serotype 1 has been found in seafoods (215, 221) and vegetables (88). Serovar 4b was associated with a cluster of human cases in Boston where raw vegetables appeared to be the source (89), while from potatoes and radishes, serovars 1/2a and 1/2 were the most frequently occurring (88). The three most prevalent serovars isolated from foods, in decreasing order, are 1/2a, 1/2b, and 4b, while from human listeriosis 4b, 1/2a, and 1/2b are the most prevalent (36).

Of serovars isolated from humans, 59% of 722 *L. monocytogenes* isolates in Britain were 4b, followed by 18%, 14%, and 4%, respectively, for 1/2a, 1/2b, and 1/2c (146). From pathological specimens throughout the world, 98% of isolates belonged to the following serovars: 1/2a, 1/2b, 1/2c, 3a, 3b, 3c, 4b, and 5 (192).

With respect to other *Listeria* spp. in foods, *L. innocua* is rather common in meats, milk, frozen seafoods, semisoft cheese, whole egg samples, and vegetables. In general, it is the most prevalent listerial species found in dairy products (132). It was found in 8 to 16% of raw milk where its presence was reported, in 46% of 57 frozen seafoods (215), and in 36% of 42 liquid whole egg samples (123). In the last study, it was the most frequently found listerial species and occurred in all 15 positive samples. In one study, this species was found in 42% of beef and poultry examined, and overall it was found two times as often as *L. monocytogenes* (199). It was found in 83% of mettwurst and 47% of pork tested in Germany (185) and in 22% of fresh salads in Britain (197). On the other hand, *L. innocua* was found in only 2% of 180 raw turkey parts (64). Although no *L. monocytogenes* could be detected in tropical fish and fishery products, 8 of 24 (33%) were positive for *L. innocua* (59). It was the next most often found in potatoes, cucumbers, mushrooms, and lettuce after *L. monocytogenes* (88).

L. welshimeri has been found in raw milk (from 0.3 to 3% of samples), meat roasts, vegetables, and turkey meat. In the latter, it was found in 16% of samples and thus was the most prevalent of listerial species (66). It was found in 24% of mettwurst and 30% of pork in Germany (185) and in frozen ground beef and deli products in France (156). The only other species of *Listeria* reported from foods are *L. grayi*, found in 89.5% of raw milk in Spain (42); also in beef and poultry; and *L. seeligeri*, found in raw milk, vegetables, cabbage, radishes, pork, and mettwurst.

Incidence

Because of the need for culture enrichments, numbers of *L. monocytogenes*/g or ml in foods are often not reported. For bulk tank raw milk in the United States (specifically in California and Ohio), the number was estimated to be about 1 cell/ml or less (136), while from a study in Canada less than 20×10^3 were found in raw milk (200). The highest number of cells found to be shed by a naturally or artificially infected cow is approximately 10^4/ml. One of 34 raw milk samples was found to contain more than 6,000/g (40). From soft cheeses studied in Britain, most contained less than 10^2/g, but in 9 of 12 imported from another country, numbers from 10^4–10^5/g were found (168). A naturally contaminated sample of brie cheese was found to contain more than 2.4×10^5/g (207), a temperature-abused sample of ricotta cheese contained 3.6×10^6/g, and in semisoft cheeses, 5.6×10^6/g were found (40). In 69 samples of imported soft cheese, Beckers et al. (10) found 10^3–10^6/g in seven samples, and another three samples were positive following enrichment. In ground beef and pork in Germany, most probable numbers (MPN) ranged between 1 and 10^3/g in one study (105), and in another 81% of positive samples contained less than 110/g by MPN (23). Turkey franks were implicated in a case of listeriosis in Texas (see below), and of two unopened packages, the number found was less than 0.3/g by MPN (219). Numbers found in meat roasts ranged from less than 10 to 10/g (98), while the numbers found in 2 of 15 samples of liquid whole eggs were 1 and 8/ml (123).

THERMAL PROPERTIES

Although *L. monocytogenes* cells were not isolated in the 1983 Massachusetts outbreak of human listeriosis in which pasteurized milk was incriminated, the adequacy of standard milk pasteurization protocols to destroy this organism was brought into question. Since 1985, a large number of studies have been reported on its thermal destruction in dairy products. *D* values have been determined on many strains of *L. monocytogenes* in whole and skim milk, cream, ice cream, and various meat products. Since this organism is an intracellular pathogen, several studies were undertaken to determine its relative heat resistance inside and outside phagocytes. Overall, standard pasteurization protocols for milk are adequate for destroying *L. monocytogenes* at levels of 10^5–10^6/ml, whether freely suspended or in intracellular state. Some of the specific findings are presented below.

Dairy Products

A summary of thermal *D* and *z* values for some *L. monocytogenes* strains is presented in Table 21-4. The *D* values indicate that the high temperature–short time (HTST) protocol for milk (71.7°C for 15 sec) is adequate to reduce normally existing numbers of this organism below detectable levels. The vat or low temperature–long time (LTLT) pasteurization protocol (62.8°C for 30 min) is even more destructive (see Chap. 14). Employing the Scott A strain (serovar 4b from the Massachusetts outbreak), *D* values ranged from 0.9 to 2.0 sec with *z* values of 6.0° to 6.5°C. The F5069 strain (serovar 4b) appeared

TABLE 21-4. Summary of Some Findings on the Thermal Destruction of *L. monocytogenes*

Strains Tested/State	Number of Cells/ml	Heating Menstrum	Heating Temperature (°C)	D Value (sec.)	z Value (°C)	Reference
Scott A, free suspension	~10⁵	Sterile skim milk	71.7	1.7	6.5	21
	~10⁵	Sterile whole milk	71.7	2.0	6.5	21
	~10⁵	Sterile whole milk	71.7	0.9	6.3	22
Scott A, intracellular	~10⁵	Whole raw milk	71.7	1.9	6.0	27
Scott A, free suspension	~10⁵	Whole raw milk	71.7	1.6	6.1	27
F5069, intracellular	~10⁶	Sterile whole milk	71.7	5.0	8.0	26
F5069, free suspension	~10⁶	Sterile whole milk	71.7	3.1	7.3	26
Scott A, free suspension	~10⁵	Ice cream mix	79.4	2.6	7.0	21
	~10⁸	pH 7.2 phos. buffer	70.0	9.0	—	20
	~10⁸	pH 5.9 meat slurry	70.0	13.8	—	20
	~10⁷	Liquid whole egg	72.0	36.0	7.1	58
Ten strains	~10⁷	Irradiated ground meats	62.0	61.0	4.92	48
Chicken/meat isolate	~10⁵	Beef	70.0	7.2		140
	~10⁵	Minced chicken	70.0	6.7		140

to be a bit more heat resistant than Scott A from these results, although Scott A was the most heat resistant of three other strains evaluated, not including F5069 (22).

The thermal resistance of *L. monocytogenes* is not affected by intracellular position (26, 134, 159). With Scott A freely suspended in whole raw milk at mean levels of 2.6×10^5 cfu/ml and heating at 71.7°C for 15 sec, no survivors could be found after five heating trials (134). In seven heating trials with Scott A engulfed in vitro by bovine phagocytes, no survivors could be detected with a mean number of 5×10^4 cfu/ml. Further, these investigators experimentally infected cows with Scott A and were still unable to find survivors following 11 pasteurization trials at 71.7°C for 15 sec with numbers of Scott A that ranged from 1.4×10^3 to 9.5×10^3 cfu/ml (134). Employing five strains of *L. monocytogenes* in whole milk, skim milk, and 11% nonfat milk solids, Donnelly and Briggs (43) found that composition did not affect heat destruction and that at 62.7°C the *D* values were 60 sec or less. The five strains employed included serotypes 1, 3, and 4. When milk that was naturally contaminated with a serotype 1 strain at around 10^4/ml was subjected to a HTST protocol at temperatures ranging from 60° to 78°C, no viable cells could be detected at processing temperatures of 69°C or above (50). In their review of the early studies on the thermal resistance of *L. monocytogenes* in milk, Mackey and Bratchell (141) concluded that normal pasteurization procedures will inactivate this organism but that the margin of safety is greater for the vat protocol (LTLT) than the HTST protocol. Their mathematical model predicted a 39 *D* for vat and a 5.2 *D* for HTST.

While the reports noted indicate that the standard pasteurization protocols are adequate to destroy *L. monocytogenes* in milk, some investigators have reported cell survival under varying conditions of heating; these, along with the specific methods employed, have been reviewed by Farber (48) and Lovett et al. (134).

Nondairy Products

For liquid whole egg and meat products, *D* values are generally higher than for milk, a fact not unpredicted considering the effect of proteins and lipids on the thermal resistance of microorganisms (discussed in Chap. 14). For one strain of *L. monocytogenes* isolated from a chicken product, *D* values at 70°C were 6.6 to 8.4 sec; they were essentially the same in beef and two poultry meats (140). In one study, viable cells could be recovered by enrichments from eight of nine samples following heating in ground beef to 70°C (20). In a study of blue crabmeat, strain Scott A at levels of about 10^7 had a *D* value of 2.61 min with *z* of 8.4°C, indicating that the crabmeat pasteurization protocol of 30 min at 85°C was adequate to render the product safe from this organism (86). Processing frankfurters to an internal temperature of 160°F (71.1°C) has been shown to effect at least a 3-log cycle reduction of strain Scott A (225). The cooking of meat products to an internal temperature of 70°C for 2 min will destroy *L. monocytogenes* (64, 140, 141).

In the sausage-type meat employed by Farber (48), noted in Table 21-4, the *D* value at 62°C was 61 sec, but when cure ingredients were added, the *D* value increased to 7.1 min, indicating some heat-protective effects of the cure compounds, which consisted of nitrite, dextrose, lactose, corn syrup, and 3% (w/v) NaCl. An approximate doubling in D value in ground beef containing 30% fat, 3.5% NaCl, 200 ppm nitrite, and 300 ppm nitrate was found by Mackey et al. (140), who attributed the increased heat resistance to the 3.5% NaCl. The destruction of strain Scott A by microwave cooking was investigated by Lund et al. (138) where more than 10^7 cells/g were placed in chicken stuffing and 10^6–10^7/g on chicken skin. By use of a home-type microwave unit, the adequacy of heating to an internal temperature of 70°C for 1 min was shown to give a 6-log reduction in numbers. The thermal destruction of *L. monocytogenes* is similar to that of most other bacteria relative to pH of suspending menstrum where resistance is higher at pH values closer to 7.0 than values in the acid range. This was demonstrated in cabbage juice where D values were higher at pH 5.6 than at 4.6 (15).

Effect of Sublethal Heating on Thermotolerance

It is unclear whether sublethal heating of *L. monocytogenes* cells renders them more resistant to subsequent thermal treatments. Some investigators have reported no effect (22, 25), and others have reported increased resistance (47, 53, 130). In one study, the heat shocking of strain Scott A at 48°C for 20 min resulted in a 2.3-fold increase in D values at 55°C (130). In another study employing Scott A in broth and ultrahigh temperature (UHT)-treated milk, an increase in heat resistance was observed following exposure to 48°C for 60 min and subsequent exposure to 60°C (53). Finally, in a study employing 10 strains at a level of about 10^7/g in a sausage mix and heat shocking at 48°C for 30 or 60 min, no significant increase in thermotolerance was observed at 62° or 64°C, but those shocked for 120 min did show an average 2.4-fold increase in D values at 64°C (47). In this study, the thermotolerance was maintained for at least 24 h when the cells were stored at 4°C. If sublethal heating does lead to greater thermoresistance, it would not pose a problem for milk that contains fewer than 10 cells/ml assuming that a two- to threefold increase in D value occurs.

VIRULENCE PROPERTIES

Of the seven listerial species, *L. monocytogenes* is the pathogen of concern for humans. Although *L. ivanovii* can multiply in the mouse model, it does so to a much less degree than *L. monocytogenes*, and up to 10^6 cells caused no infection in the mouse (92). *L. innocua*, *L. welshimeri*, and *L. seeligeri* are nonpathogens, although the last produces an hemolysin. The most significant virulence factor associated with *L. monocytogenes* is listeriolysin O.

Listeriolysin O and Ivanolysin O

In general, the pathogenic/virulent strains of *L. monocytogenes* produce β-hemolysis on blood agar (75) and acid from rhamnose but not from xylose (82). Strains whose hemolysis can be enhanced with either the prepurified exo-substance or by direct use of the culture are pathogenic (198). Regarding hemolysis, the evidence is overwhelming that all virulent strains of this species produce a specific substance that is responsible for β-hemolysis on erythrocytes and the destruction of phagocytic cells that engulf them. The substance in question has been designated **listeriolysin O (LLO)** and has been shown to be highly homologous to **streptolysin O (SLO)** and **pneumolysin (PLO)** (148, 149). It has been purified and shown to have a molecular weight of 60,000 daltons and to consist of 504 amino acids (68, 148). It is produced mainly during the exponential growth phase, with maximum levels after 8 to 10 h of growth (67). LLO has been detected in all strains of *L. monocytogenes*, including some that were nonhemolytic, but not in *L. welshimeri*, *L. grayi*, or *L. murrayi*.

L. ivanovii and *L. seeligeri* produce thiol-dependent exotoxins that are similar but not identical to LLO. Large quantities are produced by *L. ivanovii* but only small quantities by *L. seeligeri* (67). The *L. ivanovii* thiol-activated cylolysin is **ivanolysin O**. Antiserum raised to the *L. ivanovii* product cross-reacts with that from *L. monocytogenes* and SLO (117). Employing transposon-induced mutants, ILO-deficient mutants have been shown to be avirulent in mice and chick embryos (2). The gene that codes for LLO production is chromosomal and has been designated *hlyA* (148). The *hlyA* gene locus is present in all three hemolytic species, but there are differences among them (36). LLO has been cloned into the cosmid vector pHC79 and expressed in *E. coli* HG101 (213). The shuttled DNA consisted of a 40 kb (kilobase) fragment from the chromosome of serovar 1/2, and 12 of 2,000 recombinant clones produced LLO in culture media (213).

Purified LLO has been shown to share in common with SLO and PLO the following properties: activated by SH-compounds such as cysteine, inhibited by low quantities of cholesterol, and common antigenic sites as evidenced by immunological cross-reactivity. Unlike SLO, LLO is active at pH 5.5 but not at pH 7.0, suggesting the possibility of its activity in macrophage phagosomes (phagolysosomes). Its LD_{50} for mice is about 0.8 µg, and it induces an inflammatory response when injected intradermally (68).

The synthesis of LLO in serovar 1/2a strains has been shown to occur under heat shock conditions (201), and it appears to be the only major extracellular protein synthesized under these conditions (202). About 5 of 15 heat shock proteins of listeriae are coinduced with listeriolysin in all *L. monocytogenes* strains but not in other listerial species (201). Proof of the role of LLO in virulence of *L. monocytogenes* has been shown by molecular genetic analyses. A nonhemolytic mutant was created by inserting a copy of transposon Tn917 in *hlyA* of the chromosome of a hemolytic (and virulent) strain, and the mutant was avirulent in mice (37). When the avirulent strain was then transformed with a plasmid carrying only *hlyA*, hemolysin production and virulence were restored. Thus, LLO appears to be active only in maintaining intracellular survival of *L. monocytogenes* and not in entry of these cells into phagocytes (32, 120)

Indications of the role of an extracellular substance in the virulence of *L. monocytogenes* were presented many years ago. In the early 1960s, a cytolytic factor was shown to be involved in the in vitro interaction of sheep peritoneal exudate cells with *L. monocytogenes*, and it was suggested that the factor may be produced by the bacterium when inside phagocytes (157, 158). The hemolysin was characterized by several groups, including Girard et al. (71)

and Njoku-Obi et al. (158). The release by the hemolysin of β-glucuronidase and acid phosphatase from rat and rabbit lysosomes and its increased activity by reducing agents and its abatement by cholesterol were demonstrated in the 1960s (97, 114). Although the hemolysin has long been suspected of playing a key role in *L. monocytogenes* virulence, the ultimate resolution of its role was the result of a series of findings involving molecular biology and genetic techniques.

In 1986, a 26-kb conjugative transposon (Tn1545) was transferred to a hemolytic *L. monocytogenes* strain at a frequency of $1/10^8$, and the result was a loss of hemolysin production and virulence (61). The nonhemolytic mutants did not grow in mouse assays and were eliminated rapidly from spleen and liver. In related studies, the genes for LLO were shown to be inactivated by insertion of transposon Tn916 in the DNA of *L. monocytogenes* (108) and also by transposon Tn917 (37). These transposon insertions are located within *hlyA* in several avirulent mutants. The ultimate proof of the DNA locus of the *hlyA* gene and its function was demonstrated when it was cloned from a weakly hemolytic 1/2a serovar into *E. coli* strain K-12 (126).

Intracellular Invasion

When *L. monocytogenes* is contracted via the oral route, it apparently colonizes the intestinal tract by mechanisms that are poorly understood. From the intestinal tract, the organism invades tissues, including the placenta in pregnant women, and enters the bloodstream, from which it reaches other susceptible body cells. As an intracellular pathogen, it must first enter susceptible cells, and then to possess means of replicating within these cells. In the case of phagocytes, entry occurs in two steps: directly into phagosomes and from the phagosomes into the phagocyte's cytoplasm. It seems well established that LLO-producing strains can invade certain body tissues, while negative strains cannot, and yet LLO is not involved in phagocyte entry. The latter is aided by a separate extracellular invasion-associated protein designated p60 by Kuhn and Goebel (119). While mutants that lacked LLO did not survive inside phagocytes, mutants deficient in p60 did survive if their entry was otherwise aided (202). The p60 protein has been cloned in *E. coli*, sequenced, shown to consist of 484 amino acids, and to have a molecular weight of 60,000 daltons (116). In one study using hemolytic and non-hemolytic *L. monocytogenes* strains administered intragastrically (i.g.) to mice, the hemolytic strains (but not the nonhemolytic strains) were later found in mesenteric lymph nodes, spleen, and liver (178).

Further evidence that LLO is not involved in cell invasion is seen with *L. ivanovii* and *L. seeligeri*, which are hemolytic and yet are essentially avirulent and noninvasive for humans (119). Employing a human enterocytelike cell line (Caco-2, a human colon carcinoma cell line), only virulent species of *L. monocytogenes* and *L. ivanovii* induced their own phagocytosis, while *L. seeligeri*, *L. welshimeri*, and *L. innocua* did not (60). Further, the non-hemolytic mutants remained inside phagosomes and consequently did not replicate, while the virulent strains were released from the phagosomes into the cytoplasm following disruption of the vacuole membrane, presumably by

LLO (11, 60, 171). It was shown in a more recent study of noninvasive mutants of *L. monocytogenes* that lacked both LLO and an extracellular phospholipase but possessed p60 that the mutants were still deficient in their capacity to invade phagocytes (106). The extracellular phospholipase was noted by Jenkins and Watson (96) 20 years earlier. These findings suggest that the factors that are involved in the entry of *L. monocytogenes* into host cells are more complicated than once thought.

Somewhat in contrast, the role of LLO in virulence is not too clear from some studies. It has been suggested that its role in virulence may not be associated with the cytolytic function. When a mouse infection model was employed, hemolytic activity of *L. monocytogenes* was lost by transposon mutagenesis, accompanied by a loss of mouse virulence, but a direct relationship between in vitro production of hemolysin and virulence was not observed (107). In this study, levels of LLO did not correspond to infections in mice since the ability of hyperhemolytic mutants to grow in host tissues was the same as for the normal parental strains.

While the invasion-associated protein designated p60 is undoubtedly involved in the engulfment of *L. monocytogenes* cells by phagocytes, this entire process is not entirely clear. In one study where macrophages were employed in vitro, LLO and *L. monocytogenes* inhibited antigen processing and preparation by the macrophages while avirulent strains did not (32). The inhibition of antigen processing by LLO and its role in the T cell response has been shown by others (9). Both virulent and avirulent *L. monocytogenes* strains express antigenic epitopes that are recognized by protective T cells, although the avirulent strains do not persist long enough to induce cellular immune responses (110). The antigenic epitope that elicits the production of T cells has been shown to be LLO. A sublethal infection of *L. monocytogenes* elicited in a mouse a subpopulation of antilisterial T cells that reacted specifically with purified LLO (12). Heated LLO was not recognized by the antibody, but formalin-treated cells were. Berche et al. (12) speculated that the intracellular production of LLO induces macrophage damage and thus promotes a better antigenic presentation through alteration of antigen processing and persistence of immunogenic material in most tissues.

Once a virulent *L. monocytogenes* cell has been engulfed by a macrophage, it appears that LLO allows the cell to break out of the phagosome into the cell's cytosol. That LLO is active in the pH range of 5.5 suggests that it can be active in phagosomes. The ability of *L. monocytogenes* to produce catalase plays no role in their virulence in mice since transposon-induced, catalase-negative mutants grew just as well as parental catalase-positive strains (124). Once *L. monocytogenes* cells are released into the cytosol of a macrophage, they become surrounded by a coat of actin filaments within a few hours. The actin envelope reorganizes to form tails of F-actin, which facilitates intracellular movement and subsequent spread to adjacent macrophages (118, 153). Concomitant with cell-to-cell spreading, internal listerial cells carry out their lysis of macrophages with the aid of LLO. As a pore-forming SH-activated cytolysin, of which SLO is the prototype, it has been suggested that LLO binds to cholesterol-containing membranes, oligomerizes to form toxin monomers that span the membrane, and thus result in the leakage of cytoplasmic molecules that lead to cell death in a way similar to that of SLO (164, 170).

Monocytosis-Producing Activity

An interesting yet incompletely understood part of the *L. monocytogenes* cell is a lipid-containing component of the cell envelope that shares at least one property with the lipopolysaccharide (LPS) that is typical of gram-negative bacteria. In gram-negative bacteria, LPS is located in the outer membrane, but listeriae and other gram-positive bacteria do not possess outer membranes. It was shown several decades ago that phenol-water extracts of *L. monocytogenes* cells induce the production of monocytes, and it was because of this mono-cytosis-producing activity (MPA) factor that the organism was given the species name *monocytogenes*. It has been studied by a large number of investigators (63, 89, 143, 144, 174, 196, 204, 222). This LPS-like fraction accounts for about 6% of the dry weight of cells and is associated with the plasma membrane (222). It has a molecular weight of about 1,000 daltons, contains no amino acids or carbohydrates, and stimulates only mononuclear cells (62). It possesses low tissue toxicity and is serologically inactive (204), but it kills macrophages in vitro (62). It has been shown to share the following properties with LPS: pyrogenic and lethal in rabbits, produces a localized Schwartzman reaction, contains acylated hydroxy fatty acids, produces a positive reaction with the *Limulus* amoebocyte lysate (LAL) reagent, contains 2-keto-3-deoxyoctonic acid (KDO), and contains heptose (222). Regarding its LAL reactivity, 1 µg/ml was required to produce a positive reaction (196), while the same can be achieved with picogram quantities of LPS.

MPA elicits the production of immunoglobulin M (IgM) antibodies, and that prepared from a heat-killed serovar 4b strain was shown to be responsible for the transient cold agglutinin syndrome in rabbits (222). By making chloroform-methanol or saline extracts of a serotype 1 strain, another group of investigators recovered both MPA and an immunosuppressive activity fraction (67). As little as 10 µg of the MPA factor caused a ninefold increase in circulating monocytes in mice. On the other hand, Hether and Jackson (89) were unable to confirm the LPS similarity of the phenol-water extract they obtained from a serotype 1 strain, and it was the view of these investigators that lipo-teichoic acid (LTA) is the material that is extracted—material that is the characteristic gram-positive amphiphile. LTA was found by one group of investigators to constitute 8.5 to 11.5 mg/g of *L. monocytogenes* cells (210), and it is known to produce positive reactions in the LAL test. Thus, the precise role that the MPA factor plays in the overall virulence of *L. monocytogenes* remains unclear, but some fractions have been shown to be mitogenic for bone marrow cells and also to act as immunosuppressive agents. It seems inconsistent that an intracellular pathogen would elicit the production of mononuclear cells that will ultimately effect its destruction, but since MPA kills macrophages in vitro, it may have the same function in vivo.

Sphingomyelinase

L. ivanovii is known to be infectious for sheep, in which it causes abortions, and to be a prolific producer of hemolysin on sheep erythrocytes. It has been shown to possess an LLO-like hemolysin (ILO), **sphingomyelinase**, and **lecithinase** (117). Sphingomyelinase has a molecular weight of 27,000 daltons

(212). While the LLO-like agent is reponsible for the inner complete zone of hemolysis on sheep erythrocytes, the halo of incomplete hemolysis that is enhanced by *Rhodococcus equi* appears to be caused by the two enzymes noted. In one study, a mutant defective in sphingomyelinase and another protein exhibited lowered virulence than wild-type strains (2).

ANIMAL MODELS AND INFECTIOUS DOSE

The first animal model employed to test the virulence of *L. monocytogenes* was the administration of a suspension of cells into the eye of a rabbit or guinea pig (Anton's test) where 10^6 cells produced conjunctivitis (3). Chicken embryos have been studied by a large number of investigators (186, 209). Inocula of 100 cells of *L. monocytogenes* into the allantoic sac of 10-day-old embryos led to death within 2 to 5 days, and the LD_{50} was less than 6×10^2 cells for virulent strains. *L. ivanovii* is also lethal by this method. Injections of 100 to 30,000 cells/egg into the chorioallantoic membrane of 10-day-old chick embryos resulted in death within 72 h compared to about 5 days for mice (209). While Anton's test and chick enbryos may be used to assess the relative virulence of strains of listeriae, the mouse is the model of choice for the additional information that it gives relative to cellular immunity.

Not only is the mouse the most widely used laboratory animal for virulence studies of listeriae, it is widely used in studies of T cell immunity in general. This model is employed by use of normal, baby, juvenile, and adult mice, as well as a variety of specially bred strains such as athymic (T cell deficient) nude mice. Listerial cells have been administered i.p., i.v., and i.g. When normal adult mice are used, all smooth and hemolytic strains of *L. monocytogenes* at levels of 10^3–10^4/mouse multiply in the spleen (92). With many strains, inocula of 10^5–10^6 are lethal to normal adult mice, although numbers as high as 7×10^9 have been found necessary to produce an LD_{50}. A low of 50 cells for 15-g mice has been reported (see below).

While the i.p. route of injection is often used for mice, i.g. administration is employed to assess gastrointestinal behavior of listeriae. The administration of *L. monocytogenes* to 15-g mice by the i.g. route produced more rapid infection and more deaths in the first 3 days of the 6-day test than by i.p. (168). By this method, the approximate 50% lethal dose (ALD_{50}) ranged from 50 to 4.4×10^5 cells for 15 food and clinical isolates of *L. monocytogenes* (168). Six- to 8-week-old mice were given i.p. and oral challenges of a serovar 4b strain by one group to study their effect under normal and compromised states. Cells were suspended in 11% nonfat milk solids and administered to four groups of mice: normal, hydrocortisone treated, pregnant, and cimetidine treated. Minimum numbers of cells that caused a 50% infectious dose (ID_{50}) were 3.24–4.55 log cfu for normal mice, 1.91–2.74 for the cortisone treated, and 2.48 for pregnant mice (76). The ID_{50} for those administered cimetidine was similar to the normals. These investigators found no significant difference between i.p. and i.g. administration relative to ID_{50}. Employing neonatal mice (within 24 h of birth), the LD_{50} by i.p. injection of *L. monocytogenes* was 6.3×10^1 cfu, but for 6- to 8-week-old adult mice, LD_{50} was 3.2×10^6 by the same route of administration (31). The neonatal mice were protected against a lethal dose of *L. monocytogenes* when γ-interferon was injected (see below). With 15–20 g Swiss mice treated with carrageenan, LD_{50} was found to range from about 6 to 3,100 cfu (41).

When nude mice are challenged with virulent strains of *L. monocytogenes*, chronic infections follow, and for baby mice and macrophage-depleted adult mice, virulent strains are lethal. With the adult mouse model, rough strains of *L. monocytogenes* multiplied only weakly, and a weak immunity was induced; baby mice were killed, but nude mice survived (92). In nonfatal infections by virulent strains, the organisms multiply in the spleen, and protection against reinfections results regardless of the serovar used for subsequent challenge (92).

Overall, studies with the mouse model confirm the greater susceptibility to *L. monocytogenes* of animals with impaired immune systems than normal animals as is the case with humans. The correspondence of minimal infectious doses for normal adult mice to humans is more difficult. It has been suggested that levels of *L. monocytogenes* less than 10^2 cfu appear to be inconsequential to healthy hosts (76). From the nine cheeses reported by Gilbert and Pini (70) that contained 10^4–10^5/g of *L. monocytogenes*, no known human illness resulted.

INCIDENCE AND NATURE OF THE LISTERIOSIS SYNDROMES

Incidence

While *L. monocytogenes* may have been described first in 1911 by Hülphers (103), its unambiguous description was made in 1923 by Murray et al. (154). Since that time it has been shown to be a pathogen in over 50 mammals, including humans, in addition to fowls, ticks, fish, and crustaceans. The first human case of listeriosis was reported in 1929, and the disease has since been shown to occur sporadically throughout the world. *L. monocytogenes* is the etiologic agent of about 98% of human and 85% of animal cases (146). At least 3 human cases have been caused by *L. ivanovii* and one by *L. seeligeri*. There were around 60 human cases in the United Kingdom in 1981 but around 140 in 1985, along with a similar increase in animal cases (146). Between 1986 and 1988, human listeriosis increased in England and Wales by 150%, along with a 100% increase in human salmonellosis (39). The overall mortality rate for 558 human cases in the United Kingdom was 46%, with 51% and 44%, respectively, for perinatal and adult cases (146). For the period 1983–1987, 775 cases were reported in Britain, with 219 (28%) deaths, not including abortions. When the 44 abortions are added to the deaths, the fatality rate is 34% (82). Prior to 1974, 15 documented cases were seen yearly in western France, but in 1975 and 1976 there were 115 and 54, respectively (29). All but 3 of 145 strains that were serotyped were serotype 4. There were 687 cases in France in 1987 (36). In a 9-year period prior to early 1984, Lausanne, Switzerland, experienced a mean of 3 cases of human listeriosis per year, but in a 15-month period in 1983–1984, 25 cases were seen (142). Thirty-eight of 40 strains examined were serovar 4b, and 92% had the same phage type.

In a report from the U.S. Centers for Disease Control (CDC), 255 human

TABLE 21-5.   Proved and Suspected Cases of Foodborne Listeriosis

Year	Place	Synopsis of Cases	Reference
1953	Germany	Pregnant woman who drank raw milk from cow with atypical mastitis gave premature birth to stillborn twins. Same serovar recovered from milk and stillborn.	172
1958	Russia	Organism isolated from viscera of pigs slaughtered for pork on farm where cases of infectious mononucleosis-like disease occurred. Not isolated from victims.	83
1959	Sweden	4 cases (3 infants, 1 adult; 2 infant deaths). Serotype 2 recovered from all who lived within 3 blocks. Meat came from same store but no isolates were obtained.	160
1966	Halle, Germany	In 1964 and 1965, there were 23 and 15 cases, respectively, of human listeriosis and in 1966 279 cases. Although source was not documented, 97% of 144 pregnant mothers drank pasteurized milk and only 2 of 144 drank raw milk. All consumed other dairy, fruit, and vegetable products, and 68% of mothers who delivered ate a raw beef product. Of 3,000 cattle tested, only 2 had listeriosis.	162
1979	Boston	Between September and October, 23 adult cases and 87% yielded serovar 4b. Of the 20 case patients aged 46–89 years (mean 76), 3 died. 10 of 20 were immunosuppressed; 10 were not. Most were nosocomial, and most took antacids or cimetidine. Possibly raw vegetables but pasteurized milk thought to be a possible source.	89
1980	Auckland, New Zealand	Over an 11-month period, 22 cases of perinatal listeriosis seen in 3 hospitals. Most were serovar 1b. 5 fetal deaths and 1 liveborn death occurred. Cause not established, but shellfish and raw fish thought to have played a role. No cultures were obtained from foods.	127
1981	Canada, Maritime Provinces	Between March and September, 41 cases (7 adults, 34 perinatals; 2 adult and 16 infant deaths). Coleslaw was eaten by adults. Serovar 4b isolated from home of a victim and from victim's blood. Ovine listeriosis existed on farm where cabbage was fertilized with raw sheep manure.	104

TABLE 21-5.   (Continued)

Year	Place	Synopsis of Cases	Reference
1983	Boston	Between June and August, 49 cases (42 adults, 7 mother-infant pairs; 14 deaths). Pasteurized milk incriminated but no isolates were obtained. Thirty-two of 40 cultures were serovar 4b. Milk was from farm where listeriosis existed in dairy cows. Of the farms that supplied the raw milk, organism found in 15 of 124 samples and consisted of serovars 11, 3b, 4ab, and 4b. Phage type of strain from victims not identified among raw milk isolates.	57
1983–1987	Canton de Vaud, Switzerland	There were 122 human cases in this 5-year period. Vaccherin mont d'or soft cheese contained the same serovar and phage type as over 80% of victims. After product was recalled in 1987, no new cases were seen in 1988.	18
1985	Los Angeles	142 cases (93 pregnant women or their offspring, 49 nonpregnant adults; 48 deaths—30 fetuses/neonates, 18 nonpregnant adults). 98% of latter had known predisposing conditions; 82% of 105 isolates were serovar 4b; 73% were of same phage type. Mexican-style cheese was contaminated with raw milk.	129
1986–1987	Philadelphia	36 cases over 5-month period (16 deaths including 2 newborns). 32 of 36 were nonpregnant adults; 4 were newborns. Age of former ranged from 17 to 96 years (mean 67). 24 of 32 nonpregnant were immunosuppressed. 4 of remaining 8 were 80 years or older. 4 serovars found from 22 patients (8 each 4b and 1/2b and 4 of 1/2a and 2 of 3b). Organism not isolated from any food except brie cheese consumed by 1 victim. Ice cream, salami, or vegetables suspected.	188
1987	England	A nonpregnant, immunocompetent 36-year-old female contracted clinical listeriosis from a soft country cheese. Serovar 4b was recovered from patient and leftover cheese. Patient was cured with chloramphenicol.	8
1988	United Kingdom	A pregnant female had a stillbirth 5 days before flulike symptoms. Serovar 4 from chicken and fetal specimens had same phage type. Cooked-and-chilled chicken most likely source.	112

TABLE 21-5.   (Continued)

Year	Place	Synopsis of Cases	Reference
1988	United Kingdom	Miscarriage after 2-week illness. Isolates from food and fetal specimens had same phage type. 3-month-old bottle of vegetable rennet likely source.	
1988	Texas	Woman with cancer hospitalized with sepsis caused by *L. monocytogenes*. Had eaten 1 turkey frank daily heated in microwave. Serovar 1/2a isolated from patient, open and unopened franks, and from other opened foods in refrigerator. All isolates had same MEE type. Same serovar and MEE type strains found in processing plant 4 months later. Turkey franks most likely source.	209

cases were reported for the 3-year period 1967–1969 with a fatality rate of 28.2% (28). Serotype 4 accounted for 26% and serotype 1 for 45%. The incidence was highest during the first six months of the year, and about equal numbers of males and females were affected (28). For 1986–1987, the CDC estimated that there were around 1,700 human cases in the United States per year ($7.4/10^6$ population), with a mortality rate of around 28%. From six regions of the United States in 1986–1987, 154 cases were evaluated, and the fatality rate was 28% (189). One-third of the latter cases were perinatal, and two-thirds were in elderly and immunosuppressed individuals.

A synopsis of proved and suspected human foodborne listeriosis is presented in Table 21-5. The most definitive of these is the 1985 California outbreak in which Mexican-style cheese was shown to be the vehicle food. The overall fatality rate among the 142 victims was 33.8%, and raw milk in the cheese was the probable source of the serovar 4b strain, although this serovar could not be isolated from the farms that supplied the milk (see below).

Source of Pathogens

With the incidence of human foodborne listeriosis being so low and sporadic, the source of the causative strains of *L. monocytogenes* is of great interest. While the outbreaks traced to dairy products may be presumed to result from the shedding of virulent strains into milk, this is not always confirmed. In a study of 1,123 raw milk samples from the 27 farms that supplied milk to the incriminated cheese plant in California in 1985, Donnelly et al. (44) were unable to recover the responsible 4b serovar. A serotype 1 was isolated from 16 string samples from one control farm. In a review of the human cases through most of 1986, Hird (90) concluded that while the evidence was not conclusive in all cases, it nevertheless supported zoonotic transmission to some degree (zoonosis: disease transmissible under natural conditions from vertebrate animals to humans). Hird believes the healthy animal carrier is an important source of the organism, along with clinical listeriosis in livestock, but the relative degree to which each contributes to foodborne cases is uncertain.

L. *monocytogenes* was found to be shed in milk from the left forequarter of a mastitic cow, but milk from the other quarters was uninfected (73). About 10% of healthy cattle tested in the Netherlands were positive for *L. monocytogenes*, and about 5% of human fecal samples from slaughterhouse workers in Denmark contained the organism (104). The carriage rate for healthy humans seemed to be about the same regardless of their work position within food processing plants (104). Over an 18-month period in the United Kingdom, 32 of 5,000 (0.6%) fecal samples were positive (122). Cross-infection with *L. monocytogenes* from congenitally infected newborn infants to apparently healthy neonates in hospitals has been shown to occur (147). Thus, although the organism is known to be fairly common in environmental specimens, it also exists in healthy humans at rates from less than 1% to around 15%. The relative importance of environmental, animal, and human sources to foodborne episodes must await further study.

It has been suggested that the sudden increase in foodborne listeriosis may be due to coinfection by another pathogen such as *Salmonella enteritidis* (39). The incidence of human isolations of the latter organism parallels the findings of *L. monocytogenes* in humans. It has been suggested also that infection from the gastrointestinal tract is facilitated by toxins of *E. coli* (179) and possibly by the use of antacids that neutralize stomach acidity that otherwise would be inhibitory or destructive (91).

Syndromes

Listeriosis in humans is not characterized by a unique set of symptoms since the course of the disease depends on the state of the host. Nonpregnant healthy individuals who are not immunosuppressed are highly resistant to infection by *L. monocytogenes*, and there is little evidence that such individuals ever contract clinical listeriosis. However, the following conditions are known to predispose to adult listeriosis and to be significant in mortality rate: neoplasm, AIDS, alcoholism, diabetes (type 1 in particular), cardiovascular disease, renal transplant, and corticosteroid therapy. When susceptible adults contract the disease, meningitis and sepsis are the most commonly recognized symptoms. Of 641 human cases, 73% of victims had meningitis, meningoencephalitis, or encephalitis. Cervical and generalized lymphadenopathy are associated with the adult syndrome, and thus the disease may resemble infectious mononucleosis. Cerebrospinal fluid initially contains granulocytes, but in later stages monocytes predominate. Pregnant females who contract the disease (and their fetuses are often congenitally infected) may not present any symptoms, but when they do, they are typically mild and influenzalike. As may be noted from Table 21-5, abortion, premature birth, or stillbirth is often the consequence of listeriosis in pregnant females. When a newborn is infected at the time of delivery, listeriosis symptoms typically are those of meningitis, and they typically begin 1 to 4 weeks after birth, although a 4-day incubation has been recorded (Table 21-5). The usual incubation time in adults ranges from 1 to several weeks. Among the 20 case patients studied from the cluster of cases in the Boston episode, 18 had bacteremia, 8 developed meningitis, and 13 complained of vomiting, abdominal pain, and diarrhea 72 h before onset of symptoms (91).

The control of *L. monocytogenes* in the body is effected by T lymphocytes and activated macrophages, and thus any condition that adversely affects these cells will exacerbate the course of listeriosis. The most effective drugs for treatment are coumermycin, rifampicin, and ampicillin, with the last plus an aminoglycoside antibiotic being the best combination (46). Even with that regimen, antimicrobial therapy for listeriosis is not entirely satisfactory since ill patients and compromised hosts are more difficult than competent hosts.

RESISTANCE TO LISTERIOSIS

Resistance or immunity to intracellular pathogens such as viruses, animal parasites, and *L. monocytogenes* is mediated by T cells, lymphocytes that arise from bone marrow and undergo maturation in the thymus (hence *T* for thymusderived). Unlike B cells, which give rise to humoral immunity (circulating antibodies), activated T cells react directly against foreign cells. Once a pathogen is inside a host cell, it cannot be reached by circulating antibody, but the presence of the pathogen is signaled by structural changes in the parasitized cell, and T cells are involved in the destruction of this invaded host cell, which is no longer recognized as "self."

Macrophages are important to the actions of T cells, and their need for the destruction of *L. monocytogenes* and certain other intracellular pathogens was shown by Mackeness (139). First, macrophages bind and "present" *L. monocytogenes* cells to T cells in such way that they are recognized as being foreign. When T cells react with the organism, they increase in size and form clones specific for the same organism or antigen. These T cells, said to be activated, secrete interleukin 1 (IL-1). As the activated T cells multiply, they differentiate to form various subsets.

The most important subsets of T cells for resistance to listeriosis are helper or CD4 (L3T4$^+$) and cytolytic (killer) or CD8 (Lyt2$^+$) (111). The CD4 T cells react with the foreign antigen, after which they produce lymphokines (cytokines): interleukins 1 and 2 and immune or γ-interferon (150). Gamma-interferon, whose production is aided also by tumor necrosis factor (36), induces the production of IL-2 receptor expression on monocytes (84, 93). Also, IL-2 may enhance the activation of lymphokine-activated killer cells that can lyse infected macrophages. As little as 0.6 μg/mouse of exogenously administered IL-2 has been shown to strengthen mouse resistance to *L. monocytogenes* (84). Gamma-interferon activates macrophages and CD8 T cells, and the latter react with *L. monocytogenes*–infected host macrophages—and cause their lysis. Both CD4 and CD8 T cells are stimulated by *L. monocytogenes*; they activate macrophages via their production of γ-interferon and contribute to resistance to listeriosis (109). CD8 also secretes γ-interferon when exogenous IL-2 is provided, and both CD4 and CD8 can confer some passive immunity to recipient mice (109). T cell subsets and the T cell receptor for antigens have been reviewed by Fitch (56).

Some of the events that occur in murine hosts following infection with *L. monocytogenes* are presumed to be the same events that occur in humans. When macrophages engulf *L. monocytogenes*, a factor-increasing monocytopoiesis (FIM) is secreted by the macrophages at the infection site (211). FIM is transported to the bone marrow, where it stimulates the production of more macrophages. Only viable *L. monocytogenes* cells can induce the T cell response and immunity to listeriosis. Since LLO is the virulence factor of *L. monocytogenes* that elicits the T cell response, this heat-labile protein is destroyed when cells are heat killed. The CD8 T cell subset appears to be the major T cell component responsible for antilisterial immunity, for it acts by eliciting lymphokine production by macrophages; passive immunity to *L. monocytogenes* can be achieved by transfer of the CD8 T cell subset (7). The T cell response does not occur even when mice are injected with both killed cells and recombinant IL-1a (94). Neither avirulent nor killed cells induce IL-1 in vitro; viable *L. monocytogenes* cells do (152), indicating a critical role for IL-1 and γ-interferon in the initiation of the in vivo response, for it has been shown that simultaneously administered IL-1a and γ-interferon increased resistance to *L. monocytogenes* in mice better than either alone (121). The combination was not synergistic, only additive. It appears that the primary role of γ-interferon is to elicit lymphokine production rather than acting directly, and it is known to increase the production of IL-1 (31). Gamma-interferon is detectable in the bloodstream and spleen of mice only during the first 4 days after infection (155).

This synopsis of murine resistance to *L. monocytogenes* reveals some of the multifunctional roles of the lymphokines in T cell immunity and the apparent critical importance of LLO as the primary virulence factor of this organism. What makes immunocompromised hosts more susceptible to listeriosis is the dampening effect that immunosuppresive agents have on the T cell system. Possible therapy is suggested by the specific roles that some of the lymphokines play, but whether the effects are similar in humans is unclear.

PERSISTENCE OF *L. MONOCYTOGENES* IN FOODS

Since it can grow over the temperature range of about 1° to 45°C and the pH range 4.1 to around 9.6, *L. monocytogenes* may be expected to survive in foods for long periods of time, and this has been confirmed. Strains Scott A and V7, inoculated at levels of $10^4 - 10^5$/g, survived in cottage cheese for up to 28 days when held at 3°C (183). When these two strains, with two others, were inoculated into a camembert cheese formulated with levels of $10^4 - 10^5$, growth occurred during the first 18 days of ripening, and some strains attained levels of 10^6 to 10^7 after 65 days of ripening (182). With an inoculum of 5×10^2/g and storage at 4°C, *L. monocytogenes* survived in cold-pack cheese for a mean of 130 days in the presence of 0.30% sorbic acid (181). On the other hand, in manufactured and stored nonfat dry milk, a 1 to 1.5 log reduction in numbers of *L. monocytogenes* occurred during spray drying, and more than a 4-log decrease in cfu occurred within 16 weeks when held at 25°C (45).

In ground beef, an inoculum of $10^5 - 10^6$ remained unchanged through 14 days at 4°C (100), and inocula of *L. monocytogenes* of 10^3 or 10^5 remained unchanged in ground beef and liver for over 30 days although the standard plate counts (SPC) increased three- to sixfold during this time (194). When added to a Finnish sausage mix that included 120 ppm $NaNO_2$ and 3% NaCl, the initial numbers of *L. monocytogenes* decreased only by about 1 log during a 21-day fermentation period (102). When five strains of *L. monocytogenes* were added to eight processed meats that were stored at 4.4°C for up to 12 weeks, the organisms survived on all products and increased in numbers by 3 to 4 logs in most (74). Best growth occurred in chicken and turkey products, due in part to the higher initial pH of these products. In vacuum-packaged beef in a film with barrier properties of 25–30 ml/m^2/24 h/101 kPa, an inoculated strain of *L. monocytogenes* increased about 4 logs on the fatty tissue of strip loins in 16 days and by about 3 log cycles in 20 days on lean meat when held between 5° and 5.5°C (74, 77). In meat, cheese, and egg ravioli stored at 5°C, a 3×10^5 cfu/g inoculum of strain Scott A survived 14 days (14). Lettuce and lettuce juice supported the growth of *L. monocytogenes* held at 5°C for 14 days in one study, and the organism was recovered from two uninoculated lettuce samples (205).

The studies noted are typical of others and show that the overall resistance of *L. monocytogenes* in foods is consistent with their persistence in many nonfood environmental specimens.

RECOVERY FROM FOODS

Since listeriae exist in foods generally in very low numbers, selective enrichment methods are necessary to effect their recovery. Advantage is taken of the psychrotrophic nature of *L. monocytogenes* for its enrichment, which can be achieved at 4° to 5°C with incubations from 1 week up to 1 year in a variety of media. Some quantitation of original cell numbers can be achieved by enrichment methods when an MPN-type procedure is followed.

In addition to the low numbers of cells, the other challenge in the selective isolation/enumeration of listeriae is the general existence or larger numbers of lactic acid bacteria, as well as some coryneforms, both of which share properties similar to the listeriae relative to selective agents. Agents that are specifically selective for *L. monocytogenes* have not been reported.

A large number of media and methods have been proposed for recovering these organisms from foods; some are noted in Chapter 6. The various chemicals employed as selective agents have been reviewed and discussed by Klinger (115), and comparative studies of the various media have been reported by a number of investigators (133, 207).

References

1. Adams, T. J., S. Vartivarian, and R. E. Cowart. 1990. Iron acquisition systems of *Listeria monocytogenes*. *Infect. Immun.* 58:2715–2718.
2. Ade, N., S. Steinmeyer, M. J. Loessner, H. Hof, and J. Kreft (Technical University of Munich, Weihenstephan, Freising, Germany). 1991. Personal communication.
3. Anton, W. 1934. Kritisch-experimenteller Beitrag zur Biologie des *Bacterium*

monocytogenes. Mit basonderer Beruecksichtigung seiner Beziehung zur infektiosen Mononucleose des Menschen. *Zbl. Bakteriol Abt. I. Orig.* 131:89–103.

4. Audurier, A., and C. Martin. 1989. Phage typing of *Listeria monocytogenes*. *Int. J. Food Microbiol.* 8:251–257.

5. Audurier, A., R. Chatelain, F. Chalons, and M. Piéchaud. 1979. Lysotypie de 823 souches de *Listeria monocytogenes* isolées en France de 1958 à 1978. *Ann. Microbiol. (Inst. Pasteur)* 130B:179–189.

6. Bailey, J. S., D. L. Fletcher, and N. A. Cox. 1989. Recovery and serotype distribution of *Listeria monocytogenes* from broiler chickens in the southeastern United States. *J. Food Protect.* 52:148–150.

7. Baldridge, J. R., R. A. Barry, and D. J. Hinrichs. 1990. Expression of systemic protection and delayed-type hypersensitivity to *Listeria monocytogenes* is mediated by different T-cell subsets. *Infect. Immun.* 58:654–658.

8. Bannister, B. A. 1987. *Listeria monocytogenes* meningitis associated with eating soft cheese. *J. Infect.* 15:165–168.

9. Beattie, I. A., B. Swaminathan, and H. K. Ziegler. 1990. Cloning and characterization of T-cell-reactive protein antigens from *Listeria monocytogenes. Infect. Immun.* 58:2792–2803.

10. Beckers, H. J., P. S. S. Soentero, and E. H. M. Delfou-van Asch. 1987. The occurrence of *Listeria monocytogenes* in soft cheeses and raw milk and its resistance to heat. *Int. J. Food Microbiol.* 4:249–256.

11. Berche, P., J.-L. Gaillard, and S. Richard. 1988. Invasiveness and intracellular growth of *Listeria monocytogenes. Infection* 16 (Suppl. 2):145–148.

12. Berche, P., J.-L. Gaillard, C. Geoffroy, and J. E. Alouf. 1987. T cell recognition of listeriolysin O is induced during infection with *Listeria monocytogenes. J. Immunol.* 139:3813–3821.

13. *Bergey's Manual of Systematic Bacteriology.* Vol. 2. 1986. Baltimore: Williams & Wilkins.

14. Beuchat, L. R., and R. E. Brackett. 1989. Observations on survival and thermal inactivation of *Listeria monocytogenes* in ravioli. *Lett. Appl. Microbiol.* 8:173–175.

15. Beuchat, L. R., R. E. Brackett, D. Y.-Y. Hao, and D. E. Conner. 1986. Growth and thermal inactivation of *Listeria monocytogenes* in cabbage and cabbage juice. *Can. J. Microbiol.* 32:791–795.

16. Bibb, W. F., B. G. Gellin, R. Weaver, B. Schwartz, B. D. Plikaytis, M. W. Reeves, R. W. Pinner, and C. V. Broome. 1990. Analysis of clinical and food-borne isolates of *Listeria monocytogenes* in the United States by multilocus enzyme electrophoresis and application of the method to epidemiologic investigations. *Appl. Environ. Microbiol.* 56:2133–2141.

17. Bibb, W. F., B. Schwartz, B. G. Gellin, B. D. Plikaytis, and R. E. Weaver. 1989. Analysis of *Listeria monocytogenes* by multilocus enzyme electrophoresis and application of the method to epidemiologic investigations. *Int. J. Food Microbiol.* 8:233–239.

18. Bille, J. 1990. Epidemiology of human listeriosis in Europe, with special reference to the Swiss outbreak. In *Foodborne Listeriosis,* ed. A. L. Miller, J. L. Smith, and G. A. Somkuti, 71–74. Amsterdam: Elsevier.

19. Boerlin, P., J. Rocourt, and J.-C. Piffaretti. 1991. Taxonomy of the genus *Listeria* by using multilocus enzyme electrophoresis. *Int. J. System. Bacteriol.* 41:59–64.

20. Boyle, D. L., J. N. Sofos, and G. R. Schmidt. 1990. Thermal destruction of *Listeria monocytogenes* in a meat slurry and in ground beef. *J. Food Sci.* 55:327–329.

21. Bradshaw, J. G., J. T. Peeler, J. J. Corwin, J. M. Hunt, and R. M. Twedt. 1987.

Thermal resistance of *Listeria monocytogenes* in dairy products. *J. Food Protect.* 50:543–544.

22. Bradshaw, J. G., J. T. Peeler, J. J. Corwin, J. M. Hunt, J. T. Tierney, E. P. Larkin, and R. M. Twedt. 1985. Thermal resistance of *Listeria monocytogenes* in milk. *J. Food Protect.* 48:743–755.

23. Breuer, V. J., and O. Prändl. 1988. Nachweis von Listerien und deren Vorkommen in Hackfleisch und Mettwuersten in Oesterreich. *Archiv Lebensmitlhyg.* 39:28–30.

24. Buchanan, R. L., and L. A. Klawitter. 1990. Effects of temperature and oxygen on the growth of *Listeria monocytogenes* at pH 4.5. *J. Food Sci.* 55:1754–1756.

25. Bunning, V. K., R. G. Crawford, J. T. Tierney, and J. T. Peeler. 1990. Thermotolerance of *Listeria monocytogenes* and *Salmonella typhimurium* after sublethal heat shock. *Appl. Environ. Microbiol.* 56:3216–3219.

26. Bunning, V. K., C. W. Donnelly, J. T. Peeler, E. H. Briggs, J. G. Bradshaw, R. G. Crawford, C. M. Beliveau, and J. T. Tierney. 1988. Thermal inactivation of *Listeria monocytogenes* within bovine milk phagocytes. *Appl. Environ. Microbiol.* 54:364–370.

27. Bunning, V. K., R. G. Crawford, J. G. Bradshaw, J. T. Peeler, J. T. Tierney, and R. M. Twedt. 1986. Thermal resistance of intracellular *Listeria monocytogenes* cells suspended in raw bovine milk. *Appl. Environ. Microbiol.* 52:1398–1402.

28. Busch, L. A. 1971. Human listeriosis in the United States, 1967–1969. *J. Inf. Dis.* 123:328–331.

29. Carbonnelle, B., J. Cottin, F. Parvery, G. Chambreuil, S. Kouyoumdjian, M. LeLirzin, G. Cordier, and F. Vincent. 1978. Epidemie de listeriose dans l'Ouest de la France (1975–1976). *Rev. Epidem. et Santé Publ.* 26:451–467.

30. Centers for Disease Control. 1989. Listeriosis associated with consumption of turkey franks. *Morb. Mort. Wkly. Rept.* 38:267–268.

31. Chen, Y., A. Nakane, and T. Minagawa. 1989. Recombinant murine gamma interferon induces enhanced resistance to *Listeria monocytogenes* infection in neonatal mice. *Infect. Immun.* 57:2345–2349.

32. Cluff, C. W., M. Garcia, and H. K. Ziegler. 1990. Intracellular hemolysin-producing *Listeria monocytogenes* strains inhibit macrophage-mediated antigen processing. *Infect. Immun.* 58:3601–3612.

33. Colburn, K. G., C. A. Kaysner, C. Abeyta, Jr., and M. M. Wekell. 1990. *Listeria* species in a California coast estuarine environment. *Appl. Environ. Microbiol.* 56:2007–2011.

34. Cole, M. B., M. V. Jones, and C. Holyoak. 1990. The effect of pH, salt concentration and temperature on the survival and growth of *Listeria monocytogenes*. *J. Appl. Bacteriol.* 69:63–72.

35. Conner, D. E., R. E. Brackett, and L. R. Beuchat. 1986. Effect of temperature, sodium chloride, and pH on growth of *Listeria monocytogenes* in cabbage juice. *Appl. Environ. Microbiol.* 52:59–63.

36. Cossart, P., and J. Mengaud. 1989. *Listeria monocytogenes*. A model system for the molecular study of intracellular parasitism. *Mol. Biol. Med.* 6:463–474.

37. Cossart, P., M. F. Vicente, J. Mengaud, F. Baquero, J. C. Perez-Diaz, and P. Berche. 1989. Listeriolysin O is essential for virulence of *Listeria monocytogenes*: Direct evidence obtained by gene complementation. *Infect. Immun.* 57:3629–3636.

38. Courtieu, A. L., E. P. Espaze, and A. E. Reynaud. 1986. *Proceedings of the 9th International Symposium of Problems of Listeriosis*. Nantes, France: University of Nantes Press.

39. Cox, L. J. 1989. A perspective on listeriosis. *Food Technol.* 43(12):52–59.

40. Datta, A. R., B. W. Wentz, and W. E. Hill. 1988. Identification and enumeration

of beta-hemolytic *Listeria monocytogenes* in naturally contaminated dairy products. *J. Assoc. Off. Anal. Chem.* 71:673–675.

41. Del Corral, F., R. L. Buchanan, M. M. Bencivengo, and P. H. Cooke. 1990. Quantitative comparison of selected virulence associated characteristics in food and clinical isolates of *Listeria*. *J. Food Protect.* 53:1003–1009.

42. Dominguez Rodriguez, L., J. F. Fernandez Garayzabal, J. A. Vazquez Boland, E. Rodriguez Ferri, and G. Suarez Fernandez. 1985. Isolation de micro-organismes du genre *Listeria* à partir de lait cru destine à la consommation humaine. *Can. J. Microbiol.* 31:938–941.

43. Donnelly, C. W., and E. H. Briggs. 1986. Psychrotrophic growth and thermal inactivation of *Listeria monocytogenes* as a function of milk composition. *J. Food Protect.* 49:994–998.

44. Donnelly, C. W., E. H. Briggs, and G. J. Baigent. 1986. Analysis of raw milk for the epidemic serotype of *Listeria monocytogenes* linked to an outbreak of listeriosis in California. *J. Food Protect.* 49:846–847 (abstr.).

45. Doyle, M. P., L. M. Meske, and E. H. Marth. 1985. Survival of *Listeria monocytogenes* during the manufacture and storage of nonfat dry milk. *J. Food Protect.* 48:740–742.

46. Espaze, E. P., and A. E. Reynaud. 1988. Antibiotic susceptibilities of *Listeria*: In vitro studies. *Infection* 16 (Suppl. 2):160–164.

47. Farber, J. M., and B. E. Brown. 1990. Effect of prior heat shock on heat resistance of *Listeria monocytogenes* in meat. *Appl. Environ. Microbiol.* 56:1584–1587.

48. Farber, J. M. 1989. Thermal resistance of *Listeria monocytogenes*. *Int. J. Food Microbiol.* 8:285–291.

49. Farber, J. M., G. W. Sanders, and M. A. Johnston. 1989. A survey of various foods for the presence of *Listeria* species. *J. Food Protect.* 52:456–458.

50. Farber, J. M., G. W. Sanders, J. I. Speirs, J.-Y. D'Aoust, D. B. Emmons, and R. McKellar. 1988. Thermal resistance of *Listeria monocytogenes* in inoculated and naturally contaminated raw milk. *Int. J. Food Microbiol.* 7:277–286.

51. Farber, J. M., F. Tittinger, and L. Gour. 1988. Surveillance of raw-fermented (dry-cured) sausages for the presence of *Listeria* spp. 1988. *Can. Inst. Food Sci. Technol. J.* 21:430–434.

52. Farber, J. M., M. A. Johnston, U. Purvis, and A. Loit. 1987. Surveillance of soft and semi-soft cheeses for the presence of *Listeria* spp. *Int. J. Food Microbiol.* 5:157–163.

53. Fedio, W. M., and H. Jackson. 1989. Effect of tempering on the heat resistance of *Listeria monocytogenes*. *Lett. Appl. Microbiol.* 9:157–160.

54. Fenlon, D. R. 1985. Wild birds and silage as reservoirs of *Listeria* in the agricultural environment. *J. Appl. Bacteriol.* 59:537–543.

55. Fiedler, F., J. Seger, A. Schrettenbrunner, and H. P. R. Seeliger, 1984. The biochemistry of murein and cell wall teichoic acids in the genus *Listeria*. *System. Appl. Microbiol.* 5:360–376.

56. Fitch, F. W. 1986. T-cell clones and T-cell receptors. *Microbiol. Rev.* 50:50–69.

57. Fleming, D. W., S. L. Couchi, K. L. MacDonald, J. Brondum, P. S. Hayes, B. D. Plikaytis, M. B. Holmes, A. Audurier, C. V. Broome, and A. L. Reingold. 1985. Pasteurized milk as a vehicle of infection in an outbreak of listeriosis. *N. Engl. J. Med.* 312:404–407.

58. Foegeding, P. M., and N. W. Stanley. 1990. *Listeria monocytogenes* F5069 thermal death times in liquid whole egg. *J. Food Protect.* 53:6–8.

59. Fuchs, R. S., and P. K. Surendran. 1989. Incidence of *Listeria* in tropical fish and fishery products. *Lett. Appl. Microbiol.* 9:49–51.

60. Gaillard, J.-L., P. Berche, J. Mounier, S. Richard, and P. Sansonetti. 1987. In

vitro model of penetration and intracellular growth of *Listeria monocytogenes* in the human enterocyte-like cell line Caco-2. *Infect. Immun.* 55:2822–2829.

61. Gaillard, J. L., P. Berche, and P. Sansonetti. 1986. Transposon mutagenesis as a tool to study the role of hemolysin in the virulence of *Listeria monocytogenes*. *Infect. Immun.* 52:50–55.

62. Galsworthy, S. B., and D. Fewster. 1988. Comparison of responsiveness to the monocytosis-producing activity of *Listeria monocytogenes* in mice genetically susceptible or resistant to listeriosis. *Infection* 16 (Suppl. 2):118–122.

63. Galsworthy, S. B., S. M. Gurofsky, and R. G. E. Murray. 1977. Purification of a monocytosis-producing activity from *Listeria monocytogenes*. *Infect. Immun.* 15:500–509.

64. Gaze, J. E., G. B. Brown, D. E. Gaskell, and J. G. Banks. 1989. Heat resistance of *Listeria monocytogenes* in homogenates of chicken, beef, steak and carrot. *Food Microbiol.* 6:251–259.

65. Gellin, B. G., and C. V. Broome. 1989. Listeriosis. *J. Amer. Med. Assoc.* 261:1313–1320.

66. Genigeorgis, C. A., P. Oanca, and D. Dutulescu. 1990. Prevalence of *Listeria* spp. in turkey meat at the supermarket and slaughterhouse level. *J. Food Protect.* 53:282–288.

67. Geoffroy, C., J.-L. Gaillard, J. E. Alouf, and P. Berche. 1989. Production of thiol-dependent haemolysins by *Listeria monocytogenes*. *J. Gen. Microbiol.* 135:481–487.

68. Geoffroy, C., J.-L. Gaillard, J. E. Alouf, and P. Berche. 1987. Purification, characterization, and toxicity of the sulfhydryl-activated hemolysin listeriolysin O from *Listeria monocytogenes*. *Infect. Immun.* 55:1641–1646.

69. George, S. M., B. M. Lund, and T. F. Brocklehurst. 1988. The effect of pH and temperature on initiation of growth of *Listeria monocytogenes*. *Lett. Appl. Microbiol.* 6:153–156.

70. Gilbert, R. J., and P. N. Pini. 1988. Listeriosis and food-borne transmission. *Lancet* i:472–473.

71. Girard, K. F., A. J. Sbarra, and W. A. Bardawil. 1963. Serology of *Listeria monocytogenes*. I. Characteristics of the soluble hemolysin. *J. Bacteriol.* 85:349–355.

72. Gitter, M. 1976. *Listeria monocytogenes* in "oven ready" poultry. *Vet. Rec.* 99:336.

73. Gitter, M., R. Bradley, and P. H. Blampied. 1980. *Listeria monocytogenes* infection in bovine mastitis. *Vet. Rec.* 107:390–393.

74. Glass, K. A., and M. P. Doyle. 1990. Fate of *Listeria monocytogenes* in processed meat products during refrigerated storage. *Appl. Environ. Microbiol.* 55:1565–1569.

75. Goebel. W., S. Kathariou, M. Kuhn, Z. Sokolovic, J. Kreft, S. Kohler, D. Funke, T. Chakraborty, and M. Leimeister-Waechter. 1988. Hemolysin from *Listeria*—biochemistry, genetics and function in pathogenesis. *Infection* 16 (Suppl. 2): 149–156.

76. Golnazarian, C. A., C. W. Donnelly, S. J. Pintauro, and D. B. Howard. 1989. Comparison of infectious dose of *Listeria monocytogenes* F5817 as determined for normal versus compromised C57B1/6J mice. *J. Food Protect.* 52:696–701.

77. Grau, F. H., and P. B. Vanderlinde. 1990. Growth of *Listeria monocytogenes* on vacuum-packaged beef. *J. Food Protect.* 53:739–741.

78. Gray, M. L. 1963. Epidemiological aspects of listeriosis. *A. J. Pub. Hlth.* 53: 554–563.

79. Gray, M. L. 1963. *2nd Symposium on Listeric Infection*. Bozeman: Montana State College.

80. Gray, M. L., and A. H. Killinger. 1966. *Listeria monocytogenes* and listeric infections. *Bacteriol. Rev.* 30:309–382.

81. Green, S. S. 1990. *Listeria monocytogenes* in meat and poultry products. *Interim Rept. to Nat'l. Adv. Comm. Microbiol. Spec. Foods*, FSIS/USDA, Nov. 27.

82. Groves, R. D., and H. J. Welshimer. 1977. Separation of pathogenic from apathogenic *Listeria monocytogenes* by three in vitro reactions. *J. Clin. Microbiol.* 5:559–563.

83. Gudkova, E. I., K. A. Mironova, A. S. Kus'minskii, and G. O. Geine. 1958. A second outbreak of listeriotic angina in a single populated locality. *Zh. Microbiol. Epidemiol. Immunobiol.* 9:24–28 (Eng. transl.).

84. Haak-Frendscho, M., K. M. Young, and C. J. Czuprynski. 1989. Treatment of mice with human recombinant interleukin-2 augments resistance to the facultative intracellular pathogen *Listeria monocytogenes. Infect. Immun.* 57:3014–3021.

85. Hall, S. M., N. Crofts, R. J. Gilbert, P. N. Pini, A. G. Taylor, and J. McLauchlin. 1988. Epidemiology of listeriosis, England and Wales. *Lancet* ii:502–503.

86. Harrison, M. A., and Y.-W. Huang. 1990. Thermal death times for *Listeria monocytogenes* (Scott A) in crabmeat. *J. Food Protect.* 53:878–880.

87. Hayes, P. S., J. C. Feeley, L. M. Graves, G. W. Ajello, and D. W. Fleming. 1986. Isolation of *Listeria monocytogenes* from raw milk. *Appl. Environ. Microbiol.* 51:438–440.

88. Heisick, J. E., D. E. Wagner, M. L. Nierman, and J. T. Peeler. 1989. *Listeria* spp. found on fresh market product. *Appl. Environ. Microbiol.* 55:1925–1927.

89. Hether, N. W., and L. L. Jackson. 1983. Lipoteichoic acid from *Listeria monocytogenes. J. Bacteriol.* 156:809–817.

90. Hird, D. W. 1987. Review of evidence for zoonotic listeriosis. *J. Food Protect.* 50:429–433.

91. Ho, J. L., K. N. Shands, G. Friedland, P. Eckind, and D. W. Fraser. 1986. An outbreak of type 4b *Listeria monocytogenes* infection involving patients from eight Boston hospitals. *Arch. Intern. Med.* 146:520–524.

92. Hof, H., and P. Hefner. 1988. Pathogenicity of *Listeria monocytogenes* in comparison to other *Listeria* species. *Infection* 16 (Suppl. 2):141–144.

93. Holter, W., R. Grunow, H. Stockinger, and W. Knapp. 1986. Recombinant interferon-γ induces interleukin 2 receptors on human peripheral blood monocytes. *J. Immunol.* 136:2171–2175.

94. Igarashi, K.-I., M. Mitsuyama, K. Muramori, H. Tsukada, and K. Nomoto. 1990. Interleukin-1-induced promotion of T-cell differentiation in mice immunized with killed *Listeria monocytogenes. Infect. Immun.* 58:3973–3979.

95. Ivanov, I. ed. 1979. *Problems of Listeriosis.* Sofia, Bulgaria: National Agroindustrial Union.

96. Jenkins, E. M., and B. B. Watson. 1971. Extracellular antigens from *Listeria monocytogenes.* I. Purification and resolution of hemolytic and lipolytic antigens from culture filtrates of *Listeria monocytogenes. Infect. Immun.* 3:589–594.

97. Jenkins, E. M., A. N. Njoku-Obi, and E. W. Adams. 1964. Purification of the soluble hemolysins of *Listeria monocytogenes. J. Bacteriol.* 88:418–424.

98. Johnson, J. L., M. P. Doyle, and R. G. Cassens. 1990. Incidence of *Listeria* spp. in retail meat roasts. *J. Food Sci.* 55:572, 574.

99. Johnson, J. L., M. P. Doyle, and R. G. Cassens. 1990. *Listeria monocytogenes* and other Listeria spp. in meat and meat products. A review. *J. Food Protect.* 53:81–91.

100. Johnson, J. L., M. P. Doyle, and R. G. Cassens. 1988. Survival of *Listeria monocytogenes* in ground beef. *Int. J. Food Microbiol.* 6:243–247.

101. Jones, D. 1988. The place of *Listeria* among gram-positive bacteria. *Infection* 16 (Suppl. 2):85–88.

102. Junttila, J., J. Hirn, P. Hill, and E. Nurmi. 1989. Effect of different levels of nitrite and nitrate on the survival of *Listeria monocytogenes* during the manufacture of fermented sausage. *J. Food Protect.* 52:158–161.

103. Junttila, J. R., S. I. Niemela, and J. Hirn. 1988. Minimum growth temperatures of *Listeria monocytogenes* and nonhaemolytic listeria. *J. Appl. Bacteriol.* 65:321–327.

104. Kampelmacher, E. H., and L. M. Van Nooble Jansen. 1969. Isolation of *Listeria monocytogenes* from faeces of clinically healthy humans and animals. *Zentral. Bakt. Inf. Abt. Orig.* 211:353–359.

105. Karches, H., and P. Teufel. 1988. *Listeria monocytogenes* Vorkommen in Hackfleisch und Verhalten in frischer Zwiebelmettwurst. *Fleischwirtsch.* 68:1388–1392.

106. Kathariou, S., L. Pine, V. George, G. M. Carlone, and B. P. Holloway. 1990. Nonhemolytic *Listeria monocytogenes* mutants that are also noninvasive for mammalian cells in culture: Evidence for coordinate regulation of virulence. *Infect. Immun.* 58:3988–3995.

107. Kathariou, S., J. Rocourt, H. Hof, and W. Goebel. 1988. Levels of *Listeria monocytogenes* hemolysin are not directly proportional to virulence in experimental infections of mice. *Infect. Immun.* 56:534–536.

108. Kathariou, S., P. Metz, H. Hof, and W. Goebel. 1987. Tn916-induced mutations in the hemolysin determinant affecting virulence of *Listeria monocytogenes*. *J. Bacteriol.* 169:1291–1297.

109. Kaufmann, S. H. E. 1988. *Listeria monocytogenes* specific T-cell lines and clones. *Infection* 16 (Suppl. 2):128–136.

110. Kaufmann, S. H. E. 1984. Acquired resistance to facultative intracellular bacteria: Relationship between persistence, cross-reactivity at the T-cell level, and capacity to stimulate cellular immunity of different *Listeria* strains. *Infect. Immun.* 45:234–241.

111. Kaufmann, S. H. E., E. Hug, U. Vath, and I. Muller. 1985. Effective protection against *Listeria monocytogenes* and delayed-type hypersensitivity to listerial antigens depend on cooperation between specific L3T4+ and Lyt2+ T cells. *Infect. Immun.* 48:263–266.

112. Kerr, K. G., Dealler, S. F., and R. W. Lacey. 1988. Materno-fetal listeriosis from cook-chill and refrigerated food. *Lancet* ii:1133.

113. Kerr, K. G., N. A. Rotowa, P. M. Hawkey, and R. W. Lacey. 1990. Incidence of *Listeria* spp. in pre-cooked, chilled chicken products as determined by culture and enzyme-linked immunoassay (ELISA). *J. Food Protect.* 53:606–607.

114. Kingdom, G. C., and C. P. Sword. 1970. Effects of *Listeria monocytogenes* hemolysin on phagocytic cells and lysosomes. *Infect. Immun.* 1:356–362.

115. Klinger, J. D. 1988. Isolation of *Listeria*: A review of procedures and future prospects. *Infection* 16 (Suppl. 2):98–105.

116. Kohler, S., M. A. Leimeister-Waechter, T. Chakraborty, F. Lottspeich, and W. Goebel. 1990. The gene coding for protein p60 of *Listeria monocytogenes* and its use as a specific probe for *Listeria monocytogenes*. *Infect. Immun.* 58:1943–1950.

117. Kreft, J., D. Funke, A. Haas, F. Lottspeich, and W. Goebel. 1989. Production, purification and characterization of hemolysins from *Listeria ivanovii* and *Listeria monocytogenes* Sv4b. *FEMS Microbiol. Lett.* 57:197–202.

118. Kuhn, M., M.-C. Prevost, J. Mounier, and P. J. Sansonetti. 1990. A nonvirulent mutant of *Listeria monocytogenes* does not move intracellularly but still induces polymerization of actin. *Infect. Immun.* 58:3477–3486.

119. Kuhn, M., and W. Goebel. 1989. Identification of an extracellular protein of

Listeria monocytogenes possibly involved in intracellular uptake by mammalian cells. *Infect. Immun.* 57:55–61.

120. Kuhn, M., S. Kathariou, and W. Goebel. 1988. Hemolysin supports survival but not entry of the intracellular bacterium *Listeria monocytogenes*. *Infect. Immun.* 56:79–82.

121. Kurtz, R. S., K. M. Young, and C. J. Czuprynski. 1989. Separate and combined effects of recombinant interleukin-la and gamma interferon on antibacterial resistance. *Infect. Immun.* 57:553–558.

122. Kwantes, W., and M. Isaac. 1971. Listeriosis. *Brit. Med. J.* 4:296–297.

123. Leasor, S. B., and P. M. Foegeding. 1989. *Listeria* species in commercially broken raw liquid whole egg. *J. Food Protect.* 52:777–780.

124. Leblond-Francillard, M., J.-L. Gaillard, and P. Berche. 1989. Loss of catalase activity in Tn1545-induced mutants does not reduce growth of *Listeria monocytogenes* in vivo. *Infect. Immun.* 57:2569–2573.

125. Le Guilloux, M., C. Dollinger, and G. Freyburger. 1980. *Listeria monocytogenes*: Sa frequence dans les produits de charcuterie. *Bull. Soc. Vet. Prat. de France* 64(1):45–53.

126. Leimeister-Waechter, M. and T. Chakraborty. 1989. Detection of listeriolysin, the thiol-dependent hemolysin in *Listeria monocytogenes*, *Listeria ivanovii*, and *Listeria seeligeri*. *Infect. Immun.* 57:2350–2357.

127. Lennon, D., B. Lewis, C. Mantell, D. Becroft, B. Dove, K. Farmer, S. Tonkin, N. Yeates, R. Stamp, and K. Mickleson. 1984. Epidemic perinatal listeriosis. *Pediat. Inf. Dis.* 3:30–34.

128. Liewen, M. B., and M. W. Plautz. 1988. Occurrence of *Listeria monocytogenes* in raw milk in Nebraska. *J. Food Protect.* 51:840–841.

129. Linnan, M. J., L. Mascola, X. D. Lou, V. Goulet, S. May, C. Salminen, D. W. Hird, M. L. Yonekura, P. Hayes, R. Weaver, A. Audurier, B. D. Plikaytis, S. L. Fannin, A. Kleks, and C. V. Broome. 1988. Epidemic listeriosis associated with Mexican-style cheese. *N. Eng. J. Med.* 319:823–828.

130. Linton, R. H., M. D. Pierson, and J. R. Bishop. 1990. Increase in heat resistance of *Listeria monocytogenes* Scott A by sublethal heat shock. *J. Food Protect.* 53:924–927.

131. Loessner, M. J. 1991. Improved procedure for bacteriophage typing of *Listeria* strains and evaluation of new phages. *Appl. Environ. Microbiol.* 57:882–884.

132. Loessner, M. J., and M. Busse. 1990. Bacteriophage typing of *Listeria* species. *Appl. Environ. Microbiol.* 56:1912–1918.

133. Loessner, M. J., R. H. Bell, J. M. Jay, and L. A. Shelef. 1988. Comparison of seven plating media for enumeration of *Listeria* spp. *Appl. Environ. Microbiol.* 54:3003–3007.

134. Lovett, J., I. V. Wesley, M. J. Vandermaaten, J. G. Brawshaw, D. W. Francis, R. G. Crawford, C. W. Donnelly, and J. W. Messer. 1990. High-temperature short-time pasteurization inactivates *Listeria monocytogenes*. *J. Food Protect.* 53:734–738.

135. Lovett, J. 1988. Isolation and identification of *Listeria monocytogenes* in dairy products. *J. Assoc. Off. Anal. Chem.* 71:658–650.

136. Lovett, J., D. W. Francis, and J. M. Hunt. 1987. *Listeria monocytogenes* in raw milk: Detection, incidence, and pathogenicity. *J. Food Protect.* 50:188–192.

137. Ludwig, W., K.-H. Schleifer, and E. Stackebrandt. 1984. 16S rRNA analysis of *Listeria monocytogenes* and *Brochothrix thermosphacta*. *FEMS Microbiol. Lett.* 25:199–204.

138. Lund, B. M., M. R. Knox, and M. B. Cole. 1989. Destruction of *Listeria monocytogenes* during microwave cooking. *Lancet* i:218.

139. Mackeness, G. B. 1971. Resistance to intracellular infection. *J. Inf. Dis.* 123: 439–445.

140. Mackey, B. M., C. Pritchet, A. Norris, and G. C. Mead. 1990. Heat resistance of *Listeria*: Strain differences and effects of meat type and curing salts. *Lett. Appl. Microbiol.* 10:251–255.

141. Mackey, B. M., and N. Bratchell. 1989. The heat resistance of *Listeria monocytogenes*. *Lett. Appl. Microbiol.* 9:89–94.

142. Malinverni, R., J. Bille, Cl. Perret, F. Regli, F. Tanner, and M. P. Glauser. 1985. Listeriose epidemique. Observation de 25 cas en 15 mois au Centre hospitalier universitaire vaudois. *Schweiz. Med. Wschr.* 115:2–10.

143. Mara, M., J. Julak, K. Kotelko, J. Hofman, and H. Veselska. 1980. Phenol-extracted lipopeptidopolysaccharide (LPPS) complex from *Listeria monocytogenes*. *J. Hyg. Epidemiol. Microbiol. Immunol.* 24:164–176.

144. Mara, M., J. Julak, and F. Potocka. 1974. Isolation and chemical properties of biologically active fractions of the surface of *Listeria monocytogenes*. *J. Hyg. Epidemiol. Microbiol. Immunol.* 18:359–364.

145. McClain, D., and W. H. Lee. 1988. Development of USDA-FSIS method for isolation of *Listeria monocytogenes* from raw meat and poultry. *J. Assoc. Off. Anal. Chem.* 71:660–664.

146. McLauchlin, J. 1987. *Listeria monocytogenes*, recent advances in the taxonomy and epidemiology of listeriosis in humans. *J. Appl. Bacteriol.* 63:1–11.

147. McLauchlin, J., A. Audurier, and A. G. Taylor. 1986. Aspects of the epidemiology of human *Listeria monocytogenes* infections in Britain 1967–1984; The use of serotyping and phage typing. *J. Med. Microbiol.* 22:367–377.

148. Mengaud, J., M. F. Vincente, J. Chenevert, J. M. Pereira, C. Geoffroy, B. Gicquel-Sanzey, F. Baquero, J.-C. Perez-Diaz, and P. Cossart. 1988. Expression in *Escherichia coli* and sequence analysis of the listeriolysin O determinant of *Listeria monocytogenes*. *Infect. Immun.* 56:766–772.

149. Mengaud, J., J. Chenevert, C. Geoffroy, J.-L. Gaillard, and P. Cossart. 1987. Identification of the structural gene encoding the SH-activated hemolysin of *Listeria monocytogenes*: Listeriolysin O is homologous to streptolysin O and pneumolysin. *Infect. Immun.* 55:3225–3227.

150. Mielke, M., S. Ehlers, and H. Hahn. 1988. The role of T cell subpopulations in cell mediated immunity to facultative intracellular bacteria. *Infection* 16 (Suppl. 2):123–127.

151. Miller, A. L., J. L. Smith, and G. A. Somkuti, eds. 1990. *Foodborne Listeriosis*. Amsterdam: Elsevier.

152. Mitsuyama, M., K.-I. Igarashi, I. Kawamura, T. Ohmori, and K. Nomoto. 1990. Difference in the induction of macrophage interleukin-1 production between viable and killed cells of *Listeria monocytogenes*. *Infect. Immun.* 58:1254–1260.

153. Mounier, J., A. Ryter, M. Coquis-Rondon, and P. J. Sansonetti. 1990. Intracellular and cell-to-cell spread of *Listeria monocytogenes* involves interaction with F-actin in the enterocytelike cell line Caco-2. *Infect. Immun.* 58:1048–1058.

154. Murray, E. G. D., R. A. Webb, and M. B. R. Swann. 1926. A disease of rabbits characterized by large mononuclear leucocytosis caused by a hitherto undescribed bacillus *Bacterium monocytogenes* (n. sp.). *J. Pathol. Bacteriol.* 29:407–439.

155. Nakane, A., A. Numata, M. Asano, M. Kohanawa, Y. Chen, and T. Minagawa. 1990. Evidence that endogenous gamma interferon is produced early in *Listeria monocytogenes* infection. *Infect. Immun.* 58:2386–2388.

156. Nicolas, J.-A., and N. Vidaud. 1987. Contribution a l'étude des *Listeria* presentés dans les denrées d'origine animale destinées à la consommation humaine. *Rec. Med. Vet.* 163(3):283–285.

157. Njoku-Obi, A. N., E. M. Jenkins, J. C. Njoku-Obi, J. Adams, and V. Covington.

1963. Production and nature of *Listeria monocytogenes* hemolysins. *J. Bacteriol.* 86:1–8.

158. Njoku-Obi, A. N., and J. W. Osebold. 1962. Studies on mechanisms of immunity of listeriosis. I. Interaction of peritoneal exudate cells from sheep with *Listeria monocytogenes* in vitro. *J. Immunol.* 89:187–194.

159. Northolt, M. D., H. J. Beckers, U. Vecht, L. Toepoel, P. S. S. Soentoro, and H. J. Wisselink. 1988. *Listeria monocytogenes*: Heat resistance and behaviour during storage of milk and whey and making of Dutch types of cheese. *Neth. Milk Dairy J.* 42:207–219.

160. Olding, L., and L. Philipson. 1960. Two cases of listeriosis in the newborn, associated with placental infection. *Acta Pathol. Microbiol.* 48:24–30.

161. Ortel, S. 1989. Listeriocins (monocins). *Int. J. Food Microbiol.* 8:249–250.

162. Ortel, S. 1968. Bakteriologische, serologische und epidemiologische Untersuchungen wahrend einer Listeriose-Epidemie. *Deutsche Gesundheitswesen* 23: 753–759.

163. Parish, M. E., and D. P. Higgins. 1989. Survival of *Listeria monocytogenes* in low pH model broth systems. *J. Food Protect.* 52:144–147.

164. Parrisius, J., S. Bhakdi, M. Roth, J. Tranum-Jensen, W. Goebel, and H. P. R. Seeliger. 1986. Production of listeriolysin by beta-hemolytic strains of *Listeria monocytogenes*. *Infect. Immun.* 51:314–319.

165. Pearson, L. J., and E. H. Marth. 1990. *Listeria monocytogenes*—threat to a safe food supply: A review. *J. Dairy Sci.* 73:912–928.

166. Petran, R. L., E. A. Zottola, and R. B. Gravani. 1988. Incidence of *Listeria monocytogenes* in market samples of fresh and frozen vegetables. *J. Food Sci.* 53:1238–1240.

167. Piffaretti, J.-C., H. Kressebuch, M. Aeschbacher, J. Bille, E. Bannerman, J. M. Musser, R. K. Selander, and J. Rocourt. 1989. Genetic characterization of clones of the bacterium *Listeria monocytogenes* causing epidemic disease. *Proc. Natl. Acad. Sci. USA* 86:3818–3822.

168. Pine, L., G. B. Malcolm, and B. D. Plikaytis. 1990. *Listeria monocytogenes* intragastric and intraperitoneal approximate 50% lethal doses for mice are comparable, but death occurs earlier by intragastric feeding. *Infect. Immun.* 58: 2940–2945.

169. Pini, P. N., and R. J. Gilbert. 1988. The occurrence in the U.K. of *Listeria* species in raw chickens and soft cheeses. *Int. J. Food Microbiol.* 6:317–326.

170. Pinkney, M., E. Beachey, and M. Kehoe. 1989. The thiol-activated toxin streptolysin O does not require a thiol group for cytolytic activity. *Infect. Immun.* 57:2553–2558.

171. Portnoy, D. A., P. S. Jacks, and D. J. Heinrichs. 1988. Role of hemolysin for the intracellular growth of *Listeria monocytogenes*. *J. Exp. Med.* 167:1459–1471.

172. Potel, J. 1953–1954. Aetiologie der Granulomatisis Infantiseptica. *Wiss. Martin-Luther U.* 3:341–364.

173. Ralovich, B. 1984. *Listeriosis Research—Present Situation and Perspective*. Budapest: Akademiai Kiado.

174. Robinson, B. B., and A. N. Njoku-Obi. 1964. Preparation and characterization of a toxic polysaccharide from *Listeria monocytogenes*. *Bacteriol. Proc.*, M203, p. 82.

175. Rocourt, J., U. Wehmeyer, P. Cossart, and E. Stackebrandt. 1987. Proposal to retain *Listeria murrayi* and *Listeria grayi* in the genus *Listeria*. *Int. J. Syst. Bacteriol.* 37:298–300.

176. Rocourt, J., A. Audurier, A. L. Courtieu, J. Durst, S. Ortel, A. Schrettenbrunner, and A. G. Taylor. 1985. A multi-centre study on the phage typing of *Listeria monocytogenes*. *Zbl. Bakt. Hyg. A* 259:489–97.

177. Rocourt, J., F. Grimont, P. A. D. Grimont, and H. P. R. Seeliger. 1982. DNA relatedness among serovars of *Listeria monocytogenes* sensu lato. *Curr. Microbiol.* 7:383–388.

178. Roll, J. T., and C. J. Czuprynski. 1990. Hemolysin is required for extraintestinal dissemination of *Listeria monocytogenes* in intragastrically inoculated mice. *Infect. Immun.* 58:3147–3150.

179. Rolle, M., and H. Mayer. 1956. Zur Pathogenese der Listeriose. Zentral Bakteriol. *Abt. I. Orig.* 166:479–483.

180. Ruhland, G. J., and F. Fiedler. 1987. Occurrence and biochemistry of lipoteichoic acids in the genus *Listeria. System. Appl. Microbiol.* 9:40–46.

181. Ryser, E. T., and E. H. Marth. 1988. Survival of *Listeria monocytogenes* in cold-pack cheese food during refrigerated storage. *J. Food Protect.* 51:615–621.

182. Ryser, E. T., and E. H. Marth. 1987. Fate of *Listeria monocytogenes* during the manufacture and ripening of camembert cheese. *J. Food Protect.* 50:372–378.

183. Ryser, E. T., E. H. Marth, and M. P. Doyle. 1985. Survival of *Listeria monocytogenes* during manufacture and storage of cottage cheese. *J. Food Protect.* 48:746–750.

184. Schlech, W. F., III, P. M. Lavigne, R. A. Bortolussi, A. C. Allen, E. V. Haldane, A. J. Wort, A. W. Hightower, S. E. Johnson, S. H. King, E. S. Nicholls, and C. V. Broome. 1983. Epidemic listeriosis—evidence for transmission by food. *N. Engl. J. Med.* 308:203–206.

185. Schmidt, U., H. P. R. Seeliger, E. Glenn, B. Langer, and L. Leistner. 1988. Listerienfunde in rohen Fleischerzeugnissen. *Fleischwirtsch.* 68:1313–1316.

186. Schoenberg, A. 1989. Method to determine virulence of *Listeria* strains. *Int. J. Food Microbiol.* 8:281–284.

187. Schultz, G. 1967. Untersuchungen über das Vorkommen von Listerien in Rohmilch. *Monatsh. Veterinaermed.* 22:766–768.

188. Schwartz, B., D. Hexter, C. V. Broome, A. W. Hightower, R. B. Hirschhorn, J. D. Porter, P. S. Hayes, W. F. Bibb, B. Lorber, and D. G. Faris. 1989. Investigation of an outbreak of listeriosis: New hypotheses for the etiology of epidemic *Listeria monocytogenes* infections. *J. Inf. Dis.* 159:680–685.

189. Schwartz, B., C. A. Ciesielski, C. V. Broome, S. Gaventa, G. R. Brown, B. G, Gellin, A. W. Hightower, and L. Mascola. 1988. Association of sporadic listeriosis with consumption of uncooked hot dogs and undercooked chicken. *Lancet* ii: 779–782.

190. Seeliger, H. P. R., and B. Langer. 1989. Serological analysis of the genus *Listeria.* Its values and limitations. *Int. J. Food Microbiol.* 8:245–248.

191. Seeliger, H. P. R. 1961. *Listeriosis,* 2d ed. New York: Hafner

192. Seeliger, H. P. R. and K. Höhne. 1979. Serotyping of *Listeria monocytogenes* and related species. *Meth. Microbiol.* 13:31–49.

193. Selander, R. K., D. A. Caugant, H. Ochman, J. M. Musser, M. N. Gilmour, and T. S. Whittam. 1986. Methods of multilocus enzyme electrophoresis for bacterial population genetics and systematics. *Appl. Environ. Microbiol.* 51:873–884.

194. Shelef, L. A. 1989. Survival of *Listeria monocytogenes* in ground beef or liver during storage at 4° and 25°C. *J. Food Protect.* 52:379–383.

195. Shelef, L. A. 1989b. Listeriosis and its transmission by food. *Prog. Food Nutr. Sci.* 13:363–382.

196. Singh, S. P., B. L. Moore, and I. H. Siddique. 1981. Purification and further characterization of phenol extract from *Listeria monocytogenes. Am. J. Vet. Res.* 42:1266–1268.

197. Sizmur, K., and C. W. Walker. 1988. *Listeria* in prepackaged salads. *Lancet* i:1167.

198. Skalka, B., J. Smola, and K. Elischerova. 1982. Routine test for in vitro differentiation of pathogenic and apathogenic *Listeria monocytogenes* strains. *J. Clin. Microbiol.* 15:503–507.

199. Skovgaard, N., and C.-A. Morgen. 1988. Detection of *Listeria* spp. in faeces from animals, in feeds, and in raw foods of animal origin. *Int. J. Food Microbiol.* 6:229–242.

200. Slade, P. J., D. L. Collins-Thompson, and F. Fletcher. 1988. Incidence of *Listeria* species in Ontario raw milk. *Can. Inst. Food Sci. Technol. J.* 21:425–429.

201. Sokolovic, Z., A. Fuchs, and W. Goebel. 1990. Synthesis of species-specific stress proteins by virulent strains of *Listeria monocytogenes*. *Infect. Immun.* 58:3582–3587.

202. Sokolovic, Z., and W. Goebel. 1989. Synthesis of listeriolysin in *Listeria monocytogenes* under heat shock conditions. *Infect. Immun.* 57:295–298.

203. Sorrells, K. M., D. C. Enigl, and J. R. Hatfield. 1989. Effect of pH, acidulant, time, and temperature on the growth and survival of *Listeria monocytogenes*. *J. Food Protect.* 52:571–573.

204. Stanley, N. F. 1949. Studies on *Listeria monocytogenes*. I. Isolation of a monocytosis-producing agent (MPA). *Aust. J. Expt. Biol. Med.* 27:123–131.

205. Steinbruegge, E. G., R. B. Maxcy, and M. B. Liewen. 1988. Fate of *Listeria monocytogenes* on ready to serve lettuce. *J. Food Protect.* 51:596–599.

206. Stuart, S. E., and H. J. Welshimer. 1974. Taxonomic reexamination of *Listeria* Pirie and transfer of *Listeria grayi* and *Listeria murrayi* to a new genus, *Murraya*. *Int. J. Syst. Bacteriol.* 24:177–185.

207. Swaminathan, B., P. S. Hayes, V. A. Przybyszewski, and B. D. Plikaytis. 1988. Evaluation of enrichment and plating media for isolating *Listeria monocytogenes*. *J. Assoc. Off. Anal. Chem.* 71:664–668.

208. Sword, C. P., and M. J. Pickett. 1961. The isolation and characterization of bacteriophages from *Listeria monocytogenes*. *J. Gen. Microbiol.* 25:241–248.

209. Terplan, G., and S. Steinmeyer. 1989. Investigations on the pathogenicity of *Listeria* spp. by experimental infection of the chick embryo. *Int. J. Food Microbiol.* 8:277–280.

210. Uchikawa, K., I. Sekikawa, and I. Azuma. 1986. Structural studies on lipoteichoic acids from four *Listeria* strains. *J. Bacteriol.* 168:115–122.

211. van Furth, R., W. Sluiter, and J. T. van Dissel. 1988. Roles of factor increasing monocytopoiesis (FIM) and macrophage activation in host resistance to *Listeria monocytogenes*. *Infection* 16 (Suppl. 2):137–140.

212. Vazquez-Boland, J.-A., L. Dominguez, E.-F. Rodriguez-Ferri, and G. Suarez. 1989. Purification and characterization of two *Listeria ivanovii* cytolysins, a sphingomyelinase C and a thiol-activated toxin (ivanolysin O). *Infect. Immun.* 57:3928–3935.

213. Vicente, M. F., F. Baquero, and J. C. Perez-Diaz. 1985. Cloning and expression of the *Listeria monocytogenes* hemolysin in *Escherichia coli*. *FEMS Microbiol. Lett.* 30:77–79.

214. Watkins, J., and K. P. Sleath. 1981. Isolation and enumeration of *Listeria monocytogenes* from sewage, sewage sludge and river water. *J. Appl. Bacteriol.* 50:1–9.

215. Weagant, S. D., P. N. Sado, K. G. Colburn, J. D. Torkelson, F. A. Stanley, M. H. Krane, S. C. Shields, and C. F. Thayer. 1988. The incidence of *Listeria* species in frozen seafood products. *J. Food Protect.* 51:655–657.

216. Weis, J., and H. P. R. Seeliger. 1975. Incidence of *Listeria monocytogenes* in nature. *Appl. Microbiol.* 30:29–32.

217. Welshimer, H. J. 1960. Survival of *Listeria monocytogenes* in soil. *J. Bacteriol.* 80:316–320.

218. Welshimer, H. J., and J. Donker-Voet. 1971. *Listeria monocytogenes* in nature. *Appl. Microbiol.* 21:516–519.
219. Wenger, J. D., B. Swaminathan, P. S. Hayes, S. S. Green, M. Pratt, R. W. Pinner, A. Schuchat, and C. V. Broome. 1990. *Listeria monocytogenes* contamination of turkey franks: Evaluation of a production facility. *J. Food Protect.* 53:1015–1019.
220. Wesley, I. V., and F. Ashton. 1991. Restriction enzyme analysis of *Listeria monocytogenes* strains associated with food-borne epidemics. *Appl. Environ. Microbiol.* 57:969–975.
221. Wesley, I. V., R. D. Wesley, J. Heisick, F. Harrel, and D. Wagner. 1990. Restriction enzyme analysis in the epidemiology of *Listeria monocytogenes*. In: *Symposium on Cellular and Molecular Modes of Action of Selected Microbial Toxins in Foods and Feeds*, ed. J. L. Richard, 225–238. New York: Plenum.
222. Wexler, H., and J. D. Oppenheim. 1979. Isolation, characterization, and biological properties of an endotoxin-like material from the gram-positive organism *Listeria monocytogenes*. *Infect. Immun.* 23:845–857.
223. Wong, H.-C., W.-L. Chao, and S.-J. Lee. 1990. Incidence and characterization of *Listeria monocytogenes* in foods available in Taiwan. *Appl. Environ. Microbiol.* 56:3101–3104.
224. Woodbine, M., ed. 1975. *Proceedings of the 6th International Symposium on Listeriosis*. Leicester, U.K.: Leicester University Press.
225. Zaika, L. L., S. A. Palumbo, J. L. Smith, F. Del Corral, S. Bhaduri, C. O. Jones, and A. H. Kim. 1990. Destruction of *Listeria monocytogenes* during frankfurter processing. *J. Food Protect.* 53:18–21.

22

Foodborne Gastroenteritis Caused by *Salmonella*, *Shigella*, and *Escherichia*

Among the gram-negative rods that cause foodborne gastroenteritis, the most important are members of the genus *Salmonella*. This syndrome, along with those caused by *Shigella* and *Escherichia coli*, are discussed in this chapter. The general incidence of some of these organisms in a variety of foods is discussed in Chapter 4.

SALMONELLOSIS

The salmonellae are small, gram-negative, nonsporing rods that are indistinguishable from *E. coli* under the microscope or on ordinary nutrient media. They are widely distributed in nature, with humans and animals being their primary reservoirs. *Salmonella* food poisoning results from the ingestion of foods containing appropriate strains of this genus in significant numbers.

The classification of organisms within the genus *Salmonella* is under review. In the ninth edition of *Bergey's Manual*, *S. choleraesuis* is the type species, and the genus is divided into five subgenera (7). A proposal has been made (82) to redefine the genus to consist of only one species, *S. enterica* (L. adj. *enterica*, of the gut). The one-species concept would recognize the following subspecies based on DNA-DNA hybridization properties.

S. enterica subsp. *enterica*
S. enterica subsp. *salamae*
S. enterica subsp. *arizonae*
S. enterica subsp. *diarizonae*
S. enterica subsp. *houtenae*
S. enterica subsp. *bongori*
S. enterica subsp. *indica*

If this proposal is accepted, *S. enterica* would become the type species. Further, common species names of foodborne serovars such as *S. typhimurium* and *S. newport* would not be italicized but would be written as *Salmonella* Typhimurium, *Salmonella* Newport, and the like.

For epidemiologic purposes the salmonellae can be classified into three groups (19, 128).

1. Those that infect humans only: These include *S. typhi*, *S. paratyphi* A, and *S. paratyphi* C. This group includes the agents of typhoid and the paratyphoid fevers, which are the most severe of all diseases caused by salmonellae. Typhoid fever has the longest incubation time, produces the highest body temperature, and has the highest mortality rate. *S. typhi* may be isolated from blood and sometimes the urine of victims. The paratyphoid syndrome is milder than that of typhoid.
2. The host-adapted serovars (some of which are human pathogens and may be contracted from foods): Included are *S. gallinarum* (poultry), *S. dublin* (cattle), *S. abortus-equi* (horses), *S. abortus-ovis* (sheep), and *S. choleraesuis* (swine).
3. Unadapted serovars (no host preference). These are pathogenic for humans and other animals, and they include most foodborne serovars. The foodborne salmonellosis syndrome is described in a later section.

Regardless of species and subspecies classifications, the genus *Salmonella* consists of around 2,200 serotypes (serovars) based on 67 O-antigen groups and numerous H-antigens (128). Pending final resolution of the one-species proposal, *Salmonella* classification is treated in the traditional way in the remainder of this chapter and throughout this text.

Serotyping of *Salmonella*

The classification of these organisms by antigenic analysis is based on the original work of Kauffmann and White and is often referred to as the Kauffmann-White Scheme. Classification by this scheme makes use of both somatic and flagellar antigens. Somatic antigens are designated O antigens, and flagellar antigens are designated H antigens. The K antigens are capsular antigens that lie at the periphery of the cell and prevent access of anti-O agglutinins (antibodies) to their homologous somatic antigens. The K antigen differs from ordinary O antigens in being destroyed by heating for 1 h at 60°C and by dilute acids and phenol. The use of H, O, and K antigens as the basis of classification of *Salmonella* spp. is based on the fact that each antigen possesses its own genetically determined specificity.

When classification is made by use of antigenic patterns, species and serovars are placed in groups designated A, B, C, and so on, according to similarities in content of one or more O antigens. Thus, *S. hirschfeldii*, *S. choleraesuis*, *S. oranienberg*, and *S. montevideo* are placed in Group C$_1$ because they all possess O antigens 6 and 7 in common. *S. newport* is placed in Group C$_2$ due to its possession of O antigens K and 8 (Table 22-1). For further classification, the flagellar or H antigens are employed. These antigens are of two types: specific phase or phase 1, and group phase or phase 2. Phase 1 antigens are

TABLE 22-1.   Antigenic Structure of Some of Common Salmonellae

Group	Species/Serovars	O antigens[a]	H Antigens Phase 1	Phase 2
A	*S. paratyphi* A	*1*, 2, 12	a	(1,5)
B	*S. schottmuelleri*	*1*, 4, (5),12	b	1,2
	S. typhimurium	*1*, 4, (5), 12	i	1,2
C₁	*S. hirschfeldii*	6, 7, (Vi)	c	1,5
	S. choleraesuis	6, 7	(c)	1,5
	S. oranienburg	6, 7	m,t	—
	S. montevideo	6, 7	g,m,s (p)	(1,2,7)
C₂	*S. newport*	6,8	e,h	1,2
D	*S. typhi*	9, 12, (Vi)	d	—
	S. enteritidis	*1*, 9, 12	g,m	(1,7)
	S. gallinarum	*1*, 9, 12	—	—
E₁	*S. anatum*	3, 10	e,h	1,6

[a] The underlined antigens are associated with phage conversion. ( ) = May be absent.

shared with only a few other species or varieties of *Salmonella*; phase 2 may be more widely distributed among several species. Any given culture of *Salmonella* may consist of organisms in only one phase or of organisms in both flagellar phases. The H antigens of phase 1 are designated with small letters, and those of phase 2 are designated by arabic numerals. Thus, the complete antigenic analysis of *S. choleraesuis* is as follows: 6, 7, c, 1, 5, where 6 and 7 refer to O antigens, c to phase-1 flagellar antigens, and 1 and 5 to phase-2 flagellar antigens (Table 22-1). *Salmonella* subgroups of this type are referred to as serovars. With a relatively small number of O, phase 1, and phase 2 antigens, a large number of permutations are possible, allowing for the possibility of a large number of serovars.

The naming of *Salmonella* is done by international agreement. Under this system, a serovar is named after the place where it was first isolated—*S. london*, *S. miami*, *S. richmond*, and so on. Prior to the adoption of this convention, species and subtypes were named in various ways—for example, *S. typhimurium* as the cause of typhoid fever in mice.

Distribution of *Salmonella*

The primary habitat of *Salmonella* spp. is the intestinal tract of animals such as birds, reptiles, farm animals, humans, and occasionally insects. Although their primary habitat is the intestinal tract, they may be found in other parts of the body from time to time. As intestinal forms, the organisms are excreted in feces from which they may be transmitted by insects and other living creatures to a large number of places. As intestinal forms, they may also be found in water, especially polluted water. When polluted water and foods that have been contaminated by insects or by other means are consumed by humans and other animals, these organisms are once again shed through fecal matter with a continuation of the cycle. The augmentation of this cycle through the international shipment of animal products and feeds is in large part responsible for the worldwide distribution of salmonellosis and its consequent problems.

While *Salmonella* spp. have been recovered repeatedly from a large number of different animals, their incidence in various parts of animals has been shown to vary. In a study of slaughterhouse pigs, Kampelmacher (59) found these organisms in spleen, liver, bile, mesenteric and portal lymph nodes, diaphragm, and pillar, as well as in feces. A higher incidence was found in lymph nodes than in feces. The frequent occurrence of *Salmonella* spp. among susceptible animal populations is due in part to the contamination of *Salmonella*-free animals by animals within the population that are carriers of these organisms or are infected by them. A carrier is defined as a person or an animal that repeatedly sheds *Salmonella* spp., usually through feces, without showing any signs or symptoms of the disease. Upon examining poultry at slaughter, Sadler and Corstvet (102) found an intestinal carrier rate of 3 to 5%. During and immediately after slaughter, carcass contamination from fecal matter may be expected to occur.

Animal Feeds

The industry-wide incidence rate of salmonellae in animal feeds in 1989 was about 49%. Among U.S. Department of Agriculture (USDA) inspected packers and renderers, the rate was between 20 and 25%, and only 6% for pelleted animal foods (40). In a study of breeder/multiplier and broiler houses, 60% of meat and bone meal contained salmonellae, and feed was considered to be the ultimate source of salmonellae to breeder/multiplier houses (56). It has been noted that salmonellae contamination in U.S. broiler production changed little between 1969 and 1989 (56). Salmonellae contamination of rendered products is most likely due to recontamination. The primary serovars found in animal feeds are *S. senftenberg*, *S. montevideo*, and *S. cerro*. *S. enteritidis* has not been found in rendered products or finished feeds.

In an examination of the rumen contents of healthy cattle after slaughter, 45% were found to contain salmonellae (41). Some 57% of samples taken from the environment of cattle in transit to slaughter were positive for these organisms.

Food Products

Salmonellae have been found in commercially prepared and packaged foods with 17 of 247 products examined being positive (1). Among the contaminated foods were cake mixes, cookie doughs, dinner rolls, and cornbread mixes. These organisms have been found in coconut meal, salad dressing, mayonnaise, milk, and many other foods. In a study of health foods, none of plant origin yielded salmonellae, but from two of three lots of beef liver powder from the same manufacturer were isolated *S. minnesota*, *S. anatum*, and *S. derby* (121).

In a study of 40 sausage-producing plants in 1969, the overall incidence of salmonellae was 28.6% of the 566 samples examined (55). Ten years later, the overall incidence had decreased to 12.4% of 603 examined. Of the 40 matched plants, the incidence decreased in 20, increased in 13, and remained the same in 7.

In a study of the incidence of salmonellae in 69 packs of raw chicken pieces,

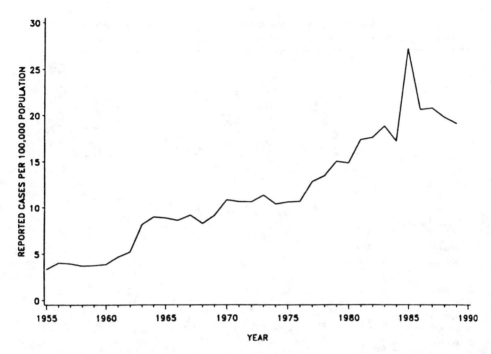

FIGURE 22-1.   Nontyphoid salmonellosis in the United States, 1955–1989.
From Morbidity and Mortality Weekly Report *38:38, 1989.*

34.8% were positive, and 11 serovars were represented, with *S. muenchen* being the most common (28). From a study in Venezuela, 41 of 45 chicken carcasses studied yielded salmonellae consisting of 11 serovars, with *S. anatum* being most frequently isolated (96).

Eggs, poultry, meat, and meat products are the most common food vehicles of salmonellosis to humans. In a study of 61 outbreaks of human salmonellosis for the period 1963 to 1965, eggs and egg products accounted for 23, chicken and turkey for 16, beef and pork for 8, ice cream for 3, potato salad for 2, and other miscellaneous foods for 9 (116). In 1967, the most common food vehicles involved in 12,836 cases of salmonellosis from 37 states were beef, turkey, eggs and egg products, and milk. Of 7,907 salmonellae isolations made by the Centers for Disease Control (CDC) in 1966, 70% were from raw and processed food sources with turkey and chicken sources accounting for 42%. Human isolations of nontyphoid salmonellae in the United States for the years 1955–1989 are presented in Figure 22-1. It may be noted that 34,766 of the cases for 1982 affected children under 5 years of age. The increase in isolations from adults is accounted for in part by the consumption of improperly pasteurized and raw milk. The specific serovar associated with raw milk consumption is *S. dublin*. Persons from whom this serovar was isolated were older and had underlying illness requiring longer periods of hospitalization.

Up to 70% of broiler carcasses have been found to be contaminated with salmonella. The organisms appear not to be normal flora of poultry but are acquired from the environment via insects, rodents, feeds, other animals, and humans (78). In a study of 418 pooled ovary samples from 32 of 42 flocks, 111

TABLE 22-2.   Five Most Prevalent *Salmonella* Serovars Reported from Different Sources, 1956–1990 (numbers indicate rank)

Serovars	Humans, U.K., 1956–1960	Foods, U.S., 1963–1965	Foods, U.S., 1974–1985	Humans, U.S., 1967	Nonhuman Sources, U.S., 1967	Humans, U.S., 1972	Humans, U.S., 1980	Raw Pork Sausage, U.S., 1979	Raw Chicken, Venezuela, 1983	Chicken Carcasses, Portugal, 1986–1987	Ovaries of U.S. Layer Hens, 1989–1990	U.S. Broiler Production/Processing System, 1989
S. adelaide												3
S. agona								3		2	2	
S. anatum				3	3			2	1			4
S. bloemfontein									2			
S. derby				4	4			1		5		
S. enteritidis	3					3	3				1	
S. hadar												5[a]
S. halmstad									4			
S. havana									5			
S. heidelberg	2	5		2	2	5	2			1		2
S. indiana												5[a]
S. infantis		1	5	5	5	4	5	4				
S. kentucky			2								5[a]	
S. mbandaka											4	
S. meunchen								5[a]				
S. montevideo		4									5[a]	
S. newport	4					2	4			3[a]		
S. oranienberg		2									3	
S. saint-paul										3[a]		
S. schwarzengrund									3			
S. senftenberg			3									
S. typhimurium	1	3		1	1	1	1	5[a]				1
S. tennessee			1									
S. thompson	5											
S. worthington			4									
References	119	116	124	19	19	122	16	55	96	73	3	56

[a] Tie.

(26.6%) were positive for salmonellae (3). The five most frequently isolated serovars among the 15 different ones found are indicated in Table 22-2.

S. typhimurium is invariably the most commonly found foodborne serovar throughout the world. In general, the incidence of salmonellae serovars in foods may be expected to parallel their incidence in humans and animals, but this is not always true (Table 22-2). The unusually high prevalence of S. enteritidis in the 1980s and early 1990s is exceptional, and this serovar is discussed further below.

Growth and Destruction of Salmonellae

These organisms are typical of other gram-negative bacteria in being able to grow on a large number of culture media and produce visible colonies well

TABLE 22-3.   Minimum pH at which Salmonellae Would Initiate Growth under Optimum Laboratory Conditions

Acid	pH
Hydrochloric	4.05
Citric	4.05
Tartaric	4.10
Gluconic	4.20
Fumaric	4.30
Malic	4.30
Lactic	4.40
Succinic	4.60
Glutaric	4.70
Adipic	5.10
Pimelic	5.10
Acetic	5.40
Propionic	5.50

Source: Chung and Goepfert (17); copyright © 1970 by Institute of Food Technologists.

Note: Tryptone-yeast extract-glucose broth was inoculated with 10^4 cells per milliliter of *Salmonella anatum, S. tennessee,* or *S. senftenberg*.

within 24 h at about 37°C. They are generally unable to ferment lactose, sucrose, or salicin, although glucose and certain other monosaccharides are fermented, with the production of gas. Although they normally utilize amino acids as N-sources, in the case of *S. typhimurium*, nitrate, nitrite, and NH_3 will serve as sole sources of nitrogen (90). While lactose fermentation is not usual for these organisms, some serovars can utilize this sugar.

The pH for optimum growth is around neutrality, with values above 9.0 and below 4.0 being bactericidal. A minimum growth pH of 4.05 has been recorded for some (with HCl and citric acids), but depending on the acid used to lower pH, the minimum may be as high as 5.5 (17). The effect of acid used to lower pH on minimum growth is presented in Table 22-3. Aeration was found to favor growth at the lower pH values. The parameters of pH, a_w, nutrient content, and temperature are all interrelated for salmonellae, as they are for most other bacteria (122). For best growth, the salmonellae require pH between 6.6 and 8.2. The lowest temperatures at which growth has been reported are 5.3°C for *S. heidelberg* nd 6.2°C for *S. typhimurium* (76). Temperatures of around 45°C have bee reported by several authors to be the upper limit for growth. Regarding available moisture, growth inhibition has been reported for a_w values below 0.94 in media with neutral pH, with higher a_w values being required as pH is decreased toward growth minima.

Unlike the staphylococci, the salmonellae are unable to tolerate high salt concentrations. Brine above 9% is reported to be bactericidal. Nitrite is effective, with the effect being greatest at the lower pH values. This suggests that the inhibitory effect of this compound is referable to the undissociated HNO_2

molecule. The survival of *Salmonella* spp. in mayonnaise was studied by Lerche (70), who found that they were destroyed in this product if the pH was below 4.0. It was found that several days may be required for destruction if the level of contamination is high, but within 24 h for low numbers of cells. *S. thompson* and *S. typhimurium* were found to be more resistant to acid destruction than *S. senftenberg*.

With respect to heat destruction, all salmonellae are readily destroyed at milk pasteurization temperatures. Thermal *D* values for the destruction of *S. senftenberg* 775 W under various conditions are given in Chapter 14. Shrimpton et al. (108) reported that *S. senftenberg* 775 W required 2.5 min for a 10^4-10^5 reduction in numbers at 54.4°C in liquid whole egg. This strain is the most heat resistant of all salmonellae serovars. This treatment of liquid whole egg has been shown to produce a *Salmonella*-free product and destroy egg alpha-amylase (see Chap. 14 for the heat *pasteurization* of egg while). It has been suggested (10) that the alpha-amylase test may be used as a means of determining the adequacy of heat pasteurization of liquid egg (compare with the pasteurization of milk and the enzyme phosphatase). In a study on the heat resistance of *S. senftenberg* 775 W, Ng et al. (87) found this strain to be more heat sensitive in the log phase than in the stationary phase of growth. These authors also found that cells grown at 44°C were more heat resistant than those grown at either 15° or 35°C.

With respect to the destruction of *Salmonella* in baked foods, Beloian and Schlosser (6) found that baked foods reaching a temperature of 160°F or higher in the slowest heating region can be considered *Salmonella* free. These authors employed *S. senftenberg* 775 W at a concentration of 7,000 to 10,000 cells/ml placed in reconstituted dried egg. With respect to the heat destruction of this strain in poultry, it is recommended that internal temperatures of at least 160°F be attained (81). Although *S. senftenberg* 775 W has been reported to be 30 times more heat resistant than *S. typhimurium* (87), the latter organism has been found to be more resistant to dry heat than the former (39). These authors tested dry heat resistance in milk chocolate.

The destruction of *S. pullorum* in turkeys was investigated by Rogers and Gunderson (98), who found that it required 4 h and 55 min to destroy an initial inoculum of 115 million in 10- to 11-lb turkeys with an internal temperature of 160°F, and for 18-lb turkeys with an initial inoculum of 320 million organisms, 6 h and 20 min were required for destruction. The salmonellae are quite sensitive to ionizing radiation, with doses of 5 to 7.5 kGy being sufficient to eliminate them from most foods and feed. The decimal reduction dose has been reported to range from 0.4 to 0.7 kGy for *Salmonella* spp. in frozen eggs. The effect of various foods on the radiosensitivity of salmonellae is shown in a study by Ley et al. (71). These investigators found that for frozen whole egg, 5 kGy gave a 10^7 reduction in the numbers of *S. typhimurium*, while 6.5 kGy was required to give a 10^5 reduction in frozen horsemeat, between 5 and 7.5 kGy for a 10^5-10^8 reduction in bone meal, and only 4.5 kGy to give a 10^3 reduction of *S. typhimurium* in desiccated coconut.

In dry foods, *S. montevideo* was found to be more resistant than *S. heidelberg* when inoculated into dry milk, cocoa powder, poultry feed, meat, and bone meal (58). Survival was greater at a_w 0.43 and 0.52 than at a_w 0.75.

The *Salmonella* Food-Poisoning Syndrome

This syndrome is caused by the ingestion of foods that contain significant numbers of non–host specific species or serotypes of the genus *Salmonella*. From the time of ingestion of food, symptoms usually develop in 12 to 14 h, although shorter and longer times have been reported. The symptoms consist of nausea, vomiting, abdominal pain (not as severe as with staphylococcal food poisoning), headache, chills, and diarrhea. These symptoms are usually accompanied by prostration, muscular weakness, faintness, moderate fever, restlessness, and drowsiness. Symptoms usually persist for 2 to 3 days. The average mortality rate is 4.1%, varying from 5.8% during the first year of life, to 2% between the first and fiftieth year, and 15% in persons over 50. Among the different species of *Salmonella*, *S. choleraesuis* has been reported to produce the highest mortality rate—21%.

While these organisms generally disappear rapidly from the intestinal tract, up to 5% of patients may become carriers of the organisms upon recovery from this disease.

Numbers of cells on the order of 10^7–10^9/g are generally necessary for salmonellosis. That outbreaks may occur in which relatively low numbers of cells are found has been noted (23). From three outbreaks, numbers of cells found were as low as 100/100 g (*S. eastbourne* in chocolate) to 15,000/g (*S. cubana* in a carmine dye solution). In general, minimum numbers for gastroenteritis range between 10^5–10^6/g for *S. bareilly* and *S. newport* to 10^9–10^{10} for *S. pullorum* (11).

Salmonella Toxins and Virulence Properties

It appears that the pathogenesis of salmonellosis may involve two toxins—an enterotoxin and a cytotoxin. The enterotoxin was first demonstrated in 1975 by Koupal and Deibel (67). Using *S. enteritidis* and the suckling mouse assay, a difficult-to-separate cell envelope–associated toxin was prepared, producing results in the suckling mouse assay similar to those elicited by the heat-stable and heat-labile enterotoxins of *E. coli* (see Chap. 7 and the section on *E. coli* in this chapter). Loop assays including the infant rabbit gave inconsistent results. Employing *S. typhimurium*, an enterotoxin was produced in brain heart infusion (BHI) broth and a 2% Casamino acids medium and measured by the rabbit ileal loop assay (107). These two media were found to be the best of several that were tried. Meanwhile, another group of researchers recovered from salmonellae a toxin that induced rabbit skin permeability and acted in a manner similar to the enterotoxins of *Vibrio cholerae* and *E. coli* (104). The toxin in question produced positive responses in the rabbit ileal loop assay. It was later shown that the vascular permeability factor could be neutralized with monospecific cholera antitoxin, and that the toxin induced elongation in Chinese hamster ovary (CHO) cells similar to the toxin of *V. cholerae* (105). In their attempts to obtain larger quantities of salmonellae enterotoxin, Koo and Peterson (62) studied the influence of nutritional factors and found that glycerol, biotin, and Mn^{2+} enhanced production, while glucose was found to be a poor carbon source. It was later determined that more enterotoxin is produced during the stationary phase of growth, at pH 6–7 or higher, at 37°C, and with increased aeration (63). The enterotoxin was found to be heat labile at 100°C, to have a molecular weight over 110,000 daltons, and an isoelectric point of approximately 4.3 to 4.8 (53). Mitomycin C added 3 h after inoculation increased the quantities in culture filtrates due to bacteriophage induction and subsequent cell lysis. The weight of the evidence suggests that the salmonellae toxin described above is an enterotoxin that acts in a manner similar to those of *E. coli* by elevating intestinal cylic adenosine monophosphate (cAMP). Unlike the *E. coli* enterotoxins, it is produced in much lower quantities and is more difficult to separate from producing cells. Further, the salmonellae enterotoxin is quite similar to cholera toxin (CT) in biologic and antigenic characteristics. Heated CT (procholeragenoid) administered parenterally protects against loop fluid responses by viable salmonellae cells (92).

While the enterotoxin described was shown to affect the adenylate cyclase system and induce fluid accumulation in animal models, the intestinal toxicity associated with salmonellosis was not explained. Although the enterotoxins of *E. coli* induce fluid accumulation, tissue toxicity that is normally associated with shigellosis and salmonellosis does not occur. In other words, the pathogenesis of salmonellosis resembles that of shigellosis more than the enterotoxin-mediated syndromes. This led Koo et al. (65) to examine salmonellae extracts for a cytotoxin. Some European workers were actually the first to show cytotoxic activity of salmonellae extracts as early as 1962. When extracts from salmonellae were added to isolated rabbit intestinal epithelial cells and to Vero cells, protein synthesis was inhibited (64, 65). These investigators presented evidence to support the view that the cellular damage that occurs to the intestinal mucosa during salmonellosis is caused by a cytotoxin. Once damaged, the intestinal mucosa is more easily invaded by the infecting organisms, which can then bring about additional tissue damage. While the pathogenesis of shigellosis and salmonellosis appears to be closely related, more tissue destruction occurs in the former than in the latter. This may be due to an additional toxin—one that is more destructive—or to other factors. That prostaglandins may play a role is suggested by the finding that both salmonellae enterotoxin and CT induced their synthesis in intestinal epithelial cells when an anti-inflammatory agent (indomethacin) was employed (27).

Salmonellae, along with yersinae, shigellae, and some *E. coli* strains, can enter and multiply in eukaryotic cells. Microfilaments appear to be involved in salmonellae uptake by cells since cytochalasins, which inhibit microfilament

TABLE 22-4.   Outbreaks, Cases, and Deaths from *S. enteritidis* in the United States, January 1985–October 1989

Year	Outbreaks/Cases/Deaths
1985	19/608/1
1986	34/1042/6
1987	50/2370/15
1988	37/956/8
1989	49/1628/13

Source: *Morb. Mort. Week. Rep.* 38: 877–880, 1990.

formation, decrease invasion (34). It is possible that microfilaments induce the synthesis of *Salmonella* proteins that are required for adherence and uptake. Several new proteins have been shown to be essential for adherence and invasion of *S. choleraesuis* and *S. typhimurium* (33). Mutants that lacked the ability to synthesize these proteins were avirulent for mice. Some of the cellular permeability proteins of the membrane of salmonellae, porins, are known to induce pathological effects (38), along with other membrane proteins.

Virulence for mice by at least three serovars—*S. dublin*, *S. enteritidis* and *S. typhimurium*—is associated with plasmids of 36 to 60 Mdal (34, 44, 86), but their precise role in virulence is unclear.

Animal models and tissue culture systems for the assay of salmonellae toxins are further described in Chapter 7.

Incidence and Vehicle Foods

The precise incidence of salmonellae food poisoning in the United States is not known since small outbreaks are often not reported to public health authorities. CDC researchers indicate that over 40,000 cases occur each year, with about 500 deaths (18). Based on certain assumptions, the actual number of cases in the United States in 1988 was estimated to be between 840,000 and 4 million (118). As is the case for staphylococcal gastroenteritis, the largest outbreaks of salmonellosis typically occur at banquets or similar functions. However, the outbreak that occurred during the spring of 1985 was exceptional and clearly the largest seen to date. Over 16,000 culture-confirmed cases were traced to two brands of pasteurized 2% milk produced by a single dairy plant, and two estimates placed the actual number of cases at 168,791 and 197,581 persons (99). *S. typhimurium*, the etiologic agent, was characterized by an unusual plasmid profile and by an antimicrobial resistance pattern. The next largest outbreak occurred in 1974 on the Navajo Indian Reservation, when 3,400 persons became ill (52). The vehicle food was potato salad served to about 11,000 individuals at a barbecue. It was prepared and stored for up to 16 h at improper holding temperatures prior to serving; the serovar isolated was *S. newport*.

Since the late 1970s, *S. enteritidis* has been the cause of a series of outbreaks in the northeastern part of the United States and in parts of Europe. The U.S. outbreaks for 1985–1989 are listed in Table 22-4, along with cases and deaths. Between January 1985 and May 1987, there were 65 foodborne outbreaks in the U.S. Northeast, with 2,119 cases and 11 deaths (115). Seventy-seven percent of the outbreaks were traced to Grade A shell eggs or foods that contained eggs (3). Throughout the United States from 1973 through 1984, 44% of *S. enteritidis* outbreaks were associated with egg-containing foods (115). During the 1970s, *S. enteritidis* accounted for 5% of all U.S. salmonellae isolations but in 1989 it accounted for 20% (118). It was second only to *S. typhimurium* in 1987 accounting for 16% of all salmonellae isolations. It was the most common serovar in eggs in Spain (91) and the most predominant cause of foodborne salmonellosis in England and Wales in 1988, where it was found in both poultry meat and eggs (48). Unlike U.S. outbreaks, those in

Europe are caused by phage type 4 strains, which are more invasive for young chicks than phage types 7, 8, or 13a (48).

Why the increased incidence of *S. enteritidis* outbreaks are associated with eggs and poultry products is unclear. The organism has been found by some investigators inside the eggs and ovaries of laying hens (91), but others have failed to find it in unbroken eggs. It was recovered from the ovaries of only 1 of 42 layer flocks located primarily in the southeast United States (3). The *S. enteritidis* outbreaks have occurred more in July and August than other months, suggesting growth of the organism in or on eggs and other poultry products. The strains in question are not heat resistant, and many of the outbreaks have occurred following the consumption of raw or undercooked eggs. In one study in which *S. enteritidis* was inoculated into the yolk of eggs from normal hens, no growth occurred at 7°C in 94 days (9). Growth in yolks at 37°C was faster from normal hens than in those from hens that were seropositive.

The salmonellae outbreaks, cases, and deaths associated with foods in the United States between 1983 and 1987 are listed in Table 22-5. The 10 leading food sources of those known for 1973 through 1987 in the United States are listed in Table 22-6. The leading food sources were beef, turkey, chicken, ice cream, and pork products.

Among the unusual cases of foodborne salmonellosis in the United States

TABLE 22-5. *Salmonella* **Outbreaks, Cases, and Deaths Traced to Foods in the United States, 1983– 1987**

Years	Outbreaks/Cases/Deaths
1983	72/2,427/7
1984	78/4,479/3
1985	79/19,660/20
1986	61/2,833/7
1987	52/1,846/2

Source: Bean et al. (5).

TABLE 22-6. **Leading Vehicle Foods Known for Salmonellosis Outbreaks in the United States, 1973–1987**

Rank	Vehicle Foods	Outbreaks	Percentage
1	Beef	77	9.7
2	Turkey	36	4.5
3	Chicken	30	3.8
4	Ice cream	28	3.5
5	Pork	25	3.2
6	Dairy products	22	2.8
7	Eggs	16	2.0
8	Bakery products	12	1.5
9	Mexican food	10	1.3
10	Fruits and vegetables	9	1.1

Source: Bean and Griffin (5).

Note: An outbreak is defined as two or more cases.

was a multistate outbreak in 1990 caused by *S. chester* that was traced to cantaloupe imported from Central America (118). Another unusual outbreak was one in California in 1985 involving 45 cases caused by chloramphenicol-resistant *S. newport*. Because of the unique plasmid profile and the antibiotic resistance of this serovar, it was traced from patients to hamburger products; to abattoirs where the animals were slaughtered; to dairies that sent the cows for slaughter; and to ill dairy cows (112). Antibiotic-resistant strains of salmonellae have been found in other outbreaks (18), and they are presumed to be the result of antibiotic usage in livestock animals, but whether from use in feeds or as the result of treatment of diseased animals is not clear. Over the years 1971 through 1983, antimicrobial-resistant salmonellae were examined from 52 outbreaks in the United States. The case fatality rate was higher for persons infected with these strains than for those with antibiotic-sensitive strains, and food animals were the source of 11 of 16 resistant and only 6 of 13 sensitive strains (51). Raw milk was the source of antibiotic-resistant strains in 4 outbreaks and beef in another. In 7 outbreaks, the resistant serovars were *S. typhimurium*, and in 2 outbreaks, they were *S. newport*. In addition to unique antibiotic resistance and plasmid profiles, other methods employed to trace salmonellae serovars include restriction endonuclease analysis (MEA), biochemical fingerprinting, phage typing, and multilocus enzyme electrophoresis (MEE). (MEA and MEE are further discussed in Chap. 21.)

Foodborne salmonellosis continues to increase. The reasons for this have been addressed by Kampelmacher (59):

- The increase in mass food preparation, which favors spread of *Salmonella*
- Inappropriate methods of storing food, which, because of modern living conditions, is sometimes accumulated in excessive amounts
- The increasing habit of eating raw or insufficiently heated foods, partly because of overreliance on food inspection
- Increasing international food trade
- Decreased resistance to infection resulting from improved standards of general hygiene

Recovery of *Salmonella* from Foods

In view of the federal regulation prohibiting the presence of *Salmonella* spp. in foods, the recovery of these organisms from food presents a difficult problem to food microbiologists and food scientists. How can one be certain that a 1,000-lb lot of powdered eggs is free of *Salmonella* if there might be only 1 organism/100 g? The problem is made more difficult by the fact that foods normally contain larger numbers of microorganisms other than *Salmonella*, such as *Proteus, Pseudomonas, Acinetobacter*, and *Alcaligenes* spp., all of which may develop on some of the media employed for the recovery of salmonellae. The problems encountered in the recovery of these organisms from foods are similar to those that one encounters in the recovery of staphylococci from foods—a generally high ratio of total numbers of other organisms to pathogens. Where the ratio of total flora to salmonellae is rather low, any one of a large number of media may be suitable for recovering these organisms. Special methods are necessary, however, when few salmonellae exist in the presence of a high total count.

In view of the generally low ratio of salmonellae to total flora, it is necessary to inhibit or kill nonsalmonellae types, while the salmonellae are allowed to increase to numbers high enough to improve the chances of finding them on plates or in broth. The fluorescent antibody technique and other salmonellae methods are discussed in Chapter 6. For a review of salmonellae methodology, see D'Aoust et al. (22) and for a comparison of rapid and conventional methods, see Flowers (36).

Prevention and Control of Salmonellosis

The intestinal tract of humans and other animals is the primary reservoir of the etiologic agents. Animal fecal matter is of greater importance than human, and animal hides may become contaminated from the fecal source. *Salmonella* spp. are maintained within an animal population by means of nonsymptomatic animal infections and in animal feeds. Both sources serve to keep slaughter animals reinfected in a cyclical manner.

Secondary contamination is another of the important sources of salmonellae in human infections. Their presence in meats, eggs, and even air makes their presence in certain foods inevitable through the agency of handlers and direct contact of noncontaminated foods with contaminated foods (49).

In view of the worldwide distribution of salmonellae, the ultimate control of foodborne salmonellosis will be achieved by freeing animals and humans of the organisms. This is obviously a difficult task but not impossible; only about 35 of the more than 2,200 serovars account for around 90% of human isolates and approximately 80% of nonhuman isolates (75).

At the consumer level, the *Salmonella* carrier is thought to play a role, but just how important this role may be is not clear. Improper preparation and handling of foods in homes and food service establishments continue to be the primary factors in outbreaks.

Competitive Exclusion to Reduce Salmonellae Carriage in Poultry

It is generally agreed that the primary source of salmonellae in poultry products is the gastrointestinal tract, including the ceca. If young chicks become colonized with salmonellae, the bacteria may be shed in feces, through which other birds become contaminated. Among the methods that may be employed to reduce or eliminate intestinal carriage is competitive exclusion (the **Nurmi concept**).

Under natural conditions where salmonellae exist when eggs hatch, young chicks develop a gastrointestinal track flora that consists of these organisms and campylobacters, in addition to a variety of nonpathogens. Once the pathogens are established, they may remain and be shed in droppings for the life of the bird. Competitive exclusion is a phenomenon whereby feces from salmonellae-free birds, or a mixed fecal culture of bacteria, are given to young chicks so that they will colonize the same intestinal sites that salmonellae employ and thus exclude the subsequent attachment of salmonellae or other entero-pathogens. This concept was advanced in the 1970s and has been studied and found to be workable by a number of investigators relative to salmonellae exclusion. (For reviews, see refs. 79 and 94.)

The enteropathogen-free flora may be administered orally to newly hatched chicks through drinking water or by spray inoculation in the hatchery. Protection is established within a few hours and generally persists throughout the life of the fowl or as long as the flora remains undisturbed. Older birds can be treated by first administering antibacterial agents to eliminate enteropathogens, and they are then administered the competitive exclusion flora. Only viable cells are effective, and both aerobic and anaerobic components of the gut flora seem to be required. Defined culture mixtures consisting of 10 to 50 or more isolates may be employed. The crop and ceca appear to be the major adherence sites, with the ceca being higher in germ-free chickens. In one study, the protective flora remained attached to cecal walls after four successive washings (114). Partial protection was achieved in 0.5 to 1.0 h, but full protection required 6 to 8 h after treatment of 1-day-old chicks (114).

Field trials in several European countries have shown the success of the competitive exclusion treatment in preventing or reducing the entry of salmonellae in broilers and adult breeder birds (79). In chicks pretreated with a cecal culture and later challenged with a *Salmonella* sp., the latter failed to multiply in the ceca over a 48-h period, while in untreated control birds, more than 10^6/g of salmonellae were colonized in the ceca (54).

The gist of competitive exclusion is that salmonellae and the native gut flora compete for the same adherence sites on gut walls. The precise nature of the bacterial adhesins is not entirely clear, although fimbriae, flagella, and pili have been suggested. In regard to the attachment of salmonellae to poultry skin, these bacterial cell structures were found not to be critical (72). Extracellular polysaccharides of a glycocalyx nature may be involved, and if so, treatment of young chicks with this material may be as effective as the use of live cultures. Although the competitive exclusion treatment seems quite feasible for large hatcheries, its practicality for small producers seems less likely.

SHIGELLOSIS

The genus *Shigella* belongs to the family Enterobacteriaceae, as do the salmonellae and escherichiae. Only four species are recognized: *S. dysenteriae*, *S. flexneri*, *S. boydii*, and *S. sonnei*. *S. dysenteriae* is a primary pathogen that causes classical bacillary dysentery; as few as 10 cfu are known to initiate infection in susceptible individuals. Although this syndrome can be contracted from foods, it is not considered to be a food-poisoning organism in the same sense as the other three species, and it is not discussed further. Unlike the salmonellae and escherichiae, the shigellae have no known nonhuman animal reservoirs. Some of the many differences among the three genera are noted in Table 22-7. The shigellae are phylogenetically closer to the escherichiae than to the salmonellae.

The three species of concern as etiologic agents of foodborne gastroenteritis are placed in separate serologic groups based on O antigens: *S. flexneri* in group B, *S. boydii* in group C, and *S. sonnei* in group D. They are nonmotile, oxidase negative, produce acid only from sugars, do not grow on citrate as sole carbon source, do not grow on KCN agar, and do not produce H_2S. In general,

TABLE 22-7.   A Comparison of *Salmonella*, *Shigella*, and *Escherichia*

Genus	Glucose	Motility	H$_2$S	Indole	Citrate	Moles 1% G + C
Escherichia	AG	+[a]	−	+[b]	−	48–52
Salmonella	AG	+[a]	+	−	+	50–53
Shigella	A	−	−	−	−	49–53

[a] Usually.
[b] Type I strains.

their growth on ordinary culture media is not as abundant as that of the escherichiae. Of shigellae isolated from humans in the United States in 1984, 64% were *S. sonnei*, 31% *S. flexneri*, 3.2% *S. boydii*, and 1.5% *S. dysenteriae* (14).

The *Shigella* species of concern are typical of most other enteric bacteria in their growth requirements, with growth reported to occur at least as low as 10°C and as high as 48°C. In one study, growth of *S. flexneri* was not observed in BHI broth at 10°C (131). It appears that *S. sonnei* can grow at lower temperatures than the other three species. Growth at pH 5.0 has been recorded, with best growth occurring in the range of 6 to 8. With *S. flexneri*, no growth occurred at pH 5.5 at 19°C in BHI broth (131). This species has been shown to be inhibited by nitrite as temperature and pH were decreased or as NaCl was increased (130). It is unclear whether they can grow at a$_w$ values below those for the salmonellae or escherichiae. Their resistance to heat

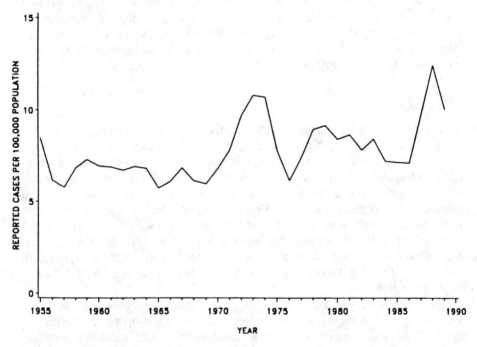

FIGURE 22-2.   Shigellosis cases in the United States, 1955–1989.
From Morbidity and Mortality Weekly Report *38:38, 1989.*

TABLE 22-8.   Outbreaks, Cases, and Deaths Associated
with Foodborne Shigellosis in the United
States, 1983–1987

Years	Outbreaks/Cases/Deaths
1983	7/1,993/0
1984	9/470/1
1985	6/241/0
1986	13/773/1
1987	9/6,494/0

Source: Bean and Griffin (5).

appears to parallel that of *E. coli* strains. Some other aspects of the growth of this genus have been reviewed (35, 110).

Foodborne Cases

The incidence of foodborne shigellosis cases in the United States for the years 1955 through 1989 is represented in Figure 22-2; outbreaks, cases, and deaths for the years 1983 through 1987 are presented in Table 22-8. For the period 1973 through 1987, foodborne shigellosis accounted for 12% of reported food poisoning cases for which an etiologic agent could be found, placing it third behind staphylococcal food poisoning (14%) and salmonellosis (45%) (4). Poor personal hygiene is a common factor in foodborne shigellosis, with shellfish, fruits and vegetables, chicken, and salads being prominent among vehicle foods. The prominence of these foods is due to the fecal-oral route of transmission. The shigellae are not as persistent in the environment as are salmonellae and escherichiae. An excellent review of the shigellae as foodborne pathogens is that of Smith (110).

Virulence Properties

Although much is known about the virulence properties of *S. dysenteriae*, much less is known about how the other three species cause disease. The pathogenic properties of *S. flexneri* are similar to enteroinvasive strains of *E. coli*, and this species appears to be more invasive than *S. sonnei* and *S. boydii*.

   S. flexneri lacks the Shiga toxin but it has four virulence-associated loci on large plasmids—three of them necessary for the expressions of adherence and invasiveness and the other for intracellular survival and multiplication. All four species invade the colonic mucosa, multiply, and elicit acute inflammatory infiltrates in the lamina propria (34). Once they have been internalized in endocytic vacuoles, their escape requires a 220-kb plasmid, especially in *S. flexneri*. This plasmid is responsible for the inhibition of protein synthesis in infected cells. The receptor-recycling inhibitor, chloroquine, inhibits invasion by both salmonellae and yersiniae but not by *S. flexneri*. Also, microfilament inhibitors, the cytochalasins, decrease invasion of *S. flexneri*, indicating the need by this organism for microfilaments for cell invasion. Congo red binding is correlated with virulence, and HeLa cells are used for some invasion studies.

E. COLI GASTROENTERITIS SYNDROMES

Although sporadic reports of *Escherichia coli*–related gastroenteritis of food origin appeared prior to the 1970s, it was the 1971 outbreak in the United States traced to imported cheese that focused attention on this organism as a foodborne pathogen. This interest coincided with that of medical microbiologists who were interested in *E. coli* as a cause of infant diarrhea. The latter work led to the development of specific and accurate in vitro and bioassay methods for assessing toxic components (see Chaps. 6 and 7) and to a better understanding of the virulence mechanisms of this organism. Although only a relatively few foodborne outbreaks have been documented, more is known about the pathogenesis of *E. coli* gastroenteritis than salmonellosis. *E. coli* as an indicator organism is discussed in Chapter 17, culture and isolation methods are covered in Chapter 5, and chemical and bioassay methods for its enterotoxins are covered in Chapters 6 and 7. The early history of this syndrome can be found in Bryan (12), Sack (100), and Mehlman et al. (80). A more recent review is that of Kornacki and Marth (66).

Strains and Distribution

The first studies on *E. coli*–associated diarrheal disease involved nursery epidemics in the mid-1940s that brought mortality rates as high as 50%. During the mid-1950s, some *E. coli* isolates were shown to produce responses in the rabbit ileal loop test similar to those of *Vibrio cholerae*. These findings led to studies of *E. coli* as a possible etiologic agent of choleralike diseases in India. The first reports of enterotoxigenic strains from young animals with diarrhea appeared in 1967, and in 1970 the production of two enterotoxins by virulent strains was reported (100).

The *E. coli* strains of importance as potential foodborne pathogens are among the fecals (see Chap. 17), and their incidence in some foods is noted in Chapter 4. In general, these strains have a wide distribution in food environments in *low* numbers. The microbiological criteria for raw/uncooked foods acknowledge this fact (see Chap. 18). As an indicator, *E. coli* in foods in sufficient numbers is taken to indicate the possibility of fecal contamination and the possible presence of other enteropathogens such as salmonellae. As a potential foodborne pathogen, the generally acceptable low numbers take on new meaning, especially when conditions permit their proliferation. While the latter can be said for coagulase-positive staphylococci, unlike fecal *E. coli*, they are not employed as indicators of sanitary quality.

E. coli strains involved in foodborne illness can be placed into five groups: enteropathogenic (EPEC), enterotoxigenic (ETEC), enteroinvasive (EIEC), enterohemorrhagic (EHEC), and facultatively enteropathogenic (FEEC) (the last strains are associated with sporadic diarrheal outbreaks and are not discussed further):

EPEC. Generally do not produce enterotoxins, although they can cause diarrhea. They produce a plasmid-mediated adherence factor that is demonstrable in HEp-2 cells and attachment-effacement (att-eff). A cytolethal distending toxin (CLDT) has been reported for some strains (8). (Possible pathogenic mechanisms are discussed below).

EIEC. Enterotoxins are not produced, and their pathogenesis is similar to that of the shigellae (see below).

ETEC. Two primary enterotoxins are produced: heat-labile (LT) containing subunits A and B (LTA and LTB) and heat-stable (STa or ST-I, and STb or ST-II). LTB is the binding site, while LTA stimulates the adenylate cyclase system. LTA and B have immunological properties similar to subunits A and B of the *V. cholerae* enterotoxin (68). LTh and LTp designate human and porcine strains, respectively. STa elicits secretory response in infant mice. STb (STp, ST-II) is primarily from swine strains. It is the most prevalent toxin associated with diarrheagenic isolates of porcine origin, and it affects only the small intestine and the ligated ileum of weaned piglets. It is trypsin sensitive and heat resistant and has a molecular weight of about 5 kdal (26). ETEC strains also produce colonizing factor antigens represented by fimbriae (or pili), which mediate binding of cells to epithelial cells.

EHEC. These strains, represented by *E. coli* 0157:H7, produce two basic toxins: Shiga-like toxin (SLT-I; verotoxin, verocytotoxin) and SLT-II. These toxins are produced by *E. coli* strains other than 0157:H7 (60, 109). SLT-I and the Shiga toxin are neutralized by anti-Shiga antisera. Both SLT-II and SLT-IIv are neutralized by antisera against SLT-II but not by anti-Shiga toxin. SLT-IIv is a variant of SLT-II that is more toxic to Vero cells than HeLa cells. It was isolated from swine with edema disease. All SLTs are cytotoxic for Vero cells and lethal for mice, and they produce positive rabbit ileal loop responses, and all except SLT-IIv are cytotoxic for HeLa cells. EHEC strains also produce plasmid-mediated fimbriae and att-eff.

Three antigens are employed to serotype these strains: O (heat-stable somatic antigens), K (heat-labile somatic antigens), and H (heat-labile flagellar antigens). About 164, 100, and 56 of O, K, and H, respectively, are known (89). Only about 30 serovars have been associated with diarrheal disease, and the first was 0111, which was isolated from children with diarrhea and constituted as high as 80 to 100% of fecal flora (100). About 1% of the cultivable fecal flora of humans consists of *E. coli* carried normally in the lower bowel, while the pathogens (except enterohemorrhagic strains) colonize the upper bowel.

With respect to serotypes/serovars, there appears to be little consistency from individual to individual. A study of this organism in the fecal flora of 13 healthy adults in England revealed wide variations in serotypes, with several individuals carrying 1 predominant type while several yielded as many as 12 different types (47). The O antigenic types carried by four to six persons were: 01, 02, 018, 068, 088, 0107, 0116, 0126, and 0132. The serovars responsible for diarrheal disease in 33 children and adults varied rather widely, with only six being recovered from 2 to 4 individuals: 06, 015, 025, 078, 0126, and 0128 (100). Some of the last types are known to be nontoxic. In a study of the incidence of pathogenic strains among 219 food handlers, 14(6.4%) were carriers of a total of eight different serovars (45). From other reports, the incidence in the feces of healthy adults and children ranges from 1.8 to 15.1%. Among EPEC strains, only 1 of 34 of serotype 014, 9 of 45 of serotype 0114,

572 Foodborne Gastroenteritis Caused by *Salmonella*, *Shigella*, and *Escherichia*

and 18 of 82 of serotype 0128 produced either enterotoxin (106). In one study, the most common serotype isolated from foods was 0149 (101).

The K88-positive strain of porcine origin has been shown to effect surface colonization in piglets and to be lethal to 50% of piglets, while the K88-negative variant was lethal to only 3% (88). The K88 antigen is a short, piluslike structure that is highly antigenic. K99 is similar but pathogenic in cows and lambs but not pigs. Although K88 produces the heat-labile toxin, it has been suggested that an immunofluorescence test for in vivo-produced pilus antigens be used for the diagnosis of ETEC infections (83). The K88 antigen is plasmid borne (111). It was observed by earlier investigators that a poor correlation exists between serotypes and ileal loop responses of *E. coli* strains. While serotyping is of value epidemiologically, it appears to be of little value as a predictor of *E. coli* virulence.

Plasmids from *E. coli* strains have been studied more than from any other bacteria. Strains from patients have been shown to contain more conjugative plasmids than those from healthy individuals. ETEC strains usually carry five or more plasmids with antibiotic resistance, enterotoxins, and adherence antigens on separate plasmids (30). Among specific plasmids in *E. coli* is ColV, which controls an iron-sequestering mechanism—possibly enterochelin or enterobactin—and serum resistance (30).

Production and Further Characteristics
of Enterotoxin

The enterotoxins are produced early in the growth phase of producing strains, with the maximum amount of ST produced after 7 h of growth in one study in a Casamino acids yeast extract medium containing 0.2% glucose (69). In a synthetic medium, ST appeared as early as 8 h, but maximal production required 24 h with aeration (13). While LT and ST appear to be produced under all conditions that allow cell growth, the release of LT from cells in enriched media was favored at pH 7.5 to 8.5 (85).

The enterotoxins, along with colonization factor antigen and hemolysin, are plasmid mediated and are encoded on transmissible Ent plasmids that also may carry genes for resistance to several antimicrobials. EIEC strains apparently carry a 140-Mdal plasmid (46), and STa is encoded within a transposon (111).

The ST toxins can withstand 100°C for 15 min, while LT is destroyed at 60°C in about 30 min. LT is acid labile; ST is not. LT is a protein with a molecular weight of about 91 kdal (21), and it possesses enzymatic activity similar to that of CT. While CT is exported from the cytoplasm to the outside of producing cells, LT is deposited into the periplasm of producing cells. Further, antisera to CT neutralize LT, and immunization with CT induces protection against both CT and LT challenges. LT producers are associated more with humans and porcines than other animals. LT is composed of two protomers: A, with a molecular weight of about 25.5 kdal, which when nicked with trypsin becomes an enzymatically active A_1 polypeptide chain of 21 kdal linked by a disulfide bond to an A_2-like chain, and B, which has a molecular weight of about 59 kdal and consists of five noncovalently linked individual polypeptide chains (29).

STa is methanol soluble and stimulates particulate intestinal guanylate cyclase, while STb is methanol insoluble. The former has a molecular weight of 1,972 daltons; the latter has a molecular weight of about 5,000 daltons. Biological activity of STa is lost upon treatment with 2-mercaptoethanol or dithiothreitol, indicating the presence and need of disulfide bridges (113). STa has been chemically synthesized (61), and it is active in neonatal piglets 1 to 3 days old and in infant mice but generally not in weaned pigs. On the other hand, STb is actived in weaned pigs and rabbit ileal loop tests but not in infant mice (13). An ST is produced by *Citrobacter freundii*, which has an amino acid sequence identical to the 18 amino acid toxin of *E. coli*. To distinguish it from the latter, it has been designated ST-Ia (42). ST-Ia–producing strains were isolated from diarrheal patients, and the toxin has a potency similar to that of *E. coli*.

Mode of Action of Enterotoxins

The *E. coli* gastroenteritis syndrome is caused by the ingestion of 10^6–10^{10} viable cells/g that must colonize the small intestines and produce enterotoxin(s). The colonizing factors are generally fimbriae or pili. The syndrome is characterized primarily by nonbloody diarrhea without inflammatory exudates in stools. The diarrhea is watery and similar to that caused by *V. cholerae*. Diarrhea results from enterotoxin activation of intestinal adenylate cyclase, which increases cyclic 3′,5′-adenosine monophosphate (cAMP), first shown for *E. coli* enterotoxins by Evans et al. (31).

With respect to LT, the B protomer mediates binding of the molecule to intestinal cells. LT binds to gangliosides, especially monosialogangliosides (GM_1) (29). CT also binds to GM_1 ganglioside, and CT and LT are known to share antigenic determinants among corresponding protomers, although they do not cross-react. Upon binding, the A polypeptide chain (of the A protomer) catalyzes ADP ribosylation of adenylate cyclase, which induces increases in intracellular cAMP.

Regarding ST, ST_a binds irreversibly to a specific high-affinity nonganglioside receptor and initiates a transmembrane signal to active particulate guanylate cyclase (37, 43). The increased levels of mucosal cGMP lead to loss of fluids and electrolytes. ST differs from CT in that only the particulate form of intestinal guanylate cyclase is stimulated by ST (37). ST_a differs from LT in that the former stimulates guanylate cyclase, while the latter and CT activate adenylate cyclase. Stimulation of guanylate cyclase by ST_a is tissue specific, and only the intestinal form of the particulate enzyme responds to ST_a. The precise role of ST_b is not understood, but it does not activate guanylate cyclase. Genes controlling its production have been genetically mapped (77) and subcloned from its plasmid and sequenced (93).

EIEC strains bind, invade, survive, and multiply within epithelial cells as do the shigellae. EPEC strains associate with intestinal mucosa in a manner known as attachment-effacement (att-eff), which leads to disruption of the epithelial cell membrane. Although plasmid-mediated factors facilitate bacterial adhesion to the small intestine, att-eff appears to be the most important virulence factor of EPEC strains (123). The mechanisms of Shiga, SLT-I, SLT-II, SLT-IIv, and the castor bean protein, ricin, are the same. They are N-glycosidases that cleave a specific adenine residue from the 28S N-glycosidases that cleave a specific adenine residue from the 28S subunit of eukaryotic rRNA, resulting in the inhibition of protein synthesis (125).

Foodborne and Related Outbreaks

Reports incriminating *E. coli* as the etiologic agent of food poisoning involving a variety of foods such as cream pie, mashed potatoes, cream puffs, and creamed fish date from around 1900. The feeding of volunteers with serotypes 055:B5, 0111:B4, and 0127:B8 at levels of 10^6–10^8 organisms was shown to produce gastroenteritis symptoms (32, 57, 127).

Two outbreaks with 76 cases and 4 deaths occurred in 1984 with beef and multiple foods as vehicles (5). One outbreak occurred in 1985 with 370 cases and no deaths; the vehicle was Mexican food. In 1986, one multiple food outbreak was recorded by CDC with 37 cases and no deaths (5). No outbreaks were recorded by CDC in 1987. The documented outbreaks prior to 1984 are summarized in Table 22-7. Not included are four to six outbreaks reported from the USSR, Eastern Europe, and Japan (12). The U.S. outbreaks in 1971 were well documented. An EIEC strain was the etiologic agent. The cheese involved was imported from France and sold under several names (brie, camembert, coulommiers), but all had been made in the same way. The serovar was 0124:B17 (74). Median time for onset of symptoms was 18 h, with a mean duration of 2 days. Common symptoms were diarrhea, fever, and nausea, in order of decreasing frequency. Less common symptoms included cramps, chills, vomiting, aches, and headaches. Among those who ate the soft-ripened cheese, the attack rate was 94%. A total of 107 outbreaks occurred in 13 states and the District of Columbia. While some of the cheese contained 10^3–10^5 coliforms/g, Enterobacteriaceae at levels of 10^6–10^7/g were found (84). Imported brie cheese was the common source of the outbreak in 1983 (Table 22-9).

Hemorrhagic Colitis

This syndrome was first recognized in 1982 in outbreaks in Michigan and Oregon affecting at least 47 individuals. It was associated with eating any of three sandwiches at fast-food restaurants of the same chain in both states, and all sandwiches contained ground beef (97). Of 43 patients studied, all had bloody diarrhea and severe abdominal cramps with 63% experiencing nausea, 49% vomiting, but only 7% fever. The mean incubation period was 3.8 to 3.9 days, and symptoms lasted for 3 to more than 7 days (97). A noninvasive, nontoxigenic *E. coli* serotype 0157:H7 was recovered from 9 of 12 stools collected within 4 days and from a frozen raw beef patty (97). This rare serotype was not recovered from stools collected 7 or more days after onset of illness (126). The only previous isolation of this serotype in the United States was from a sporadic case of hemorrhagic colitis in 1975. An outbreak of hemorrhagic colitis occurred in Ottawa, Canada, in November 1982, involving 31 of 353 residents of a home for the aged. The serotype was 0157:H7, and the apparent source was kitchen-prepared food.

The verotoxigenic strains of *E. coli* appear to be associated with bovines, and it has been found in beef and raw milk. It has been found also in water buffalo calves with diarrhea, edema disease of pigs, and human stool cultures

TABLE 22-9.   **Summary of Early Outbreaks of *E. coli* Gastroenteritis from Foods and Other Sources**

Year	Location	Vehicle Food/ Source	No. Victims/ No. at Risk	Toxin/ Strain Type	Serotype	Reference
1947	England	Salmon (apparently)	47/300		0124	50
1961	Rumania	Substitute coffee drink	10/50		086:B7;H34	20
1963	Japan	Ohagi	17/31			2
1966	Japan	Vegetables	244/435			2
1967	Japan	Sushi	835/1,736		027	103
1971	14 American states	Imported cheeses	387/?	EIEC	0124:B17	74
1980	Wisconsin	Food handler	500/>3,000	LT, ST	06:H16	120
1981	Texas	Not identified	282/3,000	LT	025:H+	129
1983	4 American states	Imported brie cheese	/?	ST	027:H20	15

in many countries. In Thailand, it was recovered from 9% of retail beef, 8 to 28% of slaughterhouse beef, and 11 to 84% of cattle fecal specimens (117). From meats in the United States and parts of Canada, this strain was recovered from 6 (3.7%) of 164 beef, 4 (1.5%) of 264 pork, 4 (1.5%) of 263 poultry, and 4 (2.0%) of 205 lamb samples (24). Although *E. coli* 0157:H7 could not be recovered from sausage in the United Kingdom, a DNA probe gave positive results on 46 (25%) of 184 samples for other EHEC strains (109). Among the EHEC strains found were 08:H25, 0100:H⁻, 0115:H10, 0146:H8. None could be found in 112 samples from 71 chickens (109).

These strains are more heat sensitive than salmonellae (D 60°C = 45 sec in ground beef), but they survived in ground beef for 9 months held at −20°C with little or no change in numbers (25). They grow poorly in EC medium if at all in the 44° to 45°C range, with no growth at 44.5°C (95) or 45.5°C (25). However, elevated temperature growth in EC medium may be achieved by reducing the medium's bile content (see Chap. 17). It is nonfluorogenic in the MUG assay (see Chap. 6).

Travelers' Diarrhea

E. coli is well established as one of the leading causes of acute watery diarrhea that often occurs among new arrivals in certain foreign countries. Among Peace Corps volunteers in rural Thailand, 57% of 35 developed the syndrome during their first 5 weeks in the country, and 50% showed evidence of infection by ETEC strains. In 1976 a shipboard outbreak of gastroenteritis was shown to be caused by serotype 025:K98:NM that produced only LT. Similar strains have been recovered from other victims of traveler's diarrhea in various countries along with EPEC and ST-producing strains.

Among other organisms associated with this syndrome are rotaviruses, Norwalk agent (virus), *Entamoeba histolytica*, *Yersinia enterocolitica*, *Giardia lamblia*, *Campylobacter jejuni/coli*, *Shigella* spp., and possibly *Aeromonas hydrophila*, *Klebsiella pneumoniae*, and *Enterobacter cloacae*.

References

1. Adinarayanan, N., V. D. Foltz, and F. McKinley. 1965. Incidence of Salmonellae in prepared and packaged foods. *J. Infect. Dis.* 115:19–26.
2. Akahane, S. 1973. Epidemiological studies on outbreaks of food poisonings due to *Escherichia coli* 0124:K72:H-. *J. Jap. Assoc. Infect. Dis.* 47:63–76.
3. Barnhart, H. M., D. W. Dreesen, R. Bastien, and O. C. Pancorbo. 1991. Prevalence of *Salmonella enteritidis* and other serovars in ovaries of layer hens at time of slaughter. *J. Food Protect.* 54:488–491.
4. Bean, N. H., and P. M. Griffin. 1990. Foodborne disease outbreaks in the United States, 1973–1987: Pathogens, vehicles, and trends. *J. Food Protect.* 53:804–817.
5. Bean, N. H., P. M. Griffin, J. S. Goulding, and C. B. Ivey. 1990. Foodborne disease outbreaks, a 5-year summary, 1983–1987. *J. Food Protect.* 53:711–728.
6. Beloian, A., and G. C. Schlosser. 1963. Adequacy of cooking procedures for the destruction of salmonellae. *Amer. J. Pub. Hlth* 53:782–791.
7. *Bergey's Manual of Systematic Bacteriology.* 1984. Vol. 1. Ed. N. R. Krieg. Baltimore: Williams & Wilkins.
8. Bouzari, S., and A. Varghese. 1990. Cytolethal distending toxin (CLDT) production by enteropathogenic *Escherichia coli* (EPEC). *FEMS Microbiol. Lett.* 71:193–198.
9. Bradshaw, J. G., D. B. Shah, E. Forney, and J. M. Madden. 1990. Growth of *Salmonella enteritidis* in yolk of shell eggs from normal and seropositive hens. *J. Food Protect.* 53:1033–1036.
10. Brooks, J. 1962. Alpha amylase in whole eggs and its sensitivity to pasteurization temperatures. *J. Hyg.* 60:145–151.
11. Bryan, F. L. 1977. Diseases transmitted by foods contaminated by wastewater. *J. Food Protect.* 40:45–56.
12. Bryan, F. L. 1979. Infections and intoxications caused by other bacteria. In *Foodborne Infections and Intoxications*, ed. H. Riemann and F. L. Bryan, 211–297. New York: Academic Press.
13. Burgess, M. N., R. J. Bywater, C. M. Cowley, N. A. Mullan, and P. M. Newsome. 1978. Biological evaluation of a methanol-soluble, heat-stable *Escherichia coli* enterotoxin in infant mice, pigs, rabbits, and calves. *Infect. Immun.* 21:526–531.
14. Centers for Disease Control. 1985. Shigellosis—United States, 1984. *Mort. Morb. Wkly. Rept.* 34:600.
15. Centers for Disease Control. 1984. Update: Gastrointestinal illness associated with imported semi-soft cheese. *Morb. Mort. Wkly. Rept.* 33:16, 22.
16. Centers for Disease Control. 1981. Human *Salmonella* isolates—United States, 1980. *Morb. Mort. Week. Rept.* 30:377–379.
17. Chung, K. C., and J. M. Goepfert. 1970. Growth of *Salmonella* at low pH. *J. Food Sci.* 35:326–328.
18. Cohen, M. L., and R. V. Tauxe. 1986. Drug-resistant *Salmonella* in the United States: An epidemiologic perspective. *Science* 234:964–969.
19. Committee on Salmonella. 1969. *An Evaluation of the Salmonella Problem.* Publ. 1683. Washington, D.C.: National Academy of Sciences.
20. Costin, I. D., D. Volculescu, and V. Gorcea. 1964. An outbreak of food poisoning in adults associated with serotype 086:B7:H34. *J. Path. Microbiol.* 27:68–78.
21. Dallas, W. S., D. M. Gill, and S. Falkow. 1979. Cistrons encoding *Escherichia coli* heat-labile toxin. *J. Bacteriol.* 139:850–858.
22. D'Aoust, J.-Y., H. J. Beckers, M. Boothroyd, A. Mates, C. R. McKee, A. B. Moran, P. Sado, G. E. Spain, W. H. Sperber, P. Vassiliadis, D. E. Wagner, and C. Wiberg. 1983. ICMSF methods studies. XIV. Comparative study on recovery

of *Salmonella* from refrigerated pre-enrichment and enrichment broth cultures. *J. Food Protect.* 46:391–399.

23. D'Aoust, J. Y., and H. Pivnick. 1976. Small infectious doses of *Salmonella*. *Lancet* 1:866.

24. Doyle, M. P., and J. L. Schoeni. 1987. Isolation of *Escherichia coli* 0157:H7 from retail fresh meats and poultry. *Appl. Environ. Microbiol.* 53:2394–2396.

25. Doyle, M. P., and J. L. Schoeni. 1984. Survival and growth characteristics of *Escherichia coli* associated with hemorrhagic colitis. *Appl. Environ. Microbiol.* 48:855–856.

26. Dubreuil, J. D., J. M. Fairbrother, R. Lallier, and S. Lariviere. 1991. Production and purification of heat-stable enterotoxin b from a porcine *Escherichia coli* strain. *Infect. Immun.* 59:198–203.

27. Duebbert, I. E., and J. W. Peterson. 1984. Enterotoxin-induced fluid accumulation during experimental salmonellosis and cholera: Requirement for prostaglandin synthesis by intestinal cells. *Bacteriol. Proc.*, 17.

28. Duitschaever, C. L. 1977. Incidence of *Salmonella* in retailed raw cut-up chicken. *J. Food Protect.* 40:191–192.

29. Eidels, L., R. L. Proia, and D. A. Hart. 1983. Membrane receptors for bacterial toxins. *Microbiol. Rev.* 47:596–620.

30. Elwell, L. P., and P. L. Shipley. 1980. Plasmid-mediated factors associated with virulence of bacteria to animals. *Ann. Rev. Microbiol.* 34:465–496.

31. Evans, D. J., Jr., L. C. Chen, G. T. Curlin, and D. G. Evans. 1972. Stimulation of adenyl cyclase by *Escherichia coli* enterotoxin. *Nature* 236:137–138.

32. Ferguson, W. W., and R. C. June. 1952. Experiments on feeding adult volunteers with *Escherichia coli* 111:B4, a coliform organism associated with infant diarrhea. *Amer. J. Hyg.* 55:155–169.

33. Finlay, B. B., F. Heffron, and S. Falkow. 1989. Epithelial cell surfaces induce *Salmonella* proteins required for bacterial adherence and invasion. *Science* 243:940–943.

34. Finlay, B. B., and S. Falkow. 1988. A comparison of microbial invasion strategies of *Salmonella*, *Shigella* and *Yersinia* species. In *Bacteria-Host Cell Interaction*, ed. M. A. Horwitz, 227–243. New York: Alan R. Liss.

35. Flowers, R. S. 1988. Bacteria associated with foodborne diseases. *Shigella*. *Food Technol.* 42(4):185–186.

36. Flowers, R. S. 1985. Comparison of rapid *Salmonella* screening methods and the conventional culture method. *Food Technol.* 39(3):103–108.

37. Frantz, J. C., L. Jaso-Friedman, and D. C. Robertson. 1984. Binding of *Escherichia coli* heat-stable enterotoxin to rat intestinal cells and brush border membranes. *Infect. Immun.* 43:622–630.

38. Galdiero, F., M. A. Tufano, M. Caldiero, S. Masiello, and M. Di Rosa. 1990. Inflammatory effects of *Salmonella typhimurium* porins. *Infect. Immun.* 58:3183–3186.

39. Goepfert, J. M., and R. A. Biggie. 1968. Heat resistance of *Salmonella typhimurium* and *Salmonella senftenberg* 775W in milk chocolate. *Appl. Microbiol.* 16:1939–1940.

40. Graber, G. 1991. Control of *Salmonella* in animal feeds. Division of Animal Feeds, Center for Veterinary Medicine, Food & Drug Administration. Report to the National Advisory Commission on Microbiological Criteria for Foods.

41. Grau, F. H., and L. E. Brownlie. 1968. Effect of some pre-slaughter treatments on the *Salmonella* population in the bovine rumen and feces. *J. Appl. Bacteriol.* 31:157–163.

42. Guarino, A., R. Giannella, and M. R. Thompson. 1989. *Citrobacter freundii* produces an 18-amino-acid heat-stable enterotoxin identical to the 18-amino-acid

Escherichia coli heat-stable enterotoxin (St Ia). *Infect. Immun.* 57:649–652.

43. Guerrant, R. L., J. M. Hughes, B. Chang, D. C. Robertson, and F. Murad. 1980. Activation of intestinal guanylate cyclase by heat-stable enterotoxin of *Escherichia coli*: Studies of tissue specificity, potential receptors, and intermediates. *J. Infect. Dis.* 142:220–228.

44. Guiney, D. G., J. Fierer, G. Chikami, P. Beninger, E. J. Heffernan, and K. Tanabe. 1988. Characterization of virulence plasmids in *Salmonella*. In *Bacteria-Host Cell Interactions*, ed. M. A. Horwitz, 329–345. New York: Alan R. Liss.

45. Hall, H. E., and G. H. Hauser. 1966. Examination of feces from food handlers for salmonellae, shigellae, enteropathogenic *Escherichia coli*, and *Clostridium perfringens*. *Appl. Microbiol.* 14:928–933.

46. Harris, J. R., I. K. Wachsmuth, B. R. Davis, and M. L. Cohen. 1982. High-molecular-weight plasmid correlates with *Escherichia coli* enteroinvasiveness. *Infect. Immun.* 37:1295–1298.

47. Hartley, C. L., H. M. Clements, and K. B. Linton. 1977. *Escherichia coli* in the faecal flora of man. *J. Appl. Bacteriol.* 43:261–269.

48. Hinton, M., E. J. Threlfall, and B. Rowe. 1990. The invasive potential of *Salmonella enteritidis* phage types for young chickens. *Lett. Appl. Microbiol.* 10:237–239.

49. Hobbs, B. C. 1961. Public health significance of *Salmonella* carriers in livestock and birds. *J. Appl. Bacteriol.* 24:340–352.

50. Hobbs, B. C., M. E. M. Thomas, and J. Taylor, 1949. School outbreak of gastroenteritis associated with a pathogenic paracolon bacillus. *Lancet* 2:530–532.

51. Holmberg, S. D., J. G. Wells, and M. L. Cohen. 1984. Animal-to-man transmission of antimicrobial-resistant *Salmonella*: Investigations of U.S. outbreaks. *Science* 225:833–835.

52. Horwitz, M. A., R. A. Pollard, M. H. Merson, and S. M. Martin. 1977. A large outbreak of foodborne salmonellosis on the Navajo Nation Indian Reservation, epidemiology and secondary transmission. *Amer. J. Pub. Hlth* 67:1071–1076.

53. Houston, C. W., C. W. Koo, and J. W. Peterson. 1981. Characterization of *Salmonella* toxin released by mitomycin C-treated cells. *Infect. Immun.* 32:916–926.

54. Impey, C. S., and G. C. Mead. 1989. Fate of salmonellas in the alimentary tract of chicks pre-treated with a mature caecal microflora to increase colonization resistance. *J. Appl. Bacteriol.* 66:469–475.

55. Johnston, R. W., S. S. Green, J. Chiu, M. Pratt, and J. Rivera. 1982. Incidence of *Salmonella* in fresh pork sausage in 1979 compared with 1969. *J. Food Sci.* 47:1369–1371.

56. Jones, F. T., R. C. Axtell, D. V. Rives, S. E. Scheideler, F. R. Tarver, Jr., R. L. Walker, and M. J. Wineland. 1991. A survey of *Salmonella* contamination in modern broiler production. *J. Food Protect.* 54:502–507.

57. June, R. C., W. W. Ferguson, and M. T. Waifel. 1953. Experiments in feeding adult volunteers with *Escherichia coli* 055:B5, a coliform organism associated with infant diarrhea. *Amer. J. Hyg.* 57:222–236.

58. Juven, B. J., N. A. Cox, J. S. Bailey, J. E. Thomson, O. W. Charles, and J. V. Shutze. 1984. Survival of *Salmonella* in dry food and feed. *J. Food Protect.* 47:445–448.

59. Kampelmacher, E. H. 1963. The role of salmonellae in foodborne diseases. In *Microbiological Quality of Foods*, ed. L. W. Slanetz et al., 84–101. New York: Academic Press.

60. Karmali, M. A. 1989. Infection by verocytotoxin-producing *Escherichia coli*. *Clin. Microbiol. Rev.* 2:15–38.

61. Klipstein, F. A., R. F. Engert, and R. A. Houghten. 1983. Properties of

synthetically produced *Escherichia coli* heat-stable enterotoxin. *Infect. Immun.* 39:117–121.

62. Koo, F. C. W., and J. W. Peterson. 1981. The influence of nutritional factors on synthesis of *Salmonella* toxin. *J. Food Safety* 3:215–232.

63. Koo, F. C. W., and J. W. Peterson. 1983. Effects of cultural conditions on the synthesis of *Salmonella* toxin. *J. Food Safety* 5:61–71.

64. Koo, F. C. W., and J. W. Peterson. 1983. Cell-free extracts of *Salmonella* inhibit protein synthesis and cause cytotoxicity in eukaryotic cells. *Toxicon* 21:309–320.

65. Koo, F. C. W., J. W. Peterson, C. W. Houston, and N. C. Molina. 1984. Pathogenesis of experimental salmonellosis: Inhibition of protein synthesis by cytotoxin. *Infect. Immun.* 43:93–100.

66. Kornacki, J. L., and E. H. Marth. 1982. Foodborne illness caused by *Escherichia coli:* A review. *J. Food Protect.* 45:1051–1067.

67. Koupal, L. P., and R. H. Deibel. 1975. Assay, characterization, and localization of an enterotoxin produced by *Salmonella. Infect. Immun.* 11:14–22.

68. Kunkel, S. L., and D. C. Robertson. 1979. Purification and chemical characterization of the heat-labile enterotoxin produced by enterotoxigenic *Escherichia coli. Infect. Immun.* 25:586–596.

69. Lallier, R., S. Lariviere, and S. St-Pierre. 1980. *Escherichia coli* heat-stable enterotoxin: Rapid method of purification and some characteristics of the toxin. *Infect. Immun.* 28:469–474.

70. Lerche, M. 1961. Zur Lebenfahigkeit von Salmonellabakterien in Mayonnaise und Fleischsalat. *Wien. tierarztl. Mschr.* 6:348–361.

71. Ley, F. J., B. M. Freeman, and B. C. Hobbs. 1963. The use of gamma radiation for the elimination of salmonellae from various foods. *J. Hyg.* 61:515–529.

72. Lillard, H. S. 1986. Role of fimbriae and flagella in the attachment of *Salmonella typhimurium* to poultry skin. *J. Food Sci.* 51:54–56, 65.

73. Machardo, J., and F. Bernardo. 1990. Prevalence of *Salmonella* in chicken carcasses in Portugal. *J. Appl. Bacteriol.* 69:477–480.

74. Marier, R., J. G. Wells, R. C. Swanson, W. Callahan, and I. J. Mehlman. 1973. An outbreak of enteropathogenic *Escherichia coli* foodborne disease traced to imported French cheese. *Lancet* 2:1376–1378.

75. Martin, W. J., and W. H. Ewing. 1969. Prevalence of serotypes of *Salmonella. Appl. Microbiol.* 17:111–117.

76. Matches, J. R., and J. Liston. 1968. Low temperature growth of *Salmonella. J. Food Sci.* 33:641–645.

77. Mazaitis, A. J., R. Maas, and W. K. Maas. 1981. Structure of a naturally occurring plasmid with genes for enterotoxin production and drug resistance. *J. Bacteriol.* 145:97–105.

78. Mead, G. C. 1982. Microbiology of poultry and game birds. In *Meat Microbiology*, ed. M. H. Brown, 67–101. London: Applied Science.

79. Mead, G. C., and P. A. Barrow. 1990. *Salmonella* control in poultry by "competitive exclusion" or immunization. *Lett. Appl. Microbiol.* 10:221–227.

80. Mehlman, I. J., M. Fishbein, S. L. Gorbach, A. C. Sanders, E. L. Eide, and J. C. Olson, Jr. 1976. Pathogenicity of *Escherichia coli* recovered from food. *J. Assoc. Off. Anal. Chem.* 59:67–80.

81. Milone, N. A., and J. A. Watson. 1970. Thermal inactivation of *Salmonella senftenberg* 775W in poultry meat. *Hlth Lab. Sci.* 7:199–225.

82. LeMinor, L., and M. Y. Popoff. 1987. Request for an opinion. Designation of *Salmonella enterica* sp. nov., nom. rev., as the type and only species of the genus *Salmonella. Int. J. Syst. Bacteriol.* 37:465–468.

83. Moon, H. W., E. M. Kohler, H. A. Schneider, and S. C. Whipp. 1980. Prevalence of pilus antigens, enterotoxin types, and enteropathogenicity among

K88-negative enterotoxigenic *Escherichia coli* from neonatal pigs. *Infect. Immun.* 27:222–230.

84. Mossell, D. A. A. 1974. Bacteriological safety of foods. *Lancet* 1:173.

85. Mundell, D. H., C. R. Anselmo, and R. M. Wishnow. 1976. Factors influencing heat-labile *Escherichia coli* enterotoxin activity. *Infect. Immun.* 14:383–388.

86. Nakamura, M., S. Sato, T. Ohya, S. Suzuki, and S. Ikeda. 1985. Possible relationship of a 36-megadalton *Salmonella enteritidis* plasmid to virulence in mice. *Infect. Immun.* 47:831–833.

87. Ng, H., H. G. Bayne, and J. A. Garibaldi. 1969. Heat resistance of *Salmonella*: The uniqueness of *Salmonella senftenberg* 775W. *Appl. Microbiol.* 17:78–82.

88. Ørskov, I., and F. Ørskov. 1966. Episome-carried surface antigen K88 of *Escherichia coli*. I. Transmission of the determinant of the K88 antigen and influence on the transfer of chromosomal markers. *J. Bacteriol.* 91:69–75.

89. Ørskov, I., F. Ørskov, B. Jann, and K. Jann. 1977. Serology, chemistry, and genetics of O and K antigens of *Escherichia coli*. *Bacteriol. Rev.* 41:667–710.

90. Page, G. V., and M. Solberg. 1980. Nitrogen assimilation by *Salmonella typhimurium* in a chemically defined minimal medium containing nitrate, nitrite, or ammonia. *J. Food Sci.* 45:75–76, 83.

91. Perales, I., and A. Audicana. 1988. *Salmonella enteritidis* and eggs. *Lancet* ii:1133.

92. Peterson, J. W. 1980. *Salmonella* toxin. *Pharm. Ther.* 11:719–724.

93. Picken, R. N., A. J. Mazaitis, W. K. Maas, M. Rey, and H. Heyneker. 1983. Nucleotide sequence of the gene for heat-stable enterotoxin II of *Escherichia coli*. *Infect. Immun.* 42:269–275.

94. Pivnick, H., and E. Nurmi. 1982. The Nurmi concept and its role in the control of salmonellae in poultry. In *Developments in Food Microbiology—1*, ed. R. Davies, 41–70. London: Applied Science Publishers.

95. Raghubeer, E. V., and J. R. Matches. 1990. Temperature range for growth of *Escherichia coli* serotype 0157:H7 and selected coliforms in *E. coli* medium. *J. Clin. Microbiol.* 28:803–805.

96. Rengel, A., and S. Mendoza. 1984. Isolation of *Salmonella* from raw chicken in Venezuela. *J. Food Protect.* 47:213–216.

97. Riley, L. W., R. S. Remis, S. D. Helgerson, H. B. McGee, J. G. Wells, B. R. Davis, R. J. Hebert, E. S. Olcott, L. M. Johnson, N. T. Hargrett, P. A. Blake, and M. L. Cohen. 1983. Hemorrhagic colitis associated with a rare *Escherichia coli* serotype. *N. Eng. J. Med.* 308:681–685.

98. Rogers, R. E., and M. F. Gunderson. 1958. Roasting of frozen stuffed turkeys. I. Survival of *Salmonella pullorum* in inoculated stuffing. *Food Res.* 23:87–95.

99. Ryan, C. A., M. K. Nickels, N. T. Hargrett-Bean, M. E. Potter, T. Endo, L. Mayer, C. W. Langkop, C. Gibson, R. C. McDonald, R. T. Kenney, N. D. Puhr, P. J. McDonnell, R. J. Martin, M. L. Cohen, and P. A. Blake. 1987. Massive outbreak of antimicrobial-resistant salmonellosis traced to pasteurized milk. *J. Amer. Med. Assoc.* 258:3269–3274.

100. Sack, R. B. 1975. Human diarrheal disease caused by enterotoxigenic *Escherichia coli*. *Ann. Rev. Microbiol.* 29:333–353.

101. Sack, R. B., D. A. Sack, I. J. Mehlman, F. Ørskov, and I. Ørskov. 1977. Enterotoxigenic *Escherichia coli* isolated from food. *J. Infect. Dis.* 135:313–317.

102. Sadler, W. W., and R. E. Corstvet. 1965. Second survey of market poultry for *Salmonella* infections. *Appl. Microbiol.* 13:348–351.

103. Sakai, S., T. Maruyama, T. Itoh, K. Saitch, and H. Zen-Yoji. 1971. An outbreak of enterocolitis ascribed to the infection of *Escherichia coli* 011:K(B):H27. *Ann. Rep. Tokyo Metropol. Res. Lab. Publ. Hlth.*

104. Sandefur, P. D., and J. W. Peterson. 1978. Isolation of skin permeability factors

from culture filtrates of *Salmonella typhimurium*. *Infect. Immun.* 14:671–679.

105. Sandefur, P. D., and J. W. Peterson. 1977. Neutralization of *Salmonella* toxin-induced elongation of Chinese hamster ovary cells by cholera antitoxin. *Infect. Immun.* 15:988–992.

106. Scotland, S. M., N. P. Day, A. Cravioto, L. V. Thomas, and B. Rowe. 1981. Production of heat-labile or heat-stable enterotoxins by strains of *Escherichia coli* belonging to serogroups 044, 0114, and 0128. *Infect. Immun.* 31:500–503.

107. Sedlock, D. M., L. R. Koupal, and R. H. Deibel. 1978. Production and partial purification of *Salmonella* enterotoxin. *Infect. Immun.* 20:375–380.

108. Shrimpton, D. H., J. B. Monsey, B. C. Hobbs, and M. E. Smith. 1962. A laboratory determination of the destruction of alpha amylase and salmonellae in whole egg by heat pasteurization. *J. Hyg.* 60:153–162.

109. Smith, H. B., T. Cheasty, D. Roberts, A. Thomas, and B. Rowe. 1991. Examination of retail chickens and sausages in Britain for vero cytotoxin-producing *Escherichia coli*. *Appl. Environ. Microbiol.* 57:2091–2093.

110. Smith, J. L. 1987. *Shigella* as a foodborne pathogen. *J. Food Protect.* 50:788–801.

111. So, M., and B. J. McCarthy. 1980. Nucleotide sequence of the bacterial transposon Tn1681 encoding a heat-stable (ST) toxin and its identification in enterotoxigenic *Escherichia coli* strains. *Proc. Natl. Acad. Sci., USA* 77:4011–4015.

112. Spika, J. S., S. H. Waterman, G. W. Soo Hoo, M. E. St. Louis, R. E. Pager, S. M. James, M. L. Bissett, L. W. Mayer, J. Y. Chiu, B. Hall, K. Greene, M. E. Potter, M. L. Cohen, and P. A. Blake. 1987. Chloramphenicol-resistant *Salmonella newport* traced through hamburger to dairy farms. *N. Eng. J. Med.* 316:566–570.

113. Staples, S. J., S. E. Asher, and R. A. Giannella. 1980. Purification and characterization of heat-stable enterotoxin produced by a strain of *E. coli* pathogenic for man. *J. Biol. Chem.* 255:4716–4721.

114. Stavric, S., T. M. Gleeson, B. Blanchfield, and H. Pivnick. 1987. Role of adhering microflora in competitive exclusion of *Salmonella* from young chicks. *J. Food Protect.* 50:928–932.

115. St. Louis, M. E., D. L. Morse, M. E. Potter, T. M. DeMelfi, J. J. Guzewich, R. V. Tauxe, P. A. Blake, and the *Salmonella enteritidis* Working Group. 1988. The emergence of Grade A eggs as a major source of *Salmonella enteritidis* infections. *J. Amer. Med. Assoc.* 259:2103–2107.

116. Steele, J. H., and M. M. Galton. 1967. Epidemiology of foodborne salmonellosis. *Hlth Lab. Sci.* 4:207–212.

117. Suthienkul, O., J. E. Brown, J. Seriwatana, S. Tienthongdee, S. Sastravaha, and P. Echeverria. 1990. Shiga-like toxin-producing *Escherichia coli* in retail meats and cattle in Thailand. *Appl. Environ. Microbiol.* 56:1135–1139.

118. Tauxe, R. V. 1991. *Salmonella*: A postmodern pathogen. *J. Food Protect.* 54:563–568.

119. Taylor, J. 1967. *Salmonella* and salmonellosis. In *Food Poisoning*, 15–32. London: Royal Society of Health.

120. Taylor, W. R., W. L. Schell, J. G. Wells, K. Choi, D. E. Kinnunen, P. T. Heiser, and A. G. Helstad. 1982. A foodborne outbreak of enterotoxigenic *Escherichia coli* diarrhea. *N. Eng. J. Med.* 306:1093–1095.

121. Thomason, B. M., W. B. Cherry, and D. J. Dodd. 1977. Salmonellae in health foods. *Appl. Environ. Microbiol.* 34:602–603.

122. Troller, J. A. 1976. *Salmonella* and *Shigella*. In *Food Microbiology: Public Health and Spoilage Aspects,* ed. M. P. deFigueiredo and D. F. Splittstoesser, 129–155. Westport, Conn.: AVI.

123. Tzioori, S., R. Gibson, and J. Montanaro. 1989. Nature and distribution of mucosal lesions associated with enteropathogenic and enterohemorrhagic *Escherichia coli* in piglets and the role of plasmid-mediated factors. *Infect. Immun.* 57:1142–1150.

124. Wagner, D. E., and S. McLaughlin. 1986. *Salmonella* surveillance by the Food and Drug Administration: A review 1974–1985. *J. Food Protect.* 49:734–738.
125. Weinstein, D. L., M. P. Jackson, L. P. Perera, R. K. Holmes, and A. D. O'Brien. 1989. In vivo formation of hybrid toxins comprising Shiga toxin and the Shiga-like toxins and role of the B subunit in localization and cytotoxic activity. *Infect. Immun.* 57:3743–3750.
126. Wells, J. G., B. R. Davis, K. Wachsmuth, L. W. Riley, R. S. Remis, R. Sokolow, and G. K. Morris. 1983. Laboratory investigation of hemorrhagic colitis outbreaks associated with a rare *Escherichia coli* serotype. *J. Clin. Microbiol.* 18:512–520.
127. Wentworth, F. H., D. W. Broek, C. S. Stulberg, and R. H. Page. 1956. Clinical bacteriological and serological observations of two human volunteers following ingestion of *E. coli* 0127:B8. *Proc. Soc. Exptl. Biol. Med.* 91:586–588.
128. WHO Expert Committee. 1988. Salmonellosis control: The role of animal and product hygiene. *Tech. Report Series 774*, Geneva: World Health Organization.
129. Wood, L. V., W. H. Wolfe, G. Ruiz-Palacios, W. S. Foshee, L. I. Gorman, F. McCleskey, J. A. Wright, and H. L. DuPont. 1983. An outbreak of gastroenteritis due to a heat-labile enterotoxin-producing strain of *Escherichia coli. Infect. Immun.* 41:931–934.
130. Zaika, L. L., A. H. Kim, and L. Ford. 1991. Effect of sodium nitrite on growth of *Shigella flexneri. J. Food Protect.* 54:424–428.
131. Zaika, L. L., L. S. Engel, A. H. Kim, and S. A. Palumbo. 1989. Effect of sodium chloride, pH and temperature on growth of *Shigella flexneri. J. Food Protect.* 52:356–359.

23

Foodborne Gastroenteritis Caused by *Vibrio*, *Yersinia*, and *Campylobacter* Species

VIBRIOSIS (*Vibrio parahaemolyticus*)

While most other known food-poisoning syndromes may be contracted from a variety of foods, *V. parahaemolyticus* gastroenteritis is contracted almost solely from seafood. When other foods are involved, they represent cross-contamination from seafood products. Another unique feature of this syndrome is the natural habit of the etiologic agent—the sea. In addition to its role in gastroenteritis, *V. parahaemolyticus* is known to cause extraintestinal infections in humans.

The genus *Vibrio* consists of at least 28 species, and 3 that are often associated with *V. parahaemolyticus* in aquatic environments and seafood are *V. vulnificus*, *V. alginolyticus*, and *V. cholerae*. Some of the distinguishing features of these species are noted in Table 23-1, and the syndromes caused by each are described below. Thorough reviews of these and related organisms have been provided by Colwell (45) and Joseph et al. (71).

V. parahaemolyticus is common in oceanic and coastal waters. Its detection is related to water temperatures, with numbers of organisms being undetectable until water temperature rises to around 19° to 20°C. A study of the Rhode River area of the Chesapeake Bay showed that the organisms survive in sediment during the winter and later are released into the water column, where they associate with the zooplankton from April to early June (73). In ocean waters, they tend to be associated more with shellfish than with other forms (90). They have been demonstrated to adsorb onto chitin particles and copepods, whereas organisms such as *Escherichia coli* and *Pseudomonas fluorescens* do not (73). This species is generally not found in the open oceans (71), and it cannot tolerate the hydrostatic pressures of ocean depths (129).

583

TABLE 23-1.   Differences between *V. parahaemolyticus* and Three Other *Vibrio* spp.

Species	*V. parahaemolyticus*	*V. alginolyticus*	*V. vulnificus*	*V. cholerae*
Lateral flagella on solid media	+	+	−	−
Rod shape	S	S	C	d
VP	−	+[a]	−	v
Growth in 10% NaCl	−	+	−	−
Growth in 6% NaCl	+	+	+	−
Swarming	−	+	−	−
Production of acetoin/diacetyl	−	+	−	+
Sucrose	−	+	−	+
Cellobiose	−	−	+	−
Utilization of putrescine	+	d	−	−
Color on TCBS agar	G	Y	G	Y

Source: *Bergey's Manual* (12).

[a] 24 h.

S = straight	G = green	d = 11 to 90% of strains positive.
C = curved	Y = yellow	v = variable; strain instability.

Growth Conditions

V. parahaemolyticus can grow in the presence of 1 to 8% NaCl, with best growth occurring in the 2 to 4% range (128). It dies off in distilled water (11). It does not grow at 4°C, but growth between 5° and 9°C has been demonstrated at pH 7.2 to 7.3 and 3% NaCl, or at pH 7.6 and 7% NaCl (Table 23-2). Its growth at 9.5° to 10°C in food products has been demonstrated (15), although the minimum for growth in open waters has been found to be 10°C (73). The upper growth temperature is 44°C, with an optimum between 30° and 35°C (129). Growth has been observed over the pH range 4.8 to 11.0, with 7.6 to 8.6 being optimum (13, 15, 129). It may be noted from Table 23-2 that the minimum growth pH is related to temperature and NaCl content, with moderate growth of one strain observed at pH 4.8 when the temperature was 30°C and NaCl content was 3%, but minimum pH was 5.2 when NaCl content was 7% (13). Similar results were found for five other strains. Under optimal conditions, this organism has a generation time of 9 to 13 min (compared to

TABLE 23-2.   Minimum pH of Growth of *V. parahaemolyticus* ATCC 107914 in TSB with 3% and 7% NaCl at Different Temperatures

Temperature (°C)	Minimum pH at NaCl Concentration	
	3%	7%
5	7.3	7.6
9	7.2	7.1
13	5.2	6.0
21	4.9	5.3
30	4.8	5.2

Source: Beuchat (13).

about 20 min for *E. coli*). Optimum a_w for growth corresponding to shortest generation time was found to be 0.992 (2.9% NaCl in tryptic soy broth). Employing the latter medium at 29°C and various solutes to control a_w, minimum values were 0.937 (glycerol), 0.945 (KCl), 0.948 (NaCl), 0.957 (sucrose), 0.983 (glucose), and 0.986 with propylene glycol (14). The organism is heat sensitive, with *D* 47°C values ranging from 0.8 to 65.1 min having been reported (16). With one strain, destruction of 500 cells/ml in shrimp homogenates was achieved at 60°C in 1 min, but with 2×10^5 cells/ml, some survived 80°C for 15 min (155). Cells are most heat resistant when grown at high temperatures in the presence of about 7% NaCl.

When the growth of *V. parahaemolyticus* was compared in estuarine water and a rich culture medium, differences were observed in cell envelope proteins and lipopolysaccharide and in alkaline phosphatase levels of K^+ and K^- strains (111). Alkaline phosphatase was slightly higher in K^- strains grown in water. Changes in cell evelope composition may be associated with the capacity of *V. parahaemolyticus* to enter a viable yet nonculturable state in waters, making its recovery from water more difficult (111).

Virulence Properties

The most widely used in vitro test of potential virulence for *V. parahaemolyticus* is the Kanagawa reaction, with most all virulent strains being positive (K^+) and most avirulent strains being negative (K^-). About 1% of sea isolates and about 100% of those from gastroenteritis patients are K^+ (125). K^+ strains produce a thermostable direct hemolysin (TDH), K^- strains produce a heat-labile hemolysin, and some strains produce both. A thermostable-related hemolysin (TRH) has been shown to be an important virulence factor for at least some *V. parahaemolyticus* strains. Of 214 clinical strains tested, 52% produced TDH only, 24% produced both TDH and TRH, and 11% produced both hemolysins (135). Of 71 environmental strains, 7% gave weak reactions to a TRH probe, but none reacted with a TDH probe. The Kanagawa reaction is determined generally by use of human red blood cells in Wagatsuma's agar medium. In addition to human red blood cells, those of the dog and rat are lysed, those of the rabbit and sheep give weak reactions, and those of the horse are not lysed (129). To determine the K reaction, the culture is surface plated, incubated at 37°C for 18 to 24 h, and read for the presence of β-hemolysis. Of 2,720 *V. parahaemolyticus* isolates from diarrheal patients, 96% were K^+, whereas only 1% of 650 fish isolates were K^+ (29). In general, isolates from waters are K^-.

The thermostable direct hemolysin has a molecular weight of 42,000 daltons and is a cardiotrophic, cytotoxic protein that is lethal to mice (67) and induces a positive response in the rabbit ileal loop assay (see Chap. 7). Its mean mouse LD_{50} by intraperitoneal (i.p.) injection is 1.5 μg, and the rabbit ileal loop dose is 200 μg (166). The hemolysin is under pH control and was found to be produced only when pH is 5.5 to 5.6 (44). That K^+ hemolysin may aid cells in obtaining iron stems from the observation that lysed erythrocyte extracts enhanced the virulence of the organism for mice (77). The membrane receptors of TDH are gangliosides G_{T1} and G_{D1a}, with the former binding hemolysin more firmly than the latter (147). The resistance of horse erythrocytes to the hemolysin apparently is due to the absence of these gangliosides (147).

A synthetic medium has been developed for the production of both the thermostable direct and the heat-labile hemolysins, and serine and glutamic acid were found to be indispensable (76). The heat stability of TDH is such that it can remain in foods after its production. In Tris buffer at pH 7.0, D 120°C and D 130°C values of 34 and 13 min, respectively, were found for semipurified toxin, whereas in shrimp D_{120} and D_{130} values were 21.9 and 10.4 min, respectively (26). Hemolysin was detected when cell counts reached 10^6/g, and its heat resistance was greater at pH 5.5 to 6.5 than at 7.0 to 8.0 (26).

The thermostable direct hemolysin gene (*tdh*) is chromosomal (148), its amino acid sequence has been determined (152), and it has been cloned in *E. coli* (75, 148). When the *tdh* gene was introduced into a K⁻ strain, it produced extracellular hemolysin (103). The nucleotide sequence of the *tdh* gene has been determined (103), and a specific *tdh* gene probe constructed, which consists of a 406-base pair (102). Employing this probe, 141 *V. parahaemolyticus* strains were tested. All K⁺ strains were gene positive—86% of them were weak positives—and 16% of K⁻ strains reacted with the probe. All gene-positive strains produced TDH as assessed by an ELISA method. Of 129 other vibrios tested with the gene probe including 19 named *Vibrio* spp., only *V. hollisae* was positive (102). The transfer of R-plasmids from *E. coli* to *V. parahaemolyticus* has been demonstrated (58).

The *V. hollisae* hemolysin is heat labile, unlike those of *V. cholerae* non-01 and *V. parahaemolyticus*, which are heat stable. The *V. hollisae* and *V. cholerae* non-01 hemolysins are immunologically related. Also, some strains of *V. cholerae* non-01, *V. hollisae*, and *V. mimicus* produce hemolysins that are similar to the TDH of *V. parahaemolyticus* (149).

At least 12 O antigens and 59 K antigens have been identified, but no correlations have been made between these and K⁺ and K⁻ strains, and the value of serotyping as an epidemiological aid has been minimal.

Since not all K⁺ strains produce positive responses in the rabbit ileal loop assay (28), and because some K⁻ strains are associated with gastroenteritis and sometimes are the only strains isolated (153), the precise virulence mechanisms are unclear. In the U.S. Pacific Northwest, K⁻ but urease-positive strains are associated with the syndrome (79). Of 45 human fecal isolates in California and Mexico, 71% of 45 were urease positive, 91% were K⁺, and the serovar was 04:K12 (1).

Adherence to epithelial cells is an important virulence property of gram-negative bacteria, and it appears that *V. parahaemolyticus* produces cell-associated hemagglutinins that correlate with adherence to intestinal mucosa (164). Pili (fimbriae) also play a role in intestinal tract colonization (101).

Gastroenteritis Syndrome and Vehicle Foods

The identity of *V. parahaemolyticus* as a foodborne gastroenteritis agent was made first by Fujino in 1951 (52). While the incidence of this illness is rather low in the United States and some European countries, it is the leading cause of food poisoning in Japan, accounting for 24% of bacterial food poisoning

TABLE 23-3.   Some Gastroenteritis Outbreaks Caused by *V. parahaemolyticus*

Year	Location/Incidence/Comments	Reference
1951	Japan. The outbreak occurred following consumption of *shirasu* (boiled and semidried young sardines). There were 272 victims, including 20 deaths.	128
1956	Japan. Salted cucumber was involved in a hospital outbreak with 120 victims.	128
1960	Japan. An explosive outbreak affecting thousands of people occurred along the Pacific coast. Horse mackerel was thought to be the vehicle.	128
1971	Maryland. Steamed crabs and crab salad were incriminated in 3 outbreaks in which 425 of about 745 people became ill. Isolations from food and victims revealed K^+ strain 04:K11 as the causative agent. This was first documented outbreak in America.	96
1974	Mexico. K^+ strains were recovered from two victims who ate raw fish.	54
1976	Guam. Some 122 cases developed from eating octopus aboard a ship.	34
1976	Louisiana. Some 100 cases resulted from eating boiled shrimp at a picnic.	34
1978	Louisiana. Boiled shrimp was the vehicle food in this outbreak involving about 67% of 1,700 persons at risk. K^+ strains were recovered from leftover shrimp and other foods as well as from the stools of 7 of 15 victims.	37

between 1965 and 1974 (19, 128). Between 1973 and 1975, three outbreaks were recorded in the United States, with 224 cases. A synopsis of some of the early outbreaks is presented in Table 23-3. Among the more than 81,000 cases reported during this period, 31 deaths were recorded (19). Of a total of 33 outbreaks recorded by the Centers for Disease Control (CDC) for the period 1973–1987, finfish and shellfish were the vehicles for 19. Only one outbreak with two cases was recorded for 1986 and none for 1987 (10).

With regard to symptomatology, findings from the Louisiana outbreak (noted in Table 23-3) illustrate the typical features. The mean incubation period in that outbreak was 16.7 h, with a range of 3 to 76 h. The symptoms lasted from 1 to 8 days, with a mean of about 4.6 days. Symptoms (along with percentage incidence of each) were diarrhea (95), cramps (92), weakness (90), nausea (72), chills (55), headache (48), and vomiting (12). Both sexes were equally affected, the age of victims ranging from 13 to 78 years.

No illness occurred among 14 volunteers who ingested more than 10^9 cells, but illness did occur in one person from the accidental ingestion of approximately 10^7 K^+ cells (128). In another study, 2×10^5 to 3×10^7 K^+ cells produced symptoms in volunteers, whereas 10^{10} cells of K^- strains did not (129, 153). Some K^- strains have been associated with outbreaks (9, 128).

Vehicle foods for outbreaks are seafood, such as oysters, shrimps, crabs, lobsters, clams, and related shellfish. Cross-contamination may lead to other foods as vehicles.

For the recovery and enumeration of *V. parahaemolyticus* from foods, an appropriate reference from Table 5-1 should be consulted.

OTHER VIBRIOS

Vibrio Cholerae

The *V. cholerae* that causes epidemic cholera belongs to serovar 0 Group 1. The strains of *V. cholerae* that are biochemically similar to the epidemic strains but that do not agglutinate in *V. cholerae* 0 Group 1 antiserum and are not associated with the epidemic disease are referred to as non-01 or non-agglutinating vibrios (NAGs). The non-01 strains have been shown to cause gastroenteritis, soft-tissue infections, and septicemia in humans. As in the case of *V. parahaemolyticus*, they are found in brackish surface waters during the warm-weather months. They are considered to be autochthonous estuarine bacteria in Chesapeake Bay (74) and also in Puerto Rican waters (116).

Among the earliest information linking non-01 *V. cholerae* to gastroenteritis in the United States are findings from 26 of 28 patients with acute diarrheal illness between 1972 and 1975. While some had systemic infections, 50% of the 28 yielded noncholera vibrios from stools and no other pathogens (69). In another retrospective study of non-01 *V. cholerae* cultures submitted to the CDC in 1979, nine were from domestically acquired cases of gastroenteritis and each patient had eaten raw oysters within 72 h of symptoms (98). One of these isolates produced a heat-labile toxin, whereas none produced heat-stable toxins.

At least five documented gastroenteritis outbreaks of non-01 *V. cholerae* occurred prior to 1981. Those in Czechoslovakia and Australia (1965 and 1973, respectively) were traced to potatoes and to egg and asparagus salads, and practically all victims experienced diarrhea. The third outbreak occurred in the Sudan, and well water was the source. Incubation periods from these three outbreaks ranged from 5 h to 4 days (18). The fourth outbreak occurred in Florida in 1979 and involved 11 persons who ate raw oysters. Eight experienced diarrheal illness within 48 h after eating oysters, and the other 3 developed symptoms 12, 15, and 30 h after eating. The fifth outbreak, which occurred in 1980 mainly among U.S. soldiers in Venice, Italy, was traced to raw oysters. Of about 50 persons at risk, 24 developed gastroenteritis. The mean incubation period was 21.5 h, with a range of 0.5 h to 5 days; and the symptoms (and percentage complaining) were diarrhea (91.7), abdominal pain (50), cramps (45.8), nausea (41.7), vomiting (29.2), and dizziness (20.8). All victims recovered in 1 to 5 days, and non-01 strains were recovered from the stools of four.

In regard to *V. cholerae* 01, 6 outbreaks with 916 cases and 12 deaths were recorded by CDC for the years 1973 through 1987 (11). Of the 6 outbreaks, 3 were traced to shellfish and 2 to finfish. A single case of 01 infection occurred in Colorado in August 1988. The victim ate about 12 raw oysters that were harvested in Louisiana and within 36 h had sudden onset of symptoms and passed 20 stools (33). *V. cholerae* 01 E1 Tor serotype Inaba was recovered from stools.

With regard to distribution, non-01 strains of *V. cholerae* have been found in the Orient and Mexico in stools of diarrheal patients along with enteropathogenic *E. coli*. *V. cholerae* non-01 was isolated from 385 persons with diarrhea in Mexico City in 1966–1967 (19). The organism causes diarrhea by an enterotoxin similar to cholera toxin (CT) and perhaps by other mechanisms (see below). A strain isolated from a patient produced a CT-like toxin, positive responses in the Chinese hamster ovary (CHO) and skin permeability assays, was invasive in rabbit ileal mucosa, but was Sereny-test negative (124).

From Chesapeake Bay, 65 non-01s were isolated in one study (74). Throughout the year, their numbers in waters were generally low, from 1 to 10 cells/l. They were found only in areas where salinity ranged between 4 and 17%. Their presence was not correlated with fecal *E. coli*, whereas the presence of the latter did correlate with *Salmonella* (74). Of those examined, 87% produced positive responses in Y-1 adrenal, rabbit ileal loop, and mouse lethality assays. Investigations conducted on waters along the Texas, Louisiana, and Florida coasts reveal that both 01 and non-01 *V. cholerae* are fairly common. Of 150 water samples collected along a Florida estuary, 57% were positive for *V. cholerae* (46). Of 753 isolates examined, 20 were 01 and 733 were non-01 types. Of the 20 01 strains, 8 were Ogawa and 12 were Inaba serovars, and they were found primarily at a sewage treatment plant. The highest numbers of both 01 and non-01 strains occurred in August and November (46). Neither the fecal coliform nor the total coliform index was an adequate indicator of the presence of *V. cholerae*, but the former was more useful than the latter. Along the Santa Cruz, California, coast, the highest numbers of non-01s occurred during the summer months and were associated with high coliform counts (80). Both 01 and non-01 strains have been recovered from aquatic birds in Colorado (105), and both types have been shown to be endemic in the Texas gulf as evidenced by antibody titers in human subjects (70).

V. cholerae 01 E1 Tor synthesizes a 82 kda preprotoxin and secretes it into culture media, where it is further processed into a 65 kda active cytolysin (163). Non-01 strains produce a cytotoxin and a hemolysin with a molecular weight of 60 kda, which is immunologically related to the hemolysin of the E1 Tor strain (165).

From a patient with travelers' diarrhea was isolated a strain of 01 from which the STa (NAG-STa) gene was cloned (104). The NAG-STa was chromosomal, and the toxin had a molecular weight of 8,815. The NAG-STa shared 50 and 46% homology to *E. coli* STh and STp, respectively (104). NAG-ST is methanol soluble, active in the infant mouse model, and similar to the ST of *Citrobacter freundii* (146). Monoclonal antibodies to NAG-ST cross-react with the ST of *Yersinia enterocolitica*. Further, *V. mimicus* ST and *Y. enterocolitica* ST are neutralized by monoclonal antibodies to NAG-ST but not *E. coli* STh or STp (146).

Vibrio vulnificus

This organism is found in seawater and seafood. It is isolated more often from oysters and clams than from crustacean shellfish products. It has been isolated from seawater from the coast of Miami, Florida, to Cape Cod, Massachusetts, with most (84%) isolated from clams (108). Upon injection into mice, 82% of tested strains were lethal. *V. vulnificus*, along with other vibrios, have been recovered from mussels, clams, and oysters in Hong Kong at rates between 6 to 9% (43). The organism is not recoverable from cold estuarine environments during winter months, and this appears to be due to the existence of viable but nonculturable forms. The latter types have been shown to contain fewer ribosomes, possess smaller cell sizes, show a decrease in C_{16} fatty acids, and to be less virulent than the comparable culturable forms (89). Like *V. aglinolyticus* (see below), *V. vulnificus* causes soft-tissue infections and primary septicemia in humans, especially in the immunocompromised and those with cirrhosis. The fatality rate for those with septicemia is more than 50%, and more than 90% among those who become hypotensive (see 162). These organisms are highly invasive, and they produce a cytotoxin with a molecular weight of about 56 kda that is toxic to CHO cells and lytic for erythrocytes. However, the cytolysin appears not to be a critical virulence factor (161). Also, an hemolysin is produced with a molecular weight of about 36 kda (162). The structural gene of *V. vulnificus* and the E1 Tor strain of *V. cholerae* 01 share areas of similarity, suggesting a common origin (163). *V. vulnificus* induces fluid accumulation in the RITARD ligated rabbit loop (see Chap. 7), suggesting the presence of an enterotoxin (137).

Infections have been seen in the United States, Japan, and Belgium. Of those seen in the United States, most occurred between May and October, and most patients were males over 40 years of age. This organism is believed to be a significant pathogen in individuals with higher than normal levels of iron (as, for example, in hepatitis and chronic cirrhosis), although its virulence is not explained entirely by its capacity to sequester iron.

Vibrio alginolyticus

This species is a normal inhabitant of seawater and has been found to cause soft-tissue and ear infections in humans. Human pathogenicity was first confirmed in 1973, although its possible role in wound infections was noted by Twedt et al. in 1969 (154). Extraintestinal infections have been reported from several countries, including the United States. Wound infections occur usually on body extremities, with most patients being males with a history of exposure to seawater.

The prevalence of *V. alginolyticus* along with that of *V. parahaemolyticus* was studied in coastal waters of the state of Washington by Baross and Liston (8), who found higher numbers of organisms in invertebrates and sediment samples than in water where the numbers were quite variable. Numbers found in oysters correlated with the temperature of overlying waters, with highest numbers associated with warmer waters. Whether this organism is of any significance as a foodborne pathogen is not clear.

For more information on *Vibrio* infections, see the reviews by Colwell (45), Joseph et al. (71), Morris and Black (97), and Klontz et al. (83).

YERSINIOSIS (*Yersinia enterocolitica*)

In the genus *Yersinia*, which belongs to the family Enterobacteriaceae, 11 species and 5 biovars are recognized, including *Y. pestis*, the cause of plague (12). The species of primary interest in foods is *Y. enterocolitica*. First isolated in New York State in 1933 by M. B. Coleman (66), this gram-negative rod is

TABLE 23-4.   Species of *Yersinia* Associated with *Y. enterocolitica* in the Environment and in Foods and Minimum Biochemical Differences

Species	VP	Sucrose	Rhamnose	Raffinose	Melibiose
Y. enterocolitica	+	+	−	−	−
Y. kristensenii	−	−	−	−	−
Y. frederiksenii	+	+	+	−	−
Y. intermedia	+	+	+	+	+
Y. bercovieri	−	+	−	−	−
Y. mollaretii	−	+	−	−	−

VP = Voges-Proskauer reaction; + = positive reaction; − = negative reaction.

somewhat unique in that it is motile below 30°C but not at 37°C. It produces colonies of 1.0 mm or less on nutrient agar, is oxidase negative, ferments glucose with little or no gas, lacks phenylalanine deaminase, is urease positive, and is unique as a pathogen in being psychrotrophic. It is often present in the environment with at least three other of the yersiniae noted in Table 23-4 (see below). General reviews relative to foods have been published by Stern and Pierson (142), Swaminathan et al. (144), and Zink et al. (169).

Growth Requirements

Growth of *Y. enterocolitica* has been observed over the temperature range of −2° to 45°C, with an optimum between 22° and 29°C. For biochemical reactions, 29°C appears to be the optimum. The upper limit for growth of some strains is 40°C, and not all grow below 4° to 5°C. Growth at 0° to 2°C in milk after 20 days has been observed. Growth at 0° to 1°C on pork and chicken has been observed (86), and three strains were found to grow on raw beef held for 10 days at 0° to 1°C (61). In milk at 4°C, *Y. enterocolitica* grew and attained up to 10^7 cells/ml in 7 days and competed well with the background flora (3). The addition of NaCl to growth media raises the minimum growth temperature. In brain heart infusion (BHI) broth containing 7% NaCl, growth did not occur at 3° or 25°C after 10 days. At pH 7.2, growth of one strain was observed at 3°C and very slight growth at pH 9.0 at the same temperature; no growth occurred at pH 4.6 and 9.6 (143). Although 7% NaCl was inhibitory at 3°C, growth occurred at 5% NaCl. With no salt, growth was observed at 3°C over the pH range 4.6 to 9.0 (143, 144). Clinical strains were less affected by these parameters than were environmental isolates. With respect to minimum growth pH, the following values were found for six strains of *Y. enterocolitica* with pH adjusted with HCl and incubated for 21 days: 4.42–4.80 at 4°C, 4.36–4.83 at 7°C, 4.26–4.50 at 10°C, and 4.18–4.36 at 20°C (27). When organic acids were used to adjust pH, the order of their effectiveness was acetic > lactic > citric. On the other hand, the order of effectiveness of organic acids in tryptic soy broth was propionic ≥ lactic ≥ acetic > citric ≥ phosphoric ≥ HCl (24).

  Y. enterocolitica is destroyed in 1 to 3 min at 60°C (60). It is rather resistant to freezing, with numbers decreasing only slightly in chicken after 90 days at −18°C (86). The calculated D 62.8°C for 21 strains in milk ranged from 0.7 to 17.8 seconds, and none survived pasteurization (51).

Distribution

Y. enterocolitica and the related species noted in Table 23-4 are widely distributed in the terrestrial environment and in lake, well, and stream waters, which are sources of the organisms to warm-blooded animals. It is more animal adapted and is found more often among human isolates than the other species in Table 23-4. Of 149 strains of human origin, 81, 12, 5.4, and 2% were, respectively, *Y. enterocolitica*, *Yersinia intermedia*, *Yersinia frederiksenii*, and *Yersinia kristensenii* (133). *Y. intermedia* and *Y. frederiksenii* are found mainly in fresh waters, fish, and foods and only occasionally are isolated from humans. *Y. kristensenii* is found mainly in soils and other environmental samples as well as in foods but rarely isolated from humans (12). Like *Y. enterocolitica*, this species produces a heat-stable enterotoxin. Many of the *Y. enterocolitica*-like isolates of Hanna et al. (62) were rhamnose-positive and consequently are classified as *Y. intermedia* and/or *Y. frederiksenii*, and all grow at 4°C. Rhamnose-positive yersiniae are not known to cause infections in humans (85).

Animals from which *Y. enterocolitica* has been isolated include cats, birds, dogs, beavers, guinea pigs, rats, camels, horses, chickens, raccoons, chinchillas, deer, cattle, swine, lambs, fish, and oysters. It is widely believed that swine constitutes the single most common source of *Y. enterocolitica* in humans. Of 43 samples of pork obtained from a slaughterhouse and examined for *Y. enterocolitica*, *Y. intermedia*, *Y. kristensenii*, and *Y. frederiksenii*, 8 were positive and all 4 species were found (64). Along with *Klebsiella pneumoniae*, *Y. enterocolitica* was recovered from crabs collected near Kodiak Island, Alaska, and was shown to be pathogenic (50).

Serovars and Biovars

The most commonly occurring *Y. enterocolitica* serovars (serotypes) in human infections are 0:3, 0:5,27, 0:8, and 0:9. Each of 49 isolates belonging to these serovars produced a positive HeLa cell response, while only 5 of 39 other serovars were positive (99). Most pathogenic strains in the United States are 0:8 (biovars 2 and 3), and except for occasional isolations in Canada, it is rarely reported from other continents. In Canada, Africa, Europe, and Japan, serovar 0:3 (biovar 4) is the most common (151). The second most common in Europe and Africa is 0:9, which has been reported also from Japan. Serovar 0:3 (biovar 4, phage type 9b) was practically the only type found in the province of Quebec, Canada, and it was predominant in Ontario (151). The

TABLE 23-5.   The Four Most Common Biovars of *Y. enterocolitica*

Substrate/Product	Biovars			
	1	2	3	4
Lipase (Tween 80)	+	−	−	−
DNAse	−	−	−	+
Indole	+	+	−	−
D-xylose	+	+	+	−

next most common were 0:5,27 and 0:6,30. From human infections in Canada, 0:3 represented 85% of 256 isolates, whereas for nonhuman sources, 0:5,27 represented 27% of 22 isolates (151). Six isolates of 0:8 recovered from porcine tongues were lethal to adult mice (48), and only 0:8 was found by Mora and Pai (99) to be Sereny positive. Employing HeLa cells, the following serovars were found to be infective: 0:1, 0:2, 0:3, 0:4, 0:5, 0:8, 0:9, and 0:21 (131). Serovar 0:8 strains are not only virulent in humans, but they possess mice lethality and invasiveness by the Sereny test. The four most common biovars of *Y. enterocolitica* are indicated in Table 23-5. It appears that only biovars 2, 3, and 4 carry the virulence plasmid (84).

Virulence Factors

Y. enterocolitica produces a heat-stable (ST) enterotoxin that survives 100°C for 20 min. It is not affected by proteases and lipases and has a molecular weight of 9,000 to 9,700 daltons, and biological activity is lost upon treatment with 2-mercaptoethanol (106, 107). When subjected to isoelectric focusing, two active fractions with PI's of 3.29 (ST-1) and 3.00 (ST-2) have been found (106). Antiserum from guinea pigs immunized with the purified ST neutralized the activity of *Y. enterocolitica* ST and *E. coli* ST (106). Like *E. coli* ST, it elicits positive responses in suckling mice and rabbit ileal loop assays and negative responses in the CHO and Y-1 adrenal cell assays (see Chap. 7). It is methanol soluble (23), and stimulates guanylate cyclase and the cAMP response in intestines but not adenylate cyclase (107, 125). It is produced only at or below 30°C (114), and its production is favored in the pH range 7 to 8. Of 46 milk isolates, only 3 produced ST in milk at 25°C and none at 4°C (34). At 25°C, more than 24 h were required for ST production. It appears to be chromosomal rather than plasmid mediated (85).

In a study of 232 human isolates, 94% produced enterotoxin, while only 32% of 44 from raw milk and 18% of 55 from other foods were enterotoxigenic (115). Of the serovars 0:3, 0:8, 0:5,27, 0:6,30, and 0:9, 97% of 196 were enterotoxigenic. Ninety percent of the rhamnose-positive strains studied by Pai et al. (115) produced enterotoxin, indicating that not all isolates were *Y. enterocolitica* and that some of the other species produce enterotoxin. It has been found that most natural waters in the United States contain rhamnose-positive strains that are either serologically untypable or react with multiple serovars (66). In another study, 43 strains of *Y. enterocolitica* from children with gastroenteritis and 18 laboratory strains were examined for ST production, and all clinical and 7 laboratory strains produced ST as assessed by the infant mouse assay, and all were negative in the Y-1 adrenal cell assay (114). Regarding the production of ST by species other than *Y. enterocolitica*, none of 21, 8, and 1 of *Y. intermedia*, *Y. frederiksenii*, and *Y. aldovae*, respectively, was positive in one study of species from raw milk, while 62.5% of *Y. enterocolitica* were ST positive (157). On the other hand, about one-third of nonenterocolitica species, including *Y. intermedia* and *Y. kristensenii*, were positive for ST in two other studies (150, 159).

Although pathogenic strains of *Y. enterocolitica* produce ST, it appears that this agent is not critical to virulence. Some evidence for the lack of importance of ST was provided by Schiemann (130), who demonstrated positive HeLa-cell and Sereny-test responses, with a 0:3 strain that did not produce enterotoxin. On the other hand, each of 49 isolates belonging to serovar 0:3 and the other 4 virulent serovars produced ST (99).

An 0:3 strain carrying a 70 kbp virulence plasmid, PYV, has been shown to bind to intestinal mucus. Mucus is thought to be an excellent nutrient base for Y. enterocolitica, and an outer membrane protein promotes adhesion to immobilized mucus (112). With this base, virulent cells may cross the intestinal epithelium and enter the underlying lymphatic tissues where they establish infection in cells of monocyte lineage. Cell invasion has been attributed to two chromosomal genes, inv and ail. By this hupothesis, all disease-causing strains carry these genes. Environmental isolates have inv but not ail. The inv gene in nonpathogenic strains is nonfunctional, epidemic strains contain both functional genes, and environmental isolates have nonfunctional inv and no ail genes (117). Whatever is the true virulence mechanism of this organism, 0:8 serovars (the most common in the United States) appear to be involved in more systemic infections than other serovars.

Virulence appears to be a result of tissue invasiveness for this organism. The latter has been shown to be mediated by a 40- to 48-Mdal plasmid (168). The 44-Mdal plasmid of Y. enterocolitica and a 47-Mdal plasmid of a Y. pestis strain have been shown to share 55% DNA sequence homology over about 80% of the plasmid genomes (118). In addition to tissue invasiveness, the 40- to 50-Mdal plasmids are responsible for calcium-dependent growth at 37°C, auto-agglutination in tissue culture medium, adult mouse lethality for serovar 0:8 strains, suckling mouse lethality, HEp-2 cell adherence, and adherence in at least three outer membrane proteins (65, 125). These plasmids appear not to be responsible for enterotoxigenicity, HEp-2 cell invasiveness, or expression of fimbrial proteins. Other plasmids exist in yersiniae ranging from 3 to 36 Mdal, but they are not virulence associated (85). Of ten strains representing six serovars that contained 42- to 44-Mdal plasmids, all were lethal to suckling mice, whereas those without these plasmids were not (6). The feeding of virulence plasmid–bearing strains to thirst-stressed mice was found to be lethal, while plasmidless strains had no effect on mice (85). The same plasmid is responsible for other virulence-associated properties of this organism, including autoagglutination (84), calcium-dependent growth (53), production of V and W antigens (31), and serum resistance (113). The V and W antigens of 0:8 (biovar 2) were immunologically identical to those of Y. pestis and Yersinia pseudotuberculosis (31). The serum resistance factors encoded for by the virulence plasmid consist of outer membrane proteins synthesized when cells are grown at 37°C but not at 25°C (94, 113). The latter cells adhered more to Henle monolayers than the 37°C-grown, thus making the latter more resistant to serum killing.

Although its role in virulence is unclear, some strains of Y. enterocolitica have been shown to produce a broad-spectrum, mannose-resistant adhesin at 20°C that agglutinated erythrocytes of at least ten animal species (92). The hemagglutination is associated with fimbriae, which were not produced when cells were cultured at 37°C. Of 21 0:3 and 0:8 serovars, 7 produced the agglutinating fimbriae, while only 46 of 115 from a variety of sources did.

To determine which of three in vitro tests best correlated with virulence, 34 strains were tested for calcium dependency, autoagglutination, and the presence of 40- to 48-Mdal plasmids. With Ca^{2+} dependency, 29 of 31 strains were positive, whereas all of 34 were positive by the other two tests (138). These authors favored autoagglutination in tissue culture medium at 35°C as being perhaps the best test. Prpic et al. (120) also found autoagglutination to be the best in vitro method, followed by Ca^{2+} dependency. In addition to *Y. enterocolitica* strains, virulent *Y. pestis* and *Y. pseudotuberculosis* auto-agglutinate in tissue culture medium, whereas avirulent strains do not (84). Schiemann and Devenish (131) have suggested that the two most important factors involved in virulence of this organism are the presence of V and W antigens and the presence of an invasive factor demonstrable by HeLa cell infectivity, but general support of this position is wanting. Since plasmid-bearing strains survived in the peritoneal cavity while plasmidless strains did not, it has been suggested that resistance to phagocytosis is the decisive virulence factor for pathogenic *Y. enterocolitica* (63). In the latter study, complement was found to play no role in virulence.

When iron-dextran was administered i.p. to mice, the median lethal dose of *Y. enterocolitica* serovars 0:3 and 0:9 was reduced by about tenfold, while Desferal (desferrioxamine B mesylate) reduced the lethal dose more than 100,000-fold (123). The 0:8 strains were less affected by these compounds, suggesting their lower requirement for iron.

Incidence of *Y. enterocolitica* in Foods

This organism has been isolated from cakes, vacuum-packaged meats, seafood, vegetables, milk, and other food products. It has been isolated also from beef, lamb, and pork (86). Of all sources, swine appears to be the major source of strains pathogenic for humans.

From 31 porcine tongues from freshly slaughtered animals, 21 strains were isolated and represented six serovars, with 0:8 the most common and 0:6,30 the second most commonly isolated (48). The other serovars recovered were 0:3, 0:13,7, 0:18, and 0:46. Of 100 milk samples examined in the United States, 12 raw and 1 pasteurized yielded *Y. enterocolitica* (100). In eastern France, 81% of 75 samples of raw milk contained *Y. enterocolitica* following enrichment, with serovar 0:5 being the most predominant (156). In Australia, 35 isolates were recovered from raw goat's milk, with 71% being rhamnose positive (68). In Brazil, 16.8% of 219 samples of raw milk and 13.7% of 280 pasteurized milk contained yersinae, with *Y. enterocolitica, Y. intermedia*, and *Y. frederiksenii* constituting 34, 65, and 2.7%, respectively (150). From raw beef and chicken in Brazil, 80% contained yersinae, 60% of ground beef and liver, and 20% of pork were also positive (159).

Gastroenteritis Syndrome and Incidence

In addition to gastroenteritis, this organism has been associated with human pseudoappendicitis, mesenteric lymphadenitis, terminal ileitis, reactive arthritis, peritonitis, colon and neck abscesses, cholecystis, and erythema nodosum. It has been recovered from urine, blood, cerebrospinal fluid, and the eyes of infected individuals. It is, of course, recovered from the stools of gastroenteritis victims. Only the gastroenteritis syndrome is addressed below.

TABLE 23-6.   Synopsis of Some Outbreaks of Yersiniosis

Year	Location	Vehicle	Characteristics of Outbreaks	Serovar	Reference
1972	Japan	Unknown	47% of 182 school children and 1 teacher infected	0:3	5
1972	Japan	Unknown	53% of 993 children and adults at a primary school affected	0:3	5
1972	Japan	Unknown	198 of 1,086 junior high school pupils infected		167
1972	North Carolina	Dog/puppies	16 of 21 persons in 4 families were infected		59
1975	North Carolina	Food handler	Two common-source outbreaks occurred in nursery schools. The children also ate snow covered with maple syrup		109
1975	Montreal	Raw milk	57 elementary school children and 1 adult infected. Serovar 0:5,27 was recovered from milk; 0:6,30 from victims		30
1976	New York state	Chocolate milk	Some 218 school children were infected. Chocolate syrup added to pasteurized milk in open vat was the apparent vehicle	0:8	18
1981	New York state	Powdered milk and chow mein	About 35% of 455 teenage campers and staff were infected from dissolved powdered milk and/or chow mein. Five had appendectomies.	0:8	134
1982	Washington state	Commercial tofu	Of the 87 victims, 56 had positive stools. Water used in processing was the apparent source of the organisms	0:8 0:Tacoma	(7, 145)
1982	Connecticut	Pasteurized milk	53 of 300 became victims with the attack rate being greatest among the 6- to 13-year-olds. Twenty of 52 stools were positive for the serovar noted.	0:8	4
1982	Tennessee, Arkansas, Mississippi	Pasteurized milk	More than 172 victims resulted from milk pasteurized in Tennessee. Seventeen patients underwent appendectomies, and 41% of victims were under age 5.	0:13,0:18	41

There is a seasonal incidence associated with this syndrome, with the fewest outbreaks occurring during the spring and the greatest number in October and November. The incidence is highest in the very young and the old. In an outbreak studied by Gutman et al. (59), the symptoms (and percentage complaining of them) were fever (87), diarrhea (69), severe abdominal pain (62), vomiting (56), pharyngitis (31), and headache (18). The outbreak led to two appendectomies and two deaths.

A synopsis of some early outbreaks in which *Y. enterocolitica* was shown or suspected as being the etiologic agent is presented in Table 23-6. Milk (raw, improperly pasteurized, or recontaminated) was the vehicle food in most. The first documented outbreak in the United States occurred in 1976 in New York State, with serovar 0:8 as the responsible strain, and chocolate milk prepared by adding chocolate syrup to previously pasteurized milk was a vehicle food (18). An outbreak of serotype 0:3 among 15 children occurred in Georgia in 1988–1989; the vehicle food was raw chitterlings (32).

Symptoms of the gastroenteritis syndrome develop several days following ingestion of contaminated foods and are characterized by abdominal pain and diarrhea. Children appear to be more susceptible than adults, and the organisms may be present in stools for up to 40 days following illness (5). A variety of systemic involvements may occur as a consequence of the gastroenteritis syndrome.

CAMPYLOBACTERIOSIS
(*Campylobacter jejuni*)

The genus *Campylobacter* consists of about 14 species, and the one of primary importance in foods is *C. jejuni* subsp. *jejuni*. Unlike *C. jejuni* subsp. *doylei*, it is resistant to cephalothin, can grow at 42°C, and can reduce nitrates. Throughout this text, *C. jejuni* subsp. *jejuni* is referred to as *C. jejuni*. The latter differs from *Campylobacter coli* in being able to hydrolyze hippurate. The campylobacters are more closely related to the genus *Arcobacter* than any other group.

Prior to the 1970s, the campylobacters were known primarily to verterinary microbiologists as organisms that caused spontaneous abortions in cattle and sheep and as the cause of other animal pathologies. They were once classified as *Vibrio* spp.

C. jejuni is a slender, spirally curved rod that possesses a single polar flagellum at one or both ends of the cell. It is oxidase and catalase positive and will not grow in the presence of 3.5% NaCl or at 25°C. It is microaerophilic, requiring small amounts of oxygen (3 to 6%) for growth. Growth is actually inhibited in 21% oxygen. Carbon dioxide (about 10%) is required for good growth. When *C. jejuni* was inoculated into vacuum-packaged processed turkey meat, cell numbers decreased, but some remained viable after 28 days at 4°C (122). Its metabolism is respiratory. In addition to *C. jejuni*, *C. coli*, *C. intestinalis*, and several other *Campylobacter* species are known to cause diarrhea in humans, but *C. jejuni* is by far the most important. More detailed information on the campylobacters can be found in Griffiths and Park (57).

Because of their small cell size, they can be separated from most other gram-negative bacteria by use of a 0.65-μm filter (21). *C. jejuni* is heat sensitive, with $D55°C$ for a composite of equal numbers of five strains being 1.09 min in peptone and 2.25 min in ground, autoclaved chicken (20). With internal heating of ground beef to 70°C, 10^7 cells/g could not be detected after about 10 min (141). It appears to be sensitive to freezing, with about 10^5 cells/ chicken carcass being greatly reduced or eliminated at $-18°C$, and for artificially contaminated hamburger meat, the numbers were reduced by 1 log cycle over a 7-day period (56).

Distribution

Unlike *Y. enterocolitica* and *V. parahaemolyticus*, *C. jejuni* is not an environmental organism but rather is one that is associated with warm-blooded animals. A large percentage of all major meat animals have been shown to contain these organisms in their feces, with poultry being prominent. A synopsis of some reports on their prevalence in some animal specimens as well as in poultry edible parts is presented in Table 23-7. Its prevalence in fecal samples often ranges from around 30 to 100%. Reports on isolations by various investigators have been summarized by Blaser (21), and the specimens and percent positive for *C. jejuni* are as follows: chicken intestinal contents (39–83), swine feces (66–87), sheep feces (up to 73), swine intestinal contents (61), sheep carcasses (24), swine carcasses (22), eviscerated chicken (72–80), and eviscerated turkey (94). The prevalence of *C. jejuni* and *C. coli* in 396 frozen and 405 fresh meats was examined. About 12% of fresh meats were positive but only 2.3% of the frozen, suggesting the lethal effects of freezing on the organisms (139). A higher percentage of chicken livers was positive (30% of fresh and 15% of frozen) than any of the other meats, which included beef, pork, and lamb livers, as well as muscle meats from these animals. Over 2,000 samples of a variety of retail-store meats were examined for *C. jejuni/coli* and *C. coli* by nine different laboratories (140). The organisms were found on 29.7% of chicken samples, 4.2% of pork sausage, 3.6% of ground beef, and about 5.1% of 1,800 red meats. Only *C. coli* was recovered from pork products. A higher incidence was noted in June and September (8.6%) than in December and March (4.5 and 3.9%, respectively).

Fecal specimens from humans with diarrhea yield *C. jejuni*, and it may be the single most common cause of acute bacterial diarrhea in humans. Of 8,097 specimens submitted to eight hospital laboratories over a 15-month period in different parts of the United States, this organism was recovered from 4.6%, salmonellae from 2.3%, and shigellae from 1% (22). The peak isolations for *C. jejuni* were in the age group 10 to 29 years. Peak isolations occur during the summer months, and it has been noted that 3 to 14% of diarrheal patients in developed countries yield stool specimens that contain *C. jejuni* (21). Peak isolations from individually caged hens occurred in October and late April–early May (47). In the latter study, 8.1% of the hens were chronic excreters of the organism, whereas 33% were negative even though they were likely exposed. The most probable source of *C. jejuni* to a duck processing farm was found to be rat and mice droppings, with 86.7% of the former being positive for this species (78).

The prevalence of *C. jejuni* on some poultry products is noted in Table 23-7. The numbers reported range from log 2.00 to 4.26/g. Once this organism is

TABLE 23-7.   Incidence of *C. jejuni* in Some Foods and Specimens

Year	Foods/Specimens	% Positive/Number Tested	Numbers/Comments	Location	Reference
1982	Feces from unweaned calves	30/30	Mean no. was $\log_{10}$ 3.26	New Zealand	55
1982	Fecal swabs from turkeys	Up to 76% pos.	Turkeys were 15–19 days old	Texas	2
1982	Cloacal swabs	41/327	82% of chicken isolates were biovar 1	Australia	132
1982	Broiler carcasses	45/40	Mean/carcass log 4.93	Australia	132
1983	Cecal cultures	All of 600	From turkeys	Colorado	91
1983	Chicken stomachs	50/20	From ready-for-market chickens	Netherlands	110
1983	Chicken ceca	92/25	Mean no. was log 4.11	Netherlands	110
1983	Chicken ceca	72/60	From 2 plants, range was log 3.72–7.25/g	California	160
1983	Chicken hearts	65/20	From ready-for-market chickens	Netherlands	110
1983	Chicken carcasses	49/120	From ready-for-market chickens	Netherlands	110
1983	Chicken livers	73/40	From ready-for-market chickens	Netherlands	110
1983	Chicken livers	69/36	Count range was log 2.00–4.15/g	California	160
1983	Chicken wings	67/36	Count range was log 2.00–4.26/g	California	160
1983	Turkey wings (fresh)	64/184	Mean count was 740/wing	California	121
1983	Turkey wings (frozen)	56/81	Mean count was 890/wing	California	121
1984	Fresh eggs	0.9/226	Only 2 eggs from fecal pos. hens contained organisms on shell	Wisconsin	47

established in a chicken house, most of the flock becomes infected over time. One study revealed that the organism appeared in all chicken inhabitants within a week once it was found among any of the inhabitants (136). In addition to poultry, the other primary source of this organism is raw milk. Some recorded outbreaks of *C. jejuni* enteritis from raw milk are noted in Table 23-8. Since the organism exists in cow feces, it is not surprising that it may be found in raw milk, and the degree of contamination would be expected to vary depending upon milking procedures. In a survey of 108 samples from bulk tanks of raw milk in Wisconsin, only 1 was positive for *C. jejuni*, whereas the feces of 64% of the cows in a grade A herd were positive (49). In the Netherlands, 22% of 904 cow fecal and 4.5% of 904 raw milk samples contained *C. jejuni* (17).

TABLE 23-8.   Synopsis of the Early Outbreaks of *Campylobacter* enteritis

Year	Location	Vehicle	Synopsis	Reference
1938	Illinois	Contaminated pasteurized milk	This presumptive outbreak involved 357 cases at 2 institutions. The outbreaks stopped after milk was boiled.	87
1978	Vermont	Water	The town's water supply was contaminated. About 2,000 of 10,000 persons were infected. Swabs from 5 of 9 victims revealed the agent.	35
1978	Colorado	Raw milk	Three of 5 family members were infected. Organism was recovered from stools of all victims as well as from cow feces.	36
1979	Iowa	Barbecued chicken	There were 8 victims of 11 who ate undercooked barbecued chicken.	38
1979	Scotland	Raw milk	There were 648 cases following an electrical failure at the dairy plant. The incubation period ranged from 1 to 13 days.	158
1980	England	Raw milk	About 75 of 300 college students were affected. The organism was found in milk samples, and 46 students had antibodies to *C. jejuni*.	126
1981	Kansas	Raw milk	There were over 264 cases. Fifty-two percent of 116 persons in households that had one or more ill family members yielded the agent.	39
1981	Oregon	Raw milk	Of 167 who drank infected milk, 77 became ill. Agent was found in stools.	40
1981	Georgia	Raw milk	There were 50 victims in 30 households but the organism was not found in milk.	119
1982	Connecticut	Cake icing	*C. jejuni* was isolated from 16 of 41 victims.	22
1984	California	Raw milk	Twelve of 35 children and adults became infected after drinking certified raw milk.	42

Virulence Properties

At least some strains of *C. jejuni* produce a heat-labile enterotoxin (CJT) that shares some common properties with the enterotoxins of *V. cholerae* (CT) and *E. coli* (LT). CJT increases cAMP levels, induces changes in CHO cells, and induces fluid accumulation in rat ileal loops (127). Maximal production of CJT in a special medium was achieved at 42°C for 24 h, and the amount produced was enhanced by polymyxin (81). The quantities produced by strains varied widely from none to about 50 ng/ml CJT protein. The amount of toxin was doubled as measured by Y-1 adrenal cell assay when cells were first exposed to lincomycin and then polymyxin (95). CJT is neutralized by CT and *E. coli* LT antisera, indicating immunological homology with these two enterotoxins (81). The *C. jejuni* LT appears to share the same cell receptors as CT and *E. coli* LT, and it contains a B subunit immunologically related to the B subunits of CT and LT of *E. coli* (82). Also, a cytotoxin is produced that is active against Vero and HeLa cells. The enterotoxin and the cytotoxin induce fluid accumulation in rat jejunal loops but not in mice, pigs, or calves. Partially purified enterotoxin contained three fractions with molecular weights of 68, 54, and 43 kda (72). Of 202 strains of *C. jejuni* and *C. coli* recovered from humans with enteritis and from healthy laying hens, 34% and 22% of the *C. jejuni* and *C. coli* strains, respectively, produced enterotoxin as determined by CHO assay (88).

C. *jejuni* enteritis appears to be caused in part by the invasive abilities of the organism. Evidence for this comes from the nature of the clinical symptoms, the rapid development of high agglutinin titers after infection, recovery of the organism from peripheral blood during the acute phase of the disease, and the finding that *C. jejuni* can penetrate HeLa cells (93). However, *C. jejuni* is not invasive by either the Sereny or the Anton assay.

Plasmids have been demonstrated in *C. jejuni* cells. Of 17 strains studied, 11 were found to carry plasmids ranging from 1.6 to 70 Mdal, but their role and function in disease is unclear (25).

A serotyping scheme has been developed for *C. jejuni*. From chickens and humans, 82 and 98%, respectively, of isolates belonged to biovar 1 (132).

Enteritis Syndrome

From the first U.S. outbreak of *C. jejuni*, traced to a water supply, in which about 2,000 individuals contracted infections, the symptoms (and percentage of individuals affected) were as follows: abdominal pain or cramps (88), diarrhea (83), malaise (76), headache (54), and fever (52). Symptoms lasted from 1 to 4 days. In the more severe cases, bloody stools may occur, and the diarrhea may resemble ulcerative colitis, while the abdominal pain may mimic acute appendicitis (21). The incubation period for enteritis is highly variable. It is usually 48 to 82 h but may be as long as 7 to 10 days or more. Diarrhea may last 2 to 7 days, and the organisms may be shed for more than 2 months after symptoms subside.

PREVENTION

V. parahaemolyticus, *Y. enterocolitica*, and *C. jejuni* are all heat-sensitive bacteria that are destroyed by milk pasteurization temperatures. The avoidance of raw seafood products and care in preventing cross-contamination with contaminated raw materials will eliminate or drastically reduce the incidence of foodborne gastroenteritis caused by *V. parahaemolyticus* and *Y. enterocolitica*. To prevent wound infections by vibrios, individuals with body nicks or abrasions should avoid entering seawaters. Yersinosis can be avoided or certainly minimized by not drinking water that has not been purified, and by avoiding raw or underprocessed milk. Campylobacteriosis can be avoided by not eating undercooked or unpasteurized foods of animal origin, especially milk.

References

1. Abbott, S. L., C. Powers, C. A. Kaysner, Y. Takeda, M. Ishibashi, S. W. Joseph, and J. M. Janda. 1989. Emergence of a restricted bioserovar of *Vibrio parahaemolyticus* as the predominant cause of *Vibrio*-associated gastroenteritis on the West Coast of the United States and Mexico. *J. Clin. Microbiol.* 27:2891–2893.
2. Acuff, G. R., C. Vanderzant, F. A. Gardner, and F. A. Colan. 1982. Examination of turkey eggs, poults and brooder house facilities for *Campylobacter jejuni*. *J. Food Protect.* 45:1279–1281.
3. Amin, M. K., and F. A. Draughon. 1987. Growth characteristics of *Yersinia enterocolitica* in pasteurized skim milk. *J. Food Protect.* 50:849–852.
4. Amsterdam, L., P. Bourbeau, and P. Checko. 1984. An outbreak of *Yersinia enterocolitica* associated with pasteurized milk in a self-contained communal group in Connecticut. *Abstracts of the American Society of Microbiology*, C-78.
5. Asakawa, Y., S. Akahane, N. Kagata, and M. Noguchi. 1973. Two community outbreaks of human infection with *Yersinia enterocolitica*. *J. Hyg.* 71:715–723.
6. Aulisio, C. C. G., W. E. Hill, J. T. Stanfield, and R. L. Sellers, Jr. 1983. Evaluation of virulence factor testing and characteristics of pathogenicity of *Yersinia enterocolitica*. *Infect. Immun.* 40:330–335.
7. Aulisio, C. C. G., J. T. Stanfield, S. D. Weagant, and W. E. Hill. 1983. Yersinosis associated with tofu consumption: Serological, biochemical and pathogenicity studies of *Yersinia enterocolitica* isolates. *J. Food Protect.* 46:226–230.
8. Baross, J., and J. Liston. 1970. Occurrence of *Vibrio parahaemolyticus* and related hemolytic vibrios in marine environments of Washington state. *Appl. Microbiol.* 20:179–186.
9. Barrow, G. I., and D. C. Miller. 1976. *Vibrio parahaemolyticus* and seafoods. In *Microbiology in Agriculture, Fisheries and Food*, ed. F. A. Skinner and J. G. Carr, 181–195. New York: Academic Press.
10. Bean, N. H., and P. M. Griffin. 1990. Foodborne disease outbreaks in the United States, 1973–1987: Pathogens, vehicles, and trends. *J. Food Protect.* 53:804–817.
11. Bean, N. H., P. M. Griffin, J. S. Goulding, and C. B. Ivey. 1990. Foodborne disease outbreaks, 5-year summary, 1983–1987. *J. Food Protect.* 53:711–728.
12. *Bergey's Manual of Systematic Bacteriology*, vol. 1. 1984. Ed. N. R. Krieg. Baltimore: Williams & Wilkins.
13. Beuchat, L. R. 1973. Interacting effects of pH, temperature, and salt concentration on growth and survival of *Vibrio parahaemolyticus*. *Appl. Microbiol.* 25:844–846.
14. Beuchat, L. R. 1974. Combined effects of water activity, solute, and temperature on the growth of *Vibrio parahaemolyticus*. *Appl. Microbiol.* 27:1075–1080.
15. Beuchat, L. R. 1975. Environmental factors affecting survival and growth of *Vibrio parahaemolyticus*. A review. *J. Milk Food Technol.* 38:476–480.
16. Beuchat, L. R., and R. E. Worthington. 1976. Relationships between heat resistance and phospholipid fatty acid composition of *Vibrio parahaemolyticus*. *Appl. Environ. Microbiol.* 31:389–394.
17. Beumer, R. R., J. J. M. Cruysen, and I. R. K. Birtantie. 1988. The occurrence of *Campylobacter jejuni* in raw cows' milk. *J. Appl. Bacteriol.* 65:93–96.
18. Black, R. E., R. J. Jackson, T. Tsai, M. Medvesky, M. Shayegani, J. C. Feeley, K. I. E. MacLeod, and A. M. Wakelee. 1978. Epidemic *Yersinia enterocolitica* infection due to contaminated chocolate milk. *N. Eng. J. Med.* 298:76–79.

19. Blake, P. A., R. E. Weaver, and D. G. Hollis. 1980. Diseases of humans (other than cholera) caused by vibrios. *Ann. Rev. Microbiol.* 34:341–367.

20. Blankenship, L. C., and S. E. Craven. 1982. *Campylobacter jejuni* survival in chicken meat as a function of temperature. *Appl. Environ. Microbiol.* 44:88–92.

21. Blaser, M. J. 1982. *Campylobacter jejuni* and food. *Food Technol.* 36(3):89–92.

22. Blaser, M. J., P. Checko, C. Bopp, A. Bruce, and J. M. Hughes. 1982. *Campylobacter* enteritis associated with foodborne transmission. *Amer. J. Epidemiol.* 116:886–894.

23. Boyce, J. M., D. J. Evans, Jr., D. G. Evans, and H. L. DuPont. 1979. Production of heat-stable, methanol-soluble enterotoxin by *Yersinia enterocolitica. Infect. Immun.* 25:532–537.

24. Brackett, R. E. 1987. Effects of various acids on growth and survival of *Yersinia enterocolitica. J. Food Protect.* 50:598–601.

25. Bradbury, W. C., M. A. Marko, J. N. Hennessy, and J. L. Penner. 1983. Occurrence of plasmid DNA in serologically defined strains of *Campylobacter jejuni* and *Campylobacter coli. Infect. Immun.* 40:460–463.

26. Bradshaw, J. G., D. B. Shah, A. J. Wehby, J. T. Peeler, and R. M. Twedt. 1984. Thermal inactivation of the Kanagawa hemolysin of *Vibrio parahaemolyticus* in buffer and shrimp. *J. Food Sci.* 49:183–187.

27. Brocklehurst, T. F., and B. M. Lund. 1990. The influence of pH, temperature and organic acids on the initiation of growth of *Yersinia enterocolitica. J. Appl. Bacteriol.* 69:390–397.

28. Brown, D. F., P. L. Spaulding, and R. M. Twedt. 1977. Enteropathogenicity of *Vibrio parahaemolyticus* in the ligated rabbit ileum. *Appl. Environ. Microbiol.* 33:10–14.

29. Bryan, F. L. 1979. Infections and intoxications caused by other bacteria. In *Foodborne Infections and Intoxications*, ed. H. Riemann and F. L. Bryan, 211–297. New York: Academic Press.

30. Canada Diseases Weekly Report. 1976. *Yersinia enterocolitica* gastroenteritis outbreak—Montreal. Report 2(11):41–44; report 2(19):73–74.

31. Carter, P. B., R. J. Zahorchak, and R. R. Brubaker. 1980. Plague virulence antigens from *Yersinia enterocolitica. Infect. Immun.* 28:638–640.

32. Centers for Disease Control. 1990. *Yersinia enterocolitica* infections during the holidays in black families—Georgia. *Morb. Mort. Week. Rept.* 39:819–821.

33. Centers for Disease Control. 1989. Toxigenic *Vibrio cholerae* O1 infection acquired in Colorado. *Morb. Mort. Week. Rept.* 38:19–20.

34. Centers for Disease Control. 1976. Foodborne and waterborne disease outbreaks. *Annual Summary*, HEW Pub. No. (CDC) 76–8185.

35. Centers for Disease Control. 1978. Waterborne *Campylobacter* gastroenteritis—Vermont. *Morb. Mort. W. Rept.* 27:207.

36. Centers for Disease Control. 1978. *Campylobacter* enteritis—Colorado. *Morb. Mort. W. Rept.* 27:226, 231.

37. Centers for Disease Control. 1978. *Vibrio parahaemolyticus* foodborne outbreak—Louisiana. *Morb. Mort. W. Rept.* 27:345–346.

38. Centers for Disease Control. 1979. *Campylobacter* enteritis—Iowa. *Morb. Mort. W. Rept.* 28:565–566.

39. Centers for Disease Control. 1981. Outbreak of *Campylobacter* enteritis associated with raw milk—Kansas. *Morb. Mort. W. Rept.* 30:218–220.

40. Centers for Disease Control. 1981. Raw-milk-associated illness—Oregon, California. *Morb. Mort. W. Rept.* 30:80–81.

41. Centers for Disease Control. 1982. Multi-state outbreak of yersiniosis. *Morb. Mort. W. Rept.* 31:505–506.

42. Centers for Disease Control. 1984. *Campylobacter* outbreak associated with certified raw milk porducts. *Morb. Mort. W. Rept.* 33:562.

43. Chan, K.-Y., M. L. Woo, L. Y. Lam, and G. L. French. 1989. *Vibrio parahaemolyticus* and other halophilic vibrios associated with seafood in Hong Kong. *J. Appl. Bacteriol.* 66:57–64.
44. Cherwonogrodzky, J. W., and A. G. Clark. 1981. Effect of pH on the production of the Kanagawa hemolysin by *Vibrio parahaemolyticus*. *Infect. Immun.* 34: 115–119.
45. Colwell, R. R., ed. 1984. *Vibrios in the Environment*. New York: Wiley.
46. DePaola, A., M. W. Presnell, R. E. Becker, M. L. Motes, Jr., S. R. Zywno, J. F. Musselman, J. Taylor, and L. Williams. 1984. Distribution of *Vibrio cholerae* in the Apalachicola (Florida) Bay estuary. *J. Food Protect.* 47:549–553.
47. Dolye, M. P. 1984. Association of *Campylobacter jejuni* with laying hens and eggs. *Appl. Environ. Microbiol.* 47:533–536.
48. Doyle, M. P., M. B. Hugdahl, and S. L. Taylor. 1981. Isolation of virulent *Yersinia enterocolitica* from porcine tongues. *Appl. Environ. Microbiol.* 42:661–666.
49. Doyle, M. P., and D. J. Roman. 1982. Prevalence and survival of *Campylobacter jejuni* in unpasteurized milk. *Appl. Environ. Microbiol.* 44:1154–1158.
50. Faghri, M. A., C. L. Pennington, L. B. Cronholm, and R. M. Atlas. 1984. Bacteria associated with crabs from cold waters, with emphasis on the occurrence of potential human pathogens. *Appl. Environ. Microbiol.* 47:1054–1061.
51. Francis, D. W., P. L. Spaulding, and J. Lovett. 1980. Enterotoxin production and thermal resistance of *Yersinia enterocolitica* in milk. *Appl. Environ. Microbiol.* 40:174–176.
52. Fujino, T., G. Sakaguchi, R. Sakazaki, and Y. Takeda. 1974. *International Symposium on Vibrio parahaemolyticus*. Tokyo: Saikon.
53. Gemski, P., J. R. Lazere, and T. Casey. 1980. Plasmid associated with pathogenicity and calcium dependency of *Yersinia enterocolitica*. *Infect. Immun.* 27:682–685.
54. Gil-Recasens, M. E., A. M. Peral-Lopez, and G. Ruiz-Reyes. 1974. Aislamiento de *Vibrio parahaemolyticus* en casos de gastroenteritis y en mariscos crudos en la ciudad de Puebla. *Rev. Lat. Amer. Microbiol.* 16:85–88.
55. Gill, C. O., and L. M. Harris. 1982. Contamination of red-meat carcasses by *Campylobacter fetus* subsp. *jejuni*. *Appl. Environ. Microbiol.* 43:977–980.
56. Gill, C. O., and L. M. Harris. 1984. Hamburgers and broiler chickens as potential sources of human *Campylobacter* enteritis. *J. Food Protect.* 47:96–99.
57. Griffiths, P. L., and R. W. A. Park. 1990. Campylobacters associated with human diarrhoeal disease. *J. Appl. Bacteriol.* 69:281–301.
58. Guerry, P., and R. R. Colwell. 1977. Isolation of cryptic plasmid deoxyribonucleic acid from Kanagawa-positive strains of *Vibrio parahaemolyticus*. *Infect. Immun.* 16:328–334.
59. Gutman, L. T., E. A. Ottesen, T. J. Quan, P. S. Noce, and S. L. Katz. 1973. An inter-familial outbreak of *Yersinia enterocolitica* enteritis. *N. Eng. J. Med.* 288: 1372–1377.
60. Hanna, M. O., J. C. Stewart, Z. L. Carpenter, and C. Vanderzant. 1977. Heat resistance of *Yersinia enterocolitica* in skim milk. *J. Food Sci.* 42:1134, 1136.
61. Hanna, M. O., J. C. Stewart, D. L. Zink, Z. L. Carpenter, and C. Vanderzant. 1977. Development of *Yersinia enterocolitica* on raw and cooked beef and pork at different temperatures. *J. Food Sci.* 42:1180–1184.
62. Hanna, M. O., D. L. Zink, Z. L. Carpenter, and C. Vanderzant. 1976. *Yersinia enterocolitica*–like organisms from vacuum-packaged beef and lamb. *J. Food Sci.* 41:1254–1256.
63. Hanski, C., M. Naumann, A. Grutzkau, G. Pluschke, B. Friedrich, H. Hahn, and E. O. Riecken. 1991. Humoral and cellular defense against intestinal murine infection with *Yersinia enterocolitica*. *Infect. Immun.* 59:1106–1111.

64. Harmon, M. C., B. Swaminathan, and J. C. Forrest. 1984. Isolation of *Yersinia enterocolitica* and related species from porcine samples obtained from an abattoir. *J. Appl. Bacteriol.* 56:421–427.

65. Heesemann, J., B. Algermissen, and R. Laufs. 1984. Genetically manipulated virulence of *Yersinia enterocolitica. Infect. Immun.* 46:105–110.

66. Highsmith, A. K., J. C. Feeley, and G. K. Morris. 1977. *Yersinia enterocolitica*: A review of the bacterium and recommended laboratory methodology. *Hlth. Lab. Sci.* 14:253–260.

67. Honda, T., K. Goshima, Y. Takeda, Y. Sugino, and T. Miwatani. 1976. Demonstration of the cardiotoxicity of the thermostable direct hemolysin (lethal toxin) produced by *Vibrio parahaemolyticus. Infect. Immun.* 13:163–171.

68. Hughes, D., and N. Jensen. 1981. *Yersinia enterocolitica* in raw goat's milk. *Appl. Environ. Microbiol.* 41:309–310.

69. Hughes, J. M., D. G. Hollis, E. J. Gangarosa, and R. E. Weaver. 1978. Noncholera vibrio infections in the United States: Clinical, epidemiologic and laboratory features. *Ann. Intern. Med.* 88:602–606.

70. Hunt, M. D., W. E. Woodward, B. H. Keswick, and H. L. Dupont. 1988. Seroepidemiology of cholera in Gulf coastal Texas. *Appl. Environ. Microbiol.* 54:1673–1677.

71. Joseph, S. W., R. R. Colwell, and J. B. Kaper. 1982. *Vibrio parahaemolyticus* and related halophilic vibrios. *CRC Crit. Rev. Microbiol.* 10:77–124.

72. Kaikoku, T., M. Kawaguchi, K. Takama, and S. Suzuki. 1990. Partial purification and characterization of the enterotoxin produced by *Campylobacter jejuni. Infect. Immun.* 58:2414–2419.

73. Kaneko, T., and R. R. Colwell. 1973. Ecology of *Vibrio parahaemolyticus* in Chesapeake Bay. *J. Bacteriol.* 113:24–32.

74. Kaper, J., H. Lockman, R. R. Colwell, and S. W. Joseph. 1979. Ecology, serology, and enterotoxin production of *Vibrio cholerae* in Chesapeake Bay. *Appl. Environ. Microbiol.* 37:91–103.

75. Kaper, J. B., R. K. Campen, R. J. Seidler, M. M. Baldini, and S. Falkow. 1984. Cloning of the thermostable direct or Kanagawa phenomenon–associated hemolysin of *Vibrio parahaemolyticus. Infect. Immun.* 45:290–292.

76. Karunsagar, I. 1981. Production of hemolysin by *Vibrio parahaemolyticus* in a chemically defined medium. *Appl. Environ. Microbiol.* 41:1274–1275.

77. Karunsagar, I., S. W. Joseph, R. M. Twedt, H. Hada, and R. R. Colwell. 1984. Enhancement of *Vibrio parahaemolyticus* virulence by lysed erythrocyte factor and iron. *Infect. Immun.* 46:141–144.

78. Kasrazadeh, M., and C. Genigeorgis. 1987. Origin and prevalence of *Campylobacter jejuni* in ducks and duck meat at the farm and processing plant level. *J. Food Protect.* 50:321–326.

79. Kelly, M. T., and E. M. Dan Stroh. 1989. Urease-positive, Kanagawa-negative *Vibrio parahaemolyticus* from patients and the environment in the Pacific Northwest. *J. Clin. Microbiol.* 27:2820–2822.

80. Kenyon, J. E., D. R. Piexoto, B. Austin, and D. C. Gillies. 1984. Seasonal variation in numbers of *Vibrio cholerae* (non-01) isolated from California coastal waters. *Appl. Environ. Microbiol.* 47:1243–1245.

81. Klipstein, F. A., and R. F. Engert. 1984. Properties of crude *Campylobacter jejuni* heat-labile enterotoxin. *Infect. Immun.* 45:314–319.

82. Klipstein, F. A., and R. F. Engert. 1985. Immunological relationship of the B subunits of *Campylobacter jejuni* and *Escherichia coli* heat-labile enterotoxins. *Infect. Immun.* 48:629–633.

83. Klontz, K. C., S. Lieb, M. Schreiber, H. T. Janowski, L. M. Baldy, and R. A. Gunn. 1988. Syndromes of *Vibrio vulnificus* infections. Clinical and epidemiologic features in Florida cases, 1981–1987. *Ann. Int. Med.* 109:318–323.

84. Laird, W. J., and D. C. Cavanaugh. 1980. Correlation of auto-agglutination and virulence of yersiniae. *J. Clin. Microbiol.* 11:430–432.

85. Lee, W. H., R. E. Smith, J. M. Damare, M. E. Harris, and R. W. Johnston. 1981. Evaluation of virulence test procedures for *Yersinia enterocolitica* recovered from foods. *J. Appl. Bacteriol.* 50:529–539.

86. Leistner, L., H. Hechelmann, and R. Albert. 1975. Nachweis von *Yersinia enterocolitica* in Faeces und Fleisch von Schweinen, Hindern und Geflugel. *Fleischwirtschaft* 55:1599–1602.

87. Levy, A. J. 1946. A gastro-enteritis outbreak probably due to a bovine strain of *Vibrio*. *Yale J. Biol. Med* 18:243–258.

88. Lindblom, G.-B., B. Kaijser, and E. Sjogren. 1989. Enterotoxin production and serogroups of *Campylobacter jejuni* and *Campylobacter coli* from patients with diarrhea and from healthy laying hens. *J. Clin. Microbiol.* 27:1272–1276.

89. Linder, K., and J. D. Oliver. 1989. Membrane fatty acid and virulence changes in the viable but nonculturable state of *Vibrio vulnificus*. *Appl. Environ. Microbiol.* 55:2837–2842.

90. Liston, J. 1973. *Vibrio parahaemolyticus*. In *Microbial Safety of Fishery Products*, ed. C. O. Chichester and H. D. Graham, 203–213. New York: Academic Press.

91. Luechtefeld, N. W., and W. L. L. Wang. 1981. *Campylobacter fetus* subsp. *jejuni* in a turkey processing plant. *J. Clin. Microbiol.* 13:266–268.

92. MacLagan, R. M., and D. C. Old. 1980. Hemagglutinins and fimbriae in different serotypes and biotypes of *Yersinia enterocolitica*. *J. Appl. Bacteriol.* 49:353–360.

93. Manninen, K. I., J. F. Prescott, and I. R. Dohoo. 1982. Pathogenicity of *Campylobacter jejuni* isolates from animals and humans. *Infect. Immun.* 38:46–52.

94. Martinez, R. J. 1983. Plasmid-mediated and temperature-regulated surface properties of *Yersinia enterocolitica*. *Infect. Immun.* 41:921–930.

95. McCardell, B. A., J. M. Madden, and E. C. Lee. 1984. *Campylobacter jejuni* and *Campylobacter coli* production of a cytotonic toxin immunologically similar to cholera toxin. *J. Food Protect.* 47:943–949.

96. Molenda, J. R., W. G. Johnson, M. Fishbein, B. Wentz, I. J. Mehlman, and T. A. Dadisman, Jr. 1972. *Vibrio parahaemolyticus* gastroenteritis in Maryland: Laboratory aspects. *Appl. Microbiol.* 24:444–448.

97. Morris, J. G., Jr., and R. E. Black. 1985. Cholera and other vibrios in the United States. *N. Eng. J. Med.* 312:343–350.

98. Morris, J. G., R. Wilson, B. R. Davis, I. K. Wachsmuth, C. F. Riddle, H. G. Wathen, R. A. Pollard, and P. A. Blake. 1981. Non-0 group 1 *Vibrio cholerae* gastroenteritis in the United States. *Ann. Intern. Med.* 94:656–658.

99. Mors, V., and C. H. Pai. 1980. Pathogenic properties of *Yersinia enterocolitica*. *Infect. Immun.* 28:292–294.

100. Moustafa, M. K., A. A.-H. Ahmed, and E. H. Marth. 1983. Occurrence of *Yersinia enterocolitica* in raw and pasteurized milk. *J. Food Protect.* 46:276–278.

101. Nakasone, N., and M. Iwanaga. 1990. Pili of *Vibrio parahaemolyticus* strain as a possible colonization factor. *Infect. Immun.* 58:61–69.

102. Nishibuchi, M., M. Ishibashi, Y. Takeda, and J. B. Kaper. 1985. Detection of the thermostable direct hemolysin gene and related DNA sequences in *Vibrio parahaemolyticus* and other *Vibrio* species by the DNA colony hybridization test. *Infect. Immun.* 49:481–486.

103. Nishibuchi, M., and J. B. Kaper. 1985. Nucleotide sequence of the thermostable direct hemolysin gene of *Vibrio parahaemolyticus*. *J. Bacteriol.* 162:558–564.

104. Ogawa, A., J.-I. Kato, H. Watanabe, B. G. Nair, and T. Takeda. 1990. Cloning and nucleotide sequence of a heat-stable enterotoxin gene from *Vibrio cholerae* non-01 isolated from a patient with traveler's diarrhea. *Infect. Immun.* 58: 3325–3329.

105. Ogg, J. E., R. A. Ryder, and H. L. Smith, Jr. 1989. Isolation of *Vibrio cholerae* from aquatic birds in Colorado and Utah. *Appl. Environ. Microbiol.* 55:95–99.

106. Okamoto, K., T. Inoue, H. Ichikawa, Y. Kawamoto, and A. Miyama. 1981. Partial purification and characterization of heat-stable enterotoxin produced by *Yersinia enterocolitica*. *Infect. Immun.* 31:554–559.

107. Okamoto, K., T. Inoue, K. Shimizu, S. Hara, and A. Miyama. 1982. Further purification and characterization of heat-stable enterotoxin produced by *Yersinia enterocolitica*. *Infect. Immun.* 35:958–964.

108. Oliver, J. D., R. A. Warner, and D. R. Cleland. 1983. Distribution of *Vibrio vulnificus* and other lactose-fermenting vibrios in the marine environment. *Appl. Environ. Microbiol.* 45:985–998.

109. Olsovsky, Z., V. Olsakova, S. Chobot, and V. Sviridov. 1975. Mass occurrence of *Yersinia enterocolitica* in two establishments of collective care of children. *J. Hyg. Epid. Microbiol. Immunol.* 19:22–29.

110. Oosterom, J., S. Notermans, H. Karman, and G. B. Engels. 1983. Origin and prevalence of *Campylobacter jejuni* in poultry processing. *J. Food Protect.* 46: 339–344.

111. Pace, J., and T.-J. Chai. 1989. Comparison of *Vibrio parahaemolyticus* grown in estuarine water and rich medium. *Appl. Environ. Microbiol.* 55:1877–1887.

112. Paerregaard, A., F. Espersen, O. M. Jensen, and M. Skurnik. 1991. Interactions between *Yersinia enterocolitica* and rabbit ileal mucus: Growth, adhesion, penetration, and subsequent changes in surface hydrophobicity and ability to adhere to ileal brush border membrane vesicles. *Infect. Immun.* 59:253–260.

113. Pai, C. H., and L. DeStephano. 1982. Serum resistance associated with virulence in *Yersinia enterocolitica*. *Infect. Immun.* 35:605–611.

114. Pai, C. H., and V. Mors. 1978. Production of enterotoxin by *Yersinia enterocolitica*. *Infect. Immun.* 19:908–911.

115. Pai, C. H., V. Mors, and S. Toma. 1978. Prevalence of enterotoxigenicity in human and nonhuman isolates of *Yersinia enterocolitica*. *Infect. Immun.* 22: 334–338.

116. Perez-Rosas, N., and T. C. Hazen. 1989. In situ survival of *Vibrio cholerae* and *Escherichia coli* in a tropical rain forest watershed. *Appl. Environ. Microbiol.* 55:495–499.

117. Pierson, D. E. and S. Falkow. 1990. Nonpathogenic isolates of *Yersinia enterocolitica* do not contain functional *inv*-homologous sequences. *Infect. Immun.* 58:1059–1064.

118. Portnoy, D. A., and S. Falkow. 1981. Virulence-associated plasmids from *Yersinia enterocolitica* and *Yersinia pestis*. *J. Bacteriol.* 148:877–883.

119. Potter, M. E., M. J. Blaser, R. K. Sikes, A. F. Kaufmann, and J. G. Wells. 1983. Human *Campylobacter* infection associated with certified raw milk. *Amer. J. Epidemiol.* 117:475–483.

120. Prpic, J. K., R. M. Robins-Browne, and R. B. Davey. 1985. In vitro assessment of virulence in *Yersinia enterocolitica* and related species. *J. Clin. Microbiol.* 22:105–110.

121. Rayes, H. M., C. A. Genigeorgis, and T. B. Farver. 1983. Prevalence of *Campylobacter jejuni* on turkey wings at the supermarket level. *J. Food Protect.* 46:292–294.

122. Reynolds, G. N., and F. A. Draughon. 1987. *Campylobacter jejuni* in vacuum packaged processed turkey. *J. Food Protect.* 50:300–304.

123. Robins-Browne, R. M., and J. K. Prpic. 1985. Effects of iron and desferrioxamine on infections with *Yersinia enterocolitica*. *Infect. Immun.* 47:774–779.

124. Robins-Browne, R. M., C. S. Still, M. Isaacson, H. J. Koornhof, P. C. Appelbaum, and J. N. Scragg. 1977. Pathogenic mechanisms of a nonagglutinable

Vibrio cholerae strain: Demonstration of invasive and enterotoxigenic properties. *Infect. Immun.* 18:542–545.

125. Robins-Browne, R. M., C. S. Still, M. D. Miliotis, and H. J. Koornhof. 1979. Mechanism of action of *Yersinia enterocolitica* enterotoxin. *Infect. Immun.* 25: 680–684.

126. Robinson, D. A., and D. M. Jones. 1981. Milk-borne campylobacter infection. *Brit. Med. J.* 282:1374–1376.

127. Ruiz-Palacios, G. M., J. Torres, E. Escamilla, B. R. Ruiz-Palacios, and J. Tamayo. 1983. Cholera-like enterotoxin produced by *Campylobacter jejuni*. *Lancet* 2:250–253.

128. Sakazaki, R. 1979. *Vibrio* infections. In *Food-borne Infections and Intoxications*, ed. H. Riemann and F. L. Bryan, 173–209. New York: Academic Press.

129. Sakazaki, R. 1983. *Vibrio parahaemolyticus* as a food-spoilage organism. In *Food Microbiology*, ed. A. H. Rose, 225–241. New York: Academic Press.

130. Schiemann, D. A. 1981. An enterotoxin-negative strain of *Yersinia enterocolitica* serotype 0:3 is capable of producing diarrhea in mice. *Infect. Immun.* 32:571–574.

131. Schiemann, D. A., and J. A. Devenish. 1982. Relationship of HeLa cell infectivity to biochemical, serological, and virulence characteristics of *Yersinia enterocolitica*. *Infect. Immun.* 35:497–506.

132. Shanker, S., J. A. Rosenfield, G. R. Davey, and T. C. Sorrell. 1982. *Campylobacter jejuni*: Incidence in processed broilers and biotype distribution in human and broiler isolates. *App. Environ. Microbiol.* 43:1219–1220.

133. Shayegani, M., I. Deforge, D. M. McGlynn, and T. Root. 1981. Characteristics of *Yersinia enterocolitica* and related species isolated from human, animal and environmental sources. *J. Clin. Microbiol.* 14:304–312.

134. Shayegani, M., D. Morse, I. DeForge, T. Root, L. M. Parsons, and P. Maupin. 1982. Foodborne outbreak of *Yersinia enterocolitica* in Sullivan County, New York with pathogenicity studies in the isolates. *Bacteriol. Proc.*, C-175.

135. Shirai, H., H. Ito, T. Hirayama, Y. Nakamoto, N. Nakabayashi, K. Kumagai, Y. Takeda, and M. Nishibuchi. 1990. Molecular epidemiologic evidence for association of thermostable direct hemolysin (TDH) and TDH-related hemolysin of *Vibrio parahaemolyticus* with gastroenteritis. *Infect. Immun.* 58:3568–3573.

136. Smitherman, R. E., C. A. Genigeorgis, and T. B. Farver. 1984. Preliminary observations on the occurrence of *Campylobacter jejuni* at four California chicken ranches. *J. Food Protect.* 47:293–298.

137. Stelma, G. N., Jr., P. L. Spaulding, A. L. Reyes, and C. H. Johnson. 1988. Production of enterotoxin by *Vibrio vulnificus* isolates. *J. Food Protect.* 51: 192–196.

138. Stern, N. J., and J. M. Damare. 1982. Comparison of selected *Yersinia enterocolitica* indicator tests for potential virulence. *J. Food Sci.* 47:582–588.

139. Stern, N. J., S. S. Green, N. Thaker, D. J. Krout, and J. Chiu. 1984. Recovery of *Campylobacter jejuni* from fresh and frozen meat and poultry collected at slaughter. *J. Food Protect.* 47:372–374.

140. Stern, N. J., M. P. Hernandez, L. Blankenship, K. E. Deibel, S. Doores, M. P. Doyle, H. Ng, M. D. Pierson, J. N. Sofos, W. H. Sveum, and D. C. Westhoff. 1985. Prevalence and distribution of *Campylobacter jejuni* and *Campylobacter coli* in retail meats. *J. Food Protect.* 48:595–599.

141. Stern, N. J., and A. W. Kotula. 1982. Survival of *Campylobacter jejuni* inoculated into ground beef. *Appl Environ. Microbiol.* 44:1150–1153.

142. Stern, N. J., and M. D. Pierson. 1979. *Yersinia enterocolitica*: A review of the psychrotrophic water and foodborne pathogen. *J. Food Sci.* 44:1736–1742.

143. Stern, N. J., M. D. Pierson, and A. W. Kotula. 1980. Effects of pH and sodium chloride on *Yersinia enterocolitica* growth at room and refrigeration temperatures. *J. Food Sci.* 45:64–67.

144. Swaminathan, B., M. C. Harmon, and I. J. Mehlman. 1982. *Yersinia enterocolitica*. *J. Appl. Bacteriol.* 52:151–183.

145. Tacket, C. O., J. Ballard, N. Harris, J. Allard, C. Nolan, T. Quan, and M. L. Cohen. 1985. An outbreak of *Yersinia enterocolitica* infections caused by contaminated tofu (soybean curd). *Amer. J. Epidemiol.* 121:705–711.

146. Takeda, T., G. B. Nair, K. Suzuki, and Y. Shimonishi. 1990. Production of a monoclonal antibody to *Vibrio cholerae* non-O1 heat-stable enterotoxin (ST) which is cross-reactive with *Yersinia enterocolitica* ST. *Infect. Immun.* 58:2755–2759.

147. Takeda, Y. 1983. Thermostable direct hemolysin of *Vibrio parahaemolyticus*. *Pharmac. Ther.* 19:123–146.

148. Taniguchi, H., H. Ohta, M. Ogawa, and Y. Mizuguchi. 1985. Cloning and expression in *Escherichia coli* of *Vibrio parahaemolyticus* thermostable direct hemolysin and thermolabile hemolysin genes. *J. Bacteriol.* 162:510–515.

149. Terai, A., H. Shirai, O. Yoshida, Y. Takeda, and M. Nishibuchi. 1990. Nucleotide sequence of the thermostable direct hemolysin gene (*tdh* gene) of *Vibrio mimicus* and its evolutionary relationship with the *tdh* genes of *Vibrio parahaemolyticus*. *FEMS Microbiol. Lett.* 71:319–324.

150. Tibana, A., M. B. Warnken, M. P. Nunes, I. D. Ricciardi, and A. L. S. Noleto. 1987. Occurrence of *Yersinia* species in raw and pasteurized milk in Rio de Janeiro, Brazil. *J. Food Protect.* 50:580–583.

151. Toma, S., and L. Lafleur. 1974. Survey on the incidence of *Yersinia enterocolitica* infection in Canada. *Appl. Microbiol.* 28:469–473.

152. Tsunasawa, S., A. Sugihara, T. Masaki, F. Sakiyama, Y. Takeda, T. Miwatani, and K. Narita. 1989. Amino acid sequence of thermostable direct hemolysin produced by *Vibrio parahaemolyticus*. *J. Biochem.* 101:111–121.

153. Twedt, R. M., J. T. Peeler, and P. L. Spaulding. 1980. Effective ileal loop dose of Kanagawa-positive *Vibrio parahaemolyticus*. *Appl. Environ. Microbiol.* 40:1012–1016.

154. Twedt, R. M., P. L. Spaulding, and H. E. Hall. 1969. Morphological, cultural, biochemical, and serological comparison of Japanese strains of *Vibrio parahaemolyticus* with related cultures isolated in the United States. *J. Bacteriol.* 98:511–518.

155. Vanderzant, C., and R. Nickelson. 1972. Survival of *Vibrio parahaemolyticus* in shrimp tissue under various environmental conditions. *Appl. Microbiol.* 23:34–37.

156. Vidon, D. J. M., and C. L. Delmas. 1981. Incidence of *Yersinia enterocolitica* in raw milk in Eastern France. *Appl. Environ. Microbiol.* 41:353–359.

157. Walker, S. J., and A. Gilmour. 1990. Production of enterotoxin by *Yersinia* species isolated from milk. *J. Food Protect.* 53:751–754.

158. Wallace, J. M. 1980. Milk-associated *Campylobacter* infection. *Hlth. Bull.* 38:57–61.

159. Warnken, M. B., M. P. Nunes, and A. L. S. Noleto. 1987. Incidence of *Yersinia* species in meat samples purchased in Rio de Janeiro, Brazil. *J. Food Protect.* 50:578–579, 583.

160. Wempe, J. M., C. A. Genigeorgis, T. B. Farver, and H. I. Yusufu. 1983. Prevalence of *Campylobacter jejuni* to two California chicken processing plants. *Appl. Environ. Microbiol.* 45:355–359.

161. Wright, A. C., and J. G. Morris, Jr. 1991. The extracellular cytolysin of *Vibrio vulnificus*: Inactivation and relationship to virulence in mice. *Infect. Immun.* 59:192–197.

162. Yamamoto, K., Y. Ichinose, H. Shinagawa, K. Makino, A. Nakata, M. Iwanaga, T. Honda, and T. Miwatani. 1990. Two-step processing for activation of the cytolysin/hemolysin of *Vibrio cholerae* O1 biotype E1 Tor: Nucleotide sequence of the structural gene (*hlyA*) and characterization of the processed products. *Infect. Immun.* 58:4106–4116.

163. Yamamoto, K., A. C. Wright, J. B. Kaper, and J. G. Morris, Jr. 1990. The cytolysin gene of *Vibrio vulnificus*: Sequence and relationship to the *Vibrio cholerae* El Tor hemolysin gene. *Infect. Immun.* 58:2706–2709.
164. Yamamoto, T., and T. Yokota. 1989. Adherence targets of *Vibrio parahaemolyticus* in human small intestines. *Infect. Immun.* 57:2410–2419.
165. Yamamoto, K., M. Al-Omani, T. Honda, Y. Takeda, and T. Miwatani. 1984. Non-01 *Vibrio cholerae* hemolysin: Purification. partial characterization, and immunological relatedness to El Tor hemolysin. *Infect. Immun.* 45:192–196.
166. Zen-Yoji, H., Y. Kudoh, H. Igarashi, K. Ohta, and K. Fukai. 1975. Further studies on characterization and biological activities of an enteropathogenic toxin of *Vibrio parahaemolyticus*. *Toxicon* 13:134–135.
167. Zen-Yoji, H., T. Maruyama, S. Sakai, S. Kimura, T. Mizuno, and T. Momose. 1973. An outbreak of enteritis due to *Yersinia enterocolitica* occurring at a junior high school. *Japan. J. Microbiol.* 17:220–222.
168. Zink, D. L., J. C. Feeley, J. G. Wells, C. Vanderzant, J. C. Vickery, W. C. Roof, and G. A. O'Donovan. 1980. Plasmid-mediated tissue invasiveness in *Yersinia enterocolitica*. *Nature* 283:224–226.
169. Zink, D. L., R. V. Lachica, and J. R. Dubel. 1982. *Yersinia enterocolitica* and *Yersinia enterocolitica*-like species: Their pathogenicity and significance in foods. *J. Food Safety* 4:223–241.

24

Foodborne Animal Parasites

The animal parasites that can be contracted by eating certain foods belong to three distinct groups: Protozoa, flatworms, and roundworms. Several of the more important members of each group of concern in human foods are examined in this chapter along with their classification.

In contrast to foodborne bacteria, animal parasites do not proliferate in foods, and their presence must be detected by direct means since they cannot grow on culture media. Because all are larger in size than bacteria, their presence can be detected rather easily by use of appropriate concentration and staining procedures. Since many are intracellular pathogens, resistance to these diseases is often by cellular phenomena similar to that for listeriosis (see Chap. 21). Finally, another significant way in which some animal parasites differ from bacteria is their requirement for more than one animal host in which to carry out their life cycles. The definitive host is the animal in which the adult parasite carries out its sexual cycle; the intermediate host is the animal where larval or juvenile forms develop. In some instances, there is only one definitive host (e.g., cryptosporiodiosis), in others more than one animal can serve as definitive host (e.g., diphyllobothriasis), and in still other cases, both larval and adult stages reside in the same host (e.g., trichinosis).

PROTOZOA

The protozoa belong to the Kingdom Protista (Protoctista), which also comprises the algae and flagellate fungi. They are the smallest and most primitive of animal forms, and the five genera of concern in foods are classified as follows:

Kingdom Protista
  Phylum Sarcomastigophora

Class Zoomastigophorea
Order Diplomonadida
Family Hexamitidae
Genus *Giardia*

Subphylum Sarcodina
Superclass Rhizopoda
Class Lobosea
Order Amoebida
Family Endamoebidae
Genus *Entamoeba*

Phylum Apicomplexa (=Sporozoa)
Class Sporozoea
Order Eucoccidiida
Family Sarcocystidae
Genus *Toxoplasma*
Genus *Sarcocystis*
Family Cryptosporidiidae
Genus *Cryptosporidium*

Giardiasis

Giardia lamblia is a flagellate protozoan that exists in environmental waters at a higher level than *Entamoeba histolytica*. The protozoal cells (trophozoites) produce cysts, which are the primary forms in water and foods. The cysts are pear shaped, with a size range of 8 to 20 μm in length and 5 to 12 μm in width. The trophozoites have eight flagella that arise on the ventral surface near the paired nuclei and give rise to "falling-leaf" motility.

Upon ingestion, *Giardia* cysts excyst in the gastrointestinal tract with the aid of stomach acidity and proteases and give rise to clinical giardiasis in some individuals. Excystation of the trophozoites occurs somewhere in the upper small intestine, and this step is regarded as being equivalent to a virulence factor (8). The trophozoites are not actively phagocytic, and they obtain their nutrients by absorption. Occasionally bile ducts are invaded, leading to cholescystitis. Compared to some of the other intestinal protozoal parasites, *Giardia* trophozoites do not penetrate deeply in parenteral tissues.

Environmental Distribution.   Water is the second most common source of giardiasis. The first recorded outbreak occurred at a ski resort in Aspen, Colo., in 1965 with 123 cases (18). Between 1965 and 1977, 23 waterborne outbreaks were recorded that affected over 7,000 persons (19). Between 1971 and 1985, 92 outbreaks were reported in the United States (18). *Giardia* cysts are generally resistant to the levels of chlorine used in the water supply. Beavers and muskrats have been shown to be the major sources of this organism in bodies of water. In a study of 220 muskrat fecal specimens collected from natural waters in southwestern New Jersey, 70% contained *Giardia* cysts (49). It is estimated that up to 15% of the U.S. population is infected with this organism.

The Syndrome, Diagnosis, and Treatment.   The incubation period for clinical giardiasis is 7 to 13 days, and cysts appear in stools after 3 to 4 weeks (67). Asymptomatic cyst passage is the most benign manifestation of *G. lamblia* infection in humans, but when clinical giardiasis occurs, symptoms may last from several months to a year or more. Up to 9.0×10^8 cysts are shed each day by patients, and they may survive as long as 3 months in sewage sludge (3). *G. lamblia* is generally noninvasive, and malabsorption often accompanies the symptomatic disease (79). Growth of the organism is favored by the high bile content in the duodenum and upper jejunum (67).

From an outbreak of giardiasis among 1,400 Americans on the Madeira Island in 1976, the symptoms, along with the percentage incidence among victims, were as follows: abdominal cramps (75%), abdominal distention (72%), nausea (70%), and weight loss (40%). The median incubation period was 4 days, and *G. lamblia* was recovered from 47% of 58 ill patients. The consumption of tapwater and the eating of ice cream or raw vegetables were significantly associated with the illness (57). The 29 victims of the 1979–1980 outbreak traced to home-canned salmon (see below) displayed the following symptoms: diarrhea (100%), fatigue (97%), abdominal cramps (83%), fever (21%), vomiting (17%), and weight loss (59%), among others (65). From another study of 183 patients, the 5 leading symptoms (and percentage complaining) were: diarrhea (92), cramps (70), nausea (58), fever (28), and vomiting (23) (79). Weight loss of about 5 lb or so is a common feature of giardiasis, and it was associated with the 1985 outbreak traced to noodle salad (66).

Giardiasis is a highly contagious disease. It has been documented in day-care centers where unsanitary conditions prevailed. The human infection rate ranges from 2.4 to 67.5% (15). The minimum infectious dose of *G. lamblia* cysts for humans is 10 or less (69).

Giardiasis is diagnosed by the demonstration of trophozoites in stool specimen by microscopic examinations using either wet mounts or stained specimens. *G. lamblia* can be grown in axenic culture, but this does not lend itself to rapid diagnosis. Effective ELISA tests have been developed. Both circulating antibodies and T lymphocytes are elicited during infection by *G. lamblia*. Since no enterotoxin has been demonstrated, diarrhea is caused by other factors (79).

The drug of choice for the treatment of giardiasis is quinacrine, an acridine derivative. Also effective are metronidazole and tinidazole (79).

Incidence in Foods and Foodborne Cases.   *Giardia* has been shown to occur in some vegetables, and it may be presumed that the organism occurs on foods that are washed with contaminated water or contaminated by unsanitary asymptomatic carriers. Of 64 heads of lettuce examined in Rome, Italy, in 1968, 48 contained *Giardia* cysts, and cysts were recovered from strawberries grown in Poland in 1981 (3).

As early as 1928, it was suggested that hospital food handlers were the likely source of protozoal infections of patients. Of 844 private patients in an urban center, 36% contracted giardiasis, and it was believed the infections were acquired by eating cyst-contaminated raw fruits and vegetables. These and some other early incidences of possible foodborne giardiasis have been discussed by Barnard and Jackson (3). Following is a list of suspected and proved foodborne giardiasis:

- Three of four members of a family who in 1960 ate Christmas pudding thought to have been contaminated by rodent feces became victims (17). *Giardia*-like cysts were found.
- In their surveillance of foodborne diseases in the United States for 1968–1969, Gangaroso and Donadio (34) recorded an outbreak of giardiasis with 19 cases for 1969 but provided no further details.
- In 1976, about 1,400 Americans on the island of Madeira contracted giardiasis. Tapwater, ice cream, and raw vegetables were the probable sources (57).
- In December 1979, 29 of 60 school employees in a rural Minnesota community contracted the disease from home-canned salmon prepared by a worker after changing the diaper of an infant later shown to have an asymptomatic *Giardia* infection (65). This was the first well-documented common-source outbreak.
- In July 1985, 13 of 16 individuals at a picnic in Connecticut met the case definition of giardiasis, and the most likely vehicle food was a noodle salad (66). Although most victims developed symptoms between 6 and 20 days after the picnic, the salad preparer became ill the day after the food was eaten by others. This was the second well-documented common-source outbreak traced to a food product.
- In 1988, 21 of 108 members of a church youth group in Albuquerque, New Mexico, were victims. Taco ingredients were the most likely vehicles from dinners prepared by parents at a church (13).

The U.S. Centers for Disease Control (CDC) recorded foodborne giardiases outbreaks in 1985 and 1986, with 1 outbreak and 13 cases in 1985 and 2 outbreaks and 28 cases in 1986 (5). The common occurrence of this organism suggests that it may be a more frequent cause of foodborne infection than is reported. The incubation period of 7 days plus could be a factor in the apparent underreporting. Another possible factor is the need to demonstrate the organism in stools and leftover foods by microscopic examination, a practice that is not routine in the microbiological examination of foods in food-borne gastroenteritis outbreaks.

The following references should be consulted for more information on giardiasis (60, 79, 88).

Amebiasis

Amebiasis (amoebic dysentery), caused by *Entamoeba histolytica*, is often transmitted by the fecal-oral route, although transmission is known to occur by water, food handlers, and foods. According to Jackson (39), there is better documentation of food transmission of amebic dysentery than for the other intestinal protozoal diseases. The organism is unusual in being anaerobic, and the trophozoites (ameba stages) lack mitochondria. It is an aerotolerant anaerobe that requires glucose or galactose as its main respiratory substrate (59). The trophozoites of *E. histolytica* range in size from 10 to 60 µm, while the cysts usually range between 10 and 20 µm. The trophozoites are motile;

the cysts are not. It is often found with *Entamoeba coli*, with which it is associated in the intestine and stools. In warm stools from a case of active dysentery, *E. histolytica* is actively motile and usually contains red blood cells that the protozoan ingests by pseudopodia. Although generally outnumbered in stools by *Entamoeba coli*, the latter never ingests red blood cells. While the trophozoites do not persist under environmental conditions, the encysted forms can survive as long as 3 months in sewage sludge (3). A person with this disease may pass up to 4.5×10^7 cysts each day (3).

The possible transmission of cysts to foods becomes a real possibility when poor personal restroom hygiene is practiced. The incidence of amebiasis varies widely, with a rate of 1.4% reported for Tacoma, Washington, to 36.4% in rural Tennessee (15). It is estimated that 10% of the world's population is infected with *E. histolytica* and that up to 100 million cases of amebic colitis or liver abscesses occur each year (89).

In its trophozoite stage, the organism induces infection in the form of abscesses in intestinal mucosal cells and ulcers in the colon. Its adherence to host cell glycoproteins is mediated by a galactose-specific lectin. It reproduces by binary fission in the large intestine. It encysts in the ileum, and cysts may occur free in the lumen. The organism produces an enterotoxic protein with a molecular weight of 35,000 to 45,000 daltons (15).

Syndrome, Diagnosis, and Treatment. The incubation period for amebiasis is 2 to 4 weeks, and symptoms may persist for several months. Its onset is often insidious, with loose stools and generally no fever. Mucus and blood are characteristic of stools from patients. Later symptoms consist of pronounced abdominal pain, fever, severe diarrhea, vomiting, and lumbago and somewhat resemble those of shigellosis. Weight loss is common, and all patients have heme-positive stools. According to Jackson (39), fulminating amebiasis with ulceration of the colon and toxicity occur in 6 to 11% of cases, especially in women stressed by pregnancy and nursing. Masses of amebae and mucus may form in the colon, leading to intestinal obstruction. Amebiasis may last in some individuals for many years in contrast to giardiasis, where disease symptoms rarely exceed 3 months (3). Under some conditions, amebiasis may result from a synergistic relationship with certain intestinal bacteria.

Amebiasis is diagnosed by demonstrating trophozoites and cysts in stools or mucosal scrapings. Immunological methods such as indirect hemagglutination, indirect immunofluorescence, latex agglutination, and ELISA are useful. The sensitivity of these tests is high with extraintestinal amebiasis, and a titer of 1:64 by indirect hemagglutination is considered significant.

This syndrome can be treated with the amebicidal drugs metronidazole and chloroquine. Resistance is mediated by cell immunity. Lymphocytes from patients in the presence of *E. histolytica* antigens have been shown to produce gamma interferon, which activates macrophages that display amebicidal properties (76).

Toxoplasmosis

This disease is caused by *Toxoplasma gondii*, a coccidian protozoan that is an obligate intracellular parasite. The generic name is based on the characteristic shape of the ameba stage of the protozoan (Gr. *toxo*, "arc"). It was first isolated in 1908 from an African rodent, the gondi—hence, its species name. In most instances, the ingestion of *T. gondii* oocysts causes no symptoms in humans, or the infection is self-limiting. In these cases, the organism encysts and becomes latent. However, when the immunocompetent state is abated, life-threatening toxoplasmosis results from the breaking out (recrudescence) of the latent infection.

Domestic and wild cats are the only definitive hosts for the intestinal or sexual phase of this organism, making them the primary sources of human toxoplasmosis. Normally, the disease is transmitted from cat to cat, but virtually all vertebrate animals are susceptible to the oocysts shed by cats. As few as 100 oocysts can produce clinical toxoplasmosis in humans, and the oocysts can survive over a year in warm, moist environments (27).

Symptoms, Diagnosis, and Treatment. In most individuals, toxoplasmosis is symptomless, but when symptoms occur, they consist of fever with rash, headache, muscle aches and pain, and swelling of the lymph nodes. The muscle pain, which is rather severe, may last up to a month or more. At times, some of the symptoms mimic infectious mononucleosis.

The disease is initiated upon the ingestion of oocysts (if from cat feces), which pass to the intestine, where digestive enzymes effect the release of the eight motile sporozoites. Oocysts are ovoid shaped, measure 10 to 12 μm in diameter, and possess a thick wall. Sporozoites are crescent shaped and measure about 3×7 μm; they cannot survive for long outside animal host tissues, nor can they survive the activities of the stomach. When freed in the intestines, these forms pass through intestinal walls and multiply rapidly in many other parts of the body, giving rise to clinical symptoms. The most rapidly multiplying forms are designated tachyzoites (Gr. *tachy*, "rapid"), and in immunocompetent individuals they eventually give rise to clusters that are surrounded by a protective wall. This is a tissue cyst, and the protozoa inside are designated bradyzoites (Gr. *bradus*, "slow"). These cysts are 10 to 200 μm in diameter, and the bradyzoites are smaller in size than the more active tachyzoites. Bradyzoites may persist in the body for the lifetime of an individual, but if the cysts are mechanically broken or break down under immunosuppression, bradyzoites are freed and begin to multiply rapidly as tachyzoites and thus bring on another active infection. The development of a cyst wall around bradyzoites coincides with the development of permanent host immunity. The cysts are normally intracellular in host cells. *T. gondii* infections are asymptomatic in the vast majority of human cases (immunocompetents), but in congenital infections and in immunocompromised hosts, such as AIDS patients, the disease is much more severe.

Unlike certain other intestinal protozoal diseases, toxoplasmosis cannot be diagnosed by demonstrating oocysts in stools since these forms occur only in cat feces. Various serologic methods are widely used to diagnose acute infection. A fourfold rise in immune globulin-G (IgG) antibody titer between acute and convalescent serum specimens is indicative of acute infection. A more rapid confirmation of acute infection can be made by the detection of immune globulin M (IgM) antibodies, which appear during the first week of infection and peak during the second to fourth weeks (68). Among other diagnostic methods are the methylene blue dye test, indirect hemagglutination, indirect immunofluorescence, and immunoelectrophoresis. With the indirect hemagglutination test, antibody titers above 1:256 are generally indicative of active infection.

Although toxoplasma infection induces protective immunity, it is in part cell mediated. In many bacterial infections where phagocytes ingest the cells, their internal granules release enzymes that destroy the bacteria. During this process, aerobic respiration gives way to anaerobic glycolysis, which results in the formation of lactic acid and the consequent lowering of pH. The latter contributes to the destruction of the ingested bacteria along with the production of superoxide, which at the acid pH, yields singlet oxygen ($^1O_2$). The latter is quite toxic. *T. gondii* tachyzoites are unusual in that once they are phagocytized, the production of H_2O_2 is not triggered, and neither do the acid pH nor the singlet oxygen events occur. Also, they reside in vacuoles of phagocytes that do not fuse with preexisting secondary lysosomes. Thus, it appears that their mode of pathogenicity involves an alteration of phagocyte membranes in such way that they fail to fuse with other endocytic or biosynthetic organelles, in addition to the other events noted (45). T cells play a role in immunity to *T. gondii*, and this has been demonstrated by use of nude rats where T cells from *T. gondii*-infected normal rats conferred to nude rats the ability to resist infection by a highly virulent strain of *T. gondii* (25).

Antimicrobial therapy for toxoplasmosis consists of sulfonamides, pyrimethamine, pyrimethamine plus clindamycin, or fluconazole. Pyrimethamine is a folic acid antagonist that inhibits dihydrofolate reductase.

Distribution of T. gondii.   Toxoplasmosis is regarded as a universal infection, with the incidence being higher in the tropics and lower in colder climes. It is estimated that 50% of Americans have circulating antibodies to *T. gondii* by the time of adulthood (68). In a study of U.S. Army recruits, 13% were positive for toxoplasma antibodies (30). In the United States, it is estimated that over 3,000 babies are infected each year with *T. gondii* because their mothers acquire the infection during pregnancy (27). Fetal infections occur in 17% of first-trimester and 65% of third-trimester cases, with the first-trimester cases being more severe (68). Among 3,000 pregnant women tested for *T. gondii* antibodies, 32.8% were positive (48).

Extensive surveys of *T. gondii* antibodies in meat animals have been reviewed by Fayer and Dubey (27) who reported that of more than 16,000 cattle surveyed, an average of 25% contained antibodies, and infectious cysts administered from cats persisted as long as 267 days, with most being found in the liver. In more than 9,000 sheep, an average of 31% had antibodies, and oocysts administered persisted 173 days, with most protozoa found in the heart. Similarly for pigs, 29% had antibodies, and oocysts persisted for 171 days, with most in the brain and heart, while for goats, oocysts persisted in the animals for 441 days, with most being found in skeletal mucles. Since the meat animals noted are herbivores, Fayer and Dubey (27) concluded that the contamination of feed and water with oocysts from cat feces must be the ultimate source of infection, aided by the practice on many farms of keeping cats to kill mice.

Food-Associated Cases.   The number of cases of toxoplasmosis that are contracted from foods is unknown, but the estimated number in the United States

TABLE 24-1.   **Estimated Number of Clinically Significant Cases of Protozoal Infections in the United States, 1985**

Infections	Cases
Amebiasis	12,000
Cryptosporiodiosis	50
Giardiasis	120,000
Toxoplasmosis[a]	2,300,000

Source: Bennett et al. (7).

[a] Excluding congenital.

from all sources for 1985 has been put at 2.3 million (Table 24-1). This estimated number far exceeds the recorded cases for the total of all other protozoal diseases.

Fresh meats may contain toxoplasma oocysts. As early as 1954, under-cooked meat was suspected to be the source of human toxoplasmosis (42). In a study in 1960 of freshly slaughtered meats, 24% of 50 porcine, 9.3% of 86 ovine, but only 1 of 60 bovine samples contained oocysts (44). *T. gondii* is more readily isolated from sheep than other meat animals (42). The following cases have been proved or suspected:

- In France in the early 1960s, 31% of 641 children in a tuberculosis hospital became seropositive for *T. gondii* after admission. When two additional meals per day of undercooked mutton were served, toxoplasmosis cases doubled. The authors concluded that the custom of this hospital to feed undercooked meat was the cause of the high number of infections (23).

  At an educational institution, 771 mothers were questioned about their preferences for meat. Of those who preferred well-done meat, 78% had toxoplasma antibody; of those who liked less well-done meats, 85% were antibody positive; and of those who ate meat rare or raw, 93% had toxoplasma antibodies (23). The authors were unable to make distinctions among beef, mutton, or horse meat. They further noted that 50% of children in France are infected with *T. gondii* before age 7 and believe this is due to the consumption of undercooked meats.
- Eleven of 35 medical students in New York City in 1968 had an increase in toxoplasma antibodies following the consumption of hamburger cooked rare at the same snack bar, and 5 contracted clinical toxoplasmosis (48).
- In 1974, a 7-month-old infant who consumed unpasteurized goat's milk developed clinical toxoplasmosis. Although *T. gondii* could not be recovered from milk, some goats in the herd had antibody titers to *T. gondii* as high as 1:512, and the child had a titer of over 1:16,000 (70).
- In 1978, 10 of 24 members of an extended family in northern California contracted toxoplasmosis after drinking raw milk from infected goats (75).
- In São Paulo, Brazil, 110 university students suffered acute toxoplasmosis after eating uncooked meat (15).

Control.   Toxoplasmosis in humans can be prevented by avoiding environ-mental contamination with cat feces and by avoiding the consumption of meat and meat products that contain viable tissue cysts. The cysts of *T. gondii* can be destroyed by heating meats above 60°C or by irradiating at a level of 30 krad (0.3 KGy) or higher (27). The organism may be destroyed by freezing, but since the results are variable, freezing should not be relied upon to inactivate oocysts.

Sarcocystosis

Of the more than 13 known species of the genus *Sarcocystis*, two are known to cause an extraintestinal disease in humans. One of these is obtained from cattle (*S. hominis*) and the other from pigs (*S. suihominis*). Humans are the definitive hosts for both species; the intermediate host for *S. hominis* is bovines and pigs for *S. suihominis*.

When humans ingest a sarcocyst, bradyzoites are released and penetrate the lamina propria of the small intestine, where sexual reproduction occurs that leads to sporocysts. The latter pass out of the bowel in feces. When sporocysts are ingested by pigs or bovines, the sporozoites are released and spread throughout the body. They multiply asexually and lead to the formation of sarcocysts in skeletal and cardiac muscles. In this stage, they are sometimes referred to as Miescher's tubules. The bradyzoite-containing sarcocysts are visible to the unaided eye and may reach 1 cm in diameter (15).

Several studies have been conducted to determine the relative infectivity of *Sarcosystis* spp. Of 20 human volunteers in five studies who ate raw beef infected with *S. hominis*, 12 became infected and shed oocysts, but only 1 had clinical illness (28). Symptoms occurred within 3 to 6 h and consisted of nausea, stomachache, and diarrhea. In 15 other volunteers who ate raw pork infected with *S. suihominis*, 14 became infected and shed oocysts, and 12 of these had clinical illness 6 to 48 h after eating the pork (28). Six who ate well-cooked pork did not contract the disease. In another study of another species of *Sarcocystis*, dogs did not become infected when fed beef cooked medium (60°C) or well done (71.1°–74.4°C), but the beef was infective when fed raw or cooked rare (37.8°–53.3°C). Dogs fed the same raw beef after storage for 1 week in a home freezer did not become infected (29). In another study, two human volunteers passed sporocysts for 40 days after eating 500 g of raw ground beef diaphragm muscle infected with *Sarcosporidia* (72).

Since bovine and porcine animals serve as intermediate hosts for these parasites, their potential as foodborne pathogens to humans is obvious.

Cryptosporiodiosis

The protozoan *Cryptosporidium parvum* was first described in 1907 in asymptomatic mice, and for decades now it has been known to be a pathogen of at least 40 mammals and varying numbers of reptiles and birds. Although the first documented human case was not recorded until 1976, this disease has a worldwide prevalance of 1 to 4% among patients with diarrhea (85), and it appears to be increasing. In England and Wales for the five years 1985–1989, the number of identified cases were 1,874, 3,694, 3,359, 2,838, and 7,769, respectively (2). This disease was the fourth most frequent cause of diarrhea during the period noted. It is estimated to cause infections in from 7 to 38% of AIDS patients in some hospitals (85). The prevalence of *C. parvum* in diarrheal stools is similar to *Giardia lamblia* (85). In humans, the disease is self-limiting in immunocompetent individuals, but it is a serious infection in the immunocompromised such as AIDS patients. The protozoan is known to be present in at least some bodies of water (see below) and thus exists the potential for food transmission. The fecal-oral route of transmission is the most important, but indirect transmission by food and milk is suspected. Thorough reviews of cryptosporiodiosis have been provided (2, 20, 21, 24, 26, 85).

C. parvum is an intracellular-extracytoplasmic coccidian protozoan that carries out its life cycle in one host. Following the ingestion of thick-walled oocysts, they excyst in the small intestine and free sporozoites that penetrate the microvillous region of host enterocytes, where sexual reproduction leads to the development of zygotes. About 80% of the zygotes form thick-walled oocysts that sporulate within host cells (20). The environmentally resistant oocysts are shed in feces, and the infection is transmitted to other hosts, when they are ingested.

The oocysts of *C. parvum* are spherical to ovoid and average 4.5 to 5.0 μm in size. Each sporulated oocyst contains four sporozoites. The oocysts are highly resistant in the natural environment and may remain viable for several months when kept cold and moist (20). They have been reported to be destroyed by treatments with 50% or more ammonia and 10% or more formalin for 30 min (20). The latter author has reported that temperatures above 60°C and below −20°C may kill *C. parvum* oocysts. Holding oocysts at 45°C for 5 to 20 min has been reported to destroy their infectivity (1). In one study, infectivity was lost after 2 months when cysts were stored in distilled water or at 15° to 20°C within 2 weeks or at 37°C in 5 days (78). In the latter study, cysts did not survive freezing even when stored in a variety of cryoprotectants. Commonly used disinfectants are ineffective against the oocysts (10), and this has been demonstrated for ozone and chlorine compounds. For a 90% or more inactivation of *C. parvum* oocysts, 1 ppm ozone required 5 min, 1.3 ppm chlorine dioxide required 60 min, and 80 ppm each of chlorine and monochloramine required about 90 min (52). The oocysts were 14 times more resistant to ClO_2 than *Giardia* cysts, and these investigators suggested that disinfection alone should not be relied upon to inactivate *C. parvum* oocysts in water.

Human cryptosporiodiosis may be acquired by at least one of five known transmission routes: zoonotic, person to person, water, nosocomial (hospital acquired), or food. Zoonotic transmission (from vertebrate animals to humans) is most likely where infected animals (such as calves) deposit fecal matter to which humans are exposed. The disease may be contracted by drinking untreated water. Oocysts at levels of 2 to 112/liter were found in 11 samples of water from four rivers in Washington and California (64). Although the minimum infectious dose for humans is not known, two of two primates became infected after the ingestion of 10 oocysts (2). The organism has been shown to be an etiologic agent of travelers' diarrhea (83).

Symptoms, Diagnosis, and Treatment. The clinical course of cryptosporiodiosis in humans depends on the immune state, with the most severe cases occurring in the immunocompromised. In immunocompetent individuals, the organism primarily parasitizes the intestinal epithelium and causes diarrhea. The disease is self-limiting, with an incubation period of 6 to 14 days, and symptoms typically last 9 to 23 days. In the immunocompromised, diarrhea is profuse and watery, with as many as 71 stools per day and up to 17 liters per day reported (26). Diarrhea is sometimes accompanied by mucus but rarely blood. Abdominal pain, nausea, vomiting and low-grade (less than 39°C) fever are less frequent than diarrhea, and symptoms may last for more than 30 days in the immunocompromised but generally less than 20 days (range of 4 to 21 days) in the immunocompetent. In an outbreak associated with a swimming pool in California in 1988, the following symptoms (and percentage affected) were given by the 44 of 60 victims: watery diarrhea (88%), abdominal cramps (86%), and fever (60%) (12). The organism was identified from stool cultures of some patients by a modified acid-fast stain. Oocysts generally persist beyond the diarrheal stage. Human and animal cases of cryptosporiodiosis cases have been summarized by Fayer and Ungar (36).

Water- and Foodborne Outbreaks.   The first demonstrated waterborne outbreak of cryptosporiodiosis occurred in Braun Station, Texas, in 1984 following the consumption of artesian well water. There were actually two outbreaks—one in May and the other in July, with 79 victims (22). A second outbreak with 13,000 victims occurred in Carrollton, Georgia, in 1987 and oocysts were found in the stools of 58 of 147 victims (36). Three separate outbreaks occurred in the United Kingdom in 1988–1989. In one, there were 500 confirmed cases that resulted from the consumption of treated water, and as many as 5,000 persons may have been affected (80). In another outbreak, 62 cases were traced to contaminated swimming pool water. Early in 1990, there was an outbreak in Scotland.

Although the fecal-oral route is known, indirect transmission by food and possibly milk is suspected and established in at least one case. The latter involved a 32-year-old man in England who contracted clinical cryptosporiodiosis from eating raw frozen bovine tripe. Oocysts were demonstrated in uneaten tripe, and this appears to be the first human infection in which cysts were demonstrated in leftovers (14).

Diagnosis and Treatment.   Diagnosis of cryptosporiodiosis requires the identification of oocysts in stools of victims. Staining methods are used, including modified acid-fast procedures, negative staining, and sugar flotation. A recently described diagnostic method is a direct immunofluorescence test used for the detection of oocysts in feces (82). The latter method employs a monoclonal antibody against an oocyst wall antigen.

Over 100 chemotherapeutic regimes have been tested and found to be ineffective (21), although spiramycin, fluconazole, and amphotericin B show some promise.

FLATWORMS

All flatworms belong to the animal phylum Platyhelminthes, and the genera discussed in this chapter belong to two classes:

Phylum Platyhelminthes
  Class Trematoda (flukes)
    Subclass Digenea
      Order Echinostomata
        Family Fasciolidae
          Genus *Fasciola*
          Genus *Fasciolopsis*
      Order Plagiorchiata
        Family Troglotrematidae
          Genus *Paragonimus*
      Order Opisthorchiata
        Family Opisthorchiidae
          Genus *Clonorchis*
  Class Cestoidea
    Subclass Eucestoda (tapeworms)
      Order Pseudophyllidea
        Family Diphyllobothriidae
          Genus *Diphyllobothrium*
      Order Cyclophyllidea
        Family Taeniidae
          Genus *Taenia*

Fascioliasis

This syndrome (also known as parasitic biliary cirrhosis and liver rot) is caused by the digenetic trematode *Fasciola hepatica*. The disease among humans is cosmopolitan in distribution, and the organism exists where sheep and cattle are raised; they, along with humans, are its principal definitive hosts.

    This parasite matures in the bile ducts, and the large operculate eggs (150 × 90 μm in size) enter the alimentary tract from bile ducts and eventually exit the host in feces. After a period of 4 to 15 days in water, the miricidium develops, enters a snail, and is transformed into a sporocyst. The sporocyst produces mother rediae, which later become daughter rediae and cercariae. When the cercariae escape from the snail, they become free swimming, attach to grasses and watercress, and encyst to form metacercariae. When ingested by a definitive host, the metacercariae excyst in the duodenum, pass through the intestinal wall, and enter the coelomic cavity. From the body cavity, they enter the liver, feed on its cells, and establish themselves in bile ducts, where they mature (15, 67).

    Fascioliasis in cattle and sheep is a serious economic problem that results in the condemnation of livers. Human cases are known, especially in France, and they are contracted from raw or improperly cooked watercress that contains attached metacercariae. Human cases are rare in the United States and are limited to the South (39). Pharyngeal fascioliasis (**halzoun**) in humans results from eating raw *Fasciola*-laded bovine liver where young flukes become attached to the buccal or pharyngeal membranes, resulting in pain, hoarseness, and coughing (15).

Symptoms, Diagnosis, and Treatment.   Symptoms develop in humans about 30 days after the infection; they consist of fever, general malaise, fatigue, loss of appetite and weight, and pain in the liver region of the body. The disease is accompanied typically by eosinophilia. Fascioliasis can be diagnosed by demonstrating eggs in stools, biliary, or duodenal fluids. Effective treatment is achieved upon the administration of praziquantel (67).

Fasciolopsiasis

Fasciolopsiasis is caused by *Fasciolopsis buski*, and the habitat of this organism is similar to *F. hepatica*. Humans serve as definitive host, several species of snails as first intermediate, and water plants (watercress nuts) as second intermediate hosts. Unlike *F. hepatica*, this parasite occurs in the duodenum and jejunum of humans and pigs, and human infection rates as high as 40% are found in parts of Thailand, where certain uncooked aquatic plants are eaten (15).

Human symptoms of fasciolopsiasis are related to the number of parasites, with no symptoms occurring when only a few parasites exist in the body. When symptoms occur, they develop within 1 to 2 months after the initial infection and consist of violent diarrhea, abdominal pain, loss of weight, and generalized weakness. Death may occur in extreme cases (67). Symptoms appear to be due to the general toxic effect of metabolic products of the flukes.

Diagnosis is made by demonstrating eggs in stools. The eggs of *F. buski* are 130 to 140 μm $\times$ 80 to 85 μm in size. Both niclosamide and praziquantel are effective in treating this disease (67).

Paragonimiasis

This parasitic disease (also known as **parasitic hemoptysis**) is caused by *Paragonimus* spp, especially *P. westermani*. It is found primarily in Asia but also in Africa and South and Central America. *P. kellicotti* is found in North and Central America. In contrast to the trematodes, *P. westermani* is a lung fluke.

The eggs of this parasite are expelled in sputum from definitive hosts (humans and other animals), and the miricida develop in 3 weeks in moist environments. A miricidium penetrates a snail (first intermediate host) and later gives rise to daughter rediae and cercariae about 78 days after entering the snail (15). The cercariae enter a second intermediate host (crab or crayfish) and encyst. The crustacean host in parts of the Orient and the Philippines are various species of freshwater crabs, where they usually form metacercarial cysts in leg and tail muscles (67). In *P. kellicotti*, cysts form in the heart region (67). When the definitive host ingests the infected crustacean, the metacercariae hatch out of their shells, bore their way as young flukes through the walls of the duodenum, and then move to the lungs, where they become enclosed in connective tissue cysts (67). The golden-brown eggs may appear in sputum 2 to 3 months later.

Symptoms, Diagnosis, and Treatment. Paragonimiasis is accompanied by severe chronic coughing and sharp chest pains. Sputum is often reddish-brown or bloody. Other nonspecific symptoms may occur when parasites lose their way to the lungs (67). Diagnosis is made by demonstrating the golden-brown eggs in sputum or stools. The eggs of *P. westermani* are 80 to 120 μm in length × 50 to 60 μm in width, Also, a complement fixation test titer of at least 1:16 is diagnostic, and ELISA tests are available. The disease can be treated with praziquantel (67).

Clonorchiasis

The Class Trematoda of the flatworms consists of parasites commonly referred to as flukes that infect the liver, lungs, or blood of mammals. *Clonorchis* (*Opisthorchis*) *sinensis* is the Chinese liver fluke that causes oriental biliary cirrhosis. Flukes typically have three hosts: two intermediate, where the larval or juvenile stage develops, and the definitive or final host, where the sexually mature adult develops. *C. sinensis* is an endoparasite whose anterior sucker surrounds the mouth. It also has a midventral sucker. Along with cats, dogs, pigs, and other vertebrates, humans may serve as definitive hosts.

When deposited in water, the eggs of *C. sinensis* hatch into ciliated larvae (miracidia), which invade the first host, usually a snail. As a larva enters the snail, it rounds up as a sporocyst and reproduces asexually to form embryos. Each embryo develops into a redia that escapes from the sporocyst and begins to feed on host tissues. Embryos within the rediae develop into cercariae, which escape from the rediae through a birth pore. A cercaria is a miniature fluke with a tail. Cercariae leave the snail and swim through water in search of their next host—usually fish, clams, and the like. They bore into the new host, shed the tail, and become surrounded by a cyst. Within the cyst, further development leads to metacercarie, which develop further in the final host, usually a vertebrate, including humans. Upon ingestion of metacercariae-containing fish, the cyst wall dissolves in the intestine, and the young flukes emerge. They then migrate through the body to their final site, the bile ducts of the liver in the case of *C. sinensis* where, among other problems, they may cause cirrhosis (see below).

Liver flukes are common in China, Korea, Japan, and parts of Southeast Asia. It is estimated that more than 20 millions persons in Asia are infested with this parasite (37). In China, it is associated often with the consumption of a raw fish dish called **ide**. Over 80 species of fish are known to be capable of harboring *C. sinensis* (37).

Symptoms, Diagnosis, Treatment, and Prevention.   Symptoms may not occur if the infection is mild, but in severe cases, damage to the liver may occur. The liver damage may lead to cirrhosis and edema, and cancer of the liver is seen occasionally (67).

Diagnosis is made by repeated microcoscopic examinations of feces and duodenal fluid for eggs. An ELISA test is useful, but cross-reactions with other trematodes may occur. Praziquantel is an effective chemotherapeutic agent.

The prevention of this syndrome is achieved by avoiding the deposition of human feces in fishing waters, but this seems unlikely in view of its wide distribution. The avoidance of raw fish products and the proper cooking of fish are more realistic alternatives. *C. sinensis* can be inactivated in fish by the same procedures as for round- and flatworms. According to Rodrick and Cheng (71), all captured fish must be considered to be potential carriers of parasites. This applies to all flukes, flat- and roundworms, and protozoa.

Diphyllobothriasis

This infection is contracted from the consumption of raw or undercooked fish, and the causative organism, *Diphyllobothrium latum*, is often referred to as the broad fish tapeworm. The definitive hosts for *D. latum* are humans and other fish-eating mammals; intermediate hosts are various freshwater fish and salmon where plerocercoid (or metacestode) larvae are formed.

When humans consume fish flesh that contains plerocercoid larvae, the larvae attach to the ileal mucosa by two adhesive grooves (bothria) on each scolex and develop in 3 to 4 weeks into mature forms. As a worm matures, its strobila, made up of proglottids, increases in length to 10 or nearly 20 m, and each worm may produce 3,000 to 4,000 proglottids that are wider than they are long (hence, broad fish tape) (see Fig. 24-1). Over 1 million eggs may be released each day into stools of victims. Eggs are more often seen in stools than proglottids, and they are not infective for humans.

When human feces are deposited in waters, the eggs hatch and release six-hooked, free-swimming larvae or coracidia (also known as oncospheres). When these forms invade small crustaceans (copepods or microcrustaceans such as *Cyclops* or *Diaphtomus*), they metamorphose into a juvenile stage designated metacestode or procercoid larvae. When a fish ingests the crustacean, the larvae migrate into its muscles and develop into plerocercoid larvae. If this fish is eaten by a larger fish, the plerocercoid migrates, but it does not undergo further development. Humans are infected when they eat fish containing these forms.

Prevalence.   A synopsis of the human history of this syndrome is presented in Table 24-2. Although the first human case was reported in 1906, it was the scattering of cases during the early 1980s that brought new attention to this disease in the United States and Canada. The cases in question resulted from the consumption of sushi, a raw fish product that has long been popular in parts of the Orient but only relatively recently has become popular in the United States. The incidence of diphyllobothriasis is high in Scandinavia and the Baltic regions of Europe. It is estimated that 5 million cases occur in Europe, 4 million in Asia, and 100,000 in North America. However, only 1 of 275 asymptomatic natives of Labrador, Canada, examined in 1977 had a positive stool culture for this organism (81).

Symptoms, Diagnosis, and Treatment.   While most cases of diphyllobothriasis are asymptomatic, victims may complain of epigastric pain, abdominal cramps, vomiting, loss of appetite, dizziness, and weight loss. Intestinal obstruction is not unknown. One of the consequences of this infestation is a vitamin B-12 deficiency, along with macrocytic anemia.

This disease is diagnosed by demonstrating eggs in stools. Treatment is the same as for taeniasis. The absence of overt symptoms does not mean the absence of the tapeworm in the intestines since the worms may persist for many years.

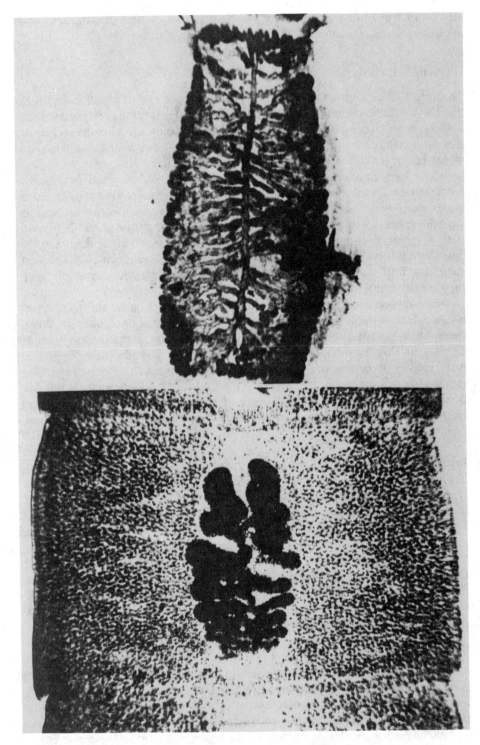

FIGURE 24-1.   Proglottid of *Taenia saginata* (top) and *Diphyllobothrium latum* (bottom), ×360.
From S. H. Abadie, J. H. Miller, L. G. Warren, J. C. Swartzwelder, and M. R. Feldman, Manual of
Clinical Microbiology, *2nd ed.; copyright © 1974 by American Society for Microbiology, used with
permission.*

TABLE 24-2.   Known Cases of Diphyllobothriasis in the United
States and on the West Coast, 1977–1981

Year	Total Cases	West Coast Cases (% of Total)
1977	121	15 (12)
1978	133	16 (12)
1979	174	17 (10)
1980	204	64 (31)
1981	203	51 (25)

Source: Ruttenbar (74).

Note: "West Coast": Alaska, California, Hawaii, Oregon, and
Washington.

Prevention.   Diphyllobothriasis can be prevented in humans by avoiding the
consumption of raw or undercooked fish. While the elimination of raw sewage
from waters will undoubtedly help to reduce the incidence, it will not break the
life cycle chain of this organism since humans are not the only definitive hosts.
Cooking fish products to an internal temperature of 60°C for 1 min or 65°C for
30 sec will destroy the organism (6), as will freezing fish to −20°C for at least
60 h (46, 47). For more information, see Bylund (9) or von Bondsdorff (87).

Cysticercosis/Taeniasis

This syndrome in humans is caused by two species of flatworms: *Taenia saginata*
(Also *Taeniarhynchus saginatus*; beef tape) and *Taenia solium* (pork tape).
They are unique among both flat- and roundworm parasites in that humans are
their definitive hosts; the adult and sexually mature stages develop in humans,
while the larval or juvenile stage develops in herbivores. These helminths have
no vascular, respiratory, or digestive systems nor do they possess a body cavity.
They depend on the digestive activities of their human hosts for all of their
nourishment. Their metabolism is primarily anaerobic.

The structure of a *T. saginata* proglottid is illustrated in Figure 24-1. The
adult worm consists of a scolex (head) that is about 1 mm in size and lacks
hooks but has four sucking discs. Behind the scolex is the generative neck,
which segments to form the strobila composed of proglottids. The latter increase
in length, with the oldest being the farthest away from the scolex. Each
proglottid has a complete set of reproductive organs, and an adult worm may
contain up to 2,000 proglottids. These organisms may live up to 25 years and
grow to a length of 4 to 6 m inside the intestinal tract. *T. saginata* sheds eight to
nine proglottids daily, each containing 80,000 eggs. The eggs are not infective
for humans.

When proglottids reach soil, they release their eggs, which are 30 to 40 µm
in diameter, contain fully developed embryos, and may survive for months.
When the eggs are ingested by herbivores, such as cattle, the embryos are
released, penetrate the intestinal wall, and are carried to striated muscles of
the tongue, heart, diaphragm, jaw, and hindquarters, where they are transformed into larval forms designated cysticerci. *Cysticercosis* is the term used

to designate the existence of these parasites in the intermediate hosts. The cysticerci usually take 2 or 3 months to develop after eggs are ingested by a herbivore. When present in large numbers, the cysticerci impart a spotted appearance to the beef tissue. Humans become infected upon the ingestion of meat that contains cysticerci.

The infection caused by the pork tape (*T. solium*) is highly similar to that described for the beef tape, but there are some significant differences. While humans are also the definitive hosts, the larval stages develop in both swine and humans. In other words, humans can serve as intermediate (cysticercosis) and definitive (taeniasis) hosts, thus making autoinfections possible. For this reason, *T. solium* infections are potentially more dangerous than those of *T. saginata*. The infection caused by larval forms of *T. solium* is sometimes designated Cysticercus cellulosae. The *T. solium* scolex has hooks rather than sucking discs, and the strobila may reach 2 to 4 m and contain only about 1,000 proglottids. Embryos of *T. solium* are carried to all tissues of the body, including the eyes and brain in contrast to *T. saginata*. While *T. saginata* exists in both the United States and many other parts of the world, *T. solium* has been eliminated in the United States. However, it does exist in Latin America, Asia, Africa, and eastern Europe. The incidence of *T. saginata* in beef in the United States is below 1% as a result of federal and local meat inspections.

Symptoms, Diagnosis, and Treatment.   Most cases of taeniasis are asymptomatic regardless of the *Taenia* species involved, but symptoms differ when humans serve as intermediate host. In these cysticercosis cases, the cysticerci develop in body tissues, including those of the central nervous system, and generally lead to eosinophilia.

Human taeniasis is diagnosed by demonstration of eggs or proglottids in stools and cysticercoisis by tissue biopsies of calcified cysticerci or by immunological methods (32). Complement fixation, indirect hemagglutination, and immunofluorescence tests are valuable diagnostic aids.

A single-dose oral treatment with niclosamide, which acts directly on the parasites, is effective in ridding the body of adult worms. This drug apparently inhibits a phosphorylation reaction in the worm's mitochondria. Another effective chemotherapeutic agent is praziquantel. With cysticercosis, surgery may be indicated.

Prevention.   The general approach in the prevention and elimination of diseases that require multiple hosts is to cut the cycle of transmission from one host to another. Since the eggs are shed in human feces, taeniasis can be eliminated by the proper disposal of sewage and human wastes, although *T. solium* infections in humans present a more complex problem. Cysticerci can be destroyed in beef and pork by cooking to a temperature of at least 60°C (43). The freezing of meats to at least −10°C for 10 to 15 days or immersion in concentrated salt solutions for up to 3 weeks will inactivate these parasites. Freezing times and temperatures necessary to ensure the death of all cysticerci from infected calves were found by one group to be as follows: 360 h at −5°C, 216 h at −10°C, and 144 h at −15°C, −20°, −25°, or −30°C (38). For more information, see Flisser et al. (31).

ROUNDWORMS

The disease-causing roundworms of primary importance in foods belong to two orders of the phylum Nematoda. The order Rhabditida includes *Turbatrix aceti* (the vinegar eel), which is not a human pathogen and is not discussed further.

Phylum Nematoda
   Class Adenophorea (=Aphasmidia)
      Order Trichinellida
         Genus *Trichinella*

Class Secernentea (=Phasmidia)
   Order Rhabditida
      Genus *Turbatrix*

   Order Ascaridida
      Genus *Ascaris*

      Subfamily Anisakinae
         Genus *Anisakis*
         Genus *Pseudoterranova* (*Phocanema*)
         Genus *Toxocara*

Trichinosis

Trichinella spiralis is the etiologic agent of trichinosis (trichinellosis), the roundworm disease of greatest concern from the standpoint of food transmission. The organism was first described in 1835 by J. Paget in London, and the first human case of trichinosis was seen in Germany in 1859 (51). While most flat- and roundworm diseases of humans are caused by parasites that require at least two different host animals, the trichinae are transmitted from host to host; no free-living stages exist. In other words, both larval and adult stages of *T. spiralis* are passed in the same host. It is contracted most often from raw or improperly cooked pork products.

The adult forms of *T. spiralis* live in the duodenal and jejunal mucosas of mammals such as swine, canines, bears, marine mammals, and humans that have consumed trichinae-infested flesh. The adult females are 3 to 4 mm long, and adult males are about half this size. While they may remain in the intestines for about a month, no symptoms are produced. The eggs hatch within female worms, and each female can produce around 1,500. These larvae, each about 0.1 mm in length, burrow through the gut wall and pass throughout the body, ultimately lodging in certain muscles. Only those that enter skeletal muscles live and grow; the others are destroyed. The specific muscles affected include those of the eye, tongue, and diaphragm. When assaying for trichinae larvae in pork, the U.S. Department of Agriculture, (USDA) employs diaphragm muscle or tongue tissues. In a recent study, the Crus muscle of the diaphragm was found to yield more larvae/g than several others (54). As the larvae burrow into muscles several weeks later, severe pain, fever, and other symptoms occur, which sometime lead to death from heart failure (see below). The larvae grow to about 1 mm in muscles and then encyst by curling up and becoming enclosed in a calcified wall some 6 to 18 months

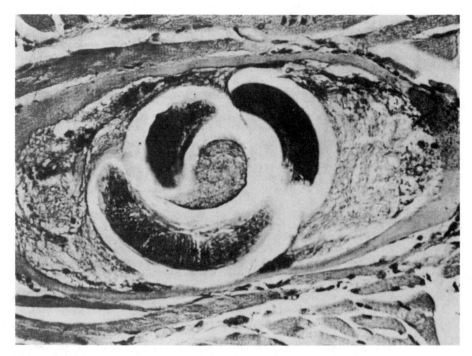

FIGURE 24-2.   *Trichinella spiralis* in muscle, ×350.
From S. H. Abadie, J. H. Miller, L. G. Warren, J. C. Swartzwelder, and M. R. Feldman, Manual of Clinical Microbiology, *2nd ed.; copyright © 1974 by American Society for Microbiology, used with permission.*

later (Fig. 24-2). The larvae develop no further until consumed by another animal (including humans), but they may remain viable for up to 10 years in a living host. When the encysted flesh is ingested by a second host, the encysted larvae are freed by the enzymatic activities in the stomach, and they mature in the lumen of the intestines.

Prevalence.   About 75 species of animals can be infected by *T. spiralis*, but avians appear to be resistant (59). During the 1930s and 1940s, about 16% of Americans were infected (59). For the period 1966–1970, 4.7% of pork contained trichinae in diaphragm muscles examined postmortem in the United States. For the five-year period 1977–1981, 686 cases with four deaths were reported in the United States (77). The CDC survey data for the years 1983–1987 show 33 outbreaks with 162 cases and 1 death, which represents a mean of around 32 cases per year for this 5-year period (5). For the 15-year period 1973–1987, the CDC recorded 128 outbreaks and 843 cases for an average of 56 per year (4). However, the actual number of cases in the United States in 1985 has been estimated to be 100,000 (7). Only 3 cases were recorded in Canada in 1982, with none in 1983 and 1984 (84). The trend in cases between 1950 and 1989 can be seen from Figure 24-3. For the 3-year period 1987–1989, fewer than 50 annual cases were reported, but 120 were reported in 1990 (14). Ninety of these occurred in Iowa among 250 immigrants from Southeast Asia who consumed raw pork sausage. An additional 15 cases occurred in Virginia; pork sausage was the vehicle food (14).

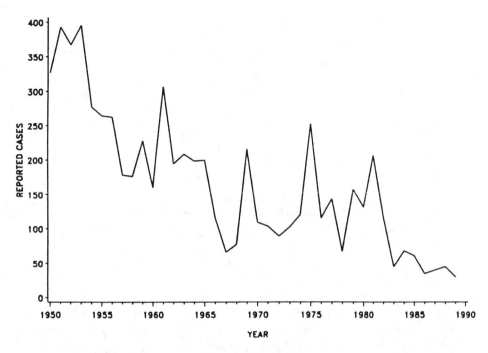

FIGURE 24-3.   Trichinosis in the United States, 1950–1989.
From Morbidity and Mortality Weekly Report *38(54):42, 1989.*

Pork was incriminated in 79% of the cases for the years 1975–1981, with bear meat in 14%, and ground beef in 7%. Studies on pork in retail ground beef have revealed that from 3 to 38% of beef samples contained pork. The presence of pork in ground beef may be deliberate on the part of some stores, or it may result from using the same grinder for both products.

Symptoms and Treatment.   One to two days after the ingestion of heavily encysted meat, trichinae penetrate the intestinal mucosa, producing nausea, abdominal pain, diarrhea, and sometimes vomiting. When only a few larvae are ingested, the incubation period may be as long as 30 days. The symptoms may persist for several days, or they may abate and be overlooked. The larvae begin to invade striated muscles about 7 to 9 days after the initial symptoms. Where 10 or fewer larvae are deposited/g of muscle tissue, there are usually no symptoms. When 100 or more per gram are deposited, symptoms of clinical trichinosis usually develop, while for 1,000 or more per gram of tissue, serious and acute consequences may occur. Muscle pain (myalgia) is the universal symptom of muscle involvement, and difficulty in breathing, chewing, and swallowing may occur (59). About 6 weeks after the initial infection, encystment occurs, accompanied by tissue pain, swelling, and fever. Resistance to reinfection develops, and it appears to be T cell mediated. Thiabendazole and mebendazole have been shown to be effective drugs for this disease.

Diagnosis.   Since the trichinae exist as coiled larvae in ovoid capsular cysts in skeletal muscles, biopsies are sometimes performed on the deltoid, biceps,

TABLE 24-3.   Required Period of Freezing at Temperatures Indicated

Temperature (°C)	Group 1 (days)	Group 2 (days)
−15	20	30
−23	10	20
−29	6	12

Source: Kotula et al. (56); copyright © 1983 by Institute of Food Technologists.

Notes: Group 1 = Less than 15.24 cm in depth; Group 2 = More than 15.24 cm in depth.
From Sec. 18.10. Regulations Governing the Meat Inspection of the United States Department of Agriculture (9 CFR 18.10, 1960).

or gastrocnemius muscles. A significant eosinophilia usually develops during the second week of the disease. Antibodies can be detected after the third week of infection; immunological methods that may be used include bentonite flocculation, cholesterol-lecithin flocculation, and latex agglutination. A bentonite titer of 1:5 is significant, but this test is not positive until at least 3 weeks after infection.

Prevention and Control.   Trichinosis can be controlled by avoiding the feeding of infected meat scraps or wild game meats to swine and by preventing the consumption of infested tissues by other animals. The feeding of uncooked garbage to swine helps to perpetuate this disease. Where only cooked garbage is fed to pigs, the incidence of trichinosis has been shown to fall sharply.

This disease can be prevented by the thorough cooking of meats such as pork or bear meat. In a study on the heat destruction of trichina larvae in pork roasts, all roasts cooked to an internal temperature of 140°F or higher were subsequently found to be free of organisms (11). Larvae were found in all roasts cooked at 130°F or lower, and in some roasts cooked at 135°F. The USDA recommendation for pork products is that the product be checked with a thermometer after standing and if any part does not attain 76.7°C (170°F), the product should be cooked further (86).

Freezing will destroy the encysted forms, but freezing times and temperatures depend upon the thickness of the product and the specific strain of *T. spiralis* (Table 24-3). The lower the temperature of freezing, the more destructive it is to *T. spiralis*, as was demonstrated in the following study. Four selected temperatures were chosen for the freezing of infected ground pork that was stuffed into casings and packed into boxes. When frozen and stored at −17.8°C, the trichinae lost infectivity between 6 and 10 days; at −12.2°C, infectivity was lost between 11 and 15 days (77). When frozen at −9.4°C, they remained infective up to 56 days and for up to 71 days when frozen at −6.7°C (92). Freezing in dry ice (−70°C) and liquid nitrogen (−193°C) destroys the larvae (53). The destruction of trichina larvae by irradiation is discussed in Chapter 12.

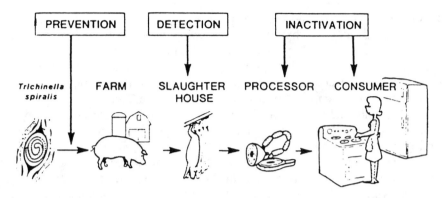

FIGURE 24-4.   Various stages in the movement of pork from the farm to consumer at which control efforts may be applied.
From Murrell (61); copyright © 1985, Institute of Food Technologists, used with permission.

The effect of curing and smoking on the viability of trichina in pork hams and shoulders was investigated by Gammon et al. (33). They employed the meat of hogs experimentally infested with *T. spiralis* as weanling pigs. After curing, the meat was hung for 30 days followed by smoking for approximately 24 h at 90° to 100°F, with subsequent aging. Live trichinae were found in both hams and shoulders 3 weeks after smoking, but none could be detected after 4 weeks. The effect of NaCl concentration, a_w, and fermentation method on viability of *T. spiralis* in genoa salami was evaluated by Childers et al. (16). Pork from experimentally infected pigs was used to prepare salami. The trichinae larvae were completely destroyed at day 30 and thereafter in salami made with 3.33% NaCl and given high-temperature (46.1°C) fermentation treatment, irrespective of product pH. No larvae were found in products made with 3.33% NaCl and given low-temperature fermentation after 30 days. In salami made with no salt, 25% of larvae were found at days 15–25, but none thereafter. A summary of the main control steps of prevention, detection, and inactivation are illustrated in Figure 24-4.

Microwave Cooking.   The efficacy of microwave ovens in destroying *T. spiralis* larvae has been investigated by several groups. In a homemaker-oriented study in which most trichina-infected pork roasts were cooked in microwave ovens by time rather than product temperature, Zimmerman and Beach (91) found that of 51 products (48 roasts and 3 pork chops) cooked in 6 different ovens, 9 remained infective. Six of the 9 did not attain a mid-roast temperature of 76.7°C, while the other 3 exceeded this temperature at some point in the cooking cycle. The authors noted that the experimentally infected pork used in the study came from pigs infected with 250,000 *T. spiralis*, which produced around 1,000 trichina/g of tissue compared to about 1 trichina/g in naturally infected pigs. While the large number of trichinae/g may have been a factor in their survival at the cooking procedures employed, the inherent unevenness of cooking in microwave ovens is of concern. In another study, while trichinae larvae were not inactivated at 77° or 82°C in microwave ovens, cooking to an internal temperature of 77°C in a conventional convection oven, flat grill, charbroiler, or deep fat did inactivate the larvae (54). Further, infected larvae survived rapid cooking that involved thawing pork chops in an industrial microwave oven followed by cooking on a charbroiler to 71° or 77°C (55).

The cooking of pork in microwave ovens is clearly a matter of concern relative to the destruction of trichinae larvae, and two factors may explain the greater efficiency of convection ovens over microwave ovens. First, microwave cooking is rapid, and herein may lie the problem. Oven heat has been shown to be more destructive to trichinae larvae in roasts when slow-cooked in conventional ovens at 200°F than when fast-cooked at 350°F (11). Second, a convection oven is more uniformly heated than some microwave ovens. This is minimized if the product is rotated in the latter-type ovens or if the oven is equipped with an automatic rotating device. Otherwise, uneven heating occurs, leading to undercooking of some parts of a roast while other parts may be overcooked. It has been shown that a set of criteria that leads to consistent doneness of pork products in microwave ovens will result in safe products (90).

Anisakiasis

This roundworm infection is caused by two closely related genera and species: *Anisakis simplex* (the herring- or whaleworm) and *Pseudoterranova decipiens* (formerly *Phocanema*; cod- or sealworm). Both of these organisms have several intermediate hosts and generally more than one definitive host. Humans are not final hosts for either, and human disease occurs as the result of humans' being accidental interlopers in the life cycles of these worms.

The definitive hosts are marine mammals—whales in the case of *A. simplex* and grey (and other) seals in the case of *P. decipiens*. Feces of these animals contain thousands of eggs, which when they enter water undergo their first molt (stage L1 to stage L2). The free-swimming larvae that result are ingested by small crustaceans (copepods), and they in turn are ingested by larger crustaceans, which serve as intermediate hosts during the second molt (stage L2 to L3). A final host may ingest L3 larvae along with the crustacean intermediate, but more often L3 is ingested by fish or squid, which may in turn be ingested by larger fish before reaching the final host. The last two molts (L3 and L4) lead to adults that mate, and these events take place in the final host. The infectious larva is L3, and it is usually found in tight, flat coils in or on fish viscera, and some larvae may occur in the belly flap muscles of fish. Because of its preference for whale hosts, *A. simplex* is found more often in fish from the northern Pacific (41, 62).

In the case of *P. decipiens*, eggs in seal feces are ingested by copepods, and L2- or early-L3-stage larvae are ingested by the first intermediate host—fish. In fish, they penetrate the stomach wall, enter the body cavity, and many burrow into fish muscles. In fish, the L3 larvae grow to 25 to 50 mm in length and are red to brown in color. The final host, seals, tends to ingest the organisms principally from smelt and other small fish.

Human infections occur upon the ingestion of fish that contain L3- and L4-stage larvae. Thus, anisakids do not mature in humans. Disease symptoms arise from the activities of the juvenile worms. *A. simplex* larvae are more harmful than those of *P. decipiens* since they often penetrate the mucosal lining, while most *P. decipiens* larvae are passed in feces or are coughed up or vomited after irritating the mucosa. *Anisakis* is most often found in cases in Japan and the Netherlands; *Pseudoterranova* is more often seen in North America.

Symptoms, Diagnosis, and Treatment. Symptoms of human anisakiasis may develop within 4 to 6 h after consumption of infested fish, and they consist of epigastric pain, nausea, and vomiting. In more severe cases, fever and bloody stools may occur within 7 days after ingesting infective fish. If the worms penetrate the mucosa, an eosinophilic granuloma may develop, or they may penetrate the gut wall and cause peritonitis. However, in the 23 North American cases through 1982, only five were caused by *Anisakis* sp., and only transient infections occurred (50). Among the four cases reported by Kliks (50), symptoms consisted of mild stomach pain and nausea from the time of ingestion up to 20 h later, and worms were coughed up or found in the mouth up to 2 weeks after consumption of the infective raw salmon.

The diagnosis of this syndrome is made difficult by the absence of eggs or other parts of the worms in feces. Larvae in the intestinal tract can be viewed by endoscopy, and surgical resection of the affected tissue can be carried out. Complement fixation and indirect immunofluorescence tests are of some diagnostic value. Thiabendazole is an effective chemotherapeutic agent for treating this disease.

Prevalence and Distribution. A synopsis of the known cases of anisakiasis is presented in Table 24-4. The first clear-cut case occurred in 1955 in the Netherlands, and between 1955 and 1965 over 149 cases were reported in that country. When freezing of herring to −20°C for 24 h was legislated, no cases were seen the following year (40). Over 1,000 cases were reported in Japan for the period 1964–1976 (63). In both instances, raw fish products such as sushi and sashimi were the vehicle foods, although lightly salted herring ("green herring") was a common vehicle in the Netherlands. Through 1976, 6 cases were recorded in the United States and 1 each in Canada, England, and Greenland (58). This disease is associated with the raw fish dish ceviche in South America, where *D. pacificum* is the usual etiologic agent (71). The first documented case in North America occurred in Boston in 1973, and through 1990 fewer than 100 cases were reported in the United States by both etiologic agents. Anisakiasis has been seen in Belgium, Britain, Chile, Denmark, France, Germany, Korea, and Taiwan. Over a 15-year period beginning in the early 1960s, some 1,200 cases were seen in Japan. Raw mackerel is probably

TABLE 24-4.   Summary of Cases of Anisakiasis

1955	First clear-cut case recorded in the Netherlands
1955–1965	About 149 cases reported in the Netherlands
1964–1976	Over 1,000 cases reported in Japan
1973	First documented case in North America (in Boston)
1980	Over 500 cases reported in Japan
1977–1981	About five cases recorded in California (two by *A. simplex* and three by *P. decipiens*)
1981–1989	Approximately 50 cases caused by *A. simplex* and 30 by *P. decipiens* seen in North America

Sources: Jackson (40), Margolis (58), Myers (63).

the most significant fish source in the Orient (71), although over 160 teleost species are believed to harbor these organisms (37).

In regard to the prevalence of anisakid larvae in fish, two extensive investigations were conducted in the late 1970s. In one, 1,010 fish belonging to 20 genera and 23 species were examined in the Washington, D.C., area and of 703 that contained parasitic nematodes, 6,547 nematodes were found, of which only 11 were *Anisakis* sp. (41). The mean content of nematode larvae/fish was 6.48, with an overall infection of 69.60%. A 2-year survey in 1974–1975 of fish and shellfish from U.S. Pacific coastal waters off Washington, Oregon, and California included 2,074 specimens (62). *Anisakis* sp. was the most frequently found, and most were found in or on fish viscera. In fish caught off the California coast, 41.6% contained anisakid larvae (59). *Anisakis* was higher than *Phocanema* (*Pseudoterranova*), due largely to the number of whales, a situation that is reversed in the eastern Canadian waters of the Gulf of St. Lawrence (62). No anisakid larvae were found in over 2,000 shellfish examined. In yet another survey in Ann Arbor, Michigan, larval densities from 63 to 91/kg of salmon tissue were found, and they were in viable form (73).

There is some controversy as to whether there is an increased incidence of anisakid larvae in commercial fish compared to decades ago. Clearly these organisms are not new in fishing waters since their presence in fish was recognized as early as 1767. A rather widely held view is that the disease in humans began to flourish when refrigeration was taken aboard fishing boats in the mid 1950s. Prior to this time, fish were caught and eviscerated, and the infective organisms were thus discarded. When fish are kept on ice for several days, some investigators believe the parasites migrate from the mesenteries to muscles following the death of fish (37, 62). On the other hand, the migration of parasites in dead fish has not been substantiated by all investigators.

Prevention.   Anisakiasis can be prevented by avoiding the consumption of raw or undercooked fish. Sushi, ceviche, and sashimi should be consumed only when properly prepared from fish that has undergone inspection for absence of infective larvae. Infective forms can be destroyed by cooking fish to an internal temperature of 60°C for 1 min or 65°C for 30 sec (6). Freezing at −20°C or below for at least 60 h is reported to render the larvae uninfective (46) although some North American species survive after 52 h at −20°C (6). Brining for 4 weeks has been found to render larvae uninfective (35).

References

1. Anderson, B. C. 1985. Moist heat inactivation of *Cryptosporidium* sp. *Am. J. Pub. Hlth.* 75:1433–1434.
2. Barer, M. R., and A. E. Wright. 1990. *Cryptosporidium* and water. *Lett. Appl. Microbiol.* 11:271–277.
3. Barnard, R. J., and G. J. Jackson. 1984. *Giardia lamblia*: The transfer of human infections by foods. In *Giardia and Giardiasis: Biology, Pathogenesis, and Epidemiology*, ed. S. L. Erlandsen and E. A. Meyer, 365–378. New York: Plenum.
4. Bean, N. H., and P. M. Griffin. 1990. Foodborne disease outbreaks in the United States, 1973–1987: Pathogens, vehicles, and trends. *J. Food Protect.* 53:804–817.
5. Bean, N. H., P. M. Griffin, J. S. Goulding, and C. B. Ivey. 1990. Foodborne

disease outbreaks, 5-year summary, 1983–1987. *J. Food Protect.* 53:711–728.

6. Bier, J. W. 1976. Experimental anisakiasis: Cultivation and temperature tolerance determinations. *J. Milk Fd. Technol.* 39:132–137.

7. Bennett, J. V., S. D. Holmberg, M. F. Rogers, and S. L. Solomon. 1987. Infectious and parasitic diseases. In *The Burden of Unnecessary Illness*, ed. R. W. Amler and H. B. Dull, 102–114. New York: Oxford University Press.

8. Boucher, S.-E. M., and F. D. Gillin. 1990. Excystation of in vitro–derived *Giardia lamblia* cysts. *Infect. Immun.* 58:3516–3522.

9. Bylund, B. G. 1982. Diphyllobothriasis. In: *CRC Handbook Series in Zoonoses*, ed. L. Jacobs and P. Arambulo III, 217. Boca Raton, Fla.: CRC Press.

10. Campbell, L., S. Tzipori, G. Hutchison, and K. W. Angus. 1982. Effect of disinfectants on survival of cryptosporidium oocysts. *Vet. Rec.* 111:414–415.

11. Carlin, A. F., C. Mott, D. Cash, and W. Zimmerman. 1969. Destruction of trichina larvae in cooked pork roasts. *J. Food Sci.* 34:210–212.

12. Centers for Disease Control. 1990. Swimming-associated cryptosporidiosis—Los Angeles County. *Morb. Mort. Wkly. Repts.* 39:343–345.

13. Centers for Disease Control. 1989. Common-source outbreak of giardiasis—New Mexico. *Morb. Mort. Wkly. Repts.* 38:405–407.

14. Centers for Disease Control (U. S.). 1991. *Trichinella spiralis* infection—United States, 1990. *Morb. Mort. Wkly. Rept.* 40:57–60.

15. Cheng, T. C. 1986. *General Parasitology*, 2d ed. New York: Academic Press.

16. Childers, A. B., R. N. Terrell, T. M. Craig, T. J. Kayfus, and G. C. Smith. 1982. Effect of sodium chloride concentration, water activity, fermentation method and drying time on the viability of *Trichinella spiralis* in Genoa salami. *J. Food Protect.* 45:816–819.

17. Conroy, D. A. 1960. A note on the occurrence of *Giardia* sp. in a Christmas pudding. *Rev. Iber. Parasitol.* 20:567–571.

18. Craun, G. 1988. Surface water supplies and health. *J. Am. Water Works Assoc.* 80:40–52.

19. Craun, G. F. 1979. Waterborne giardiasis in the United States: A review. *Amer. J. Pub. Hlth.* 69:817–819.

20. Current, W. L. 1988. The biology of Cryptosporidium. *ASM News.* 54(11):605–611.

21. Current, W. L. 1987. *Cryptosporium*: Its biology and potential for environmental transmission. *CRC Crit. Rev. Environ. Cont.* 17:21–51.

22. D'Antonio, R. G., R. E. Winn, J. P. Taylor, T. L. Gustafson, W. L. Current, M. M. Rhodes, G. W. Gary, Jr., and R. A. Zajac. 1985. A waterborne outbreak of cryptosporidiosis in normal hosts. *Ann. Int. Med.* 103:886–888.

23. Desmonts, G., J. Couvreur, F. Alison, J. Baudelot, J. Gerbeaux, and M. Lelong. 1965. Etude epidemiologique sur la toxoplasmose: De l'influence de la cuisson des viandes de boucherie sur la frequence de l'infection humaine. *Rev. Franc. Etudes Clin. Biol.* 10:952–958.

24. Dubey, J. P., C. A. Speer, and R. Fayer. 1990. *Cryptosporidiosis of Man and Animals*. Boca Raton, Fla.: CRC Press.

25. Duquesne, V., C. Auriault, F. Darcy, J.-P. Decavel, and A. Capron. 1990. Protection of nude rats against *Toxoplasma* infection by excreted-secreted antigen-specific helper T cells. *Infect. Immun.* 58:2120–2126.

26. Fayer, R., and B. L. P. Ungar. 1986. *Cryptosporidium* spp. and cryptosporiodiosis. *Microbiol. Rev.* 50:458–483.

27. Fayer, R., and J. P. Dubey. 1985. Methods for controlling transmission of protozoan parasites from meat to man. *Food Technol.* 39(3):57–60.

28. Fayer, R. 1982. Other protozoa: *Eimeria, Isospora, Cystoisospora, Besnoitia, Hammondia, Frenkelia, Sarcocystis, Cryptosporidium, Encephalitozoon,* and

Nosema. In *CRC Handbook Series in Zoonosis*, ed. J. H. Steele, 187–197. Boca Raton: CRC Press.

29. Fayer, R. 1975. Effects of refrigeration, cooking and freezing on *Sarcocystis* in beef from retail food stores. *Proc. Helm. Soc. Wash.* 42:138–140.

30. Feldman, H. A., and L. T. Miller. 1956. Serological study of toxoplasmosis prevalence. *Amer. J. Hyg.* 64:320–335.

31. Flisser, A., K. Willms, J. P. Laclette, C. Larralde, C. Ridaura, and F. Beltran. 1982. *Cysticercosis: Present State of Knowledge and Perspectives*. New York: Academic Press.

32. Flisser, A. 1985. Cysticercosis: A major threat to human health and livestock production. *Food Technol.* 39(3):61–64.

33. Gammon, D. L., J. D. Kemp, J. M. Edney, and W. Y. Varney. 1968. Salt, moisture and aging time effects on the viability of *Trichinella spiralis* in pork hams and shoulders. *J. Food Sci.* 33:417–419.

34. Gangarosa, E. J., and J. A. Donadio. 1970. Surveillance of foodborne disease in the United States. *J. Inf. Dis.* 62:354–358.

35. Grabda, J., and J. W. Bier. 1988. Cultivation as an estimate for infectivity of larval *Anisakis simplex* from processed herring. *J. Food Protect.* 51:734–736.

36. Hayes, E. B., T. D. Matte, T. R. O'Brien, T. W. McKinley, G. S. Logsdon, J. B. Rose, B. L. P. Ungar, D. M. Word, P. F. Pinsky, M. L. Cummings, M. A. Wilson, E. G. Long, E. S. Hurwitz, and D. J. Juranek. 1989. Large community outbreak of cryptosporiodiosis due to contamination of a filtered public water supply. *N. Engl. J. Med.* 320:1372–1376.

37. Higashi, G. I. 1985. Foodborne parasites transmitted to man from fish and other aquatic foods. *Food Technol.* 39(3):69–74, 111.

38. Hilwig, R. W., J. D. Cramer, and K. S. Forsyth. 1978. Freezing times and temperatures required to kill cysticerci of *Taenia saginata* in beef. *Vet. Parasitol.* 4:215–219.

39. Jackson, G. J. 1990. Parasitic protozoa and worms relevant to the U.S. *Food Technol.* 44(5):106–112.

40. Jackson, G. J. 1975. The "new disease" status of human anisakiasis and North American cases: A. review. *J. Milk Fd. Technol.* 38:769–773.

41. Jackson, G. J., J. W. Bier, W. L. Payne, T. A. Gerding, and W. G. Knollenberg. 1978. Nematodes in fresh market fish of the Washington, D.C. area. *J. Food Protect.* 41:613–620.

42. Jackson, M. H., and W. M. Hutchison. 1989. The prevalence and source of *Toxoplasma* infection in the environment. *Adv. Parasitol.* 28:55–105.

43. Jacobs, L. 1962. Parasites in food. In *Chemical and Biological Hazards in Food*, ed. J. C. Ayres et al., 248–266. Ames: Iowa State University Press.

44. Jacobs, L., J. S. Remington, and M. L. Melton. 1960. A survey of meat samples from swine, cattle, and sheep for the presence of encysted *Toxoplasma*. *J. Parasitol.* 46:23–28.

45. Joiner, K. A., S. A. Furhman, H. M. Miettinen, L. H. Kasper, I. Mellman. 1990. *Toxoplasma gondii*: Fusion competence of parasitophorous vacuoles in Fc receptor-transfected fibroblasts. *Science* 249:641–646.

46. Karl, H. 1988. Vorkommen von Nematoden in Konsumfischen. *Verfah. z. Feststel. u. Abtoetung. Rundsch. Fleisch. Lebensmittelhyg.* 40:198–199.

47. Karl, H., and M. Leinemann. 1989. Ueber lebensfaehigkeit von Nematodenlarven (*Anisakis* sp.) in gefrosteten Heringen. *Arch. Lebensmittelhyg.* 40:14–16.

48. Kean, B. H., A. C. Kimball, and W. N. Christenson. 1969. An epidemic of acute toxoplasmosis. *J. Amer. Med. Assoc.* 208:1002–1004.

49. Kirkpatrick, C. E., and C. E. Benson. 1987. Presence of *Giardia* spp. and absence of *Salmonella* spp. in New Jersey muskrats (*Ondatra zibethicus*). *Appl. Environ. Microbiol.* 53:1790–1792.

50. Kliks, M. M. 1983. Anisakiasis in the western United States: Four new case reports from California. *Am. J. Trop. Med. Hyg.* 32:526–532.

51. Kolata, G. 1985. Testing for trichinosis. *Science* 227:621, 624.

52. Korich, D. G., J. R. Mead, M. S. Madore, N. A. Sinclair, and C. R. Sterling. 1990. Effects of ozone, chlorine dioxide, chlorine, and monochloramine on *Cryptosporidium parvum* oocyst viability. *Appl. Environ. Microbiol.* 56:1423–1428.

53. Kotula, A. W., A. K. Sharar, E. Paroczay, H. R. Gamble, K. D. Murrell, and L. Douglass. 1990. Infectivity of *Trichinella spiralis* from frozen pork. *J. Food Protect.* 53:571–573.

54. Kotula, A. W., P. J. Rothenberg, J. R. Burge, and M. B. Solomon. 1988. Distribution of *Trichinella spiralis* in the diaphragm of experimentally infected swine. *J. Food Protect.* 51:691–695.

55. Kotula, A. W. 1983. Postslaughter control of *Trichinella spiralis*. *Food Technol.* 37(3):91–94.

56. Kotula, A. W., K. D. Murrell, L. Acosta-Stein, L. Lamb, and L. Douglass. 1983. Destruction of *Trichinella spiralis* during cooking. *J. Food Sci.* 48:765–768.

57. Lopez, C. E., D. D. Juranek, S. P. Sinclair, and M. G. Schultz. 1978. Giardiasis in American travelers to Madeira Island, Portugal. *Am. J. Trop. Med. Hyg.* 27:1128–1132.

58. Margolis, L. 1977. Public health aspects of "codworm" infection: A review. *J. Fish. Res. Bd. Canada* 34:887–898.

59. Marquardt, W. C., and R. S. Demaree. 1985. *Parasitology*. New York: Macmillan.

60. Meyer, E. A., ed. 1990. *Giardiasis*. Amsterdam: Elsevier.

61. Murrell, K. D. 1985. Strategies for the control of human trichinosis transmitted by pork. *Food Technol.* 39(3):65–68, 110.

62. Myers, B. J. 1979. Anisakine nematodes in fresh commercial fish from waters along the Washington, Oregon and California coasts. *J. Food Protect.* 42:380–384.

63. Myers, B. J. 1976. Research then and now on the anisakidae nematodes. *Trans. Amer. Microscop. Soc.* 95:137–142.

64. Ongerth, J. E., and H. H. Stibbs. 1987. Identification of *Cryptosporidium* oocysts in river water. *Appl. Environ. Microbiol.* 53:672–676.

65. Osterholm, M. T., J. C. Forfang, T. L. Ristinen, A. G. Daan, J. W. Washburn, J. R. Godes, R. A. Rude, and J. G. McCullough. 1981. An outbreak of foodborne giardiasis. *N. Engl. J. Med.* 304:24–28.

66. Petersen, L. R., M. L. Cartter, and J. L. Hadler. 1988. A foodborne outbreak of *Giardia lamblia*. *J. Inf. Dis.* 157:846–848.

67. Piekarski, G. 1989. *Medical Parasitology*. New York: Springer-Verlag.

68. Plorde, J. J. 1984. Sporozoan infections. In *Medical Microbiology: An Introduction to Infectious Diseases*, ed. J. C. Sherris, K. J. Ryan, C. G. Ray, J. J. Plorde, L. Corey, and J. Spizizen, 469–483. New York: Elsevier.

69. Rendtorff, R. C. 1954. The experimental transmission of human intestinal protozoan parasites. II. *Giardia lamblia* cysts given in capsules. *Am. J. Hyg.* 59:209–220.

70. Riemann, H. P., M. E. Meyer, J. H. Theis, G. Kelso, and D. E. Behymer. 1975. Toxoplasmosis in an infant fed unpasteurized goat milk. *J. Pediat.* 87:573–576.

71. Rodrick, G. E., and T. C. Cheng. 1989. Parasites: Occurrence and significance in marine animals. *Food Technol.* 43(11):98–102.

72. Rommel, M. and A.-O. Heydorn. 1972. Beiträge zum Lebenszyklus der Sarkosporidien. III. *Isospora hominis* (Railliet and Lucet, 1891) Wenyon, 1923, eine Dauerform der Sarkosporidien des Rindes und des Schweins. *Berl. Munchen. Tierarztl. Wochens.* 85:143–145.

73. Rosset, J. S., K. D. McClatchey, G. I. Higashi, and A. S. Knisely. 1982. *Anisakis* larval type I in fresh salmon. *Am. J. Clin. Pathol.* 78:54–57.

74. Ruttenber, A. J., B. G. Weniger, F. Sorvillo, R. A. Murray, and S. L. Ford. 1984.

Diphyllobothriasis associated with salmon consumption in Pacific coast states. *Am. J. Trop. Med. Hyg.* 33(3):455–459.

75. Sacks, J. J., R. R. Roberto, and N. F. Brooks. 1982. Toxoplasmosis infection associated with raw goat's milk. *J. Amer. Med. Assoc.* 248:1728–1732.

76. Salata, R. A., A. Martinez-Palomo, L. Canales, H. W. Murray, N. Trevino, and J. I. Ravdin. 1990. Suppression of T-lymphocyte responses to *Entamoeba histolytica* antigen by immune sera. *Infect. Immun.* 58:3941–3946.

77. Schantz, P. M. 1983. Trichinosis in the United States—1947–1981. *Food Technol.* 37(3):83–86.

78. Sherwood, D., K. W. Angus, D. R. Snodgrass, and S. Tzipori. 1982. Experimental cryptosporiodiosis in laboratory mice. *Infect. Immun.* 38:471–475.

79. Smith, P. D. 1989. *Giardia lamblia.* In *Parasitic Infections in the Compromised Host*, ed. P. D. Walzer and R. M. Genta, 343–384. New York: Marcel Dekker.

80. Smith, H. V., R. W. A. Girdwood, W. J. Patterson, R. Hardie, L. A. Green, C. Benton, W. Tulloch, J. C. M. Sharp, and G. I. Forbes. 1988. Waterborne outbreak of cryptosporiodiosis. *Lancet* ii:1484.

81. Sole, T. D., and N. A. Croll. 1980. Intestinal parasites in man in Labrador, Canada. *Am. J. Trop. Med. Hyg.* 29(3):364–368.

82. Sterling, C. R., and M. J. Arrowood. 1986. Detection of *Cryptosporidium* sp. infections using a direct immunofluorescent assay. *Ped. Inf. Dis.* 5:139–142.

83. Sterling, C. R., K. Seegar, and N. A. Sinclair. 1986. *Cryptosporidium* as a causative agent of traveler's diarrhea. *J. Inf. Dis.* 153:380–381.

84. Todd, E. C. D. 1989. Foodborne and waterborne disease in Canada—1983 annual summary. *J. Food Protect.* 52:436–442.

85. Tzipori, S. 1988. Cryptosporidiosis in perspective. *Adv. Parasitol.* 27:63–129.

86. U.S. Department of Agriculture. 1982. USDA advises cooking pork to 170 degrees Fahrenheit throughout. News release, USDA, Washington, D.C.

87. Von Bonsdorff, B. 1977. *Diphyllobothriasis in Man.* New York: Academic Press.

88. Wallis, P. M., and B. R. Hammond, eds. 1988. *Advances in Giardia Research.* Calgary, Canada: University of Calgary Press.

89. Walsh, J. A. 1986. Problems in recognition and diagnosis of amebiasis: Estimation of the global magnitude of morbidity and mortality. *Rev. Infect. Dis.* 8:228–238.

90. Zimmermann, W. J. 1983. An approach to safe microwave cooking of pork roasts containing *Trichinella spiralis. J. Food Sci.* 48:1715–1718, 1722.

91. Zimmermann, W. J., and P. J. Beach. 1982. Efficacy of microwave cooking for devitalizing trichinae in pork roasts and chops. *J. Food Protect.* 45:405–409.

92. Zimmermann, W. J., D. G. Olson, A. Sandoval, and R. E. Rust. 1985. Efficacy of freezing in eliminating infectivity of *Trichinella spiralis* in boxed pork products. *J. Food Protect.* 48:196–199.

25

Other Proved and Suspected Foodborne Agents: Mycotoxins, Viruses, *Aeromonas*, *Plesiomonas*, Scombroid, and Paralytic Shellfish Poisonings

MYCOTOXINS

A very large number of molds have been demonstrated to produce toxic substances designated mycotoxins. Some are mutagenic and carcinogenic, some display specific organ toxicity, and some are toxic by other mechanisms. While the clear-cut toxicity of many mycotoxins for humans has not been demonstrated, the effect of these compounds on experimental animals and their effect in in vitro assay systems leaves little doubt about their real and potential toxicity for humans. At least 14 mycotoxins are carcinogens, with the aflatoxins being the most potent (131). It is generally accepted that about 93% of mutagenic compounds are carcinogens. With mycotoxins, microbial assay systems reveal an 85% level of correlation between carcinogenicity and mutagenesis (131).

Mycotoxins are produced as secondary metabolites. The primary metabolites of fungi as well as for other organisms are those compounds that are essential for growth. Secondary metabolites are formed during the end of the exponential growth phase and have no apparent significance to the producing organism relative to growth or metabolism. In general, it appears that they are formed when large pools of primary metabolic precursors such as amino acids, acetate, pyruvate, and so on accumulate, and the synthesis of mycotoxins represents one way the fungus has of reducing the pool of metabolic precursors that its metabolic needs no longer require. In the case of aflatoxin synthesis, it begins at the onset of the stationary phase of growth and occurs along with that of lipid synthesis (125). The biosynthetic pathways of the aflatoxins have been reviewed by Magoon et al. (93). Of the three primary routes of secondary metabolism in fungi (polyketide, terpenoid, and that utilizing essential amino acids), mycotoxins are synthesized via the polyketide route.

Presented below are some of the mycotoxins demonstrated to occur in foods. More extensive treatments of this subject have been presented by Davis

641

and Diener (46), Stark (131), and Bullerman (16). Methods for detecting mycotoxins have been presented by Bullerman (19). For immunoassay methods, see Chu (38) and Chapter 6.

Aflatoxins

These are clearly the most widely studied of all mycotoxins. Knowledge of their existence dates from 1960, when more than 100,000 turkey poults died in England after eating peanut meal imported from Africa and South America. From the poisonous feed were isolated *Aspergillus flavus* and a toxin produced by this organism that was designated aflatoxin (*Aspergillus flavus* toxin—A-fla-toxin). Studies on the nature of the toxic substances revealed the following four components:

It was later determined that *A. parasiticus* produces aflatoxins. These compounds are highly substituted coumarins, and at least 18 closely related toxins are known. Aflatoxin B_1 (AFB$_1$) is produced by all aflatoxin-positive strains, and it is the most potent of all. AFM$_1$ is a hydroxylated product of AFB$_1$ and appears in milk, urine, and feces as metabolic products (52). AFL, AFLH$_1$, AFQ$_1$, and AFP$_1$ are all derived from AFB$_1$. AFB$_2$ is the 2,3-dehydro form of AFB$_1$, and AFG$_2$ is the 2,3-dihydro form of AFG$_1$. The toxicity of the six most potent aflatoxins decreases in the following order: $B_1 > M_1 > G_1 > B_2 > M_2 \neq G_2$ (7). When viewed under ultraviolet (UV) light, six of the toxins fluoresce as noted:

B_1 and B_2—blue
G_1—green
G_2—green-blue
M_1—blue-violet
M_2—violet

The proposed partial pathway for AFB_1 synthesis is as follows (see 113): Acetate > norsolorinic acid > averantin > averufanin > averufin > versiconal hemiacetal acetate > versicolorin A > sterigmatocystin > O-methylsterigmato-cystin > AFB_1. Versicolorin A is the first in the pathway to contain the essential C_2-C_3 double bond (113).

Requirements for Growth and Toxin Production.   No aflatoxins were produced by 25 isolates of *A. flavus/parasiticus* on wort agar at 2°, 7°, 41°, or 46°C within 8 days, and none was produced under 7.5 or over 40°C even under otherwise favorable conditions (120). In another study employing Sabouraud's agar, maximal growth of *A. flavus* and *A. parasiticus* occurred at 33°C when pH was 5.0 and a_w was 0.99 (71). At 15°C, growth occurred at a_w 0.95 but not at 0.90, while at 27° and 33°C, slight growth was observed at an a_w of 0.85. The optimum temperature for toxin production has been found by many to be between 24° and 28°C. In one study, maximal growth of *A. parasiticus* was 35°C, but the highest level of toxin was produced at 25°C (124).

The limiting moisture content for AFB_1 and AFB_2 on corn was 17.5% at a temperature of 24°C or higher, with up to 50 ng/g being produced (139). No toxin was produced at 13°C. Overall, toxin production has been observed over the a_w range of 0.93 to 0.98, with limiting values variously reported as being 0.71 to 0.94 (94). In another study, no detectable quantities of AFB_1 were

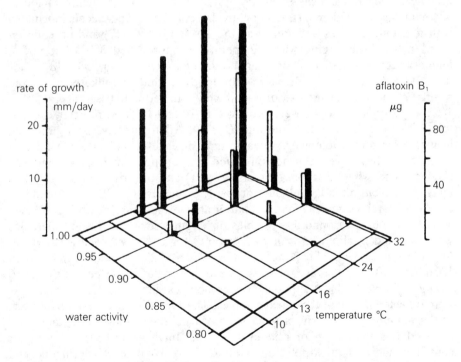

FIGURE 25-1.   Growth and aflatoxin B_1 production on malt extract-glycerine agar at various water activity values and temperatures. White columns: rate of growth; black columns: average AFB_1 production.
From Northolt et al. (105); copyright © 1976 by International Association of Milk, Food and Environmental Sanitarians.

formed by *A. parasiticus* at a_w of 0.83 and 10°C (105). The optimum temperature at a_w 0.94 was 24°C (Fig. 25-1). Growth without demonstrable toxin appeared possible at a_w 0.83 on malt agar–containing sucrose. It has been observed by several investigators that rice supports the production of high levels of aflatoxins at favorable temperatures but none is produced at 5°C on either rice or cheddar cheese (108).

Overall, the minimal and maximal parameters that control growth and toxin production by these eukaryotic organisms are not easy to define, in part because of their diverse habitats in nature and in part because of their eukaryotic status. It seems clear that growth can occur without toxin production.

It has been reported that AFG_1 is produced at lower growth temperatures than AFB_1, and while some investigators have found more AFB_1 than AFG_1 at around 30°C, others have found equal production. With regard to *A. flavus* and *A. parasiticus*, the former generally produces only AFB and AFG (46). Aeration favors aflatoxin production, and amounts of 2 mg/g can be produced on natural substrates such as rice, corn, soybeans, and the like (46). Up to 200 to 300 mg/liter can be produced in broth containing appropriate levels of Zn^{2+}. Aflatoxin synthesis can be inhibited by caffeine, which apparently inhibits the uptake of carbohydrates (12). The release of AFB_1 by *A. flavus* appears to involve an energy-dependent transport system (2).

Production and Occurrence in Foods. With respect to production in foods, aflatoxin has been demonstrated on fresh beef, ham, and bacon inoculated with toxigenic cultures and stored at 15°, 20°, and 30°C (17) and on country-cured hams during aging when temperatures approached 30°C, but not at temperatures less than 15°C or relative humidity (R.H.) over 75% (18). They have been found in a wide variety of foods, including milk, beer, cocoa, raisins, soybean meal, and so on (see below). In fermented sausage at 25°C, 160 and 426 ppm of AFG_1 were produced in 10 and 18 days, respectively, and ten times more AFG_1 was found than B_1 (90). Aflatoxins have been produced in whole-rye and whole-wheat breads, in tilsit cheese, and in apple juice at 22°C (55). They have been demonstrated in the upper layer of 3-month-old cheddar cheese held at room temperature (91) and on brick cheese at 12.8°C by *A. parasiticus* after 1 week but not for *A. flavus* (124). AFB_1 was found in 3 of 63 commercial samples of peanut butter at levels less than 5 ppb (112). From a 5-year survey of around 500 samples of Virginia corn and wheat, aflatoxins were detected in about 25% of corn samples for every crop year, with 18 to 61% of samples containing 20 ng/g or more and 5 to 29% containing more than 100 ng/g (126). The average quantity detected over the 5-year period was 21 to 137 ng/g (Table 25-1). Neither aflatoxins nor zearalenone and ochratoxin A were detected in any of the wheat samples. The 1988 drought led to an increase in the amount of aflatoxin produced in corn in some midwestern states that received less than 2 in of rain in June and July. About 30% of samples contained more than 20 ppb compared to 2 to 3 ppb levels during normal rainfall (130).

The effect of temperature cycling between 5° and 25°C on production in rice and cheese has been investigated. *A. parasiticus* produced more toxin under cycling temperatures than at 15°, 18°, or 25°C, while *A. flavus* produced less

TABLE 25-1. Aflatoxin Levels in Dent Corn Grown in Virginia, 1976–1980

Total Aflatoxin, ng/g	Collected from Trucks by Federal Grain Inspection Service (FGIS)													Collected at Harvest by Statistical Reporting Service (SRS)			
	1976		1977		1978		1979		1980					1978		1979	
	Number of Samples	(%)	Number of Samples	(%)	Number of Samples	(%)	Number of Samples	(%)	Number of Samples	(%)				Number of Samples	(%)	Number of Samples	(%)
ND[a]	77	(63)	52	(51)	63	(64)	81	(71)	18	(18)				79	(88)	93	(79)
<20	13	(10)	17	(17)	10	(10)	13	(11)	20	(20)				2	(2)	9	(8)
20–100	21	(17)	18	(18)	21	(21)	8	(7)	32	(32)				5	(6)	7	(6)
101–500	9	(7)	10	(10)	5	(5)	10	(9)	26	(26)				4	(4)	7	(6)
501–1000	1	(1)	1	(1)	—	—	2	(2)	1	(1)				—	—	—	—
>1000	2	(2)	3	(3)	—	—	—	—	2	(2)				—	—	1	(1)
Total	123		101		99		114		99					90		117	
% Incidence	37		49		36		29		82					12		21	
% ≥ 20 ng/g	27		32		26		18		61					10		13	
% > 100 ng/g	10		14		5		11		29					4		7	
Average level (ng/g), all samples	48		91		21		34		137					13		36	
Average level (ng/g), positive samples	130		187		58		118		167					110		176	

Source: Shotwell and Hesseltine (126); copyright © 1983 by Association of Official Analytical Chemists.

[a]ND = not detected.

under these conditions (108). On cheddar cheese, however, less aflatoxin was produced than at 25°C, and these investigators noted that cheese is not a good substrate for aflatoxin production if it is held much below the optimum temperature.

Aflatoxin production has been demonstrated to occur on an endless number of food products in addition to those noted. Under optimal conditions of growth, some toxin can be detected within 24 h—otherwise within 4 to 10 days (37). On peanuts, Hesseltine (70) has made the following observations:

- Growth and formation of aflatoxin occur mostly during the curing of peanuts after removal from soil.
- In a toxic lot of peanuts, only comparatively few kernels contain toxin, and success in detecting the toxin depends on collecting a relatively large sample, such as 1 kg, for assay.
- The toxin will vary greatly in amount even within a single kernel.
- The two most important factors affecting aflatoxin formation are moisture and temperature.

Aflatoxin production on corn by *A. parasiticus* has been shown to be inhibited by sodium bicarbonate (99), while in culture media, its synthesis was increased by ascorbic acid (109). A large number of other compounds that affect the synthesis of aflatoxins has been compiled and reviewed (152).

The U.S. Food and Drug Administration (FDA) has established allowable action levels of aflatoxins in foods as follows: 20 ppb for food, feeds, Brazil nuts, peanuts, peanut products, and pistachio nuts and 0.5 ppb for milk (see 87 for further information on regulations).

Relative Toxicity and Mode of Action.   For the expression of mutagenicity, mammalian metabolizing systems are essential for aflatoxins, especially AFB_1. Also essential is their binding with nucleic acids, especially DNA. While nuclear DNA is normally affected, AFB_1 has been shown to bind covalently to liver mitochondrial DNA preferential to nuclear DNA (103). Cellular macromolecules other than nucleic acids are possible sites for aflatoxins. The site of the aflatoxin molecule responsible for mutagenicity is the C_2-C_3 double bond in the dihydrofurofuran moiety. Its reduction to the 2,3-dihydro (AFB_2) form reduces mutagenicity by 200 to 500-fold (131). Following binding to DNA, point mutations are the predominant genetic lesions induced by aflatoxins, although frameshift mutations are known to occur. The mutagenesis of AFB_1 has been shown to be potentiated twofold by BHA and BHT and much less by propyl gallate employing the Ames assay, but whether potentiation occurs in animal systems is unclear (122).

The LD_{50} of AFB_1 for rats by the oral route is 1.2 mg/kg, and 1.5 to 2.0 mg/kg for AFG_1 (22). The relative susceptibility of various animal species to aflatoxins is presented in Table 25-2. Young ducklings and young trout are among the most sensitive, followed by rats and other species. Most species of susceptible animals die within 3 days after administration of toxins and show gross liver damage, which, upon postmortem examination, reveals the aflatoxins to be hepatocarcinogens (149). The toxicity is higher for young animals and males than for older animals and females, and the toxic effects are enhanced by low protein or cirrhogenic diets.

Circumstantial evidence suggests that aflatoxins are carcinogenic to humans.

TABLE 25-2.   Comparative Lethality of Single Doses of Aflatoxin B$_1$

Animal	Age (or Weight)	Sex	Route	LD$_{50}$ mg/kg
Duckling	1 day	M	PO	0.37
	1 day	M	PO	0.56
Rat	1 day	M-F	PO	1.0
	21 days	M	PO	5.5
	21 days	F	PO	7.4
	100 g	M	PO	7.2
	100 g	M	IP	6.0
	150 g	F	PO	17.9
Hamster	30 days	M	PO	10.2
Guinea pig	Adult	M	IP	ca. 1
Rabbit	Weanling	M-F	IP	ca. 0.5
Dog	Adult	M-F	IP	ca. 1
	Adult	M-F	PO	ca. 0.5
Trout	100 g	M-F	PO	ca. 0.5

Source: Wogan (149).

Notes: PO = oral; IP = intraperitoneal.

Among conditions believed to result from aflatoxins is the EFDV syndrome of Thailand, Reye's syndrome of Thailand and New Zealand (23), and an acute hepatoma condition in a Ugandan child. In the last a fatal case of acute hepatic disease revealed histological changes in the liver identical to those observed in monkeys treated with aflatoxins, and an aflatoxin etiology was strongly suggested by the findings (121). Two researchers who worked with purified aflatoxin developed colon carcinoma (49). On the other hand, it has been noted that no mycotoxin has been linked with a specific cancer in humans in the absence of chronic infection with hepatitis B virus (134). Although some mycotoxins are extremely toxic to the young of many animal species, the view exists that their toxicity for humans is overstated.

Degradation.   AFB$_1$ and AFB$_2$ can be reduced in corn by bisulfite. When dried figs were spiked with 250 ppb of AFB$_1$ and subjected to several treatments, 1% sodium bisulfite effected a 28.2% reduction in 72 h; 0.2% H_2O_2 (added 10 min before sodium bisulfite) effected 65.5% reduction; heating at 45° to 65°C for 1 h 68.4%; and UV radiation effected a 45.7% reduction (3). Aflatoxin-contaminated cottonseed treated with ammonia and fed to cows led to lower levels of AFB$_1$ and AFM$_1$ in milk than nontreated product (74). When yellow dent corn naturally contaminated with 1,600 ppm aflatoxin was treated with 3% NaOH at 100°C for 4 min, further processed and fried, 99% of the aflatoxin was destroyed (24).

Alternaria Toxins

Several species of *Alternaria* (including *A. citri*, *A. alternata*, *A. solani*, and *A. tenuissima*) produce toxic substances that have been found in apples, tomatoes, blueberries, and others. The toxins produced include alternariol, alternariol monomethyl ether, altenuene, tenuazonic acid, and altertoxin-I. On slices of apples, tomatoes, or crushed blueberries incubated for 21 days at 21°C, several *Alternaria* produced each of the toxins noted at levels up to 137 mg/100 g (132). In another study, tenuazonic acid was the main toxin produced in tomatoes, with levels as high as 13.9 mg/100 g; on oranges and lemons, *A. citri* produced tenuazonic acid, alternariol, and alternariol monomethyl ether at a mean concentration of 1.15 to 2.66 mg/100 g (133). The fruits were incubated at room temperature for 21 to 28 days. More recently, an *A. alternata* strain has been shown to produce **stemphyltoxin III**, which was found to be mutagenic by the Ames assay (48). More information on the alternaria toxins can be obtained from (79).

Citrinin

This mycotoxin is produced by *Penicillium citrinum*, *P. viridicatum*, and other fungi. It has been recovered from polished rice, moldy bread, countrycured

Citrinin

hams, wheat, oats, rye, and other similar products. Under long-wave UV light, it fluoresces lemon yellow. It is a known carcinogen. Of seven strains of *P. viridicatum* recovered from country-cured hams, all produced citrinin in potato dextrose broth and on country-cured hams in 14 days at 20° to 30°C but not at 10°C (150). Growth was found to be poor at 10°C. While citrinin-producing organisms are found on cocoa and coffee beans, this mycotoxin as well as others is not found to the extent of growth. The apparent reason is the inhibition of citrinin in *P. citrinum* by caffeine. The inhibition of citrinin appears to be rather specific, since only a small decrease in growth of the organims occurs (13).

Ochratoxins

The ochratoxins consist of a group of at least seven structurally related secondary metabolites of which ochratoxin A (OA) is the best known and the most toxic. OB is dechlorinated OA and along with OC, it may not occur naturally. OA is produced by a large number of storage fungi, including *A. ochraceus*, *A. alliaceus*, *A. ostianus*, *A. mellus*, and other species of aspergilli. Among penicillia that produce OA are *P. viridicatum*, *P. cyclopium*, *P. variable*, and others.

OA is produced maximally at around 30°C and a_w 0.95 (39). The minimum a_w supporting OA production by *A. ochraceus* at 30°C in poultry feed is 0.85

Ochratoxin A

(9). Its oral LD_{50} in rats is 20 to 22 mg/kg, and it is both hepatotoxic and neprotoxic.

This mycotoxin has been found in corn, dried beans, cocoa beans, soybeans, oats, barley, citrus fruits, Brazil nuts, moldy tobacco, country-cured hams, peanuts, coffee beans, and other similar products. Two strains of *A. ochraceus* isolated from country-cured hams produced OA and OB on rice, defatted peanut meal, and when inoculated into country-cured hams (53). Two-thirds of the toxin penetrated to a distance of 0.5 cm after 21 days, with the other one-third located in the mycelial mat. Of six strains of *P. viridicatum* recovered from country-cured hams, none produced ochratoxins. From a study of four chemical inhibitors of both growth and OA production by two OA producers at pH 4.5, the results were potassium sorbate > sodium propionate > methyl paraben > sodium bisulfite; while at a pH of 5.5, the most effective two were methyl paraben and potassium sorbate (138). Like most other mycotoxins, OA is heat stable. In one study, the highest rate of destruction achieved by cooking faba beans was 20%, and the investigators concluded that OA could not be destroyed by normal cooking procedures (51). Under UV light, OA fluoresces greenish, while OB emits blue fluorescence. It induces abnormal mitosis in monkey kidney cells.

Patulin

Patulin (clavicin, expansin) is produced by a large number of penicillia, including *P. claviforme*, *P. expansum*, *P. patulum*; by some aspergilli (*A. clavatus*, *A. terreus* and others), and by *Byssochlamys nivea* and *B. fulva*.

Patulin

Its biological properties are similar to those of penicillic acid. Some patulin-producing fungi can produce the compound below 2°C (7). This mycotoxin has been found in moldy bread, sausage, fruits (including bananas, pears, pineapples, grapes, and peaches), apple juice, cider, and other products. In apple juice, levels as high as 440 μg/liter have been found, and in cider, levels up to 45 ppm.

Minimum a_w for growth of *P. expansum* and *P. patulum* has been reported to be 0.83 and 0.81, respectively. In potato dextrose broth incubated at 12°C, patulin was produced after 10 days by *P. patulum* and *P. roqueforti*, with the former organism producing up to 1,033 ppm (17). Patulin was produced in apple juice also at 12°C by *B. nivea*, but the highest concentration was attained after 20 days at 21°C after a 9-day lag (115). The next highest amount was produced at 30°C, with much less at 37°C. These investigators confirmed that patulin production is favored at temperatures below the growth optimum, as was previously found by Sommer et al. (129). The latter investigators used *P. expansum* and found production over the range 5° to −20°C, with only small amounts produced at 30°C. In five commercial samples in Georgia, patulin levels from 244 to 3993 µg/L were found, with a mean of 1902 µg/L (146). The overall incidence of patulin in apple juice has been reviewed (66). Atmospheres of CO_2 and N_2 reduced production compared to that in air. To inhibit production, SO_2 was found more effective than potassium sorbate or sodium benzoate (115).

The LD_{50} for patulin in rats by the subcutaneous route has been reported to be 15 to 25 mg/kg, and it induces subcutaneous sarcomas in some animals. Both patulin and penicillic acid bind to —SH and —NH_2 groups, forming covalently linked adducts that appear to abate their toxicities. Patulin causes chromosomal aberrations in animal and plant cells and is a carcinogen.

Penicillic Acid

This mycotoxin has biological properties similar to patulin. It is produced by a large number of fungi, including many penicillia (*P. puberulum*, for example),

Penicillic acid

as well as members of the *A. ochraceus* group. One of the best producers is *P. cyclopium*. It has been found in corn, beans, and other field crops and has been produced experimentally on swiss cheese. Its LD_{50} in mice by subcutaneous route is 100 to 300 mg/kg, and it is a proved carcinogen.

Of 346 penicillia cultures isolated from salami, about 10% produced penicillic acid in liquid culture media, but 5 that were inoculated into sausage failed to produce toxin after 70 days (40). In another study, some 183 molds were isolated from swiss cheese; 87% were penicillia, 93% of which were able to grow at 5°C. Thirty-five percent of penicillia extracts were toxic to chick embryos, and from 5.5% of the toxic mixtures were recovered penicillic acid as well as patulin and aflatoxins (14). Penicillic acid was produced at 5°C in 6 weeks by 4 of 33 fungal strains.

Sterigmatocystin

These mycotoxins are structurally and biologically related to the aflatoxins, and like the latter, they cause hepatocarcinogenic activity in animals. At least eight derivatives are known. Among the organisms that produce them are *Aspergillus versicolor*, *A. nidulans*, *A. rugulosus*, and others. The LD_{50} for rats by intraperitoneal injection of 60 to 65 mg/kg. Under UV light, the toxin fluoresces dark brick-red. Although not often found in natural products, they have been found in wheat, oats, Dutch cheese, and coffee beans. While related to the aflatoxins, they are not as potent. They act by inhibiting DNA synthesis.

Zearalenone

There are at least five naturally occurring zearalenones, and they are produced by *Fusarium* spp., mainly *F. graminearum* (formerly *F. roseum*, = *Gibberella*

Zearalenone

zeae) and *F. tricinctum*. Associated with corn, these organisms invade field corn at the silking stage, especially during heavy rainfall. If the moisture levels remain high enough following harvesting, the fungi grow and produce toxin. Other crops, such as wheat, oats, barley, and sesame, may be affected in addition to corn.

The toxins fluoresce blue-green under long-wave UV and greenish under short-wave UV. They possess estrogenic properties and promote estrus in mice and hyperestrogenism in swine. While they are nonmutagenic in the Ames assay, they produce a positive response in the *Bacillus subtilis* Rec assay (131). It was not detectable in 12 commercial milk samples assayed by an immunoaffinity-ELISA method (8).

VIRUSES

Much less is known about the incidence of viruses in foods than about bacteria and fungi, for several reasons. First, being obligate parasites, viruses do not grow on culture media as do bacteria and fungi. The usual methods for their cultivation consist of tissue culture and chick embryo techniques. Second, since viruses do not replicate in foods, their numbers may be expected to be low relative to bacteria, and extraction and concentration methods are necessary for their recovery. Although much research has been devoted to this methodology, it is difficult to effect more than about a 50% recovery of virus particles from products such as ground beef. Third, laboratory virological techniques are not practiced in many food microbiology laboratories. Finally, not all viruses of potential interest to food microbiologists can be cultured by existing methods (the Norwalk virus is one example). In spite of these reasons, there does exist a growing body of information on viruses in foods.

Since it has been demonstrated that any intestinal bacterial pathogen may under unsanitary conditions be found in foods, the same may be presumed for intestinal viruses, even though they may not proliferate in foods. Cliver et al. (42) have noted that virtually any food can serve as a vehicle for virus transmission, and they have stressed the importance of the anal-oral mode of transmission, especially for viral hepatitis of food origin. Just as nonintestinal bacteria of human origin are sometimes found in foods, the same may be true for viruses, but because of their tissue affinities, foods would serve as vehicles only for the intestinal or enteroviruses. These agents may be accumulated by some shellfish up to 900-fold (58). Viral gastroenteritis is believed to be second only to the common cold in frequency (45).

A grouping of the viruses that may be found in foods is presented in Table 25-3. Other than hepatitis A and rotaviruses, those of primary concern are 28 to 35 nm in diameter and include the Norwalk, calici-, and astroviruses, and they are discussed under Norwalk agents below. For further general information, see reviews by Gerba et al. (58, 60), Cliver (42, 44) and a Centers for Disease Control (CDC) report (27). Methods for virus recovery from foods have been outlined by Cliver et al. (43), Sobsey et al. (128), and the CDC (27), and a thorough review of virus uptake by shellfish is that of Richards (114).

Incidence in Foods and the Environment.   The most common food source of gastroenteritis-causing viruses is shellfish. While crustaceans do not concentrate viruses, molluscan shellfish do because they are filter feeders. When poliovirus 1 was added to waters, blue crabs were contaminated, but they did not concentrate the virus (68). Shucked oysters artificially contaminated with 10^4 pfu (plaque-forming units) of a poliovirus retained viruses during refrigeration for 30 to 90 days with a survival rate of 10 to 13% (50). It has been reported that the uptake of enteroviruses by oysters and clams is not likely when viruses in the water column are less than 0.01 pfu/ml (88). The recovery method employed by the latter authors was capable of detecting 1.5 to 2.0 pfu/shellfish.

Although the coliform index is of proved value as an indicator of intestinal

TABLE 25-3.   **Human Intestinal Viruses with High Potential as Food Contaminants**

1. Picornaviruses
    Polioviruses 1-3
    Coxsackievirus A 1-24
    Coxsackievirus B 1-6
    ECHOvirus 1-34
    Enterovirus 68-71
    Probably hepatitis A
2. Reoviruses
    Reovirus 1-3
    Rotaviruses
3. Parvoviruses
    Human gastrointestinal viruses
4. Papovaviruses
    Human BK and JC viruses
5. Adenoviruses
    Human adenoviruses types 1-33

Source: Larkin (89); copyright © 1981 by International Association of Milk, Food and Environmental Sanitarians.

bacterial pathogens in waters, it appears to be inadequate for enteroviruses, which are more resistant to adverse environmental conditions than bacterial pathogens (111). In a study of more than 150 samples of recreational waters from the upper Texas Gulf, enteroviruses were detected 43% of the time when by coliform index the samples were judged acceptable, and 44% of the time when judged acceptable by fecal coliform standards (59). In the same study, enteroviruses were found 35% of the time in waters that met acceptable standards for shellfish harvesting, and the investigators concluded that the coliform standard for waters does not reflect the presence of viruses. From a study of hard-shell clams off the coast of North Carolina, enteric viruses were found in those from open and closed beds (144). (Closed waters are those not open to commercial shellfishing because of coliform counts.) From open beds, 3 of 13 100-g samples were positive for viruses, while all 13 were negative for salmonellae, shigellae, or yersiniae. From closed waters, 6 of 15 were positive for salmonellae, and all were negative for shigellae and yersiniae (144). The latter investigators found no correlation between enteric viruses and total coliforms or fecal coliforms in shellfish waters, or total coliforms, fecal coliforms, "fecal streptococci," or aerobic plate counts (APC) in clams. Although enteric viruses may be found in shellfish from open waters, less than 1% of shellfish samples examined by the FDA contained viruses (89). (See Chap. 17 for further discussion of safety indicators and viruses.)

With respect to the capacity of certain viruses to persist in foods, it has been shown that enteroviruses persisted in ground beef up to 8 days at 23° or 24°C and were not affected by the growth of spoilage bacteria (69). In a study of 14 vegetable samples for the existence of naturally occurring viruses, none were found, but coxsackievirus B5 inoculated onto vegetables did survive at 4°C for 5 days (82). In an earlier study, these investigators showed that coxsackievirus B5 showed no loss of activity when added to lettuce and stored at 4°C under moist conditions for 16 days. Several enteric viruses failed to survive on the surfaces of fruits, and no naturally occurring viruses were found in nine fruits examined (83). ECHO virus 4 and poliovirus 1 were found in 1 each of 17 samples of raw oysters by Fugate et al. (57), and poliovirus 3 was found in 1 of 24 samples of oysters. Of seven food-processing plants surveyed for human viruses, none was found in a vegetable-processing plant or in three that processed animal products (84). The latter investigators examined 60 samples of market foods but were unable to detect viruses in any. They concluded that viruses in the U.S. food supply are very low.

Destruction in Foods.   The survival of the hog cholera (HCV) and African swine fever (ASFV) viruses in processed meats was studied by McKercher et al. (96). From pigs infected with these viruses, partly cooked canned hams and dried pepperoni and salami sausages were prepared; while virus was not recovered from the partly cooked canned hams, they were recovered from hams after brining but not after heating. The ASFV retained viability in the two sausage products following addition of curing ingredients and starters but were negative after 30 days. HCV also survived the addition of curing ingredients and starter and retained viability 22 days later.

The effect of heating on destruction of the foot-and-mouth virus was evaluated by Blackwell et al. (11). When ground beef was contaminated with virus-infected lymph node tissue and processed to an internal temperature of 93.3°C, the virus was destroyed. However, in cattle lymph node tissue, virus survived for 15 but not 30 min at 90°C. The boiling of crabs was found sufficient to inactivate 99.9% of poliovirus 1, and a rotavirus and an ECHOvirus was destroyed within 8 min (68). A poliovirus was found to survive stewing, frying, baking, and steaming of oysters (50). In broiled hamburgers, enteric viruses could be recovered from 8 of 24 patties cooked rare (to 60°C internally) if the patties were cooled immediately to 23°C (135). No viruses were detected if the patties were allowed to cool for 3 min at room temperature before testing.

Hepatitis A Virus

There are more documented outbreaks of hepatitis A traced to foods than any other viral infection. The virus belongs to the family Picornaviridae, as do the polio, ECHO, and coxsackie viruses, and all have single-stranded RNA genomes. The incubation period for infectious hepatitis ranges from 15 to 45 days, and lifetime immunity usually occurs after an attack. The fecal-oral route is the mode of transmission, and raw or partially cooked shellfish from polluted waters is the most common vehicle food.

In the United States in 1973, 1974, and 1975, there were five, six, and three outbreaks respectively, with 425, 282, and 173 cases. The 1975 outbreaks were traced to salad, sandwiches, and glazed doughnuts served in restaurants. The recorded outbreaks and cases in the United States for 1983 through 1987 are presented in Table 25-4. According to the CDC, hepatitis A increased in the United States between 1983 and 1989 by 58%—from 9.2 to 14.5 per 100,000 persons (28). Of the cases in 1988, 7.3% were either food- or waterborne (28). Synopses of some of the early outbreaks are presented in Table 25-5.

Norwalk and Related Viruses

These viruses are referred to as *small, round structural viruses (SRSVs)* and also as Norwalk agents. They belong to a large and heterogenous family, and there is some evidence that they belong to the family Caliciviridae. They are identified by the name of the locale of their initial isolation. The Norwalk virus was first recognized in a school outbreak in Norwalk, Ohio, in 1968, and water was suspected but not proved as the source. It is the prototype of SRSVs.

TABLE 25-4.   Outbreaks, Cases, and Deaths Associated with Viral Foodborne Gastroenteritis in the United States, 1983–1987

Years	Outbreaks/Cases/Deaths		
	Hepatitis A	Norwalk Agent	Other Viruses
1983	10/530/1	1/20/0	—
1984	2/29/0	1/137/0	1/444/0
1985	5/118/0	4/179/0	1/114/0
1986	3/203/0	3/463/0	—
1987	9/187/0	1/365/0	—
Totals	29/1,067/1	10/1,164/0	2/558/0

Source: N. H. Bean, P. M. Griffin, J. S. Goulding, and C. B. Ivey. *J. Food Protect.* 53:711–728, 1990.

TABLE 25-5.   Synopsis of Some Foodborne Viral Infections

Virus	Vehicle Foods/Source/Comments	Reference
Polio	10 outbreaks with 161 cases, 1914–1949. Raw and pasteurized milk were the vehicles in 6 and 2 outbreaks, respectively.	41
Polio 3	Oysters from the Louisiana coast.	57
ECHO 9	Oysters from the New Hampshire coast.	97
ECHO 4	Oysters from the Texas Gulf waters.	57
Enteroviruses	Oysters from the Texas Gulf.	63
Coxsackie B-4	Oysters.	97
Hepatitis A	Over 1,600 cases 1945–1966 from a variety of foods.	41
Hepatitis A	About 1,600 cases from oysters and clams, 1955–1966.	41
Hepatitis A	A total of 3 outbreaks in 1975 (173 cases) from glazed donuts, sandwiches, and salad/sandwiches in New York, Oklahoma, and Oregon.	31
Hepatitis A	Eighteen of about 580 persons at risk contracted the disease from sandwiches in Pennsylvania.	32
Hepatitis A	Oysters.	111
Hepatitis A	From 3 outbreaks in New York state, there were 10 victims of about 238 at risk. Vehicle food was raw clams.	34
Hepatitis A	Fifty-six cases were contracted in a restaurant that employed a food handler with the disease in New Jersey.	33
Hepatitis A	Some 203 persons contracted the disease at a drive-in restaurant in Oklahoma in 1983.	35
Hepatitis A	From a salad bar–type restaurant in Texas, 123 cases occurred in 1983. A food handler was suspected.	35
Norwalk virus	Oysters. About 2,000 persons were infected in Australia in 1978.	101
Norwalk virus	Green salad (lettuce). Possibly cross-contamination from seafoods. About 71% of 190 persons became ill in New Jersey in 1979.	64
Norwalk virus	Raw oysters were involved in this outbreak in Florida in 1980. Six of 13 persons at risk contracted the disease. The agent was identified by RIA.	65
Norwalk virus	Of 6 outbreaks in the United Kingdom 1976–1980 where the incubation period was between 24 and 48 h, about 45% of 2,590 persons became ill after consuming a variety of foods including seafoods.	5
Norwalk virus	Coleslaw, potato, fruit, and tossed salads. In Minnesota in 1982, 57% of 383 persons became ill after eating above foods prepared by one ill worker and another worker following recovery from gastroenteritis.	147
Norwalk virus	Cake/frosting. A food handler with gastroenteritis infected up to 3,000 persons in Minnesota in 1982.	86
Norwalk virus	Raw clams/oysters. Between May and December 1982, 103 documented gastroenteritis outbreaks with 1,017 cases occurred in New York State. Norwalk agent was identified in 71% of 7 outbreaks.	100
Norwalk virus	Ice used in soft drinks. Over 200 university students and football team members in Philadelphia exhibited symptoms typical of Norwalk gastroenteritis, and a 27-nm virus was identified in this 1987 outbreak.	30

The Norwalk and closely related agents are 28 to 38 nm in size and are placed in three groups.

1. The SRSVs, which have amorphous surfaces. In addition to Norwalk, they consist of Hawaii, Snow Mountain, Taunton, and others.

2. **Caliciviruses** contain surface hollows. They resemble the picornaviruses and cause vomiting and diarrhea in 1 to 3 days in children. Antibodies do not cross-react with the Norwalk agent.
3. **Astroviruses** contain a five- or six-pointed surface star. They are known to cause occasional gastroenteritis in children and adults, although children under age 7 are most susceptible. Symptoms develop within 24 to 36 h and consist of vomiting, diarrhea, and fever.

Some strains of calici- and astroviruses have been cultured in vitro. Only the Norwalk agent is discussed in the remainder of this section.

The Norwalk virus is infective primarily for older children and adults, and about 67% of adults in the United States possess serum antibodies (45). Of 74 acute nonbacterial outbreaks reported to the U.S. CDC for 1976 through 1980, 42% were attributed to the Norwalk virus based on a fourfold rise in antibody titer (76). It is known to be associated with travelers' diarrhea, and polluted water is an obvious source.

The reported food-associated outbreaks and cases in the United States for 1983 through 1987 are presented in Table 25-4, and synopses of some food-associated outbreaks are presented in Table 25-5. Of 430 foodborne outbreaks in the United States in 1979, 4% displayed the pattern of Norwalk gastroenteritis (77). This virus was thought to be the cause of more foodborne gastroenteritis than any other single bacterium in the state of Minnesota in 1985 (85).

Among the earliest reported outbreaks that involved SRSVs is one that occurred in 1976 in England. Between December 21, 1976, and January 10, 1977, 33 outbreaks and 797 cases occurred; cockles were incriminated (6). The incubation period was 24 to 30 h, and from 12 of 14 stool samples, small, round virus particles measuring 25 to 26 nm in diameter were demonstrated, but they were not found in cockles. Although the investigators believed that the agents were neither Norwalk nor Hawaii, these outbreaks are regarded by some as Norwalk virus outbreaks. The 1978 outbreak in Australia that involved at least 2,000 persons was well documented, and the vehicle food was oysters (101). The virus was found in 39% of fecal specimens examined by electron microscopy, and antibody responses were demonstrated in 75% of paired sera tested. The incubation period ranged from 18 to 48 h, with most cases occurring in 34 to 38 h. Nausea was the first symptom, usually accompanied by vomiting, nonbloody diarrhea, and abdominal cramps, with symptoms lasting 2 to 3 days. In the Philadelphia outbreak noted in Table 25-5, 99% of victims reported nausea as one symptom (30). Another outbreak in Australia was traced to bottled oysters and symptoms occurred in 24 to 48 h (54). The oysters had an aerobic plate count (APC) of 2.2×10^4/g and a fecal coliform count of 500/100 g. The first documented outbreaks in the United States are those that occurred in New Jersey in 1979, where lettuce was the vehicle food, and the Florida outbreak in 1980 that was traced to raw oysters (see Table 25-5). In the latter, the agent was identified by a radioimmunoassay method.

The Norwalk agent is more resistant to destruction by chlorine than other enteric viruses. In volunteers, 3.75 ppm chlorine in drinking water failed to inactivate the virus, while poliovirus type 1 and human and simian rotaviruses were inactivated (78). Some Norwalk viruses remained infective at residual chlorine levels of 5 to 6 ppm. Hepatitis A viruses are not as resistant as Norwalk, but both are clearly more resistant to chlorine than the rotaviruses.

Rotaviruses

The first demonstration of these viruses occurred in 1973 in Australia, and they were first propagated in the laboratory in 1981. Six groups have been identified, and three are known to be infectious for humans. Group A is the most commonly encountered among infants and young children throughout the world. Group B causes diarrhea in adults, and they have been seen only in China. Rotaviruses belong to the family Reoviridae, they are about 70 nm in diameter, are nonenveloped, and contain dsRNA. The fecal-oral route is the primary mode of transmission.

Rotaviruses cause an estimated one-third of all hospitalizations for diarrhea in children below age 5, and the peak season for infection occurs during the winter months (25, 27). Most susceptible are children between the ages of 6 months and 2 years, and virtually every child in the United States is infected by age 4 (27). Although most persons are immune by age 4, high inoculum or lowered states of immunity can lead to milder illness among older children and adults (27). They are known to be transmitted among children in day-care centers and by water. A community waterborne outbreak occurred in Eagle-Vail, Colorado, in 1981, and 44% of 128 persons, most of them adults, became ill (72). They are believed to be only infrequent causes of foodborne gastroenteritis (44).

The incubation period for rotavirus gastroenteritis is 2 days. Vomiting occurs for 3 days accompanied by watery diarrhea for 3 to 8 days and often by abdominal pain and fever (27). They are known to be associated with travelers' diarrhea (123, 127).

For the 23-month period between January 1989 and November 1990, 48,035 stool specimens were examined in the United States, with 9,639 (20%) being positive for rotavirus (25). The highest percentage of positive stools occurred in February (36%) and the lowest in October (6%). Between 1979 and 1985, an annual average of 500 children died from diarrheal illness in the United States, and 20% were caused by rotavirus infections (25).

The host cell receptor protein for rotavirus also serves as the beta-adrenergic receptor. Once inside cells, they are transported to lysosomes where uncoating occurs.

Rotaviral infections can be diagnosed by immunoelectron microscopy, ELISA, and latex agglutination methods.

AEROMONAS, PLESIOMONAS, AND OTHER GRAM-NEGATIVE BACTERIA

Aeromonas

This genus consists of several species that are often found in gastrointestinal specimens. Among these are *A. caviae*, *A. eucrenophila*, *A. schubertii*, *A. sobria*, *A. veronii*, and *A. hydrophila*. An enterotoxin has been identified in *A.*

caviae (102) and *A. hydrophila* (see below), and the other species noted are associated with diarrhea. Since *A. hydrophila* has received the most study, the discussion that follows is based on this species. The aeromonads are basically aquatic forms that are often associated with diarrhea, but their precise role in the etiology of gastrointestinal syndromes is not clear.

A. hydrophila is an aquatic bacterium found more in salt waters than in fresh waters. It is a significant pathogen to fish, turtles, frogs, snails, and alligators and a human pathogen, especially in compromised hosts. It is a common member of the bacterial flora of pigs. Diarrhea, endocarditis, meningitis, soft tissue infections, and bacteremia are caused by *A. hydrophila* (47).

Virulent strains of *A. hydrophila* produce a 52 kda single polypeptide that possesses enterotoxic, cytotoxic, and hemolytic activities (92, 116, 143). This multifunctional molecule displays immunological cross-reactivity with the cholera toxin (116), and it resembles **aerolysin**, a thiol-activated toxin analogous to listeriolysin O discussed in Chapter 21. Cytotonic activity has been associated with a *A. hydrophila* toxin, which induced rounding and steroidogenesis in Y-1 adrenal cells. Also, positive responses in the rabbit ileal loop, suckling mouse, and CHO assays have been reported for a cytotonic toxin (36).

A large number of studies have been conducted on *A. hydrophila* isolates from various sources. In one study, 66 of 96 (69%) isolates produced cytotoxins, while 32 (80%) of 40 isolates from diarrheal disease victims were toxigenic, with only 41% of nondiarrheal isolates being positive for cytotoxin production. Most enterotoxigenic strains are VP (Voges-Proskauer test) and hemolysin positive and arabinose negative (20), and produce positive responses in the suckling mouse, Y-1 adrenal cell, and rabbit ileal loop assays. In a study of 147 isolates from patients with diarrhea, 91% were enterotoxigenic, while only 70% of 94 environmental strains produced enterotoxin as assessed by the suckling mouse assay (21). All but 4 of the clinical isolates produced hemolysis of rabbit red blood cells. Of 116 isolates from the Chesapeake Bay, 71% were toxic by the Y-1 adrenal cell assay and toxicity correlated with lysine decarboxylase and VP reactions (75). In yet another study, 48 of 51 cultures from humans, animals, water, and sewage produced positive responses in rabbit ileal loop assays with 10^3 or more cells, and cell-free extracts from all were loop positive (4).

Isolates from meat and meat products possessed biochemical markers that are generally associated with toxic strains of other species, with the mouse LD_{50} being log 8–9 cfu for most strains tested (107). The latter authors suggested the possibility that immunosuppressive states are important factors in food-associated infections by this organism, a suggestion that could explain the difficulty of establishing this organism as the sole etiologic agent of foodborne gastroenteritis.

With regard to growth temperature and habitat, 7 of 13 strains displayed growth at 0° to 5°C, 4 of 13 at 10°C, and 1 at a minimum of 15°C (117). The psychrotrophs had optimum growth between 15 and 20°C. The maximum growth temperature for some strains was 40 to 45 with optimum at 35°C (67). Regarding distribution, the organism was found in all but 12 of 147 lotic and lentic habitats (67). Four of those habitats that did not yield the organism were either hypersaline lakes or geothermal springs. Some waters contained up to 9,000/ml. An ecologic study of *A. hydrophila* in the Chesapeake Bay revealed numbers ranging from less than 0.3/liter to 5×10^3/ml in the water column, and about 4.6×10^2/g of sediment (75). The presence of this organism correlated with total, aerobic, viable, and heterotrophic bacterial counts, and its presence was inversely related to dissolved O_2 and salinity, with the upper salt level being about 15%. Fewer were found during the winter than during the summer months.

Plesiomonas

P. shigelloides is found in surface waters and soil and has been recovered from fish, shellfish, other aquatic animals, as well as from terrestrial meat animals. It differs from *A. hydrophila* in having a G + C content of DNA of 51%, versus 58 to 62% for *A. hydrophila*. It has been isolated by many investigators from patients with diarrhea and is associated with other general infections in humans. It produces a heat-stable enterotoxin, and serogroup $0:17$ strains react with *Shigella* group D antisera (1). In a study of 16 strains from humans with intestinal illness, *P. shigelloides* did not always bind Congo red, the strains were noninvasive in HEp-2 cells, and they did not produce Shiga-like toxin on Vero cells (1). Although a low-level cytolysin was produced consistently, the mean LD_{50} for outbred Swiss mice was 3.5×10^8 cfu. Heat-stable enterotoxin was not produced by either of the 16 strains, and it was the conclusion of these investigators that this organism possesses a low pathogenic potential (1).

P. shigelloides* was recovered by Zajc-Satler et al. (153) from the stools of six diarrheal patients. It was believed to be the etiologic agent, although salmonellae were recovered from two patients. Two outbreaks of acute diarrheal disease occurred in Osaka, Japan, in 1973 and 1974, and the only bacterial pathogen recovered from stools was *P. shigelloides*. In the 1973 outbreak, 978 of 2,141 persons became ill, with 88% complaining of diarrhea, 82% of abdominal pain, 22% of fever, and 13% of headaches (140). Symptoms lasted 2 to 3 days. Of 124 stools examined, 21 yielded *P. shigelloides* 017:H2. The same serovar was recovered from tapwater. In the 1974 outbreak, 24 of 35 persons became ill with symptoms similar to those noted. *P. shigelloides* serovar 024:H5 was recovered from three of eight stools "virtually in pure culture" (140). The organism was recovered from 39% of 342 water and mud samples, as well as from fish, shellfish, and newts.

A 15-year-old female contracted gastroenteritis, and 6 h after she took one tablet of trimethoprim-sulfadiazine, *P. shigelloides* could be recovered from her blood (110). The latter authors noted that 10 of the previously known 12 cases of *P. shigelloides* bacteremia were in patients who were either immuno-compromised or presented with other similar conditions. The 15-year-old had a temperature of 39°C and passed up to 10 watery stools daily. The isolated strain reacted with *S. dysenteriae* serotype 7 antiserum, placing it in O group 22 of *P. shigelloides* (110).

Growth of *P. shigelloides* has been observed at 10°C (117), and 59% of 59 fish from Zaire waters contained the organism (142). In the latter study, the organism was found more in river fish than lake fish. It appeared not to produce an enterotoxin since only 4 of 29 isolates produced positive responses in rabbit ileal loops (118). Foodborne cases have not been documented, but the organism has been incriminated in at least two outbreaks (98).

Regarding other gram-negative bacteria, heat-stable enterotoxins have been shown for *Klebsiella pneumoniae* (80) and for *Enterobacter cloacae* (81). The STs were similar and resembled *E. coli* ST. Since LT and ST are plasmid-mediated toxins and since the Enterobacteriaceae are common inhabitants of the gastrointestinal tract, it is not unreasonable to expect that, through plasmid transfer, some non–*E. coli* organisms may acquire and express these toxin-synthesizing genes. The potential significance of some of these organisms in foods has been noted (141).

HISTAMINE-ASSOCIATED (SCOMBROID) POISONING

Illness contracted from eating scombroid fish or fish products containing high levels of histamine is often referred to as scombroid poisoning. Among the scombroid fishes are tuna, mackerel, bonito, and others. The histamine is produced by bacterial decarboxylation of the generally large quantities of histidine in the muscles of this group. Sufficient levels of histamine may be produced without the product's being organoleptically unacceptable, with the result that scombroid poisoning may be contracted from both fresh and organoleptically spoiled fish. The history of this syndrome has been reviewed by Hudson and Brown (73), who questioned the etiologic role of histamine. This is discussed further below.

The bacteria most often associated with this syndrome are *Morganella* spp., especially *M. morganii*, of which all strains appear to produce histamine at levels up to 400 mg%. Among other bacteria known to produce histidine decarboxylase are *K. pneumoniae*, *Hafnia alvei*, *Citrobacter freundii*, *Clostridium perfringens*, *Enterobacter aerogenes*, *Vibrio alginolyticus*, and *Proteus* spp. From room-temperature spoiled skipjack tuna, 31% of 470 bacterial isolates produced from 100 to 400 mg% of histamine in broth (106). The strong histamine formers were *M. morganii*, *Proteus* spp., and a *Klebsiella* sp., while weak formers included *H. alvei* and *Proteus* spp. Skipjack tuna spoiled in seawater at 38°C contained *C. perfringens* and *V. alginolyticus* among other histidine decarboxylase producers (151). From an outbreak of scombroid poisoning associated with tuna sashimi, *K. pneumoniae* was recovered and shown to produce 442 mg% histamine in a tuna fish infusion broth (136). This syndrome has been associated with foods other than scombroid fish, particularly cheeses, including swiss cheese, which in one case contained 187 mg% histamine; the symptoms associated with the outbreak occurred in 30 min to 1 h after ingestion (137).

The number of outbreaks reported to the CDC for the years 1972 through 1986 were 178 with 1,096 cases but no deaths (29). The largest outbreaks of 51, 29, and 24 occurred in Hawaii, California, and New York, respectively. The three most common vehicle foods were mahi mahi (66 outbreaks), tuna (42 outbreaks), and bluefish (19 outbreaks). While fresh fish normally contains 1 mg% of histamine, some may contain up to 20 mg%, a level that may lead to symptoms in some individuals. The FDA hazardous level for tuna is 50 mg% (29). The cooking of toxic fish may not lead to safe products.

Histamine content of stored skipjack tuna can be estimated if incubation times and temperatures of storage are known. Frank et al. (56) found that

100 mg% formed in 46 h at 70°F, in 23 h at 90°F, and in 17 h at 100°F. A nomograph was constructed over the temperature range of 70° to 100°F, underscoring the importance of low temperatures in preventing or delaying histamine formation. Vacuum packaging is less effective than low-temperature storage in controlling histamine production (145). The culture medium of choice for detecting histamine-producing bacteria is that of Niven et al. (104).

Histamine production is favored by low pH, but it occurs more when products are stored above the refrigerator range. The lowest temperature for production of significant levels was found to be 30°C for *H. alvei*, *C. freundii*, and *E. coli*; 15°C for two strains of *M. morganii*; and 7°C for *K. pneumoniae* (10).

The syndrome is contracted by eating fresh or processed fish of the type noted; symptoms occur within minutes and for up to 3 h after ingestion of toxic food, with most cases occurring within 1 h. Typical symptoms consist of a flushing of the face and neck accompanied by a feeling of intense heat and general discomfort, and diarrhea. Subsequent facial and neck rashes are common. The flush is followed by an intense, throbbing headache tapering to a continuous dull ache. Other symptoms include dizziness, itching, faintness, burning of the mouth and throat, and the inability to swallow (see 73). The minimum level of histamine thought necessary to cause symptoms is 100 mg%. Large numbers of *M. morganii* in fish of the type incriminated in this syndrome and a level of histamine more than 10 mg% is considered significant relative to product quality.

The first 50 incidents in Great Britain occurred between 1976 and 1979, with all but 19 occurring in 1979. Canned and smoked mackerel was the most common vehicle, with bonita, sprats, and pilchards involved in one outbreak each. The most common symptom among the 196 cases was diarrhea (61).

Regarding etiology, Hudson and Brown (73) believe the evidence does not favor histamine per se as the agent responsible for the syndrome. They suggest a synergistic relationship involving histamine and other as yet unidentified agents such as other amines or factors that influence histamine absorption. This view is based on the inability of large oral doses of histamine or histamine-spiked fish to produce symptoms in volunteers. On the other hand, the suddenness of onset of symptoms is consistent with histamine reaction, and the association of the syndrome with scombroid fish containing high numbers of histidine decarboxylase-producing bacteria cannot be ignored. While the precise etiology may yet be in question, bacteria do play a significant if not indispensable role.

PARALYTIC SHELLFISH POISONING

This syndrome is contracted by eating toxic mussels, clams, oysters, scallops, or cockles. These bivalves become toxic after feeding on certain dinoflagellates of which *Gonyaulax catenella* is representative of the U.S. Pacific Coast flora. Along the North Atlantic coast of the United States and over to northern Europe, *G. tamarensis* is found, and its poison is more toxic than that of *G. catenella*. *G. acatenella* is found along the coast of British Columbia. Masses or blooms of these toxic dinoflagellates give rise to the **red tide** condition of seas.

The paralytic shellfish poison (PSP) is **saxitoxin**, and its structural formula is as follows:

Saxitoxin exerts its effect in humans through cardiovascular collapse and respiratory failure (148). It blocks the propagation of nerve impulses without depolarization (119), and there is no known antidote. It is heat stable, water soluble, and generally not destroyed by cooking. It can be destroyed by boiling 3 to 4 h at pH 3.0. A D value at 250°F of 71.4 min in soft-shell clams has been reported (62).

Symptoms of PSP develop within 2 h after ingestion of toxic mollusks, and they are characterized by paresthesia (tingling, numbness, or burning), which begins about the mouth, lips, and tongue and later spreads over the face, scalp, and neck and to the fingertips and toes. The mortality rate is variously reported to range from 1 to 22%.

Between 1793 and 1958, some 792 cases were recorded, with 173 (22%) deaths (95). In the 15-year period 1973–1987, 19 outbreaks (with a mean of 8 cases) were reported by state health departments to the CDC. In 1990, there were 19 cases from two outbreaks in the states of Massachusetts and Alaska alone. In the former, 6 fishermen became ill after eating boiled mussels that contained 4,280 µg/100 g saxitoxin (26). The raw mussels contained 24,400 µg/ 100 g. The 13 cases in Alaska resulted in one death, and gastric contents from the victim who died contained 370 µg/100 g of PSP toxin, while a sample of the butterclam that was consumed contained 2,650 µg/100 g (26). The maximum safe level of PSP toxin is 80 µg/100 g (26).

Outbreaks of PSP seem to occur between the months of May and October on the U.S. West Coast and between August and October on the East Coast. Mollusks may become toxic in the absence of red tides. Detoxification of mollusks can be achieved by their transfer to clean waters, and a month of more may be required.

References

1. Abbott, S. L., R. P. Kokka, and J. M. Janda. 1991. Laboratory investigations on the low pathogenic potential of *Plesiomonas shigelloides*. *J. Clin. Microbiol.* 29:148–153.
2. Achmoody, J. B., and J. R. Chipley. 1978. The influence of metabolic inhibitors and incubation time on aflatoxin release from *Aspergillus flavus*. *Mycologia* 70:313–320.
3. Altug, T., A. E. Yousef, and E. H. Marth. 1990. Degradation of aflatoxin B_1 in dried figs by sodium bisulfite with or without heat, ultraviolet energy or hydrogen peroxide. *J. Food Protect.* 53:581–582.

4. Annapurna, E., and S. C. Sanyal. 1977. Enterotoxicity of *Aeromonas hydrophila*. *J. Med. Microbiol.* 10:317–323.

5. Appleton, H., S. R. Palmer, and R. J. Gilbert. 1981. Foodborne gastroenteritis of unknown aetiology: A virus infection? *Brit. Med. J.* 282:1801–1802.

6. Appleton, H., and M. S. Pereira. 1977. A possible virus aetiology in outbreaks of food-poisoning from cockles. *Lancet* 1:780–781.

7. Ayres, J. C., J. O. Mundt, and W. E. Sandine. 1980. *Microbiology of Foods*, 658–683. San Francisco: Freeman.

8. Azcona, J. I., M. M. Abouzied, and J. J. Pestka. 1990. Detection of zearalenone by tandem immunoaffinity-enzyme-linked immunosorbent assay and its application to milk. *J. Food Protect.* 53:577–580.

9. Bacon, C. W., J. G. Sweeney, J. D. Hobbins, and D. Burdick. 1973. Production of penicillic acid and ochratoxin A on poultry feed by *Aspergillus ochraceus*: Temperature and moisture requirements. *Appl. Microbiol.* 26:155–160.

10. Behling, A. R., and S. L. Taylor. 1982. Bacterial histamine production as a function of temperature and time of incubation. *J. Food Sci.* 47:1311–1314, 1317.

11. Blackwell, J. H., D. Rickansrud, P. D. McKercher, and J. W. McVicar. 1982. Effect of thermal processing on the survival of foot-and-mouth disease virus in ground meat. *J. Food Sci.* 47:388–392.

12. Buchanan, R. L., and D. F. Lewis. 1984. Caffeine inhibition of aflatoxin synthesis: Probable site of action. *Appl. Environ. Microbiol.* 47:1216–1220.

13. Buchanan, R. L., G. Tice, and D. Marino. 1982. Caffeine inhibition of ochratoxin A production. *J. Food Sci.* 47:319–321.

14. Bullerman, L. B. 1976. Examination of Swiss cheese for incidence of mycotoxin producing molds. *J. Food Sci.* 41:26–28.

15. Bullerman, L. B. 1984. Effects of potassium sorbate on growth and patulin production ,by *Penicillium patulum* and *Penicillium roqueforti*. *J. Food Protect.* 47:312–316.

16. Bullerman, L. B. 1984. Formation and control of mycotoxins in food. *J. Food Protect.* 47:637–646.

17. Bullerman, L. B., P. A. Hartman, and J. C. Ayres. 1969. Aflatoxin production in meats. I. Stored meats. *Appl. Microbiol.* 18:714–717.

18. Bullerman, L. B., P. A. Hartman, and J. C. Ayres. 1969. Aflatoxin production in meats. II. Aged dry salamis and aged country cured hams. *Appl. Microbiol.* 18:718–722.

19. Bullerman, L. B. 1987. Methods for detecting mycotoxins in foods and beverages. In: *Food and Beverage Mycology*, ed. L. R. Beuchat, 2d ed., 571–598. New York: Van Nostrand Reinhold.

20. Burke, V., J. Robinson, H. M. Atkinson, and M. Gracey. 1982. Biochemical characteristics of enterotoxigenic *Aeromonas* spp. *J. Clin. Microbiol.* 15:48–52.

21. Burke, V., J. Robinson, M. Cooper, J. Beamon, K. Partridge, D. Peterson, and M. Gracey. 1984. Biotyping and virulence factors in clinical and environmental isolates of *Aeromonas* species. *Appl. Environ. Microbiol.* 47:1146–1149.

22. Busby, W. F., Jr., and G. N. Wogan. 1979. Food-borne mycotoxins and alimentary mycotoxicoses. In *Food-borne Infections and Intoxications*, ed. H. Riemann and F. L. Bryan, 519–610. New York: Academic Press.

23. Butler, W. H. 1974. Aflatoxin. In *Mycotoxins*, ed. I. F. H. Purchase, 1–28. New York: Elsevier.

24. Camou-Arriola, J. P., and R. L. Price. 1989. Destruction of aflatoxin and reduction of mutagenicity of naturally-contaminated corn during production of a corn snack. *J. Food Protect.* 52:814–817.

25. Centers for Disease Control. 1991. Rotavirus surveillance—United States, 1989–1990. *Morb. Mort. Wkly. Rept.* 40:80–81, 87.

26. Centers for Disease Control. 1991. Paralytic shellfish poisoning—Massachusetts and Alaska, 1990. *Mort. Morb. Wkly. Rept.* 40:157–161.
27. Centers for Disease Control. 1990. Viral agents of gastroenteritis. Public health importance and outbreak management. *Morb. Mort. Week. Rept.* 39:1–23.
28. Centers for Disease Control. 1990. Foodborne hepatitis A—Alaska, Florida, North Carolina, Washington. *Morb. Mort. Week. Rept.* 39:228–232.
29. Centers for Disease Control. 1989. Scombroid fish poisoning—Illinois, South Carolina. *Morb. Mort. Week. Rept.* 38:140–142, 147.
30. Centers for Disease Control. 1987. Outbreak of viral gastroenteritis—Pennsylvania and Delaware. *Morb. Mort. Week. Rept.* 36:709–711.
31. Centers for Disease Control. 1976. Foodborne and waterborne disease outbreaks. *Annual Summary,* HEW Publ. No. (CDC) 76-8185.
32. Centers for Disease Control. 1977. Foodborne outbreak of hepatitis A—Pennsylvania. *Morb. Mort. W. Rept.* 26:247–248.
33. Centers for Disease Control. 1982. Outbreak of food-borne hepatitis A—New Jersey. *Morb. Mort. W. Rept.* 31:150–152.
34. Centers for Disease Control. 1982. Enteric illness associated with raw clam consumption—New York. *Morb. Mort. W. Rept.* 31:449–451.
35. Centers for Disease Control. 1983. Food-borne hepatitis A—Oklahoma, Texas. *Morb. Mort. W. Rept.* 32:652–654, 659.
36. Chakraborty, T., M. A. Montenegro, S. C. Sanyal, R. Helmuth, E. Bulling, and K. N. Timmis. 1984. Cloning of enterotoxin gene from *Aeromonas hydrophila* provides conclusive evidence of production of a cytotonic enterotoxin. *Infect Immun.* 46:435–441.
37. Christensen, C. M. 1971. Mycotoxins, *CRC Crit. Rev. Environ. Cont.* 2:57–80.
38. Chu, F. S. 1984. Immunoassays for analysis of mycotoxins. *J. Food Protect.* 47:562–569.
39. Chu, F. S. 1974. Studies on ochratoxins. *CRC Crit. Rev. Toxicol.* 2:499–524.
40. Ciegler, A., H.-J. Mintzlaff, D. Weisleder, and L. Leistner. 1972. Potential production and detoxification of penicillic acid in mold-fermented sausage (salami). *Appl. Microbiol.* 24:114–119.
41. Cliver, D. O. 1967. Food-associated viruses. *Hlth Lab. Sci.* 4:213–221.
42. Cliver, D. O., R. D. Ellender, and M. D. Sobsey. 1983. Methods for detecting viruses in foods: Background and general principles. *J. Food Protect.* 46:248–259.
43. Cliver, D. O., R. D. Ellender, and M. D. Sobsey. 1983. Methods to detect viruses in foods: Reading and interpretation of results. *J. Food. Protect.* 46:345–357.
44. Cliver, D. O. (and the IFT Expert Panel on Food Safety & Nutrition). 1988. Virus transmission via foods. *Food Technol.* 42(10):241–248.
45. Cukor, G., and N. R. Blacklow. 1984. Human viral gastroenteritis. *Microbiol. Rev.* 48:157–179.
46. Davis, N. D., and U. L. Diener. 1987. Mycotoxins. In *Food and Beverage Mycology*, 2nd ed., ed. L. R. Beuchat, 517–570. New York: Van Nostrand Reinhold.
47. Davis, W. A., II, J. G. Kane, and V. F. Garagusi. 1978. Human *Aeromonas* infections. A review of the literature and a case report of endocarditis. *Med.* 57:267–277.
48. Davis, V. M., and M. E. Stack. 1991. Mutagenicity of stemphyltoxin III, a metabolite of *Alternaria alternata*. *Appl. Environ. Microbiol.* 57:180–182.
49. Deger, G. E. 1976. Aflatoxin—human colon carcinogenesis? *Ann. Intern. Med.* 85:204.
50. DiGirolamo, R., J. Liston, and J. R. Matches. 1970. Survival of virus in chilled, frozen, and processed oysters. *Appl. Microbiol.* 20:58–63.
51. El-Banna, A. A., and P. M. Scott. 1984. Fate of mycotoxins during processing of foodstuffs. III. Ochratoxin A during cooking of faba beans (*Vicia faba*) and polished wheat. *J. Food Protect.* 47:189–192.

52. Enomoto, M., and M. Saito. 1972. Carcinogens produced by fungi. *Ann. Rev. Microbiol.* 26:279–312.

53. Escher, F. E., P. E. Koehler, and J. C. Ayres. 1973. Production of ochratoxins A and B on country cured ham. *Appl. Microbiol.* 26:27–30.

54. Eyles, M. J., G. R. Davey, and E. J. Huntley. 1981. Demonstration of viral contamination of oysters responsible for an outbreak of viral gastroenteritis. *J. Food Protect.* 44:294–296.

55. Frank, H. A. 1968. Diffusion of aflatoxins in foodstuffs. *J. Food Sci.* 33:98–100.

56. Frank, H. A., D. H. Yoshinaga, and I.-P. Wu. 1983. Nomograph for estimating histamine formation in skipjack tuna at elevated temperatures. *Mar. Fish. Rev.* 45:40–44.

57. Fugate, K. J., D. O. Cliver, and M. T. Hatch. 1975. Enteroviruses and potential bacterial indicators in Gulf Coast oysters. *J. Milk. Food Technol.* 38:100–104.

58. Gerba, C. P., and S. M. Goyal. 1978. Detection and occurrence of enteric viruses in shellfish: A review. *J. Food Protect.* 41:743–754.

59. Gerba, C. P., S. M. Goyal, R. L. LaBelle, I. Cech, and G. F. Bodgan. 1979. Failure of indicator bacteria to reflect the occurrence of enteroviruses in marine waters. *Amer. J. Pub. Hlth.* 69:1116–1119.

60. Gerba, C. P., J. B. Rose, and S. N. Singh. 1985. Waterborne gastroenteritis and viral hepatitis. *CRC Crit. Rev. Environ. Cont.* 15:213–236.

61. Gilbert, R. J., G. Hobbs, G. K. Murray, J. G. Cruickshank, and S. E. J. Young. 1980. Scombrotoxic fish poisoning: Features of the first 50 incidents to be reported in Britain (1976–1979). *Brit. Med. J.* 281:71–72.

62. Gill, T. A., J. W. Thompson, and S. Gould. 1985. Thermal resistance of paralytic shellfish poison in soft-shell clams. *J. Food Protect.* 48:659–662.

63. Goyal, S. M., C. P. Gerba, and J. L. Melnick. 1979. Human enteroviruses in oysters and their overlying waters. *Appl. Environ. Microbiol.* 37:572–581.

64. Griffin, M. R., J. J. Surowiec, D. I. McCloskey, B. Capuano, B. Pierzynski, M. Quinn, R. Wojnarski, W. E. Parkin, H. Greenberg, and G. W. Gary. 1982. Foodborne Norwalk virus. *Amer. J. Epidemiol.* 115:178–184.

65. Gunn, R. A., H. T. Janowski, S. Lieb, E. C. Prather, and H. B. Greenberg. 1982. Norwalk virus gastroenteritis following raw oyster consumption. *Amer. J. Epidemiol.* 115:348–351.

66. Harrison, M. A. 1989. Presence and stability of patulin in apple products: A review. *J. Food Saf.* 9:147–153.

67. Hazen, T. C., C. B. Fliermans, R. P. Hirsch, and G. W. Esch. 1978. Prevalence and distribution of *Aeromonas hydrophila* in the United States. *Appl. Environ. Microbiol.* 36:731–738.

68. Hejkal, T. W., and C. P. Gerba. 1981. Uptake and survival of enteric viruses in the blue crab, *Callinectes sapidus*. *Appl. Environ. Microbiol.* 41:207–211.

69. Herrmann, J. E., and D. O. Cliver. 1973. Enterovirus persistence in sausage and ground beef. *J. Milk Food Technol.* 36:426–428.

70. Hesseltine, C. W. 1967. Aflatoxins and other mycotoxins. *Hlth Lab. Sci.* 4:222–228.

71. Holmquist, G. U., H. W. Walker, and H. M. Stahr. 1983. Influence of temperature, pH, water activity and antifungal agents on growth of *Aspergillus flavus* and *A. parasiticus*. *J. Food Sci.* 48:778–782.

72. Hopkins, R. S., G. B. Gaspard, F. P. Williams, Jr., R. J. Karlin, G. Cukor, and N. R. Blacklow. 1984. A community waterborne gastroenteritis outbreak: Evidence for rotavirus as the agent. *Amer. J. Pub. Hlth.* 74:263–265.

73. Hudson, S. H., and W. D. Brown. 1978. Histamine (?) toxicity from fish products. *Adv. Food Res.* 24:113–154.

74. Jorgenssen, K. V., D. L. Park, S. M. Rua, Jr., and R. L. Price. 1990. Reduction of mutagenic potentials in milk: Effects of ammonia treatment on aflatoxin-

contaminated cottonseed. *J. Food Protect.* 53:777–778.

75. Kaper, J. B., H. Lockman, R. R. Colwell, and S. W. Joseph. 1981. *Aeromonas hydrophila*: Ecology and toxigenicity on isolates from an estuary. *J. Appl. Bacteriol.* 50:359–377.

76. Kaplan, J. E., G. W. Gary, R. C. Baron, N. Singh, L. B. Schonberger, R. Feldman, and H. B. Greenberg. 1982. Epidemiology of Norwalk gastroenteritis and the role of Norwalk virus in outbreaks of acute nonbacterial gastroenteritis. *Ann. Intern. Med.* 96:756–761.

77. Kaplan, J. E., R. Feldman, D. S. Campbell, C. Lookabaugh, and G. W. Gary. 1982. The frequency of a Norwalk-like pattern of illness in outbreaks of acute gastroenteritis. *Amer. J. Pub. Hlth.* 72:1329–1332.

78. Keswick, B. H., T. K. Satterwhite, P. C. Johnson, H. L. DuPont, S. L. Secor, J. A. Bitsura, G. W. Gary, and J. C. Hoff. 1985. Inactivation of Norwalk virus in drinking water by chlorine. *Appl. Environ. Microbiol.* 50:261–264.

79. King, A. D., Jr., and J. E. Schade. 1984. *Alternaria* toxins and their importance in food. *J. Food Protect.* 47:886–901.

80. Klipstein, F. A., and R. F. Engert. 1976. Purification and properties of *Klebsiella pneumoniae* heat-stable enterotoxin. *Infect. Immun.* 13:373–381.

81. Klipstein, F. A., and R. F. Engert. 1976. Partial purification and properties of *Enterobacter cloacae* heat-stable enterotoxin. *Infect. Immun.* 13:1307–1314.

82. Konowalchuk, J., and J. I. Speirs. 1975. Survival of enteric viruses on fresh vegetables. *J. Milk Food Technol.* 38:469–472.

83. Konowalchuk, J., and J. I. Speirs. 1975. Survial of enteric viruses on fresh fruit. *J. Milk Food Technol.* 38:598–600.

84. Kostenbader, K. D., Jr., and D. O. Cliver. 1977. Quest for viruses associated with our food supply. *J. Food Sci.* 42:1253–1257, 1268.

85. Kuritsky, J. N., M. T. Osterholm, J. A. Korlath, K. E. White, and J. E. Kaplan. 1985. A statewide assessment of the role of Norwalk virus in outbreaks of food-borne gastroenteritis. *J. Inf. Dis.* 151:568.

86. Kuritsky, J. N., M. T. Osterholm. H. B. Greenberg, J. A. Korlath, J. R. Godes, C. W. Hedberg, J. C. Forfang, A. Z. Kapikian, J. C. McCullough, and K. E. White. 1984. Norwalk gastroenteritis: A community outbreak associated with bakery product consumption. *Ann. Int. Med.* 100:519–521.

87. Labuza, T. P. 1983. Regulation of mycotoxins in food. *J. Food Protect.* 46:260–265.

88. Landry, E. F., J. M. Vaughn, T. J. Vicale, and R. Mann. 1982. Inefficient accumulation of low levels of monodispersed and feces-associated poliovirus in oysters. *Appl. Environ. Microbiol.* 44:1362–1369.

89. Larkin, E. P. 1981. Food contaminants—viruses. *J. Food Protect.* 44:320–325.

90. Leistner, L., and F. Tauchmann. 1979. Aflatoxinbildung in Rohwurst durch verschiedene *Aspergillus flavus*-Stämme und einer *Aspergillus parasiticus*-Stamm. *Fleischwirtschaft* 50:965–966.

91. Lie, J. L., and E. H. Marth. 1967. Formation of aflatoxin in Cheddar cheese by *Aspergillus flavus* and *Aspergillus parasiticus*. *J. Dairy Sci.* 50:1708–1710.

92. Ljungh, Å., and T. Wadström. 1982. *Aeromonas* toxins. *Pharmac. Ther.* 15:339–354.

93. Maggon, K. K., S. K. Gupta, and T. A. Venkitasubramanian. 1977. Biosynthesis of aflatoxins. *Bacteriol. Rev.* 41:822–855.

94. Marth, E. H., and B. G. Calanog. 1976. Toxigenic fungi. In *Food microbiology: Public Health and Spoilage Aspects*, ed. M. P. deFigueiredo and D. F. Splittstoesser, 210–256. Westport, Conn.: AVI.

95. McFarren, E. F., M. L. Shafer, J. E. Campbell, K. H. Lewis, E. T. Jensen, and E. J. Schantz. 1960. Public health significance of paralytic shellfish poison. *Adv. Food Res.* 10:135–179.

96. McKercher, P. D., W. R. Hess, and F. Hamdy. 1978. Residual viruses in pork products. *Appl. Environ. Microbiol.* 35:142–145.

97. Metcalf, T. G., and W. C. Stiles. 1965. The accumulation of enteric viruses by the oyster, *Crassostrea virginica*. *J. Infect. Dis.* 115:68–76.

98. Miller, M. L., and J. A. Koburger. 1985. *Plesiomonas shigelloides*: An opportunistic food and waterborne pathogen. *J. Food Protect.* 48:449–457.

99. Montville, T. J., and P. K. Goldstein. 1989. Sodium bicarbonate inhibition of aflatoxigenesis in corn. *J. Food Protect.* 52:45–48.

100. Morse, D. L., J. J. Guzewich, J. P. Hanrahan, R. Stricof, M. Shayegani, R. Deibel, J. C. Grabau, N. A. Nowak, J. E. Herrmann, G. Cukor, and N. R. Blacklow. 1986. Widespread outbreaks of clam- and oyster-associated gastroenteritis. Role of Norwalk virus. *New Engl. J. Med.* 314:678–681.

101. Murphy, A. M., G. S. Grobmann, P. J. Christopher, W. A. Lopez, G. R. Davey, and R. H. Millsom. 1979. An Australia-wide outbreak of gastroenteritis from oysters caused by Norwalk virus. *Med. J. Austr.* 2:329–333.

102. Namdari, H., and E. J. Bottone. 1990. Cytotoxin and enterotoxin production as factors delineating enteropathogenicity of *Aeromonas caviae*. *J. Clin. Microbiol.* 28:1796–1798.

103. Niranjan, B. G., N. K. Bhat, and N. G. Avadhani. 1982. Preferential attack of mitochondrial DNA by aflatoxin B_1 during hepatocarcinogenesis. *Science* 215:73–75.

104. Niven, C. F., Jr., M. B. Jeffrey, and D. A. Corlett, Jr. 1981. Differential plating medium for quantitative detection of histamine-producing bacteria. *Appl. Environ. Microbiol.* 41:321–322.

105. Northolt, M. D., C. A. H. Verhulsdonk, P. S. S. Soentoro, and W. E. Paulsch. 1976. Effect of water activity and temperature on aflatoxin production by *Aspergillus parasiticus*. *J. Milk Food Technol.* 39:170–174.

106. Omura, Y., R. J. Price, and H. S. Olcott. 1978. Histamine-forming bacteria isolated from spoiled skipjack tuna and jack mackerel. *J. Food Sci.* 43:1779–1781.

107. Palumbo, S. A., M. M. Bencivengo, B. Del Corral, A. C. Williams, and R. L. Buchanan. 1989. Characterization of the *Aeromonas hydrophila* group isolated from retail foods of animal origin. *J. Clin. Microbiol.* 27:854–859.

108. Park, K. Y., and L. B. Bullerman. 1983. Effect of cycling temperatures on aflatoxin production by *Aspergillus parasiticus* and *Aspergillus flavus* in rice and cheddar cheese. *J. Food Sci.* 48:889–896.

109. Patel, U. M., S. R. Bapat, and P. J. Dave. 1990. Induction of aflatoxin biosynthesis in *Aspergillus parasiticus* by ascorbic acid–mediated lipid peroxidation. *Curr. Microbiol.* 20:159–164.

110. Paul, R., A. Siitonen, and P. Karkkainen. 1990. *Plesiomonas shigelloides* bacteremia in a healthy girl with mild gastroenteritis. *J. Clin. Microbiol.* 28:1445–1446.

111. Portnoy, B. L., P. A. Mackowiak, C. T. Caraway, J. A. Walker, T. W. McKinley, and C. A. Klein. 1975. Oyster-associated hepatitis: Failure of shellfish certification programs to prevent outbreaks. *J. Amer. Med. Assoc.* 233:1065–1068.

112. Ram, B. P., P. Hart, R. J. Cole, and J. J. Pestka. 1986. Application of ELISA to retail survey of aflatoxin B_1 in peanut butter. *J. Food. Protect.* 49:792–795.

113. Reynolds, G., and J. J. Pestka. 1991. Enzyme-linked immunosorbent assay of versicolorin A and related aflatoxin biosynthetic precursors. *J. Food Protect.* 54:105–108.

114. Richards, G. P. 1988. Microbial purification of shellfish: A review of depuration and relaying. *J. Food Protect.* 51:218–251.

115. Roland, J. O., and L. R. Beuchat. 1984. Biomass and patulin production by *Byssochlamys nivea* in apple juice as affected by sorbate, benzoate, SO_2 and

temperature. *J. Food Sci.* 49:402–406.

116. Rose, J. M., C. W. Houston, D. H. Coppenhaver, J. D. Dixon, and A. Kurosky. 1989. Purification and chemical characterization of a cholera toxin-cross-reactive cytolytic enterotoxin produced by a human isolate of *Aeromonas hydrophila*. *Infect. Immun.* 57:1165–1169.

117. Rouf, M. A., and M. M. Rigney. 1971. Growth temperatures and temperature characteristics of *Aeromonas*. *Appl. Microbiol.* 22:503–506.

118. Sanyal, S. C., S. J. Singh, and P. C. Sen. 1975. Enteropathogenicity of *Aeromonas hydrophila* and *Plesiomonas shigelloides*. *J. Med. Microbiol.* 8:195–198.

119. Schantz, E. J. 1973. Some toxins occurring naturally in marine organisms. In *Microbial Safety of Fishery Products,* ed. C. O. Chichester and H. D. Graham, 151–62. New York: Academic Press.

120. Schindler, A. F. 1977. Temperature limits for production of aflatoxin by twenty-five isolates of *Aspergillus flavus* and *Aspergillus parasiticus*. *J. Food. Protec.* 40:39–40.

121. Serck-Hanssen, A. 1970. Aflatoxin-induced fatal hepatitis? *Arch. Environ. Hlth.* 20:729–731.

122. Shelef, L. A., and B. Chin. 1980. Effect of phenolic antioxidants on the mutagenicity of aflatoxin B_1. *Appl. Environ. Microbiol.* 40:1039–1043.

123. Sheridan, J. F., L. Aurelian, G. Barbour, M. Santosham, R. B. Sack, and R. W. Ryder. 1981. Traveler's diarrhea associated with rotavirus infection: Analysis of virus-specific immunoglobulin classes. *Infect. Immun.* 31:419–429.

124. Shih, C. N., and E. H. Marth. 1972. Experimental production of aflatoxin on brick cheese. *J. Milk Food Technol.* 35:585–587.

125. Shih, C. N., and E. H. Marth. 1974. Some cultural conditions that control biosynthesis of lipid and aflatoxin by *Aspergillus parasiticus*. *Appl. Microbiol.* 27:452–456.

126. Shotwell, O. L., and C. W. Hesseltine. 1983. Five-year study of mycotoxins in Virginia wheat and dent corn. *J. Assoc. Off. Anal. Chem.* 66:1466–1469.

127. Smith, G. C., L. Aurelian, M. Santosham, and R. B. Sack. 1983. Rotavirus-associated traveler's diarrhea: Neutralizing antibody in asymptomatic infections. *Infect. Immun.* 41:829–833.

128. Sobsey, M. D., R. J. Carrick, and H. R. Jensen. 1978. Improved methods for detecting enteric viruses in oysters. *Appl. Environ. Microbiol.* 36:121–280.

129. Sommer, N. F., J. R. Buchanan, and R. J. Fortlage. 1974. Production of patulin by *Penicillium expansum*. *Appl. Microbiol.* 28:589–593.

130. Stahr, H. M., R. L. Pfeiffer, P. J. Imerman, B. Bork, and C. Hurburgh. 1990. Aflatoxins—the 1988 outbreak. *Dairy Food Environ. Sanit.* 10:15–17.

131. Stark, A.-A. 1980. Mutagenicity and carcinogenicity of mycotoxins: DNA binding as a possible mode of action. *Ann. Rev. Microbiol.* 34:235–262.

132. Stinson, E. E., D. D. Bills, S. F. Osman, J. Siciliano, M. J. Ceponis, and E. G. Heisler. 1980. Mycotoxin production by *Alternaria* species grown on apples, tomatoes, and blueberries. *J. Agric. Food Chem.* 28:960–963.

133. Stinson, E. E., S. F. Osman, E. G. Beisler, J. Siciliano, and D. D. Bills. 1981. Mycotoxin production in whole tomatoes, apples, oranges, and lemons. *J. Agric. Food Chem.* 29:790–792.

134. Stoloff, L. 1987. Carcinogenicity of aflatoxins. Science 237:1283–1284 (letter to editor with two responses).

135. Sullivan, R., R. M. Marnell, E. P. Larkin, and R. B. Read, Jr. 1975. Inactivation of poliovirus 1 and coxsackievirus B-2 in broiled hamburgers. *J. Milk Food Technol.* 38:473–475.

136. Taylor, S. L., L. S. Guthertz, M. Leatherwood, and E. R. Lieber. 1979. Histamine

production by *Klebsiella pneumoniae* and an incident of scombroid fish poisoning. *Appl. Environ. Microbiol.* 37:274–278.

137. Taylor, S. L., T. J. Keefe, E. S. Windham, and J. F. Howell. 1982. Outbreak of histamine poisoning associated with consumption of Swiss cheese. *J. Food Protect.* 45:455–457.

138. Tong, C.-H., and F. A. Draughon. 1985. Inhibition by antimicrobial food additives of ochratoxin A production by *Aspergillus sulphureus* and *Penicillium viridicatum*. *Appl. Environ. Microbiol.* 49:1407–1411.

139. Trenk, H. L., and P. A. Hartman. 1970. Effects of moisture content and temperature on aflatoxin production in corn. *Appl. Microbiol.* 19:781–784.

140. Tsukamoto, T., Y. Konoshita, T. Shimada, and R. Sakazaki. 1978. Two epidemics of diarrhoeal disease possibly caused by *Plesiomonas shigelloides*. *J. Hyg.* 80:275–280.

141. Twedt, R. M., and B. K. Boutin. 1979. Potential public health significance of non-*Escherichia coli* coliforms in food. *J. Food Protect.* 42:161–163.

142. Van Damme, L. R., and J. Vandepitte. 1980. Frequent isolation of *Edwardsiella tarda* and *Plesiomonas shigelloides* from healthy Zairese freshwater fish: A possible source of sporadic diarrhea in the tropics. *Appl. Environ. Microbiol.* 39:475–479.

143. Wadström, T., Å. Ljungh, and B. Wretlind. 1976. Enterotoxin, haemolysin and cytotoxic protein in *Aeromonas hydrophila* from human infections. *Acta Path. Microbiol. Scand. Sect. B* 84:112–114.

144. Wait, D. A., C. R. Hackney, R. J. Carrick, G. Lovelace, and M. D. Sobsey. 1983. Enteric bacterial and viral pathogens and indicator bacteria in hard shell clams. *J. Food Protect.* 46:493–496.

145. Wei, C. I., C.-M. Chen. J. A. Koburger, W. S. Otwell, and M. R. Marshall. 1990. Bacterial growth and histamine production on vacuum packaged tuna. *J. Food Sci.* 55:59–63.

146. Wheeler, J. L., M. A. Harrison, and P. E. Koehler. 1987. Presence and stability of patulin in pasteurized apple cider. *J. Food Sci.* 52:479–480.

147. White, K. E., M. T. Osterholm, J. A. Mariotti, J. A. Korlath, D. H. Lawrence, T. L. Ristinen, and H. B. Greenberg. 1986. A foodborne outbreak of Norwalk virus gastroenteritis. *Amer. J. Epidemiol.* 124:120–126.

148. Wills, J. H., Jr. 1966. Seafood toxins. In *Toxicants Occurring Naturally in Foods*, 147–163. Pub. No. 1354. Washington, D.C.: National Academy of Science.

149. Wogan, G. N. 1966. Chemical nature and biological effects of the aflatoxins. *Bacteriol. Rev.* 30:460–470.

150. Wu, M. T., J. C. Ayres, and P. E. Koehler. 1974. Production of citrinin by *Penicillium viridicatum* on country-cured ham. *Appl. Microbiol.* 27:427–428.

151. Yoshinaga, D. R., and H. A. Frank. 1982. Histamine-producing bacteria in decomposing skipjack tuna (*Katsuwonus pelamis*). *Appl. Environ. Microbiol.* 44:447–452.

152. Zaika, L. L., and R. E. Buchanan. 1987. Review of compounds affecting the biosynthesis or bioregulation of aflatoxins. *J. Food Protect.* 50:691–708.

153. Zajc-Satler, J., A. Z. Dragav, and M. Kumelj. 1972. Morphological and biochemical studies of 6 strains of *Plesiomonas shigelloides* isolated from clinical sources. *Zbt. Baktr. Hyg. Abt. Orig. A.* 219:514–521.

APPENDIX

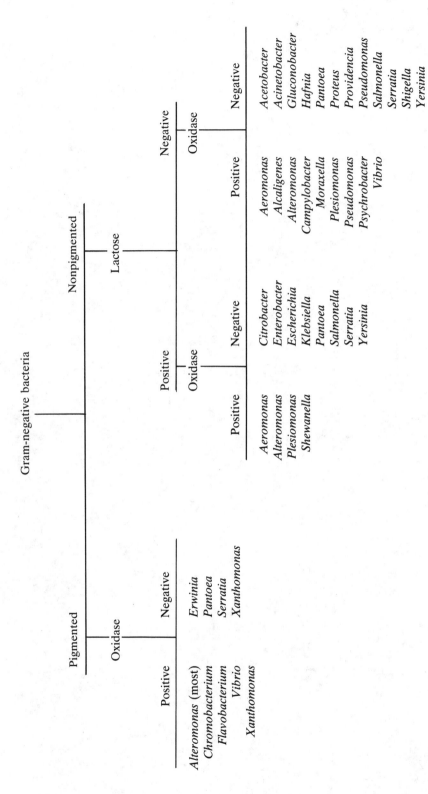

A. Relationships of common foodborne genera of gram-negative bacteria. For details, consult *Bergey's Manual*.

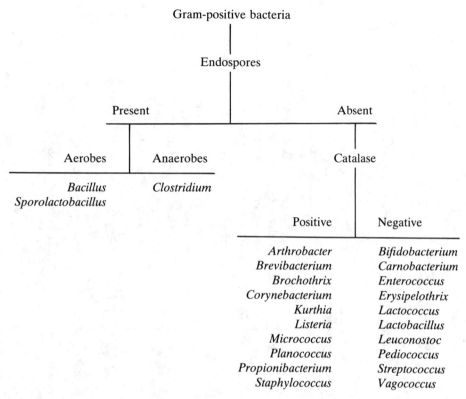

B.   Relationship of common foodborne genera of gram-positive bacteria. For details, consult *Bergey's Manual*.

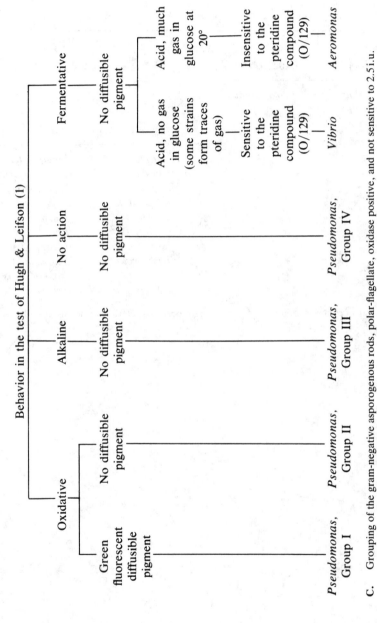

Behavior in the test of Hugh & Leifson (1)

- **Oxidative**
  - Green fluorescent diffusible pigment → *Pseudomonas*, Group I
  - No diffusible pigment → *Pseudomonas*, Group II
- **Alkaline**
  - No diffusible pigment → *Pseudomonas*, Group III
- **No action**
  - No diffusible pigment → *Pseudomonas*, Group IV
- **Fermentative**
  - No diffusible pigment
    - Acid, no gas in glucose (some strains form traces of gas) — Sensitive to the pteridine compound (O/129) → *Vibrio*
    - Acid, much gas in glucose at 20° — Insensitive to the pteridine compound (O/129) → *Aeromonas*

C. Grouping of the gram-negative asporogenous rods, polar-flagellate, oxidase positive, and not sensitive to 2.5 i.u. penicillin, on the results of four other tests.

After J. M. Shewan, G. Hobbs, and W. Hodgkiss. 1960. A determinative scheme for the identification of certain genera of gram-negative bacteria, with special reference to the Pseudomonadadeae. J. Appl. Bacteriol. 23:379–390.

INDEX

675